International Conference on Electrical Engineering, Control and Robotics (EECR 2018)

MATEC Web of Conferences Volume 160 (2018)

Chengdu, China
12 – 14 January 2018

Editors:

X. Lei
S.P. Kozaitis
C.-C. Ku

ISBN: 978-1-5108-6105-3

Printed from e-media with permission by:

Curran Associates, Inc.
57 Morehouse Lane
Red Hook, NY 12571

Some format issues inherent in the e-media version may also appear in this print version.

Printed by Curran Associates, Inc. (2018)

For additional information, please contact EDP Sciences – Web of Conferences at the address below.

EDP Sciences – Web of Conferences
17, Avenue du Hoggar
Parc d'Activité de Courtabœuf
BP 112
F-91944 Les Ulis Cedex A
France

Phone: +33 (0) 1 69 18 75 75
Fax: +33 (0) 1 69 28 84 91

contact-edps@webofconferences.org

Additional copies of this publication are available from:

Curran Associates, Inc.
57 Morehouse Lane
Red Hook, NY 12571 USA
Phone: 845-758-0400
Fax: 845-758-2633
Email: curran@proceedings.com
Web: www.proceedings.com

TABLE OF CONTENTS

Preface

2018 International Conference on Electrical Engineering, Control and Robotics (EECR 2018) was held in Chengdu, China, from January 12 to 14, 2018, co-organized by University of Electronic Science and Technology of China, Sichuan Institute of Electronics, and Southwest Minzu University.

The goal of this conference is to foster pure and applied researches in recognition of academic and industry, and to encourage interdisciplinary and international collaboration for the participants. Accepted and presented papers will be published in MATEC Web of Conferences Proceedings (ISSN: 2261-236X). The conference proceedings will be indexed by Ei Compendex, SCOPUS, etc.

There were five keynotes in EECR 2018, first given by Dr. May Huang, a professor and department chair of Electrical and Computer Engineering of International Technological University, USA, with topic: Trend of AI Chips; followed by Associate Professor, Dr. Quang Ha, from University of Technology Sydney, Australia, with topic: Smart City Infrastructure: Energy Management and Environment Monitoring; Prof. Chun-Yi Su from Concordia University, Canada, gave topic of Modeling and Control of Hysteresis Nonlinearities in Smart Actuators; Prof. Xingwen Liu from Southwest Minzu University, China, gave topic of Covering Method in Stability Analysis and Controller Design of Switched Systems with Delays; and Assoc. Prof. Haiqing Li, University of Electronic Science and Technology of China, gave topic of Service Intelligent Robots Change People's Lives.

Total 63 papers are included in Proceeding of EECR 2018, presenting research works and results in fields of Control Theory and Control System, Smart Grid and Power Electronics Technology, Robot Design and Intelligent Control, as well as Information Engineering and Image Processing.

The coordinators of various sessions have devoted a considerable time and energy in soliciting papers from relevant researchers on presentations at the conference. The chairpersons of the sessions played important role in conducting the presentations of the sessions effectively.

On behalf of the conference committee, we express sincere appreciation to session chairs, presenters and coordinators. We also appreciate all the reviewers of the manuscripts for theirvaluable comments within the time allotted to them.

Thanks to everyone who was involved in EECR 2018 for making it successful! We expect that Proceedings of EECR 2018could provide new ideas in areas of electrical engineering, control and robotics.

Prof. May Huang
International Technological University, USA

edp sciences

MATEC Web of Conferences

Statement of Peer review

In submitting conference proceedings to MATEC Web of Conferences, I certify to the Publisher that I adhere to the **Policy on Publishing Integrity** of the journal in order to safeguard good scientific practice in publishing.

1. All articles have been subjected to peer review administered by the proceedings editors.
2. Reviews have been conducted by expert referees, who have been requested to provide unbiased and constructive comments aimed, whenever possible, at improving the work.
3. Proceedings editors have taken all reasonable steps to ensure the quality of the materials they publish and their decision to accept or reject a paper for publication has been based only on the merits of the work and the relevance to the journal.

Title, date and place of the conference

2018 International Conference on Electrical Engineering, Control and Robotics (EECR 2018)

Proceedings editor(s):

Xuelin Lei

East China University of Science and Technology, China

Date and editor's signature

22nd. March

edp sciences

MATEC Web of Conferences

Statement of Peer review

In submitting conference proceedings to MATEC Web of Conferences, I certify to the Publisher that I adhere to the **Policy on Publishing Integrity** of the journal in order to safeguard good scientific practice in publishing.

1. All articles have been subjected to peer review administered by the proceedings editors.
2. Reviews have been conducted by expert referees, who have been requested to provide unbiased and constructive comments aimed, whenever possible, at improving the work.
3. Proceedings editors have taken all reasonable steps to ensure the quality of the materials they publish and their decision to accept or reject a paper for publication has been based only on the merits of the work and the relevance to the journal.

Title, date and place of the conference

2018 International Conference on Electrical Engineering, Control and Robotics (EECR 2018)
Jan.12-14, 2018
Chengdu, China

Proceedings editor(s): S. Kozaitis

Date and editor's signature 3/22/18

17, avenue du Hoggar - PA de Courtabœuf – BP 112 - 91944 Les Ulis Cedex A (France)
Tél. : 33 (0)1 69 18 75 75 - Fax : 33(0)1 69 07 45 17 – www.edpsciences.org

Statement of Peer review

In submitting conference proceedings to MATEC Web of Conferences, I certify to the Publisher that I adhere to the **Policy on Publishing Integrity** of the journal in order to safeguard good scientific practice in publishing.

1. All articles have been subjected to peer review administered by the proceedings editors.
2. Reviews have been conducted by expert referees, who have been requested to provide unbiased and constructive comments aimed, whenever possible, at improving the work.
3. Proceedings editors have taken all reasonable steps to ensure the quality of the materials they publish and their decision to accept or reject a paper for publication has been based only on the merits of the work and the relevance to the journal.

Title, date and place of the conference

2018 International Conference on Electrical Engineering, Control and Robotics (EECR 2018)
Jan.12-14, 2018
Chengdu, China

Proceedings editor(s):

Date and editor's signature Cheung-Chieh Ku 2018.03.22

17, avenue du Hoggar - PA de Courtabœuf – BP 112 - 91944 Les Ulis Cedex A (France)
Tél. : 33 (0)1 69 18 75 75 - Fax : 33(0)1 69 07 45 17 – www.edpsciences.org

Online Inductance and Capacitance Identification Based on Variable Forgetting Factor Recursive Least-Squares Algorithm for Boost Converter

Chen Chen[1], Run Min[1], Qiaoling Tong[1], Shifei Tao[2], Dian Lyu[1], and Linkai Li[1]

[1]*School of Optical and Electronic Information, Huazhong University of Science and Technology, Wuhan 430074, China*
[2] *Wuhan Marine Machinery Co., Ltd., Wuhan 430074, China*

Abstract. The control performance of boost converter suffers from the variations of important component parameters, such as inductance and capacitance. In this paper, an online inductance and capacitance identification based on variable forgetting factor recursive least-squares (VFF-RLS) algorithm for boost converter is proposed. First, accurate inductance and capacitance identification models and the RLS algorithm are introduced. In order to balance the steady-state identification accuracy and parameter tracking ability, a forgetting factor control technique is investigated. By recovering system noise in the error signal of the algorithm, the value of forgetting factor is dynamically calculated. In addition, since the sampling rate is much lower than the existing identification methods, the proposed algorithm is practical for low-cost applications. Finally, the effectiveness of the proposed algorithm is verified by experiment. The experiment results show that the algorithm has good performance in tracking inductance and capacitance variations.

1 Introduction

Recently, digitally controlled pulsewidth modulated dc-dc converters have become a hot research topic because of their lower sensitivity to system noises and component parameter variations, programmability, ability to implement complex control strategies, and possibilities to update controller online [1-5]. The main control techniques of digital dc-dc converter can be classified into voltage mode control and current mode control. The former as a single-loop control can be easily designed and analysed. But it has low response speed to line voltage and load. Compared with the voltage mode control, the current mode control has faster response speed and larger loop gain bandwidth since the inner current loop control is added [6]. The control performance of the two methods depends on accurate converter modeling. As two important parameters in the converter model, the inductance of inductor and capacitance of output capacitor vary over time in practice due to temperature variation, aging effect and load level. Therefore, in order to improve the control performance of dc-dc converter system, it is necessary to identify the inductance and capacitance online for timely tracking their variations.

Several methods for online parameter identification of dc-dc converter have been studied [7-12]. For the load estimation of boost converter, an algebraic parameter identification and classical asymptotic observers are compared in [7]. Different from asymptotic observers, the algebraic identification method does not suffer from stability and convergence issues. Further, Jesús et al. proposes an algebraic identification algorithm for unknown time-varying parameters, such as load resistance and input voltage supply value. On this basis, the generalized proportional–integral based adaptive controller can be tuned online. In [9], a real-time detection technology is studied for capacitance and equivalent resistance (ESR) of output capacitor. With the sampling of capacitor current and ripple voltage, the capacitor parameters can be obtained by Kalman filter. However, the sampling rate is 16 times than the switching frequency, which leads to high requirements for hardware. By analyzing the characteristics of capacitor voltage ripple, a current-sensorless capacitance and ESR identification approach for buck converter is investigated [10]. The output voltage only needs to be sampled twice in a switching cycle. But, the identification accuracy will be affected by inductance variation. In [11], an online identification algorithm for capacitance, ESR, inductance, and load is proposed. Based on the sampling information of MOSFET gate signal, inductor current and output voltage, the algorithm can estimate the above four parameter values through different data processing methods, such as the least mean squares algorithm. In [12], inductance and capacitance can be reconstructed by an observer with adaptive feedback law. For the two methods, since the sampling frequency is much higher than switching frequency and large amounts of data need to be processed, their demands for hardware, such as ADC and digital signal processor, are high, which is not suitable for low-cost applications. In addition, they are difficult to implement in converters with high switching frequency. The above literatures address the online identification of load, input voltage, capacitance, ESR, and inductance. However, none of these algorithms can simultaneously estimate inductance and capacitance with low requirements for system hardware.

In this paper, an inductance and capacitance identification based on variable forgetting factor recursive least-squares (VFF-RLS) algorithm for boost converter is proposed. The value of forgetting factor is derived by recovering system noise in the error signal of the algorithm. Based on the proposed accurate inductance and capacitance identification model, inductance and capacitance variations can be accurately and rapidly tracked. Since the proposed algorithm is simple and its maximum sampling rate is only two times of switching frequency, it is practical for the low-cost applications. The experiment results fully verify the feasibility of the proposed online parameter identification for boost converter.

2 Online inductance and capacitance identification based on RLS

The schematic diagram of boost converter with online inductance and capacitance identification algorithm based on RLS is shown in Fig. 1, where $\hat{L}$ and $\hat{C}$ are the identification values of inductance and capacitance, respectively. The digital control of boost converter, illustrated in shaded area, is used to guarantee the stable operation of the system by designing a suitable digital controller. Any kind of constant frequency digital control method can be utilized, such as the classical voltage mode control and current model control.

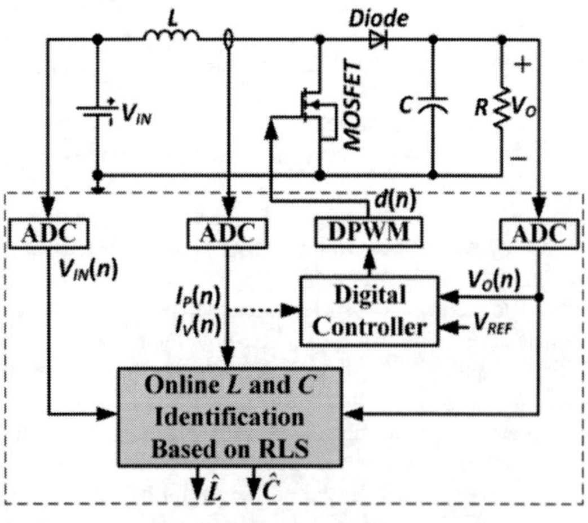

Figure 1. The structure of boost converter with online L and C identification algorithm based on FF-RLS

Since the identification results of inductance and capacitance do not converge in steady state, the identification should be carried out in a dynamic process. First, the dynamic process is created by slightly changing the voltage reference V_{REF}, i.e., a small pulse signal is injected into V_{REF}. Then, according to the sampled data of input voltage, output voltage, and inductor current, inductance and capacitance are identified through the RLS algorithm.

2.1 Discrete-time modeling of boost converter

The accuracy of parameter identification is built on the basis of accurate converter modelling. In order to accurately identify the inductance and capacitance, the conventional average state equation of boost converter is first improved.

By considering the inductor current ripple and combining it with the used leading-edge pulse width modulation method, the improved discrete-time average state equation is obtained.

$$\begin{aligned} L\frac{I_P(k+1)-I_P(k)}{T} &= V_{IN}(n)-D'(n)V_O(n) \\ C\frac{V_O(n+1)-V_O(n)}{T} &= D'(n)\bar{I}_L(n)\,|\,D'-\frac{V_O(n)}{R} \end{aligned} \quad (1)$$

where $V_O(n)$ and $I_P(n)$ are the sampled output voltage and inductor current at the very beginning of the nth switching cycle, respectively. $\bar{I}_L(n)\,|\,D'$ is the average inductor current during the switching off period. It can be expressed as

$$\bar{I}_L(n)\,|\,D' = \frac{I_P(n)+I_V(n)}{2} \quad (2)$$

where $I_V(n)$ is the valley inductor current of the nth switching cycle, which is sampled in the switch-on moment.

From (1), the inductance and capacitance identification models, which are suitable for the standard RLS method, are given by

$$\theta_L\left[V_{IN}(n)-D'(k)V_O(n)\right]=I_P(n+1)-I_P(n) \quad (3)$$

$$\theta_C\left[D'(k)\bar{I}_L(n)\,|\,D'-\frac{V_O(n)}{R}\right]=V_O(n+1)-V_O(n) \quad (4)$$

where $\theta_L = T/L$, $\theta_C = T/C$ and they are the parameters to be identified. As can be seen from (3) and (4), the effect of the boost converter's nonlinearity on parameter identification is also eliminated in models. By applying RLS method with (3) and (4), θ_L and θ_C can be calculated, then the inductance and capacitance are estimated.

2.2 RLS for parameter identification

Recursive least square algorithm is a widely used adaptive filter. It has the characteristics of fast calculation speed, less calculation, fast convergence and low sensitivity to noise. Therefore, it is commonly applied in the field of parameter identification, especially for online parameter identification.

Recursive formula of RLS algorithm is shown as follows:

$$e(n) = y(n) - \hat{\theta}^T(n-1)\varphi(n) \quad (5)$$

$$K(n) = \frac{P(n-1)\varphi(n)}{\lambda + \varphi^T(n)P(n-1)\varphi(n)} \quad (6)$$

$$\hat{\theta}(n)=\hat{\theta}(n-1)+K(n)e(n) \tag{7}$$

$$P(n)=\frac{1}{\lambda}\left[I-K(n)\varphi^{T}(n)\right]P(n-1) \tag{8}$$

where $\lambda\in[0,1]$ is the forgetting factor. $e(n)$, $y(n)$, and $\varphi^{T}(n)$ are the error signal, desired output of adaptive filter, and input signal vector, respectively. $\hat{\theta}(n-1)$ is parameter estimation value obtained after (*n*-1)*th* recursion. $K(n)\in R^{N}$ is the Kalman gain, $P(n)\in R^{N\times N}$ is the error covariance matrix, and N is the number of parameters to be identified. The initial value of $P(n)$, i.e., $P(0)$, is selected by the designer with experience.

By applying the RLS method to (3) and (4), respectively, the RLS-based online inductance and capacitance identification is obtained. For inductance identification, $y(n)$, $\varphi(n)$, and $\hat{\theta}(n-1)$ are written as $y(n)=I_P(n+1)-I_P(n)$, $\varphi(n)=V_{IN}(n)-D'(k)V_O(n)$, and $\hat{\theta}(n-1)=\hat{\theta}_L(n-1)$, while for capacitance identification, $y(n)$, $\varphi(n)$, and $\hat{\theta}(n-1)$ are expressed as $y(n)=V_O(n+1)-V_O(n)$, $\varphi(n)=D'(k)\bar{I}_L(n)\,|\,D'-V_O(n)/R$ $\hat{\theta}(n-1)=\hat{\theta}_C(n-1)$. The identification algorithm is recursively calculated once per switching cycle. According to the duty cycle, the sampled inductor current and voltages, $\varphi(n)$ and $y(n)$ can be obtained. Then Kalman gain $K(n)$ and error signal $e(n)$ is calculated to correct the parameter identification value of last switching cycle, $\hat{\theta}(n-1)$. Thus, the new observation data is constantly introduced in each switching cycle to update the identification value until an expected accuracy is achieved. Finally, the inductance and capacitance can be estimated from the optimal $\hat{\theta}_L(n)$ and $\hat{\theta}_C(n)$.

3 VFF-RLS algorithm for inductance and capacitance identification

In the classical RLS algorithm, λ is set to a fixed value. The effect of past observation data is decreased at a fixed rate. If λ is close to 1, the algorithm has high precision and low parameter tracking ability. Decreasing λ improves tracking performance, but the steady-state accuracy is also lowered at the same time. In order to satisfy these two contradictory demands, a variable forgetting factor RLS algorithm is investigated for inductance and capacitance identification.

Since the error signal in (5) is a prior error computed with the parameter identification value of the (*n*-1)*th* cycle, it is a prior error. The posteriori error can be expressed as

$$\varepsilon(n)=y(n)-\hat{\theta}^{T}(n)\varphi(n) \tag{9}$$

From (5), (7), and (9), $\varepsilon(n)$ can be rewritten as

$$\varepsilon(n)=e(n)\left[1-\varphi^{T}(n)K(n)\right] \tag{10}$$

The value of forgetting factor is designed by recovering the system noise in the error signal of the adaptive filter, i.e., the forgetting factor $\lambda(n)$ can be adjusted according to (11).

$$E\{e(n)\varepsilon(n)\}=E\{v^{2}(n)\} \tag{11}$$

where $E\{v^2(n)\}=\sigma_v^2$ is the power of system noise.

Substituting (6) and (10) into (11), with the assumption that the input and error signals are uncorrelated, (12) can be obtained.

$$E\left\{\frac{\lambda(n)}{\lambda(n)+q(n)}\right\}=\frac{\sigma_v^2}{\sigma_e^2(n)} \tag{12}$$

where $q(n)=\varphi^{T}(n)P(n-1)\varphi(n)$, $E\{e^2(n)\}=\sigma_e^2(n)$ is the power of a priori error signal. The assumption is true when the identified parameters have started to converge to the actual value. By solving (12), the variable forgetting factor is given by

$$\lambda(n)=\frac{\sigma_v^2\sigma_q^2(n)}{\sigma_e^2(n)-\sigma_v^2} \tag{13}$$

where $E\{q^2(n)\}=\sigma_q^2(n)$. In real application, $\sigma_e^2(n)$ and $\sigma_q^2(n)$ can be estimated by

$$\hat{\sigma}_e^2(n)=\alpha\hat{\sigma}_e^2(n-1)+(1-\alpha)e^2(n) \tag{14}$$

$$\hat{\sigma}_q^2(n)=\alpha\hat{\sigma}_q^2(n-1)+(1-\alpha)q^2(n) \tag{15}$$

where $\alpha\in[0,1]$ is the weighting factor.

Since the value of λ must be in the range of [0,1], the forgetting factor of the proposed algorithm is given by

$$\lambda(n)=\min\left\{\frac{\sigma_v^2\hat{\sigma}_q^2(n)}{\xi+\hat{\sigma}_e^2(n)-\sigma_v^2},\lambda_{\max}\right\} \tag{16}$$

where ξ is a small positive constant that prevents the division by zero. Before the identification results converge or when an abrupt change occurs in the system, $\hat{\sigma}_e^2(n)$ is larger than σ_v^2, hence $\lambda(n)$ goes to lower values, which makes the algorithm achieves fast convergence and tracking. When the results converge to the steady-state value, $\lambda(n)$ is set to $\lambda_{\max}$, then the algorithm can have high identification accuracy.

4 Experiment results

To validate the performance of the proposed identification method, a prototype is employed for testing the steady-state accuracy and tracking ability of the proposed identification. Experimental system hardware consists of a control stage and power stage. The core component of control stage is Texas Instruments TMS320F2812 DSP system, which is used to implement the control and parameter identification algorithms. The power stage contains a non-synchronous boost converter and signal conversion circuits. The specifications of the boost converter are shown in Table 1. The nominal inductance is obtained by connecting two inductors (L_1=28μH and L_2=100μH) in parallel. Similarly, the nominal capacitance is got by connecting two capacitors (C_1=56μF and C_2=10μF) in series. The digital peak current control is utilized to ensure the system stable operation.

Table 1. Specifications of the tested boost converter.

Inductance of the power inductor	22μH
Capacitance of the output capacitor	66μF
Input voltage	6V
Output voltage	12V
Rated output current (ROC)	1.2A
Voltage ripple under ROC	1%
Switching frequency	100kHz

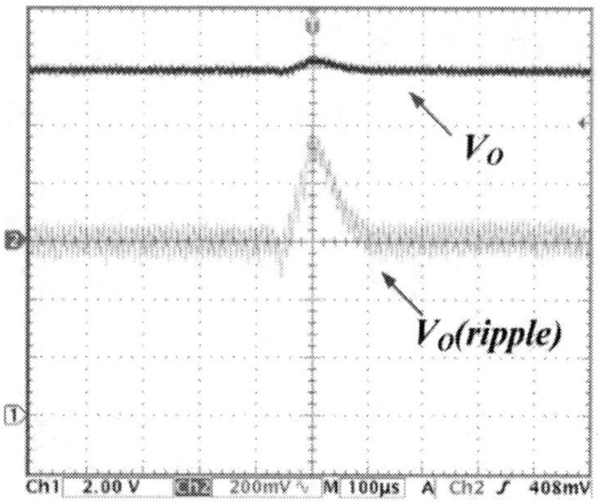

Figure 2. Output voltage waveform when small pulse signal is injected

Fig. 2 shows the output voltage waveform during the small pulse injection. The injected pulse signal lasted 20μs. Its amplitude is 3% of expected voltage reference. As can be seen from Fig. 2, after the voltage reference changes, there is a transient response in the output voltage. With the recovery of voltage reference, the output voltage then returns to the steady-state. The whole process lasts 200μs. The maximum derivation of output voltage is 0.32V, which is less than 3% of the steady-state output voltage. Therefore, the injection effect on output voltage is small. The boost converter can be regarded as always working in steady state.

During the transient state, the input voltage, output voltage, and inductor current are sampled by AD converters, providing the data for estimating inductance and capacitance. In order to verify the robustness of the proposed algorithm, the parameter identification is executed with different input voltages and the initial values of inductance and capacitance are set to 20μH and 56μH, which derivate from the actual values. The identification results are shown in Fig. 3 and Fig. 4, respectively. As Fig. 3 shows, under the rated input voltage, the identification result of inductance converges to 20.95μH after 14 iterations. The identification error is only 4.77% compared with the actual value (22μH). When the input voltage decreased to 5V, the identified inductance is 21.07μH. The estimated error is still low. As can be seen from Fig. 4, under the rated input voltage, the identification result of capacitance converges to 67.64μF after 14 iterations with 2.48% estimation error. When the input voltage is 5V, the estimated capacitance is 67.69μF, of which the estimation error is still very small. Therefore, the proposed algorithm can accurately identify the inductance and capacitance and will not affected by changes in input voltage.

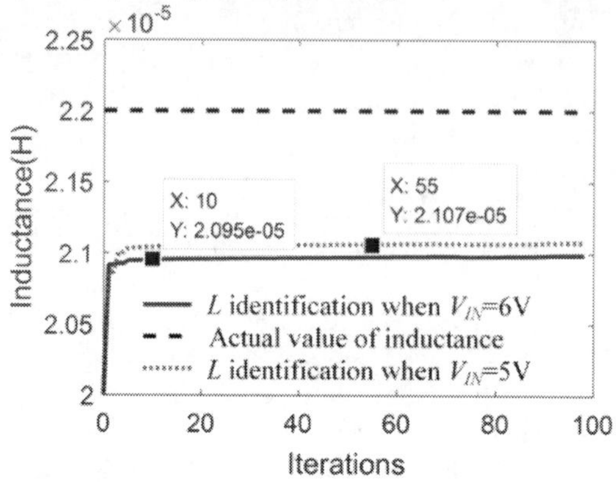

Figure 3. Identification results of inductance under different input voltages

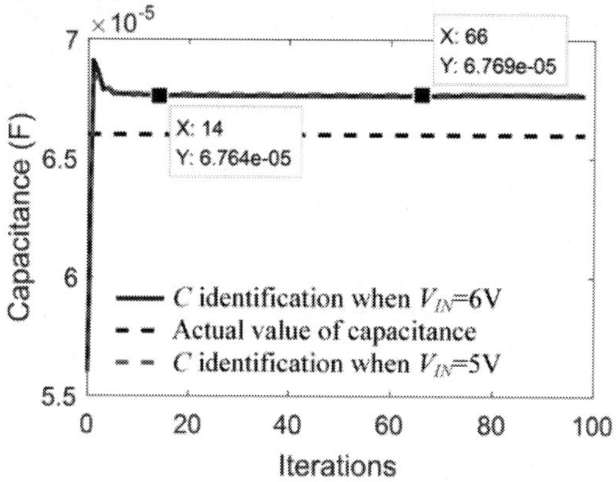

Figure 4. Identification results of capacitance under different input voltages

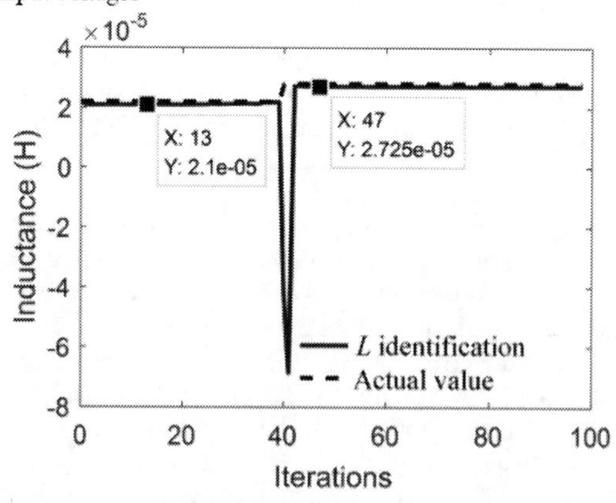

Figure 5. Identification results of inductance under inductance and capacitance variations

In order to verify the tracking capabilities and steady state accuracy of the proposed algorithm, step changes in inductance and capacitance are applied. To simulate a sudden change in inductance and capacitance, the inductor L_2 and capacitor C_2 are simultaneously and transiently disconneted from the boost converter. The identification results are shown in Fig. 5 and Fig. 6, respectively. As can be seen from Fig. 5 and Fig. 6, when inductance and capacitance step change, the estimated inductance and capacitance converge to the steady-state after 7 and 10 iterations. The identification results are 27.25μH and 56.83μF, respectively, of which the estimation errors are less than 2.7%.

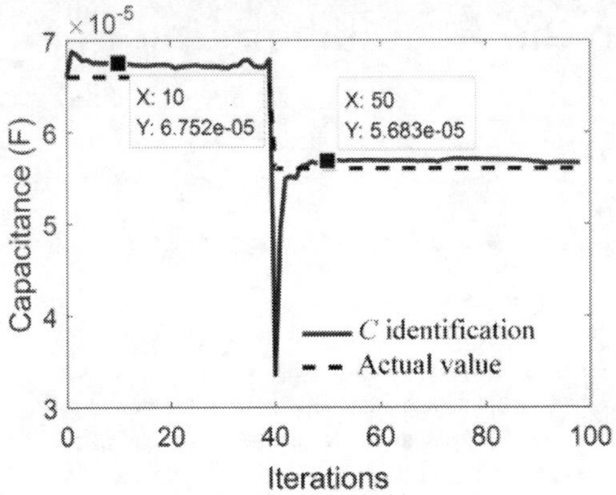

Figure 6. Identification results of capacitance under inductance and capacitance variations

As can be seen from the experiment results, even if there are sudden changes in inductance and capacitance, the errors between the identified values and actual values are always less than 5%. The proposed algorithm can track the actual inductance and capacitance quickly and precisely.

5 Conclusion

An online inductance and capacitance identification based on VFF-RLS algorithm is proposed in this paper. With the consideration of inductor current ripple, the inductance and capacitance identification models for standard RLS method are derived. By recovering the system noise in the error signal of the algorithm, the value of forgetting factor in RLS algorithm is dynamically calculated, which solves the problem of the classical RLS algorithm that cannot balance the steady-state accuracy and parameter tracking ability. The experiment results indicate that the proposed algorithm can accurately and rapidly track the inductance and capacitance variations. In addition, the algorithm is simple and easy to implement in practice, which is suitable for low-cost applications. Since the proposed algorithm is applied for boost converters, a further investigation of the application on other topologies and the effect of parasitic parameters on converter modeling, still need to be considered in future works.

Acknowledgement

The financial support of this research by the National Science Foundation of China, under Grant No. 61202469, 61640311 is greatly appreciated.

References

1. G. Feng, E. Meyer, Y.F. Liu. A new digital control algorithm to achieve optimal dynamic performance in DC-to-DC converters. IEEE Trans. Power. Electron., **22**, 4 (2007)
2. Y.S. Lai and C.A. Yeh, Predictive digital-controlled converter with peak current-mode control and leading-edge modulation. IEEE Trans. Ind. Electron., **56**, 6 (2009)
3. M. Veerachary and R. Saxena, Design of robust digital stabilizing controller for fourth-order boost DC–DC converter: A quantitative feedback theory approach, IEEE Trans. Ind. Electron., **59**, 2 (2012)
4. Q. Tong, Q. Zhang, R. Min, X. Zou, Z. Liu, and Z. Chen. Sensorless predictive peak current control for boost converter using comprehensive compensation strategy. IEEE Trans. Ind. Electron., **61**, 6 (2014)
5. Y. Qiu, H. Liu, and X. Chen, Digital average current-mode control of PWM DC–DC converters without current sensors. IEEE Trans. Ind. Electron., **57**, 5 (2010)
6. Q. Zhang, R. Min, Q. Tong, X. Zou, Z. Liu, and A. Shen. Sensorless predictive current controller DC-DC converter with a self-correction differential current observer. IEEE Trans. Ind. Electron., **61**, 12 (2014)
7. Gensior A, Weber J, Rudolph J, and H. Güldner. Algebraic Parameter Identification and Asymptotic Estimation of the Load of a Boost Converter. IEEE Trans. Ind. Electron., **55**, 9 (2008)
8. L.F. Jesús, H.M. Arturo, G.R. Carlos, and H.S. Ramírez. Robust nonlinear adaptive control of a "boost" converter via algebraic parameter identification. IEEE Trans. Ind. Electron., **61**, 8 (2014)
9. K. Abdennadher, P. Venet, G. Rojat, J.M. Rétif, and C. Rosset. A real-time predictive-maintenance system of aluminum electrolytic capacitors used in uninterrupted power supplies. IEEE Trans. Ind. Appl., **46**, 4 (2010)
10. K. Yao,W. Tang,W. Hu, and J. Lyu, A current-sensorless online ESR and C identification method for output capacitor of buck converter. IEEE Trans. Power Electron., **30**, 12 (2015)
11. G. Buiatti, A. Amaral, and A. Cardoso, An online technique for estimating the parameters of passive components in non-isolated DC/DC converters. *In Proc. IEEE Int. Symp. Ind. Electron.*, 606–610 (2007)
12. Z. Cen and P. Stewart, Condition parameter estimation for photovoltaic buck converters based on adaptive model observers. IEEE Trans. Rel., **66**, 1 (2017)

Application Analysis of AD8479 in the Acquisition Unit of Electronic Current Transformer

Xi Liu, Jianhua Yang, Bifu Qian, Lei Zhang

State Grid Wenzhou Electric Power Supply Company, Wenzhou, China

Abstract. Electronic current transformer acquisition unit was installed outdoors and near the circuit breaker. The poor environments, including high common mode voltage, electromagnetic interference and other factors were a serious of disadvantages impacts on the acquisition unit. The paper introduced the parameter characteristics and functions of Rogowski coil, the differential amplifier chip AD8479, and filter circuit in detail. The application circuit was designed based on proper parameters. Lastly, the interference was inhibited effectively, and the electronic current transformer acquisition unit run more reliably resulted from theoretical analysis and field test.

1 Introduction

Electronic current transformer is one of the most important equipment in intelligent power substation. The output signal of acquisition Unit could be processed by the acquisition unit of it, such as conditioning effect, integral and sample output. The performance of acquisition unit determines the overall accuracy of the transformer, also determines its transient response and reliability. [1-4]

The VFTO (Very Fast Transient Over-voltage, VFTO) stemming from switch operations and ground faults could seriously affect the normal operation of the collection unit, which is the important guarantee of power grid security.

With the development of modern smart power grid and smart substation, electronic current transformers based on Rogowski coil are increasingly installed in GIS (Gas Insulated Switchgear) equipment. The VFTO and TEV (Transient Enclosure Voltage, TEV) and electromagnetic interference generated when the isolating switches are operated. Although the Rogowski coil of acquisition unit is equipped with the protection and filter circuit, it still be interfered by high common mode interference voltage and wide temperature. The efficiency of collecting device is greatly reduced because of the interference, affecting the operation condition of GIS equipment. [5-8]

To solve this key problem, this paper focuses on the working principle of ADI's differential amplifier AD8479, and the design of the corresponding filter circuit and power supply are stated. The circuit designed in this paper can effectively suppress interference and ensure the efficient and stable operation of the equipment.

2 Electronic current transformer based on Rogowski coil

Current transformer uses Rogowski coil as a current sensing unit, which is hollow coils of a special structure. Because it has lots of good features such as large measuring electric current pulse amplitude, frequency band width, no core saturation phenomenon, high linearity and good electrical insulation performance, it has been widely applied in the condition of power system fault transient and pulse power technology application field.

The Rogowski coil is coiled evenly around the ring section of nonmagnetic skeleton. The hollow inductor coil is made. It can efficiently isolate high voltage circuits. Its working principle is shown as Fig.1. the e(t) is the secondary side induction voltage, that is the differential voltage of the output signal. The output voltage is proportional to the rate of change of the measured current. [9-13]

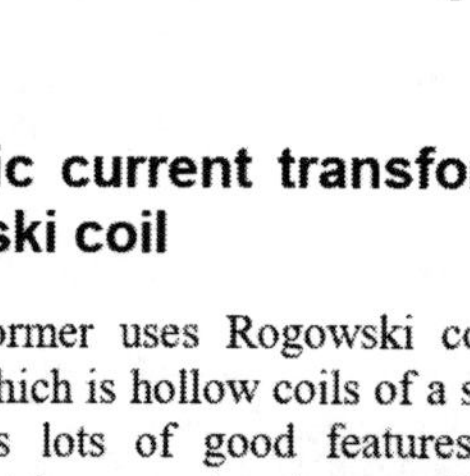

Figure 1. Principle of Rogowski coil

$$e(t) = -M\frac{di}{dt} \tag{1}$$

In (1), M is the mutual inductance between the bus and the coil, the 'i' is the bus-bar current

3 Differential amplifier AD8479

Extracting the weak signal from the large common mode noise source, instrument amplifier are used making of two or three operational amplifiers, although the instrument amplifier has excellent common mode rejection ratio, but it requires that the input voltage range is always smaller than the power supply voltage, and cannot cope with the condition of the signal source greater than the power supply voltage or the signal superimposed on the high common mode voltage. And high common mode voltage precision differential amplifier AD8479 Apply to it, it is a precision differential amplifier, it has high input common-mode voltage range, accurate measurement difference signal can be obtained in the highest ±600V high common-mode voltage, it also could provide the highest ±600V to input common-mode or difference of transient voltage protection. The following characteristics are its advantages: low dissonance voltage, low offset voltage drift, low gain drift, low co-mode rejection, and excellent CMRR (Common Mode Rejection Ratio, CMRR) in a wide frequency range. The internal principle of the AD8479 chip is shown below Fig.2.

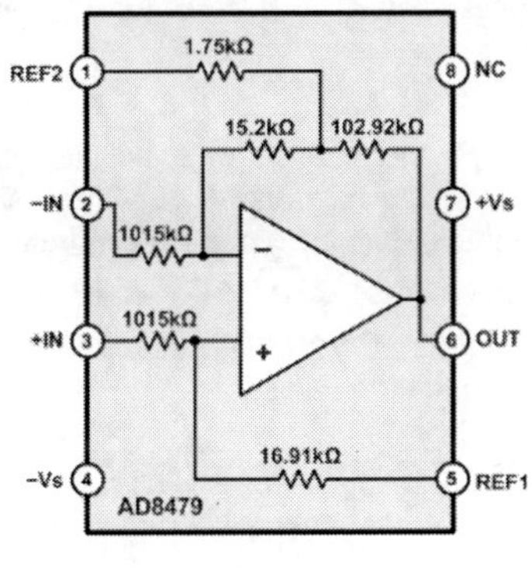

Figure 2. Schematic diagram of AD8479

4 CMRR of AD847

When the isolation switch in GIS equipment is operated, the VFTO has a high value and can even reach the base value of 3 times. The Wave-front steep time is low to several ns, frequency is very high. TEV and electromagnetic interference can affect the measurement of the output signal of Rogowski coil

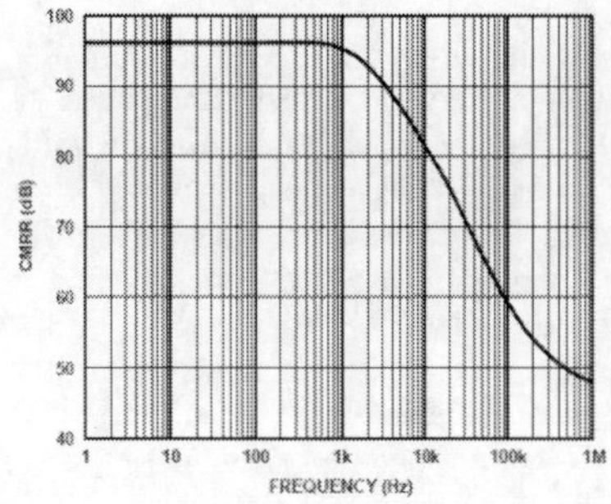

Figure 3. The relationship between CMRR and frequency of AD8479

In the high frequency region of the Fig.3, the performance of the CMRR of AD8479 decreases sharply with the increase of frequency. At this time, the CMRR performance of AD8479 can be significantly improved by the co-mode choke coil in the secondary output side of the Rogowski coil. The common mode choke can suppress the common mode noise and inhibit the noise. In order to further restrict the bandwidth of the differential amplifier signals including difference mode and common mode, the RC filter circuit is built between the tube feet of 2 and 3 in AD8479, as shown in figure 4 below.

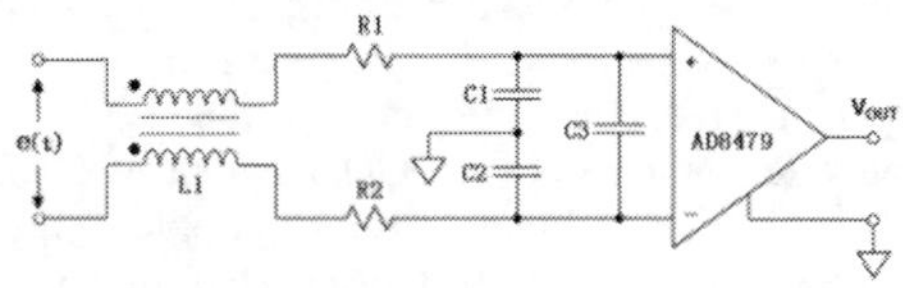

Figure 4. RC filter circuit

In the Fig.4, L1 is the common mode choke coil of RF. For the equilibrium of the differential arm of AD8479, R1= R2, C1=C2 in general. So -3db Common mode bandwidth BW_{CM} is:

$$BW_{CM} = \frac{1}{2\pi * R1 * C1} \tag{2}$$

-3db Difference bandwidth BW_{DIFF} is:

$$BW_{DIFF} = \frac{1}{2\pi * R1(2C3 + C1)} \tag{3}$$

Need of special note is, RF common-mode choke coil at high frequencies, such as in a few MHz to produce apparent impedance, the effect of L1 on the bandwidth of RC filter is not considering in the above (2) and (3). The precise resistances with a temperature coefficient of 15ppm and 0.1% accuracy are selected to be R1 and R2In the wide temperature range, because the deviation of the differential output of AD8479 is caused by the non-synchronous drift of R1 and R2, to ensure the consistency of R1 and R2, R1 and R2 are encapsulated together as a module. The capacitor parameters and the stable NP0 patch capacitors are selected to be C1, C2 and C3.

5 AD8479 power system design

The design of the differential amplifier AD8479 power supply system is very important. If the design is not properly designed, the input common mode voltage will be reduced, the output signal bandwidth and the pendulum will be affected, and the performance will decrease. As shown in fig.5, the common mode voltage of AD8479 can reach 600V with the two power supplies of ±15V. In the case of plus or minus 5V, the common mode voltage is less than 300V. When only +5V power is supplied, its common mode voltage is less than 230V.

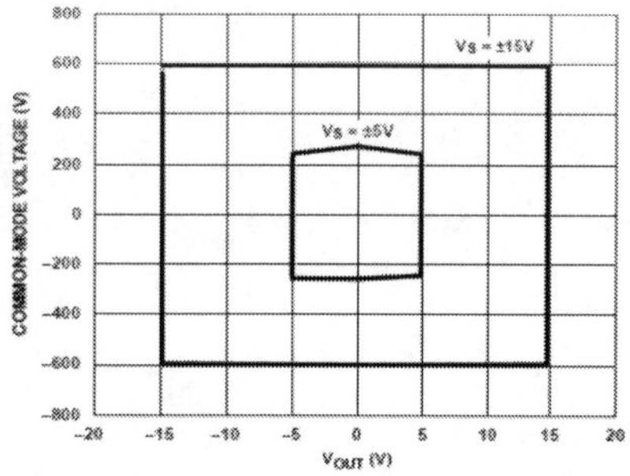

Figure 5. The relationship between the common mode voltage and the output voltage

The power pin effecting on the input terminal of AD8479 could not be ignored. Any noise and coupling that exists on the power line may have an effect on the output signal. As shown in figure 6, when PSRR (Power Supply Rejection Ratio, PSRR) is equal to 20dB and 300k Hz. PSRR is almost zero. The ripple or noise on the power supply is directly reflected in the output terminal. Therefore, in this paper, the application of plus or minus 15V DC/DC power supply chip is done. And the low dropout regulator is designed after the output end. It can provide AD8479 with relatively clean power supply.

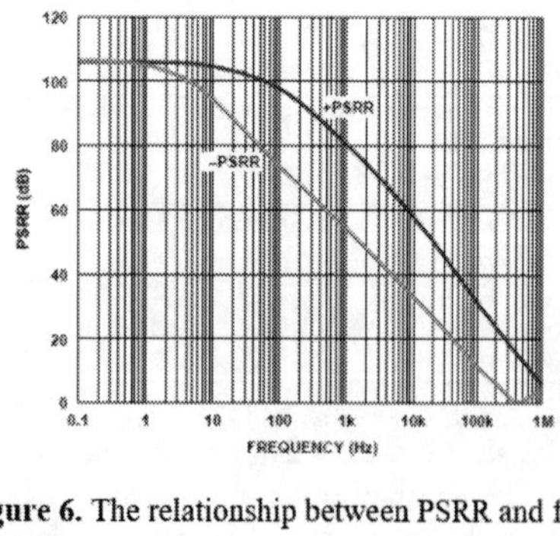

Figure 6. The relationship between PSRR and frequency

6 Noise analysis

A In the high common mode voltage, AD8479 detects the small signal of the electronic current transformer output based on Rogowski coil. There is a two-order low-pass passive filter between the output of AD8479 and the ADC Successive approximation chip AD7606 of 16-bit in Fig.7. [14-18]

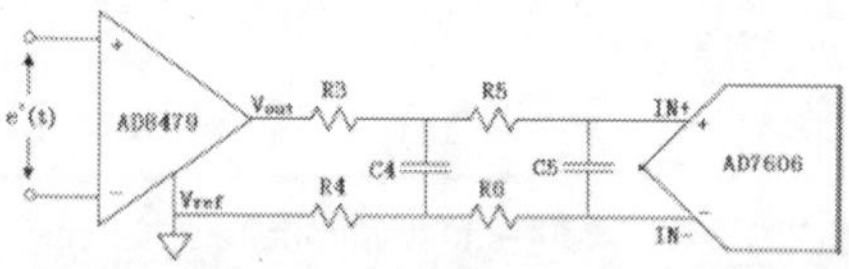

Figure 7. AD7606 Drive circuit diagram

In practical application, the equivalent relationship between the -3db cut-off frequency of the filter f_{-3dB} and the effective noise bandwidth of the loop BW_n can be obtained by the proper selection of the components of R3, R4, R5, R6, and C4 and C5, showed as below:

$$BW_n = f_{-3dB} * k_n \tag{4}$$

k_n is the noise bandwidth conversion factor.

f_{-3dB}=2kHz; k_n=1.22 because the filter has 2 poles.

The noise analysis of the AD8479 application circuit consists of three parts: voltage noise, current noise and resistance noise. The calculation formula is as follows:

$$E_{n\_in} = \sqrt{E_{n\_v}^2 + E_{n\_i}^2 + E_{n\_r}^2} \tag{5}$$

E_{n_in} is input noise;

E_{n_v} is voltage noise;

E_{n_i} is current noise;

E_{n_r} is resistance noise.

Since the Gain of AD8479 is fixed to 1, the RMS noise at the output terminal is $E_{n\_out} = E_{n\_in}$:

$$E_{n\_out} = E_{n\_in} * Gain \tag{6}$$

The estimated peak noise of the output about is $E_{n\_out\_pp}$:

$$E_{n\_out\_pp} = E_{n\_out} * 6 \tag{7}$$

Through the analysis of Fig.4 and Fig.7, combined with the voltage spectrum density curve shown in Fig.8, the formula (4), (5), (6) and (7) can be used to estimate the output $E_{n\_out\_pp}$ of AD8479 about is 30 uV_{p-p}. The application of electronic current transformer can be fully satisfied in theory.

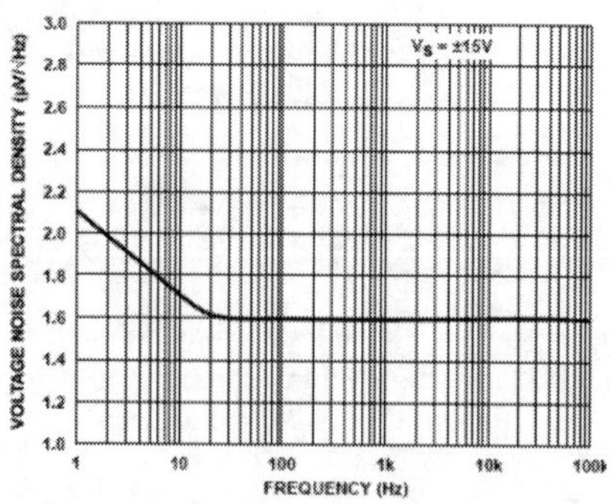

Figure 8. The relationship between voltage spectrum density and frequency

7 Experiment demonstrates

Taking the product of Guodian Nanjing Automation CO., Ltd in Xi 'an High Voltage Apparatus Research institute

Co,. Ltd, testing differential amplifier chip AD8479 in 363 kV high voltage intelligent GIS disconnector opening bus charging current test process, although acquisition cell interference signal port voltage was up to 10 kV, frequency reaches 1-10 MHz, acquisition unit could still work normally.

From the Fig.9, the recording of charging current test waveforms showed the reliability. Channel 1 and 2 represented the bus voltage in the closed isolation switch, the voltage waveform was the step shape, which was the typical VFTO waveform; Channel 3 and 4 were the waveform of protection current (that was, the output of Rogowski coil). The noise was effectively suppressed, and the signal waveform collected by AD8479 was not distorted and has no large value, which conforms to the requirements of the test specification. Channel 5 mean measuring coil current waveform.

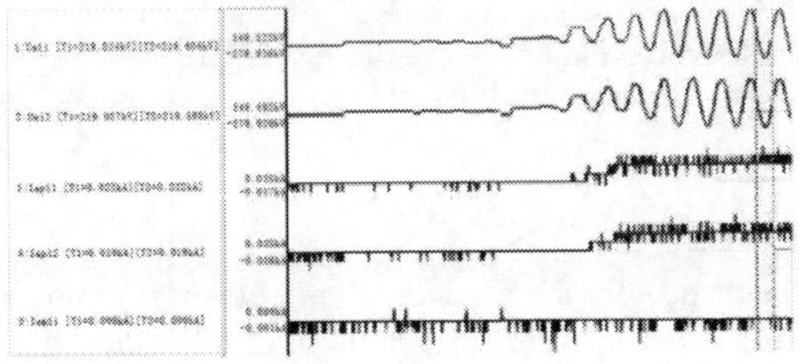

Figure 9. Testing waveforms

8 Conclusion

As the core part of electronic current transformer, the normal working state of acquisition unit was the basic guarantee of stable and reliable operation of the whole power system. The method of using the differential amplifier AD8479 chip to extract the small signal from the secondary side of the differential amplifier AD8479 chip was adopted in the paper. It proved that this kind of application completely meets the requirements of electromagnetic interference, precision, transient response and so on.

References

1. C. Y. Zhang, C. X. Zhang, X. X. Wang, et al. Signal processing system for digital closed-loop fiberoptic current sensor[J]. Proceedings of the CSEE. 2009, 29(30), pp, 42-46.
2. J. Q Gao, X. G Guo. Discussion on intelligent electric power equipment[J]. High Voltage Apparatus, 2010, 46(12), pp, 1-4.
3. Z. H. Xiao. Study and comment of the optical transformers in power system[J]. Power System Protection and Control, 2014, 42(12): 148-154.
4. H. H. Ye, P. Xu, H. L. Zong, et al. The configuration and switch over of voltage transformer in digital substation[J]. Automation of Electric Power Systems, 2008, 32(24), pp,93-95.
5. Y. M. Li, T. F. Zheng, S. B. Zhou, et al. A realization solution of key links for merging unit in smart substation[J]. Automation of Electric Power Systems. 2013, 37(11), pp, 93-98.
6. H. Z. Wang. Research on the factors of disturbance magnetic field of improved air-core coil current transformer[J]. Development & Innovation of Machinery & Electrical Products. 2009, 22(3), pp, 166-168.
7. H. L. Zhang. Study and desgin on the Rogowski coil[D]. Qinhuangdao: Yanshan University, 2006.
8. W. B. Li, C. X. Mao, J. M. Lu, et al. Study of the influence of the distributed capacitance on dynamic property of Rogowski coil[J]. Transactions of China Electrotechnical Society, 2004, 19(6), pp, 12-17.
9. R. Wu, Makinwa, K. A. Huijsing. A chopper current feedback instrumentation amplifier with a 1MHz noise corner and an AC-coupled ripple reduction loop[J]. IEEE Journal of Solid-State Circuits. 2009, 44(12), pp, 3232-3243.
10. Y. H. Lam, W. H. Ki. CMOS bandgap references with self-biased symmetrically matched current–voltage mirror and extension of sub-1-V design[J]. IEEE Transactions on Very Large Scale Integration (VLSI) Systems, 2010, 18(6), pp, 857-865.
11. S. B. Liang. Measuring methods of composlte errors on current transformer[J]. Relay, 2005, 33(18), pp, 79-83.
12. K. Xiang. Study on digital output interface of electronic transformer[D]. Wuhan, China, Huazhong Universtity of Science and Technology, 2006.
13. E. Qiao, Z. P. An, C. M. Luo. Rogowski coil used in optical current transformer[J]. Transformer, 2000, 37(5), pp, 7-22.
14. B. P. Lathi. Linear system and signals[M]. Carmicheal, California, USA : Berkeley-Cambridge Press, 1992.
15. W. C. Xie, Y. X. Dai. Technology research of integrator for Rogowski electronic current transducer[J]. Electrical Measurement & Instrumentation, 2011, 48(5), pp, 10-13.
16. H. Wen, Z. S. Teng, S. Y. Guo. Triangular self-convolution window with desirable sidelobe behaviors for harmonic analysis of power system[J]. IEEE Transactions on Instrumentation and Measurement, 2010, 59(3), pp, 543-552.
17. H. M. Wang, S. X. Zheng, F. Liu. Optimization design for Rogowski coil of high voltage electronic current transducer[J]. High Voltage Engineering, 2005, 31(9), pp, 5-7.
18. Z. L. Long. Design and performance of miniature Rogowski coil for heavy current measurement[J]. High Voltage Engineering, 2007, 33(7), pp, 79-84.

Prescribed Performance Control for Two-axis Optronic Stabilized Platform

WANG Yexing[1], LUO Changxin[1], ZHANG Tao[1], LIU Yaning[1] and XIAO Chuhan[1]

[1]Air and Missile Defense College, Air Force Engineering University, 710051 Xi'an, China

Abstract. Aiming at improving the tracking and stabilizing performance of two-axis optronic stabilized platform with Stribeck friction and uncertain velocity disturbance, a prescribed performance control strategy with unknown initial errors is designed. By designing a new performance function, the limit of traditional prescribed control that the initial error has to be known accurately is broken through. The strategy possesses strong robustness against unknown disturbance, and the state error is restrained to a predefined arbitrary small residual. It is guaranteed that the closed-loop system is uniformly ultimately bounded. The simulation results demonstrate the effectiveness of proposed strategy.

1 Introduction

Optronic stabilized platform(OSP) is a kind of high-precision servo tracking system, which is usually mounted on a mobile carrier for stable tracking of moving targets. Optical tracking system is widely used in robot self-balance system[1], unmanned aerial vehicle[2], airborne aiming system [3] and non-strapdown seeker servo system[4].The control object of the optical tracking system is the optical tracking platform. The main factors that affect the tracking performance are the carrier speed interference and the friction torque between the shaft systems[5][6].Two difficulties when dealing with the improvement of the system performance. One is the interference suppression in order to ensure tracking stability and the other is the system response and accuracy.How to effectively suppress interference in order to ensure the stability of tracking, and how to ensure that the system quickly and accurately track the target are two important aspects to improve system performance.

To realize the system performance with fast dynamic response and high-precision control of OSP, researches have tried varieties of approaches. For classical control structure, the proportional-integral-derivative control has been used to many systems [7], however, the robustness can't be guaranteed. To enhance the robustness of the system, H_∞ control is introduced to eliminate the sensitivity for disturbance, however, the robustness conflicts with its performance sharply[8]. Furthermore, sliding mode control is also used to deal with the non-linearity of OSP systems [9], it enables all system states with arbitrary values to converge to user-specified surface, while its chattering problem is an unavoidable drawback. Observer-based control is also used to eliminate disturbance and gets better performance [10]. Disturbance observer achieves high-precision estimation of interference, while the parameters have to be adjusted intricately. In reference [11], the adaptive RBFNN was designed to estimate and compensate for disturbances to improve the control performance of OSP. However, its control performance was easily affected by the selected upper bound of residual approximation error, and the problem of calculating delay have not been solved yet.

A common characteristic of the aforementioned researches is that the transient and steady performance cannot be prescribed, methods mentioned can only guarantee the convergence of error to a residual set. In this case, a prescribed performance control(PPC) is proposed in[12],based on PPC, it's easy to design a prescribed performance bound to acquire desired transient and steady performance[13]. So far, there has not been application of PPC to OSP.

In this paper, PPC is utilized to improve the performance of OSP in the presence of unknown model perturbation and external disturbance. A new performance function is designed to break through the limitation of knowing the initial errors accurately. The designed scheme does not reside to the information of disturbance, which simplify the controller design and increase the practicality of the algorithm.

The paper is organised as follows:

In Section 2, the OSP model and preliminaries are presented. In Section 3, prescribed performance controller is designed along with the stability analysis. In Section 4, simulations are carried out to demonstrate the effectiveness.

2 Problem formulation

2.1 Constitution of two-axis optronic stabilized platform

Fig.1 shows the schematic diagram of two-axis stabilized platform. We can see that stabilized platform consists of two gimbals, which are pitch gimbal and yaw gimbal. From Fig.1 we can see the relationships between two gimbals: gyroscopes measuring the angular rate of pitch and yaw gimbal, angel sensors measuring the angle of pitch and yaw gimbal, moment sensors measuring the moment of pitch and yaw motor.

The system is driven by two servo motors, the sensor such as imaging sensor is placed in the inner frame.

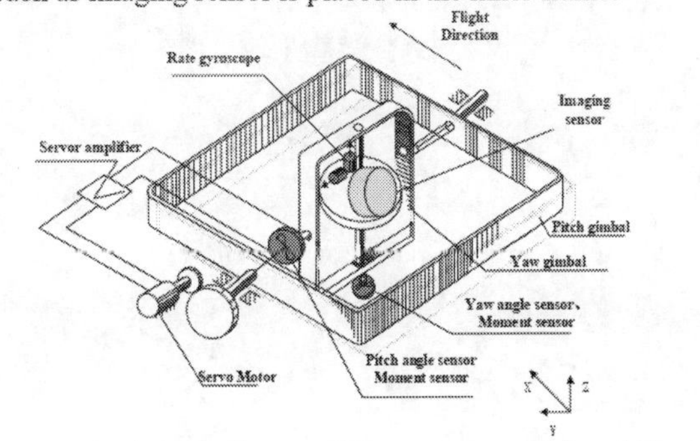

Figure 1. Schematic diagram of the two-axis optronic stabilized platform

2.2 Dynamic model of the plant

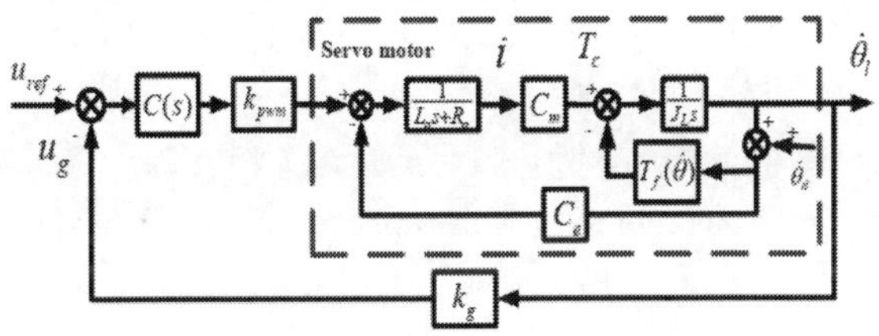

Figure 2. Stabilized-loop block diagram

Fig.2 shows the pitch control block of stabilized platform. Where the block within the red imaginary line stands for the servo motor; $C(s)$ is the controller of stabilized loop; k_{PWM} is the power amplifier coefficient; $T_l(\dot{\theta})$ is the moment of friction; $\dot{\theta}_l$ is the angular rate of stabilized platform in the inertial space; $\dot{\theta}_d$ is the perturbation of angular rate; $T_f(\dot{\theta})$ is the friction moment; k_g is the simplified transfer function of rate gyroscope, cascade coefficient $1/k_g$ makes it a unit feedback system; T_c is the moment output of servo motor.

Combined with the dynamic equation of stable platform and the dynamic equation of motor, mathematical model is acquired as follows [14]:

$$\begin{cases} \dot{x}_1 = \dfrac{x_2}{J_L} - \dfrac{F_f(x_1+\dot{\theta}_d)}{J_L} \\ \dot{x}_2 = -\dfrac{C_m C_e}{L_a}x_1 - \dfrac{R_a}{L_a}x_2 + \dfrac{C_m k_{PWM}}{L_a}u - \dfrac{C_e C_m \dot{\theta}_d}{L_a} \end{cases} \tag{1}$$

Where x_1 , x_2 are state variables, they represent $\dot{\theta}_l$ and T_c respectively. u is the input of controller.

The friction moment is modified as Stribeck friction model as Fig.3[15].

$$F_f(v) = [F_c + (F_s - F_c)e^{-(\frac{v}{v_s})^2}]\mathrm{sgn}(v) + B_v v \tag{2}$$

Where F_c , F_s are coulomb friction and static friction force respectively.

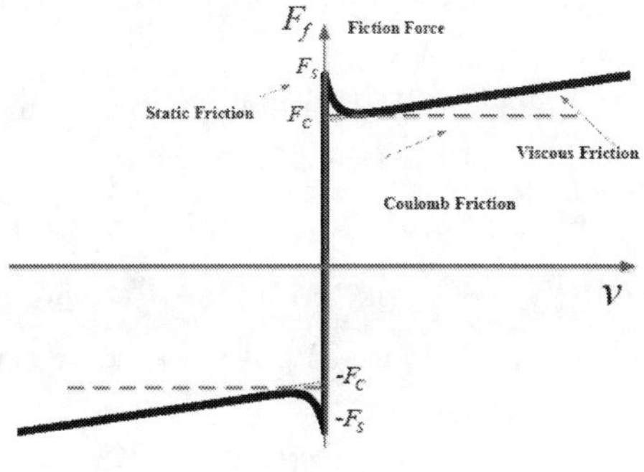

Figure 3. Friction force with Stribeck effect

2.3 Control problems for the stabilized platform

There are some troublesome characteristics in the stabilized platform:

1) It is hard to establish accurate mathematical model of the stabilized platform due to its coefficient uncertainty and perturbation;
2) The strong nonlinearity of friction model at velocity zero point may cause dead-area phenomenon of velocity tracking;
3) When there exists unknown external velocity disturbances and internal friction disturbances, it is hard to guarantee the transient and steady state performance.

3 Controller design

3.1 Prescribed performance

In this subsection, some preliminaries of prescribed performance are summarized. And a new performance function is designed.

Definition 1. [16] A smooth function $\hbar(t): \mathbb{R}_+ \to \mathbb{R}_+$ will be called a performance function if it satisfies:

(1) $\hbar(t)$ is positive and decreasing;

(2) $\lim_{t\to+\infty} \hbar(t) = \hbar_\infty > 0$.

According to Definition 1, a novel performance function is formulated as:

$$\hbar_i(t) = \operatorname{csch}(\kappa_i t + \theta_i) + \hbar_{i\infty} \quad (3)$$

With κ_i, θ_i and $\hbar_{i\infty}$ chosen positive constants appropriately, where $i = 1, 2, \ldots N$. Then $\hbar_i(t)$ satisfies:

(1) $\hbar_i(t) = \operatorname{csch}(\kappa_i t + \theta_i) + \hbar_{i\infty} = \dfrac{2}{e^{t_i} - 1} + \hbar_{i\infty} > \hbar_{i\infty}$

(2) $\lim_{\theta_i \to +\infty} \hbar_i(0) \to +\infty$; (3) $\lim_{t \to +\infty} \hbar_i(t) = \hbar_{i\infty}$.

Consider tracking error $e_i(t)$, prescribed performance is achieved if the following inequality is satisfied.

$$-\underline{\delta}_i \hbar_i(t) < e_i(t) < \overline{\delta}_i \hbar_i(t) \quad (4)$$

Where $\underline{\delta}_i, \overline{\delta}_i$ are positive constants associated with $e_i(t)$, $i = 1, 2, \ldots, N$.

The performance bound is constrained by(4), $e_i(t)$ is shown in Fig.4. When θ_i is chosen adequately small, and $\underline{\delta}_i, \overline{\delta}_i$ are positive, we can easily get $-\underline{\delta}_i \hbar_i(0) \to -\infty$ and $\overline{\delta}_i \hbar_i(0) \to +\infty$. So, for a bounded $e_i(0)$ and an adequately small θ_i, the following inequality is acquired.

$$-\underline{\delta}_i \hbar_i(0) < e_i(0) < \overline{\delta}_i \hbar_i(0) \quad (5)$$

Appropriately, with the aforesaid proposed performance function(3), there is no need to get accurate initial error $e_i(0)$.

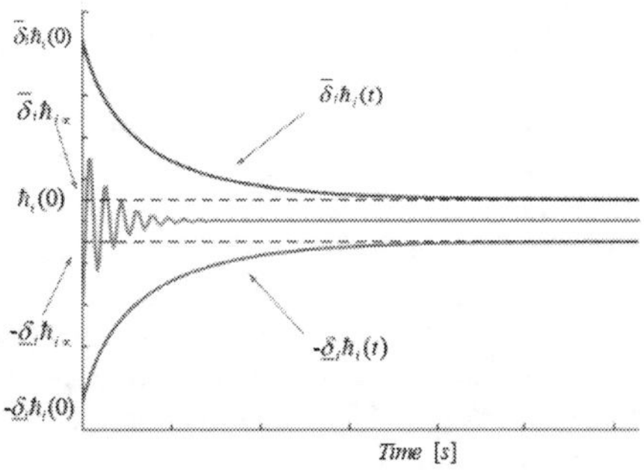

Figure 4. The PPB on the error $e_i(t)$

Remark1

1. Provided that $-\underline{\delta}_i \hbar_i(0) < e_i(0) < \overline{\delta}_i \hbar_i(0)$, $\underline{\delta}_i \hbar_{i\infty}$, $\overline{\delta}_i \hbar_{i\infty}$ determine the maximum allowable size of the error and the steady state. The parameter κ_i domains the decreasing rate of $\hbar_i(t)$, which will domain the decreasing rate bound of $e_i(t)$ when the overshoot is no more than $\underline{\delta}_i \hbar_i(0)$, $\overline{\delta}_i \hbar_i(0)$. All parameters mentioned above can be arbitrary regulated.
2. The performance function applied in [17,18] is imposed by $\rho(t)$, the tracking error $e(t)$ is given as follows:

$$\begin{cases} -\delta_1 \hbar_i(t) < e(t) < 1, & if\ e(0) > 0 \\ -1 < e(t) < \delta_2 \hbar_i(t), & if\ e(0) < 0 \end{cases} \quad (6)$$

with $0 \le \delta_1 \le 1, 0 \le \delta_2 \le 1$. Appropriately, the behavior bounds are different with different initial value of the error. We can also see that the controller design and stabilized analysis are supposed to differ because of different initial value. Additionally, acquirement of the $e(0)$ adds the difficulty of algorithm applying.

Considering that it is unable to devise controller directly based on q.(4), an error transfer function $\mu(\varepsilon_i)$ is introduced to transfer the "constrained" system into equivalent "unconstrained" system.

$$e_i(t) = \mu_i(\varepsilon_i)\hbar_i(t) \quad (7)$$

With

$$\mu(\varepsilon_i) = \frac{\overline{\delta}_i e^{\varepsilon_i} - \underline{\delta}_i e^{-\varepsilon_i}}{e^{\varepsilon_i} + e^{-\varepsilon_i}} \quad (8)$$

Where $\varepsilon_i, i = 1, 2, \ldots, N$ donates transfer error.

$\mu(\varepsilon_i)$ satisfies:

(1) $\mu(\varepsilon_i)$ is smooth and strictly increasing;

(2) $\lim_{\varepsilon_i \to -\infty} \mu_i(\varepsilon_i) = -\underline{\delta}_i$, $\lim_{\varepsilon_i \to +\infty} \mu_i(\varepsilon_i) = \overline{\delta}_i$.

Based on the properties of $\mu(\varepsilon_i)$, original error is transferred into

$$-\underline{\delta}_i < \mu_i(t) < \overline{\delta}_i \quad (9)$$

If we multiply the above formulas, considering $\hbar_i(t)$ is positive, we can get

$$-\underline{\delta}_i \hbar_i(t) < e_i(t) < \overline{\delta}_i \hbar_i(t) \quad (10)$$

Appropriately, (4)is equivalent illustrated by Eq. (7).

Furthermore, since $\mu_i(\varepsilon_i)$ is strictly increasing as well as $\hbar_i(t) \ge \hbar_{i\infty} > 0$, we can get the inverse transformation (7).

$$\varepsilon_i(t) = \mu_i^{-1}(\varepsilon_i) = \frac{1}{2}\ln\left(\frac{e_i / \hbar_i + \underline{\delta}_i}{\overline{\delta}_i - e_i / \hbar_i}\right) \quad (11)$$

And

$$\dot{\varepsilon}_i(t) = r_i\left(\dot{e}_i - \frac{\dot{\hbar}_i}{\hbar_i} e_i\right) \quad (12)$$

With

$$r_i = \frac{1}{2}\left(\frac{1}{e_i / \hbar_i + \underline{\delta}_i} - \frac{1}{e_i / \hbar_i - \overline{\delta}_i}\right) > 0 \quad (13)$$

$$\dot{\hbar}_i(t) = \kappa_i - \kappa_i[\coth(\kappa_i t + \theta_i)]^2$$

3.2 Design of prescribed performance controller

Assumption 1. The desired trajectory $x_d(t)$ is limited and unknown, and $\dot{x}_d(t)$ is limited.

The following control scheme is defined

$$\begin{aligned}
&\mu_2(t)=\frac{\tilde{x}_1}{\hbar_2(t)}\\
&a_2=-k_1\varepsilon_1\\
&\mu_1(t)=\frac{\tilde{x}_2}{\hbar_1(t)}\\
&a_2=-k_2\varepsilon_2\\
&u=a_3
\end{aligned} \tag{14}$$

with

$$\begin{cases}
\tilde{x}_1=\tilde{\theta}_l=\dot{\theta}_l-\dot{\theta}_d=x_1-x_d\\
\tilde{x}_2=\tilde{T}_c=T_c-T_d=x_2-a_2\\
\varepsilon_1=\ln(\dfrac{\mu_1(t)+\underline{\delta_1}}{\overline{\delta}_1-\mu_1(t)})\\
\varepsilon_2=\ln(\dfrac{\mu_2(t)+\underline{\delta_2}}{\overline{\delta}_2-\mu_2(t)})\\
\hbar_1(t)=\operatorname{csch}(k_3t+\theta_1)+\hbar_{1\infty}\\
\hbar_2(t)=\operatorname{csch}(k_4t+\theta_2)+\hbar_{\infty}
\end{cases} \tag{15}$$

Where $k_1>0$, $k_2>0$ are design constants of proportion controlling coefficient, u is the controller input.

Through inspection of Eqs. (14)-(15), it is easy to find that the control law doesn't need any derivatives of x_i and a_i, so the controller design and implementation are greatly simplified.

3.3 Stabilized analysis

Assumption 2. There is a bounded constant B_f ,satisfying

$$\frac{\partial F_f(x_1)}{\partial x_1}\ge B_f$$

Assumption 3. The velocity disturbance Δ is limited.

Theorem 1. Consider the closed-loop system consisting of stabilized platform with controllerEq.(14), all the signals involved are uniformly ultimately bounded.

Choose the following Lyapunov function as

$$V=V_1+V_2$$

with

$$V_1=\frac{1}{2}\varepsilon_1^2,\quad V_2=\frac{1}{2}\varepsilon_2^2$$

Substituting time derivative of V into Eq. (11), we can get

$$\begin{aligned}
\dot{V}&=\dot{V}_1+\dot{V}_2\\
&=\varepsilon_1\dot{\varepsilon}_1+\varepsilon_2\dot{\varepsilon}_2\\
&=\varepsilon_1\cdot\frac{2\dot{\mu}_1(t)}{1-\mu_1^2(t)}+\varepsilon_2\cdot\frac{2\dot{\mu}_2(t)}{1-\mu_2^2(t)}
\end{aligned} \tag{16}$$

where

$$\dot{\mu}_1(t)=\frac{\dot{\tilde{x}}_1\cdot\hbar_1(t)-\tilde{x}_1\cdot\dot{\hbar}_1(t)}{\hbar_1^2(t)} \tag{17}$$

Combining with Eq.(1), we can get

$$\dot{\mu}_1(t)=\frac{x_2/J_L-F_f(x_1+\Delta)-\dot{x}_{1d}-\dot{\hbar}_1(t)\cdot\mu_1(t)}{\hbar_1(t)} \tag{18}$$

Furthermore, with (18) we have

$$\begin{aligned}
\dot{V}_1=&\frac{2\varepsilon_1(t)((-k_1\varepsilon_1+\hbar_2(t)\cdot\mu_2(t))/J_L)}{\hbar_1(t)(1-\mu_1^2(t))}\\
&\frac{-F_f(x_1+\Delta)-\dot{x}_{1d}-\dot{\hbar}_1(t)\cdot\mu_1(t)}{\hbar_1(t)(1-\mu_1^2(t))}
\end{aligned} \tag{19}$$

According to Assumption 3, and within the metabolic range of $x_1, F_f(x_1+\Delta)$ is limited. $\dot{x}_{1d}, \dot{\hbar}_1(t), \hbar_2(t), \mu_2(t)$ are limited too.

Define a limited constant F_1 , it is obvious that

$$\left|\hbar_2(t)\cdot\mu_2(t)/J_L\text{-}F_f(x_1+\Delta)-\dot{x}_{1d}-\dot{\hbar}_1(t)\cdot\mu_1(t)\right|\le F_1 \tag{20}$$

Moreover,

$$\begin{aligned}
\dot{V}_1\le&\frac{2|\varepsilon_1(t)|\cdot\bar{F}_1}{\hbar_1(t)(1-\mu_1^2(t))}\text{-}\frac{2\varepsilon_1^2(t)\cdot k_1}{\hbar_1(t)(1-\mu_1^2(t))}\\
=&\frac{2|\varepsilon_1(t)|(\bar{F}_1-k_1|\varepsilon_1(t)|/J_l)}{\hbar_1(t)(1-\mu_1^2(t))}
\end{aligned} \tag{21}$$

Define

$$\Omega_{\varepsilon_1}=\left\{\varepsilon_1\,\middle|\,|\varepsilon_1(t)|\le\frac{J_l\bar{F}_1}{k_1}\right\} \tag{22}$$

It is obvious that $\dot{V}_1$ will be negative if $\varepsilon_1\notin\Omega_{\varepsilon_1}$.

Thus, e_1 is bounded.

Similarly, combined with Eq. (1) and Eq. (14),the following equation is acquired.

$$\begin{aligned}
\dot{\mu}_2(t)=&\frac{-\dfrac{C_mC_e}{L_a}x_1-\dfrac{R_a}{L_a}x_2+\dfrac{C_mk_{PWM}}{L_a}u}{\hbar_1(t)}\\
&\frac{-\dfrac{C_eC_m}{L_a}\Delta-\dot{x}_{2d}-\dot{\hbar}_2(t)\cdot\mu_2(t)}{\hbar_1(t)}
\end{aligned} \tag{23}$$

Combining with Eq. (14) and Eq. (23),We can get

$$\begin{aligned}
\dot{V}_2=&\frac{2\varepsilon_2(t)(-\dfrac{C_eC_m}{L_a}x_1-\dfrac{R_a}{L_a}x_2-\dfrac{C_eC_m}{L_a}\Delta}{\hbar_2(t)(1-\mu_2^2(t))}\\
&\frac{-k_1\dot{\varepsilon}_1-\dfrac{C_mk_{PWM}}{L_a}k_2\varepsilon_2-\dot{\hbar}_2(t)\cdot\mu_2(t))}{\hbar_2(t)(1-\mu_2^2(t))}
\end{aligned} \tag{24}$$

Table 1. Model Parameter.

Symbol	J_L	C_m	L_a	R_a	k_{PWM}	C_e	F_C	F_S	B_V	v_s
Parameter	0.0012	0.625	0.062	5.1	3.75	0.75	0.2	0.3	0.017	0.2

According to Eq.(21), x_1 is limited. Based on the Singular value theory, x_2 is limited. Considering that Δ , $\dot{\hbar}_2(t)$ are limited, invoking Eq. (12), $\dot{\varepsilon}_1$ is limited.

Define a limited constant F_2 , it is obvious that

$$\left|-\frac{C_eC_m}{L_a}x_1-\frac{R_a}{L_a}x_2-\frac{C_eC_m}{L_a}-k_1\dot{\varepsilon}_1-\Delta\dot{\hbar}_2(t)\cdot\mu_2(t)\right|\le F_2 \quad (25)$$

Moreover,

$$\dot{V}_1\le\frac{2|\varepsilon_2(t)|\cdot\bar{F}_2}{\hbar_2(t)(1-\mu_2^2(t))}-\frac{2C_mk_{PWM}\cdot k_2\varepsilon_2^2(t)}{L_a\hbar_2(t)(1-\mu_2^2(t))}$$
$$=\frac{2|\varepsilon_2(t)|(\bar{F}_2-|\varepsilon_2(t)|C_mk_{PWM}\cdot k_2/L_a)}{\hbar_2(t)(1-\mu_2^2(t))} \quad (26)$$

$$\text{Define }\Omega_{\varepsilon_2}=\left\{\varepsilon_2\,\middle|\,|\varepsilon_2(t)|\le\frac{L_a\bar{F}_2}{C_mk_{PWM}\cdot k_2}\right\} \quad (27)$$

It is obvious that $\dot{V}_2$ will be negative if $\varepsilon_2\notin\Omega_{\varepsilon_2}$.

Thus, e_2 is bounded.

Thus, all the signals involved are uniformly ultimately bounded. This is the end of proof.

Remark2. By Theorem 1, the bound of ε_1 and ε_2 are achieved. So we can conclude that the velocity tracking error and the moment tracking error can converge to the prescribed performance bound.

4 Simulation results

In this section, simulations are carried out to verify the effectiveness of the proposed control scheme for stabilized platform. The plant is stabilized platform of seeker.

The simulation step size is chosen as $T=0.001s$.

The parameters of the control system are given in Table 1.

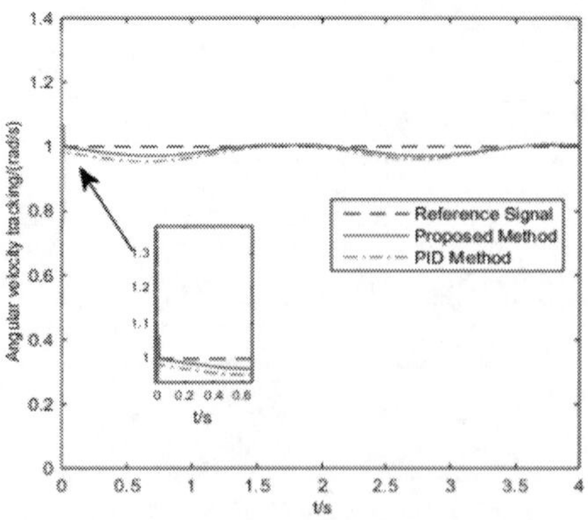

Figure 5. Angular velocity tracking response

Choosing step signal as the reference signal, Fig.5 shows the response of angular velocity tracking with the velocity disturbance chosen as $\sin(3t-0.5)(rad/s)$. From Fig.5, we can see that the proposed method responds quickly, and it has smaller overshoot and smaller tracking error than PID method.

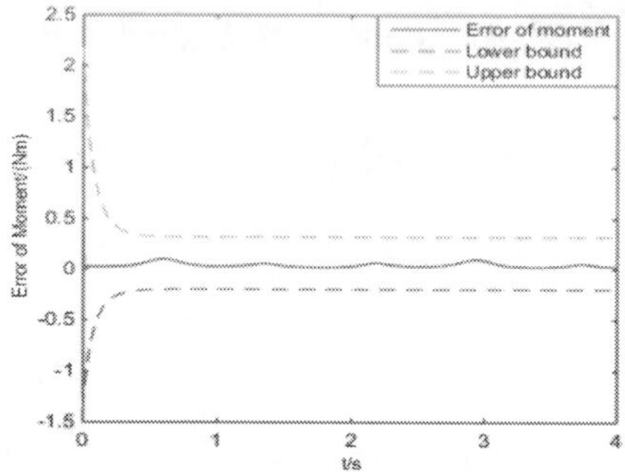

(a) Output error of moment with parameter perturbation

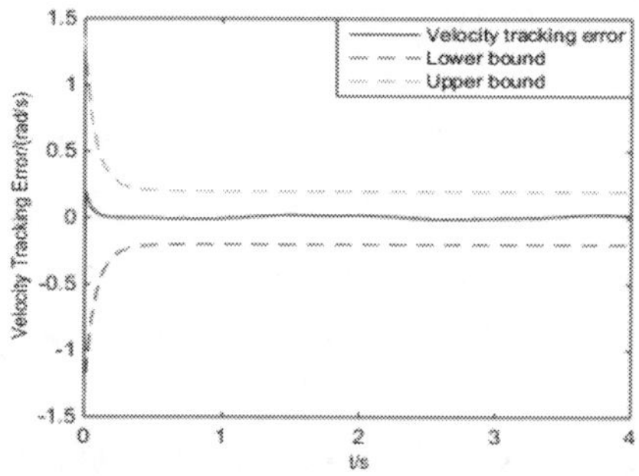

(b) Output error velocity with parameter perturbation

Figure 6. Response of system state variety with step signal

To verify the robustness of the proposed method, 50%, perturbation is applied to coefficient $\frac{1}{J_L},-\frac{C_mC_e}{L_a},-\frac{R_a}{L_a}$.

From Fig.6, we can see that the error of states converges to zero with acceptable accuracy, Under the condition of intense parameter perturbation and angular velocity disturbance, the steady deviation of the response is no more than 3%.

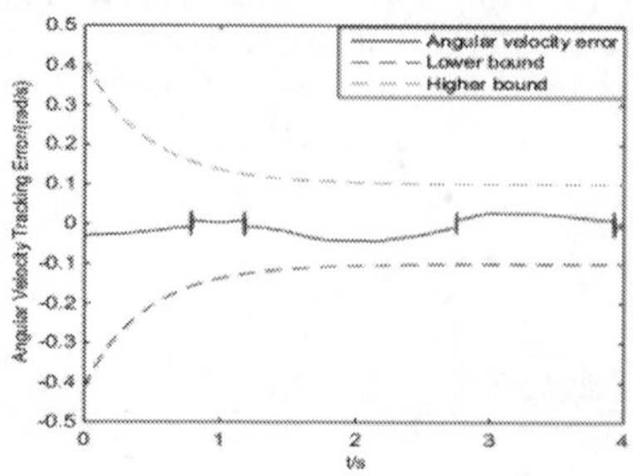

Figure 7. Response of system state variety of sine signal

To verify the control effect of the plant's angular velocity crossing the zero point frequently, sine signal is chosen as the reference signal, with the disturbance chosen as $\cos(3t)(rad/s)$, similarly, 50% parameter perturbation is introduced.

We can see from Fig.7 that the stabilized platform is capable of achieving prescribed bounds on the transient and steady state output error performance, through observation of Fig.7, it is easy to find that the system achieves more than 95% isolation of internal and external disturbance.

*Remark*3. The proposed controller doesn't reside to the construction of friction model or special analytical friction expressions, it has promoting value in engineering practice with its easy design structure and relaxed application condition.

5 Conclusion

In this paper, a prescribed performance controller, which allows transient performance and steady value of moment and angular velocity tracking errors to achieve prescribed bound, is addressed for stabilized platform with Stribeck friction. The designed performance function takes satisfying effect without acquiring the initial error. It's verified that the plant using the proposed method responses rapidly. And it achieves high-precision tracking with parameter perturbation and unknown disturbance. Besides, the phenomenon of speed dead-area is eliminated by the proposed method. Finally, because of its simple design structure and its loose applying condition, it has promoting value in engineering practice.

References

1. Haarhoff D, Kolditz M, Abel D, et al. Actuator design for stabilizing single tendon platforms. 23rd International Conference on Robotics, 46,371-378.(2017)
2. Stepaniak, MJ, de Haag, MU,Van Graas, F, Field programmable gate array-based attitude stabilization. Journal of aerospace computing information and communication. 6(7),451-463. (2009)
3. Song XR, Chen H,Xue YG, Stabilization precision control methods of photoelectric aim-stabilized system. 351,115-120. (2015)
4. Liu F, Wang H, Fuzzy PID controller for two-axis airborne optoelectronic stabilized platform. International journal of computing information and control, 13(4), 1307-1322.(2017)
5. Kennedy P J, Kennedy R L, Direct versus indirect line of sight stabilization. IEEE Transactions on Control System Technology, 11(8), 1293-1299. (2003)
6. Sun, H., Zhang, S. M.,Computation model and error budget for image motion of aerial imaging system. Electronics Optics& Control, 20(11), 2492-2499. (2012)
7. Zhang YS., Yang T., Li CY., Liu SS., Du CC., Liu M., et al. Fuzzy-PID control for the position loop of aerial inertially stabilized platform. Aerospace Sci Technology, 36, 21–26.(2014)
8. Martin R, Zdenek H, Structured MIMO design for dual-stage inertial stabilization: case study for HIFOO and Hinfstruct solvers. Mechatronics ,23(8),1084–1093.(2013)
9. Zou Y., Lei XS., A compound control method based on the adaptive neural network and sliding mode control for inertial stable platform. Neurocomputing,155,286–294.(2015)
10. ZhouXY., Yuan JQ., Cai TT., Dual-rate-loop control based on disturbance observer of angular acceleration for a three-axis aerial inertially stabilized platform. ISA Transactions, 63, 288-298. (2016)
11. Ekstrand B, Equations of motion for a two-axes gimbal system. IEEE TransAerospace Electron System, 37(3),1083–1091.(2001)
12. Bechlioulis C, Rovithakis G, Robust adaptive control of feedback linearizable MIMO nonlinear systems prescribed performance. IEEE Trans. Automation Control,53(9),2090-2099.(2008)
13. Zhao K, Song Y D, Ma T D, et al. Prescribed performance of uncertain Euler-Lagrange system subject to full-state constraints. IEEE Transactions on Neural Networks and Learning System, To be punished, doi: 10.1109/TNNLS.2017.2727223.
14. Zhang Z Y, Li Z Q, Zhou Q K, Application in prestiction friction compensation for angular velocity loop of inertially stabilized platforms. Chinese Journal of Aeronautics, 27(3),655-662. (2014)
15. Wang Y S, Wang Q J, Shi F H, Development of a set of Stribeck curves for conformal contacts of rough surfaces. Tribology& Lubrication Technology, 63(4),32-40.(2007)
16. Bechlioulis, Charalampos P, Rovithakis, George A, Prescribed performance adaptive control for multi-input multi-output affine in the control nonlinear systems. IEEE Transactions on automatic control, 55(5),1220-1226.(2010)
17. Li, D., Liu G P., Hua C C., Modeling and prescribed performance control for cold rolling mill HAGC system with measurement delay, Proceedings of the 36th Chinese Control Conference, 726-731. (2017)
18. A. Theodorakopoulos and G. Rovithakis, Prescribed performance control of strict-feedback system with hysteresis input nonlinearity. 2015 European Control Conference,3220-3225.(2015)

Application of Copper Cladding Aluminum Composites in UHV Portable Earthing and Short-circuiting Wires

Jianjun ZHU[1], Guangkai YU[2], TING Liu[2], KAI Liu[2], Xinglie LEI[2], Ziming SU[2], Linyue ZHU[3]

[1]State Grid Corporation of China，100031 Beijing，China.
[2]China Electric Power Research Institute，430074 Wuhan，China.
[3]Beijing NO.161 High School, Beijing 100031, China.

Abstract. Aiming at the heavy weight and inconvenience when carrying and installing copper earthing wires on the UHV transmission lines, in this paper, we present the use of copper clad aluminum(CCA) composite materials as a lightweight method for UHV earthing wire conductor. Theoretical calculations and tests of the fusing current in a short time for copper and CCA material are conducted. The results show that the theoretical value of the earthing wire conductor's fusing current corresponds with the test value on condition of the conductor cross section greater than $4mm^2$ as well as fusing time less than 1.5s. The CCA-10 earthing wires get 36.2% weight reduction compared with copper wires.

1 Introduction

UHV power transmission has the advantages of large transmission capacity, long distance of power transmission, low wire loss and less land occupation and other outstanding advantages. It is a kind of advanced power transmission technology with resource saving and environment friendliness. With the rapid development of China's UHV power grid, it has higher requirements for the practicality, convenience and safety of UHV special safety tools [1-2]. As a "save line" in the electrical maintenance operation, the earthing wire need to ensure the reliable connection between the maintenance equipment and the earth during the operation at the same time, and it should also be easy to carry and install by the operating personnel[3-4]. Conventional earthing wire conductors use the conventional soft copper stranded wire[5]. While UHV system has high voltage level, if it use the traditional material, the size and weight of UHV earthing wire would double increase compared with the conventional earthing wire. So in field use, it could not be carried and installed by the single person and it would require more manpower and take longer time to install a group of earthing wire, which greatly reduce the maintenance efficiency. Therefore, it is of great significance to develop a light ground lead suitable for overhauling of UHV lines.

In order to guarantee that the earthing wire would not fuse when the circuit is misoperated and electrified, the rated short-circuit current value of the earthing wire at the rated short-circuit time should be less than its fusing current value and the appropriate margin should be left[6]. In regards to the calculation of the wire fuse current, Onderdonk formula of fusing current calculation of the copper material is provided by the domestic standard SSD332-89 [7] and the American standard ASTM F855-15[8]; in IEC standard 61230[9], the relation curve between the current density, time and temperature rise of copper, aluminum, silver, steel and other different materials is provided; in the document[10-12], the electric heating fusing in rapid temperature rise and slow temperature rise of the wire is studied theoretically and experimentally. For the copper wire with diameter of 1.6 mm, when the current duration is less than 100s, the law of the fusing current and time is basically in accordance with the calculation model of the adiabatic process; in the document[13], the change law of the conductor resistance in the conductor fusing process is studied.

In regards to the material of the earthing wire conductor, the US standard ASTM F855-15[8] only stipulates to use the copper wire. The domestic standard DL/T879-2004[14] mentions that the aluminum wire could be used, but any performance parameter requirement is not given. IEC standard 61230[9] stipulates that the earthing wire could use copper, aluminum or aluminum alloy conductor and the aluminum wire should be careful to use and save. Copper clad aluminum (CCA) is a new type of material that enabled a strong metallurgical bonding of copper layer and aluminum layer at the interface. It not only has high electrical conductivity and thermal conductivity of copper, but also has the low density and low price of aluminum[15-16]. CCA has been widely used in high-frequency signal transmission, power transmission as a transmission conductor [17-18].

Table 1. Characteristic comparison of soft pure copper, pure aluminum and CCA.

Wire type		Annealed copper wire	Annealed aluminum wire	Annealed CCA		
Wire model[20-22]		TR	LR	CCA-10A	CCA-15A	CCA-20A
Structure parameter	Volume ratio of copper layer(%)	100	0	10	15	20
	Minimum wiredrawing diameter(mm)	0.02	0.3	0.08	0.08	0.08
Physical property	Density(g/cm^3)	8.93	2.70	3.32	3.63	3.95
	Melting point(℃)	1083	660	1083/660	1083/660	1083/660
	Specific heat ($J{\cdot}kg^{-1}{\cdot}℃^{-1}$)	385.2	889.5	754.0	703.6	661.3
Electrical properties	Specific resistance below 20℃ (Ω·m)	1.724×10-8	2.83×10-8	2.66×10-8	2.58×10-8	2.51×10-8
	Temperature coefficient of resistance (1/℃)	3.93×10-3	4.29×10-3	4.23×10-3	4.21×10-3	4.19×10-3
Mechanical property	Strength of extension (MPa)	220-270	70-100	130-170	130-170	130-170
	Elongation at break (%)	30-45	23-25	25-30	25-30	25-30
	Elastic modulus (MPa)	120000	63000	69000	72000	75000

This paper proposed light scheme of UHV earthing wire where copper clad aluminum stranded wire replaced the existing copper wire. The dimension parameters of earthing wire of copper and copper clad aluminum materials are determined and the weight loss effect of copper-clad aluminum composite material is verified quantitatively.

2 Characteristics of copper clad aluminum composite wire

Copper clad aluminum composite wire is made of longitudinal seam welding through copper strips, blank drawing and annealing process[19] after the aluminum rod covered the pure copper strip. Now soft copper clad aluminum wire characteristics parameter after the annealing process is compared with the soft copper wire and soft aluminum wire as shown in Table 1. It could be seen from the table that the characteristics of the copper clad aluminum composite wire is between the copper and aluminum and is related with the content ratio of copper and aluminum. When the volume ratio of copper layer increases, the conductivity and mechanical strength of the composite wire is higher, while the weight also increases accordingly.

3 Calculation and test of instantaneous fusing current of the earthing wire conductor material

3.1 Promotion algorithm of Onderdonk formula

3.1.1 Derivation of onderdonk formula for copper material

The use of copper clad aluminum composite wire as UHV earthing wire conductor has small density compared to the pure copper wire while it has higher electrical and mechanical property than the pure aluminum wire. In addition, the current copper earthing wire is usually made of monofilament stranded wire with diameter of 0.1-0.2mm, which has good flexibility and is easy to coil and bend when being stored and carried; while the pure aluminum has poor flexibility, and the diameter of monofilament under the existing drawing process is no less than 0.3mm, and the strand is easy to break during stranding process, which affects the overall performance of the strand. After the use of aluminum alloy and modification, the flexibility and mechanical properties could be increased, the copper clad aluminum base after clad-welding could be made into soft state monofilament with diameter of 0.08mm through the process of drawing and annealing and it could reach the flexibility the same as the copper stranded wire. The copper and aluminum atoms of the well processed copper clad aluminum composite wire penetrate each other in the copper and aluminum interface to form a solid compound layer[23], and even in the high temperature or greater pull force, it is not easy to have the interface separation, which has a stable structural characteristics.

According to the domestic standard SSD332-89[7] and the American standard ASTM F855-15[8], the fusing current of copper material carrying short-circuit earthing wire could be calculated by using Onderdonk formula:

$$I_m = A\sqrt{\frac{\log(\frac{T_m - T_A}{234 + T_A} + 1)}{33t}} \tag{1}$$

In the formula (1), I_m is the melting current value, A; A is the cross section of the earthing wire, the circular mil; t is the time to withstand the current, s; T_m is the melting temperature of the metal, °C; T_A is the ambient temperature, °C.

The formula is derived as follows:

As the duration of the short-circuit current is very short, the earthing wire and the external environment have very little heat exchange, which could be approximated as an adiabatic process. The Power loss generated by the short-circuit current all changed to thermal energy, and it leads to the temperature increase of the earthing wire. The first order differential equation could be listed between time t and the unit length of the earthing wire:

$$c\rho A \cdot d\theta(t) = \rho_A[1+\alpha_A \cdot \theta(t)](I^2/A) \cdot dt \quad (2)$$

In the formula (2), c is the specific heat capacity, ρ is the mass density, A is the section area of the earthing wire, ρ_A is the conductor resistivity of the initial ambient temperature T_A . α_A is the resistance temperature coefficient, and I is the effective value of the short-circuit current . After (2) is simplified, it could be gained:

$$\frac{d\theta(t)}{1+\alpha_A \cdot \theta(t)} = \frac{\rho_A I^2}{c\rho A^2}dt \quad (3)$$

Integral both sides, after solving formula (3), it could be gained (make t=0, $\theta(0)$ =0; when t=t, $\theta(t)=\theta$):

$$\ln(1+\alpha_A \cdot \theta) = \frac{\alpha_A \rho_A I^2}{c\rho A^2} \cdot t \quad (4)$$

Formula (4) could be transformed to:

$$(\frac{\alpha_A \rho_A}{c\rho})\frac{I^2}{A^2} \cdot t = \ln(1+\frac{\theta}{1/\alpha_A}) \quad (5)$$

It is approximately to consider that when the original temperature is in certain range (for example 0-100°C), $\alpha_A \rho_A$ is the constant value:

$$\alpha_A \rho_A = \alpha_{20}\rho_{20} \quad (6)$$

In formula (6), α_{20} 、 ρ_{20} are the temperature coefficient of the conductor resistance and the specific resistance at 20 °C respectively. For copper materials, they are:

$$\begin{aligned} &\alpha_{20} = 0.00393(1/°C) \\ &\rho_{20} = 1.724 \times 10^{-8}(\Omega \cdot m) \\ &c = 385.2[J/(kg \cdot °C)] \\ &\rho = 8930(kg/m^3) \end{aligned} \quad (7)$$

The value of formula (7) is put the left side of the formula (5), and it could be gained:

$$1.97 \times 10^{-17}\frac{I^2}{A^2} \cdot t = \ln(1+\frac{\theta}{1/\alpha_A}) \quad (8)$$

The section area unit is converted to circular mil from square meter (1m2=1.98×109circ-miles), and is changed to the usual logarithmic base, and it could be gained:

$$33\frac{I^2}{A^2} \cdot t = \log(1+\frac{\theta}{1/\alpha_A}) \quad (9)$$

In addition, it could be gained from formula (6)、(7) that:

$$\begin{aligned} 1/\alpha_A &= \rho_A/(\alpha_{20}\rho_{20}) = [\rho_{20}+\rho_{20} \cdot \alpha_{20}(T_A - T_{20})]/(\alpha_{20}\rho_{20}) \\ &= \frac{1}{\alpha_{20}} + T_A - T_{20} = \frac{1}{0.00393} + T_A - 20 = 234 + T_A \end{aligned} \quad (10)$$

When formula (10) is put to formula (9), it could be gained that:

$$33\frac{I^2}{A^2} \cdot t = \log(1+\frac{\theta}{234+T_A}) \quad (11)$$

When the wire temperature reaches the melting point T_m, the temperature rise is $\theta = T_m - T_A$. At this time, I is the maximum short-circuit current I_m, and it is taken into the formula (11), which could be gained that:

$$33\frac{I_m^2}{A^2} \cdot t = \log(1+\frac{T_m - T_A}{234+T_A}) \quad (12)$$

After formula (12) is transformed, formula (1) could be gained, ie the equation of Onderdonk. When using the copper earthing wire, take T_m=1083°C (melting point of copper), T_A=40°C, and A (circular mil) is converted to S (mm^2) , and the above formula could be simplified as:

$$I_m = 283.7S_{Cu}/\sqrt{t} \quad (13)$$

3.1.2 Calculation of fusing current of copper clad aluminum composite material

For pure aluminum material, it could be gained:

$$\begin{aligned} &\alpha_{20} = 0.00429(1/°C) \\ &\rho_{20} = 2.87 \times 10^{-8}(\Omega \cdot m) \\ &c = 889.5[J/(kg \cdot °C)] \\ &\rho = 2703(kg/m^3) \end{aligned} \quad (14)$$

According to the above method, the calculation formula of melting current of aluminum wire could be derived as follows:

$$I_{m(Al)} = A\sqrt{\frac{\log(\frac{T_m - T_A}{213+T_A}+1)}{87.2t}} \quad (15)$$

Take aluminum wire T_m=660°C, (melting point of aluminum), T_A=40°C and convert A (circular mil) to S (mm^2), and the above formula could be simplified as:

$$I_m = 155.5S_{Al}/\sqrt{t} \quad (16)$$

For CCA composite wire, since the compound layer at the interface of copper and aluminum is usually no more than 20μm[23], when calculating the fusing current, the impact of the interface layer and the thermal conductivity between the copper, aluminum layer during transient process could be ignored. For CCA wire whose copper layer volume ration is 15%, ie, the copper, aluminum section area ratio of the wire is $\frac{S_{Cu}}{S_{Al}} = \frac{15}{85} = 0.176$. The skin effect is ignored and it is considered that the current and resistance flew through the copper layer and aluminum layer at the same time had inversed proportion: $\frac{I_{Cu}}{I_{Al}} = \frac{R_{Al}}{R_{Cu}} = \frac{\rho'_{Al}/S_{Al}}{\rho'_{Cu}/S_{Cu}} = \frac{1.724}{2.87} \div 0.176 = 3.4$. If the total current through the wire is I, then:

$$I_{Cu} = 0.23I \quad (17)$$

$$I_{Al} = 0.77I \quad (18)$$

Put formula (18) to formula (16), and the section area of aluminum layer is gained as followings:

$$S_{Al} = \frac{0.77I\sqrt{t}}{155.5} = \frac{I\sqrt{t}}{201.9} \quad (19)$$

Then the section area of corresponding copper layer is:

$$S_{Cu} = S_{Al} \cdot 0.176 = \frac{I\sqrt{t}}{1147} \quad (20)$$

In addition, put formula (17) into the formula (13), and the flowing current could be gained to ensure the minimum section area of no fusing copper material:

$$S'_{Cu} = \frac{0.23I\sqrt{t}}{283.7} = \frac{I\sqrt{t}}{1233} \quad (21)$$

After comparing (20)、(21), it could be gained that $S'_{Cu} < S_{Cu}$. So when the short circuit current flew through CCA, the aluminum material would fuse firstly. Put formula (18) and $S_{Al} = 0.85S$ to formula (16), the copper layer volume ration could be gained, which is the relationship between 15% CCA fusing current and section area: $0.77I_m = 155.5 \times 0.85 S_{CCA\text{-}15} / \sqrt{t}$:

$$I_m = 171.6 S_{CCA\text{-}15} / \sqrt{t} \quad (22)$$

Similarly, fusing current calculation formula of the CCA wire with copper volume ratio of 10% and 20% could be derived:

$$I_m = 165.8 S_{CCA\text{-}10} / \sqrt{t}$$
$$I_m = 176.2 S_{CCA\text{-}20} / \sqrt{t} \quad (23)$$

The reduction coefficient K (value is 1.08-1.15) [5] is introduced, and the rated short-circuit current of the earthing wire is:

$$I_r = I_m / K \quad (24)$$

The unit length weight m of the earthing wire conductor is:

$$m = \rho S \quad (25)$$

Put formula (24), (25) into the formula (13), (22) and (23), the rated short-circuit current Ir could be gained (unit: A). When the rated short-circuit time is t (unit: s), the unit length weight of copper and CCA earthing wire is:

$$m_{Cu} = 3.15 \times 10^{-5} K I_r \sqrt{t} (kg/m)$$
$$m_{CCA\text{-}10} = 2.01 \times 10^{-5} K I_r \sqrt{t} (kg/m) \quad (26)$$
$$m_{CCA\text{-}15} = 2.12 \times 10^{-5} K I_r \sqrt{t} (kg/m)$$
$$m_{CCA\text{-}20} = 2.24 \times 10^{-5} K I_r \sqrt{t} (kg/m)$$

It could be seen from the above formula that when the rated short-circuit current of the earthing wire is the same with the short circuit time, although the section area needed by CCA earthing wire is greater than the pure copper wire, the overall weight of earthing wire conductor decreases because CCA material density is much smaller than copper; the smaller the proportion of copper layer is, the more the weight reduces. The theoretical calculation showes that the section area of CCA wire with copper layer volume ratio of 10% is 71.1% larger than that of copper wire, but the weight could be reduced by 36.2%.

3.2 Test and analysis of fusing current

3.2.1 Test arrangement

In order to further verify the fuse characteristics of CCA material in the instantaneous current and compare with the copper material, the copper of the different section and CCA-10 copper clad aluminum small strand samples are selected to carry out the fusing current test, and the test layout is shown below. The tested wire samples are applied the power frequency current through the high current generator of the power frequency, and the oscilloscope collected the voltage across of the shunt (micro-European rating resistor) thus to obtain the wire current waveform and fuse time and use the high-speed camera to record the whole process of wire sample fuse, and the test electrode spacing is 100mm.

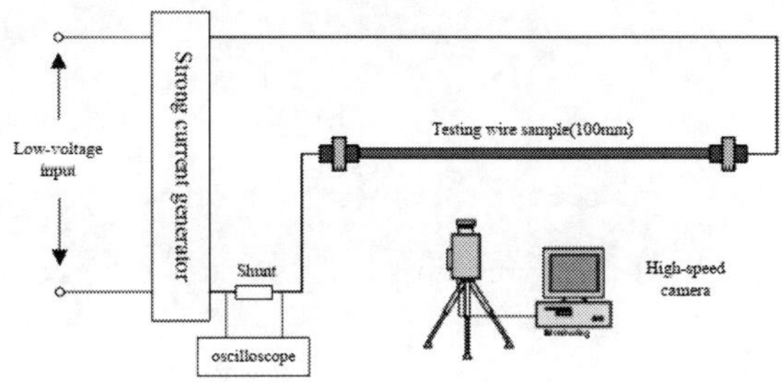

Figure 1. Layout diagram of fusing current test of earthing wire conductor material.

3.2.2 Analysis of test results

Table 2. Fusing current test results of copper and CCA stranded wire samples.

Sample Specification	Effective value of test current (A)	Fusing time(s)	
		Test value	Calculation value
2.5mm² copper stranded wire	396	4.02	3.21
	528	2.35	1.80
	556	2.05	1.63
4mm² copper stranded wire	424	7.24	7.16
	575	4.12	3.89
	650	3.24	3.05
2.5mm² CCA stranded wire	100	Not fuse for10s	17.18
	236	3.98	3.08
	330	2.25	1.58
4mm² copper stranded wire	377	3.56	3.09
	490	2.10	1.83
	556	1.54	1.42

The fusing current test of 2.5mm² and 4mm² pure copper and CCA-10 copper clad aluminum composite wire is carried out respectively. The two wires are both stranded by the soft monofilament with diameter of

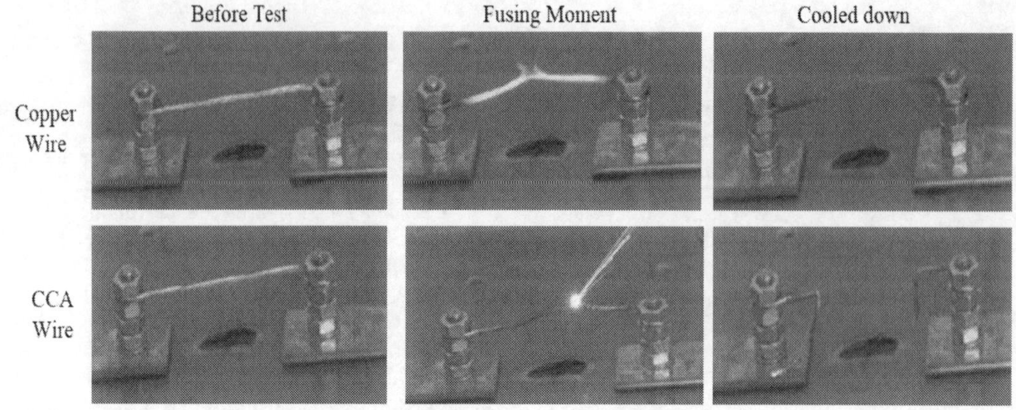

Figure 2. Layout diagram of fusing current test of earthing wire conductor material.

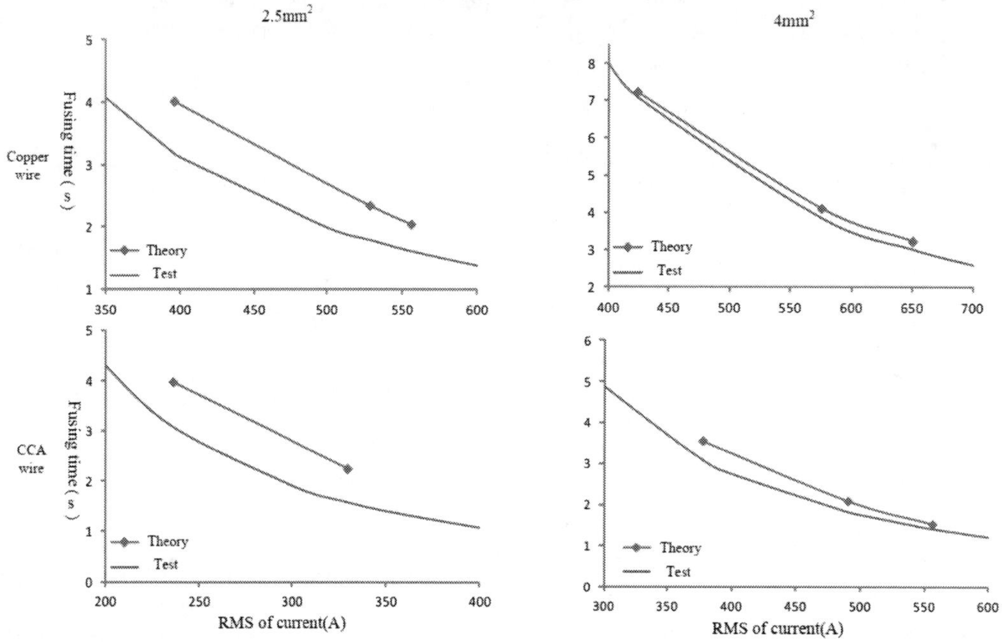

Figure 3. Comparison of fusing current test and theoretical calculation.

0.2mm. By adjusting the output current, recording the fusing time in different currents, the test process and the results are shown in figure 2, table 2.

The experimental results are compared with the theoretical results in section 3.1, and the relation curve between the fusing time and the effective value of fusing current is shown in figure3. It could be seen from the figure that:

(1) If the effective values of the current which flew through the samples with the same material and section are the same, the fuse time obtained by the test is greater than the theoretical calculation value, which is because the heat exchange between the wire and the external environment is ignored and is regarded as an adiabatic process. While during the actual test, the insulation measures are not taken and part of the energy dissipated to the surrounding environment.

(2) The larger the section of the test sample, the smaller the average heat dissipation area, so the closer the test value is to the calculated value.

(3) The greater the test current is, the shorter the fusing time is, the smaller the dissipation heat is, so the closer the test value is to the calculated value.

(4) When the wire section is greater than 4mm2 and the fusing time is less than 1.5s, the error between the test result and the calculated value is less than 10%, and the influence of heat exchange with the surrounding environment could be neglected.

4 Conclusion

This paper put forward the use of copper clad aluminum(CCA) composite materials as a lightweight method for UHV earthing wire conductor. Theoretical calculations and tests of the fusing current in a short time for copper and CCA material are conducted. The following conclusions are drawn:

1)Copper clad aluminum composite material has the advantages of both high conductivity of copper and low density of aluminum. It is proposed to use copper clad aluminum to replace the traditional copper materials as high voltage earthing wire conductor.

2)By theoretical calculation, compared with copper earthing wire, the section area of CCA with copper layer volume ration of 10% increases 71.1%, but the weight reduces 36.2%.

3)Under the same time, the fusing current test result of the copper and CCA small stranded sample is higher than the theoretical calculated value under adiabatic conditions. When the wire section is larger than $4mm^2$ and the fusing time is less than 1.5s, the error between the test result and calculation value is less than 10% and the heat exchange effect with the surrounding environment could be ignored.

References

1. LIU Zhenya. Innovation of UHVAC transmission technology in China[J]. Power System Technology, 2013, 37(3): 1-8.
2. LIU Zhenya. Ultra-high voltage AC & DC grid[M]. Beijing, China: China Electric Power Press, 2013.
3. ZHAO Zhijun. Earthing and short-circuiting wires - the survive lines in electrical maintenance.[J]Heilongjiang Science and Technology Information, 2008, (14): 30.
4. DAI Keming. Correct use of portable earthing and short-circuiting wires.[J]. Electric Safety Technology, 2011, 13(11): 20-24.
5. LI Jinxiong. Application of earthing and short-circuiting devices[M]. Beijing, China: China Electric Power Press, 2008.
6. ZHU Zhongming. The discussion of rated short-circuit current for portable earthing and short-circuiting wires[J]. East China Electric Power, 1994, (10): 23-25.
7. SSD332-89. Technical standard of portable earthing and short-circuiting wires[S]. Beijing, China: China Standard Press, 1989.
8. ASTM F855-15. Standard specifications for temporary protective grouds to be used on de-energized electric power lines and equip-ment[S]. 2015.
9. IEC 61230:2008. Live working - Portable equipment for earthing and short-circuiting[S]. 2008.
10. Babrauskas V, Wichman I S. Fusing of wires by electrical current[C]. Fire and Materials 12th International Conference and Exhibition, San Francisco, USA: Interscience Communications, 2011: 769-778.
11. Choi C S, Kim H K, Kim D W, et al. Analysis of dispersive characteristics and structures of copper wire melted by overcurrent[J], IEEJ Trans, 2005, 125(12): 1327-1331.
12. Wright A, Newbery P G. Electric fuses[M], 3rd ed. Stevenage, UK: Institution of Electrical Engineers, 2004: 191-194.
13. Tanaka T, Yamasaki M. Modeling of fuses for melting time and fusing current analysis[C]. Telecommunications Energy Confer-ence(INTLEC), Chicago, USA: IEEE, 2004: 671-675.
14. DL/T 879-2004. Portable equipment for earthing and short-circuiting for live working[S]. Beijing, China: China Standard Press, 2004.
15. Gibson A. The economics of copper clad aluminum bometallic cables[J]. Wire & Cable Technology International, 2005, 33(4): 82-83.
16. ZHANG Jianyu, YAO Jinjin, ZENG Xiangyong, et al. Research progress of copper cladding aluminum composites[J]. The Chinese Journal of Nonferrous Metals, 2014, 24(5): 1275-1284.
17. SHOU Weichun, LI Xiangli. The application of CCA wire in coaxial cable[J]. Optical Fiber & Electric Cable, 2000(2): 40-45.
18. DONG Yongjun, WANG Jianwei, HUANG Chunde, et al. Ther-modynamic equilibrium calculation of copper clad aluminum composite[J]. Chinese Journal of Rare Metals.
19. ZHENG Chuantao, DAI Yakang. The characteristics of copper clad aluminum wire effect on the service performance of copper clad aluminum cable[J]. Electic Wire & Cable, 2016, (6): 4-7.
20. GB/T 3953-2009. Round copper wire for electrical purposes[S] Beijing, China: China Standard Press, 2009.
21. GB/T 3955-2009. Round aluminum wire for electrical purposes[S] Beijing, China: China Standard Press, 2009.
22. GB/T 29197-2012. Copper-clad aluminum wire[S] Beijing, China: China Standard Press, 2012.
23. ZHAO Zhengshu, DAI Yakang. Study of copper clad aluminum wire annealing technology[J] Electic Wire & Cable, 2011, (6): 6-8, 23.

Full Wave Simulation and Design of Film Bulk Acoustic Wave Filters

Du Bo[1,2], Lin Jian[2], Ou Li[2], Shi Yu[1] and MA Jin-yi[2]

[1]*State Key Lab.of Electronic Thin Films and Integrated Devices, University of Electronic Science and Technology of China, 610054 Chengdu,China*

[2]*26th Institude of China Electronics Technology Group Corporation, 400060 Chongqing, China*

Abstract. In order to simulate the frequency response of thin film bulk acoustic wave filter accurately, the MBVD model of film bulk acoustic wave filter chip and three dimension electro-magnetic model of package were respectively established. And then, the collaborative simulation of acoustic and electro-magnetic components was realized in ADS software. Finally, thin film bulk acoustic wave filter with working frequency of 2.825GHz and 5.6 GHz were designed and fabricated by this method. Insertion loss of 2.825GHz FBAR filter is less than 1.5dB, and rejection is more than 40dB. Insertion loss of 5.6GHz FBAR filter is less than 1.7dB, and rejection is more than 30dB. The simulation and experimental results show that the proposed method has good design precision and can simulate the influence of package on the performance of the filter accurately.

1 Introduction

Thin film bulk acoustic wave resonator (FBAR) filter has the advantages of higher operating frequency, smaller size, higher Q value and better power handling capability, comparing surface acoustic wave (SAW) filter, LC filter and ceramic filter [1]. Furthermore, its process is compatible with integration circuit (IC) process [2]. So, FBAR filter has been used widely in mobile communication market.

FBAR utilizes the longitudinal wave resonate characteristics of piezoelectric materials such as AlN or ZnO to realize electrical frequency selection [3,4]. Acoustic waves are about five orders of magnitude shorter than electromagnetic waves, resulting in significantly small size. FBAR filter is combined by multiple FBAR resonators in a certain circuit structure, including ladder structure and bridge structure [5]. Ladder structure is often used in single ended circuit, and the bridge structure is used in balance ended circuit structure.

The frequency of FBAR is mainly determined by the acoustic velocity and thickness of piezoelectric film. The acoustic velocity of AlN is about 11000m/s, and the thickness of AlN can be controlled between 0.1 micros to 2 micros. So, FBAR can work at the frequency of 1 to 10 GHz [6]. At this high working frequency, the parasitic parameters of the package have significant influence on the frequency response of the filter. Therefore, in order to realize more accuracy design, the effect of package should be considered in the FBAR filter simulation model.

The FBAR filter is made by MEMS process, and it is based on Microwave acoustic theory. While the ceramic package is made of HTCC process and its influence on the filter is based on electro-magnetic wave theory. How to realize co-simulation of acoustic and electro-magnetic components in the design process is difficult.

Traditional design method is to fit the equivalent circuit parameters of package by measuring the frequency response of the packaged filter, and then use the equivalent circuit for the filter design [7]. The drawback of this approach is that in order to get accurate parameters, it needs to produce and measuring the device multi-times. Meanwhile, if the package or bonding method is changed, it needs to fit the measurement results again. So, its efficiency is low.

This paper used the ADS software to set up a FBAR MBVD (Modified Butterworth-Van Dyke) equivalent circuit model for the simulation of filter chip. The three dimension (3D) electro-magnetic model of ceramic package was established with HFSS software. And then, the S parameters of package was achieved and added into FBAR filter MBVD model. This method combined electro-magnetic and acoustic simulation, called full wave simulation. Using this method, FBAR filters working at 2.825GHz and 5.6GHz was designed, produced and measured.

2 Full wave simulations

2.1 FBAR equivalent circuit model

In order to realize the fast and accurate design of FBAR filter, MBVD model is usually adopted. The equivalent circuit model use lumped elements to describe FBAR resonator, so it is conducive to the rapid optimization of filter frequency response [8].

Fig. 1 is MBVD model of FBAR resonator, comprising 6 lumped parameter components. The dynamic impedance is on behalf of electrical and mechanical response of piezoelectric materials characterization, including dynamic capacitance C_m, dynamic inductance L_m and resistance R_m. Capacitance C_0 represents the dielectric properties of piezoelectric material. R_0 represents the dielectric loss of piezoelectric materials and related mechanical vibration loss. R_s is from the ohmic loss of electrodes.

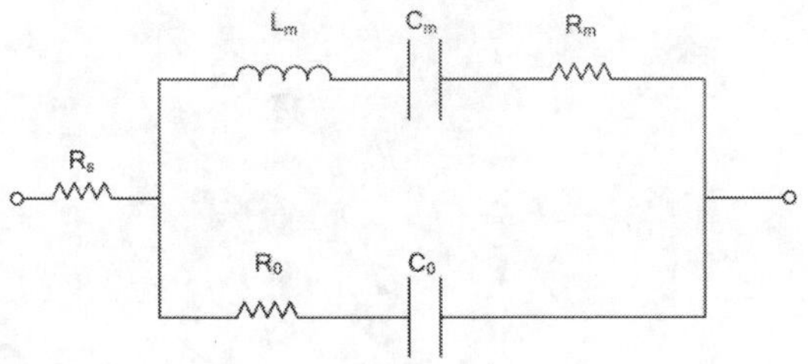

Figure 1. MBVD equivalent circuit model.

FBAR resonator needs be fabricated and measured, and the MBVD model parameters of FBAR resonator can be accurately extracted by fitting measured S parameters. The measurement and fitted Smith circle of FBAR are shown in Fig.2. The MBVD model fits well with measured data above resonator frequency (fs). However, because of parasitic resonances caused by Rayleigh-Lamb (RL) modes, it can't fit the measured data below fs. Nevertheless, the precision of MBVD model is good enough for filter simulation.

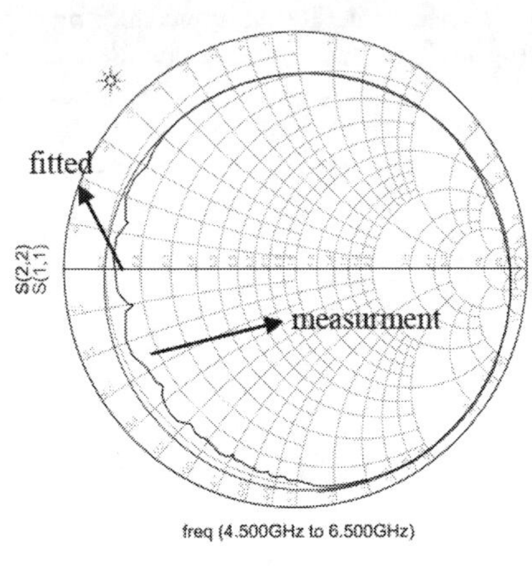

Figure 2. Measured and fitted Smith circle of FBAR.

According to the ladder or bridge circuit structure, as shown in Fig.3, the MBVD models of each resonator are combined, and then the equivalent circuit simulation model of the filter is constructed. In order to realize a flat passband, the anti-resonant frequency of parallel resonators is approximately equal to the resonant frequency of series resonators. Generally, the frequency of parallel resonators is 6%~8% lower than that of series resonators.

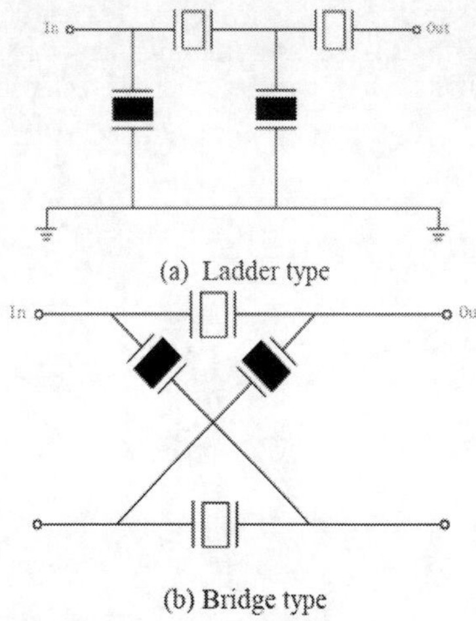

Figure 3. FBAR filter circuit structures.

The ladder structure is single ended structure, which is widely used. It can realize good rectangle coefficient, but its rejection is poor, so it is necessary to introduce inductors in the parallel resonators to improve the out-of-band rejection characteristics. Fig.4 is the admittance of resonator series with and without inductor. As shown, the introduction of inductor can cause the series resonant frequency of resonator to move to lower frequency. Therefore, different inductors can produce several zero points, and the rejection of filter can be improved.

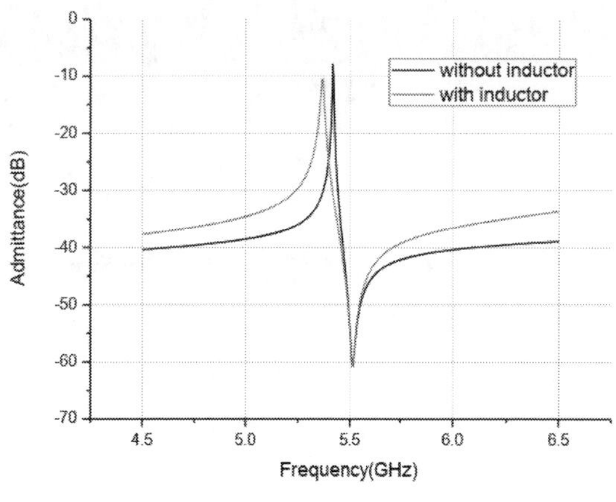

Figure 4. The admittance of resonator series with and without inductor.

2.2 The influence of package to filter frequency response

FBAR filter has high working frequency between 1GHz to 10GHz. At this high frequency, the inductors series with the parallel resonators are usually realized by parasitic inductance of package, such as bonding wire,

 https://doi.org/10.1051/matecconf/201816001005

embedded inductor and so on. At the same time, the parasitic capacitance of the package also has significant influence on the out-of-band rejection of the filter.

Fig. 5 and Fig. 6 show the measured frequency responses of FBAR filters working at 1.56GHz, packaged with different ceramic envelopes. It's clearly that the frequency responses of these filters are different, especially for out-of-band rejection characteristics.

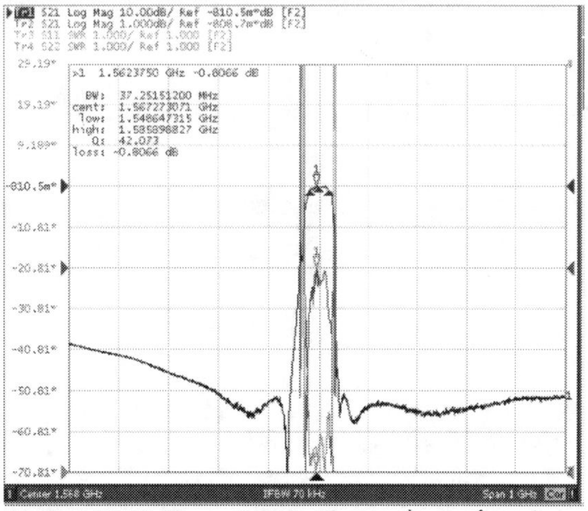

Figure 5. FBAR filter packaged with ceramic envelope A.

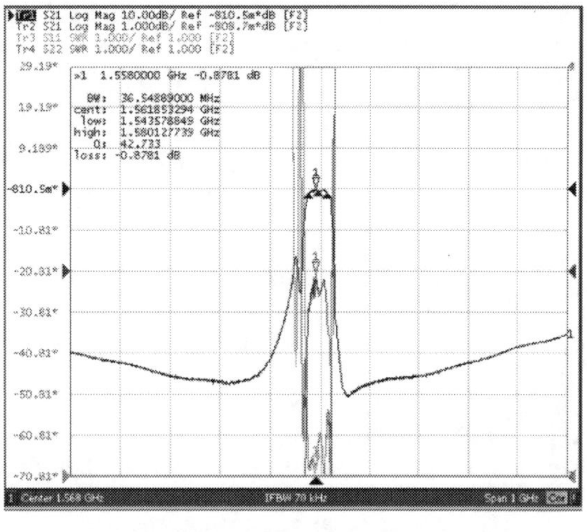

Figure 6. FBAR filter packaged with ceramic envelope B.

During the process of design FBAR filer chip, if not considering the influence of package, the measured results will be different with simulations.

2.3 3D electromagnetism model of FBAR package

In order to extract the electromagnetic parameters of package accurately, 3D electro-magnetic model of ceramic package was established by electro-magnetic simulation software HFSS, as shown in Figure 7. And the 8 ports S parameters were calculated and extracted. Inside the model, two ports were set as input and output ports of package, and the other six ports were set as interconnecting ports of the package and the filter chip.

The model includes ceramic package, bonding wire, chip substrate, bonding pad and so on. The size of ceramic package is 3mm × 3mm × 1.1mm, the bonding wires are gold wire. The chip substrate material is high resistance silicon, and the bonding pad material is gold film. The chip size is 1mm × 1mm × 0.35mm.

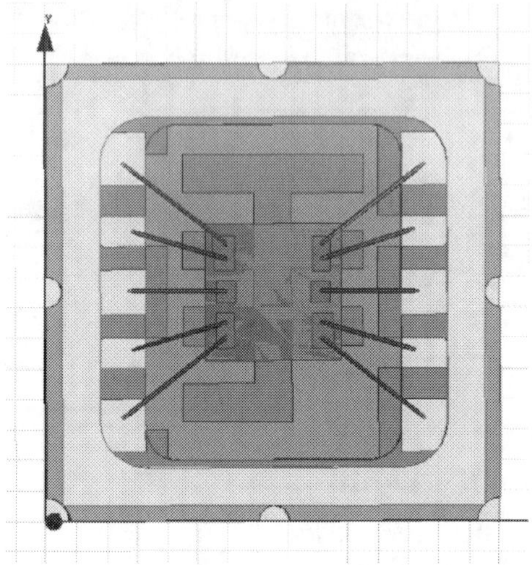

Figure 7. 3D electromagnetism model of FBAR package.

2.4 Full wave simulation of FBAR filter

ADS software was used to realize full wave simulation of FBAR filter. S parameters of package was introduced into ADS, and connected with MBVD model of FBAR filter chip. Input and output ports of package were connected with terminals of S-parameter simulators, which are used to define the impedance and location of the ports, as shown in Fig 8. After building the model, the resonator areas of filter were tuned to optimize the filter frequency response. If the frequency response can't meet the specification requirements, it is necessary to refine the bonding method or replace the package.

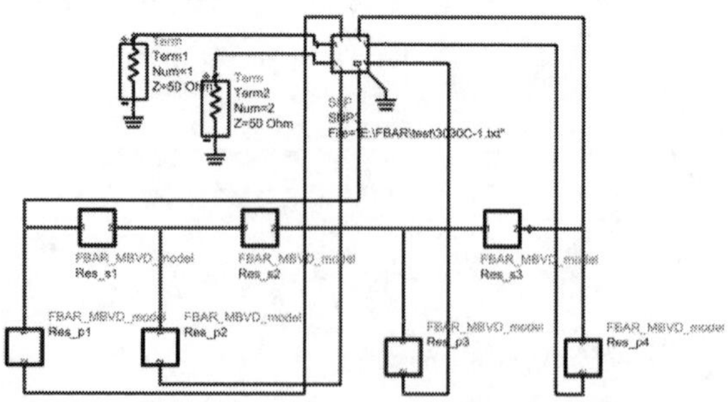

Figure 8. Full wave simulation model of FBAR filter.

3 Fabrication and measurement of FBAR filters

Using full wave simulation method, the FBAR filters working at S band and C band are optimized and fabricated and measured.

3.1. S band FBAR filter

Fig. 9 is S band FBAR filter chip photograph, and Fig. 10 is comparison of simulation result of S band FBAR filter with measurement result. The central frequency of the device is 2.825 GHz, the insertion loss is 1.5 dB, and the out of band rejection is more than 40dB.

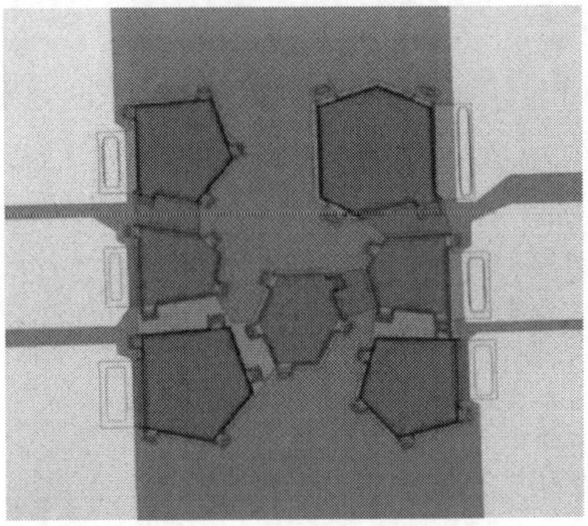

Figure 9. S band FBAR filter chip photograph.

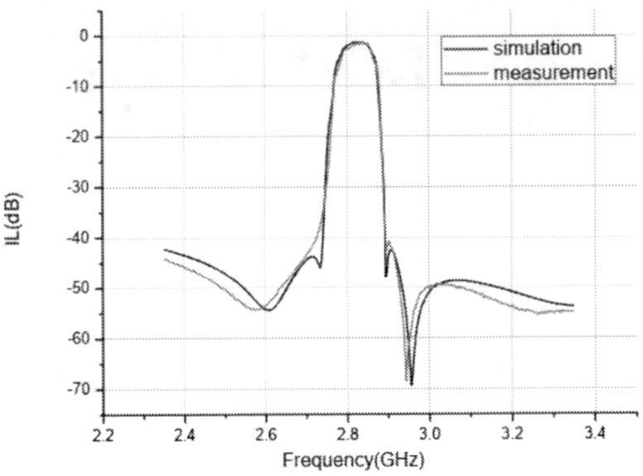

Figure 10. Comparison of simulation result of S band FBAR filter with measurement result.

3.2 C band FBAR filter

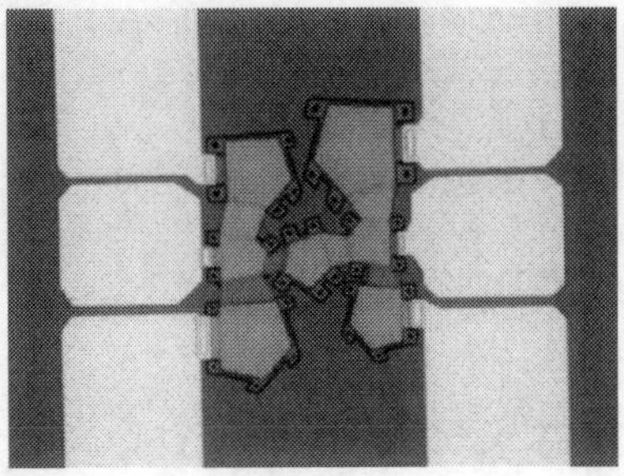

Figure 11. C band FBAR filter chip photograph.

Fig. 11 is C band FBAR filter chip photograph, and Fig. 12 is comparison of simulation result of C band FBAR filter with measurement result. The central frequency of the device is 5.6 GHz, the insertion loss is 1.7 dB, and the out of band rejection is more than 30dB.

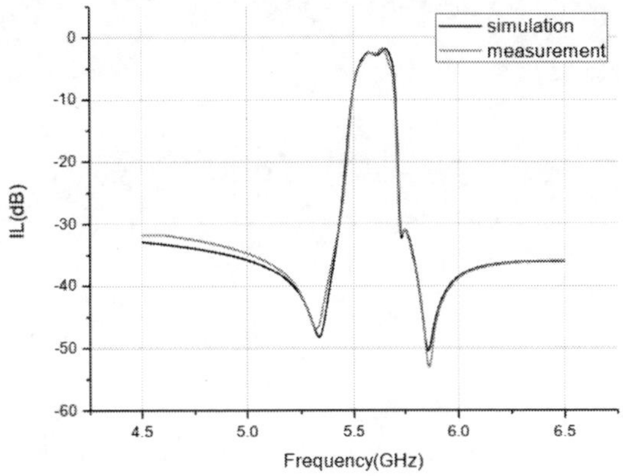

Figure 12. Comparison of simulation result of C band FBAR filter with measurement result.

All those results show that the simulation results of S and C band filters are in good agreement with the measured results.

4 Conclusions

The FBAR filter operates at very high frequencies and covers 1-10GHz. In such a high frequency, the influence of parasitic parameters of bonding wire and packages to the frequency responses of FBAR filter is significant. If not considering the influence of package in the filter design process, it is likely to lead to deterioration in the frequency responses of FBAR filter after filter chip packaged.

However, how to realize co-simulation of acoustic and electro-magnetic components in the design process is difficult. This paper used full wave design method, combined electro-magnetic and acoustic models.

With this design method, the optimized design of FBAR filter with 2.825GHz and 5.6GHz working frequency were completed. Insertion loss of 2.825GHz FBAR filter is less than 1.5dB, and rejection is more than 40dB, and insertion loss of 5.6GHz FBAR filter is less than 1.7dB, and rejection is more than 30dB.

The simulation and measurement results show that this design method has good simulation accuracy, and can be used for FBAR design.

References

1. J. S. Shin, Y.K. Park, Y.I. Kim et al, Proceedings of the 40th European Microwave Conference, 1257-1260 (2010)
2. M. Aissi, E. Tournier, M.A. Dubois, et al, IEEE International Solid-State Circuits Conference, 17.5 (2006)
3. F. Martin, M.-E. Jan, B. Belgacem, M.-A. Dubois, P. Muralt, Thin Solid Films, 514, 341-343 (2006)
4. L. Mai, J.Y. Lee, V.S. Pham, and G.Y, IEEE Microwave and Wireless Components Letters, 17, 12, (2007)
5. Uzunov, D. Gajdajiev, MIXDES 2011, 18th International Conference "Mixed Design of Integrated Circuits and Systems", 16-18 (2011)
6. M. Hara, T. Yokoyama, M. Ueda and Y. Satoh, IEEE Ultrasonic Symposium, 1152-1155 (2007)
7. F. M. Pitschi, J. E. Kiwitt, B. Bader, and K. C. Wagner, IEEE International Ultrasonics, Ferroelectrics, and Frequency Control Joint 50th Anniversary Conference, 1525-1528 (2004)
8. J. D. Larson, SM, P.D. Bradley, et al, IEEE Ultrasonic Symposium, 863-868 (2000)

Magnetic field computation and simulation of the coil systems using Comsol software

Ivan Zatonov[1], Pavel Baranov[1] and Andrey Kolomeycev[1]

[1] *Department of Electronic Engineering, National Research Tomsk Polytechnic University, 634050 Tomsk, Russia*

Abstract. The article considers the calibration system computation with the inhomogeneity of magnetic field less than 0.1 %. Method of calculation is described. The numerical simulation using finite-element analyze was made for such system as: Helmholtz coils, improved Helmholtz coils, four and six coils system. Authors made the calculation of the magnetic field homogeneity toward axial direction. Based on procure results, authors analyzed magnetic field homogeneity and compared different types of coil systems.

1 Introduction

The measuring of magnetic flux density plays the key role in such brunches of since and technology like: task of orientation, navigation, stabilization, flaw detection, nondestructive testing, magnetic tomographic measurement, shielding action for quantum computer, etc.

Magnetometer gains currency for magnetic field measuring [1]. The stormy developing of robotic engineering complexes of surface-mounted, underwater and space system appropriation requires magnetometers evolution that will provide opportunity for weak magnetic field measuring. The theme of magnetic field measuring was reviewed by following scientists [2,3].

Permanent field source of magnetic flux density is used for magnetometers calibration and graduation. The inaccuracy of magnetic field source must be three times less than the inaccuracy of the calibrated sensor. For instance, for the measuring device with inaccuracy near of 1 %, it is necessary to provide homogeneity magnetic field with inaccuracy 0.1- 0.3 %.

Some of the most ordinary ways for homogeneity magnetic field creation is using of Helmholtz coil. There are a lot of works devoted to the calculation of Helmholtz coils magnetic field [4-8]. One of the others way for uniform magnetic field generation is to use spheroidal coils[9]. The system with bigger number of coils is used to create magnetic field with higher [10], such as: Improved Helmholtz coils, four and six coils system. In this article we made the comparison of magnetic homogeneity in axial direction for four types of coils with similar radius.

2 Methods of magnetic field's axial component computation for the coil system

According to Biot-Savart law [11], flux density on a distance 'z' from the center of coordinate frame created by circular coil can be calculated by the following formula:

$$B(z) = \frac{\mu_0 NIR^2}{2\left[R^2 + (z-h)^2\right]^{\frac{2}{3}}}, \qquad (1)$$

where μ_0 is vacuum permeapility equals to $4\pi\times10^{-7}$ H/m; N is the number of windings in coil; I is current value in the conductor, A; R is coil radius, m; h is distance between coil center and central point of the coordinate system, m.

Helmholtz coils is formed by two circular coils with the similar radius and equal distance from the coordinate frame (Figure 1).

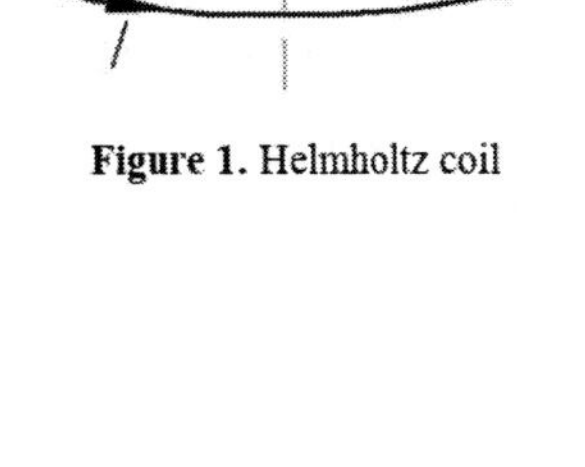

Figure 1. Helmholtz coil

The magnetic flux density in the axial direction for the Helmholtz coil is calculated by:

$$B(z)=\frac{\mu_0 NIR^2}{2}\left\{\left[(z+h)^2+R^2\right]^{-\frac{3}{2}}+\left[(z-h)^2+R^2\right]^{-\frac{3}{2}}\right\}. \quad (2)$$

To calculate the 'h' it is necessary and appropriate to solve criterion (3). The second member of Taylor series in the condition of z=0 must be equal to zero:

$$B^{(2)}(0)=0. \quad (3)$$

Considering equations (2) and (3) we can have the following formula:

$$\frac{\mu_0 NIR^2}{2}\left[\frac{30h^2}{\left(R^2+h^2\right)^{\frac{7}{2}}}-\frac{6}{\left(R^2+h^2\right)^{\frac{5}{2}}}\right]=0. \quad (4)$$

Solution of the equation (4) demonstrates that 'h' equals to '$R/2$'. As it can be seen from the above the final formula for the calculation of magnetic flux density in the axial direction for the Helmholtz coil could be written as:

$$B(z)=\frac{\mu_0 NIR^2}{2}\left\{\left[\left(z+\frac{R}{2}\right)^2+R^2\right]^{-\frac{3}{2}}+\left[\left(z-\frac{R}{2}\right)^2+R^2\right]^{-\frac{3}{2}}\right\}. \quad (5)$$

Improved Helmholtz coils (Figure 2) consists of three coils with equal distance between each other and the same radius of the coil. In paper [7] authors made a research for three coil system. The Magnetic flux density for that coil system is defined by the 6th equal.

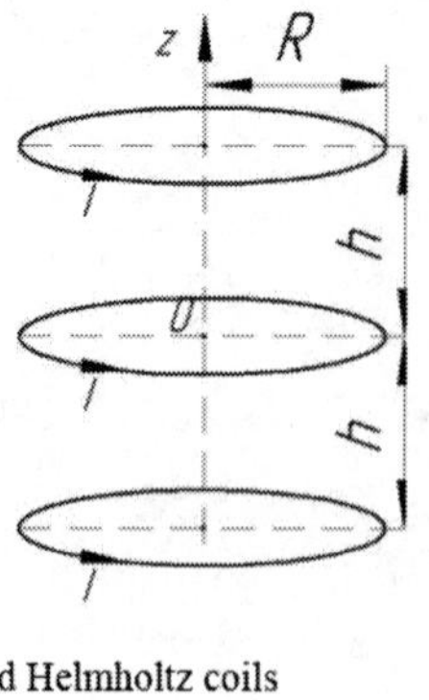

Figure 2. Improved Helmholtz coils

$$B(z)=\frac{\mu_0 NIR^2}{2}\left\{k\left(z^2+R^2\right)^{-\frac{3}{2}}+\left[(z+h)^2+R^2\right]^{-\frac{3}{2}}+\left[(z-h)^2+R^2\right]^{-\frac{3}{2}}\right\}, \quad (6)$$

where 'k' is the coefficient that defines the number of windings in the central coil.

Homogeneity magnetic field created by the improved Helmholtz coils must fulfill the following condition:

$$\begin{matrix} B^{(2)}(0)=0 \\ B^{(4)}(0)=0 \end{matrix}. \quad (7)$$

The set of equations (7) is also appropriate for the three coils system of Maxwell.

Computation of the magnetic flux density in the axial direction for the multicoil system (Figure 3) is performed by the following equation:

$$B(z)=\frac{\mu_0 NIR^2}{2}\left\{\left(z^2+R^2\right)^{-\frac{3}{2}}+k_1\left[\left(z+h_1\right)^2+R^2\right]^{-\frac{3}{2}}+k_1\left[\left(z-h_1\right)^2+R^2\right]^{-\frac{3}{2}}+\ldots+k_n\left[\left(z+h_n\right)^2+R^2\right]^{-\frac{3}{2}}+k_n\left[\left(z-h_n\right)^2+R^2\right]^{-\frac{3}{2}}\right\}, \quad (8)$$

Where k_i is the coeffitient that defines the number of windings in ith coil; h_i is the distance between the coil center and the starting of coordinate system.

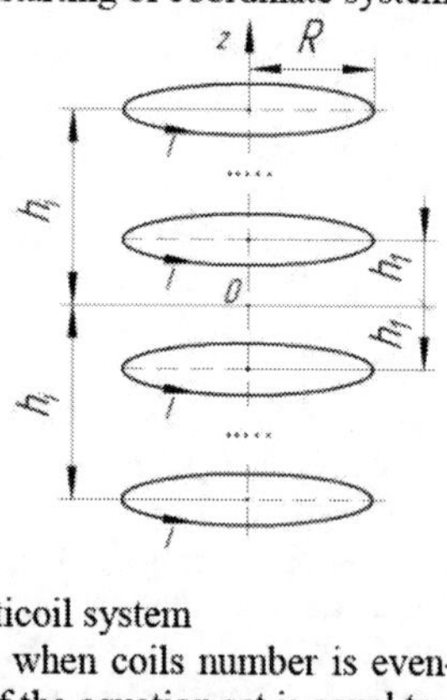

Figure 3. Multicoil system

In the case when coils number is even-numbered the first member of the equation set is equal to zero.

Homogeneity magnetic field created by the system with n/2 pairs of coils must fulfill the constraint condition(9):

$$\begin{cases} B^{(2)}(0)=0 \\ \ldots \\ B^{(2(n-1))}(0)=0 \end{cases}, \quad (9)$$

Each of equations allows the defining of the magnetic flux density value on the coordinate system center possible. Due to the value of magnetic flux density in the coordinate frame center and equations that are defining the changing character of magnetic field in the axial direction, we receive expression for relative inhomogeneity of magnetic flux density in the direction 'z' for each analyzed types of coil system:

$$\delta_z = \frac{B(z) - B(0)}{B(0)} \cdot 100\,\%, \tag{10}$$

It is important to take into account that the calculation is carried out for coils with the cross section aiming to zero. Real constructions have some volume. the finite element analysis was carried out in to take into account of the cubic content.

3 Finite element analysis

Authors used the program Comsol to conduct the finite element analysis. Every coil system was provided with its own task. The geometry of Helmholtz coils is presented on the figure 4. The size of coils section was chosen according to the number of windings in coil.

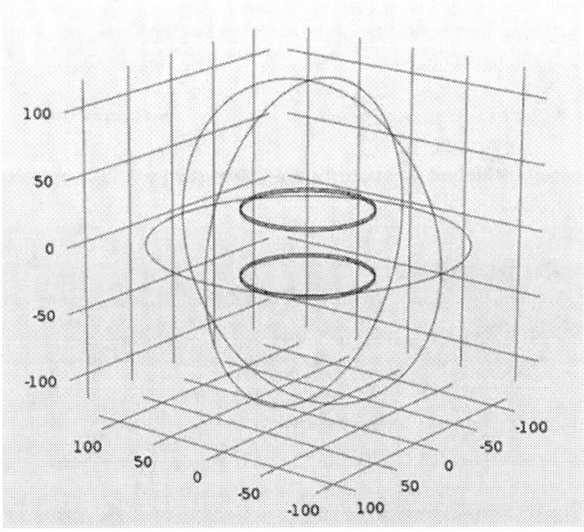

Figure 4. The geometry of Helmholts coils in Comsol

The calculation was carried in module "Magnetic field". For the windings parameters setting the authors used Multi-turn coil section. The following parameters were set: the number of windings, the current in a conductor and the movement direction. The last parameter was set through the reference edge. It should be noted that the current movement direction in both coils match.

The block with manually setting mesh was added to the model in order to receive precise measuring data avoiding fluctuation. It helped to receive data with interval lesser than 1mm. The set mesh for the improved Helmholtz coils is presented on the figure 5.

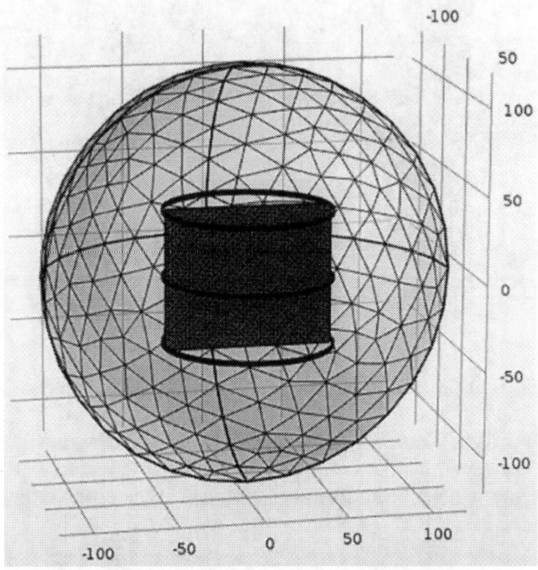

Figure 5. Finite element mesh for the improved Helmholtz coils

The stationary type of the task solution was chosen for each of the models as well as iterative solver.

The calculated coils parameters values are persented in the table 1.

Table.1. Coils parameters

i	Parameter value	
	h,m	N
Helmholtz coils		
-	0.025	10
Improved Helmholtz coils		
Central position	0	5
Edge position	0.037	10
4 coils system		
1	0.01	10
2	0.045	26
6 coils system		
1	0.009	10
2	0.029	12
3	0.0615	31
8 coils system		
1	0.0135	10
2	0.0425	12
3	0.0805	17
4	0.146	34

4 Calculation results

The figure below presents results of the magnetic field calculation of the six coils system in the axial direction with the DC value of 1 mA.

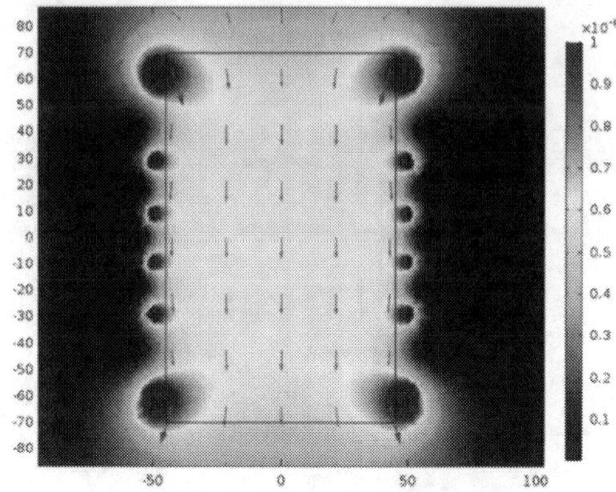

Figure 6. The distribution of six coil system magnetic field

The figure 7 shows the heterogeneity for four coil system in two-dimensional space. The maximum value of the magnetic flux density heterogeneity is observed near coils surface.

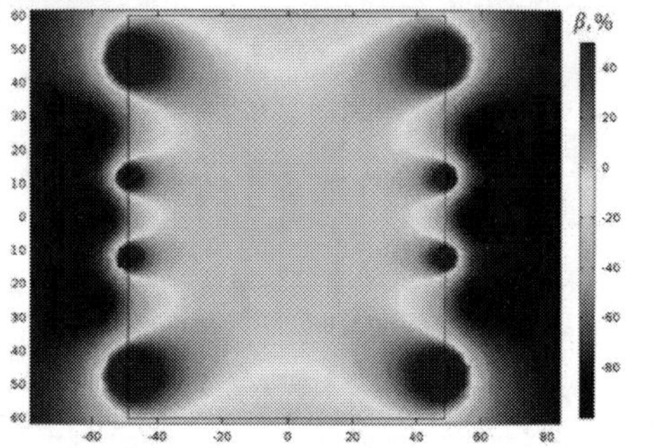

Figure 7. The heterogeneity of the magnetic field for the four coil system

The figure 8 presents the heterogeneity of the magnetic field in the axial direction for the considered types of coil systems.

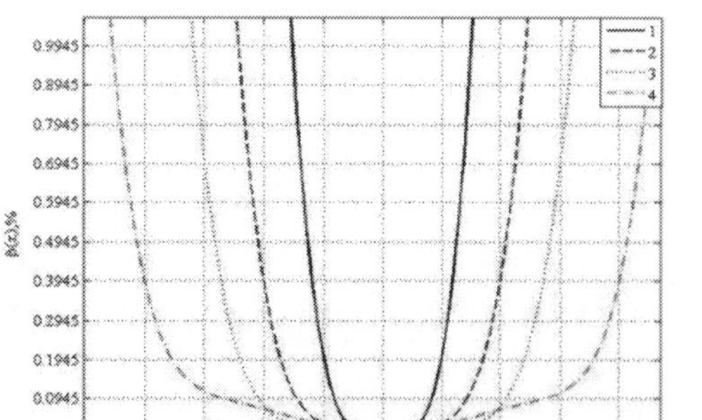

Figure 8. The heterogeneity of the magnetic field in the axial direction, where 1 – the Helmholtz coil, 2 – Improved Helmholtz coil, 3 – four coils system, 4 – six coils system.

The changing of one of the parameters presented in the table 1 leads to the magnetic field inhomogeneity change.

The figure 9 shows that the variation of the magnetic field inhomogeneity parameter in the case of the fourth coils pairs depends on the value of 'h'.

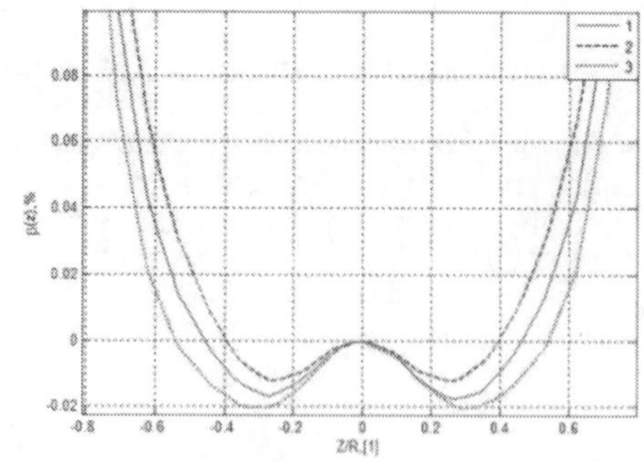

Figure 9. The heterogeneity of the magnetic field for the 8 coils system in axial direction, where 1 – $h_4 = 146\,mm$; 2 – $h_4 = 145.5\,mm$; 3 – $h_4 = 146.5\,mm$

This figure demonstrates, that a slight change in the parameters leads to increase or decrease of the field inhomogeneity. Increasing of the 'h' for the fourth pair coils allows to expand the area of the desired inhomogeneity.

Acknowledgments: The research is funded from Russian Science Foundation (RSF), Grant Number 17-79-10083.

Conclusion

The conducted comparison showed the needed heterogeneity of the magnetic field less than 0.1 % is provided by the four and six coils system. The increase of the coils number allows to enlarge the homogeneity of the magnetic field in the axial direction. Thus, four and six coils systems could be used to create a homogeneous magnetic field.

References

1. B. Ando, S. Baglio, A.R. Bulsara, C. Trigona, Sen. &Act. Phys.**151**,145(2009)
2. T. Charubin, M. Nowicki, R. Szewczyk, Mechatron 2017, **644**,613-618(2017)
3. A. Tejera,C.Carretero, J.T. Boys, IEEE Trans. Pow. Electron, **32**(11),8349-8359(2016)
4. J. Wang, S. She, S. Zhang, Rev. of Scen. Instr.**75**, 2175(2002)
5. V. Baranova, P.Baranov, Mech.&Mash.(Dyn.),1-4(2014)
6. S. Magdaleno-Adame, JC. Olivares-Galvan, E. Campero-Littlewood, R. Escarela Perez, E. Blanco-Brisset, *In Excerpt from the proceedings of the COMSOL conference* (Comsol, Massachusetts,2010)
7. R. Beiravand. IEEE Trans. Ind. Electron.,**64**,7120-7131(2017)
8. M. Hosoya and E. Goto, Rev. Sci. Instrum. **62**(10), 2472–2475 (1991)
9. Y. Öztürk, B. Aktaş, Rev. Scien. Instr.,**87**,106103(2016)
10. V. Ogay, Mater. Sci. Eng.,**6**,1-6(1014)
11. D. Jiles, *Introduction to Magnetism and Magnetic Materials, Third Edition* (CRC Press,2015).

A Single-Phase Grounding Fault Judgment Method Based on Mutation Current Logic Matrix

Wu Xiaobin[1], Xu Shihua[1], Zhao Hui[1] and Zhang Wei [2]

[1] *R & D Center, Fujian Automation Electric Power Technology Co.,Ltd., 350012 Fuzhou, Fujian, P. R. China*
[2] *Power Grid & Distribution Department, Jicheng Electronics Co., Ltd,250100 Jinan, Shandong, P. R. China*

Abstract. In order to solve the problem of single-phase grounding fault judgment in non-solid-earthed distribution network, the power flow in non-solid-earthed distribution is analyzed. A single-phase grounding fault judgment method based on mutation current logic matrix is proposed. The minimum fault judgment area model is constructed by the parent and child nodes. The feeder mutation current matrix is generated by the feeder real-time current matrix and the feeder history current matrix. The feeder mutation current matrix is transformed into the feeder mutation current logic matrix by the capacitance current threshold, and the non-zero elements of the feeder mutation current logic matrix are extracted into the mutation current logic sequence list. Then the single-phase grounding fault can be determined in the minimum fault judgment area of the last element of the mutation current logic sequence list as the parent node. The detailed calculation formulas are given. This method is also applicable to the non-solid-earthed distribution network containing distributed generation. The simulation results show that the method proposed in this paper has a good adaptability to the permanent grounding fault and it is worth popularizing.

1 Introduction

FA (Feeder Automation), as an important means to improve the reliability of power supply, has been widely deployed in the current DMS (Distribution Management System), and it plays an important role in distribution fault location. However, at present, the feeder automation system can only deal with the short-circuit fault location, but there is no better way to deal with the single-phase grounding fault judgment in non-solid-earthed distribution network [1-5].

The neutral grounding mode of 10kV distribution network in China mainly includes two kinds of neutral unearthed and neutral earthed via Peterson coil. When the single-phase grounding fault occurs, these grounding modes can limit the short-circuit current to a smaller value. But because of the step voltage, it often leads to human and livestock casualties, so the fault must be isolated as soon as possible. If the neutral grounding mode is changed into neutral earthed through small resistance, when the single-phase grounding fault occurs, the substation outlet circuit breaker will trip. Although this method can isolate the fault, the reliability of power supply is seriously reduced due to frequent tripping of the outlet switch. Therefore, the non-solid-earthed operation mode is desirable in most areas of our country, but it also needs to be removed as soon as possible [6-10].

Passive method and active method are two commonly used methods for single-phase grounding fault location in non-solid-earthed distribution [11-13]. However, the installation of special terminal equipment is required for both the active and passive methods to meet the special sampling requirements. The current distribution network has installed a large number of sampling terminal equipment, such as FTU(Feeder Terminal Unit), DTU(Distribution Terminal Unit), fault indicator, etc., most of these devices do not have a special sampling and determination requirements. If a special sampling decision device is installed, most of the equipment needs to be installed under the condition of power failure, and the reliability of power supply is further reduced. Therefore, the best possible use of existing equipment and sampling data to solve the problem of single-phase grounding fault judgment, try not to install special equipment, to avoid duplication of waste of resources, these practices can be better to build a conservation-oriented society.

Based on the characteristics of single-phase grounding fault location, this paper analyzes the power flow in non-solid-earthed distribution, and presents a single-phase grounding fault judgment method based on mutation current logic matrix.

2 Steady state characteristic analysis of single-phase grounding fault

In the non-solid-earthed distribution, whether the grounding transformer is grounded via Peterson coil or not, its influence on distribution network is the same, which is to reduce the fault short-circuit current. Fig. 1 is the distribution of capacitance current and load current in the non-solid-earthed distribution network. Wherein, the G1 is the equivalent power supply of the system, #1~#4 is the bus, and the ABC are the three phases of bus #3.TRA1 is Y/Y/D type three winding transformer, and rated voltage is 110kV/35kV/10kV. TRA2 is Y/Y type two winding transformer, rated voltage is 10kV/400V, and it can be used as grounding transformer. Feeder1 and Feeder2 are two feeders. Load1 and Load2 are equivalent loads. FP is the fault point.

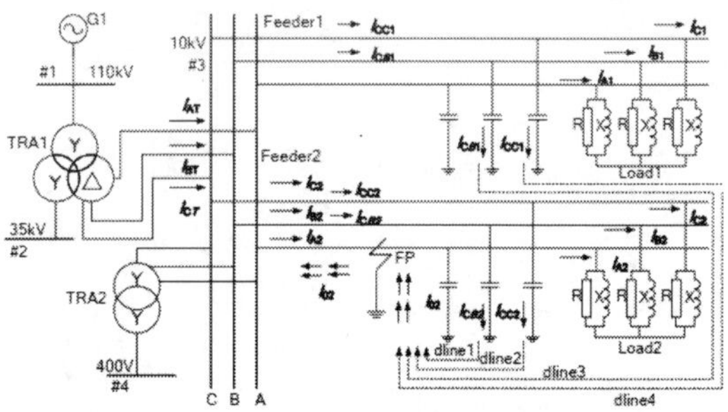

Figure 1. The non-solid-earthed distribution power flow.

When the A phase of the feeder Feeder2 is single-phase grounding fault at the FP point, the B phase and C phase capacitive current phasors I_{CB1}, I_{CC1} and I_{CB2}, I_{CC2} in the non-fault line Feeder1 and the fault line Feeder2 flow from the bus to the feeder and flow to the fault point. As shown in figure dline1~dline4. TRA1 transformer low side because it is triangular wiring, so the winding between the ground capacitance is small, negligible, and the capacitor current between the windings have been incorporated into the load current, as shown phasors I_{AT}, I_{BT}, I_{CT} in Figure 1. TRA2 ground transformer due to neutral point is not grounded, so the capacitance current is also negligible. The total capacitance current flowing through the fault point is the zero sequence current I_{02}, which is the sum of the four capacitor currents whose direction flows from the fault point to the bus.

In the feeder Feeder1 and Feeder2, the load current of each phase is denoted by I_{A1}, I_{B1}, I_{C1}, I_{A2}, I_{B2}, I_{C2} respectively, as shown in the figure. When the fault occurs, due to symmetry, the load current of each phase does not change.

As can be seen from Fig. 1, the current I_{AL2} flowing through the fault phase A in the upstream region of the fault point FP is a superposition of the load current I_{A2} and the zero sequence current I_{02}. And the current flowing through the downstream of the fault point is only the load current I_{A2}. The current flowing through the BC phase of the fault line is the same as before and after the fault, and the value is the load current. The current of ABC three-phase flow in non fault line Feeder1 has no change before and after the fault, and its value is the load current.

For cable lines or cable and overhead hybrid lines, the line capacitance current will be much larger than the overhead line capacitance current, and the substation 10kV feeder is generally more than a dozen, in addition feeder branch more effective length. Therefore, in the city or suburban distribution network with cable line and cable overhead hybrid line, the ground current flowing through the fault point will not be neglected relative to the load current when the single-phase grounding fault occurs.

3 A single-phase grounding fault judgment method

3.1. Minimum fault judgment area

According to the above analysis, the fault location can be done by the mutation of the steady-state current before and after the single-phase grounding fault.

The real-time steady-state current value of the switch needs to be obtained by installing intelligent terminals such as FTU, DTU and fault indicator; The current value of the distribution transformer should be transmitted through the marketing power acquisition system, or through the TTU(Transformer Terminal Unit), intelligent distribution transformer terminal and other equipment acquisition; The current value of substation outlet switch can be collected by intelligent terminal or transmitted through EMS (Energy Management System) system. The switch or distribution equipment without data acquisition will not play a role in the single-phase grounding fault judgment. The switches discussed in this paper have the function of measurement acquisition, and the measurement value is accurate and reliable.

It is stipulated that a switch and all of its sub-node devices (switches and distributors) form a minimum fault judgment area from the start of the substation outlet, the direction of the trenches, to the distribution transformers. The minimum fault judgment area is the minimum range that can be used for fault location. The details are:

$$SMA = \left[MA_1, MA_2, \cdots, MA_i, \cdots, MA_n\right] \quad (1)$$

$$MA_i = \left[a_{i1}, b_{i1}, b_{i2}, \cdots, b_{ij}, \cdots b_{ih}\right] \quad (2)$$

Wherein, SMA is the set of minimum fault judgment area for the feeder. MA_i is the i minimum fault judgment area, i=1,2,3,…,n; n is the number of minimum fault judgment areas; a_{i1} is the parent node device of the i minimum fault judgment area; b_{ij} is the j sub-node device with the a_{i1} as the parent node in the i minimum fault judgment area, j=1,2,3,⋯,h; h is the number of all child nodes with a_{i1} as the parent node.

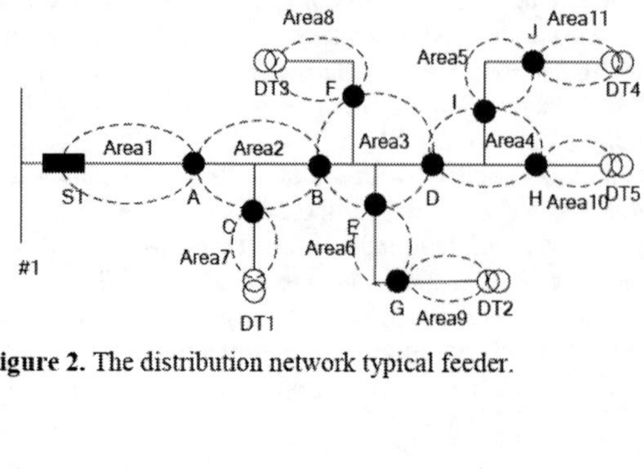

Figure 2. The distribution network typical feeder.

Figure 2 is the distribution network typical feeder; # 1 is the substation bus; S1 is substation outlet switch; A~J are sectional switches which is installed with FTU, DTU, fault indicator and other acquisition devices. DT1 ~ DT5 are distribution transformers.

By definition, the sub-node device with switch S1 as the parent node is switch A. Then the switches S1, A constitute the minimum fault judgment area 1, as shown in Figure 2 dotted line. Similarly, the feeder can be broken down into 11 minimum fault judgment areas, as shown in Table 1.

Table 1. The minimum fault judgment area.

Area	Area 1	Area 2	Area3	Area4	Area5	Area 6
MA_i	S1,A	A,B,C	B,D,E,F	D,H,I	I,J	E,G
Area	**Area 7**	**Area 8**	**Area9**	**Area10**	**Area11**	
MA_i	C,DT1	F,DT3	G,DT2	H,DT5	J,DT4	

Because the switches C, F, G, H, J sub-nodes are distribution transformers, and the distribution transformer has data acquisition function, so the area 7~11 contains the distribution transformer.

3.2 Establishment of mutation current logic matrix model

Because the single-phase grounding fault can cause the current mutation of the upstream fault phase of the fault point, the fault point can be determined by analyzing the current changes before and after the fault of each feeder.

The real-time current matrix *RCM* is defined to describe the three-phase (A,B,C) real-time operation state of all switches and distribution transformers in a distribution feeder. The details are as follows:

$$RCM = \begin{bmatrix} ra_1 & ra_2 & \cdots & ra_k & \cdots & ra_m \\ rb_1 & rb_2 & \cdots & rb_k & \cdots & rb_m \\ rc_1 & rc_2 & \cdots & rc_k & \cdots & rc_m \end{bmatrix} \quad (3)$$

Wherein, ra_k is the A-phase real-time current value of the k switch or distribution transformer included in the feeder; rb_k is the B-phase real-time current value of the k switch or distribution transformer included in the feeder; rc_k is the C-phase real-time current value of the k switch or distribution transformer included in the feeder; k=1,2,...,m, M is the total number of switches and distribution transformers included in the feeder.

The history current matrix HCM is defined to describe the three-phase (A,B,C) historical operation state of all switches and distribution transformers in a distribution feeder. The details are as follows:

$$HCM = \begin{bmatrix} ha_1 & ha_2 & \cdots & ha_k & \cdots & ha_m \\ hb_1 & hb_2 & \cdots & hb_k & \cdots & hb_m \\ hc_1 & hc_2 & \cdots & hc_k & \cdots & hc_m \end{bmatrix} \quad (4)$$

Wherein, ha_k, hb_k and hc_k are the A-phase, B-phase and C-phase historical current value of the k switch or distribution transformer included in the feeder. The history time can be set according to the terminal type. For high-precision real-time acquisition devices with short sampling intervals such as FTU and DTU, the historical time can be set to 2 to 5 minutes before the current time; For the fault indicator, TTU and other sampling interval longer acquisition device, the historical time can be set to 5 to 10 minutes before the current time; For FTU, DTU, fault indicator mixed configuration, the historical time can be set to 5 to 10 minutes before the current time; For EMS and marketing power information acquisition system to transmit data, the historical time can be set to 5 to 10 minutes before the current time. The purpose of the above time setting is to ensure that all data is sent to the DMS.

3.3 Fault judgment method

When the single-phase grounding fault occurs, the real-time current matrix and the history current matrix before the fault can be used to calculate the steady-state current changes caused by the fault. The details are as follows:

$$ACM = RCM - HCM \quad (5)$$

Wherein, *ACM* is the feeder mutation current matrix, which is also a 3 × m matrix.

For the convenience of calculation, the sudden change of feeder current is represented by the mutation current logic matrix *ALM*. The details are as follows:

$$ALM = \begin{bmatrix} la_1 & la_2 & \cdots & la_k & \cdots & la_m \\ lb_1 & lb_2 & \cdots & lb_k & \cdots & lb_m \\ lc_1 & lc_2 & \cdots & lc_k & \cdots & lc_m \end{bmatrix} \quad (6)$$

$$la_k = \begin{cases} 1 & aa_k \ge \alpha MV \\ 0 & aa_k < \alpha MV \end{cases} lb_k = \begin{cases} 1 & ab_k \ge \alpha MV \\ 0 & ab_k < \alpha MV \end{cases} \quad (7)$$
$$lc_k = \begin{cases} 1 & ac_k \ge \alpha MV \\ 0 & ac_k < \alpha MV \end{cases}$$

Wherein, la_k is the A-phase mutation current logic value of the k device of the feeder; lb_k is the B-phase mutation current logic value of the k device of the feeder; lc_k is the C-phase mutation current logic value of the k device of the feeder; aa_k, ab_k, ac_k are the A-phase, B-phase, and C-phase current changes of the k device in the *ACM*. α is the threshold coefficient, the value range is (0,1).

MV is the capacitance current threshold, its value with the line type, line length, substation outlet circuit number are related. The ground capacitance current of the cable feeder will be greater than the ground capacitance current of the cable overhead feeder, and is greater than the ground capacitance current of the overhead feeder. In engineering practice, the value can be determined according to the maximum value of the steady-state current variation of all outgoing lines of the substation after the grounding fault in a year, and proper adjustment is made according to the operation conditions.

If the mutation current logic matrix *ALM* is zero, then the feeder does not occur single-phase grounding fault. If the *ALM* is a non-zero matrix, it indicates that the feeder has a single-phase grounding fault, and the non-zero element line indicates that the grounding fault occurs.

If the f (f=1,2,3) row elements in the mutation current logic matrix *ALM* are non-zero, all non-zero elements are extracted to form the mutation current logic table *ACL*. The details are as follows:

$$ACL = \left[cl_1, cl_2, \cdots, cl_e, \cdots, cl_g\right] \quad (8)$$

Wherein, cl_e is the device number represented by the e logical value 1 in line f of *ALM*, e=1, 2, 3, ..., g, g is the total number of non-zero elements in line f of *ALM*.

Obviously, *ACL* is an unordered table. According to the parent-child relationship of the device, the *ACL* is reordered from the parent to the child, and the logic sequence table *ACSL* of the mutated current is obtained, the details are as follows:

$$ACSL = \left[cs_1, cs_2, \cdots, cs_w, \cdots, cs_g\right] \quad (9)$$

Wherein, cs_w is the device number, w=1,2,3,... g, cs_{w-1} is the parent node device number of the device cs_w, and cs_{w+1} is the sub node device number of the device cs_w.

Combined with the above analysis, the grounding fault occurs in a minimum fault judgment area, and the last element cs_g of the logic sequence table *ACSL* is the parent node of the area.

For the non-solid-earthed distribution network with distributed generation, if the distributed generation is non-earthed running, when the grounding fault occurs, the distributed generation on the system grounding current distribution is similar to the general load [14-15]. Therefore, the grounding fault judgment method described in this paper is also applicable to the non-solid-earthed distribution network where the distributed generation is non-earthed.

4 Case analysis

As shown in Figure 2, it is assumed that a permanent fault of A-phase occurs between the switches D, H and I. Through the PSCAD simulation system, the three-phase simulation current waveform of substation outlet switch S1 is shown in Figure 3. The horizontal axis is time, the unit is s; the longitudinal axis is current, and the unit is A.

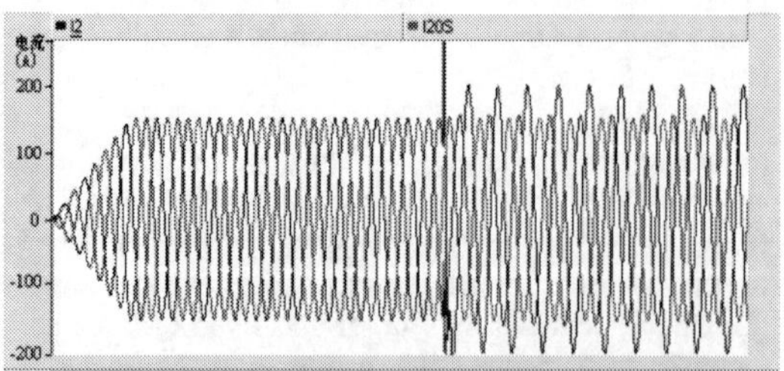

Figure 3. The three-phase current simulation waveform.

The real-time measurement data RD_a, RD_b, RD_c of the equipment A, B and C after the fault occurred, and the historical measurement data HD_a, HD_b and HD_c 5 minutes before the fault occurred are shown in Table 2.

Table. 2 The measured values of equipment (A).

	S1	A	B	C	D	E	F	G
RD RD_a	105.6	105.6	95.5	10.1	61.5	19.5	14.5	19.4
HD_a	94.2	94.2	85.1	9.1	52.1	18.5	14.5	18.5
RD_b	94.4	94.4	84.3	10.1	50.3	19.5	14.5	19.4
HD_b	94.2	94.2	85.2	9.2	52.1	18.5	14.5	18.5
RD_c	94.7	94.7	84.6	10.1	50.6	19.7	14.3	19.7
HD_c	94.2	94.2	85.2	9.2	52.1	18.5	14.5	18.5
	H	I	J	DT1	DT2	DT3	DT4	DT5
RD_a	23.1	27.2	27.2	10.1	19.5	14.2	27.1	23.0
HD_a	23.9	28.2	28.2	9.1	18.5	14.2	28.2	23.9
RD_b	23.1	27.2	27.2	10.1	19.5	14.2	27.1	23.0
HD_b	23.9	28.2	28.2	9.1	18.5	14.2	28.2	23.9
RD_c	23.3	27.3	27.3	10.1	19.7	14.3	27.3	23.3
HD_c	23.9	28.2	28.2	9.1	18.5	14.2	28.2	23.9

According to the method described in Section 2, *MV* is 10, α is 0.6, then the mutation current logic matrix *ALM* is:

$$ALM = \begin{bmatrix} 1 & 1 & 1 & 0 & 1 & 0 & 0 & 0 & 0 & 0 & 0 & 0 & 0 & 0 & 0 & 0 \\ 0 & 0 & 0 & 0 & 0 & 0 & 0 & 0 & 0 & 0 & 0 & 0 & 0 & 0 & 0 & 0 \\ 0 & 0 & 0 & 0 & 0 & 0 & 0 & 0 & 0 & 0 & 0 & 0 & 0 & 0 & 0 & 0 \end{bmatrix}$$

Because only the first element of *ALM* has non-zero value, the grounding fault occurs in the A-phase of the feeder. According to formula (8) and (9), the logic sequence table *ACSL* of the sudden change current can be obtained:

$$ACSL = \left[S1, A, B, D\right]$$

It can be seen that the fault occurs in the minimum failure determination area where the switch D is the parent node. According to the formulas (1), (2) and Table.1, it can be seen that the area where the ground fault occurs is the minimum fault judgment area Area4.

5 Conclusion

According to the above analysis, it can be seen that in the non-solid-earthed distribution network, when the single-phase grounding fault occurs, the current in the upstream fault phase is abruptly changed due to the superposition of the normal load current and the capacitance current, and the current in the downstream fault phase will keep the load current constant. For other non-fault phase and non-fault feeder current in the upstream or downstream of the fault point, the steady-state current values are load current without mutation.

A single-phase grounding fault judgment method based on mutation current logic matrix is proposed. Then the single-phase grounding fault can be determined in the minimum fault judgment area of the last element of the mutation current logic sequence list as the parent node. This method is also applicable to the non-solid-earthed distribution network containing distributed generation. The simulation results show that the method proposed in this paper has a good adaptability to the permanent grounding fault and it is worth popularizing.

References

1. Zhang Guping, Jiang Chao, Li Gang. *Method of Fault Area & Section Location for Non-solidly Earthed Distribution System*. Proceedings of the CSEE. **32** (2012)
2. Chen Bobo, Qu Weifeng, Yang Hongyu, etal. *Research on single phase grounding arc model and line selection for neutral ineffectively grounding system*. Power System Protection and Control. **44** (2016)
3. Lai Ping, Zhou Xiangling, Qiu Dan. *Research on transient-current frequency analysis and faulty line detecting method in indirectly grounding power system*. Power System Protection and Control.**43** (2015)
4. Chen Kui, Wei Xiaoguang, Chen Jingbo,etal. *Fault Line Detection Using Sampled Data Processing and adaboost for Small Current Grounding System*. Proceedings of the CSEE.**34** (2014)
5. Yang Jianxiang, Chen Zhongxue, Dou Minna. *A novel frequency band self-adaptive selection method of fault line selection in distribution network*. Power System Protection and Control.**43** (2015)
6. Yan Feng, Yang Qixun, Qi Zheng, etal.Study onfaultlocation for distribution network based on travelling wavetheory. Proceedings ofthe CSEE.24 (2004)
7. Wu Lepeng, Huang Chun, Lin Dabin,etal. *Faulty line selection based on transient wavelet energyfor non-solid . earthed network* . Electric Power Automation Equipment.**33** (2013)
8. Liu Mouhai, Fang Tao, Jiang Yun,etal . *A new correlation analysis approach to fault line selection based on transient main-frequency components*. Power System Protection and Control. **44** (2016)
9. Ma Shicong, Xu Bingy in,Gao Houlei. *An earth fault locating method in feeder automation system by examining correlation oftransient zero mode currents*. Automation of Electric Power Systems. **32** (2008)
10. Xue Yongduan, Li Juan, Xu Bingyin. *Transient Equivalent Circuit and Transient Analysis of Single-phase Earth Fault in Arc Suppression Coil Grounded System*. Proceedings of the CSEE. **35** (2015)
11. Gao Zhipeng, Zhang Huifen, Sun Xuna. *A method of fault line selection and fault point location with halFwave DC injection in distribution network*. Power System Protection and Control. **41** (2013)
12. Zhang Huifen, Pan Zhencun, Sang Zaizhong. *Injecting current based method for fault location in neutral isolated power system*. Automation ofElectric Power Systems. **28** (2004)
13. Wu Jun , Lu Gang . *Virtual grounding identification technology in neutral point non-effectively grouned system* . Power System Protection and Control. **38** (2010)
14. Sun Yongchao, Tai Nengling, Zheng Xiaodong. *Research on single-phase grounding fault location of distribution network with DG*. Journal of Electric Power Science and Technology. **31** (2016)
15. Guo Liwei, Xue Yongduan, Zhang Linli,etal. *Analysis of Single-phase Earth Fault in Low Resistance Grounded Distribution Network Containing DG*. Automation of Electric Power Systems. **39** (2015)

Research on grounding grid corrosion classification method based on convolutional neural network

Jingyi Du, Liqian Yan ,Haixia Wang,and Qiong Huang

College of Electrical and Control Engineering, Xi'an University of Science and Technology ,Xi'an, China

Abstract. Aiming at the problem that the traditional detection methods can not accurately classify the corrosion degree of grounding grids. The corrosion image is taken as the research object , the convolution neural network is used as the algorithm firstly to classify the corrosion degree. Firstly, the corrosion simulation experiment was carried out, and the sample library was established by using the corrosion image collected in different stages. Then, according to the LeNet-5 model , the traditional CNN and improved CNN models were designed for corrosion classification of grounding grid. Simulation experiments were carried out in the preprocessed samples. Finally, the experimental results of Soft-max and SVM classifier are compared and analyzed. The results show: the classification results of the two models were better than those of the original samples, and the classification performance of SVM is better than that of Soft-max. The improved model can improve classification accuracy. This study fills the blank of detecting the corrosion degree of grounding grid by image method, and it is significant to quickly grasp the corrosion degree to avoid faults or accidents.

1 Introduction

Grounding grid is the necessary facilities for substation grounding, and it is the key to ensure the electrical equipment, power system and personal safety. The grounding material of domestic commonly used steel, burying in soil medium underground complex, corrosion will be gradually deepened, eventually led to the loss of grounding function even ruptured, causing a series of grounding accidents. Therefore, accurate understanding of the corrosion of grounding grid, timely detection of problems and take corresponding measures, is very important for the safe operation of the power system.

In order to ensure the normal operation of the grounding grid, it is necessary to take effective measures to detect the corrosion state of the grounding grid. But the grounding conductor is buried in the soil all the year round, so it is very expensive and difficult to operate when the corrosion condition is checked[1-3]. Based on this, this paper first proposes the grounding grid corrosion image as the research object, and uses convolutional neural network as the research algorithm to classify the corrosion degree of the grounding grid.

2 Classification basis of grounding grid corrosion

2.1 Classification criteria of corrosion degree

This paper, the corrosion grade of grounding grid is evaluated by subjectively according to the national standard GB/T6461-2002 "rating of specimens and specimens after corrosion test on metal substrate and other inorganic covering layer". Table 1 is the corrosion appearance rating corresponding to the corrosion appearance features given in the standard.

Table 1. Representation of appearance rating.

rating	Change of surface appearance of test sample
A	very few corrosive spots
B	Slight discoloration
C	Obvious discoloration and loss of light
D	Slight corrosion occurred
E	The corrosion products or pitting corrosion on the local surface of the sample is thin
F	Severe discoloration and corrosion products or pitting distribution on the surface of the sample
G	There is a thick corrosion product layer or pitting corrosion on the whole surface
H	There is very thick corrosion product layer or pitting corrosion on the whole surface, and there is very deep pitting corrosion
I	Matrix metal corrosion occurs

The subjective evaluation of the extent of corrosion damage in the standard can be divided into four categories: VS is very mild, S is mild, M is moderate, X is severe, and the appearance rating number from A to I indicates that the degree of corrosion is from light to heavy. According to the features of corrosion degrees in the standard on the corrosion image samples plus artificial participation classification label, the appearance ratings in the A and B as a very mild corrosion, C, D, E as mild corrosion, F, G as moderate corrosion, H, I as severe corrosion.

Very mild and mild corrosion condition, grounding grid can work normally only routine monitoring; moderate to increase monitoring efforts, and make timely rectification; severe corrosion case, watch out for the grounding accidents, to immediately repair [7].

2.2 Establishment of corrosion image sample library

Carbon steel is widely used in grounding grid materials in China, and the corrosion problem is relatively serious. In the simulation experiment of grounding grid corrosion, a carbon steel plate is buried in the soil with depth of 0.8m, and the relatively complex environment. The corrosion image information is collected at different time. There is not a relatively complete system of grounding grid corrosion image database in the existing literature, and create a real grounding grid corrosion image database needs years of research and accumulation.

The image cutting method is used to expand the sample of corrosion image data[4].The specific cutting method is the use of PhotoShop image processing software, each pretreatment of the corrosion image size normalization, become a number of 32*32 small square image as the input of the network, and the principle of shear is maximum range preserving image corrosion feature information. Through the above treatment, we can build a sample library containing four kinds of image with different corrosion degree, each of 2500 kinds. The pretreatment samples are shown in Figure 1.

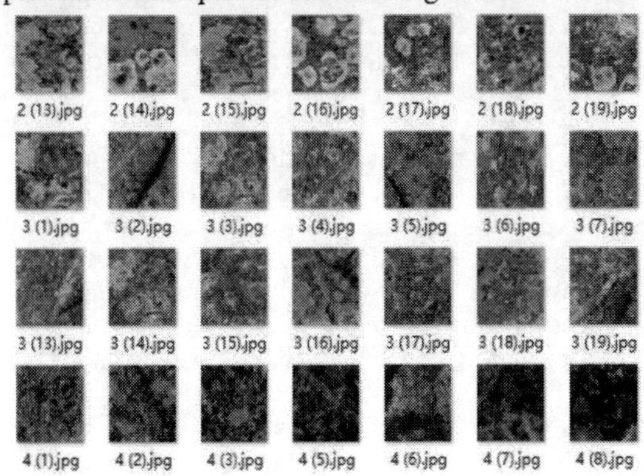

Figure 1. Sample after pretreatment.

3 Design of traditional CNN model

3.1. Brief introduction of CNN

Convolutional neural network contains four basic structure layers, which are input layer, output layer, convolution layer and sampling layer [5]. In CNN, the convolutional layer and the sampling layer are alternately connected, and the feature is extracted by the convolution layer, reducing the dimension by the sampling layer , in order to reduce the complexity of the model and improve the convergence speed of the network.

3.2 Model design

1) Input and output layer design

In the traditional CNN model, the size of each image in the sample library is 32*32, stored in the pixel form of the whole picture 32*32=1024 in the data file, and the whole pixel is used as the input of the network.

According to the standard GB/T6461-2002, the corrosion images can be divided into four categories: very mild corrosion, mild corrosion, moderate corrosion, severe corrosion, so the output layer of the network is set to 4 neuron nodes.

2) Intermediate layer design

The traditional CNN classification model is to improve the existing problems of LeNet-5 network, mainly from the following aspects:

a) The maximum sampling method is selected. Based on the nonlinear principle of maximum sampling, the sampled images are divided into non overlapping regions. The pixel is the basic unit for segmentation, and the output value is the largest pixel value in each rectangle area.

b) Using Soft-max and SVM[6] instead of radial basis functions. This paper chooses Soft-max and SVM as the classifier model CNN to achieve better classification effect.

Soft-max regression has achieved good results in solving multi classification problems. A training sample $\{(x^{(1)},y^{(1)}),(x^{(2)},y^{(2)})\ldots(x^{(m)},y^{(m)})\}$ is made up of M images with class labels, this paper, k=4 is used to classify the corrosion degree of grounding grids.

For training sample x, the probability value $p(y=j/x)$ is estimated by using the assumed function for each category j That is, the probability that the sample x is assigned to each classification result is estimated. Suppose that the function $h_\theta(x)$ can be expressed as

$$h_\theta(x^{(i)})=\begin{bmatrix} p(y^{(i)}=1\mid x^{(i)};\theta) \\ p(y^{(i)}=2\mid x^{(i)};\theta) \\ \vdots \\ p(y^{(i)}=k\mid x^{(i)};\theta) \end{bmatrix}=\frac{1}{\sum_{j=1}^{k} e^{\theta_j^T x^{(i)}}}\begin{bmatrix} e^{\theta_1^T x^{(i)}} \\ e^{\theta_2^T x^{(i)}} \\ \vdots \\ e^{\theta_k^T x^{(i)}} \end{bmatrix} \quad (1)$$

Among them,$\theta_1,\theta_2,\ldots,\theta_k \in R^{n+1}$, are the parameters of the model, the fractional part is normalized.

The cost function can be expressed as:

$$J(\theta)=-\frac{1}{m}\left[\sum_{i=1}^{m}\sum_{j=1}^{k}1\{y^{(i)}=j\}\log\frac{e^{\theta_j^T x^{(i)}}}{\sum_{l=1}^{k}e^{\theta_l^T x^{(i)}}}\right] \quad (2)$$

In Soft-max regression, the probability of classifying sample X into category j is:

$$p(y^{(i)}=j\mid x^{(i)};\theta)=\frac{e^{\theta_j^T x^{(i)}}}{\sum_{l=1}^{k}e^{\theta_l^T x^{(i)}}} \quad (3)$$

The support vector machine (SVM) method is based on the limited sample information to find the best compromise between the complexity of the model and learning ability[7].

To construct hyperplane in high dimensional feature space, and then transform nonlinear classification into linear classification problem, that is, the following optimization problems:

$$\min\left(\frac{1}{2}\|w\|^2 + C\sum_{i=1}^{l}\xi_i\right) \tag{4}$$

Constraint as:

$$y_i(<w,\phi(x_j)>+b) \geq 1-\xi_i, i=1,\cdots,l \tag{5}$$

The Lagrange function is introduced to solve convex optimization problems.

$$L=\frac{1}{2}\|w\|^2 + C\sum_{i=1}^{l}\xi_i - \sum_{i=1}^{l}\alpha_i(y_i(<w,\phi(x_i)>+b)-1+\xi_i - \sum_{i=1}^{l}\gamma_i\xi_i \tag{6}$$

The $\alpha_i,\gamma_i \geq 0, i=1,\ldots,l$, α_i,γ_i are Lagrange multipliers.The extreme value of function L should satisfy the condition:

$$\frac{\partial}{\partial w}L=0, \frac{\partial}{\partial b}L=0, \frac{\partial}{\partial \xi_i}=0 \tag{7}$$

Finally, the discriminant function is obtained:

$$f(x)=\text{sgn}(\sum_{i=1}^{l}\alpha_i y_i K(x_i,x)+b) \tag{8}$$

c) Increase the number of convolution kernels of the stacking layer. In addition, the next layer feature map with convolution kernel size determined by a convolution kernel is $w \times w$, stride=k filter, to a new image ($n_h \times n_w$) convolution processing, new feature map size $(n_h-w+k)k \times (n_w-w+k)k$.

d) Adding Dropout to F5 layer. The basic idea of Dropout is that the activation value of some neurons is output to 0 by the probability P, and the P is set to 0.5.

3.3 Model establishment

Through the above improvements, two traditional CNN models, CNN-Soft-max and CNN-SVM are formed, and the model structure is shown in Figure 2.

3.4 Model training

a) Training process

The CNN training method is supervised training. When the model is convergent, the optimal input and output functions can be obtained by preserving the optimal parameters learned. The training process of the network is shown in figure 3.

The CNN training process is the same as the traditional BP, including two stages. The first stage is the forward propagation stage:

(a) Select a tagged image sample (x, t) from the training set and send the image into the network.

(b) The corresponding actual output y is calculated.

In the first stage, the tagged image starts from the input and transforms layers by layer to extract the feature and reduce the dimension. Then the final extracted feature is sent to the classification layer to output.This process is the forward calculation of each layer weight and activation value:

$$O_p = F_n(\ldots(F_2(F_1(X_p W^{(1)})W^{(2)})\ldots)W^{(n)}) \tag{9}$$

The second stage, back propagation stage:

(a) Calculate the difference between the output y of the model and the corresponding ideal output, that is, the label value t.

(b) The error is transmitted back, and the update weight parameter is adjusted according to the minimization error

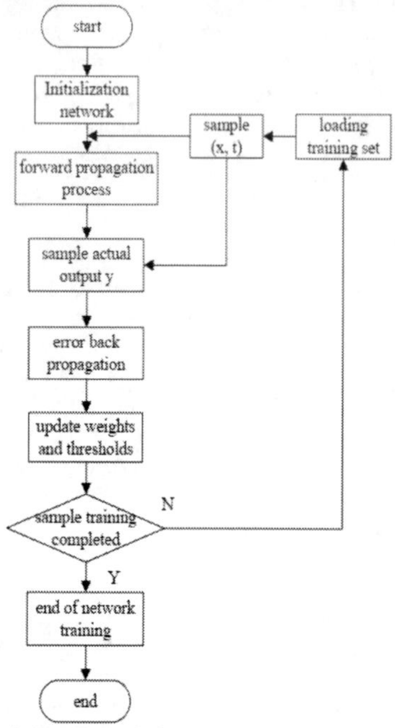

Figure 3. Training flow chart

b) Setting training parameters

In this paper, the implementation of convolutional neural network is mainly based on the MATLAB Deep Learning toolbox DeepLearn Toolbox.

Network initialization. CNN network initialization is the initialization of the relevant parameters, the main work is to roll layer, output layer weights and bias of the initialization parameter settings, DeepLearnToolbox toolbox for the random initialization of weights.

Learning rate.The learning rate (learning_rate, represented by alpha) is proportional to the training speed of the network. When the alpha is larger, the training speed of the network can be accelerated, but the convergence is poor. On the contrary, the network stability is good, but it will fall into local optimization to a great extent. The setting of learning rate is usually adjusted by experience and adjusted according to the actual training situation.

(c) Convergence condition.The iterative method is used to train the network, and the training model is stopped when the error is not obvious after the continuous epoch times batchsize (batch training), and the model parameters are preserved when the minimum classification error is obtained. The number of epoch batch training can be set according to the actual situation.

(d) Parameter setting in simulation process. The parameter settings in the MATLAB simulation process are as follows:

options.epochs =50,options.batchsize = 20,
options.alpha = 1e-1,options.momentum =0.95,

The selection of training method is to minimize the error function by using the random gradient descent algorithm of the driving quantity, and the momentum is set to 0.95. The weights of the convolution layer and the output layer are randomly initialized, and the bias is initialized to 0.

4 Design of improved CNN model

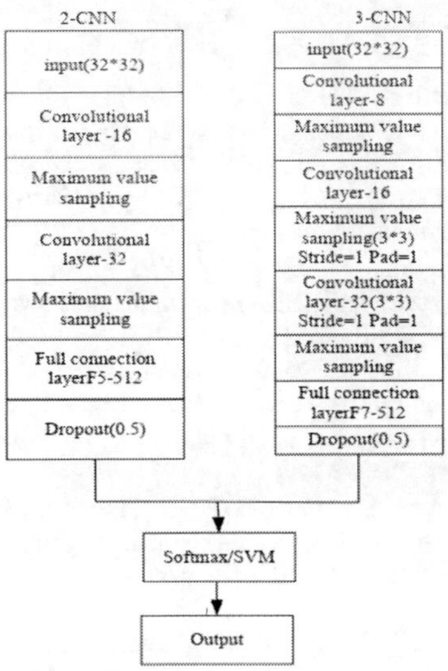

Figure 4. Schematic diagram of improved CNN model

The improved CNN model as shown in Figure 4, the left is a traditional CNN model (2-CNN), the difference is the improved model of ReLU using the activation function, ReLU function is an activation function in recent years deep learning fields commonly used the mathematical form of $f(x)=\max(0,x)$. On the right is a 3 layer network volume (3-CNN), Pad=1 said in a circle around the map feature extraction to ensure zero, more features, the model contains two different topologies of CNN parameters, save two CNN minimum error after the end of the model training.Two kinds of different network topology can learn and extract the corrosion characteristics of more abundant, the better generalization performance.

Model training, the training image input the 2-CNN and 3-CNN, in the connection layer respectively output 512 one-dimensional feature vector, the feature vector stitching two get 512 dimensional a 1024 dimensional feature vector as the classification model of Soft-max and SVM input, to get a training class of higher performance, generalization ability the stronger CNN classification model.

The training set and test set are also generated by MATLAB code of traditional model, and the training method and initialization parameter setting are the same as the traditional model.

5 Model simulation experiment

5.1 Training set and test set generation

MATLAB loads the input image from the mat format database, so first the image sample set is converted into mat data format. Input conversion code to convert.Jpg format image data in the database to mat format files in the file editor, and add to the image classification label, and random selection of 80% of the total sample as the training set and test set as 20%.Table 2 gives partial image samples of.Mat format generated by MATLAB, and adds classified labels. The classified label values are set to [0 0 0 1], [0 0 1 0], [0 1 0 0], [1 0 0 0], which represent four kinds of image data, very mild corrosion, mild corrosion, moderate corrosion, and severe corrosion, respectively.

5.2 Simulation experiment

The overall block diagram of algorithm simulation is shown in Figure 5. The simulation process mainly consists of two parts: training phase and testing phase, the former loading training set update parameters through the back-propagation algorithm, the end of the training save model parameters; test stage test data will be fed to the trained model, the results is given, and record the accuracy.

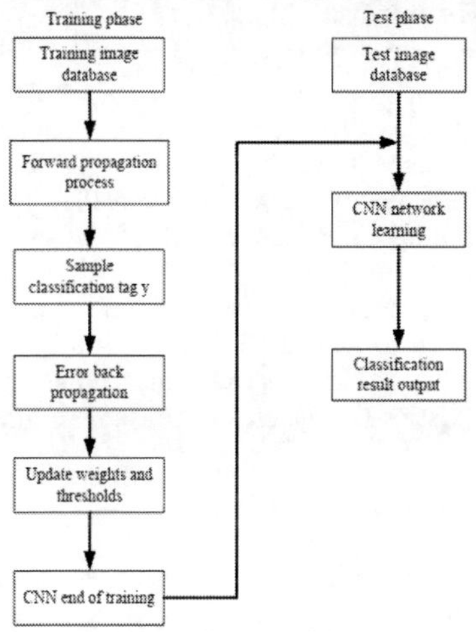

Figure 5. The overall block diagram of algorithm simulation

6 Analysis of experimental results

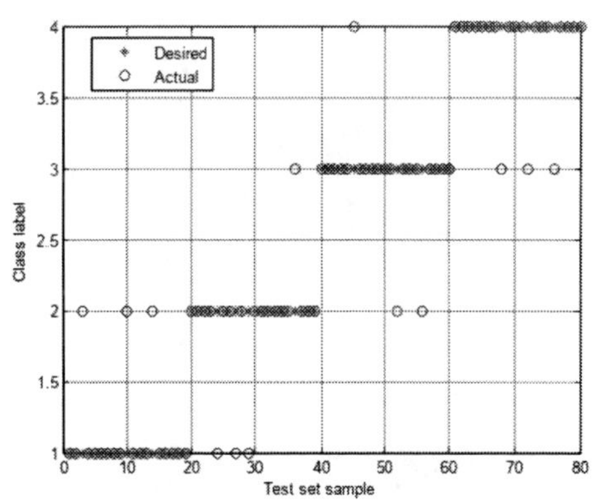

Figure 6. Traditional CNN-SVM test results

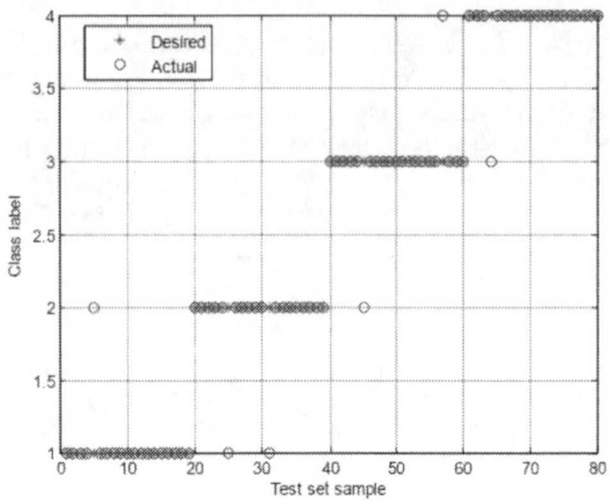

Figure 7. Improved CNN-SVM test results

Table 2. Comparison of Recognition Rates of Two CNN Models (%).

Algorithm	Preprocessing sample bank	
	Training set recognition rate	Test set recognition rate
CNN-Softmax	83.14	79.26
CNN-SVM	85.21	81.55
Improved CNN-Softmax	87.58	84.09
Improved CNN-SVM	91.25	89.37

Figure 6 is the test results of the traditional CNN-SVM on some test sets, in which the abscissa is the sample number of the test set, and the ordinate is the corresponding classification label (drawing 1, 2, 3, 4 as the class labels).In order to observe the classification results, experiment selected 80 test samples of the mapping, it is shown that, the 80 test samples, the traditional model and improved model in the pretreatment samples were 13, 6 sample classification error.

7 Conclusion

This paper, the relative advanced depth learning convolutional neural network is applied to image recognition of grounding grid corrosion degree. The improved CNN model has higher classification accuracy than the traditional CNN model, and the improved CNN-SVM model has better classification effect.Compared with the traditional detection methods of corrosion, corrosion of the image recording is more realistic, and this study fills the blank of detecting the corrosion degree of grounding grid by image method, and provides a new research idea for the research and application of related fields. At the same time, it is of great significance to quickly grasp the corrosion degree of grounding grid to avoid faults or accidents.

References

1. Guemes JA, Hernando FE.Method for calculating the ground Resistance of grounding grids using FEM[J]. IEEE Transactions on Power Delivery, 2004, 19(2): 595.
2. 2.Sverak JG. Progress in step and touch voltage equations of ANSI/IEEE Std 80-historical perspective. IEEE Transactions on Power Delivery[J]. 1998, 13(3): 762.
3. Dawalibi FP,Ma J, Southey RD. Behavior of grounding systems in multiplayer soils: a parametric analysis [J]. IEEE Transactions on Power Delivery, 1994, 9(1): 334.
4. Nguyen A, Yosinski J, Clune J. Deep neural networks are easily fooled: High confidence predictions for unrecognizable images[J]. 2014:427-436.
5. Hubel D H, Wiesel T N. Receptive fields, binocular interaction and functional architecture in the cat's visual cortex.[J]. Journal of Physiology, 1962, 160(1):106.
6. Abdel-Rahman E M, Mutanga O, Adam E, et al. Detecting Sirex noctilio, grey-attacked and lightning-struck pine trees using airborne hyperspectral data, random forest and support vector machines classifiers[J]. Isprs Journal of Photogrammetry & Remote Sensing, 2014, 88(2):48–59.
7. Srivastava N, Hinton G, Krizhevsky A, et al. Dropout: a simple way to prevent neural networks from overfitting[J]. Journal of Machine Learning Research, 2014, 15(1):1929-1958

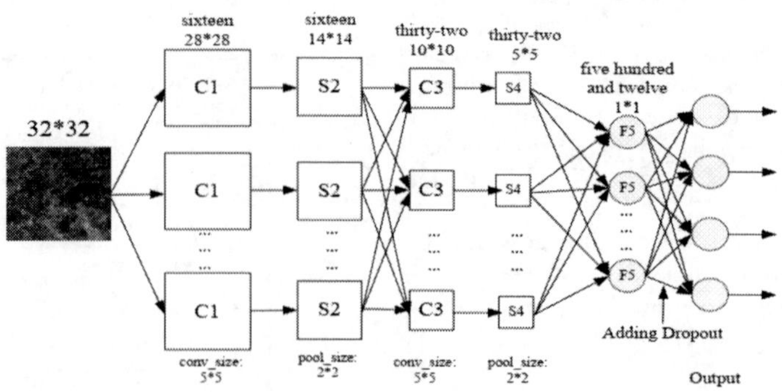

Figure 2. Traditional CNN model structure

MATEC Web of Conferences **160**, 01009 (2018) https://doi.org/10.1051/matecconf/201816001009
EECR 2018

SLG(Single-Line-to-Ground) Fault Location in NUGS(Neutral Un-effectively Grounded System)

Wenhai Zhang[1], Meimei He[2], Jie Ren[1], Yongke Wen[1], Zheng Zhang[1], Zhou Pu[1], Rongjian Zhang[1]

[1]*State Grid Sichuan Maintenance Company, Chengdu, Sichuan Province, China*
[2]*Sichuan Electric Power Design & Consulting Co,. Ltd., Chengdu, Sichuan Province, China*

Abstract. This paper reviews the SLG(Single-Line-to-Ground) fault location methods in NUGS(Neutral Un-effectively Grounded System), including ungrounded system, resonant grounded system and high-resistance grounded system which are widely used in Northern Europe and China. This type of fault is hard to detect and location because fault current is the sum of capacitance current of the system which is always small(about tens of amperes). The characteristics of SLG fault in NUGS and the fault location methods are introduced in the paper.

1 Introduction

In medium voltage, 10kV~35kV, distribution system in China and Nordic countries are mainly NUGS (Neutral Un-effectively Grounded System). The SLG (single-line-to-ground) fault is the most common fault type in NUGS which is also called earth fault by some scholars. It is reported that about 50-80% of all faults are of this type[1]. This type of fault is different from the line-to-line fault in NUGS and the SLG fault in neutral effective grounded system, the load is not affected by the fault and the fault current is always tens of Amperes which is determined by the capacitive charging current of the system. But the healthy phases' voltage would arise to line-to-line voltage which may cause insulation breakdown and lead to fault expands. The SLG fault may lead to the fire hazard and electric shock. So it is necessary to locate the SLG fault as soon as possible, then isolate it and repair such a fault.

Research on SLG fault location has been conducted for decades. There are many methods have been proposed and can be divided into single measurement based and multiple measurements based, steady signal based and transient signal based, time domain based and frequency based, and so on. However, there is no one could solve the problem very well for many reasons, especially the small fault current and complex system structure. All the methods will be introduced in the paper.

2 The Characteristics of SLG Fault and Problem Definition

The typical SLG fault model in NUGS can be depicted as in Fig 1. The system is neutral ungrounded system when breaker K is open, and it is resonant grounded system when K is closed. Considering the fault resistance R_f and fault distance D is zero, the fault phase voltage will drop to zero and two healthy phase to ground voltages will increase to times of phase voltage. There is no large fault current for the fault path is high impedance. Fault current If is the sum of all line to ground capacitive current.

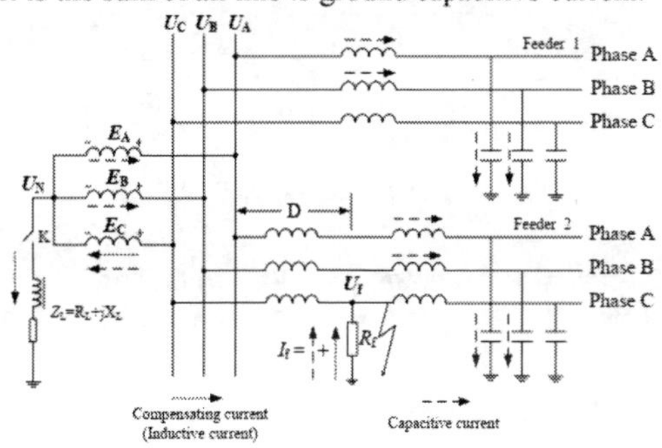

Figure 1. The SLG fault model in neutral ungrounded and resonant grounded system

The transient is obvious when SLG fault happened because large number of energy-storage elements(inductance and capacitance) contained in the system which is introduced in [2] in detail. It has larger amplitude than the steady state fault current. The typical three phase voltage and three phase current waveforms of fault feeder are shown in Fig. 2. When a SLG fault happens, three different components can be distinguished in the transient signal[3]. The discharge transient is initiated when the voltage of the faulty phase falls and the charge stored in its earth capacitances is removed. Because of the voltage rise of the two sound phases, another component, called charge transient, is created. The interline compensating components equalize the voltages of parallel lines at their substation terminals. In compensated networks there is, in addition, a decaying DC-transient of the suppression coil circuit. This component is usually at its highest, when the fault takes place close to voltage zero. If the coil is saturated, the current may also include harmonics.

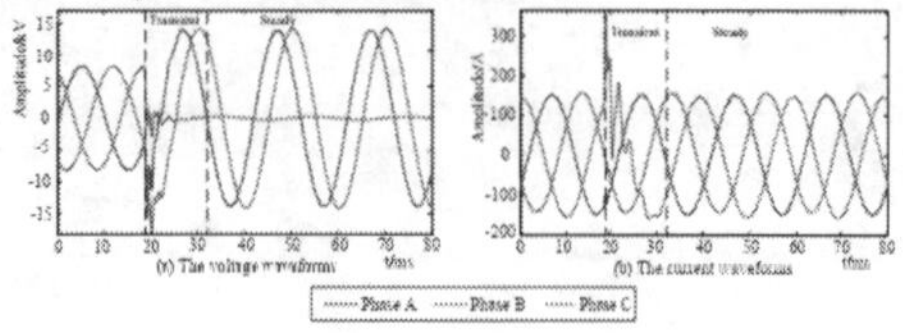

Figure 2 The typical waveforms of SLG fault in NUGS(fault in phase C)

3 Review of SLG Fault Location Methods in NUGS

3.1 The Classification of the Methods

In the past two decades, many algorithms have been proposed to locate the SLG fault in NUGS. Those methods can be divided into many groups (see Fig 3), which are introduced in this chapter respectively. The passive methods utilize the features cause by fault, and the active methods need some other device to produce disturbance for accurate fault location. The passive methods can be subdivided into single-ended methods and multiple measurements methods based on the number of required measurement devices.

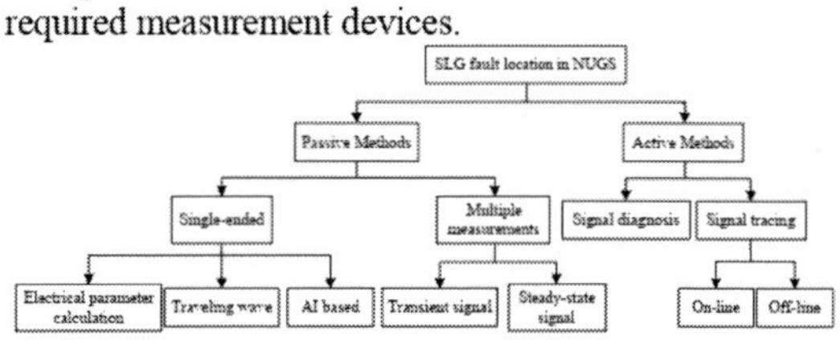

Figure 3. SLG fault location methods in NUGS

3.2 Electrical Parameter Calculation Based Methods

3.2.1 Fundamental Based Methods

This types method mainly used in effectively grounded system. Because the fault would cause large fault current which can be used for fault distance calculation. It also can be classified into phase analysis based method[4], sequence analysis based method[5].

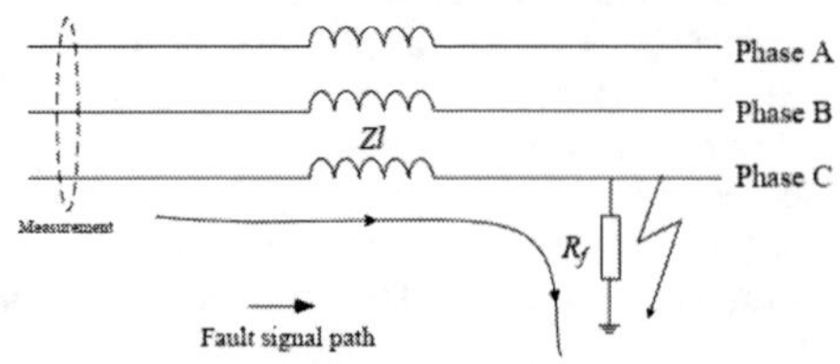

Figure 4. Fault signal path

3.2.2 Transient Based Methods

The transient based methods calculate the fault distance by calculating the inductance L of fault path. There are three methods to calculate the L, including differential equation method, Fourier transform based method and wavelet based method.

(1) Differential equation methods

The simplified equivalent circuit of fault path is considered as one order R-L circuit. Then the differential equation algorithm calculates the fault distance by numerical solving the differential equation as follows:

$$u(t) = R \cdot i(t) + L \cdot \frac{di(t)}{dt} \tag{1}$$

where u(t) is the instantaneous voltage; i(t) is the instantaneous current; R is the resistance of equivalent R-L circuit; L is the inductance of equivalent R-L circuit.
The equation can be solved with a numerical integration and the inductance of the fault path can be calculated if three equally spaced pairs of phase currents and voltage samples are available[6].

$$L = \frac{\Delta t}{2}\left[\frac{(i_{k+1}+i_k)(v_{k+2}+v_{k+1})-(i_{k+2}+i_{k+1})(v_{k+1}+v_k)}{(i_{k+1}+i_k)(i_{k+2}-i_{k+1})-(i_{k+2}+i_{k+1})(i_{k+1}-i_k)}\right] \tag{2}$$

(2) Fourier transform based methods

In ref[7], a method using the Fourier-transform to calculated the line impedance in frequency domain. The reactance of the faulty line length is obtained directly as the imaginary part of the impedance calculated from the corresponding frequency spectrum components of the voltage and current. Ref[8] use this type method in the cable fault location. But the high frequency signal contained in transient signal is an exponential decaying signal which could cause error using FFT analysis.

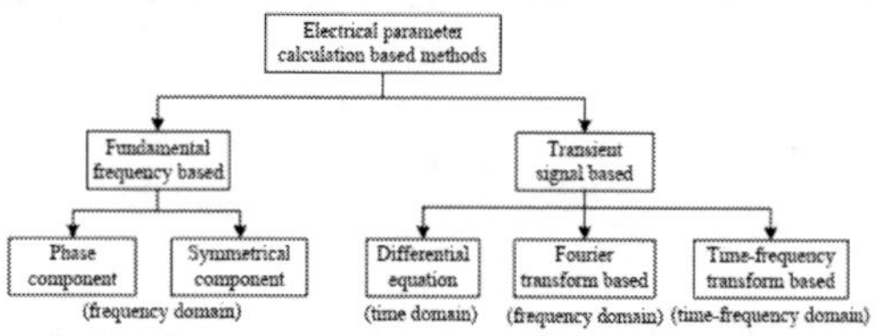

Figure 5. Electrical parameter calculation based methods

(3) Time-frequency transform based methods

This type of method uses time-frequency transform to extract transient voltage and current to calculate the line impedance. It is based on one specified damped exponential signal for electrical parameter calculation. The wavelet transform is the typical method to extract time variant signal. Ref[9] use continuous wavelet transform to extract the specified signal to calculate fault path inductance.

$$L = \frac{1}{\omega} imag\left[\frac{U_w(k\Delta t, f)}{I_w(k\Delta t, f)}\right] \tag{3}$$

Where ω is angular frequency; L is inductance of fault path; Uw and Iw are the wavelet coefficient of voltage and frequency separately; f is frequency for the distance calculation.

The electrical parameter calculation based methods estimate the fault distance by calculating the inductance of transmission line from substation to fault position. These methods are mainly single ended, and can be divided into fundamental signal based and transient signal based as shown in Fig 5. The fault distance equation can be formulated in time domain and frequency domain.

3.3 Traveling Wave Based Methods

This type of method is not affected by fault type, because it is not based on the fault current analysis. Traveling waves(also called surges) on power lines arise from a number of causes, of which the most common are faults, switching operations, and lighting. Surges on overhead power lines travel at the speed of light, approximately 3×10^8 m/sec, and consist of a voltage wave and a current wave related through the surge impedance of the line[10]. There are two types of traveling wave based methods, first one utilizes the traveling wave propagates time from fault point to substation combine with traveling wave speed for calculation; another one is based on the traveling wave frequency for calculation.

3.3.1 Time-distance based methods

The traditional traveling wave based method in distribution system is the single-ended method using the time difference between the first arrival of an incident traveling wave generated by a fault and corresponding reflected wave reflected from point. We nominated this type method time-distance based traveling wave method in this paper. The principle of the method is shown in Fig 6. Knowing this time delay and traveling wave velocity the place of fault is determined. In three-phase networks, the traveling waves are mutually coupled which means a single traveling wave velocity does not exist. Therefore, the phase domain signals are first decomposed into their modal components by means of the modal transformation matrices and each node is treated separately.

Considering the fault is detected at end S, the single-ended fault distance calculation is as follows[11]:

$$D=\frac{v\cdot(t_{s3}-t_{s1})}{2} \tag{7}$$

where D is fault distance, v is traveling wave velocity; t0 is fault time, then the fault generate arrives at end S at time t_{s1}, and t_{s3} is the time traveling wave reflect from fault point. It is necessary to distinguish between the reflected wave from the fault point and that from the remote bus-bar.

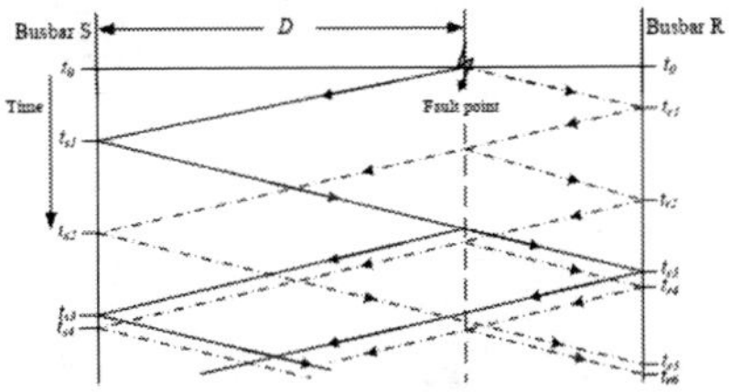

Figure 6 The reflection of traveling wave in the system

In this method, high-frequency voltage signals in the frequency range of 1 to 10MHz are captured for fault location purpose[11]. In order to apply the method in three-phase systems, modal transform are used to extract the aerial mode and ground mode signals. It has been proven that the fault-generated high-frequency transient signal is not affected by the variation of fault type or fault path impedance.

The most critical issue in these methods is detecting and extracting desired part of the signal, which indicates the traveling of the waves. There are many methods have been proposed to identify the reflected wave, including cross-correlation[12], wavelet transform and so on.

3.3.2 Frequency-distance based methods

Another type of traveling wave based method is emerged in recent years which utilize the relationship between specific paths in the network and traveling wave frequency. Fault-originated traveling waves propagate along the network and reflect at line terminations, junctions between feeders, and the fault location. The characteristic frequency of the path is determined by the length and structure of the system[13].

Ref[14] introduced the relationship between characteristic frequency, traveling wave speed and signal path. Assuming the network topology and the traveling wave speeds of the various propagation modes are known, frequency of mode through path can be evaluated a priori as:

$$f_{p,i}=\frac{v_i}{n_p L_p} \tag{8}$$

where v_i is the traveling speed of the ith propagation mode, L_p is the length of the pth path and $n_p(\in \mathrm{N})$ is the number of times needed for a given traveling wave to propagates along path p before attain again the same polarity. p-1 values are used to identify the faulted section and the remaining one to identify the fault distance between observation point m and the fault location. The method of associating a characteristic frequency to each path was introduced in the paper.

3.4 Multiple Measurement Based Methods

The multiple measurements based methods is a kind of fault section identification method. The fault section is identified by comparing the current or some other features measured along the distribution line. It can be recognized as expansion of fault feeder identification. The methods can be subdivided into transient based and steady state based. Most of these methods are on-line fault location, and there are some requirements for the measurement device, such as sample frequency, synchronization, tele-communication and so on. The comparison of the different methods are shown in Table 1.

3.5 Signal Tracing Based Method

This method is a typical active method which locates the fault by tracing the injected signal path by signal detector, including handheld detector and fixed detector along the line. The signal injection methods have been used for the fault detection and fault feeder identification. The fault location can be recognized as the expansion of these methods which need to trace the injected signal path.

The injected signal would pass through the fault point and back to the injected point through the ground. It means that the fault is in the downstream of measure point if the injected signal can be detected in this point.

The requirements for the injected signal are that: 1) it is distinguishable; 2) the signal and its source do not adversely affect the performance of the equipment. There are three typical research team have used this type of method as shown in Fig. 7. Then the three methods and modified would be introduced separately.

3.6 Signal Diagnosis Methods

The signal diagnosis algorithm is based on transfer functions and frequency spectrum analysis, using the response of the applied impulse in the frequency domain to construct criteria for the fault location[27]. It is a type of active method which needs signal injection. The method has been proposed for several years but it is verified just by the computer simulation and hasn't been used in field test.

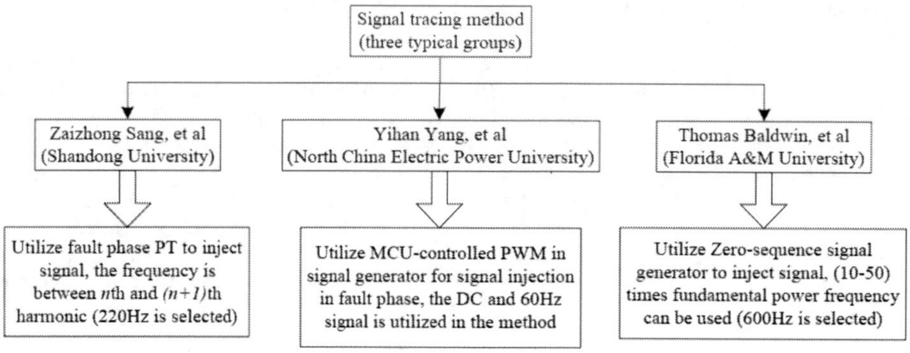

Figure 7. The comparison of three typical signal tracing based methods

Table 1. The comparison of the different multiple measurements based method

Features		Ref.	Measurements	General principle
Steady state based methods (using fundamental signal)	Amplitude of I_0	14	I_0 along the line and magnetic field detection	Upload the I_0 along the line to primary substation for fault section identification, then detect the fault position by detecting the magnetic field of fundamental frequency signals
	The phase difference between U_0 of bus and I_0 of the line	15	U_0 of bus and I_0 along the line	The voltage and current need to be synchronous measurement. Measures current using portable current measurement based on magnetic field detection.
	The phase difference between the I_0 of two adjacent nodes	16	I_0 along the line	The fault section is identified based on the absolute different between the approximate zero-sequence angles of two adjacent nodes.
	Direction of zero sequence power	17	U_0 of bus and I_0 along the line	Measure the zero sequence current using magnetic detection, use the synchronized zero sequence voltage of bus to calculate zero sequence power
	5th harmonic	18	electric and magnetic field	Detect the fault point by the portable and non-contact voltage and current measurement, the amplitude and phase are considered simultaneously
	Composite method	19	U_0 of bus and I_0 along the line	Combine the magnitude and phase comparison implemented by using the mobile zero sequence CT and wide area phase measurement
	The increment of I_0	20	U_0 of bus and I_0 along the line	Detect zero sequence current variation after adjust Peterson coil
Transient signal based methods	Transient reactive power direction	21	The electric field and magnetic field	The reactive power direction calculated using the transient signals in the selected frequency band is used for the fault section identification
	Polarity of a specific frequency band power	22	Residual voltage and residual current	Fault section identification is based on the polarity of a specific frequency band power computed by multiplying the DWT detail coefficient
	The waveform similarity of two adjacent nodes	23	The waveform of I_0	Calculate the correlation of transient zero current for fault section identification.
	The relationship between phase current and I_0	24	Three phase currents	Utilize the difference between phase current and zero sequence current
	The product of transient line-to-line voltage and I_0	25	Line-to-line voltage and I_0 along the line	The product of transient line-to-line voltage after Hilbert transformation and transient zero-mode current is defined as the parameter of fault direction.
	Composite	26	Line-to-line voltage and I_0 along the line	Upload the fault data by FTU, and identified the fault section combine transient reactive power direction method and transient current similarity location method

Where U_0 is zero sequence voltage and I_0 is zero sequence current.

3.7 AI Based Methods

The AI(artificial intelligence) based method is proposed with the development of computer and power system simulation. It is also used by some scholars for the SLG fault location, including ANN(artificial neural network) based method proposed by [28] and Fuzzy neural network based method proposed by [29]. All the fault features including transient and steady fault features could be used in the type methods. But it needs training and retraining when the structure of the system changed. The large number of actual fault data or simulation results is need for the training.

4 Conclusion

Although there are many methods have been proposed, there is no one is practical and low-cost which can be spreaded based on investigation. There are some methods have been in field test, such as transient signal based parameter calculation methods, multi-measurements based methods and signal injection methods. But the transient signal based parameter calculation method is low robustness for the amplitude of transient signal is affected by fault resistance and fault inception angle; multi-measurements based methods are high-investigation maintenance workload and the reliability is also affected by fault resistance. The signal injection based methods are time consuming and the operation is complex. Based on the features of SLG fault and the existing location methods, some advices are proposed as follows:

(1) Considering the complexity of distribution system, increase accurate measurement points in the system for accurate calculation is important;

(2) The signal injection based methods are the most reliable, but most of them are off-line and time consuming. The idea of active method can be considered to enhance the reliability of location for the SLG fault features are too weak;

(3)There are big differences between the developments of distribution systems; the measurements in the systems are different. Then the different fault location methods should be proposed to adapt the measurement condition of system.

References

1. Seppo Hanninen, and Matti Lehtonen. "Characteristics of earth faults in electrical distribution networks with high impedance earthing." Electric Power Systems Research 44.3 (1998): 155-161.
2. Druml, Gernot, Andreas Kugi, and Olaf Seifert. "A new directional transient relay for high ohmic earth faults." CIRED 17th International Conference on Electricity Distribution, Barcelona, Spain. 2003.
3. Seppo Hanninen. Single phase earth faults in high impedance grounded networks: characteristics, indication and location. VTT Technical Research Centre of Finland, 2001.
4. Seppo Hanninen, and Matti Lehtonen. "Earth fault distance computation with fundamental frequency signals based on measurements in substation supply bay." VTT TIEDOTTEITA (2002).
5. Soon-Ryul Nam, et al. "Ground-fault location algorithm for ungrounded radial distribution systems." Electrical Engineering 89.6 (2007): 503-508.
6. Hanninen, S., et al. "Comparison of wavelet and differential equation algorithms in earth fault distance computation." Proceedings of Conference on Power System Computation, Norway. 1999.
7. Igel, M., H - J. Koglin, and P. Schegner. "New algorithms for earthfault distance protection in insulated and compensated networks." European Transactions on Electrical Power 1.5 (1991): 253-259.
8. Ke Jia, David Thomas, and Mark Sumner. "A new single-ended fault-location scheme for utilization in an integrated power system." Power Delivery, IEEE Transactions on 28.1 (2013): 38-46.
9. Imriš, Peter. Transient based earth fault location in 110 kV subtransmission networks. Helsinki University of Technology, 2006.
10. Zhang, Fan, et al. "New Algorithm Based on Traveling Wave for Location of Single Phase to Ground Fault in Tree Type Distribution Network." Zhongguo Dianji Gongcheng Xuebao(Proceedings of the Chinese Society of Electrical Engineering). Vol. 27. No. 28. 2007.
11. YU, Sheng-nan, Hai BAO, and Yi-han YANG. "Practicalization of Fault Location in Distribution Lines [J]." Proceedings of the CSEE 28 (2008): 016.
12. Borghetti, Alberto, et al. "On the use of continuous-wavelet transform for fault location in distribution power systems." International Journal of Electrical Power & Energy Systems 28.9 (2006): 608-617.
13. Ye, Lei, et al. "An improved fault-location method for distribution system using wavelets and support vector regression." International Journal of Electrical Power & Energy Systems 55 (2014): 467-472.
14. ZHANG, Li, et al. "Distribution Network Fault Location Based on the Zero Sequence Current and Magnetic Field Detection Spot." Automation of Electric Power Systems 14 (2008): 017.
15. ZHANG, Li, et al. "Method of mobile phase-comparison for fault location of distribution network." Proceedings of the CSEE 7 (2009): 015.
16. Nam, Soon-Ryul, et al. "Single line-to-ground fault location based on unsynchronized phasors in automated ungrounded distribution systems."Electric Power Systems Research 86 (2012): 151-157.
17. Li, Zhang, et al. "Online fault location of neutral point ungrounded distribution network based on zero-sequence power direction." Automation of Electric Power Systems 32.17 (2008): 79-82.

18. LI, Meng-qiu, et al. "A new approach on detecting the single-to ground fault location on power system with neutral unearthed [J]." Proceedings of the Csee10 (2001)
19. Yang Yihan, et al. " Research on fault location online for neutral point ungrounded system" Journal of North China Electric University, 35.6 (2008): 1-6.
20. Wang, Yumei, Hao Guo, and Shengda Hui. "Fault Location Approach for Resonant Grounded System Based on Remnant Current Increment Method."Automation of Electric Power Systems 2 (2011): 015.
21. Sun, Bo, et al. "Single phase to ground fault section location based on transient signals in non-solidly earthed network." Autom Electr Power Syst 32.3 (2008): 52-55.
22. Elkalashy, Nagy I., et al. "DWT-based detection and transient power direction-based location of high-impedance faults due to leaning trees in unearthed MV networks." Power Delivery, IEEE Transactions on 23.1 (2008): 94-101.
23. Ma, Shicong, et al. "An earth fault locating method in feeder automation system by examining correlation of transient zero mode currents." Autom Electric Power Syst 32.7 (2008): 48-52.
24. Sun, Bo, et al. "Earth fault location based on transient phase current in non-solidly earthed network." Power System Protection and Control 40.18 (2012): 69-74.
25. Linli, ZHANG., et al. "Transient Fault Locating Method Based on Line Voltage and Zero-mode Current in Non-solidly Earthed Network." Proceedings of CSEE 32.013 (2012): I0015-I0015.
26. Xue Yongduan, et al. "Small-current grounding fault location based on transient signals of distribution automation system" Electric Power Automation Equipment, 33.12(2013)
27. Yi Guiye, Yang Xuechang, Wu Zhensheng. "Transfer function algorithm for locating ground faults in power distribution networks" J Tsinghua Univ(Sci & Tech), 2000, 40(7): 31-34.
28. Hanninen, S., and M. Lehtonen. "Earth fault distance computation with artificial neural network trained by neutral voltage transients." Power Engineering Society Summer Meeting, 2001. Vol. 2. IEEE, 2001.
29. Chunju, Fan, et al. "Application of wavelet fuzzy neural network in locating single line to ground fault (SLG) in distribution lines." International Journal of Electrical Power & Energy Systems 29.6 (2007): 497-503.

Research on lightning stroke model and characteristics of intelligent transformer

Mu LI[1], Tao LI[1], Jia Jie WANG[1] and Ling Hong TAO[1]

[1]Wuhan NARI Limited Liability Company, State Grid Electric Power Research Institute, Wuhan 430074, China

Abstract. In order to improve the reliability of power supply, it is necessary to study the intelligent transformer of the key equipment in the smart substation. Firstly, the transformer model based on the magnetic coupling mechanism is given according to the magnetic coupling characteristics of the transformer. Then, considering the static coupling characteristics of the transformer under high frequency conditions, an improved transformer model based on wide frequency mechanism is obtained. Finally, the effectiveness of the proposed model is verified by comparing the simulation data with the actual measured data for an actual object.

1 Introduction

With the development and application of sensor technology and information technology, the transformation from traditional substation to smart substation has gradually become a reality. Transformer is one of the key equipment in smart substation.

To ensure the safe and reliable operation of the transformer, it is necessary to comprehensively and accurately grasp the characteristics of the over voltage surge in lightning wave shows, according to the characteristics of lightning and risk to develop the corresponding protective measures and strategies. Now we focus on analyzing the lightning stroke model of transformer and its lightning characteristics[1-3].

2 Model of intelligent transformer

Compared with the transformer in conventional substation, there is no special design for the transformer body structure in the smart substation. It only collects the measured values of various sensors directly into the merging unit near the transformer, and then transforms it to the optical signal and then uploads it to second-order system equipment of the main control room through the optical cable.

In order to analyze the influence of lightning on the intelligent transformer more accurately, we extend the simplified transformer model, build a more refined transformer model, and build a lightning stroke model of intelligent transformer combined with sensors[4-5].

In order to obtain the wide frequency characteristic of transformer, it is necessary not only to analyze the magnetic coupling between transformer windings in low frequency, but also to consider the static coupling between transformer conductors under high frequency. In fact, with the increase of frequency, the distributed capacitance parameters which are closely related to the structure size of the transformer will affect the operating characteristics of the transformer, and interact with the power electronic devices on both sides, therefore, the distributed capacitance of the transformer can not be ignored. Therefore, the traditional low frequency transformer model is no longer applicable because it can not take into account the capacitance effect in the transformer. Figure. 1 is the circuit diagram of a single phase double winding transformer.

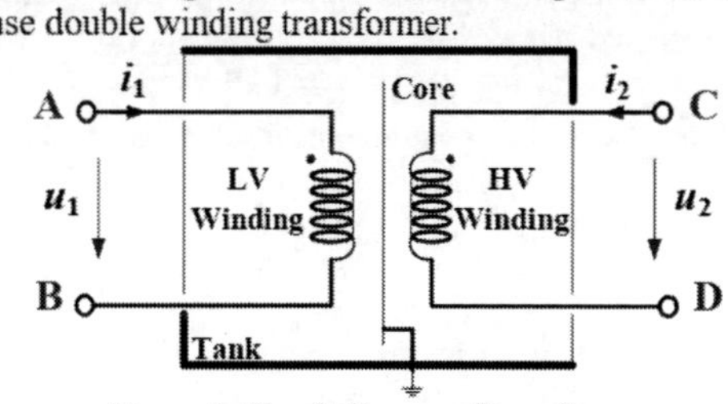

Figure 1. Circuit diagram of transformer

With the development of flexible DC converter system to high voltage and large capacity, the capacity and voltage level of transformer will be improved, the turn number of transformer winding becomes larger, and the oil immersed type is needed to meet the insulation demand in the high-power field. In addition, in order to improve the power density and reduce the loss of the transformer, some magnetic materials with low loss density and high permeability, such as ultra-thin silicon steel sheet and nanocrystalline alloy, are also used as transformer cores, and these core materials are also conductors. Therefore, it is necessary to analyze the capacitance effect between transformer winding, magnetic core and fuel tank.

The electrical characteristics of transformer at low frequency mainly depend on the magnetic coupling between transformer windings, but in the high frequency section, the electrostatic coupling between transformer winding, magnetic core and oil tank can not be ignored. Therefore, in order to establish the broadband mechanism model of transformer, it is necessary to deeply analyze the magnetic coupling and electrostatic coupling characteristics of the transformer.

2.1 Model based on magnetic coupling mechanism

Because we focus on the external characteristics of the transformer, it is considered that the magnetic core of the transformer operates in the coupled linear region. The coupling inductance model of the transformer is shown in figure 2(a).

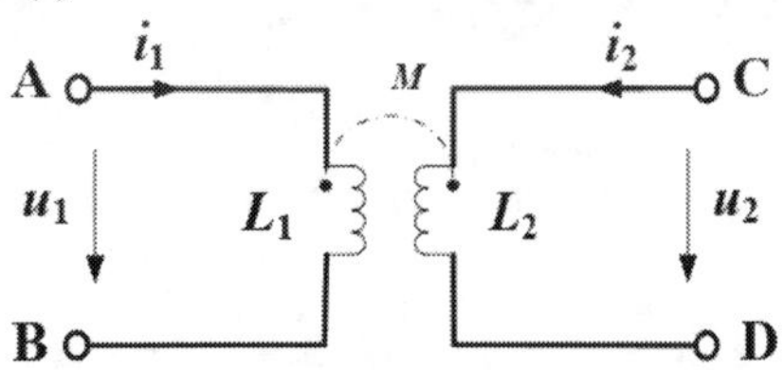

(a) Model with coupled inductors

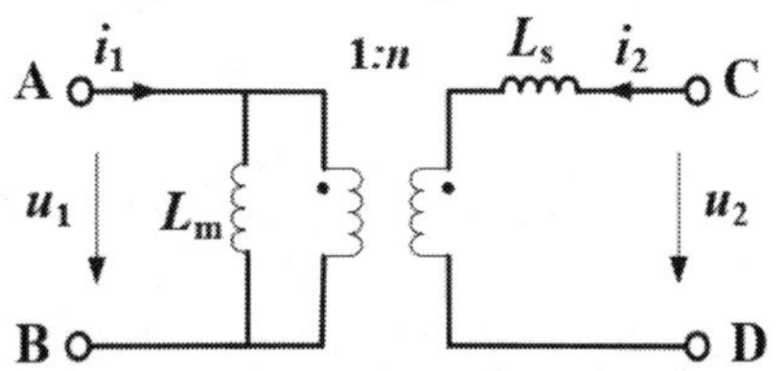

(b) Equivalent model with ideal transformer

Figure 2. Transformer model based on Lossless magnetic mechanism

The inductance matrix of transformer winding can be fully expressed by self inductance and mutual inductance, and the leakage inductance parameter from C-D side can be expressed as:

$$L_s = L_2 - M^2/L_1 = (1-k^2)L_2 \tag{1}$$

Here, k is the coupling coefficient of transformer windings. Due to the existence of the magnetic core of the transformer, the k is usually close to 1.

Because the leakage inductance is represented as the difference between the two large numbers, the leakage inductance obtained by equation 1 may not be accurate. Considering the leakage inductance will affect the impedance characteristics of the transformer in high frequency, we adopt an equivalent model with ideal transformer, as shown in Figure 2 (b). Lm is the excitation inductance from the A-B side, and the n is the ideal transformer ratio:

$$L_m = L_1 \tag{2}$$

$$\mathrm{n} = \mathrm{k}\sqrt{\frac{L_2}{L_1}} \tag{3}$$

The equivalent relation of the two transformer models in Figure 2 can be proved by the Kirchhoff current law of the primary side of transformer and the Kirchhoff voltage law of the secondaly side of the transformer. Using the model with ideal transformer as shown in Figure 2 (b), the leakage inductance can be taken as an independent variable of the model, and can be directly extracted by magnetic field analysis method. At the same time, the mutual inductance is eliminated so that the circuit analysis of the transformer can be easily carried out.

For an actual transformer, winding and core losses need to be considered to improve the accuracy of the model. On the basis of the model shown in Figure 2 (b), two winding resistances Rs1 and Rs2 are added, and the magnetic core loss equivalent resistance Rm is parallel to the excitation branch, and finally the final magnetic mechanism model is shown in Figure 3. The model is a low frequency model of traditional transformer, and has been widely used in the field of power system and motor. However, the model is not suitable for overvoltage analysis of transformer because it does not consider the capacitance effect between winding, magnetic core and tank.

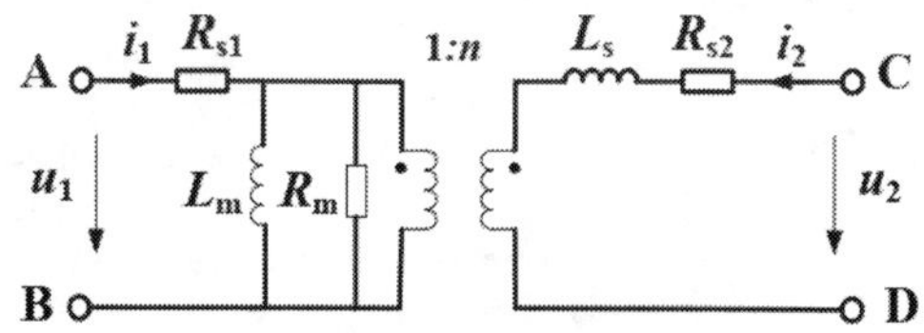

Figure 3. Transformer model based on magnetic mechanism

2.2 Model based on electrostatic coupling mechanism

The distributed capacitance is closely related to the electrostatic coupling between the conductors in the transformer. The conductors in transformer include low voltage winding conductor, high voltage winding conductor, magnetic core and oil tank. Due to the voltage drop between the conductors of the transformer, there is a coupling capacitance between these conductors. Considering the low-voltage winding voltage, high voltage winding voltage and winding voltage, the static characteristics of the winding can be described by six capacitors, as shown in Figure 4. However, this model does not consider the electrostatic coupling between winding, magnetic core and fuel tank.

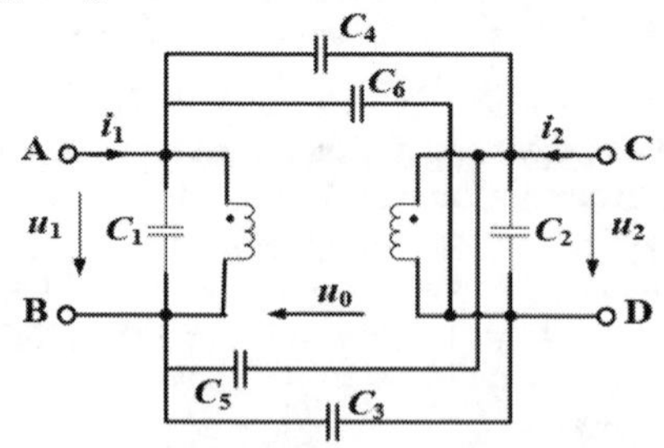

Figure 4. Model with six capacitor

So we propose a transformer model considering the electrostatic coupling mechanism of capacitance, as shown in figure 5. The core is connected to the tank and connected to the ground.

In Figure 5, in addition to six capacitors representing the electrostatic characteristics of the transformer windings, new four capacitors C7, C8, C9 and C10, are also introduced between the winding terminals and the core and the fuel tank to reflect the capacitance effect between the winding, the core and the tank. By adding these capacitors, the electrostatic coupling characteristics of the transformer are fully considered.

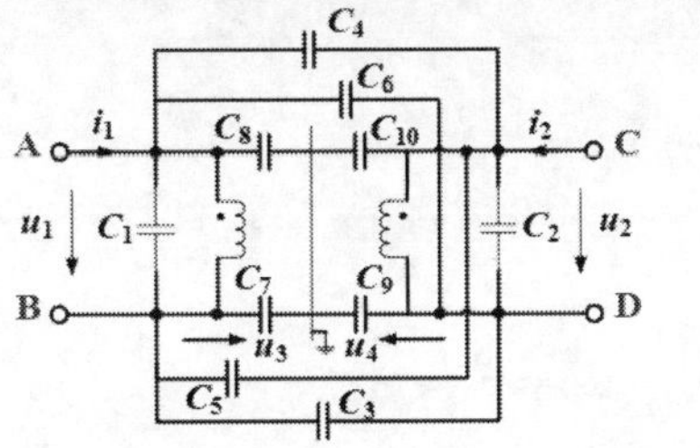

Figure 5. Model of adding new capacitance

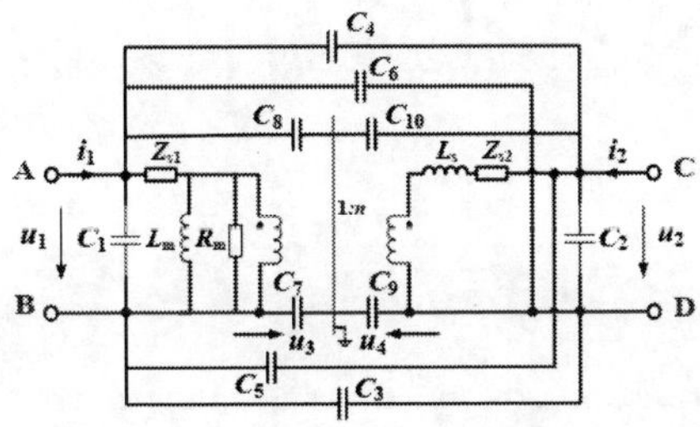

Figure 6. Broadband mechanism model of transformer

Combining the magnetic mechanism model of transformer with its capacitance mechanism model, the broadband mechanism model of transformer is obtained, as shown in Figure 6. It is noted that, compared with the magnetic mechanism model shown in Figure 2, the broadband mechanism model shown in Figure 6 replaces the winding resistance for internal impedance, which can reflect the skin effect of the winding at high frequency.

3 Analysis of model characteristics

In order to verify the effectiveness of the broadband model of transformer, we use Agilent 4294A impedance analyzer to measure the impedance characteristics of a 500kV/220kV single-phase transformer under different conditions, such as no-load, short circuit and load (with 125k ohm resistive load and 40uH inductive load). At the same time, the circuit simulation of the transformer is also made by using the broadband mechanism model presented previously. Here, the impedance analyzer is connected with the low-voltage winding and the high voltage winding, respectively expressed as "1" and "2"; "OC", "SC", "R" and "L" respectively represent the open circuit, the short circuit, the resistive load and the state of inductive load of the other winding.

The experimental and simulation results are shown in Figure 7 and Figure 8. The thick lines show the measurement results of impedance analyzer. The fine dotted lines represent the simulation results of the magnetic mechanism model of the transformer (the low frequency model of the traditional transformer). The coarse dotted lines represent the simulation results of the broadband mechanism model of the transformer.

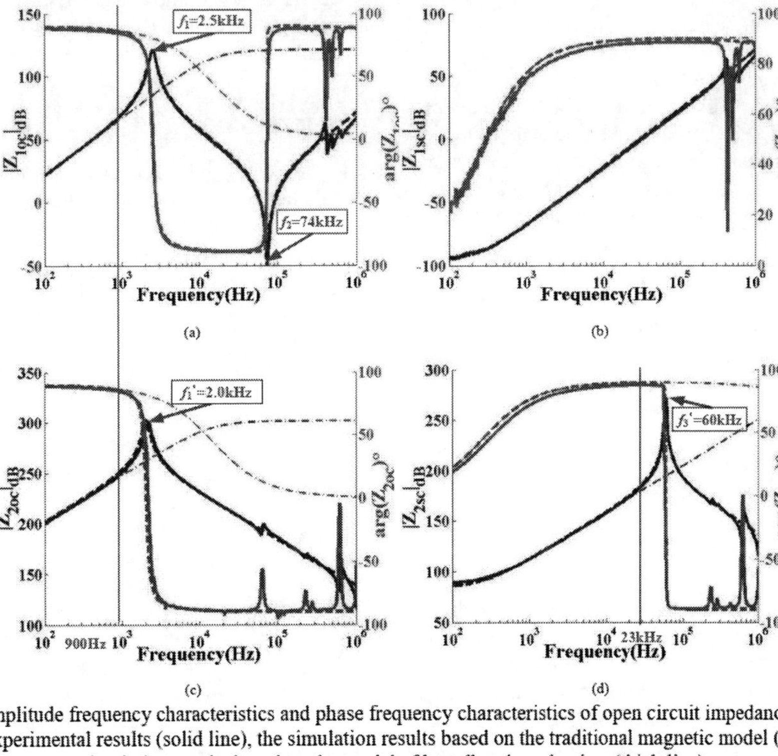

Figure 7. The amplitude frequency characteristics and phase frequency characteristics of open circuit impedance and short circuit impedance : experimental results (solid line), the simulation results based on the traditional magnetic model (dotted line), the simulation results based on the model of broadband mechanism (thick line).
(a) high voltage winding open circuit (b) high voltage winding short circuit
(c) low voltage winding open circuit (d) low voltage winding short circuit

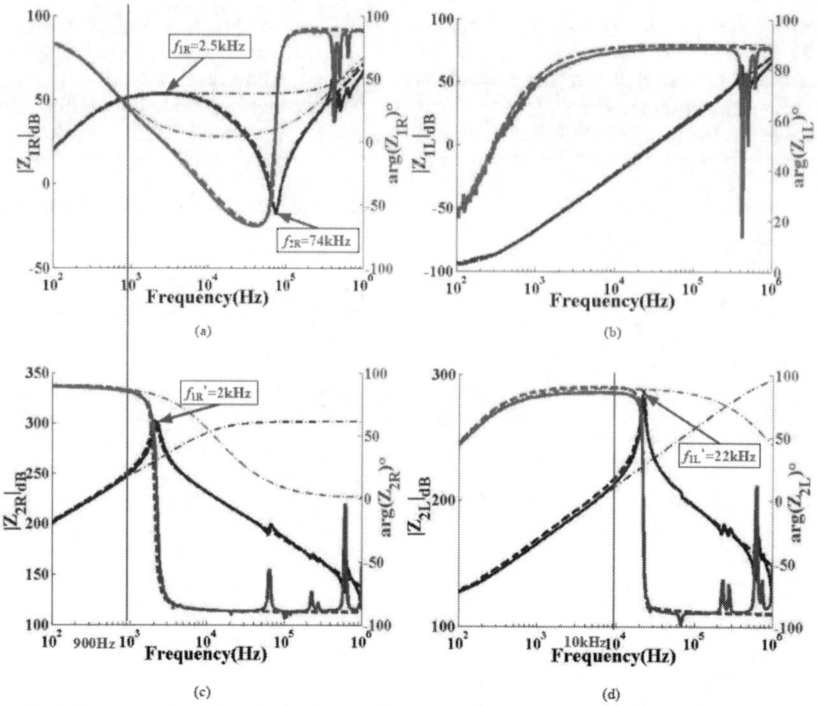

Figure 8. The amplitude frequency characteristics and phase frequency characteristics of open circuit impedance and short circuit impedance : experimental results (solid line), the simulation results based on the traditional magnetic model (dotted line), the simulation results based on the model of broadband mechanism (thick line).
(a) high voltage winding connected with a resistive load (b) high voltage winding connected with a inductive load
(c) low voltage winding connected with a resistive load (d) low voltage winding connected with a inductive load

From Figure 7 and Figure 8, it can be seen that the simulation results of the traditional transformer model are only consistent with the actual measurement results in the low frequency band, and the simulation results are obviously deviated with the increase of the frequency. The frequency at which the simulation results deviate from the experimental results at 20% is defined as the critical frequency, as shown in the vertical lines in Figure 7 and figure 8. It can be seen that the critical frequency of the traditional transformer model is about 900Hz for the open circuit impedance characteristics of the transformer, which is far lower than the working frequency of the transformer 20kHz. This indicates that the capacitance effect between transformer winding, magnetic core and tank is necessary to be considered. Because some of the capacitors in the model are shorted in the short-circuit state of the transformer, the critical frequency of the Z1sc exceeds 1MHz, and the critical frequency of Z2sc is 23kHz.

4 Conclusion

The analysis results show that compared with the traditional transformer model, regardless of the open circuit state or the load state, the simulation results of the transformer model based on wide band mechanism are in good agreement with the actual measurement results, and the critical frequency is as high as 300kHz. The transformer model based on broadband mechanism can accurately reflect the characteristics of magnetic coupling and static coupling of transformer, and also verify the effectiveness of transformer parameter extraction method.

References

1. W.Xiong, Z.Wenliang, Technical problems of electromagnetic compatibility in power system, High voltage technology, 23,53-58 (1997)
2. V. S. Rashkes, L. D. Ziles, Very high frequency overvoltages at air EHV substations during disconnect switch operations, IEEE on PD, 3,1618-1623(1996)
3. S.Zhusen, Z.Yufang, Z.Guangzhou, et al. Research on electromagnetic disturbance and protection measures in 500kV substation (1), high voltage technology,26,16-18(2000)
4. C.Dingxin, Q.Xuedi, Calculation method of magnetic field produced by line current in "wire earth" loop, Proceedings of the Symposium on electromagnetic environment and electromagnetic compatibility of power system, Guangzhou, 27-29(2002)
5. W.Xiong, Electromagnetic disturbance and level of 500kV substation, Symposium on electromagnetic environment and electromagnetic compatibility of power system, Guangzhou,5-8(2002)

Study of Non-contact Voltage Detector of 1000kV UHV AC Based on MEMS Electric Field Sensor

Jianjun ZHU[1], Xinglie LEI [2], Ziming SU[2], Ting LIU[2], Kai LIU[2], Guangkai YU[2], Linyue ZHU[3]

[1]*State Grid Corporation of China, Beijing 100031, China.*
[2]*State Key Laboratory of Power Grid Environmental Protection, China Electric Power Research Institute, Wuhan 430074.*
[3]*Beijing NO.161 High School, Beijing 100031, China.*

Abstract. This paper expounds the working principle of non-contact voltage detector and the electric field sensor which made by the technology of microelectromechanical systems (MEMS), determined the requirements of the detecting area and the threshold of the detector through simulation, and correct the threshold by carrying out experiments, which shows that by setting the threshold of the detector at 25kV/m, when the detector is 10-15m away from the conductors, or 2m away from the tower toward the conductor at the tower body whereas the same high as the conductor or on the crossarm, the detector can effectively identify whether the conductor is power on or off. The MEMS no-contact voltage detector and the method put forward by this paper can solve the problem of voltage detecting on UHV transmission lines, which has wide popularization and application value.

1 Introduction

Voltage detecting is an important content of operation on transmission line, at present the main method of voltage detecting used on HV transmission lines is by capacitive type voltage detector. Study of voltage detecting on 500kV and the following voltage transmission project are quite a lot, and related products and technologies are mature and reliable and has been popularized and applied [1-5]. But for the UHV transmission line, contact voltage detector needs very long insulation rod, so it is difficult to use [6-7].

For the UHV transmission lines, non-contact voltage detecting is a fast and convenient method. At present, research on non-contact voltage detecting method includes ultraviolet pulse [8-10], ultra high frequency partial discharge detection [11-13] and electric field detection near the ground [14-15]. Ultraviolet pulse and ultra-high frequency partial discharge detection can easily be affected by weather conditions and equipment status, and have poor accuracy and high cost; electric field detection near the ground can be influenced by the environment of the ground, the tower structure, the wire layout and other factors, it has poor anti-interference, complex operation, and there are false detecting and misinformation.

In this article, according to the operation characteristics of UHV transmission lines and non-contact voltage detecting problems, based on the MEMS electric field sensor, we determined the method and threshold of non-contact voltage detecting through simulation and corrected it by testing, developed two types of non-contact voltage detectors.

2 Principle of MEMS electric field sensor and voltage detecting method

The MEMS electric field sensor is based on the principle that the conductor induces charge in the electric field, as shown in figure 1. The shielding electrode and induction electrode of the sensor are conductive polysilicon material, when the shielding electrode vibrates, the induced charge on the induction electrode will change, therefore produces induced current. By using the differential detection circuit, the induced current can be detected, so as to realize the detection of electric field.

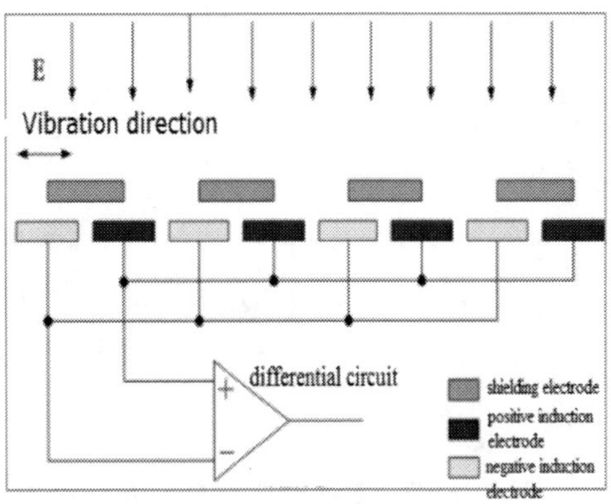

Figure 1. Working principle diagram of MEMS sensor.

Assume that the frequency of the sensor is ω_1 (2kHz~4kHz), and the initial phase is φ_1. If the electric field E is the power frequency electric field, the input electric field is:

$$E=E_0\sin(\omega_e t+\varphi_e) \quad (1)$$

E_0 is the electric field amplitude, ω_e is the signal frequency (50Hz), and φ_e is the initial phase.

As the shielding electrode vibrates periodically, the sensor formats induced current periodic, and the voltage output converted by the I/V will be:

$$V_s=\frac{dQ}{dt}\cdot R_s=K_1\cdot E_0\cdot\sin(\omega_1 t+\varphi_1)\cos(\omega_e t+\varphi_e)$$
$$+K_2\cdot E_0\cdot\cos(\omega_1 t+\varphi_1)\sin(\omega_e t+\varphi_e) \quad (2)$$

K_1 and K_2 are constants.

After the output signal is demodulated, the amplitude or the effective value is extracted, and then the signal is amplified and processed, and transmitted to the alarm circuit so to realize the function of voltage detecting. Functional diagram block of the non-contact voltage detector based on MEMS sensor is as shown in Figure 2.

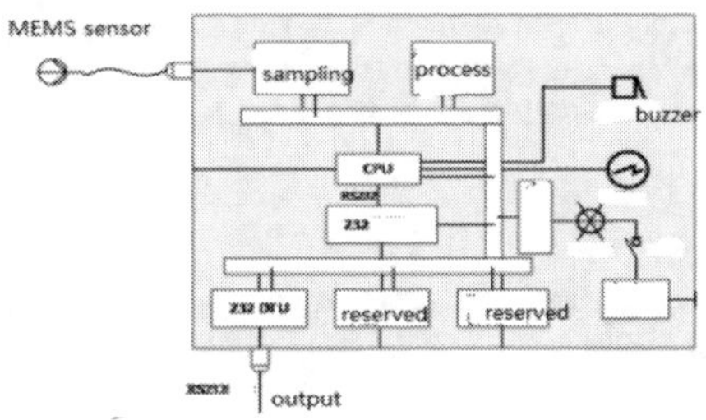

Figure 2. Functional diagram block of non-contact voltage detector.

By using the MEMS electric field sensor, the electrified condition can be determined by detecting the electric field around the UHV line. The existing research shows that the distribution of electric field around the UHV transmission line is greatly affected by the environment, and the distortion of electric field can be caused by different terrain structures, surface buildings and vegetation, etc.[16-19]. UHV double circuit lines can induce charge when one circuit is out, and there will be electric field around it; and other UHV voltage lines on the same tower will also affect the distribution of the electric field around the UHV transmission line [20-22].

In order to avoid the influence of environmental factors and induction, we can detect the electric field near the circuit. In the premise of ensuring safety, we can put the voltage detector as near as possible to the circuit. The closer to the wire, the greater the electric field intensity, the detection will be more reliable; at the same time, the closer to the wire, the more open of the surrounding, the smaller the impact of the environment.

So the following two ways can be taken into consideration: 1) Using UAV carrying the voltage detector to the certain area near the circuit; 2) the workers climb up the tower, send the voltage detector to the certain area by using insulating rod. The UAV can achieve precise positioning, ranging and automatic flight and other functions, which can fully meet the detecting requirements; the workers can carry a portable insulating rod and the voltage detector to climb the tower, which is simple and safe. The two methods not only can avoid the influence of the ground environment factor, but also can avoid the inductive voltage of other lines by setting the threshold of the detector.

3 Simulation and analysis of threshold of the non-contact voltage detector

According to the principle of MEMS electric field sensor, this article built the UHV transmission line simulation models through ANSYS, and calculated the electric field distribution, so as to determine the threshold of non-contact voltage detector. According to the structural characteristics of UHV transmission line and the detecting methods, electric field distributions around the mid-span conductors and around the tower body were calculated.

3.1 Threshold of detecting near mid-span conductors

Mid-span refers to that the detecting position is more than twice the height of the tower away from the tower, under this condition the influence of tower can be ignored. Distribution of electric field around the conductor can be calculated using two-dimensional calculation method, according to the typical design of UHV transmission lines, the calculation models are as shown in Figure 3.

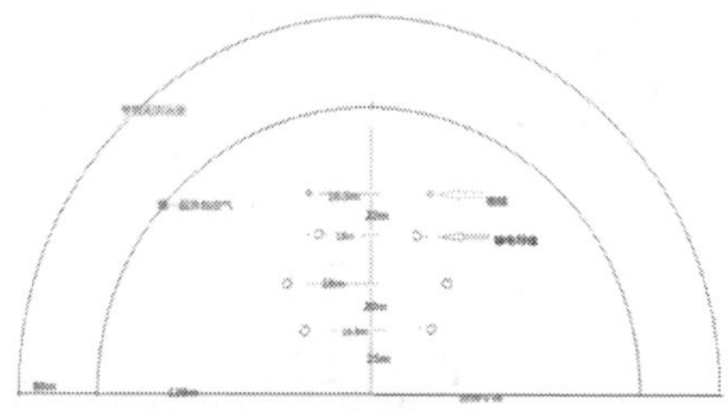

(a) Two-dimensional model

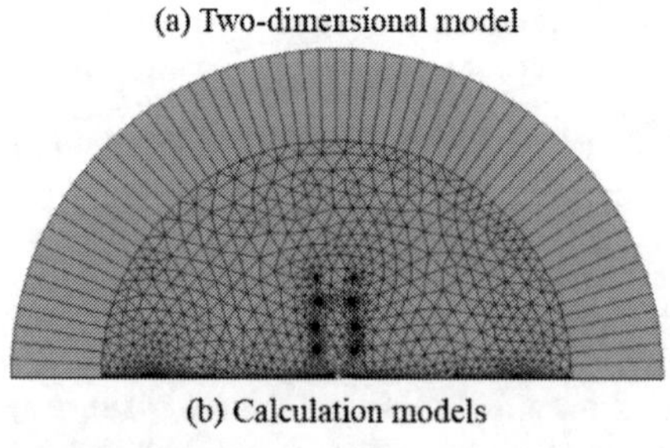

(b) Calculation models

Figure 3. Models of the conductors.

(1) Operation with both circuits

Assume that the double circuit of UHV transmission line is running at rated voltage, and the electric field distribution of the vertical plane is calculated, as shown in Figure 4.

Figure 4. Electric field around the conductors when both circuits are charged.

Taking the wire as the center, the electric field lines are as shown in Figure 5.

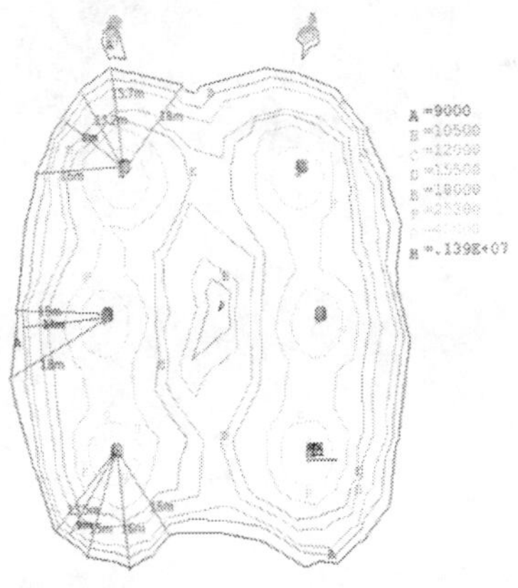

Figure 5. Electric field lines around the conductors when both circuits are charged.

(2) Operation with only one circuit

UHV double circuit lines can operate with only one circuit, according to the actual operating parameters, the electric field is calculated and is as shown in Figure 6 and Figure 7.

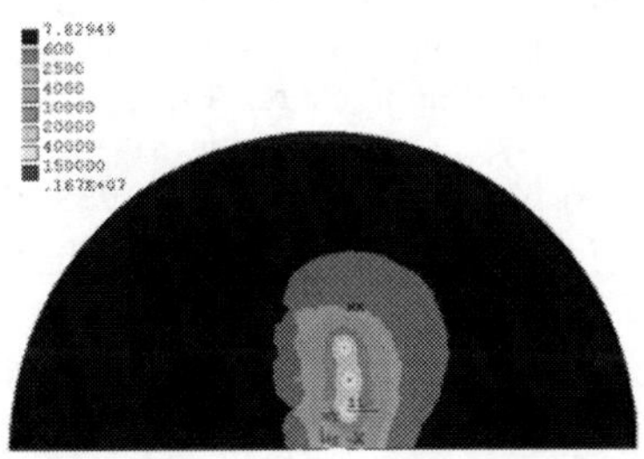

Figure 6. Electric field around the conductors when only one circuit is charged.

The results show that when the double circuit are both charged, the electric field generated around the two circuit transmission lines on both sides are the same, the closer to the conductor, the stronger the electric field intensity.

When only one circuit is running, the distribution of the electric field around the running line is nearly the same as that of the double circuits. The electric field over 2m away from the un-changed circuit is less than 7kV/m. The maximum and minimum electric field 15m away from the left and right wire on the vertical path are as shown in table 1.

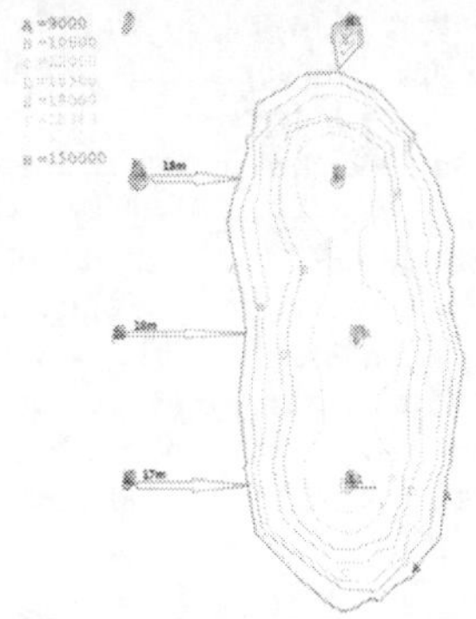

Figure 7. Electric field lines around the conductors when only one circuit is charged.

Table 1. Eigenvalue of the electric field (kV/m).

	Un-charged side	Charged side
Maximum value	0.88	20.04
Minimum value	0.46	7.82

Considering the safety of voltage detecting, chose the outside aera that 2m-15m away from the conductor as the detection aera of the non-contact voltage detector, and set the threshold as 7kV/m, then it can accurately tell the voltage condition.

3.2 Threshold of detecting on the tower

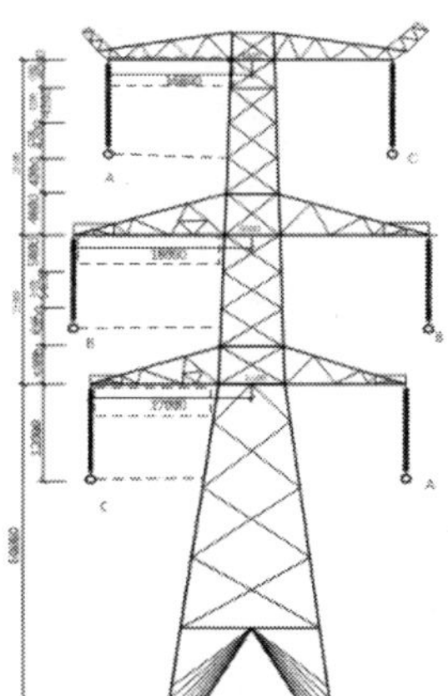

Figure 8. Structure of straight tower of the UHV transmission lines.

Ttypical tower structure of UHV double circuit lines is as shown in Figure 8, in this situation the electric field around the tower body is calculated including the situation of both circuits are charged and with only one circuit is charged.

(1) Operation with both circuits

Assume that the double circuit of UHV transmission line is running at rated voltage, and the electric field distribution of the vertical plane around the tower body is calculated, as shown in figure 9.

In the same way, the electric field distribution lines is as shown in figure 10.

The simulation results show that the electric field are larger at the tower body where is as high as the conductor and at the cross arm than the other positions.

(2) Operation with only one circuit

When one circuit is charged and the other circuit is off, the electric field distribution and contour around the tower are calculated. The results are as shown in Figure 11 and Figure 12.

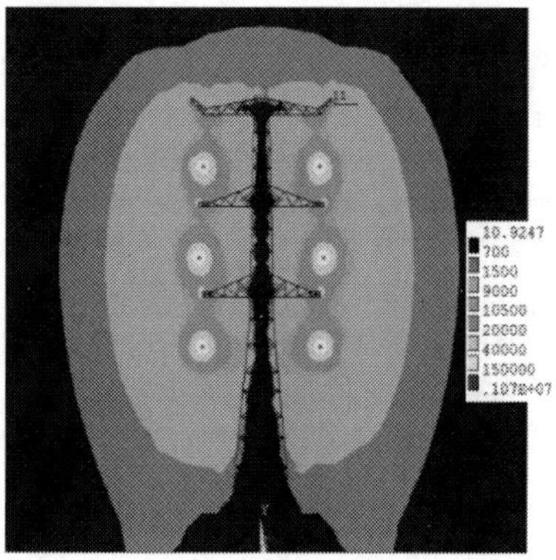

Figure 9. Electric field around the tower body when both circuits are charged.

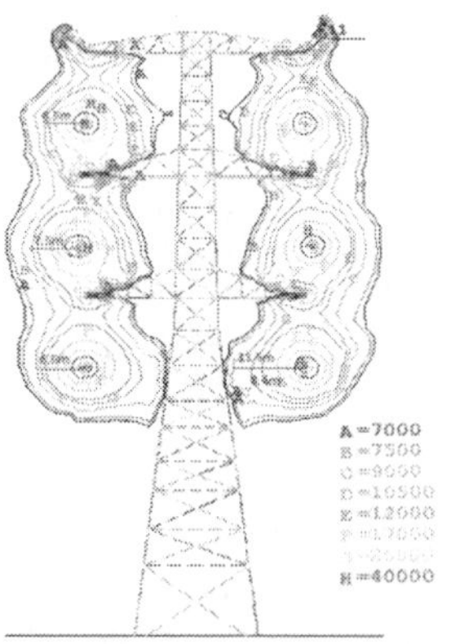

Figure 10. Electric field lines around the tower body when both circuits are charged.

It can be concluded from the calculation results that, when both circuits are charged, the electric field distribution on both side of the tower are the same, when only one circuit is charged, the electric field distribution on the live side is almost the same as that of both circus are charged, and the electric field distribution electric field distribution electric field distribution on the other side is almost shielded by the tower.

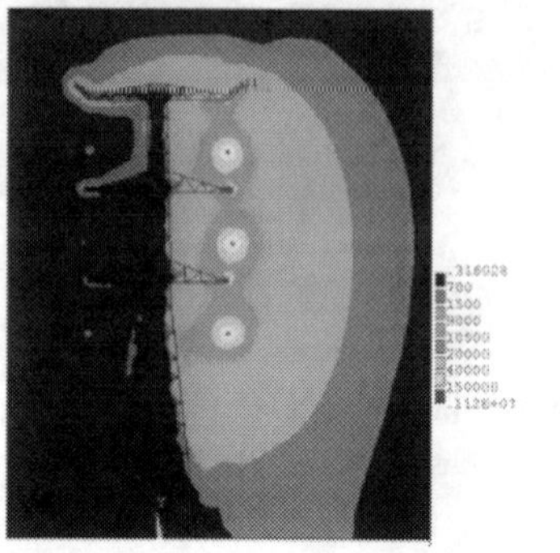

Figure 11. Electric field around the tower body when only one circuit is charged.

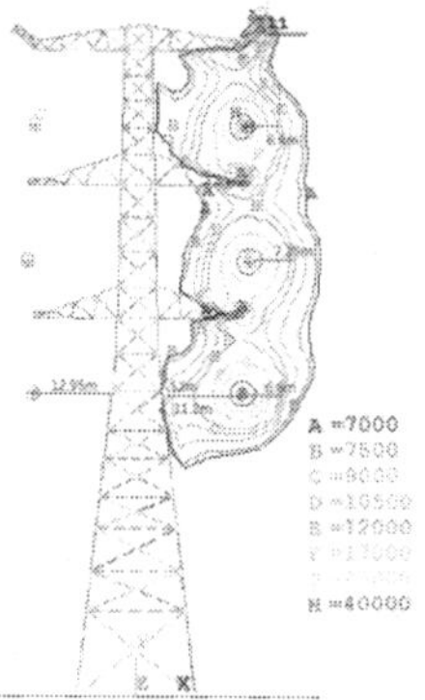

Figure 12. Electric field lines around the tower body when only one circuit is charged.

When the is circuit is charged, at the hight as the conductor on the tower body, the electric field intensity is no less than 9.8kV/m where 2m away from the tower body toward the conductor , and electric field intensity is no less than 12kV/m on the cross arms. When the circuit is off (with the other one charged), the electric field intensity less than 1kV/m from the tower and 10m away toward the conductor, and it is is less than 1kV/m where is 5m away from the conductor on the cross arms.

For the safety of voltage detecting, the threshold of the detector can be is set as 9kV/m, and fix the non-contact voltage detector to a 2m-long insulating rod, the operator can hold the insulating rod, and detect the conductor on the tower body or on the cross arms, which can accurately tell the voltage condition.

4 Calibration test of the threshold

In the actual detecting process, due to the introduction of the detector and auxiliary equipment, electric field will be distort. To determine the final threshold of non-contact voltage detector, prototype of non-contact voltage detector based on MEMS sensor is made(Figure 13). The prototype combines the sensor, signal processing circuit, alarming circuit and a display module, the sensor probe is at top of the detector, and there are start switch, liquid crystal display, parameter setting buttons and charging port on the outside of it.

In order to study the quantitative effect of the detector to the electric field distortion, this paper carried out calibration tes in the laboratory. The test uses a standard electric field electrode with an electrode spacing of 1m, and the power supply is adjustable from 0-50kV, which is shown in Figure 14. By changing the arrangement of the detector, including vertical, lateral and 60 degree angle with vertical direction, adjusting the voltage and recording electric field intensity, the result is as shown in Figure 15.

Figure 13. MEMS non-contact voltage detector.

From the results, the measurement of electric field have a good linear relationship with the power supply voltage (standard electric field). In the vertical situation, the measurement result is about 12 times of the standard electric field strength, in the lateral situation the result is smaller than inthe vertical situation, which is about 4 times of the standard electric field strength, and in the 60 degrees angle situation, the result is about 7 times. According to the test results, when the senor is right toward the the high voltage electrode, the measured value of electric field distortion is the largest.

Figure 14. Arrangement of electric field measurement correction test.

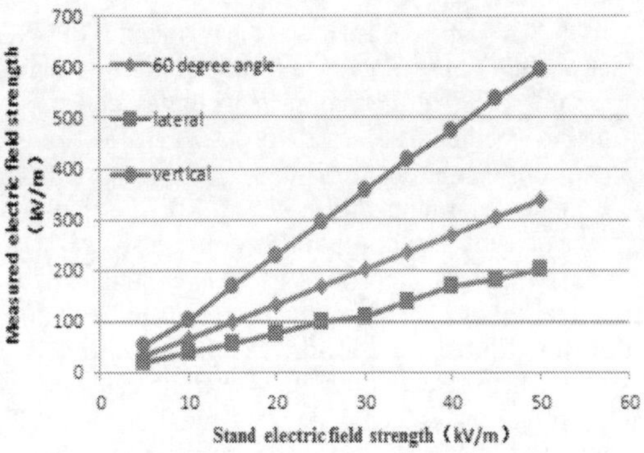

Figure 15. Results of electric field measurement correction test.

According to the simulation results of Chapter 2, when detecting at mid-span conductors, the maximum value of the electric field in the test area of the outage circuit is about 0.88kV/m, and the minimum value of the electric field is about 7.82 kV/m in the test area of the charged circuit. According to the influence of non-contact voltage detector, the alarming threshold should be set in the range of [0.88 * 12,7.82 * 4], namely [10.56,31.28]. Similarly, when detecting on the tower, the alarming threshold should be set in the range of [12,39.2].

Considering the other lines on the same tower or in the same corridor, mutiply the threshold by 2(for that the induced voltage is no more than the operation voltage), then the alarming threshold can be set to 25kV/m, which can judge the charging condition of UHV transmission lines by the methods put foward in chapter 2.

5 Applications

According to the calculation and test results, the applications are carried on actual UHV lines by using the non-contact voltage detector and methos proposed by this article, the applications are as shown in Figure 16.

Figure16. Application of the MEMS non-contact voltage detector.

The application results show that according to the detecting methods and the threshold and the the test area , interference of environmental factors and effective other transmission lines can be avoid, and can judge the charging condition of the UHV transmission lines correctly, which can solve the voltage detecting problem of UHV lines.

6 Conclusion

In this paper, based on the principle of MEMS electric field sensor, study and application of alarming threshold of the non-contact voltage detector and detecting methods are carried out, and the following conclusions can be acquired:

1)the alarming threshold of MEMS non-contact voltage detector can be set to 25kV/m.

2) when detecting at mid-span conductors, it can effectively judge the charging condition inthe aera of 2-15m away from the conductor.

3) when detecting on the tower, it can effectively judge the charging condition by extending the MEMS non-contact voltage detector toward the conductor about 2m,whether at the tower body or on the cross arm.

4) the non-contact voltage detector and the method is simple and feasible, which can meet the requirements of voltage detecting on UHV transmission lines.

Reference

1. Khan T H, Kabir S M L, Hussain S, et al. Design and implementation of a low cost electricity meter testing bench[C]// Industrial Electronics & Applications (ISIEA), 2010 IEEE Symposium on. IEEE, 2010:34 - 39.
2. Jiang T, Huang X, Yang Y, et al. Study on anti-interference of high voltage ultraviolet electroscope[C]// Automation Congress, 2008. Wac. IEEE, 2008:1-4.
3. FENG Kun, RUAN Jiangjun, DUN Shenglei, et al. A Fault Analysis and Sensitivity Improvement of Capacitive High Voltage Electroscope[J]. Electromechanical Components, 2014, 34(3):42-45.
4. LI He, DENG Zhixiang, WANG Chuang, et al. Study on Electrical Inspection's Dead Zone of Capacitive Detector[J]. Electrical & Energy Management Technology, 2014(15):10-13.
5. DL/T 740-2014 capacitive type voltage detectors[M]. China Power Press, 2014.
6. WANG Peijun, ZHANG Ping, WANG Liyong. Problems and Countermeasures in the use of capacitive type electroscope[J]. Electric Safety Technology, 2006, 8(8):54-55.
7. WANG Yong. Improvement of high voltage transmission line electroscope[J]. 2006, 8: 29.
8. DAI Libo. Application of UV imaging technology in live inspection of high voltage equipment[J]. Autormation of Electric Power Systems, 2003, 27(20).
9. JIANGYi. Application of ultraviolet corona detector in corona discharge detection[J]. East China Electric Power.Vol.32,No.8,Aug.2004:34-35.
10. Liu Jungang. Research of Noncontact Electroscope of Ultra-High Voltage Based on Ultraviolet Detection[D]. Chongqing University.
11. XU Yang. Optical Pulse Method for Partial Discharge Measurement and the Comparison with Electrical Current Method[J]. High Voltage Engineering, 2001,27(4):3-5.
12. Kurrer R, Feser K. The application of ultra-high-frequency partial discharge measurements to gas-insulated substations [J]. IEEE Trans. on Power Delivery, 1998, 13(3): 777-782.
13. ZHANG Haifeng, PANG QIchang, CHEN Xiuchun. The characteristics of High-Voltage corona and its detection[J].Electrical measurement and instrumentation .2006.2.Vol.43,No.482:6-9.
14. XU Yi-sheng , LI Mei-jun. HV Live Displaying Device of Field Intensity Type[J]. High Voltage Apparatus,2005, 41(5): 383-384.
15. Gao Guihua. The Design of Non-contact Electroscope of Ultra-high Voltage Based On Field Intensity Distribution[J]. XIAN University
16. LUO Chao, FANG Shubo, ZHA Zhiming and etc. research on distribution characteristics of electromagnetic field and attenuation law of pure sound for UHV AC transmission line[J]. High Voltage Apparatus,2017(2017-08-03).
17. Xiao Dongping. Research on Three Dimension Calculation Models and Shielding Measures for Electromagnetic Field of Ultra-High Voltage AC Transmission Lines[D]. Chongqing University.
18. Luo Xiao. Research on Electromagnetic Field of UHV Transmission Lines and Substations[D]. Chongqing University.
19. ZOU Anxin XU Luwen. Simulation Analysis on Effect of Phase Sequence on Power Frequency Electromagnetic Field Under UHV AC Transmission Lines[J]. High Voltage Apparatus, 2015,03(51),82-87.
20. IN Xin , LI Xue-bin , XU Jian-yuan. Simulation and Analysis of Induced Voltage and Current for UHV Double Circuit Transmission Lines on the Same Tower[J]. High Voltage Engineering, 2010,36(09):2193-2198.
21. HE Yuwei, LIU Yakun, YAN Nannan and etc。Evaluation of Induced Voltage and Current on Low-Voltage Distribution Lines Under EHV/UHV AC and DC Transmission Lines[J]. Power System Technology, 2015,39(09):2640-2646.
22. ZHAO Kangwei,ZHANG Chao,PENG Long,CAI Yufei,KE Qinghua,WEI Ran.Research on Induced Voltage and Current of UHV Circuits Produced by Parallel Lines[J].Electric power, 2017,50(02):46-51+87.

Linear Motor Motion Control Experiment System Design Based on LabVIEW

He Cuixian[1,2], Li Jianqi[1,2], Yang Feng[2] and Li MingHao[2]

[1]*Cooperative Innovation Center for The Construction & Development of Dongting Lake Ecological Economic Zone, Hunan University of Arts & Science, Changde, Hunan, China*
[2] *College of Electrical Engineering, Hunan University of Arts & Science, Changde, Hunan, China*

Abstract. In order to meet the needs of experimental training of electrical information industry, a linear motor motion experiment system based on LabVIEW was developed. This system is based on the STM32F103ZET6 system processor controller, a state signal when the motor moves through the grating encoder feedback controller to form a closed loop, through the RS232 serial port communication with the host computer, the host computer is designed in the LabVIEW interactive environment monitoring software. Combined with the modular design concept proposed overall program, given the detailed hardware circuit, targeted for the software function design, to achieve man-machine interface. The system control of high accuracy, good stability, meet the training requirements for laboratory equipment, but also as a reference embodiment of the linear motor monitoring system.

1 Introduction

With the rapid development of intelligent manufacturing technology, the manufacturing industry has put forward higher requirements for the control accuracy, which makes the linear motor widely researched and developed[1]. There are many advantages like simple structure, low noise, fast speed, high accuracy to use linear motor as driving device. It greatly improve the level of manufacturing[2]. Recently many electrical and information majors in related universities have set up experimental training courses for industrial movement control in order to better train students' ability and carry out the training policy of applied talents. These courses are different in contents and forms among colleges. But traditional linear control system of rotating motor drive is used in common experimental devices, which can not meet the requirements for talent training in existing industrial enterprises[11]. The virtual instrument used as experimental development equipment, has powerful function, easy debugging and many advantages like this[3-4]. Therefore, this paper takes Labview as development platform and proposes an experimental equipment system for monitoring and debugging the linear motor.

2 System overall design

The block diagram of linear motor monitoring system is shown in Figure 1. This controller uses linear actuator to control linear motor driving and dragging load. State signal while motor running obtained by linear actuator detection, feeds back to the controller after grating encoder processing and forms a closed loop system[12]. The controller controls the linear motor moving to the specified position according to the deviation value produced by the target value and the feedback value. The controller communicates with the host computer by RS232 serial port mode. The controller receives the target value sent by the host computer, and then adjusts the movement mode of the linear motor according to the instruction, in order to make the motor move to the new position. The controller can also send the motor state signal obtained by the feedback circuit to the host computer.

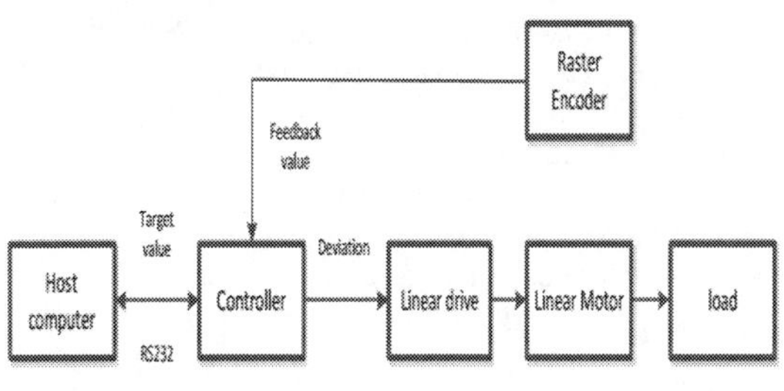

Figure 1. The block diagram of linear motor monitoring system.

3 System hardware design

The system uses STM32F103ZET6 (hereinafter referred to as STM32) as the main controller, which mainly includes power module, control module, detection and feedback module and serial communication module [5-6], as shown in Figure 2.

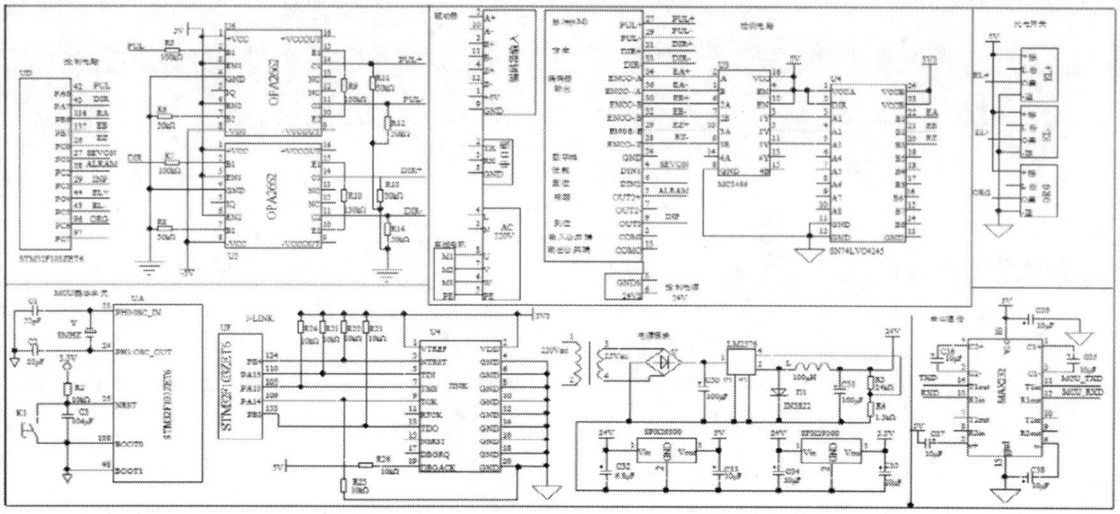

Figure 2. Hardware circuit.

3.1. Power module

The 220V voltage after step-down transform, rectification and filtration, then through the voltage regulator LM2576 will be output to 24V voltage, then use SPX29300 rise 3.3V to 5V power supply to various parts of the system.

3.2 Control module

In STM32, PWM generated from CH1 channel of timer TIME3 is taken as PUL. PWM generated from CH2 channel of timer TIME3 is taken as DIR. The PUL pulse and the DIR pulse are converted into differential control signals through the operational amplifier OPA2662, and output to the linear motor driver. The driver controls the motion of the linear motor according to the number of pulses.

3.3 Detection and feedback module

The encoder of the linear driver can automatically detect the motion state of the motor and produce EA+ and EA-, EB+ and EB-, EZ+ and EZ- differential state signals[14]. After the grating encoder MC3486, those state signals will be converted into two orthogonal coded pulse signals EA, EB and one reference point pulse signal EZ and feed back to STM32[7]. STM32 uses a timer TIME4 encoder mode to determine and count the direction of the feedback pulse signal. After the converter, the SN74LVC4245 converts the 5V pulse to 3.3V pulses so that the STM32 is easy to collect samples[8].

3.4 Serial communication module

The communication module adopts serial port RS232 mode which can realize communication within the distance of 20 meters and can adapt to the teaching conditions in most environments. The transform of communication power level between the host computer and the lower computer is realized by MAX232.

4 System software design

4.1 Communication protocol

In order to ensure the normal communication between the host computer and the lower computer, communication protocol is defined. The communication protocol mainly includes communication start, communication enable, communication enable close, positive direction, negative position and other functions. Each function is defined by the corresponding command format [9]. For example, when the system receives the "communication start" command, open communication channel to send and receive data. When the system needs to be shut down, then send "communication close" command to close and set memory free, so that the second time to update data. The specific agreement as shown in Table 1:

Table 1. Communication protocol.

Functions	Protocol commands	Remarks
Communication start	communication start	
Motor enable	communication enable 1	
Motor enable close	communication enable 0	
Positive direction	communication direction 0	
Negative direction	communication	

	direction 1	
Distance	communication distance x	X stands for distance value
Origin motion	communication backmove	
Output port	communication output y	Y is hexadecimal character string, represents the output port
Input port	communication input z speed v position s d	Z, v, d are all hexadecimal character string, respectively represents input port, speed and position
Communication close	communication close	

4.2 Design of main program

The main program flow chart of the system is shown in Figure 3. the system first completes the clock, serial port, timer initialization, then automatically starts the detection. Then it will send the detected speed and location information to the host computer every 0.5 second. The system will start the control program when receiving the start command from the host computer[10].

In the control program, the controller processes the feedback data, then generates a control pulse signal according to the target value and the feedback value, and then transmits the signal to the driver. The number of pulses transmitted by the controller determines the moving distance of the overload in the linear motor. That is to say, the moving distance is equal to the width of a single pulse multiplied by the number of pulses. The movement direction is controlled by a I/O port. The high power level is at positive direction, and the low power level is at the negative direction.

In the detection program, the STM32 universal timer TIM4 encoder mode is set. The main idea is to judge whether the EA phase is in advance of the EB phase. If advanced, the output pulse will increase. If not, then reduce. If the EZ phase comes a pulse, it will count to zero[8].

4.3 Host computer monitor program

The program mainly adopts event structure to write. The host computer receives the data according to the receiving protocol "communication input z exp_position p speed v position s dir d /n". Then it will separate character string z, p, v, s and d by using 6 regular expressions. And then display data represented by these strings into the interactive interface. At the same time, a data function is needed to create and a shift register is needed to add to store last time date. If <Motor enable> is "ENABLE", then the host computer will send command "communication enable 1" to the lower computer. If <Motor enable> is "DISABLE", then the host computer will send command "communication enable 0" to the lower computer. If the moving direction is positive, then the host computer will send "communication direction 0" to the lower computer. If the moving direction is negative, then the host computer will send "communication direction 1" to the lower computer[15]. Once the "communication start" value is set, the data type of distance on the interactive interface is default. The value of the form of digital conversion into a decimal string, and put the string to connect, then the host computer will send " communication distance x " to the lower computer. Once the " origin motion " value is set, the host computer will send "communication back move" to the lower computer[13]. Once the "I/O port" is set and the "Actual level" is high level, the Enable will keep steady; While the "Actual level" is low level, then first untie all the I/O port value and reverse and then tie up. If the data type of new value of event branch is hexadecimal, it will send command "communication output y" to the lower computer. If "Stop" value is set, it will send command "communication close" to the lower computer. It will delay 50ms through a tiled sequential structure and turn off the VISA resource name, and then the lower computer will close the communication protocol.

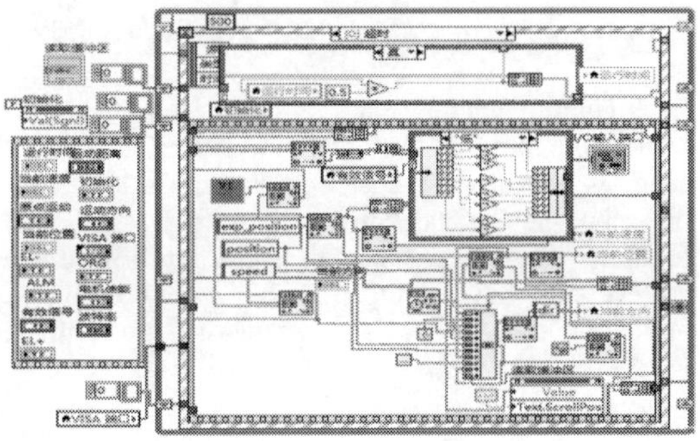

Figure 3. Read/Write program.

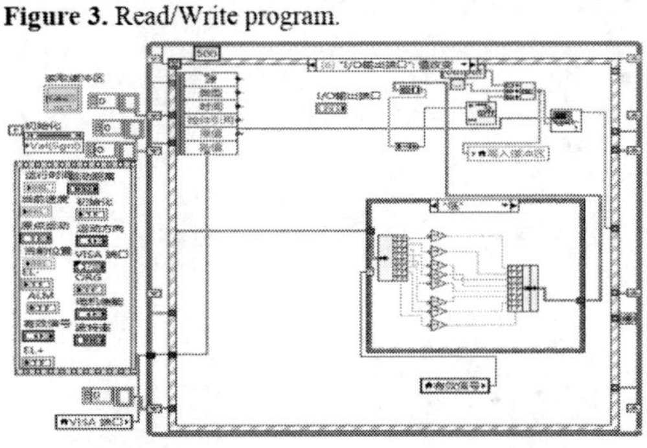

Figure 4. I/O port control program.

4.4 Interactive interface design of the host computer

The "motion control interface" is designed on the LabVIEW platform, as shown in Figure 5. It uses VISA to write serial drive. VISA is a multi type function library, which supports multi platform work and multi interface control. When the external device changes, you only need to replace several program modules[10]. In LabVIEW, the basic steps of serial communication are divided into 3 steps: First, serial initialization, it uses VI SA Configure Serial Port.vi nodes to set the serial port number, baud rate, stop bits, parity bits, data bits. Second, read and write serial port. It uses VISA Read node and VISA Write node to read and write the serial port. Third, close the serial port, which is to close all the read or write operations.

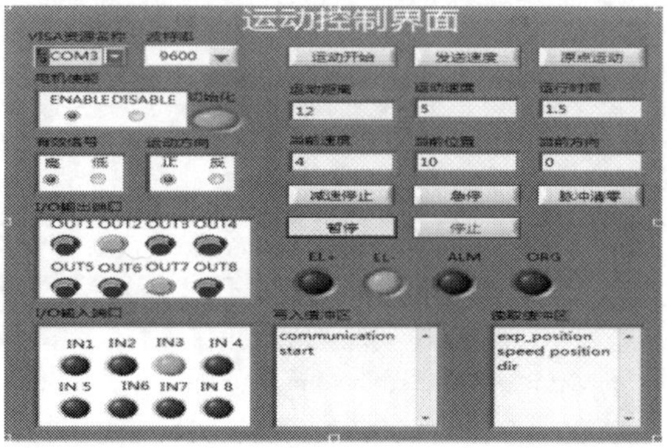

Figure 5. Interactive interface.

5 System test analysis

Each control function is tested one by one through the interaction interface, and the state value of the motor, such as the current speed, current position, current direction, can be monitored in real time, as shown in Figure 6. The lower computer can responds in time and correctly, as shown in Figure 6. When the position is set at the minimum interval, the moving distance of the load is 26 μ m, which meets the requirements of the experimental expect accuracy.

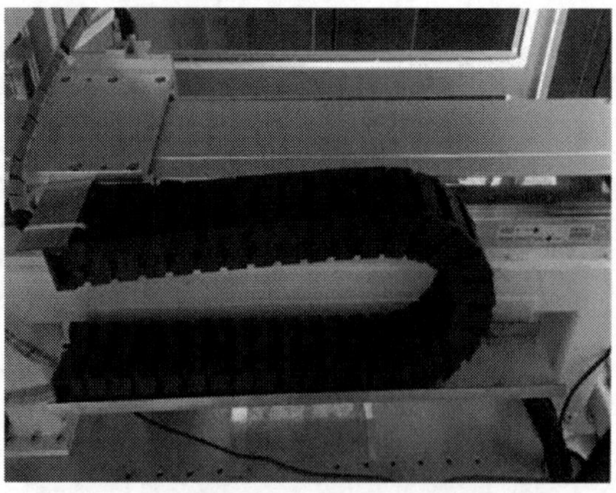

Figure 6. Interactive interface.

6 Results

The equipment system composed of the host computer, controller, linear drive and linear motor realizes the monitoring of linear motor movement. The closed-loop control circuit ensures the stability of the system. The host computer and controller communicates in accordance with the protocol through RS232 serial port. The host computer monitoring program is designed under the LabVIEW environment and the interactive property is good. According to the test result show in table 2 and figure 6, the accuracy of control and monitoring is higher. The function of the system has reached the expected design requirements, and the cost is low. It can be used as a practical teaching equipment, and it also provides a reference for the linear motor monitoring system program.

Table 2. Test data.

Expected distance(mm)	-50	-30	-20	-10	0	10	20	30	50
Actual distance(mm)	-49.91	-29.92	-19.96	-9.99	-0.01	9.98	19.94	29.92	49.92
Error(mm)	-0.09	-0.08	-0.04	-0.01	0.01	0.02	0.06	0.08	0.08

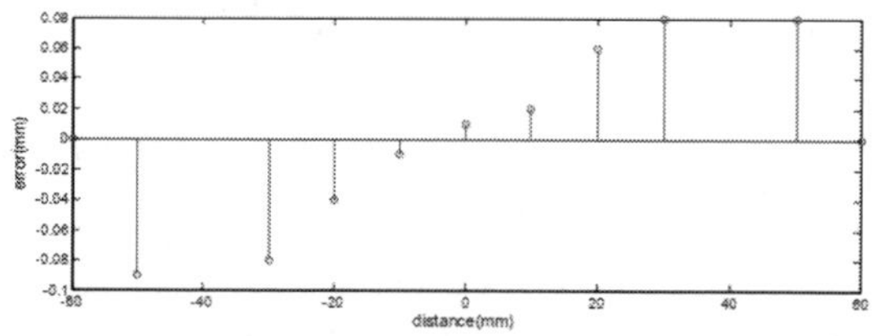

Figure 6. The relation between error and distance.

Acknowledgements

The author would like to thank all the anonymous reviewers for their valuable comments and thoughtful suggestions that improved the quality of the presented work. This work is partially supported by the National Natural Science Foundation of China (Grant No. 61403136， 61703157),and Changde City Science and Technology Bureau project, China (Grant No. 2015SK18).

References

1. Wang Huiyong, Man Zhonglei, Li Xiangnan, Jiang Shouhe, Dong Youli, Wang Bin. Application of the linear motor and the development trend in the manufacturing [J]. Motor and control applications, 2013,40 (03): 1-4.
2. BinYao,LiXu.Adaptive Robust Motion Controlof Linear Motors for Precision Manufacturing[J]. Mechatronics, 2002,12(4),595-616.
3. Lei Zhenshan, Wei Li, Zhao Chenguang, Tang Xiaojiao. LabVIEW senior programming and virtual instrument engineering application ［M］. Beijing: China Railway Press.2009: (81-85).
4. Cao Rongmin, Zhou Huixing, Su ang, Wu Yingnian, Hou Ming. Design of innovative experimental platform for linear motor based on virtual instrument

[J]. Experimental technology and management, 2013,30 (09): 83-87+96.
5. i QingPeng, Cao Rongmin. Design and implementation of linear motor driver based on DRV8301 driver IC [J].Automation of manufacturing industry, 2013, (24): 126-130.
6. Jia YiChong, Bao Guangqing, Yang Qiaoling. Control of permanent magnet synchronous linear motor based on state error [J].. Automation and instrumentation, 2016, (02): 78-82.
7. PCH, Cheng Pengfei, Wang Yan, Yin Tianming. Detection on position and speed of Switched Reluctance Motor Based on grating encoder [J]. Electric automation, 2013, (35): 31-33.
8. Shen Jiaoxiao. Research of permanent magnet synchronous linear motor control system based on [D]. STM32.Southeast University, 2016.
9. Li Ruixian. Software design and implementation of motor control experiment system based on LabVIEW [J]. Journal of Shandong University of Technology: Natural Science Edition, 2006,20 (4): 52-56.
10. Wang Haixia, Yan Guiding, Li Baohui, Liu Xiaoyan, Wang Yi. Software design and implementation of linear motor motion control system [J]. Journal of electronic measurement and instrumentation, 2013, (03): 264-269.
11. Wang Weijin. Overview of the development and application of linear motors [J]. micro motors (servo technology), 2004, (01): 45-46+50.
12. Liu Shuai. Introduction and application of grating encoder [J]. special equipment for electronics industry, 2014,43 (09): 42-45.
13. Shen Baoshan, Ji Changying, Guo Yuping, Zhang Jile. Design of [J]. mechanical and electronic data acquisition system based on LabVIEW 2009 (04): 76-78.
14. Xie noritada. Linear motors in the electronics industry [J]. encoder electronic industry special equipment, 2007, (05): 69-73.
15. Li Da, Wei Xuezhe, Sun Zechang. Design and implementation of.LabVIEW data acquisition system [J]. China Instrument and meter, 2007, (01): 49-52.

Design of Ultra-low-end Controllers for Efficient Stepper Motor Control

Mordecai Raji[1],Akeem Shokanbi[3] and Happy Monday[2]

[1,a]University of Electronic Science and Tech. of China, Electronics Department,611731West Hi-Tech Zone,Chengdu,Sichuan,China
[2]University of Electronic Science and Tech. of China, Electronics Department,611731 West Hi-Tech Zone, Chengdu,Sichuan, China
[3]University of Electronic Science and Tech. of China, Automation Department,611731 West Hi-Tech Zone, Chengdu,Sichuan, China

Abstract. This paper lays emphasis on thedevelopment of low cost controllers for stepper motors in contrast to its resource limitations such as memory size, few I/O pins and computing power compared to High-end designs. The microchip AVR ATtiny45 microcontroller was employed alongside a redesigned (reduced-input pin count) pulse distribution circuit for two H-Bridge drivers. The motor can rotate in both directions as well as possible speed control. The concept of the motor control signals was modeled in Matlab/Simulink, firmware was written in Atmel AVRStudio development environment, while theoverall design was carried out in Proteus software followed by Hardware implementation. Total material cost is about $5 which would be less in commercial production cases.

1Introduction

Since the advent of stepper motors, it has gained widespread use in different applications including medical equipment, robotics, antenna positioning, printers, scanners, industrial applications, disk drives including micro-surgical operations, etc. The widespread acceptance is due to its advantage over conventional DC/AC motors such as precise motion, position,and speed, more so, compared to conventional motors, stepper motors are more flexible to control [1][2][3]. Stepper motors are brushless motors, so they can only be commutated (in the right sequence) by digital signals to achieve motion. For a bipolar stepper motor, signals from the controller are coupled to the coils with two H-Bridge as shown below.

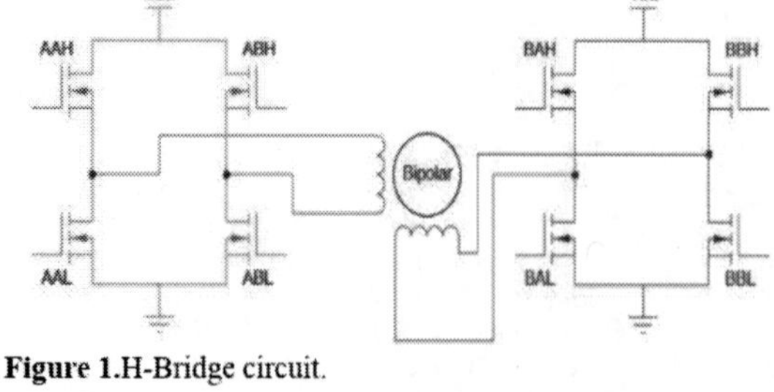

Figure 1.H-Bridge circuit.

The applied signal can either make the motor move or hold a position. Motion can be afull step, half step, and micro-step to higher resolutions of micro-steps, therefore the performance of stepper motors depends largely on the controller. Most stepper motor controllers are based on (mid-end) MCUs, ARMs, CPLDs, DSPs and FPGAs[3] which usually include four kinds of systems; they are microcontroller (MCU) system, professional motion control and programmable logic controller (PLC) system, PC and motion control system respectively. All the above control methods require the user to further develop Windows program [3].

The purpose of this paper is to design and implement a very simple and cost-effective stepper motor controller with micro stepping capabilities. Experimental setup features the Microchip ATtiny45 and a custom driver circuit to achieve comparable features and results as that of high-end controllers. For thefurther sake of simplicity, the control buttons for motor direction of rotations are also used for step functions (and speed control). Arguably, interfacing the ATtiny45 AVR microcontroller (or other 8 pin microcontroller) with many stepper motor driver ICs will leave room for little or no other I/O function because it has only 6 I/O pins. An example; the L298 dual H-bridge monolithic stepper motor driver has 6 control pins.

2Theories of operation

2.1 Concept of PWM and microstepping

Stepper motors can be operated in full step and half step. The need for higher resolution, less vibration,and smoother movement steps gave birth to the concept of microstepping. Therefore to achieve microstepping, the technique of Sinusoidal Pulse Width Modulation (PWM) is usually employed. SPWM is the sinusoidal variation of the Pulse Width of an output Signal. It is typically generated from the output of a comparator; fed in one of it input a carrier signal and the other a modulating signal. Figure 2 depicts AVR PWM timer in Fast PWM mode using sawtooth waveform as the carrier signal and the compare (reference) signal for modulation. So, SPWM can simply be achieved by sinusoidal variation of the reference signal level using the Sine Look-Up Table stored in the EEPROM of the microcontroller.

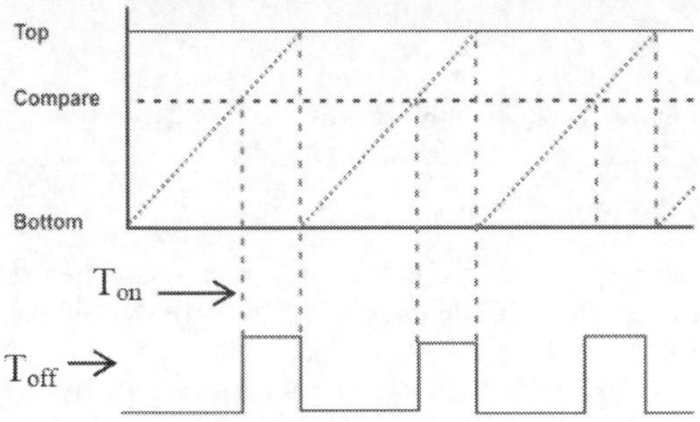

Figure 2. Pulse width modulation.

The Average value of the output waveform is given by [4]

$$\bar{y} = \frac{1}{T} \int_0^T f(t)\, dt. \tag{1}$$

where T is the period and f(t) is the pulse waveformThe above expression is simplified as

$$\bar{y} = D \cdot y_{\max} \tag{2}$$

where D is the Duty Cycle and $Y_{min} = 0$.
Bipolar stepper motors require two sine output phased 90 degrees to each other i.e. a sine and a cosine wave[5] as shown in figure 3.

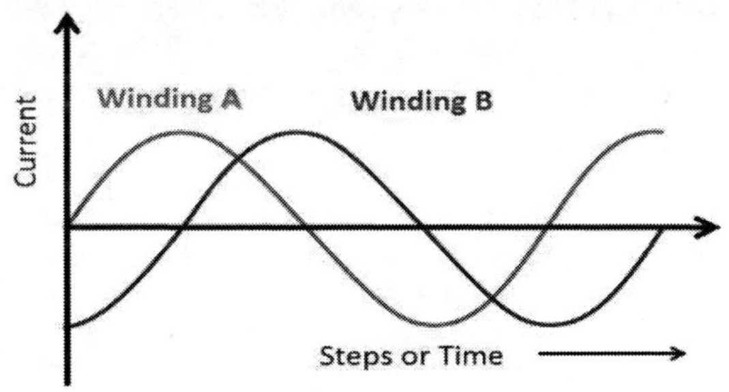

Figure 3. 90-degree current phase difference (for microstepping) through the winding of a stepper motor.

2.2 Control hardware and software

The ATtiny45 is an 8-bit AVR RISC based microcontroller, combines 4KB ISP flash memory, 256-Bytes EEPROM, 256 SRAM and 6 general purpose I/O lines. The figure below reveals the function assigned to each pin. The firmware & Hardware was carefully designed to accommodate all basic functions.

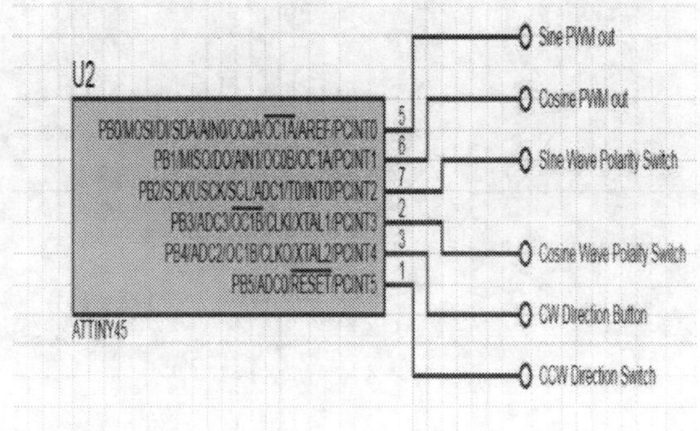

Figure4. Pin-out assignments for the Atiny 45 microcontroller operation.

2.3Control algorithm

The algorithm is optimized for a bipolar stepper motor. The PWM is generated simply from one sine Look-up table(LUT) using two table pointers phased 90 degrees to each other as opposed to using two LUTs which will use up more memory. The outputs of the pointers are sent to the OC0A and OC0B pins (PB0 and PB1 respectively) in fast PWM mode. The direction of the table pointers depends on the button pressed.

In any instance where any of the table pointer indexes is zero, the polarity of the affiliated switch pin will be inverted as to generate anegative sine or negative cosine wave for the required duration. This is essential for the two H-Bridge. At every 90 degrees, the polarity pin changes state: four times in one full rotation. In other to further understand and verify this theory, a model of this concept was created and simulated using Matlab/Simulink. Figure 3 shows the model component setup and the result. The result verifies the appropriate gating signals for the H-Bridge.

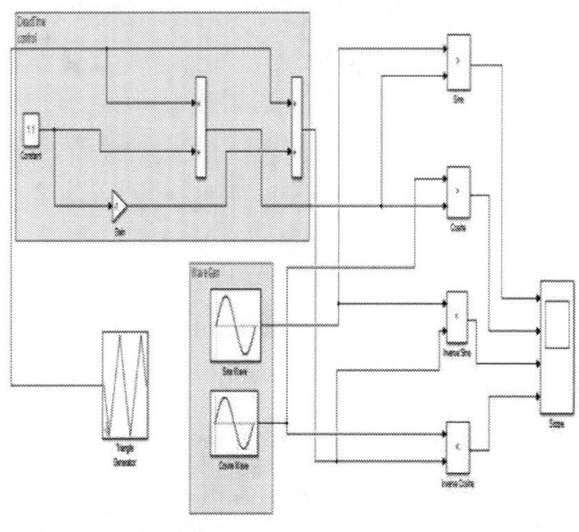

Figure 5. Simulink model.

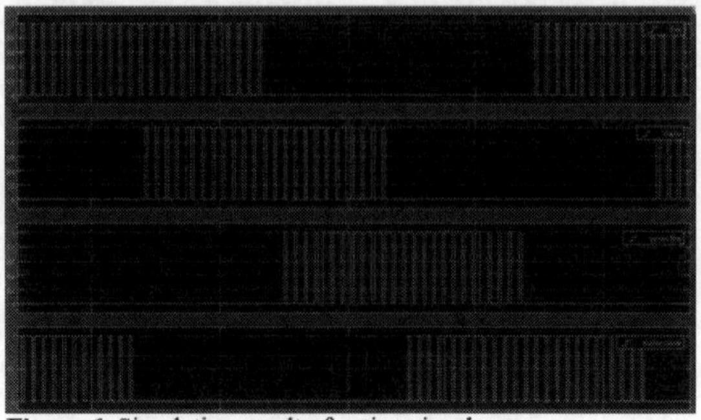

Figure 6. Simulation result of gating signals.

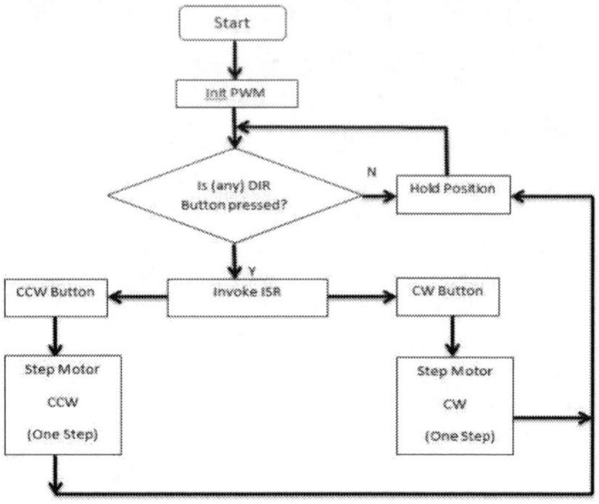

Figure 7. Flowchart.

```
ISR (TIMER0_COMPA_vect)
{
    if((PINB & 0b010000)==0) //if Right button is pressed. Forward direction
    {
        i++; //sine index pointer increment
        j++; //cosine index pointer increment
    }

    if((PINB & 0b100000)==0) //if left button is pressed. Reverse direction
    {
        i--; //sine index pointer decrement
        j--; //cosine index pointer decrement
    }

    if(i==0) // check sine PWM phase angle (180 degrees or 0 degrees)
    {
        PORTB^=(1<<2); //invert polarity
    }

    if(j==0) //check cosine PWM phase angle (180 degrees or 0 degrees)
    {
        PORTB^=(1<<3); //invert polarity
    }

    OCR0A = pgm_read_byte(&sinewave[i]); //update sine PWM output
    OCR0B = pgm_read_byte(&sinewave[j]); //update cosine PWM output
}
```

Figure 8. Control algorithm of the stepper motor.

2.4 H-Bridge driver and control circuitry

Due to the very limited number of GPIO pins of the ATtiny45 (and other 8-pin microcontrollers), driving the two H-Bridge circuits directly would leave little or no room for other control functions. As a result, 'two pre-stage circuits were developed which will serve as adaptors as to reduce the pin count. The circuit is a simple combination of logic gates and also acts as pulse output switch (distribution) pin to the two H-Bridge.

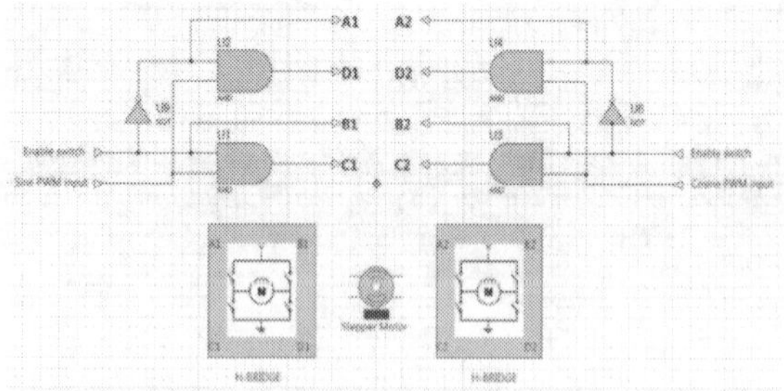

Figure 9. H-Bridge driver circuit.

2.5 Stepper motor

Stepper Motors are classified into Variable Reluctance Motor (VRM), Permanent Magnet (PM) Motor and Hybrid Motors (HM). PM has relatively lower torque than VRM [6] while Hybrid Motors hasan advantages over the other two such as high torque and ability to achieve higher speed. Therefore for the experimental setup, a hybrid stepper motor was selected. The Table below shows the Motor specifications from the manufacturer (NEMA).

Item	Specifications
Step Angle	1.8°
Step Angle Accuracy	± 5% (full step, no load)
Resistance Accuracy	± 10%
Inductance Accuracy	± 20%
Temperature Rise	80°C Max.(rated current,2 phase on)
Ambient Temperature	-20°C~+50°C
Insulation Resistance	100M Ω Min. ,500VDC
Dielectric Strength	500VAC for one minute
Shaft Radial Play	0.02Max. (450 g-load)
Shaft Axial Play	0.08Max. (450 g-load)
Max. radial force	28N (20mm from the flange)
Max. axial force	10N

Figure 10. Stepper motor specifications.

3 Experimental setup, verification and result

For experimental purposes, a high torque hybrid stepper motor is used with the controller. Initially, the setup was put into motor free running mode for easy determination of the controller's efficacy. In this mode, the speed depends on the microcontroller's clock frequency, number of values in the LUT and the number of clock cycle it takes to serve the interrupt each time. For an 8-bit timer, interrupt occurs every 256 clock cycle [7], theClock frequency is 8MHz, no pre-scaling.

Stepper motors have their specific problems; low-frequency oscillations around the synchronism frequency at high speeds [8]; running the motor at higher frequencies (speed) resulted into step loss; primarily due to the drastic current drop in each phase of the motor. A chopper driver should be employed.

At the final test, the pushbuttons were installed; each corresponding to the motor's direction of rotation. The microcontroller was reprogramed to respond to the pushbuttons'low-level signals. The pushbuttons are on

https://doi.org/10.1051/matecconf/201816002003

PB4 and PB5 for forward and reverse direction respectively. PB5 serves the same function as the factory-configured reset pin; therefore the Microcontroller's corresponding reset fuse should be disabled for that pin so the pin could function as a typical IO. These pushbuttons could be replaced with a host controller; the host controller can easily determine the motor speed and position by pulse rate and pulse counting respectively without the need for a rotary encoder. However, positioning and tracking would only be efficient by ensuring the host controller is configured to deliver the right pulse length per count and prevention of step loss because feedback is not provided.

The position of the motor shaft is determined by the torque phasor due to the currents applied to the motor coils [8], the torque phasor is determined as desired by the microcontroller by incrementing or decrementing the index pointer value to the LUT to generate sine wave signals phased 90 degrees to each other. The relationship between the electrical position and the current per phase is given by [9]:

$$\begin{cases} i_A = I_0 \cdot cos\,\theta_e \\ i_B = I_0 \cdot sin\,\theta_e \end{cases}$$

where Θ_e is the electrical position and I_o is the current per phase.

For practical purpose, the popular L298 dual H-bridge Bipolar Stepper Motor driver is used. It has 6 control input pins, therefore, the microcontroller pins need to be adapted to the driver IC. Figure 9 is an excerpt of the above H-Bridge driver circuit reduced to two NOT gates. The L298 accepts TTL inputs. For the experiment, gate ICs was not used; two transistors were wired as inverters to form NOT gates. The L298 dual H-Bridge driver circuit was chosen because of thehigh current capability of 2A DC operation and 2.5A at 80% duty cycle [10] and it is a flexible driver IC. In most stepper motor configurations, The Enable pins are used for activating or deactivating the H-Bridge but in this design, the Sine PWM output and the Cosine PWM output are fed to the EnA (pin 6) and EnB (pin 11) respectively while the pulse distribution is achieved with the In1-In4 (pin 5, 7, 10, and 12).

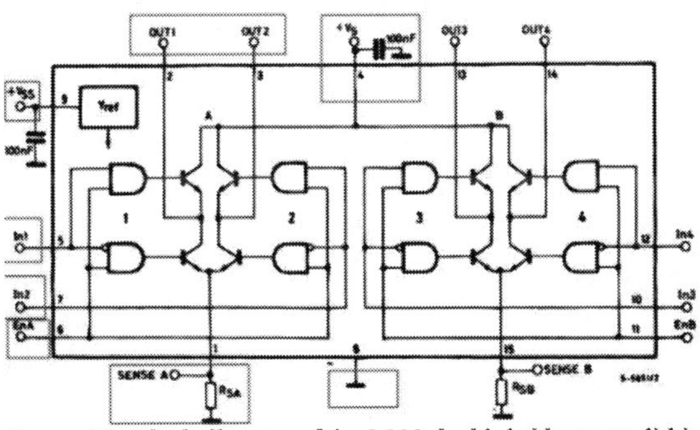

Figure 11. Block diagram of the L298 dual h-bridge monolithic driver.

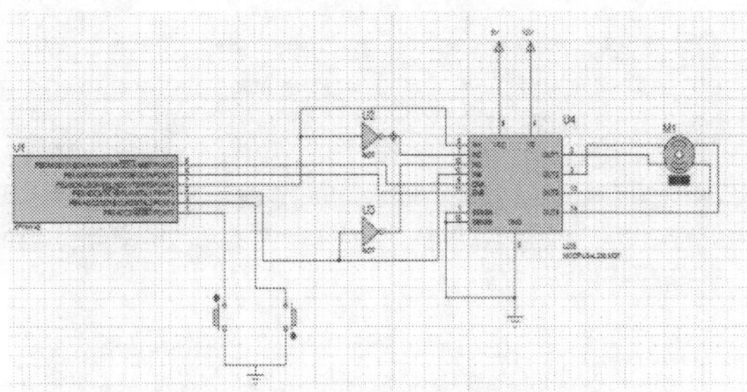

Figure 12. Using with the L298 dual h-bridge driver IC.

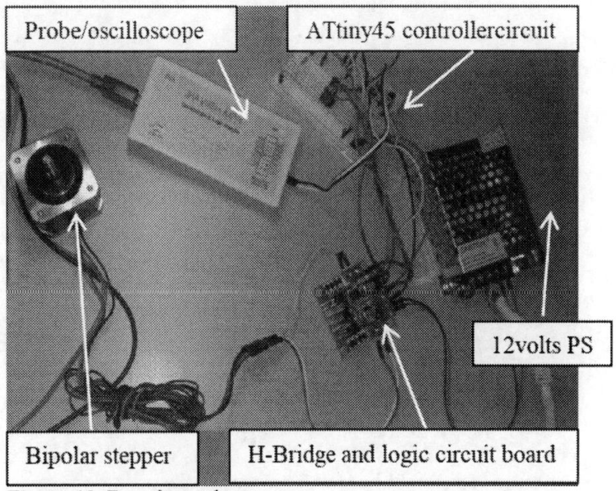

Figure 13. Experimental setup.

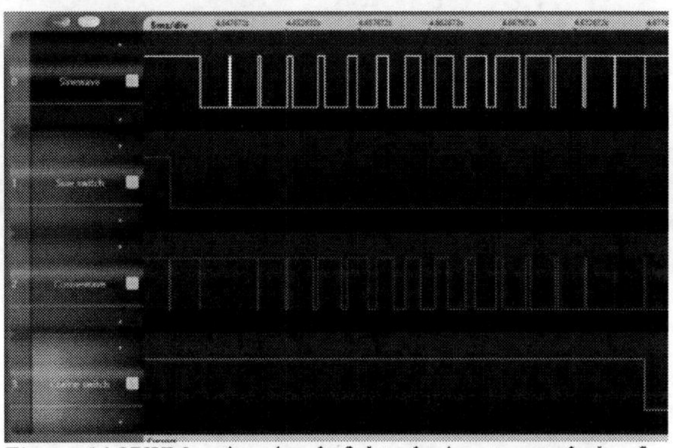

Figure 14.SPWM gating signals fed to the input control pin of the L298 driver IC.

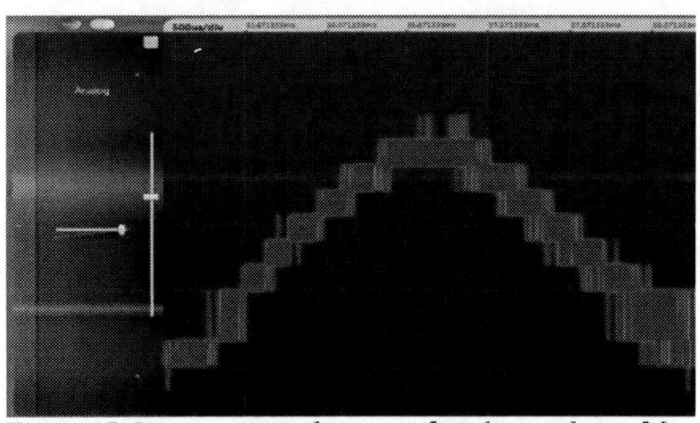

Figure 15. Stepper motor voltage waveform in one phase of the stepper motor (16 micro steps).

4Conclusion

This paper aims to prove that microcontrollers could be put into more effective use (pushing its boundary) in control systems and other embedded system domains by skillful management of its (limited) resources and that design Engineers should not be quick into resorting to costly controllers (DSPs, FPGAs) without fully exploring its capabilities.

Basic control of a stepper motor has been discussed with simple implementations aimed at achieving results comparable to high-end controllers. Some techniques have been presented to make the circuit as simple as possible with the least possible footprint and cost implementation. Vibration and step loss have been reduced with a considerable increase in torque by either adjusting the controller's clock frequency or the number of samples in the LUT at the least. Resonance is minimal, step resolution is high and setup is fairly easy. In a commercial purpose, due to the simplicity of the system, the microcontroller and the H-Bridge driver could be integrated to a single chip as a mixed-signal Integrated circuit or on a single circuit board.

Future work would be to squeeze more functions into a controller of the same level without compromising efficiency such as running the pushbuttons on a single pin of the microcontroller by exploring the inbuilt ADC. Thetwo pushbuttons (or even more) will be identified by different voltage levels.This would leave a pin vacant which could be used for other functions such as sensing current through the stepper motor or for positioning.

References

1. M. Y. Tarnini, "Fast and Cheap Stepper Motor Drive," *4th International Conf. Renew. Energy Res. Appl.*, **vol. 5**, no. 2, pp. 689–693 (2015)
2. W. Ruifeng, W. Zhe, and W. Liying, "Stepper Motor Control based on AT89S51 Microcontroller," *2015 8th Int. Conf. Intell. Comput. Technol. Autom.*, pp. 1–4 (2015)
3. G. L. ZHANG Benhua, LI Chenghua, SUN Shiming, "Design on a Unipolar and Unidirectional Stepper Motor Circuit," *Int. Conf. Electron. Mech. Eng. Inf. Technol.*, pp. 1795–1797 (2011)
4. https://en.wikipedia.org/wiki/Pulse-width_modulation
5. A. Nk, D. Krishnan, S. Moorthi, and M. P. Selvan, "FPGA Based Microstepping Scheme for Stepper Motor in Space-Based Solar Power Systems," *2012 IEEE 7th Int. Conf. Ind. Inf. Syst.*, pp. 6–10(2012)
6. Ł. Przeniosło and M. Hołub, "Development of microprocessor,timeoptimized stepper motor driving algorithm," *2017 22nd Int. Conf. Methods Model. Autom. Robot.*, pp. 174–179 (2017)
7. "PWM Sine Wave Generation," https://web.csulb.edu/~hill/ee470/Lab%202d%20-%20Sine_Wave_Generator.pdf.
8. G. Baluta, "Microstepping Mode for Stepper Motor Control," *2007 Int. Symp. Signals, Circuits Syst.*, pp. 1–4(2007)
9. Acarnley, "Stepping Motors: a Guide to Modern Theory and Practice, "4th ed., IEE Control Engineering Series 63, ISBN: 0-85296-029-8, . T E. Michael Faraday House pp.48-51 (2002)
10. http://www.st.com/content/ccc/resource/technical/document/datasheet/82/cc/3f/39/0a/29/4d/f0/CD00000240.pdf/files/CD00000240.pdf/jcr:content/translations/en.CD00000240.pdf

Efficient Voltage Regulation with Modified Hybrid SEPIC DC-DC-Converter

Md. Ali Azam Khan[1] and Mohammad Ali Choudhury[2]

[1]Department of EECE, MIST, Dhaka, Bangladesh
[2]Department ofEEE,BUET,Dhaka, Bangladesh

Abstract. Switch mode dc-dc converters are attractive for their small size, ease of control and efficient power conversion. Output voltage is regulated by duty cycle control of semiconductor switch of switch mode dc-dc converters. The voltage gain and efficiency of practical switching regulators deviate from ideal values at extreme duty cycles. Also, desired gain /attenuation is not achievable at high/low duty cycles. In applications where high gain or high attenuation of voltage is desired with acceptable energy conversion efficiency, hybrid dc-dc switching converters are used. Hybrid dc-dc converters are combination of voltage multiplier/division circuit with appropriate SMPS circuits. By incorporating voltage multiplier/division cell with conventional SEPIC converters, desired voltage gain (either very low or very high) may be achieved at acceptable energy conversion efficiency. In the present work with an aim to attain very high voltage gain by conventional SEPIC topologies, a new voltage multiplier cell consisting of multiple inductors and diodes is proposed.

1. Introduction

Modern power electronic switches can operate at high frequencies. The high frequency DC-DC chopper-converters are known as switch mode power supplies (SMPS). SMPS are extensively used as power supplies of computers and electronic equipment. According to the type of voltage regulation (step-down/step-up/step-up-down) SMPS has three basic (Buck, Boost and Buck-Boost/Ĉuk/Sepic) and many derived configurations. The voltage regulation is achieved by duty cycle (D) control of the switching signal of the semiconductor switch of the SMPS. In a practical SMPS, the efficiency and the voltage-gain are function of the duty cycle D. Each type of SMPS has high value of efficiency for limited range of voltage gain (within a range of duty cycle D). In case of extreme gains (either high or low), hybrid SMPS converters may be used to maintain high conversion efficiency. Hybrid SMPS converters are combination of voltage multiplier/divider circuits with appropriate SMPS circuits. In this Thesis work the SEPIC, the hybrid boost dc-dc converter with two inductors, the hybrid proposed SEPIC converter with two inductors and the proposed hybrid SEPIC dc-dc converter with three inductors are studied by simulation at variable duty cycles.

2 Multiple inductor multiplier cell

Proposed two inductor cell for SEPIC dc-dc hybrid converter is shown in Figure-1. The circuit of Figure-2 takes the form of Figure-a when the switch is ON. Currents flow through the inductors according to current division rule and rises with time. Instantaneous voltage across the inductor L_1 and L_2 are $v_{L1} = L_1 \frac{di_{L1}}{dt}$ and $v_{L2} = L_2 \frac{di_{l2}}{dt}$ respectively. If the two inductors have half the value of the inductor of single inductor conventional dc-dc converter, then i_{L_2} and i_{L_1} will be larger than i_L of the L of conventional dc-dc converter. As a result v_{L_2} and v_{L_1} will be higher than v_L of conventional dc-dc converter. When the switch of the proposed cell turns OFF, V_{dc}+ v_{L_1} or V_{dc}+ v_{L_2} appear across the switch providing the boost voltage to the next stage of the converter. Since v_{L_1} or v_{L_2} stage of the converter,

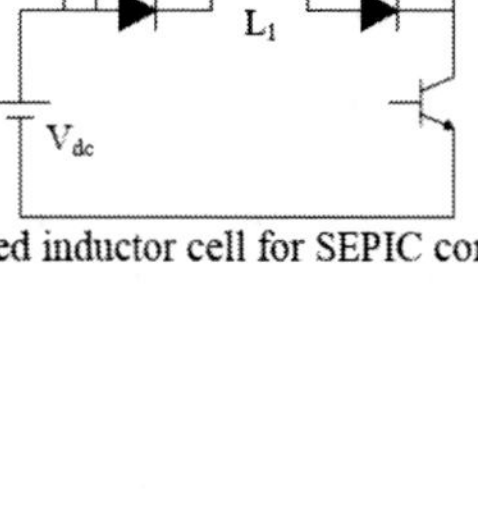

Figure1 . Proposed inductor cell for SEPIC converter.

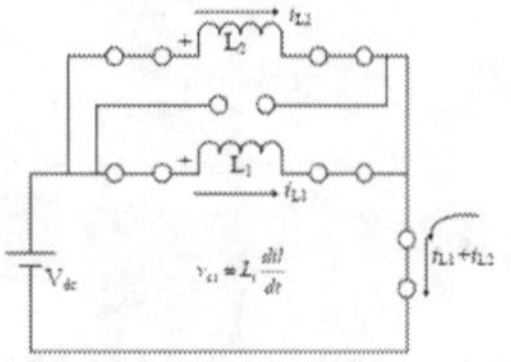

(a)Circuit for Figure1.(when switch is ON)

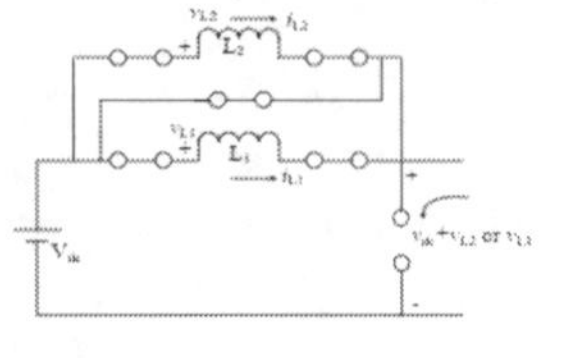

(b) Circuit for Figure1.(when switch is OFF)

Figure 2 Circuit of Figure 1 when is turns ON and OFFis higher than v_1 of conventional dc-dc converter, the proposed cell will provide higher Boost voltage to the next stage of the dc-dc converter at a lower duty cycle then the conventional dc-dc SEPIC converter. The proposed cell is different than the SEPIC cell of Figure 2 and 3 made of two inductors of reference [1].It can be multiplied several times as shown in Figure 4 and 5.

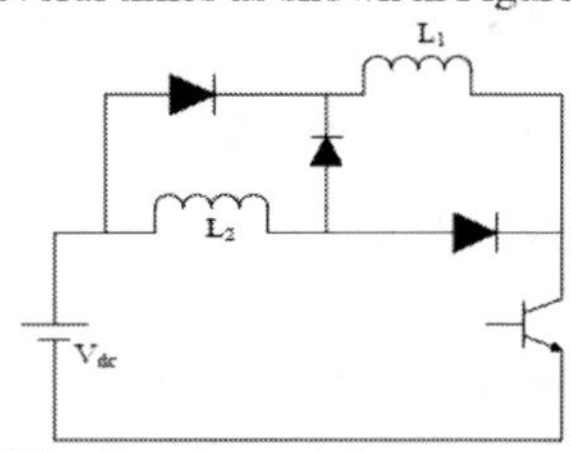

Figure 2. SEPIC cell made two inductors as reported in reference [1].

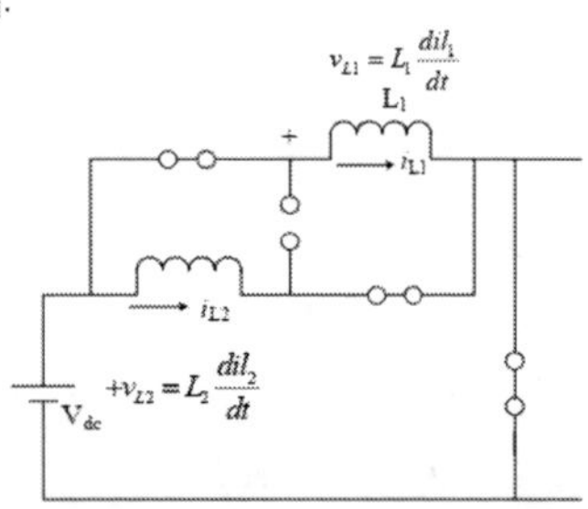

(a) SEPIC cell of Figure 2.(when switch is ON)

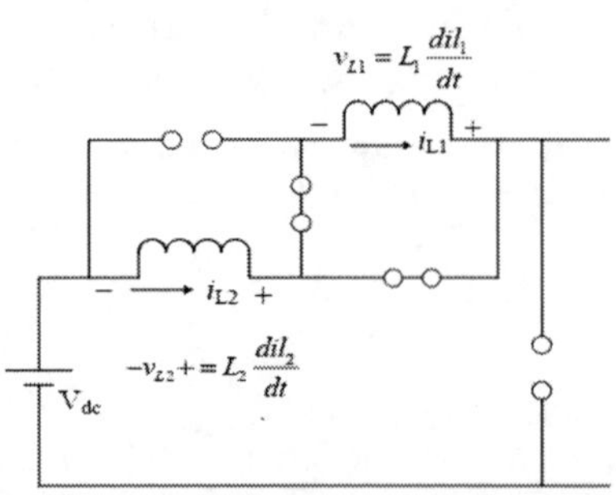

(b) SEPIC cell of Figure 2.(when switch is OFF)

Figure 3. SEPIC cell made of two inductors as reported in reference [1] with switch ON and OFF. Proposed three and four inductor cells

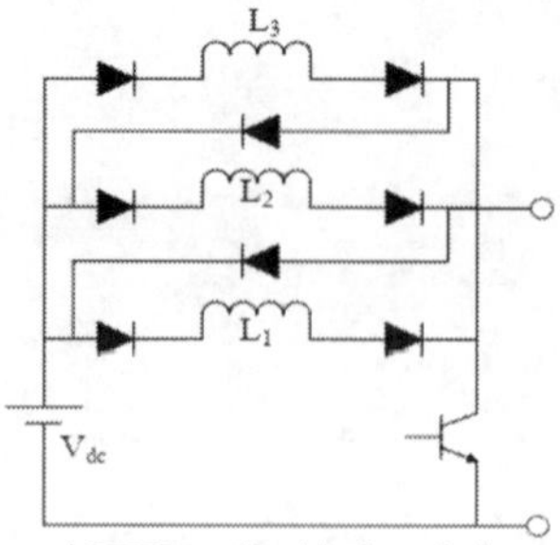

Figure 4. Proposed SEPIC cell with three inductors.

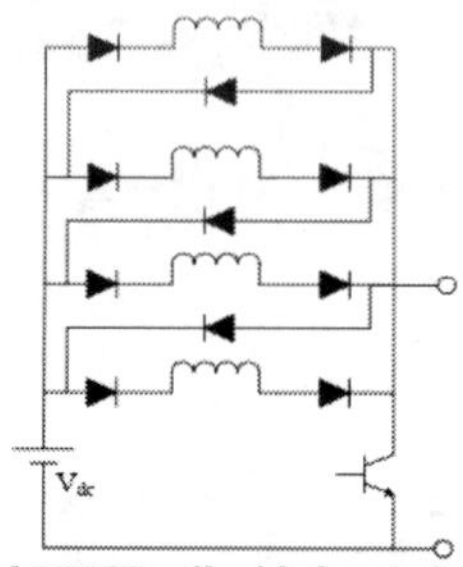

Figure 5. Proposed SEPIC cell with four inductors.

2 Results and discussions

The SEPIC dc-dc convertor, the hybrid sepic[1]dc-dc converter with two inductors, the hybrid proposed sepic converter with two inductors and the proposed hybridsepic dc-dc converter with three inductors are studied by simulation at variable duty cycles. The total inductors in all the circuits are kept at 1 mH for the purpose of comparison on the same component value.The conventional SEPIC circuit is shown in figure 6 and its typical waveforms are shown in figure 7. The SEPIC hybrid boost circuit of reference [1] is shown in figure 10 and its typical waveforms is shown in Figure 11. The proposed SEPIC hybrid circuit with two and three inductor cells are shown in Figures 14 and 18 respectively. Typical waveforms of Figures 14 and 18 are shown in Figures 15 and 19 respectively. The input voltage, output voltage, voltage gain, input power, output power and efficiency are tabulated in tables 2-5 and depicted in graphs of Figure 8-9, Figure 12-13, Figure 16-17, Figure 20-21respectively. Table 1.A compares the results of four configurations for voltage gain of 5.1 in the first row, which indicate that for 1mH total inductor the voltage gain 5.1 takes place at 0.8, 0.5, 0.6 and .4 duty cycle for normal SEPIC, hybrid SEPIC [1], proposed hybrid SEPIC of 2 inductors and proposed hybrid of 3 inductor circuit. The efficiency of the [1]SEPIC hybrid circuit is higher than normal, proposed hybrid circuit for two inductors and also for 3 inductor circuit.

Also in the second row of the table 1.A it is evident that voltage gain of the multiple inductors can be made higher than any other topology. The increase of inductor number keeping the total inductor value same is not possible in the hybrid circuit of the paper[1].

Table 1.A comparison of normal SEPIC, SEPIC hybrid topologyof[1], proposed sepic hybrid topology with two inductor, proposed topology with three inductor

Normal SEPIC			SEPIC Hybrid of [1] with 2 Inductor		
Duty cycle	Voltage Gain	Efficiency %	Duty cycle	Voltage Gain	Efficiency %
0.8	5.1	90	0.5	5.145	96.11
.875	6.4	79.78	.875	10.32	67.01

Table 1.A (Continued)

Sepic Hybrid of Proposed Topology with 2 Inductor			SEPIC Hybrid of Proposed Topology with 3 Inductor		
Duty cycle	Voltage Gain	Efficiency %	Duty cycle	Voltage Gain	Efficiency %
0.6	5.1	85	0.4	5.1	85
0.875	9.24	60.39	0.875	12.12	34.47

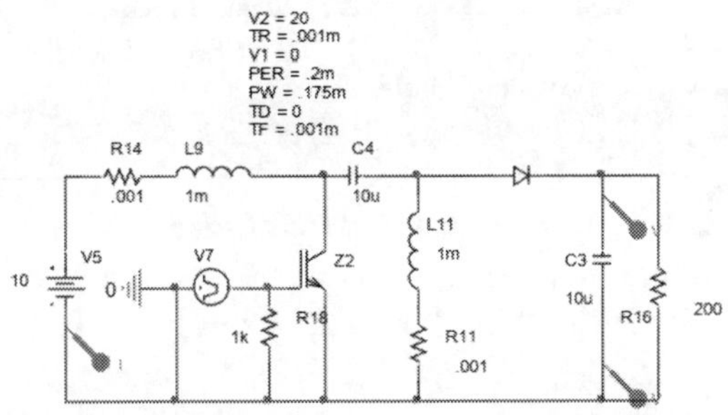

Figure 6. SEPIC DC-DC converter.

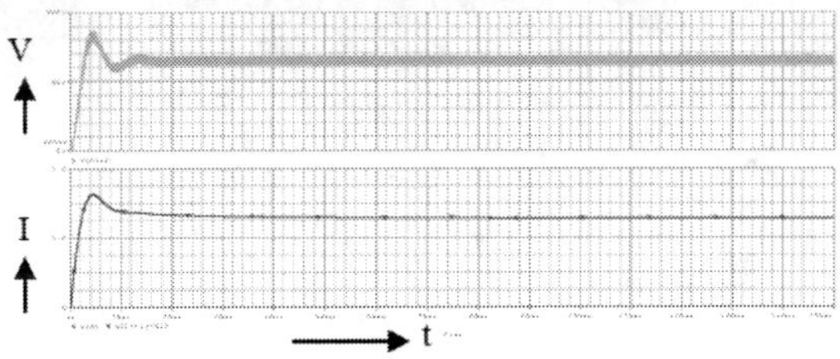

Figure 7. Typical Waveform Sepic DC-DC Converter of Figure 6.

Table 2. Performance of Sepic DC-DC converter circuit of figure 6.

Duty Cycle	V_{in} (V)	I_{in} (A)	Vout (V)	Iout (A)	Pin (W)	Pout (W)	Voltage gain	Efficiency in percentage
0.125	10	0.04	8.99	0.04	0.41	0.39	.89	95.12
0.25	10	0.15	17.17	0.08	1.55	1.52	1.71	98.06
0.375	10	0.29	23.50	0.11	2.94	2.74	2.35	93.19
0.50	10	0.52	31.17	0.15	5.22	4.83	3.11	92.52
0.625	10	0.82	39.53	0.19	8.21	7.51	3.95	91.47
0.75	10	1.21	46.78	0.23	12.10	10.89	4.67	90.00
0.875	10	2.57	64.25	0.32	25.77	20.56	6.42	79.78

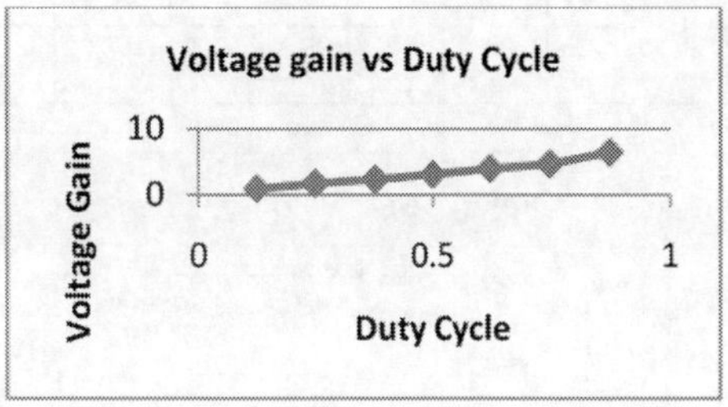

Figure 8. Voltage gain vs duty cycle curve of Sepic DC-DC converter of figure 6.

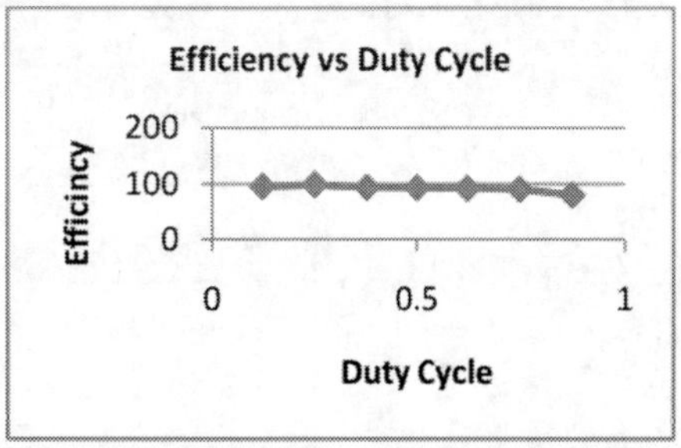

Figure 9. Efficiency vsduty cycle curve of Sepic DC-DC converter of figure6.

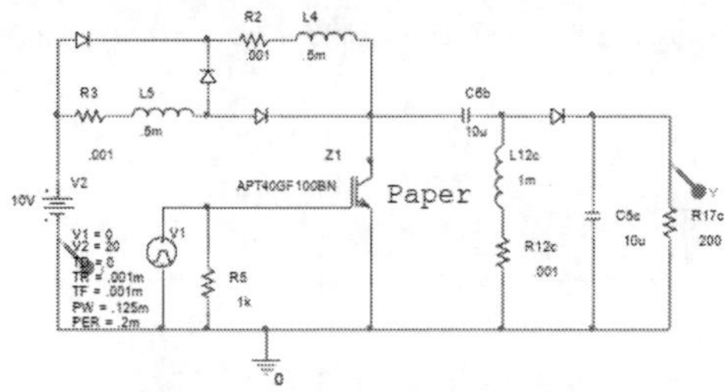

Figure 10. SEPIC hybrid DC-DC converter [1].

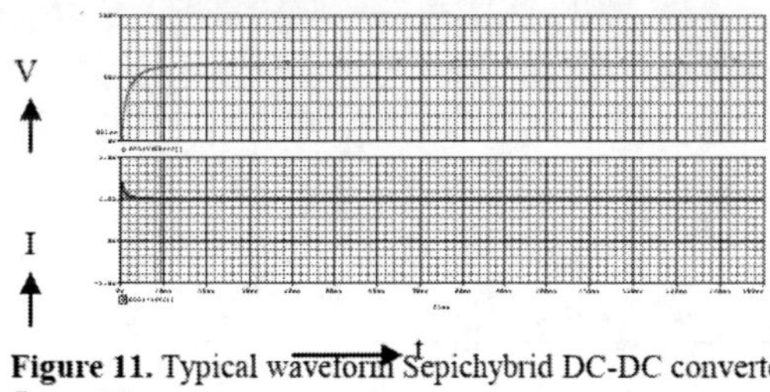

Figure 11. Typical waveform Sepichybrid DC-DC converter of figure 10.

Table 3. Performance of sepic hybrid DC-DC converter of figure10.

Duty Cycle	V_{in} (V)	I_{in} (A)	V_{out} (V)	I_{out}(A)	P_{in} (W)	P_{out} (W)	Voltage gain	Efficiency in percentage
0.125	10	0.14	16.08	0.08	1.4	1.29	1.60	92.34
0.25	10	0.38	27.09	0.13	3.83	3.66	2.70	95.80
0.375	10	0.79	39.6	0.19	7.91	7.84	3.96	99.12
0.5	10	1.37	51.45	0.25	13.77	13.23	5.14	96.11
0.62	10	2.06	62.8	0.31	20.6	19.71	6.28	95.72

0.75	10	2.98	72.33	0.36	29.8	26.15	7.23	87.77
0.875	10	7.95	103.22	0.51	79.5	53.27	10.32	67.01

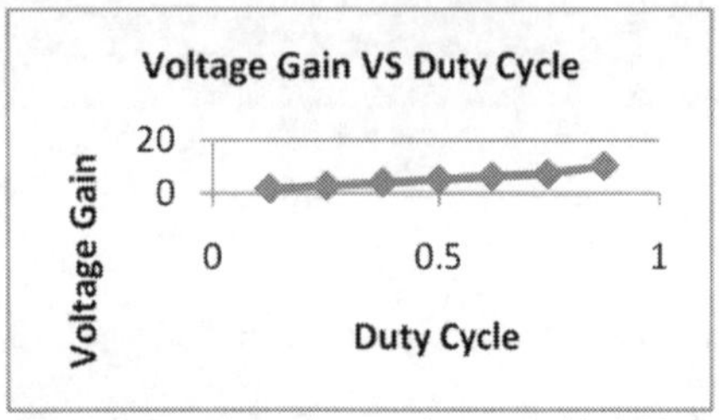

Figure 12. Voltage gain vs duty cycle curve of hybrid sepicDC-DC converter of figure 10.

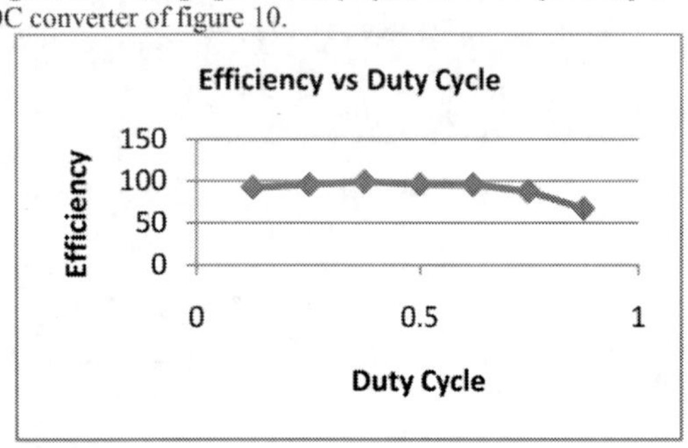

Figure 13. Efficiency vsduty cycle curve of hybridsepic DC-DC converter of figure10.

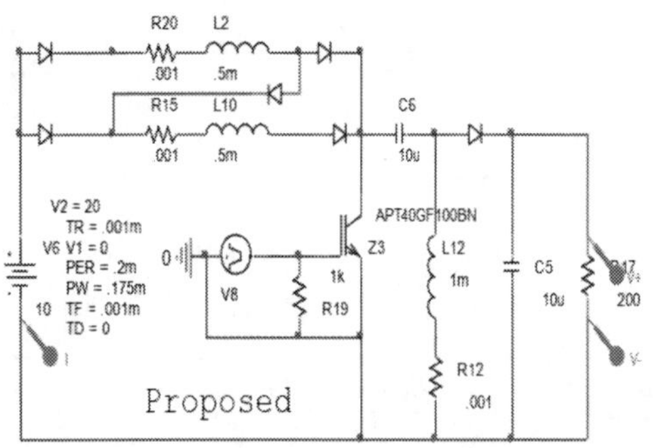

Figure 14. SEPIC hybrid DC-DC converter (proposed).

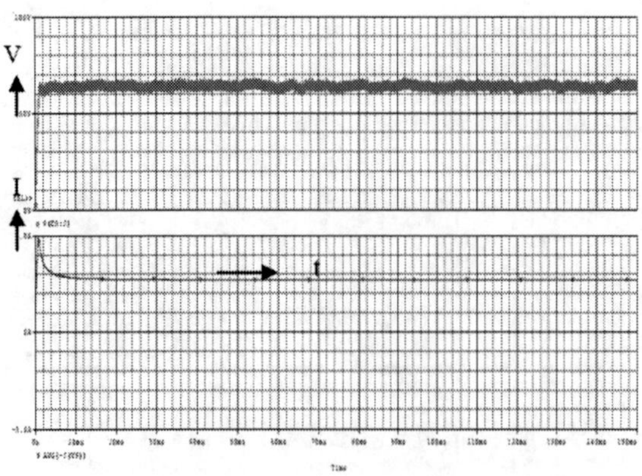

Figure 15. Typical waveform SEPIC hybrid DC-DC converter of figure-14.

Table 4. Performance of Sepichybrid DC-DC converter of figure-14.

Duty Cycle	V_{in} (V)	I_{in} (A)	V_{out} (V)	I_{out} (A)	P_{in} (W)	P_{out} (W)	Voltage gain	Efficiency in percentage
0.125	10	0.10	14.04	0.07	1.02	0.98	1.40	90.42
0.25	10	0.31	22.96	0.11	3.17	2.63	2.29	83.14
0.375	10	0.69	35.29	0.17	6.99	6.22	3.52	89.08
0.5	10	1.18	45.15	0.22	11.8	10.19	4.51	86.37
0.625	10	1.86	56.13	0.28	18.6	15.75	5.61	84.69
0.75	10	2.84	64.14	0.32	28.4	20.56	6.41	72.42
0.875	10	7.08	92.48	0.46	70.8	42.76	9.24	60.39

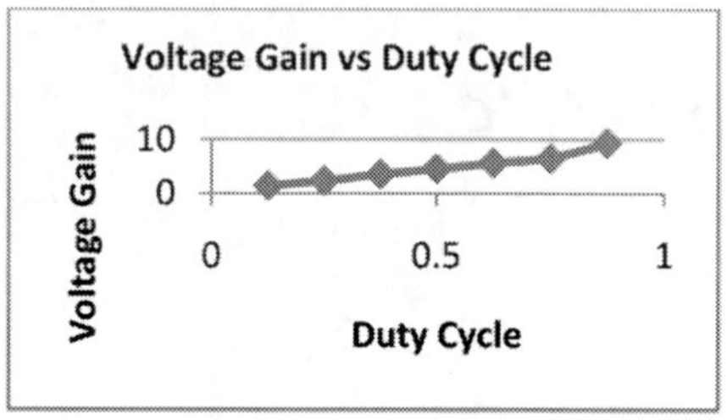

Figure 16. Voltage gain vs duty cycle curve of hybridSepic DC-DC converter of figure14.

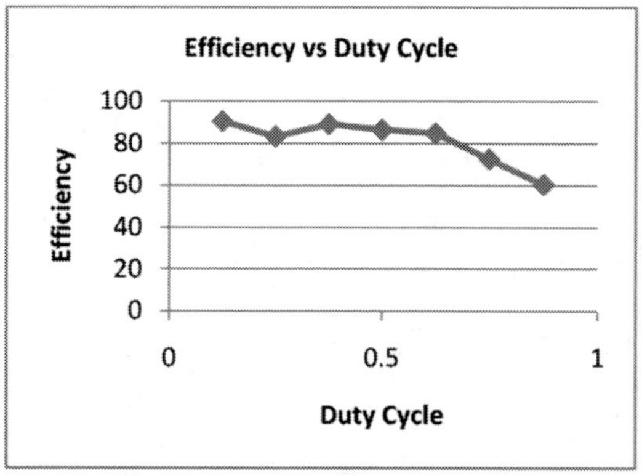

Figure 17. Efficiency vsduty cycle curve of hybridSepic DC-DC converter of figure14.

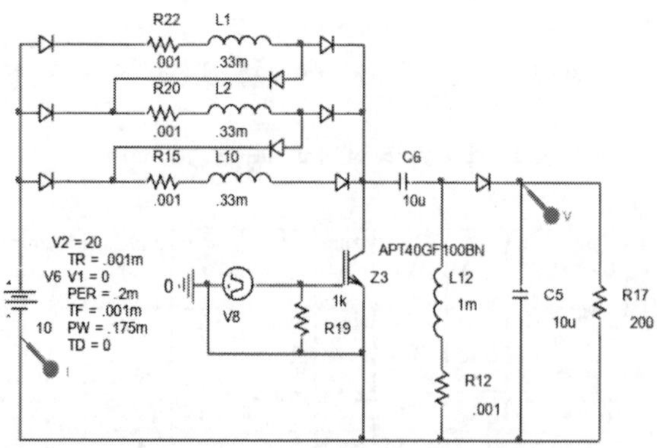

Figure 18. SEPIC hybrid DC-DC converter.

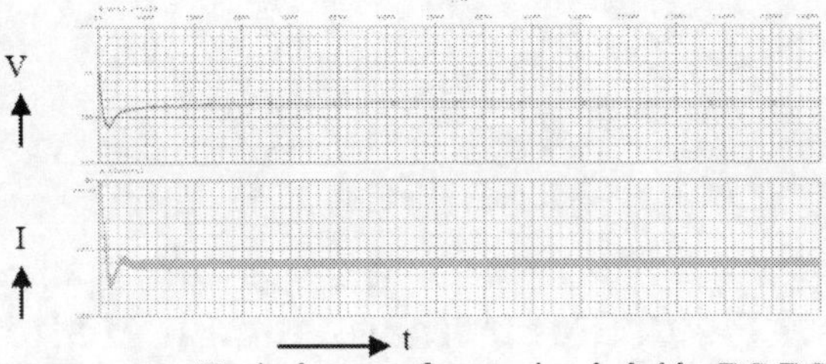

Figure 19. Typical wavefromsepic hybrid DC-DC converter of figure18.

Table 5. Performance ofsepichybrid DC-DC converter of figure 18.

Duty Cycle	Vin (V)	Iin (A)	Vout (V)	Iout (A)	Pin (W)	Pout (W)	Voltage gain	Efficiency in percentage
0.125	10	0.24	19.30	0.09	2.44	1.85	1.93	75.81
0.25	10	0.68	31.50	0.15	6.80	4.94	3.15	80.45
0.375	10	1.33	48.44	0.242	13.30	11.72	4.84	88.12
0.50	10	2.34	61.06	0.30	23.44	18.62	6.10	79.43
0.625	10	3.73	74.27	0.37	37.33	27.55	7.42	73.80
0.75	10	6.95	85.84	0.42	69.55	36.82	8.58	52.94
0.875	10	20.70	121.22	0.60	207.07	73.45	12.12	34.47

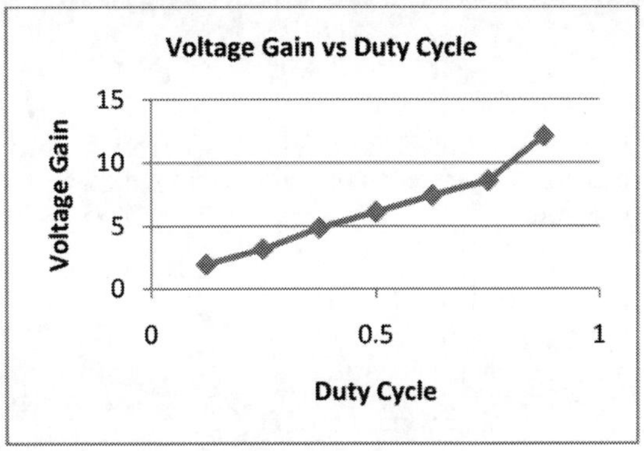

Figure 20. Voltage gain vs duty cycle curve of hybrid SEPIC DC-DC converter of figure18.

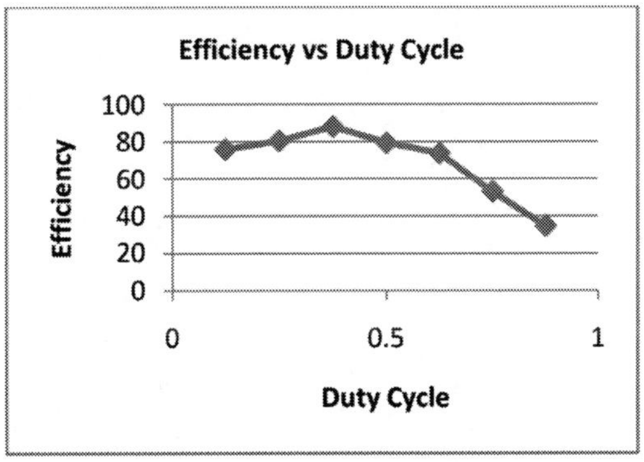

Figure 21. Efficiency vsduty cycle curve of hybrid Sepic DC-DC converter of figure18.

3 Conclusions

The proposed cellis investigated with conventional SEPIC dc-dc converters topologies with expected result of attaining same voltage gain than conventional dc-dc converter counterpart with higher efficiencies at lower duty cycle. The dc-dc converters were found to behave similarly for voltage gain and efficiency at higher duty cycles. It was also observed that the voltage gain of these converters in hybrid mode operations can be varied widely to higher gain levels than conventional converters.

References

1. B. Axelrod, Y. Berkovich, A. Ionovici: "Switched-capacitor/ switched-inductor structures for getting transformerless hybrid DC– DC PWM converters", IEEE Transactions on Circuits and Systems I. vol. **55**, no. 2, pp. 687–696. (2008)
2. [2] A. Kishon Robert, C. Baskaran "Single switch high stepup dc-dc converters with voltage multiplier cell", International Journal of Advanced Trends in Computer Science and Engineering, Vol.2, No.2, Pages : 95-100 (2013) Special Issue of NCRTECE 2013 - Held during 8-9 February, 2013 in SMK Fomra Institute of Technology, OMR, Thaiyur, Kelambakkam, Chennai pp 95-100, (2013)
3. A. Gopi and R. Saravanakumar, "High Boost Isolated DC-DC Converter with Controller."World Applied Sciences Journal 23 (4): pp. 486-494, 2013 ISSN 1818-4952, (2013)
4. Daniel W Hart., "Power Electronics", McGraw Hill, (2010)
5. R. W. Erickson, "Fundamental of Power Electronics", Chapman and Hall, USA, (1997)
6. M. H. Rashid, "Power Electronics Handbook", Academic Press, USA, (2001)
7. Timothy Skvarenina, "The Power Electronics Handbook", CRC Press (2002)
8. Fang Lin Luo and Hong Ye, "Advanced Multi-Quadrant DC/DC Converters", CRC Press, (2006)
9. Keng C Wu., "Switch Mode Power Converter", Elsevier Academic Press, (2006)
10. Dorin O. Neaesu, "Power Switching Converters", CRC Press (2006)

Numerical Simulation of Internal Flow Field, Electromagnetic Field and Temperature Field in Deep-Water Motor

Qi Zhang , Zengliang Li, Ran Yu, Zhaocheng Sun and Le Zhang

College of Mechanical Electronic Engineering, China University of Petroleum, Qingdao 266580 China

Abstract. In this paper, considering the heat exchange between the ends of stator&rotor and lubricating oil, the internal flow field model is established to analyse the flow state. The variation law of the pressure drops in the internal flow field and the friction loss of the rotor surface are specified. The operation characteristic of the motor and the variation law of the magnetic field loss are obtained through the analysis of motor magnetic field. Based on the results of internal flow field and magnetic field, the temperature distribution law of motor is obtained. Compared with the result of the temperature simulation in which only air gap path is considered, the stator temperature shows totally different distribution law, and the stator temperature increases from 32°C to 42°C. Compared with the result of the temperature simulation without considering heat source loss change, the maximum temperature of the motor decreases from 71°C to 49°C. The results show that the ends of stator&rotor and the variation of heat source loss can influence the temperature simulation results.

1 Introduction

As oil filled asynchronous motors, Deep-water motors are not only meeting the requirement of power, but also have stronger ability to adapt complex work environment. With the development of marine equipment, the unit capacity of Deep-water motor is increasing. The motor get more loss, which makes the temperature rise. The service life of motor will be shortened as the motor is overheating. Due to the special working conditions, the maintenance cost of the motor is high. Accordingly, it is important to obtain the rules of motor temperature distribution accurately.

There are many analytical methods of the temperature field [1-5]. Finite volume method is widely used as a method of coupling flow field, magnetic field and temperature field [6-10]. The establishment of temperature field model and the calculation of heat source loss are the basis of numerical simulation. The traditional modelling approach only considers the influences of air gaps path on the temperature distribution, and ignores the influence of the end of stator&rotor on temperature [11-12]. The calculation result of the heat source loss is always a constant, and ignores the influences of the temperature rise on viscosity and resistance [13-14]. These factors have resulted in the difference between thermal field simulation and actual temperature.

In this paper, the internal flow field, electromagnetic field and temperature field of the motor are established by using ANSYS. The variation law of heat source loss and the temperature distribution law of the motor are obtained.

2 Internal flow field analysis of Deep-water motor

2.1 Internal flow field model

According to the fluid dynamics theory, the governing equations of lubricating oil flow in the motor are mass conservation equation and momentum conservation equation.

$$\partial u / \partial x + \partial v / \partial y + \partial w / \partial z = 0 \tag{1}$$

$$\frac{\partial(\rho u_x)}{\partial t} + \frac{\partial(\rho u_x u_x)}{\partial x} + \frac{\partial(\rho u_y u_x)}{\partial y} + \frac{\partial(\rho u_z u_x)}{\partial z} = \frac{\partial}{\partial x}\left(\mu \frac{\partial u_x}{\partial x}\right) + \frac{\partial}{\partial y}\left(\mu \frac{\partial u_x}{\partial y}\right) + \frac{\partial}{\partial z}\left(\mu \frac{\partial u_x}{\partial z}\right) - \frac{\partial p}{\partial x} + S_u$$

$$\frac{\partial(\rho u_y)}{\partial t} + \frac{\partial(\rho u_x u_y)}{\partial x} + \frac{\partial(\rho u_y u_y)}{\partial y} + \frac{\partial(\rho u_z u_y)}{\partial z} = \frac{\partial}{\partial x}\left(\mu \frac{\partial u_y}{\partial x}\right) + \frac{\partial}{\partial y}\left(\mu \frac{\partial u_y}{\partial y}\right) + \frac{\partial}{\partial z}\left(\mu \frac{\partial u_y}{\partial z}\right) - \frac{\partial p}{\partial x} + S_v$$

$$\frac{\partial(\rho u_z)}{\partial t}+\frac{\partial(\rho u_x u_z)}{\partial x}+\frac{\partial(\rho u_y u_z)}{\partial y}+\frac{\partial(\rho u_z u_z)}{\partial z}=$$
$$\frac{\partial}{\partial x}\left(\mu\frac{\partial u_z}{\partial x}\right)+\frac{\partial}{\partial y}\left(\mu\frac{\partial u_z}{\partial y}\right)+\frac{\partial}{\partial z}\left(\mu\frac{\partial u_z}{\partial z}\right)-\frac{\partial p}{\partial x}+S_w \tag{2}$$

where, p is the pressure on the fluid micro body. ρ is fluid density. u_x、u_y and u_z are the components of velocity vector u in the direction of x、y and z.

The basic parameters of the Deep-water motor studied in this paper are as follows. The load power is 60kW. The rated voltage is 1140V. The inner and outer diameters of stator are 120mm and 260mm. The air gap length is 0.6mm.The core length is 300mm. The slot ratio of stator and rotor is 36/26. The extreme logarithm is 2. The winding form is single layer concentric type. The number of turns per slot is 35. The rated flow rate and pressure of single-stage impeller are $1m^3/h$ and 0.2MPa. Considering the heat exchange between the ends of stator&rotor and lubricating oil, the internal flow field model of the motor is established based on the single tooth and single groove of stator&rotor as shown in Fig. 1.

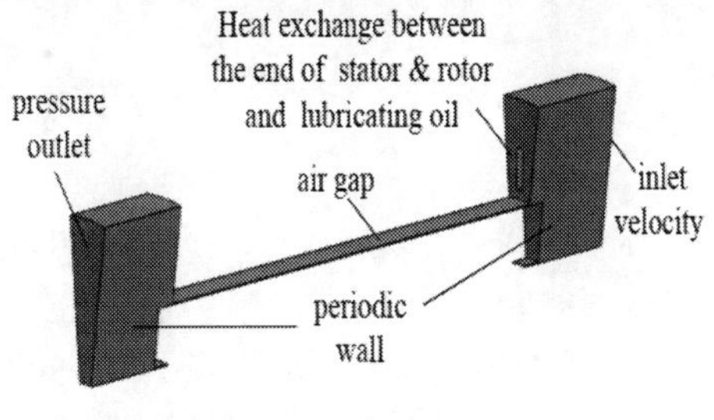

Figure 1. Internal flow field model.

2.2 Internal flow field result analysis

The boundary conditions of the internal flow field are set in Fig. 1. According to the rated flow of lubricating oil, inlet velocity is 0.004m/s. The viscosity is 0.0971kg/ms at the temperature of 24°C. The flow state in the air gap is laminar by Reynolds number. The rotor line speed is 9.19m/s. The internal flow field center cross-section traces and radial velocity distribution of the air gap are obtained as shown in Fig.2. In the air gap, the lubricating oil flow is dominated by swirling flow, and it velocity distribution decreases from rotor side to stator side. The maximum speed is 9.2m/s.

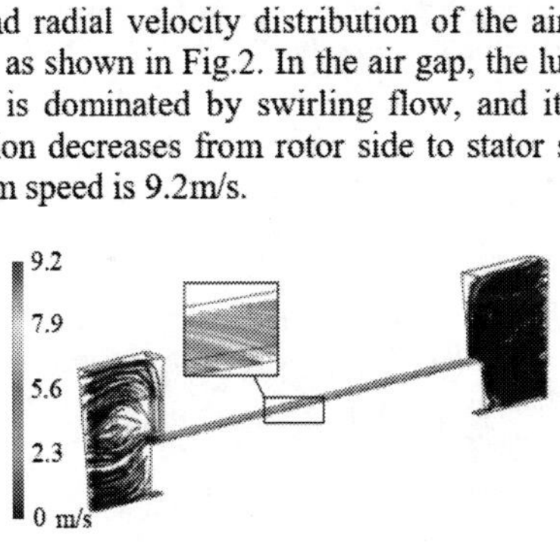

(a)Internal flow field center cross-section traces

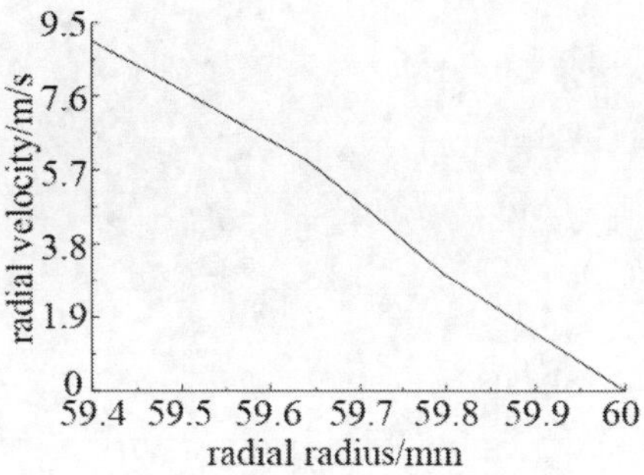

(b)Radial velocity distribution of the air gap

Figure 2. Internal flow field simulation result.

The motive power for lubricating oil flow comes from the impeller at the bottom of the motor. Considering the relationship between temperature and viscosity, the resistance of lubricating oil will change, which causes the impeller to provide pressure changes. According to the impeller characteristic curve, the change of pressure will make the corresponding flow change. In order to ensure the impeller work in the high-efficiency section, the pressure drop of the internal flow field under different viscosities and inlet velocities is given in Fig.3. As the viscosity and inlet velocity decrease, the pressure drop of the internal flow field decreases gradually. The results provide the basis for selecting the number of impellers in series.

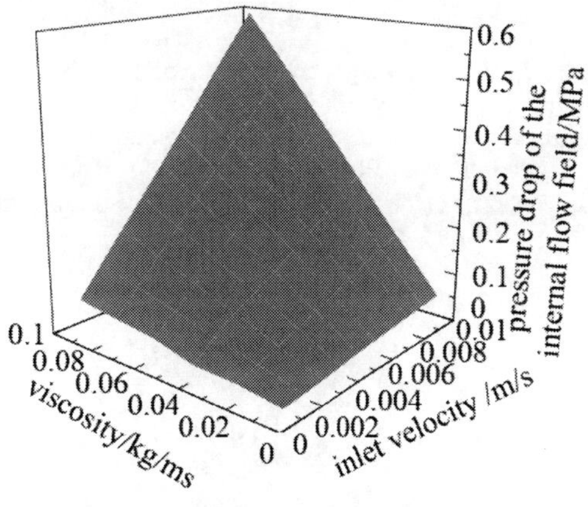

Figure 3. Pressure drop of the internal flow field under different viscosities and inlet velocities.

According to the analysis of the internal flow field, the friction loss of the rotor surface is related to the viscosity and the inlet velocity. As shown in Fig.4, the friction loss of the rotor surface increases with the increase of the viscosity, but basically unchanged with the increase of the inlet velocity.

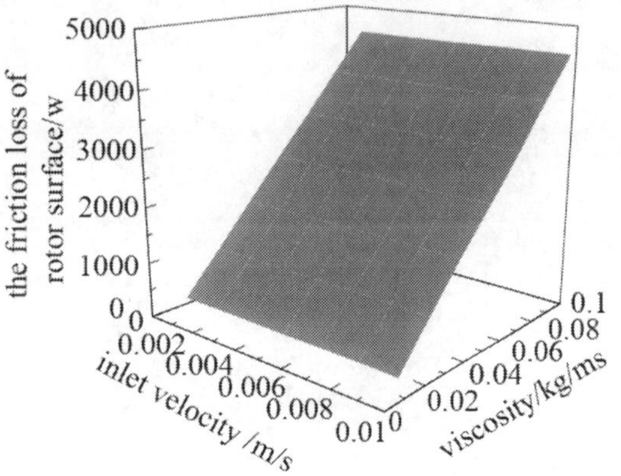

Figure 4. The friction loss of rotor surface under different viscosities and inlet velocities.

3 Electromagnetic field analysis of Deep-water motor

The Maxwell equation is used to represent the variation law of time varying electromagnetic field.

$$\begin{cases} \nabla \times H = J + \dfrac{\partial D}{\partial t} \\ \nabla \times E = -\dfrac{\partial B}{\partial t} \\ \nabla \cdot D = \eta \\ \nabla \cdot B = 0 \end{cases} \tag{3}$$

where, H is the magnetic field strength. B is magnetic induction intensity. E is field strength. D is electric displacement. J is current density. η is charge density.

According to the size of stator and rotor, two-dimensional electromagnetic field model is established as shown in Fig. 5.

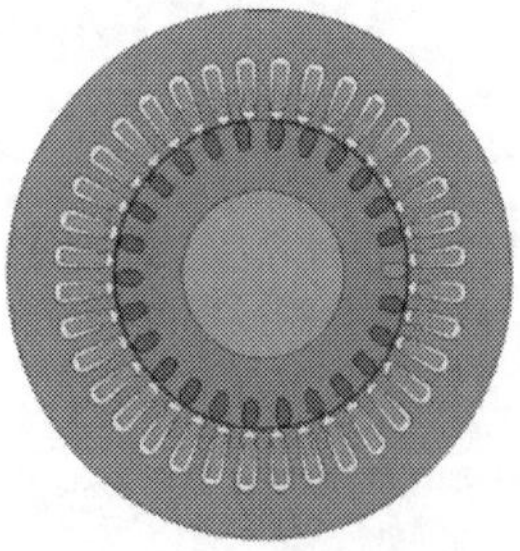

Figure 5. Two-dimensional electromagnetic field model.

The parallel boundary conditions of magnetic lines are applied on the outer circumference of stator. Constant power load is applied. Voltage source excitation is added. The simulation results of the electromagnetic field in the starting time of 0.2s are obtained.

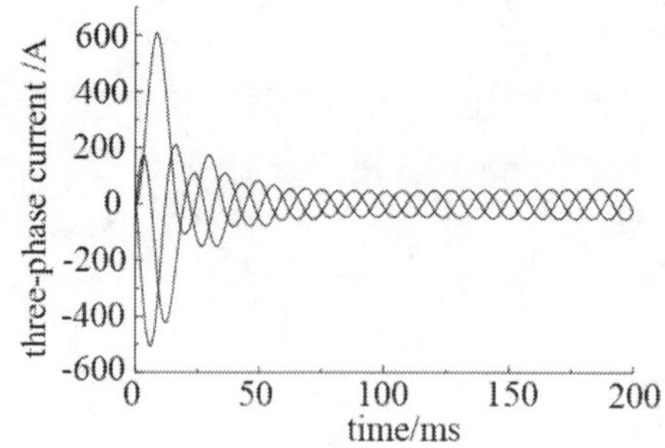

Figure 6. Three-phase current of the winding varies with time.

The three-phase current of the winding varies with time as shown in Fig.6. The three-phase current is sinusoidal alternating current. The current is very unstable at the initial time. The maximum amplitude is far more than the stable current. The stator winding current tends to be stable after 75ms.

The single-phase current changes with motor speed as shown in Fig. 7. When motor speed is 0, the current is about 288A. With the increase of motor speed, the current decreases gradually. When the rated speed is reached, the current is about 33.6A.

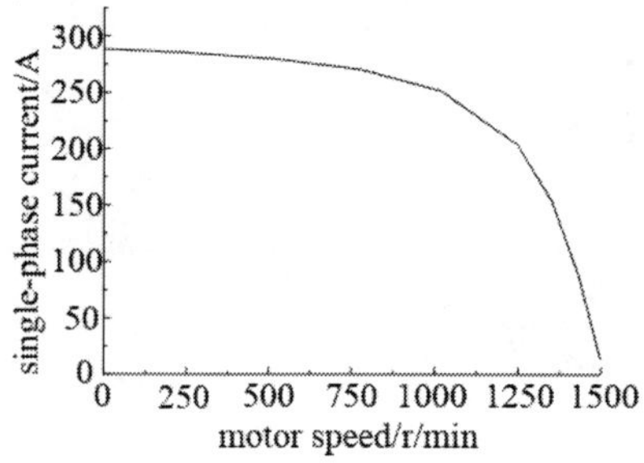

Figure 7. The single-phase current changes with motor speed.

The magnetic flux distribution in the motor at different times is shown in Fig.8. At 0.002s, the magnetic flux of rotor is mainly distributed on the surface of rotor tooth. The magnetic flux density of rotor yoke is small. At 0.01s, the motor operation gradually stabilized. The magnetic flux concentrated on the surface of rotor teeth gradually dispersed and transferred to rotor teeth. At 0.04s, the magnetic flux of rotor passes through the cogging, which is mainly distributed in the yoke, while almost no magnetic flux passes through rotor tooth surface.

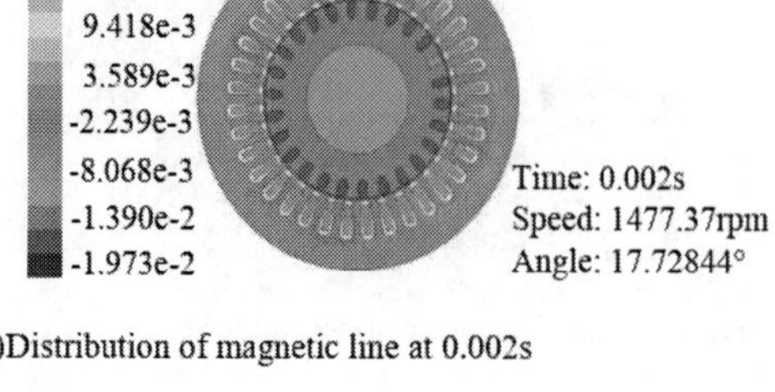

(a)Distribution of magnetic line at 0.002s

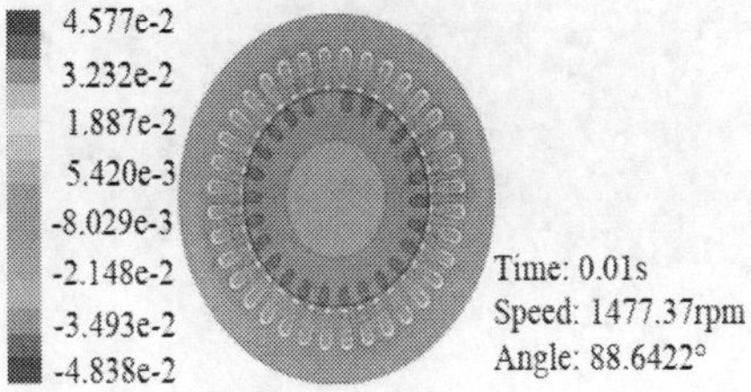

(b)Distribution of magnetic line at 0.01s

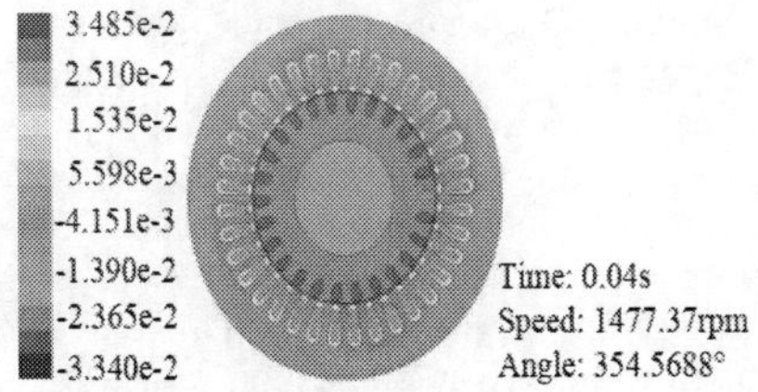

(c)Distribution of magnetic line at 0.04s

Figure 8. Magnetic flux distribution in the motor at different times.

According to the above analysis, the radial and tangential flux density curves of in different regions include tooth top, tooth and yoke in a stable power frequency period are shown in Fig.9. The radial and tangential flux density curves fluctuate obviously in the tooth top region. The flux density curve in the tooth region mainly fluctuates in the radial direction. In the yoke area, the radial flux density curve has less fluctuation, and the tangential flux density curve is smooth

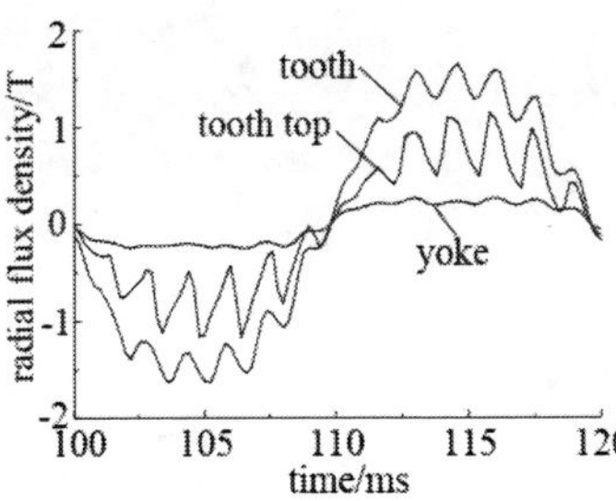

(a)Radial flux density curve

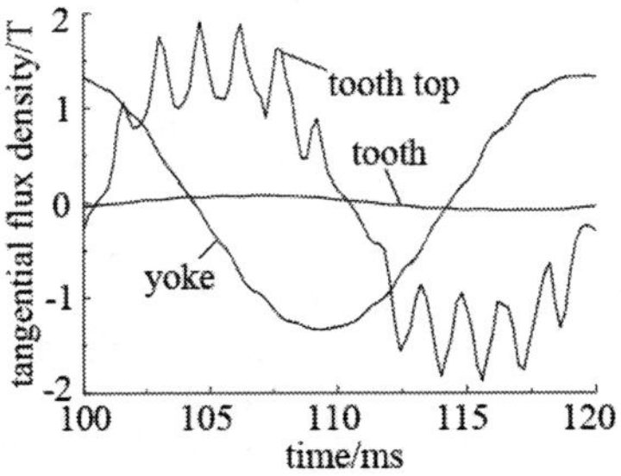

(b)Tangential flux density curve

Figure 9. Radial and tangential magnetic flux density curves in different regions.

In the actual motor magnetic field, due to the existence of harmonic magnetic field, it will make the radial and tangential magnetic fluxes produce non-positive rotation changes. The magnetization density of harmonic number of different harmonic times is introduced and calculated by Fourier decomposition method.

$$P=\sigma_h\sum_{k=1}^{N}f_k\left(\left(B_{r\max}\right)_k^2+\left(B_{t\max}\right)_k^2\right)+\sigma_e\sum_{k=1}^{N}f_k^2\left(\left(B_{r\max}\right)_k^2+\left(B_{t\max}\right)_k^2\right)$$
$$+\sigma_a\sum_{k=1}^{N}f_k^{1.5}\left(\left(B_{r\max}\right)_k^{1.5}+\left(B_{t\max}\right)_k^{1.5}\right) \quad (4)$$

where σ_h and α are the hysteresis loss coefficient. σ_e is the eddy current loss coefficient. σ_a is the abnormal loss coefficient. B_{rmax} and B_{tmax} are the radial and tangential fluxes of the different harmonic times. f_k is the different harmonic frequency.

The iron loss density distribution is obtained according to the equation (4) as shown in Fig. 10.

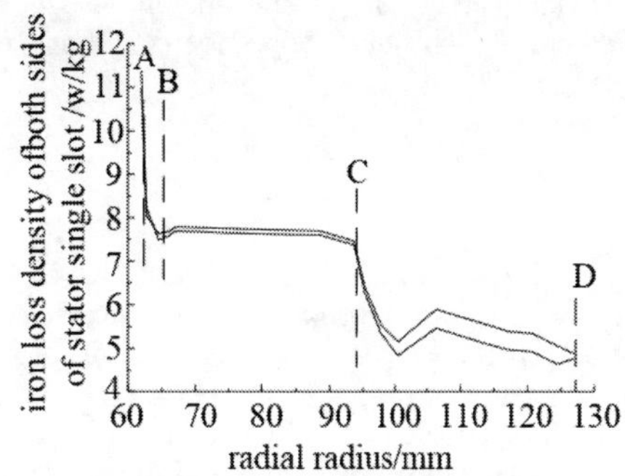

Figure 10. Iron loss density distribution.

In Fig. 10, AB is tooth top region. BC is tooth region. CD is yoke region. The distribution regularity of iron loss density both sides of stator single slot has the same trend. In the tooth top region, iron loss density decreases sharply with the increase of the radius. The iron loss density in the tooth region is basically unchanged. Iron loss density in the yoke region first decreases, then increases and then decreases with the increase of the radius, which is related to the direction of the magnetic field lines in the yoke.

The current loss in bar-conductor and winding is called as copper loss. It is calculated according to Joule-Lenz's law.

$$P_1=3I_x^2R_x \quad (5)$$

where, I_x is the excitation current. R_x is the equivalent resistance converted to the reference operating temperature.

The change of copper loss under different viscosity is obtained by using viscosity instead of temperature. As shown in Fig.11, copper loss decreases with the increase of viscosity.

 https://doi.org/10.1051/matecconf/201816002005

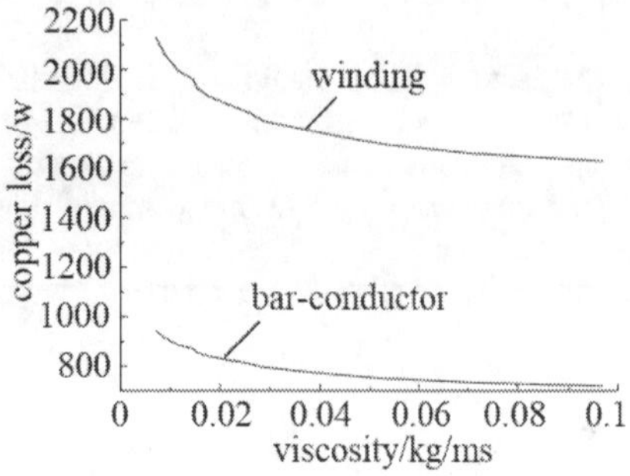

Figure 11. Variation law of copper loss.

4 Temperature field analysis of Deep-water motor

According to heat transfer theory, the fluid-solid coupling interface of motor has the same or the conservation of temperature T and heat flux density q, and its expression is:

$$T_{\text{soild}}=T_{\text{fluid}} \tag{6}$$

$$q_{\text{soild}}=q_{\text{fluid}} \tag{7}$$

The lubricating oil in the motor is given inlet temperature T_f. The third boundary conditions of heat transfer are selected.

$$\left(-\lambda\frac{\partial T}{\partial n}\right)_{soild}=\left(\alpha\left(T-T_f\right)\right)_{fluid} \tag{8}$$

The heat transfer process in the fluid-solid coupling region is considered as a whole heat transfer process by the finite volume method.

The three-dimensional temperature field model of motor is established by using SolidWorks, which is shown in Fig. 12(a). The model is divided into 9 sub-blocks such as stator block and rotor block by using ICEM. On the basis of dividing the hexahedral grids in each sub-block, the sub-blocks are spliced as shown in Fig.12(b). The meshes of the interfaces between the sub-blocks should be consistent.

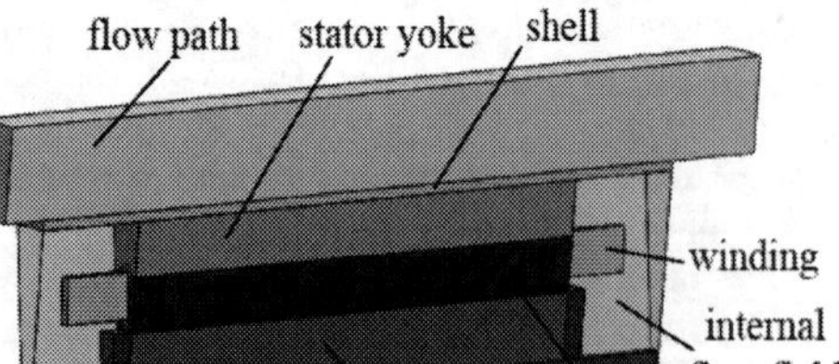

(a)Three-dimensional temperature field model

(b)Meshing

Figure 12. Three-dimensional temperature field model and meshing.

According to the analysis results of internal flow field and electromagnetic field, the UDF command is used to realize the variation law of rotor surface friction loss and copper loss with viscosity. The iron loss density is applied in different regions of the stator. The pressure inlet and the pressure outlet are set in the outer flow path. The other settings are consistent with the analysis of the internal flow field. The simulation results of the motor temperature are obtained as shown in Fig. 13.

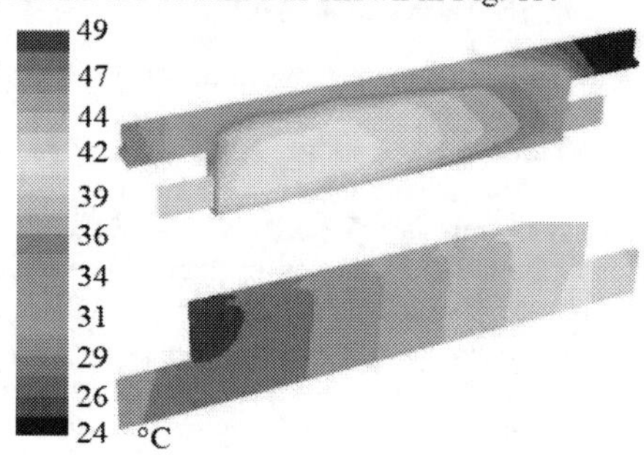

(a)Temperature distribution

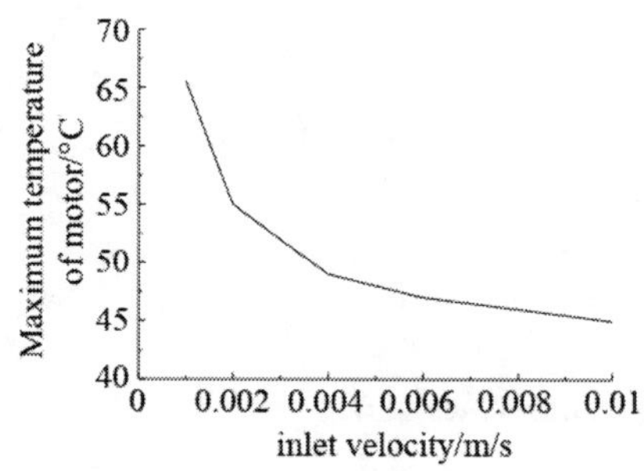

(b)Maximum temperature at different inlet velocities

Figure 13. Temperature field simulation result.

In Fig. 13, the rotor temperature is distributed along the axial direction, and the maximum temperature is 49°C at the air gap outlet because the rotor heat is mainly taken away by the air gap lubricating oil. Stator temperature is significantly lower than rotor temperature. Shell temperature is the lowest. The ends temperature of the stator and rotor is greater at the air gap outlet side than at the air gap inlet side. With the increase of the inlet velocity, the maximum temperature of the motor gradually decreases. The variation of the motor temperature after 0.004m/s is small.

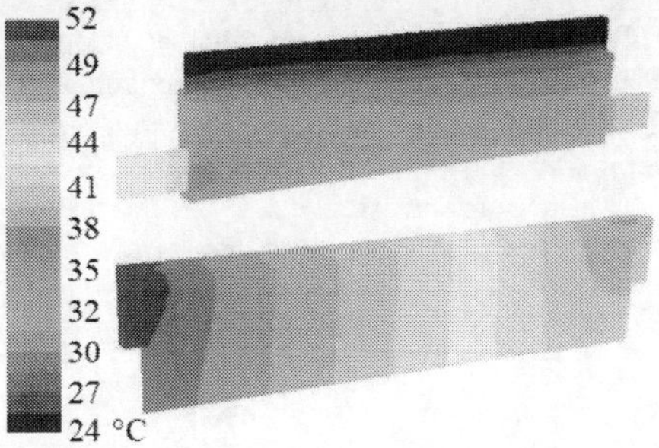

Figure 14. Temperature simulation considering only the air gap path.

Compared with the simulation results in Fig.13 and Fig. 14, the maximum temperature of the motor slightly increases. The tendency of the stator temperature is greatly different. This is mainly because the heating of the motor increases the temperature of the lubricating oil on the outlet side of the air gap. Due to the heat exchange between the end of the stator and rotor and the lubricating oil, the stator temperature in Fig.13 is obviously higher than that in Fig. 14. The comparison shows that the ends of the stator and rotor and heat source loss changes have a certain influence on the motor temperature analysis result.

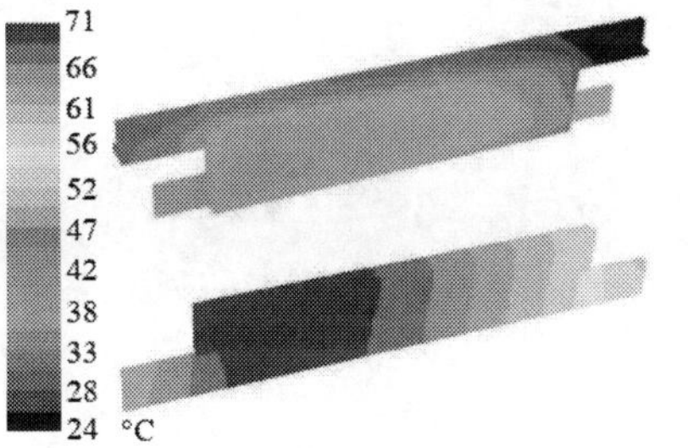

Figure 15. Temperature simulation without considering heat source loss change.

5 Conclusions

In this paper, the internal flow field, electromagnetic field and temperature field of Deep-water motor are numerically simulated. The following conclusions are drawn.

(1) In the air gap, the oil flow is dominated by swirling flow, and it velocity distribution decreases from the rotor side to the stator side. The pressure drop in the inner flow field decreases with the decrease of viscosity and inlet velocity. The friction loss of the rotor surface increases with the increase of the viscosity, but basically unchanged with the increase of the inlet velocity.

(2) After the start-up, the winding current fluctuates greatly and reaches a steady state after 75ms. The current decreases with the increase of rotor speed. After the motor is stable, the magnetic lines are evenly distributed on the stator yoke core. The iron loss density decreases with the increase of radial radius. The copper loss decreases with the increase of viscosity.

(3) Based on the results of internal flow field and magnetic field analysis, the law of motor temperature distribution is obtained. The maximum temperature of the motor appears at the air gap outlet. Compared with the temperature simulation considering only the air gap path, the stator temperature distribution law is obviously different. Compared with the result of temperature without considering heat source loss change, the maximum temperature of the motor decreased from 71°C to 49°C

This research is sponsored by high technology ship research project of ministry of Industry and Information Technology, Grant No.2014020024.

References

1. Fu Xinghe, Lin Mingyao, Xu Da, et al. *Computation and Analysis of 3D-Transient Temperature Field for a Permanent Magnet-Induction Hybrid Excitation Generator[J]*. Transactions of China Electrotechnical Society, **28**, 3 (2013): 107-113.
2. Chen Yiguang, Zheng Jun, Wei Juan, et al. *Design of PMSM for Actuator and Its Temperature Field Analysis[J]*. Transactions of China Electrotechnical Society, **30**, 14 (2015): 94-99.
3. Qi Wenjuan, Zou Jiming, Hao Guiqing, et al. Thermal *Analysis of Underwater Oil-filled BLDC Motors[C]*. ICEMS, **24**, 1 (2011): 1-4.
4. Li Ming, Tang Renyuan, Chen Lixiang. *Screw* Pump *Permanent Magnet AC Servo Temperature of the Motor Calculation and Analysis[J]*. Electrical Engineering, 9 (2012): 5-8.
5. Staton D.A, Cavagnino A. *Convection Heat* Transfer *and Flow Calculations Suitable for Electric Machines Thermal Models[J]*. IEEE Transactions on Industrial Electronics, **55**, 10 (2008): 3509-3516.
6. Yujiao Z, Jiangjun R, Tao H, et al. *Calculation of* Temperature *Rise in Air-cooled Induction Motors through 3-D Coupled Electromagnetic Fluid-dynamical and Thermal Finite-element Analysis[J]*. IEEE Transactions on Magnetics, **48**, 2, (2012): 1047-1050.
7. Li Weili, Li Yong, Yang Xuefeng, et al. *Temperature and Fluid Flow Field Calculation and Analysis of Stator End of Air Cooled Turbo-generator[J]*. Proceedings of the CSEE, **29**, 36, (2009): 80-87.
8. Zhang Yujiao, Ruan Jiangjun, Huang Tao, et al. Calculation *of temperature rise in air-cooled induction motors through 3-D coupled electromagnetic fluid-dynamical and thermal finite-element analysis[J]*. IEEE Transactions on Magnetics, **48**, 2, (2012): 1047-1050.
9. Yulei L. *Calculation and Analysis of 3D* Temperature *Field in Steady State of the Submersible Motor[C]*. ICECE, (2010): 3427-3430.

10. Vong P K, Rodger D. *Coupled electromagnetic-*thermal *modeling of electrical achines[J]*. IEEE Transactions on Magnetics, **39**, 3, (2003): 1614-1617.
11. Meng Dawei, Liu Yulei, Zhang Qingjun, et al. *Calculation and analysis of 3D temperature of the submersible motor based on coupled field[J]*. Electric Machines and Control, **14**, 1, (2010): 52-55.
12. Meng Dawei, Liu Yu, Xu Yongming. *Analysis and* Calculation *of 3D Temperature Field of Submersible Motor Rotor[J]*. Electric Machines and Control, **13**, 3, (2009): 367-376.
13. Shehab Ahmed, Hamid A.Toliyat. *Coupled Field Analysis Needs in the Design of Submersible Electric Motors[C]*. IEEE Electric Ship Technologies Symposium. (2007): 231-237.
14. Zou Jibin, Qi Wenjuan, Xu Yongxiang, et al. *Design of Deep Sea Oil-Filled Brushless DC Motors Considering the High Pressure Effect[J]*. IEEE Transactions on Magnetics, **48**, 11, (2012): 4220-4223.

Research on Speed Sensorless Vector Control System of Asynchronous Motor Based on MRAS

Zili Liao, Qijin Zhao, Xinxi Zhang and Luming Chen

Department of Arms and Control Engineering, Academy of Army Armored Force, Beijing, China

Abstract. This paper analysed the basic principle of speed sensorless vector control system. Based on speed and current closed loop vector control, combined with a simple and feasible current hysteresis control strategy, the whole speed sensorless system of asynchronous motor is simulated in MATLAB/Simulink. The method uses the Model Reference Adaptive System (MRAS). The observation and analysis of waveform shows that the system has good static performance and robustness. The control effects are also as similar as the vector control system which contains speed sensor.

1 Introduction

At present, AC induction motors have a wide range of applications in the production and life. To achieve better control results, accurate speed feedback is required. Due to the limited reliability and high cost of the current speed sensors (Most of them are photoelectric encoders, resolver, etc.) in harsh environments, it is very important to study the control system of high performance speed sensorless[1-2].

In the sensorless control technology, speed identification is the key. The main methods having been used are direct calculation, Extended Kalman Filter (EKF), model reference adaptive method (MRAS), high frequency injection and intelligent control method[3-4]. By comparing these methods, MRAS is suitable for medium and high speed operation and is easier to implement in practical application[5]. In this paper, on the basis of vector control, build the motor simulation model, and the MRAS based speed estimation method is applied to the model. By observing the simulation results, the motor speed control and speed identification performance have been analyzed.

2 Vector control strategy based on current hysteresis loop

2.1 Current hysteresis PWM control

Hysteresis current pulse width modulation(PWM) technology is a method of converting the DC voltage into a voltage pulse sequence by using the power switching device. It can achieve the purpose of frequency conversion and voltage variation by controlling the voltage pulse width or period[6]. In the voltage source inverter, the hysteresis current tracking control provides a method to control the transient current output. The basic idea is using output current as the command signal and the actual current as feedback signal, then determine the power switch of the inverter by comparing the instantaneous value of the two switches so that the actual output tracking command signal. Hysteresis current PWM modulation control system is simple relatively. And it is different from the Space vector pulse width modulation(SVPWM) which needing Combination and calculation of complex voltage space vectors[7-8]. It controls the current directly, and the control precision is due to the switching frequency of inverter components and hysteresis width. By setting reasonable hysteresis width, simple direct control of current can be carried out.

In figure 1, the principle of single phase half bridge inverter circuit for PWM current tracking control is presented.

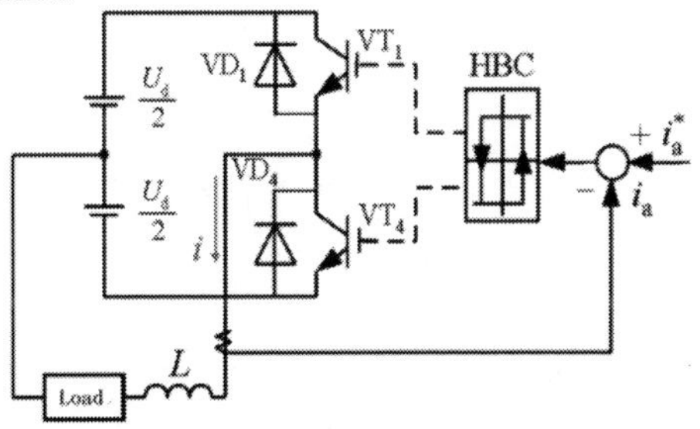

Figure 1. Schematic diagram of hysteresis current tracking control

As shown in figure 1, the deviation Δi_a between the given current i_a^* and the output current i_a is taken as the input of the comparator with hysteresis characteristics. The on-off of power device VT_1 and VT_4 is controlled by its output. When Δi_a is positive, the VT_1 is switched on, and then i_a increases and, when the VD_4 freewheeling, i_a decreases. When Δi_a is negative, the VD_4 is switched on, and then the absolute value of i_a increases, and when the VD_1 freewheeling, i_a the absolute value of i_a decrease. In this way, the current command is tracked in zigzag range in the range of $i_a + \Delta I$ and $i_a - \Delta I$ by controlling the hysteresis comparator with a ring width of $2\Delta I$ [9]. The current tracking waveform is shown in figure 2.

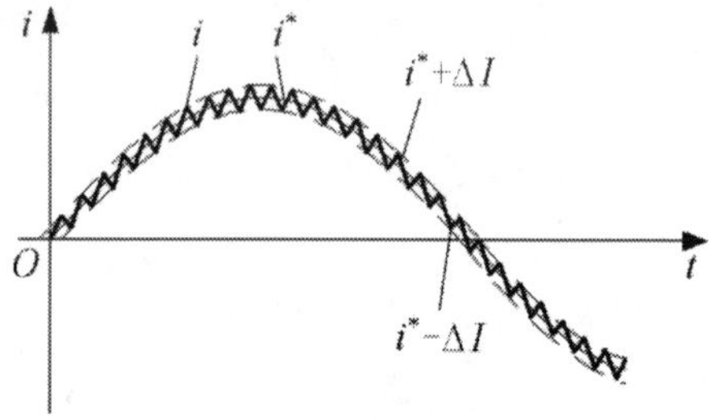

Figure 2. Current waveform in current hysteresis loop tracking control

2.2 Vector control principle and mathematical model of asynchronous motor

Vector control decompose the stator current vector of AC motor into two DC components (magnetic flux component and torque component) through coordinate transformation, so as to realize the decoupling of torque control and flux linkage control of induction motor. In the two-phase system, the d-axis can be regarded as the field winding of the DC motor, i_d is equivalent to the excitation current while the q-axis can be regarded as the armature winding, i_q is equivalent to the armature current which is proportional to the electromagnetic torque[10]. So that control the d-axis current and q-axis current, so as to achieve the desired control effect and avoid the nonlinear and strong coupling problems.

The equations of the motor in the d-q coordinate system are obtained by Clarke and Park transformation based on the motor model equation in the three-phase stationary coordinate system. The stator and rotor voltage equations are:

$$\begin{cases} u_{sd} = R_s i_{sd} + \dfrac{d\psi_{sd}}{dt} - \psi_{sq}\dfrac{d\theta_1}{dt} \\ u_{sq} = R_s i_{sq} + \dfrac{d\psi_{sq}}{dt} + \psi_{sd}\dfrac{d\theta_1}{dt} \\ u_{rd} = R_r i_{rd} + \dfrac{d\psi_{rd}}{dt} - \psi_{rq}\dfrac{d\theta_2}{dt} \\ u_{rq} = R_r i_{rq} + \dfrac{d\psi_{rq}}{dt} + \psi_{rd}\dfrac{d\theta_2}{dt} \end{cases} \tag{1}$$

The stator and rotor flux equations are:

$$\begin{cases} \psi_{sd} = L_s i_{sd} + L_m i_{rd} \\ \psi_{sq} = L_s i_{sq} + L_m i_{rq} \\ \psi_{rd} = L_r i_{rd} + L_m i_{sd} \\ \psi_{rq} = L_r i_{rq} + L_m i_{sq} \end{cases} \tag{2}$$

The electromagnetic torque equations is:

$$T_e = \frac{P_m L_m}{L_r}(i_{sq}\psi_{rd} - i_{sd}\psi_{rq}) \tag{3}$$

Where $u_{sd}, u_{sq}, u_{rd}, u_{rq}$ are the d and q axis voltage; i_{sd}, i_{sq}, i_{rd}, i_{rq} are the d and q axis current; $\psi_{sd}, \psi_{sq}, \psi_{rd}, \psi_{rq}$ are the d and q axis current flux; R_s, R_r represent stator and rotor resistance; L_s, L_r, L_m stator and rotor self-inductance and mutual inductance between them; θ_1 and θ_2 are represent the angle between the d and q axis and the α axis of the stationary two phase coordinate system respectively; T_e is electromagnetic torque; P_m is polar logarithm.

3 Speed sensorless control method based on MRAS

Speed sensorless control strategy eliminates the speed sensor with shortcomings such as low reliability. It obtains the motor speed through a certain algorithm using the motor stator voltage, current and other electrical signals which easily measured. The key to the problem is how to estimate the speed information quickly, and maintain high control accuracy, which can meet the requirements of real-time control.

Figure3 is the basic structure of Model Reference Adaptive System(MRAS). As the most widely used speed estimation method, the basic idea of MRAS is as follows. It designs the adaptive control system using Popov stability theory in the two phase static coordinate asynchronous motor system where the rotor flux voltage model without speed as the reference model and the current model of rotor flux with speed as the adjustable model[11].

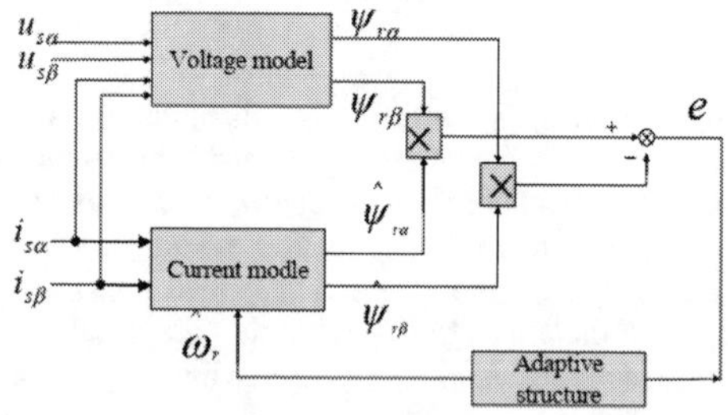

Figure 3. the basic structure of motor MRAS

The key to MRAS is the selection of adaptive law. In general, the speed is time-varying. But because of the inertia of the transmission system, in the small range of instantaneous speed change, it can be assumed that all the parameters are constant except the rotational speed as the variable to be identified. The estimated value of the current model is the equation(4).

 https://doi.org/10.1051/matecconf/201816002006

$$p\begin{bmatrix}\hat{\psi}_{r\alpha}\\ \hat{\psi}_{r\beta}\end{bmatrix}=\begin{bmatrix}-T_r & -\hat{\omega}_r\\ -\hat{\omega}_r & -T_r\end{bmatrix}\begin{bmatrix}\hat{\psi}_{r\beta}\\ \hat{\psi}_{r\alpha}\end{bmatrix}+L_mT_r\begin{bmatrix}i_{s\alpha}\\ i_{s\beta}\end{bmatrix} \quad (4)$$

Define state error $e=\psi_r-\hat{\psi}_r$ and We can obtain error equation (5) from the difference between the actual model and the estimated model.

$$p\begin{bmatrix}e_{r\alpha}\\ e_{r\beta}\end{bmatrix}=\begin{bmatrix}-T_r & -\omega_r\\ -\omega_r & -T_r\end{bmatrix}\begin{bmatrix}e_{r\alpha}\\ e_{r\beta}\end{bmatrix}-(\hat{\omega}_r-\omega_r)\begin{bmatrix}0 & -1\\ -1 & 0_r\end{bmatrix}\begin{bmatrix}\hat{\psi}_{r\alpha}\\ \hat{\psi}_{r\beta}\end{bmatrix} \quad (5)$$

Put it another form:

$$\dot{e}=A_m e-\omega \quad (6)$$

According to the Popov inequality, the identification model of rotational speed can be obtained under the premise that $\eta(0,t)\geq-\gamma^2$ is established:

$$\hat{\omega}_r=\int_0^t K_1(\hat{\psi}_{r\alpha}\psi_{r\beta}-\psi_{r\alpha}\hat{\psi}_{r\beta})dt+K_p(\hat{\psi}_{r\alpha}\psi_{r\beta}-\psi_{r\alpha}\hat{\psi}_{r\beta})$$
$$+\hat{\omega}(0)=(K_p+K_1/s)(\hat{\psi}_{r\alpha}\psi_{r\beta}-\psi_{r\alpha}\hat{\psi}_{r\beta}) \quad (7)$$

4 Simulation analysis based on MATLAB/Simulink

In order to verify the feasibility of induction motor speed sensorless vector control system based on MRAS, the simulation experiment is carried out by using the mentioned methods in MATLAB/Simulink. The main parameters of the asynchronous motor used in the experiment are listed in Table 1:

Table 1. motor parameters

Parameters	Values
Line voltage U	380V
Rated frequency f	50Hz
Number of pole pairs P_n	4
Stator resistance R_s	0.253Ω
Rotor resistance R_r	0.531Ω
Self inductance of Stator L_s	0.102mH
Self inductance of rotor L_r	0.052mH
Mutual inductance L_m	0.053mH
Inertia J	0.22kg·m^2

4.1 Simulation model

On the basis of analyzing the mathematical model of three-phase asynchronous motor, a simulation model of current hysteresis vector control system based on speed sensorless is established. The system adopts double closed loop structure, and the speed, torque and flux control adopt PI regulator with integral and output limiting, and the current is controlled by hysteresis. The system model includes rotor flux observation subsystem, speed estimation subsystem (figure 4), current hysteresis PWM subsystem (figure 5) and so on. The whole control system is shown in figure6. The whole system is simulated by continuous control system.

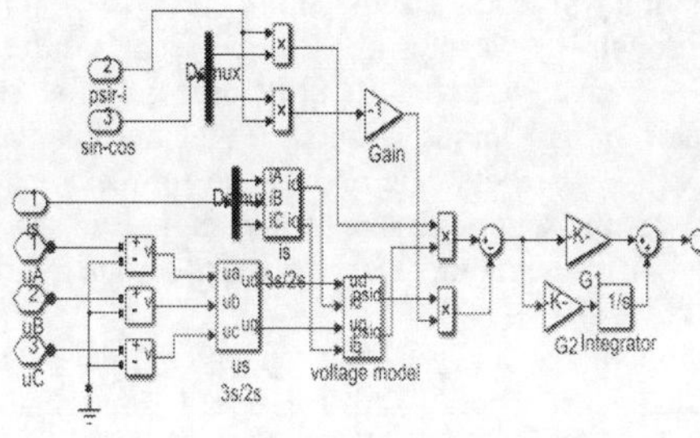

Figure 4. Speed estimation module

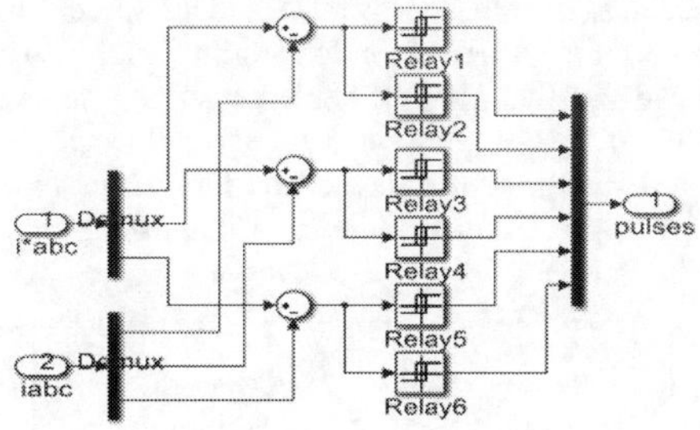

Figure 5. Hysteresis-band current PWM output module

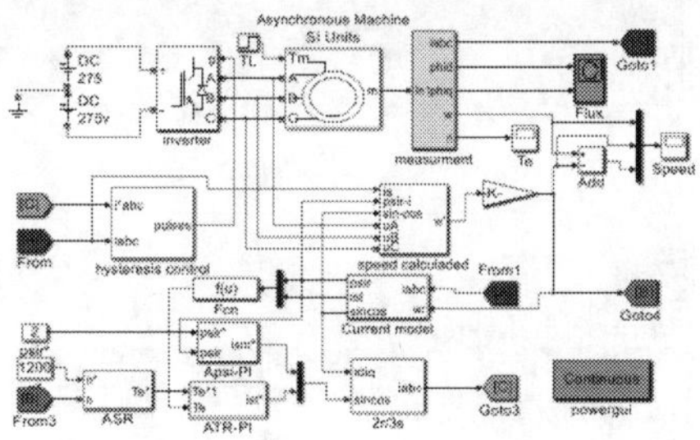

Figure 6. Speed sensorless vector control system

The parameter settings of the PI regulator are shown in Table2.

Table 2. regulator parameters

PI Regulator	K_p	K_i	Regulator clipping	
			Upper limit	*lower limit*
Speed regulator	12	3.2	80	-80
Torque regulator	5.5	18	55	-55
Flux regulator	3	125	16	-16
Speed estimation	1.28	385	—	—

4.2 Simulation analysis

In order to verify the static and dynamic performance of the system, the current and speed change of the motor under sudden load are given, and the accuracy of the speed estimation is also observed. In the simulation, the initial speed is set to 1200rpm, the motor is unloaded at startup, and the load torque is added to 45N·m after 0.6s. The simulation waveform is shown in figure 7.

Through the waveform, it can be seen that the motor reaches a given speed at about 0.40second after starting, and the no-load torque is about 0 and the fluctuation is smaller after steady state. When the torque is suddenly increased, the speed is basically kept at a given speed, and the torque of the motor reaches the load torque rapidly and tends to be stable.

Through the analysis, the vector control system has good response speed, and the speed control is more accurate. In the estimation of speed, it can be seen from figure 7(d) that the actual speed is almost the same as the estimated speed after steady state, and the speed estimation is still accurate after sudden load torque. In the early stage of speed rise there will be some deviation, this is mainly because of the changing current in the early, and There is a time delay in the adjustable model, which makes the current change with certain delay and is amplified in its rapid change

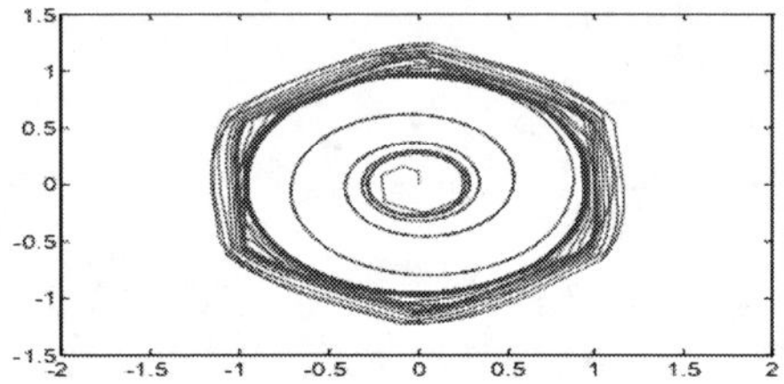

(a)stator flux linkage

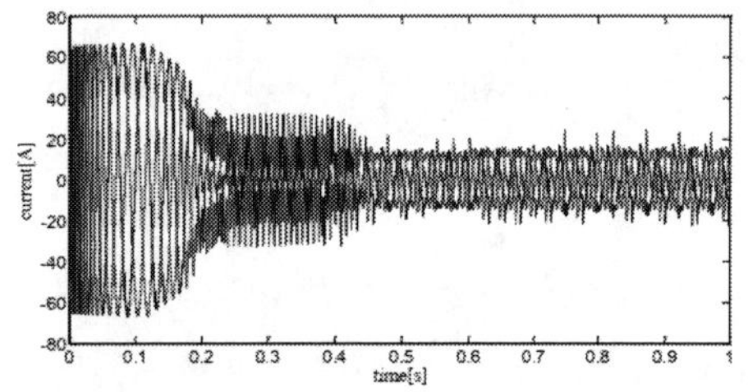

(b)three-phase stator current

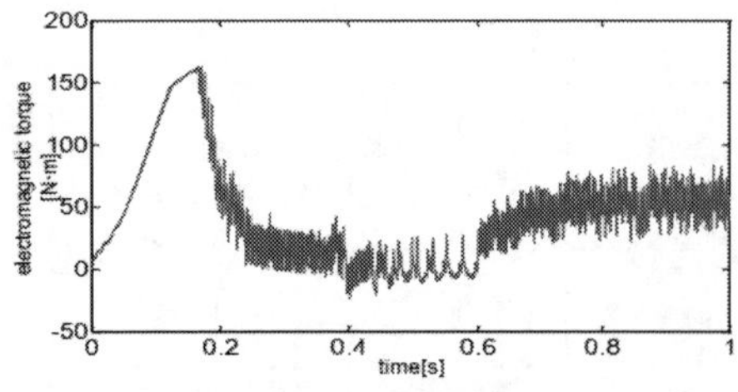

(c)electromagnetic torque

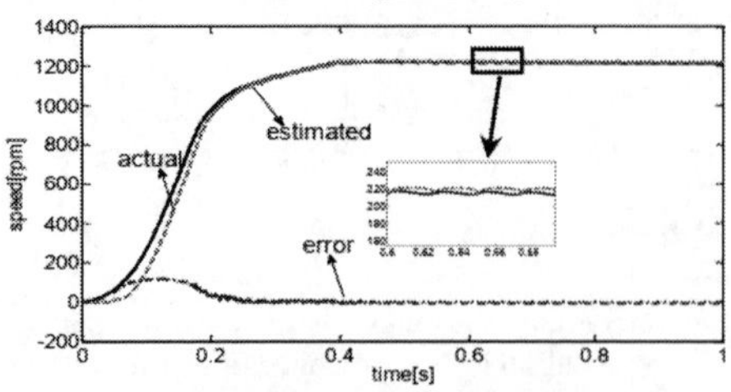

(d)actual and estimated speed

Figure 7. Performance under load of MRAS system

Figure8 is the speed sensorless vector control Counter-starting and steering process. Figure8(a) is the flux linkage curve which showing a well circle, indicating that the system pulsation is small. From 0 to 0.15 second is the Counter-starting process of the motor. In the starting process, the speed calculated by the speed observer has a certain deviation from the speed measured by the measuring module. From 0.15 to 0.3 second is the reverse stable phase, and at the time 0.3second, the system switched to positive 500rpm, the motor speed began to decline, and quickly entered the positive start stage. As shown in figure 8(c), the motor always keeps the maximum torque until the start is completed. Looking at figure8(d), there is almost no difference between the observed rotational speed and the steady speed in the two stable stages.

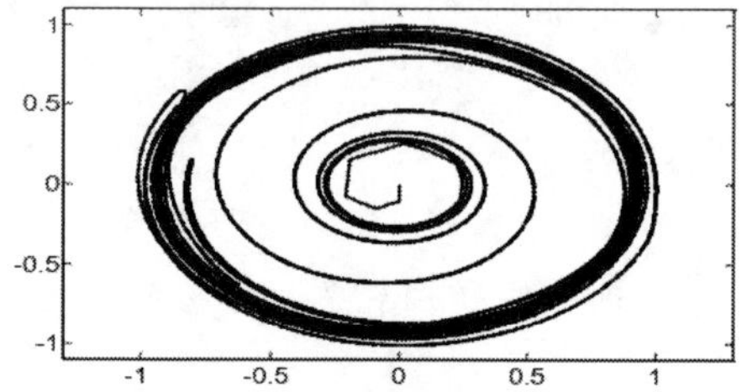

(a) stator flux linkage

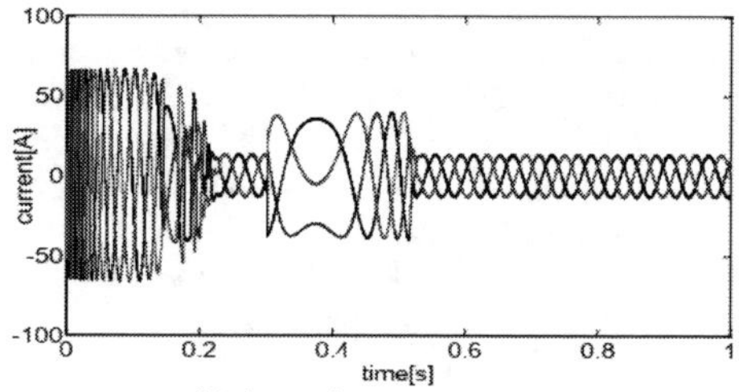

(b) three-phase stator current

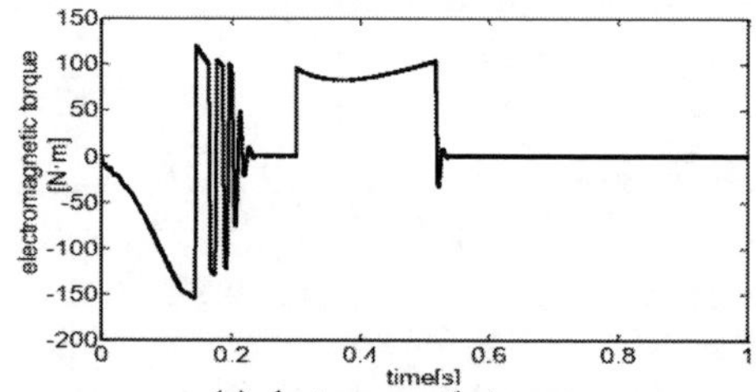

(c) electromagnetic torque

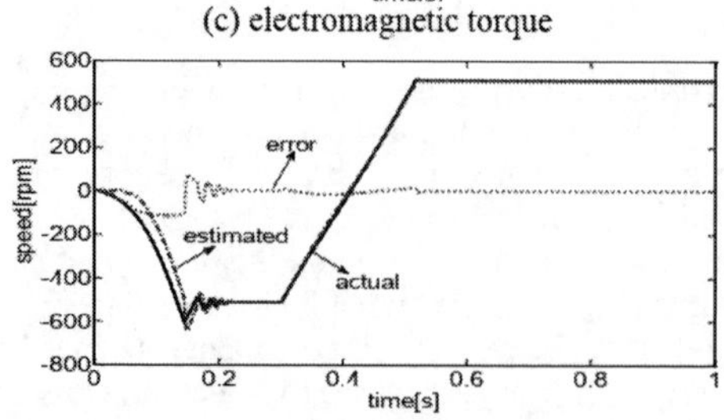

(d) actual and estimated speed

Figure 8. Positive and negative adjustment process of MRAS system

5 Conclusion

In this paper, the basic principle of vector control is analyzed. Based on MRAS speed identification method, a speed sensorless vector control system is built in MATLAB/Simulink. By analyzing the simulation waveform, it is proved that the system has good dynamic and static performance, and the speed identification is feasible. Compared with the installation of speed controller, this control method can not only improve the reliability of the system, but also save a part of the cost, which has certain value for practical engineering application.

References

1. A. Mecke, I. Lee, J.R. Baker jr., M.M. BanaszakHoll, B.G. Orr, Eur. Phys. J. E **14**, 7 (2004)
1. P.F. Zheng, T. Fukao. Robust speed identification for speed-sensorless vector control of induction motors. IEEE Transactions on Industry Applications.J. **30**,5:1234-1240(1994)
2. T.N. Ohtani, Takada, K. Tanaka. Vector control of induction motor without shaft encoder. Industry Applications IEEE Transactions.J. **28**, 1:157-164(1992)
3. Y.R. Kim, S. K. Sul, M. H. Park. Speed sensorless vector control of induction motor using extended Kalman filter. Industry Applications IEEE Transactions. J. **30**, 5:1225-1233(1994)
4. Yao, Qing Hua, M. Cao. Research of Vector Control for Asynchronous Motor System Based on MRAS.Key Engineering Materials. J.:769-772 (2013)
5. E. D.Mitronikas, A. N. Safacas. An improved sensorless vector-control method for an induction motor drive. IEEE Transactions on Industrial Electronics.J.**52**,6:1660-1668(2005)
6. Dwivedi, Anurag, A. N. Tiwari. Analysis of three-phase PWM rectifiers using hysteresis current control techniques: a survey. International Journal of Power Electronics.J.**8**,4: 349(2017)
7. V. Neenu. Adaptive Modified Hysteresis Current Controlled Grid Connected Photovoltaic Inverter. International Journal of Engineering & Technical Research.J. **6**, 4(2017).
8. L.Zhen, L. Xu. Sensorless field orientation control of induction machines based on a mutual MRAS scheme. Industrial Electronics IEEE Transactions.J.**45**,5 : 824-831(1998)
9. E. Levi, R. Bojoi, F. Profumo. Multiphase induction motor drives - a technology status review. Iet Electric Power Applications.J.**4**,1: 489-516 (2007)
10. M. Cirrincione, M. Pucci. An MRAS-based sensorless high-performance induction motor drive with a predictive adaptive model. IEEE Transactions on Industrial Electronics.J.**52**.2:532-551(2005)
11. M. Jain, M. Singh, A. Chandra, S.S. Williamson, Sensorless control of permanent magnet synchronous motor using ANFIS based MRAS.Electric Machines & Drives ConferenceIEEE, C. :599-606 (2011)

Research on lightning stroke model and characteristics of electronic transformer

Mu LI[1], Lian Lian LI[1], Shu Bo FANG[1] and Jing Jing WANG[1]

[1]Wuhan NARI Limited Liability Company, State Grid Electric Power Research Institute, Wuhan 430074, China

Abstract. In order to improve the reliability of power supply, a large number of electronic voltage and current transformers are used in digital substations. In this paper, the mathematical model of the electronic transformer is analyzed firstly, and its circuit model is given. According to the difference of working characteristics between voltage transformer and current transformer, the circuit model of voltage type electronic transformer and current type electronic transformer is given respectively. By analyzing their broadband transmission characteristics, the accuracy of the model is verified, and their lightning analysis models are obtained.

1 Introduction

In order to improve the reliability of power supply, a large number of electronic transformers are used instead of traditional transformers in the digital substation, so as to avoid the differential saturation, ferroresonance and other risks. At the same time, it also solves the problem of heavy equipment, inconvenient installation and transportation under the high voltage level.

To ensure the safe and reliable operation of the electronic transformer, it is necessary to comprehensively and accurately grasp the characteristics of the over voltage surge in lightning wave shows, according to the characteristics of lightning and risk to develop the corresponding protective measures and strategies. Now we focus on analyzing the lightning stroke model of electronic voltage and current transformer and its lightning characteristics[1-3].

2 Model of electronic transformer

The mathematical model of electronic transformer is using the network analyzer and other measuring instruments to measure the wide band voltage transmission characteristics of the electronic transformer equipment, broadband current transmission characteristics and scattering parameters, and then convert the measured data and rational approximation, and get their rational approximation expression. On the basis of the expression of rational approximation, the circuit model is established, and the broadband transmission model of the electronic transformer device or the T equivalent circuit model are obtained. The method of establishing wide band model of electronic transformer is explained here. Firstly, taking voltage transformer as an example, it explains the method of establishing wide band mathematical model of transformer[4-5].

2.1 Mathematical model of electronic transformer

The broadband transmission characteristics of the measured electronic transformer, whether it is voltage transformer or current transformer, can be mathematically approximated by rational fitting method. Its typical description form is:

$$f(s) = \sum_{n=1}^{N} \frac{c_n}{s - a_n} + d \tag{1}$$

Here, the pole an and the corresponding cn residue can be real, can also be a complex conjugate; d is a real number, N is the total number of poles.It is assumed that there are k pairs of conjugate complex poles and N-2k real poles in the left half plane, and K pairs of conjugate complex poles are:

$$\left.\begin{aligned} a_{2n-1} &= -p_{rn} + jp_{in} \\ a_{2n} &= -p_{rn} + jp_{in} \end{aligned}\right\} n = 1,2,\cdots,K \tag{2}$$

Here, p_m>0. Then the residue corresponding to complex conjugate poles:

$$\left.\begin{aligned} c_{2n-1} &= -c_{rn} + jc_{in} \\ c_{2n} &= -c_{rn} + jc_{in} \end{aligned}\right\} n = 1,2,\cdots,K \tag{3}$$

The other N-2k real poles are:

$$a_n < 0, n = 2K + 1, \cdots, N \tag{4}$$

The corresponding residue:

$$c_n < 0, n = 2K + 1, \cdots, N \tag{5}$$

Thus it can be obtained:

$$f(s)=\sum_{n=1}^{K}\left(\frac{2c_{rn}s+2c_{rn}p_{rn}-2p_{in}c_{in}}{s^2+2p_{rn}s+p_{rn}^2+p_{in}^2}\right)+\sum_{n=2K+1}^{N}\frac{c_n}{s-a_n}+d \tag{6}$$

$$f(s)=\sum_{n=1}^{K}f_{1n}(s)+\sum_{n=1}^{K}f_{2n}(s)+\sum_{n=2K+1}^{N}f_{3n}(s)+f_4(s) \tag{7}$$

Among them:

$$f_{1n}(s)=\left(\frac{2c_{rn}p_{rn}-2p_{in}c_{in}}{s^2+2p_{rn}s+p_{rn}^2+p_{in}^2}\right) \tag{8}$$

$$f_{2n}(s)=\left(\frac{2c_{rn}s}{s^2+2p_{rn}s+p_{rn}^2+p_{in}^2}\right) \tag{9}$$

$$f_{3n}(s)=\frac{c_n}{s-a_n} \tag{10}$$

$$f_4(s)=d \tag{11}$$

Using the mathematical approximation expressions, the circuit model corresponding to f1n (s), f2n (s), f3n (s) and f4n(s) can be obtained.

From equation 8, f1n(s)can be rewritten as:

$$f_{1n}(s)=H_{1n}\frac{1}{X_{1n}s^2+Y_{1n}s+1} \tag{12}$$

Among them:

$$H_{1n}=2\frac{c_{rn}p_{rn}-p_{in}c_{in}}{p_{rn}^2+c_{in}^2} \tag{13}$$

$$X_{1n}=\frac{1}{p_{rn}^2+P_{in}^2}>0 \tag{14}$$

$$Y_{1n}=\frac{2p_{rn}}{p_{rn}^2+p_{in}^2}>0 \tag{15}$$

The circuit diagram corresponding to equation 12 is shown in Figure 1.

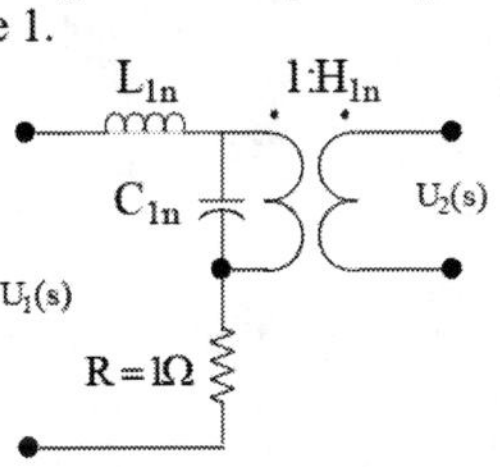

Figure 1. Circuit model corresponding to f1n (s).

In Figure 1, the calculation of the parameters of each component can be given by the following formula:

$$C_{1n}=Y_{1n}=\frac{2p_{rn}}{p_{rn}^2+p_{in}^2},\quad L_{1n}=\frac{X_{1n}}{C_{1n}}=\frac{1}{2p_{rn}} \tag{16}$$

Similarly, from equation 9, f2n(s)can be rewritten as:

$$f_{2n}(s)=H_{2n}\cdot\frac{Y_{2n}s}{X_{2n}s^2+Y_{2n}s+1} \tag{17}$$

The circuit diagram corresponding to equation 17 is shown in Figure 2.

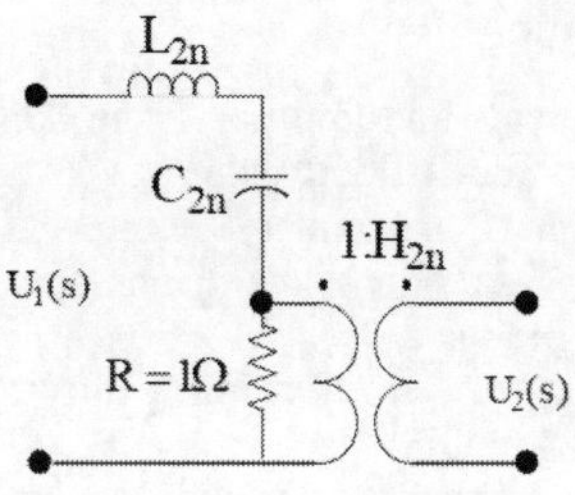

Figure 2. Circuit model corresponding to f2n (s).

From equation 10, f3n(s) and f4n(s) can be rewritten as:

$$f_{3n}(s)=\frac{c_n}{s-a_n}=H_{3n}\frac{1}{s/p_n+1} \tag{18}$$

The equivalent circuit model of f3n (s) and f4n(s) is as follows:

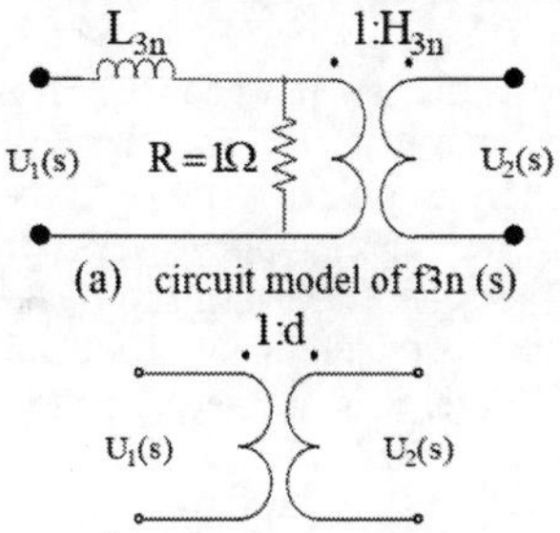

(a) circuit model of f3n (s)

(b)circuit model of f4n (s)

Figure 3. Equivalent circuit model of f3n (s) and f4n(s).

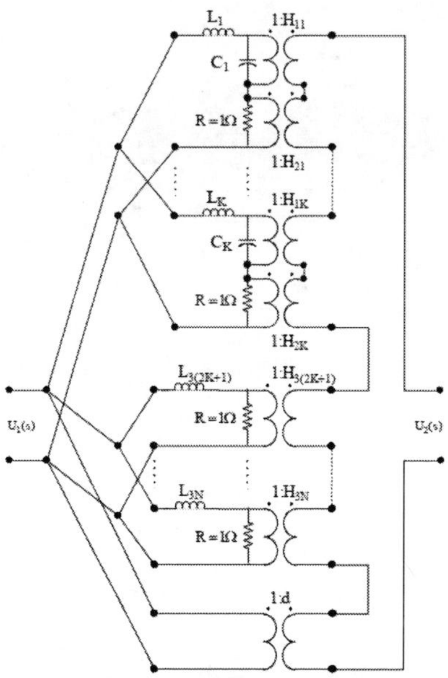

Figure 4. Circuit model corresponding to f (s).

By combining the Figure 1, Figure 2 and Figure 3 according to equation 7, the equivalent circuit model corresponding to the voltage transfer function f (s) can be obtained as shown in Figure 4, in which all the transformers are ideal transformers.

2.2 Circuit model of electronic voltage and current transformer

The instrument transformer is the key transmission channel for transmitting lightning overvoltage along overhead lines or primary equipment to the two equipment. There are three main types of electronic transformer products.

The first is an electronic current and voltage transformer for GIS. Each phase needs to install two transformers, which are installed on both sides of the switch fracture. Each transformer consists of two independent groups of sensors and their corresponding sensing modules. Each sensor consists of a capacitive divider, a low power iron core coil and an air core coil.

The second is an open type independent installation of electronic current and voltage transformers, which consists of two independent groups of sensors and their corresponding sensing modules. Each sensor consists of a capacitive divider, a low power iron core coil and an air core coil.

The third is the low power transformer (LPCT). The economy and advantage of the electronic voltage and current transformer are proportional to the voltage level. Because of the higher voltage level transformer, its saturation characteristics, insulation complexity, large volume, high cost and other shortcomings are more obvious.Therefore, it is not necessary to popularize and apply electronic transformers blindly in each substation voltage level. We believe that the application of electronic transformers is unnecessary and uneconomical for systems below 110kV voltage, especially for 10-35kV systems, and the use of LPCT is a realistic and economical solution.

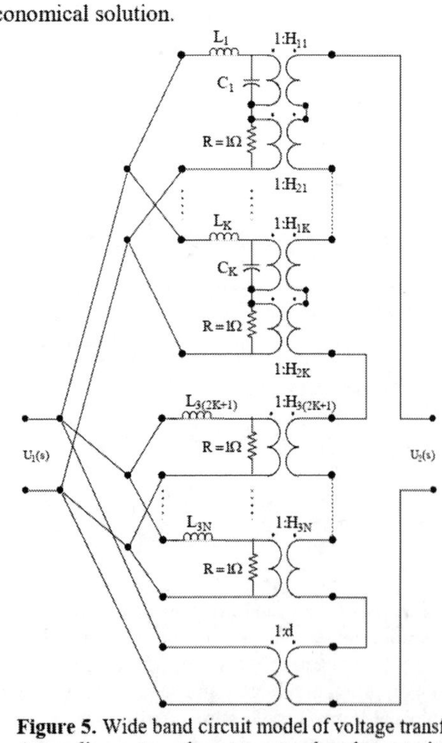

Figure 5. Wide band circuit model of voltage transformer.

According to the structural characteristics of electronic voltage and current transformer and the mathematical model of transformer, the broadband circuit model of electronic voltage transformer and electronic current transformer is obtained, as shown in Figure 5 and Figure 6.

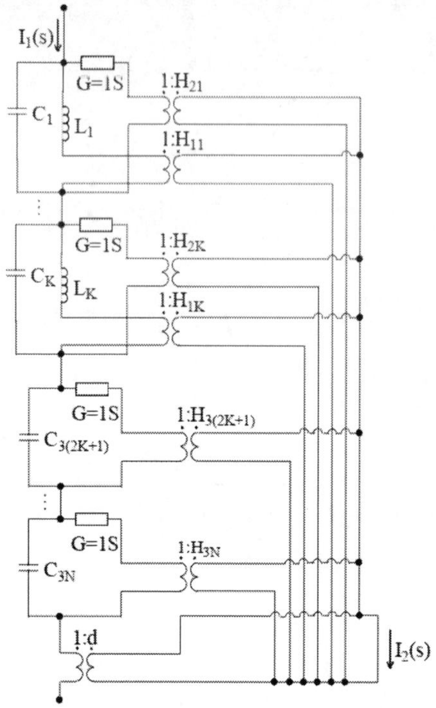

Figure 6. Wide band circuit model of current transformer.

3 Analysis of model characteristics

Here, we give the model of voltage transformer and current transformer based on test results, and study their characteristics.

3.1. Voltage transformer

The scattering parameters of a 500kV capacitor voltage transformer are measured. The model of the voltage transformer is $\mathrm{TYD}500/\sqrt{3}-0.005\mathrm{H}$, and the rated voltage ratio is $5000/\sqrt{3}kV/0.1/\sqrt{3}\,kV$.

On the basis of the measurement of scattering parameters, the mathematical data are approximated by rational fitting method. The approximation result has 40 poles of the left half plane. Using the proposed modeling method, we can obtain the wide band transmission model of the voltage transformer.

Component parameters in the model are shown in Table 1 and Table 2, and the constant term corresponding to the ideal transformer for 1:d=1:5.8647 * 10-2. The voltage transmission characteristic of the wide band transmission model of the transformer is shown in the virtual curve in Figure 7. As shown in Figure 7, the measured voltage transmission characteristics are in good agreement with the voltage transmission characteristics of the circuit model. Figure 7 shows the voltage transfer characteristics of the voltage transformer in the lower frequency range. It can be seen that when the ratio between the voltage of the low voltage side of the transformer and the voltage of the high voltage side is below 20kHz, the value of the amplitude frequency characteristic of the voltage transmission is basically kept at 0.02, which shows that the voltage ratio of the voltage transformer in the 20kHz is kept better. This shows that the voltage ratio of the voltage transformer is kept better in 20kHz. When the frequency is higher than 20kHz, the voltage transfer characteristics of the transformer are complex, and some resonance points appear.

Table 1. Component parameters of circuit models corresponding to conjugate complex poles.

Pole number	L_n(mH)	C_n(μF)	H_{1n}	H_{2n}
1	7.66E-02	2.09E-01	-2.47E-03	-2.61E-03
2	3.93E-02	1.29E-01	1.21E-03	6.69E-03
3	2.70E-01	1.53E-02	-3.59E-03	3.07E-01
4	1.48E-02	1.99E-01	2.17E-03	5.81E-02
5	5.33E-03	2.52E-01	-3.76E-02	2.87E-02
6	1.22E-03	2.20E-02	-5.47E-03	2.17E-02
7	4.85E-04	4.36E-03	-6.84E-03	6.88E-03
8	3.42E-04	3.89E-03	-6.18E-03	3.37E-02
9	2.02E-03	4.83E-04	-2.89E-04	-5.63E-03
10	4.99E-04	1.84E-03	-7.15E-03	5.54E-02
11	1.47E-03	4.77E-04	6.76E-04	5.49E-02
12	2.39E-04	2.35E-03	8.99E-03	1.24E-01
13	2.10E-03	2.26E-04	1.96E-04	-1.42E-02
14	5.96E-04	6.32E-04	2.09E-04	-2.60E-02
15	1.35E-04	2.73E-03	1.65E-02	1.81E-01
16	9.82E-02	3.71E-06	2.21E-04	-3.59E-02
17	2.08E-03	1.45E-04	-3.34E-04	1.49E-02
18	1.05E-03	2.85E-04	8.10E-04	2.50E-02
19	2.21E-04	5.41E-04	-2.37E-03	4.79E-02

Table 2. Component parameter of circuit model corresponding to real number pole.

Pole number	L3n (mH)	H3n
1	2.71E+02	-6.44E-03
2	1.76E-02	4.04E-03

Figure 8 shows the voltage transfer characteristics at high frequencies, and the measured voltage transmission characteristics at high frequencies coincide with the voltage transmission characteristics of the circuit model. If we want to obtain better approximation results, we can increase the number of poles in the data fitting, but this will lead to the increase of the order of rational approximation expression and the number of circuit elements.

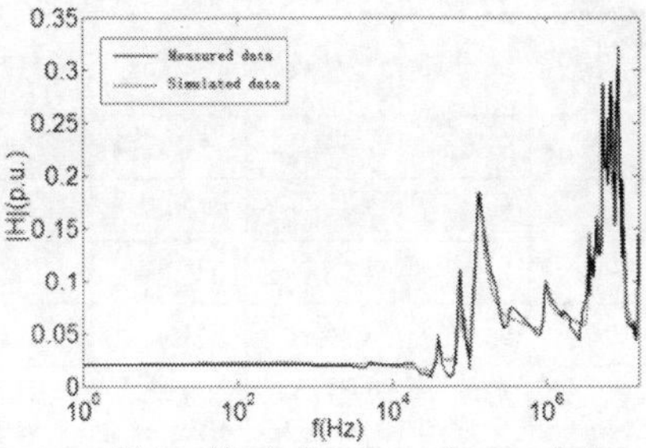

(a)Amplitude frequency characteristic

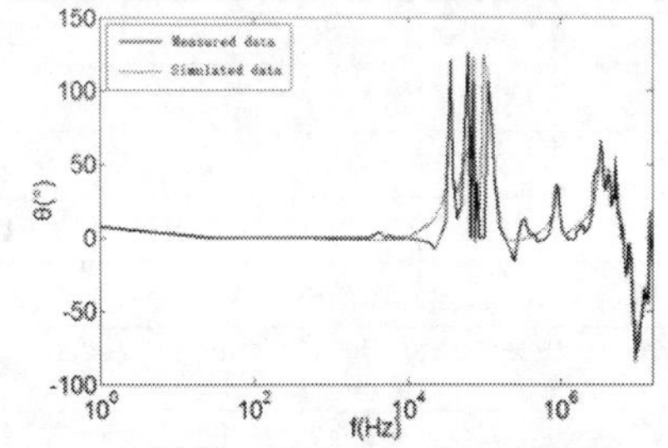

(b) Phase frequency characteristic

Figure 7. Wide frequency voltage transmission characteristics of voltage transformer.

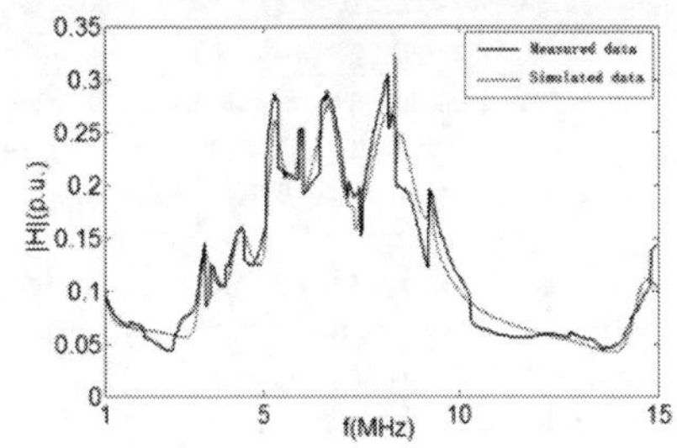

(a) Amplitude frequency characteristic

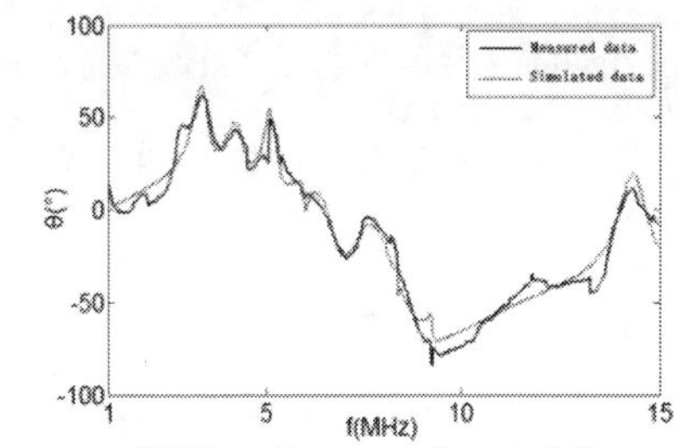

(b) Phase frequency characteristic

Figure 8. Voltage transmission characteristics of voltage transformer with frequency higher than 1MHz.

3.2 Current transformer

The scattering parameters of a 500kV current transformer are measured. The model of the current transformer is LB1-500, and the rated current ratio is 1250A/1A.

The mathematical data are approximated by the rational fitting method, and the fitting results of 24 poles are obtained. Using the proposed modeling method, we can obtain the wide band transmission model of the current transformer. Component parameters in the model are shown in Table 3 and Table 4, and the current ratio of ideal transformer corresponding to F4 (s) =d is 1:d = 1:0.014069.

Table 3. Component parameters of circuit models corresponding to conjugate complex poles.

Pole number	Ln(mH)	Cn(μF)	H1n	H2n
1	6.38E-4	1.59E+4	-3.39E-4	-2.58E-3
2	1.14E-6	7.43E+1	9.20E-3	-7.65E-1
3	1.58E-6	4.98E+1	1.39E-1	-8.26E+0
4	1.80E-6	4.26E+1	1.36E-1	1.01E+1
5	2.05E-6	9.97E+0	8.53E-4	-3.64E-2
6	1.45E-5	1.17E-1	5.39E-1	2.46E-1
7	1.20E-6	4.00E-1	-1.27E-2	-1.63E-1
8	4.09E-8	4.83E+0	1.45E-3	-1.38E-1
9	3.43E-6	5.67E-2	-3.21E-1	-5.3E-1
10	7.53E-8	2.33E+0	3.54E-3	-1.94E-1
11	4.08E-9	3.25E+1	-4.16E-4	1.65E-2

Table 4. Component parameter of circuit model corresponding to real number pole.

Pole number	L3n (mH)	H3n
1	1.15E+05	-3.58E-01
2	7.92E+03	-1.44E-01

The current transmission characteristics of the circuit model are shown in the dashed line curves of Figure 9 and Figure 10, and they are consistent with the measured data. Figure 9 shows the current transmission characteristics of the current transformer under low frequency. Figure 10 shows the current transmission characteristics of the current transformer at higher frequency. Under the 20kHz frequency, the current transfer characteristic of the current transformer is relatively stable, and the current ratio is 10A/5A=2. When the frequency is more than 20kHz, the change of current transmission characteristics of current transformer is large, and there is a resonance point occurred near 580kHz.

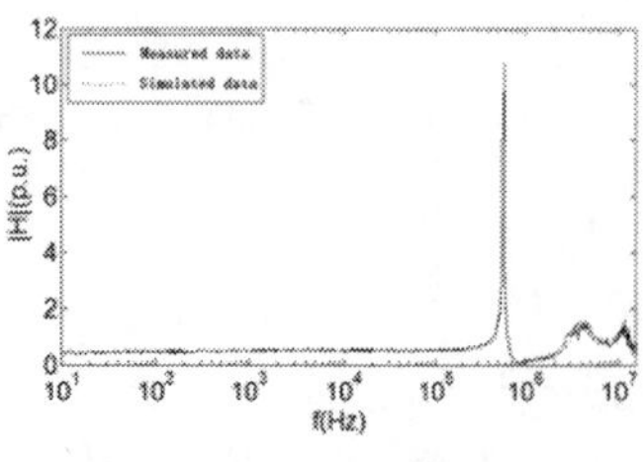

(a)Amplitude frequency characteristic

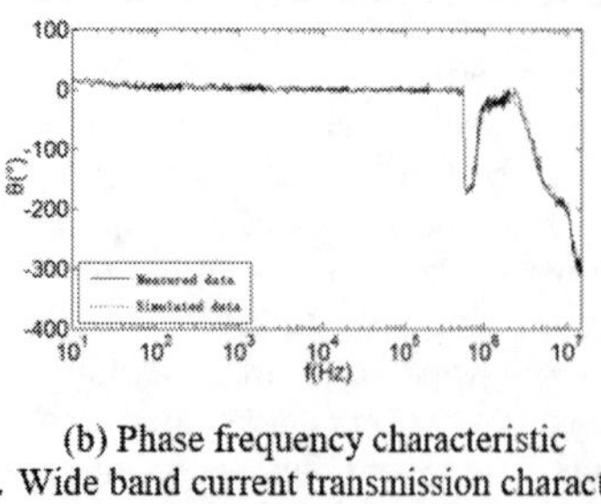

(b) Phase frequency characteristic

Figur 9. Wide band current transmission characteristics of current transformer.

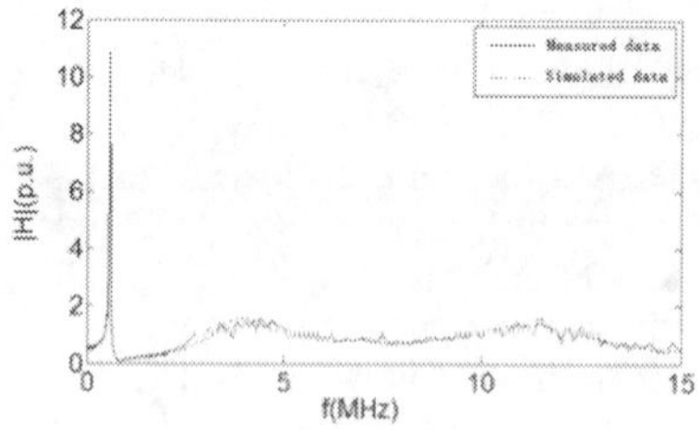

(a)Amplitude frequency characteristic

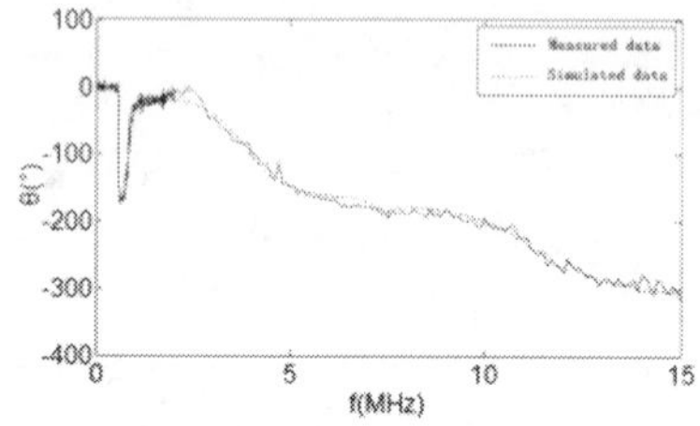

(b) Phase frequency characteristic

Figure 10. Current transmission characteristics of current transformer under higher frequency.

4 Conclusion

Based on the above analysis results, it can be seen that the electronic transformer model has good accuracy in high frequency range, but the model is complex, and it is not easy to use. Therefore, if we only need to consider the transformer model suitable for analyzing the lightning overvoltage, it is only necessary to ensure the accuracy of the model in the 1MHz band. Therefore, for the voltage transformer mentioned above, only the first 6-order parameters need to be considered, and the current transformer only needs to consider the 1-order model.

References

1. W.Xiong, Z.Wenliang, Technical problems of electromagnetic compatibility in power system, High voltage technology, 23,53-58 (1997)
2. V. S. Rashkes, L. D. Ziles, Very high frequency overvoltages at air EHV substations during disconnect switch operations, IEEE on PD, 3,1618-1623(1996)
3. S.Zhusen, Z.Yufang, Z.Guangzhou, et al. Research on electromagnetic disturbance and protection measures in 500kV substation (1), high voltage technology,26,16-18(2000)
4. C.Dingxin, Q.Xuedi, Calculation method of magnetic field produced by line current in "wire earth" loop, Proceedings of the Symposium on electromagnetic environment and electromagnetic compatibility of power system, Guangzhou, 27-29(2002)
5. W.Xiong, Electromagnetic disturbance and level of 500kV substation, Symposium on electromagnetic environment and electromagnetic compatibility of power system, Guangzhou,5-8(2002)

Study on Maneuvering Target On-axis Tracking Algorithm of Modified Current Statistical Model

Xiong Zhenkai [1,2], Li Fanying [1] , Zhang Lei [1]
[1]China Shipbuilding Industry Corporation 713 Research Institute Zhengzhou, China
[2]The School of Information Engineering of Zhengzhou UniversityZhengzhou, China

Abstract. Aiming at the model adaptability and the filter precision on the maneuvering target on-axis tracking,The paper put forward a filter algorithm based on modified current statistical model. The algorithm can enhance the model adaptability to the weak and non-maneuvering maneuvering target. The method uses Unscented Kalman Filter to obtain the importance density function of each particle, improves the Particle Filter estimation performance.By applying the proposed algorithm to the on-axis tracking system,the simulation results demonstrate that algorithm can effectively improve filter performance and tracking precision.

1 Introduction

Maneuvering target tracking technology is widely used in scientific research, aviation, aerospace and other fields, and put forward higher requirements. In particular, in the aspect of photoelectric tracking, due to the requirement of tracking precision, the requirements of maneuvering target tracking are higher. For this reason, the researchers propose methods such as composite axis tracking, equivalent composite tracking, on-axis tracking and multiple composite axis tracking. With the development of computer technology, on-axis tracking to solve the high precision optical tracking provides a good way[1]. One of the core of on-axis tracking is the filtering prediction algorithm, which is to predict the accurate target position information and velocity information for the control system[2,3,4,5]. The exact prediction premise of the maneuvering target prediction is that the target motion model can match the actual. However, because it is necessary to fully describe the actual state of the target motion, it is necessary to pass the high-order Taylor expansion, which is not desirable in the specific mathematical processing.

The establishment of the target model can not only facilitate the late mathematical calculations, but also similar to the target movement state. The "current" statistical model is a model that is more reasonable to describe the maneuvering target state[6,7]. The "current" probability density of the maneuvering acceleration is described by the modified Rayleigh distribution. The model is characterized by nonzero mean and modified Rayleigh distribution Maneuvering acceleration characteristics, and thus more realistic[8,9,10,11,12,13]. However, for weak and non-maneuvering maneuvering target, the current statistical model can not be fully described[14].

The key to the prediction of maneuvering target prediction is the measurement equation.The detection of sensor noise , the external uncertain disturbance and other nonlinear factors are the main aspects of the filter precision. The best way to solve the nonlinear factor is to use nonlinear filtering Prediction algorithm[15]. Based on these considerations, this paper designs an improved maneuvering target on-axis tracking filter algorithm for the current statistical model, which satisfies the fusion algorithm of Unscented Kalman Filter (UKF) and Particle Filter (PF) Nonlinear filter tracking requirements.

2 On-axis Tracking Principle

On-axis tracking is the target position information and speed information by computer technology to provide input for the servo system control. The servo system and the detection system are relatively independent[16,17]. The servo system is composed of speed loop and position loop. The detection system is composed of the detector and filter unit.After the detector output and the servo feedback information fusion, the parameter is used as filter input.Filter output is used as the servo system input[18].

The servo bandwidth should be wide for control precision and control optimization to reduce the detection noise interference[19,20,21,22], the detection system bandwidth should be narrow. On-axis tracking schematic diagram shown in Figure 1.

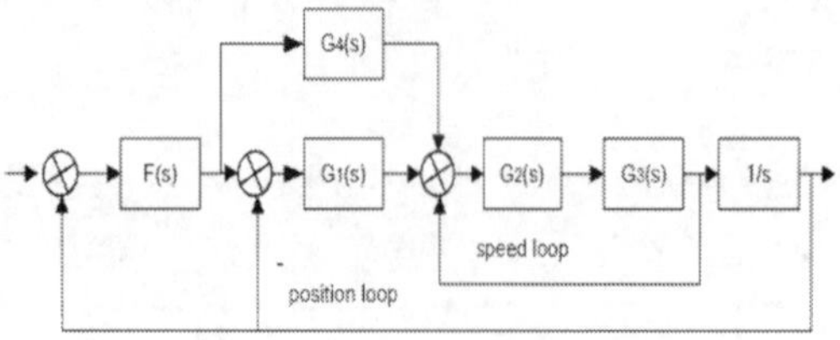

Figure 1. On-axis tracking principle figure.

In the figure, $\Delta\theta$ is the target error distance information by the sensor , $F(s)$ is the filtering predictive function, $G_1(s)$ is the position loop control function, $G_2(s)$ is the speed loop control function, $G_3(s)$ is the speed loop transfer function, and $G_4(s)$ is the feedforward control function.

3 Modified Current Statistical Model

The discrete state equation and the measurement equation of the one-dimensional current statistical model are:

$$X(k+1)=F(k+1)X(k)+G(k)\bar{\omega}(k)+M(k) \quad (1)$$

$$Z(k+1)=H(k)X(k)+N(k) \quad (2)$$

$X(k)$ is System state vector,

$X(k)=[x(k) \quad \dot{x}(k) \quad \ddot{x}(k)]^T$,

$F(k+1)$ is State transition matrix,

$$F(k+1)=\begin{bmatrix} 1 & T & \dfrac{e^{-\alpha T}+\alpha T-1}{\alpha^2} \\ 0 & 1 & \dfrac{1-e^{-\alpha T}}{\alpha} \\ 0 & 0 & e^{-\alpha T} \end{bmatrix}$$

$G(k)$ is input matrix,

$$G(k)=\begin{bmatrix} \dfrac{1}{\alpha}(\dfrac{\alpha T^2}{2}-T+\dfrac{1-e^{-\alpha T}}{\alpha}) \\ T-\dfrac{1-e^{-\alpha T}}{\alpha} \\ 1-e^{-\alpha T} \end{bmatrix}$$

$\bar{\omega}(k)$ is the acceleration mean, $Z(k+1)$ is the measurement vector, $H(k)$ is the measurement matrix, $M(k)$ is the so-called noise, the covariance

$$Q(k)=2\alpha\delta_a^2\begin{bmatrix} q_{11} & q_{12} & q_{13} \\ q_{21} & q_{22} & q_{23} \\ q_{31} & q_{32} & q_{33} \end{bmatrix},$$

$R(k)$ is the covariance of the measurement noise $N(k)$.

$$q_{11}=\frac{[1-e^{-2\alpha T}+2\alpha T-2\alpha^2T^2+\dfrac{2\alpha^3T^3}{3}-4\alpha Te^{-\alpha T}]}{2\alpha^5},$$

$$q_{12}=\frac{[e^{-2\alpha T}+1-2e^{-\alpha T}+2\alpha Te^{-\alpha T}+\dfrac{2\alpha^3T^3}{3}-2\alpha T+\alpha^2T^2]}{2\alpha^4},$$

$$q_{13}=\frac{(1-e^{-2\alpha T}-2\alpha Te^{-\alpha T})}{2\alpha^3},$$

$$q_{22}=\frac{(1+e^{-2\alpha T}-2e^{-\alpha T})}{2\alpha^3},$$

$$q_{33}=\frac{(1-e^{-2\alpha T})}{2\alpha}$$

δ_ω^2 is the acceleration variance, α is the maneuvering frequency, T is the discrete time of the sampling period.

$$\delta_\omega^2=\begin{cases} \dfrac{4-\pi}{\pi}(\omega_{max}-\bar{\omega}_k)^2 & \bar{\omega}_k \ge 0 \\ \dfrac{4-\pi}{\pi}(\omega_{-max}+\bar{\omega}_k)^2 & \bar{\omega}_k < 0 \end{cases} \quad (3)$$

ω_{max} is the maximum maneuvering acceleration, ω_{-max} is the minimum maneuvering acceleration.

Due to the modified Rayleigh distribution , the target model $\frac{4-\pi}{4}\omega_{-max} \le \bar{\omega}_k \le \frac{4-\pi}{4}\omega_{max}$, the acceleration distribution description is not available, thus resulting in poor precision for weak and non-maneuvering tracking. Therefore, it is necessary to re-describe the acceleration variance[23,24]. In the Gaussian distribution , the probability of random variables beyond 3ω is 0.0027, the acceleration mean is zero for the weak and non-maneuvering targets, and then,

$$\left|\frac{4-\pi}{4}\omega_{max}\right| \le 3\delta_\omega \quad (4)$$

$$\delta_\omega^2=\frac{\left[(4-\pi)\omega_{max}/4\right]}{9} \quad (5)$$

thus

$$\delta_\omega^2=\begin{cases} \dfrac{4-\pi}{\pi}(\omega_{max}-\bar{\omega}_k)^2 & \bar{\omega}_k > \dfrac{4-\pi}{4}\omega_{max} \\ \dfrac{4-\pi}{\pi}(\omega_{-max}+\bar{\omega}_k)^2 & \bar{\omega}_k < \dfrac{4-\pi}{4}\omega_{-max} \\ \dfrac{\left[(4-\pi)\omega_{max}/4\right]}{9} & \dfrac{4-\pi}{4}\omega_{-max} \le \bar{\omega}_k \le \dfrac{4-\pi}{4}\omega_{max} \end{cases} \quad (6)$$

4 PF and UKF fusion filter

UKF through Unscented transformation, the calculation sampling point through the non-linear propagation, its posterior statistical value use linear regression to obtain posterior statistics, the Gaussian random variable Taylor expansion to the third order, is a filter precision higher nonlinear filter[25].

PF is a posterior probability density based on Monte Carlo random sampling particle weight and de-approximation states[26]. It is very suitable for nonlinear filtering due to the sample form rather than function form. However, since the particle filter usually uses the state transition density as an important density function, the filter is liable to fail when the new measurement present to the tail of prior distribution or is found to be more sharp with respect to the prior distribution.At the same time, The repeat sampling link, so that the diversity of particles lost, filter performance decreased.

Therefore, it is necessary to generate the importance density function by a suboptimal filtering method. For this reason, UKF is used to generate the importance density function, and the latest measurement information is used in the particle sampling process to improve the sampling particle validity of approximation the distribution[27].

The specific steps of the UKF and PF fusion filter:

(1)Initialization:k=0 , random sampling points are drawn from prior distributions .

$\hbar_0^i, \hbar_0^i \sim p(\hbar_0)(i=1,2,\cdots N)$;

$\omega_0^i = \frac{1}{n}$ is the weight, n is the particle number.

(2) Importance sampling: for each particle $\hbar_{k-1}^i$, the mean $\bar{\hbar}_k^i$ and covariance P_k^i is calculated by UKF, and get the importance density function $N(\bar{\hbar}_k^i, P_k^i)$ [28].

$$q(\hbar_k^i|\hbar_{k-1}^i, z_k) = N(\bar{\hbar}_k^i, P_k^i) \tag{7}$$

(3) Weight update: extract N particles, calculate weights

$$\omega_k^i = \omega_{k-1}^i \frac{p(z_k|\hbar_k^i)p(\hbar_k^i|\hbar_{k-1}^i)}{q(\hbar_k^i|\hbar_{k-1}^i, z_k)} \tag{8}$$

normalizing weights

$$\bar{\omega}_k^i = \omega_k^i \Big/ \sum_{i=1}^{n} \omega_k^i \tag{9}$$

(4) Resampling[29]:

Resampling judgment: when $1 \Big/ \sum_{i=1}^{n} \omega_i^2 > N_o$, nonresampling; otherwise resampling, N_o is the sampling threshold .

(a)Rank n particles by weight, the former n_1 particle were selected and normalized by equation (9).

(b) Calculate the filter density for selected particles

$$p(\hbar_k|z_k) = \sum_{i=1}^{n_1} \omega_k^i \delta(\hbar_k - \hbar_k^i) \tag{10}$$

(c) The weight of the selected n_1 particle are restored to the pre-normalized weight by equation (9).

(d) All n particles are normalized by equation (9).

(5)Estimate calculate:

$$\hat{\hbar}_k = \sum_{i=0}^{n} \omega_k^i \hbar_k^{i*} \tag{11}$$

$$P_k = \sum_{i=0}^{2n} W_i^c [\chi_k - \hat{\hbar}_k][\chi_k - \hat{\hbar}_k]^T \tag{12}$$

W_i^c is the sigma weight, χ_k is the sample point.

(6)Jump calculating

K=k+1,go to step 2.

5 Simulation and Experiment analysis

Aiming at the TV tracking system, the system simulation model is established. transfer function

$$G_1(s) = \frac{37.5(0.01s+1)}{(0.34s^2+0.05s+1)}$$

$$G_2(s)G_3(s) = \frac{37.5}{(0.01s+1)(0.34s^2+0.05s+1)}$$

Based on the origin of the coordinates,if the target model is constant velocity model,the equation (1) and (2) can be showed:

$$X(k+1) = \begin{bmatrix} 1 & T & 0 & 0 \\ 0 & 1 & 0 & 0 \\ 0 & 0 & 1 & T \\ 0 & 0 & & 1 \end{bmatrix} X(k) + \begin{bmatrix} T^2/2 \\ T \\ T^2/2 \\ T \end{bmatrix} M(k)$$

$$Z(k+1) = \begin{bmatrix} \sqrt{(x_k^2+y_k^2)^2} \\ ac\tan\frac{y_k}{x_k} \end{bmatrix} + N(k)$$

$X(k)$ is System state vector,

$X(k) = [x_k \quad v_{x_k} \quad y_k \quad v_{y_k}]^T$,

it show the displacement and velocity components in x and y directions. $M(k)$ is the gaussian white noise, the covariance $Q(k) = \begin{bmatrix} 0.5^2 & 0 \\ 0 & 0.5^2 \end{bmatrix}$. $N(k)$ is also the gaussian white noise, the covariance

$$R(k) = \begin{bmatrix} 20^2 & 0 \\ 0 & 0.05^2 \end{bmatrix}$$

The period of filtering $T = 0.1s$,the initial value $X(0) = [1000 \quad 150 \quad 1000 \quad 30]^T$,the particle number is 100. One hundred Monte Carlo simulation is performed by Intel core i7-5500U computer. The simulation applied the MCS-PF、MCS-UKF and MCS-UPF. Position error curve shown in Figure2,velocity error curve shown in Figure3, single performance period shown in table1.

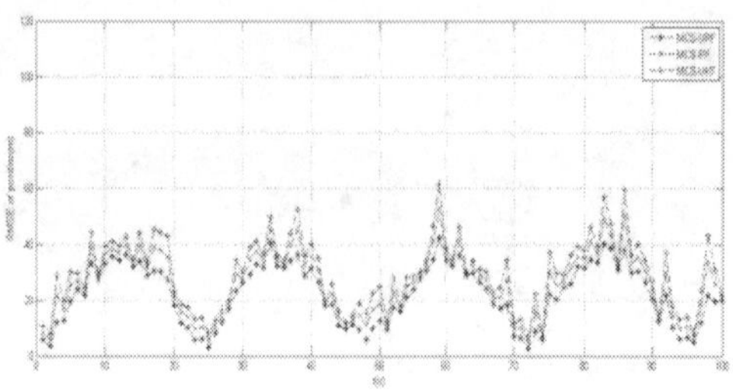

Figure 2. Position error curve figure.

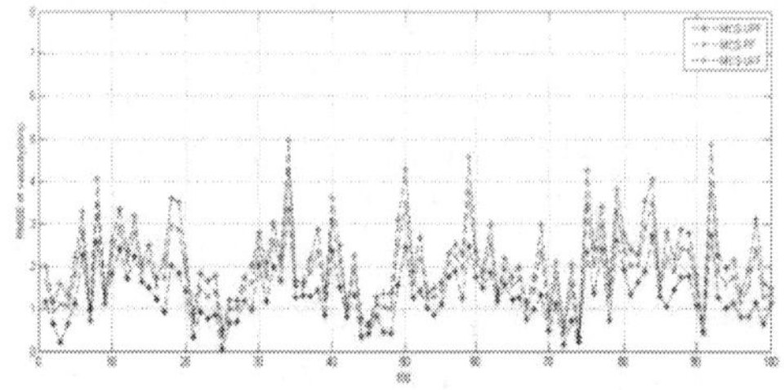

Figure 3. Velocity error curve figure.

Table 1. Time comparision.

Filter method	Single time(s)
MCS-UKF	0.65
MCS-PF	1.43
MCS-UPF	3.19

6 Conclusions

Aiming at the problem of poor adaptability for filtering models and the problem of low filter precision, the paper put forward a filter algorithm based on modified current statistical model for improving the precision of on-axis tracking system.The algorithm enhances the model's adaptive ability to weak and non-maneuvering maneuvering target.The algorithm increases the estimated capacity of PF algorithm by UKF to calculate the importance density function.The algorithm is applied to the on-axis tracking system simulation. It can effectively improve the filtering performance, enhance the tracking precision and stability. The algorithm improves the filtering performance, but increases the computational complexity. The next step study must focus on the problem to reduce calculation and shorten the calculation time.

References

1. Michanil N P. Signal Processing.87,7(2007)
2. O.Cappe, S.Godsill, E.Moulines.Proceedings of IEEE. 95 ,5(2007)
3. Z K Xiong,F Y Li..2017 IEEE 2nd Advanced Information Technology, Electronic and Automation Control Conference(2017)
4. Lu Y,Wu T,Chun Zhu S. Proceedings of the IEEE Conference on Computer Vision and Pattern Recognition(2014)
5. Merwe v.d.R, Douce A.Cambridge University Enginnering Department, England Cambridge University press(2000)
6. J.Simon, K. U. Jeffrey. Proceedings of IEEE.92, 3(2004)
7. H R ZHOU.Acta Aeronautica Et Astronautuca Sinica.4,1(1983)
8. Y Qing, L Ming.2017 International Conference on Smart Grid and Electrical Automation(2017)
9. Z K XIONG, T F CHEN. High Power Laser and Particle Beams.24 ,6 (2012)
10. X Wang, Z Q Hou. IEEE Transactions on Cybernetics, 99(2017)
11. X J FANG, F LIU, Y QIN. Acta Electronica Sinica. 34, 6 (2006)
12. Nadine M. Moacdieh, Nadine Sarter.IEEE Transactions on Human-Machine Systems.47,6(2017)
13. J LIU, Y J WANG, J ZHANG. IEEE Conf on Decision and Control (2014)
14. F H Liu, C Gong, T Zhou,K Fu.IEEE Transactions on Multimedia. 19,12(2017)
15. Y SHI, C Z HAN. Journal of Acta Automatica Sinica. 37(2011)
16. Seon Heo,Hyung Il Koo,Nam Ik Cho. IEEE Transactions on Circuits and Systems for Video Technology.27,11(2017)
17. H X YU, C K FU. Proc. of the 2nd International conference on Signal Processing Systems(2010)
18. T F Wu,Y Lu, S C Zhu. IEEE Transactions on Pattern Analysis and Machine Intelligence.39, 12 (2017)
19. Pavani S K,Delgado D,Frangi A F. Pattern Recognition.43,1(2010)
20. N Wang, S Lv; Meng Joo Er, W H Chen. IEEE Transactions on Intelligent Vehicles. 1, 3(2017)
21. Danelljan M,Hager G,Khan F.Proceedings of the Conference on British Machine Vision Conference(2014)
22. X D Zhao, P Shi. IEEE Transactions on Neural Networks and Learning Systems.27, 9(2016)
23. K Mechrotra, P R Mahapatra. IEEE Trans. On Aerospace and Electronic Systems.33, 4(1997)
24. Lasitha Mekkayil; Hariharan Ramasangu. 2017 IEEE Region 10 Symposium(2017)
25. J G MA, T TANG. Infrared and Laser Engineering.42, 1(2013)
26. Ma C,Huang J B,Yang X. Proceeding of the IEEE International Conference on Computer on Computer Vision(2015)
27. Lin M,Zhang C,Zhang Z. Neurocomputing.159(2015)
28. G G Rigatos. IEEE Transactions on Industrial Electronics.59, 10(2012)
29. Cehovin L,Kristan M, Leonardis A. IEEE transactions on pattern analysis and machine intelligence.35,4(2015)

The Electric Vehicle Time-of-Use Price Optimization Model Considering the Demand Response

Zejing Shi[1,2], Ninghui Zhu[1,2] and Jinsong Yu[1,2]

[1]State Grid Electric Power Research Institute, NARI Group Corporation, 211000,Nanjing,China
[2]China EPRI&Science Technology Co. Ltd., 102200, Beijing, China

Abstract. A large number of electric vehicles connecting to the distribution grids usually introduce significant fluctuations to the grid and the loads. To solve the problem, guiding the users coordinated charging is proposed. Firstly, the uncontrolled charging power prediction models of electric vehicles are established, and the Monte Carlo method is adopted to simulate the power demands of different electric vehicles, and the influences on the load peak-valley ratios and the voltages and losses of the grid are all analyzed. Then the vehicle responses model considering the time-of-use price is analyzed, and the vehicle response ratios are obtained under different time-of-use prices. Finally the multi-objective optimization model is constructed including the minimum peak-valley ratio, maximum consumption satisfaction index and cost satisfaction index. In the procedure, vehicles and the grid are both taken into account. The results indicate the proposed method could guide the users coordinated charging, and the peak shaving and valley filling is also achieved, and the operation of the distribution grid is improved.

1.Introduction

As the increasing development of electric vehicles and storage systems, the electric vehicles will develop continuously and steadily under the policy encouragement and support. Experts predict that in China, the total number of electric vehicles is about 5,000,000, and simultaneously, 12,000 new charging stations and 4,800,000 charging poles will be built in 2020.

The extensive use of electric vehicles is closely related to the planning, constructions and operations of the distribution grid [1-4]. To prevent the grid fluctuations induced by electric vehicles charging, lots of methods are proposed and categorize into directly control method and indirectly guide method. To achieve minimum loss of the grid, reference [5] constructs an optimal model considering the charging power, and iterated revision andTaylor expansion methods are adopted. The minimum load fluctuations and minimum charging costs are both taken into account in [6], and linear weighting method is utilized to normalize the two objective functions. According to the pricing sensitivity model, reference [7] constructs the minimum mean squared error function, and adopts particle swarm optimization method to optimize the time-of-use price. Three kinds of electric vehicles are taken into account in [8], and the charging models are established, and the optimization peak and valley time price project is derived. Reference [9] builds optimization models to minimize the peak-valley ratio, and also constructs the starting moment model of charging considering time-of-use prices, and finally uses the genetic algorithmfor optimization.

Taking private electric vehicles as examples, this paper simulates the influences to the gird during different power electric vehicles charging. Then the vehicle responses to the time-of-use price are analyzed, and the multi-objective optimization model is established depending on the vehicles and the grid. Finally, NSGA-((Non-Dominated Sorting in Genetic Algorithms) II is adopted to solve the model equation.

2. The fluctuations to the grid induced by electric vehicles uncontrolled charging

2.1 The simulation of demanded power during the electric vehiclescharging

The electric vehicle E150 of Baic Motor is considered here, and the parameters are shown in Table 1.

Table 1. The simulation parameters of electric vehicle

Battery Capacity/Kwh	Charging Power /Kw	Maximum daily mileage/Km
25.6	3.5	130

(1) The charging time and daily mileage

According to the statistical data of domestic vehicles by US Department of Transportationin 2001, reference [10]derives the daily mileage and returning time of electric vehiclesaccord withthe lognormal distribution and normal distribution respectively through data fitting.

The electric vehicledaily mileaged（Km）comply with lognormal distributionμ_d of σ_d and, the probability density functionis:

$$f_d = \frac{1}{d\sigma_d\sqrt{2\pi}}\exp\left[-\frac{(\ln d-\mu_d)^2}{2\sigma_d^2}\right] \quad (1)$$

where $\mu_{t_s} = 19.5$; $\sigma_{t_s} = 0.5$

The charging time t_s (Assuming returning time of electric vehicles is the charging home) satisfies the normal distribution, and the function is:

$$f_{t_s} = \frac{1}{\sigma_{t_s}\sqrt{2\pi}}\exp\left[-\frac{(t_s-\mu_{t_s})^2}{2\sigma_{t_s}^2}\right] \quad (2)$$

where $\mu_{t_s} = 19.5$;

(2)The charging power of electric vehicles

Monte Carlo method is utilized to derive daily mileagedepending on (1), and consequently, the initial SOC (state of charge) of electric vehicle before charging is also deduced according to (3)

$$SOC = 1-\frac{d}{d_m} \quad (3)$$

According to the initial SOC of electric vehicle,the required time for the electric vehicle fully charged is calculated as:

$$t_c = \frac{(1-SOC)\times C_n}{P_n} \quad (4)$$

Then the charging power of a single electric vehicle is shown in (5)

$$P_c = \begin{cases} P_n & t_s \le t \le t_s + t_c \\ 0 & \text{others} \end{cases} \quad (5)$$

(3)The charging power simulation of multiple electricvehicles.

Penetration ratiois defined asthe ratio ofthe charging loads and in distribution grid. The total power of active loads in 33-node distribution network is 3715 kW, if the penetration coefficientis 20%, and consequently, the number of electric vehicles is 212. The simulation flow chart according to Monte Carlo method is shown in Figure1.

2.2 The influence analysis on the distribution grid during electric vehicle charging

By simulations, the loads curves of distribution grid when different power electric vehicles are connected are shown in Figure2.

Figure2 shows as the adding of the electric vehiclepentration ratios, the maximum load increases simultaneously. At the same time, thepeak-valley ratios also increase, resulting inlow efficiency of grid equipment and large grid peaking pressure.

IEEE33-nodesdistribution network is taken as an example here, and the influences on the node voltages and grid losses are both analyzed when large power electric vehicles are connected. The voltage base value is 10kV, and thepower base value is10MVA.Assuming the charging loads is evenly distributed.

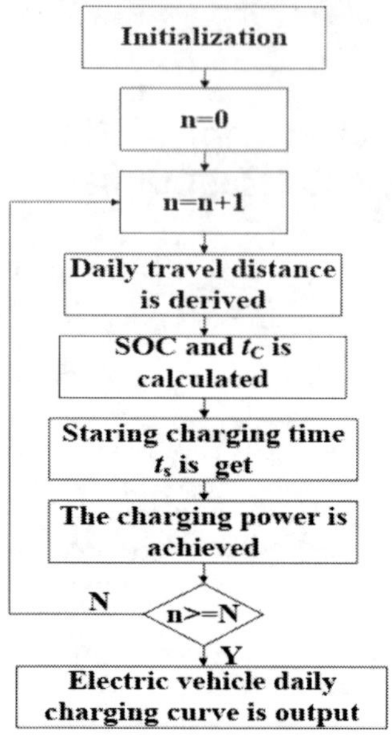

Figure 1. The simulation flowchart by Monte Carlo method.

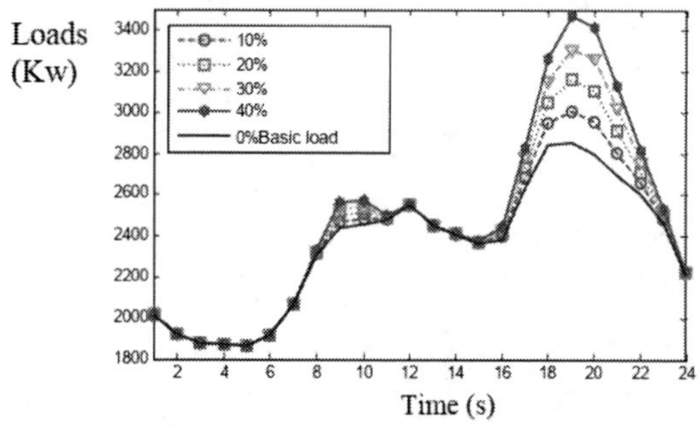

Figure 2. The loads curves of distribution grid with different penetration ratios.

Network loss rate is defined as the ratio of daily network loss and the total load power, and is shown as follows.

$$R_{pl} = \frac{\sum_{T=1}^{24} P_{loss}^T}{\sum_{T=1}^{24}\left(P_{l+ev}^T + P_{loss}^T\right)} \times 100\% \quad (6)$$

where R_{pl} is the network loss rate, and P_{loss}^T is the power losses of all networksat the moment T , and $P_{l+ev}^T + P_{loss}^T$ is the active powerreceived from the source node at the moment*T*.

Figure3 indicates curves of the voltage per unit value of distribution nodes at 19:00. When the penetration ratios are 0% and 10%,the voltage amplitudes of all nodes are between 0.93 and 1.05, and consequently, the voltage deviation is allowable. When the penetration ratio is 40%, the voltage amplitudes of nodes 12-18 and nodes 30-33are all lower than0.93, transcending the limit.

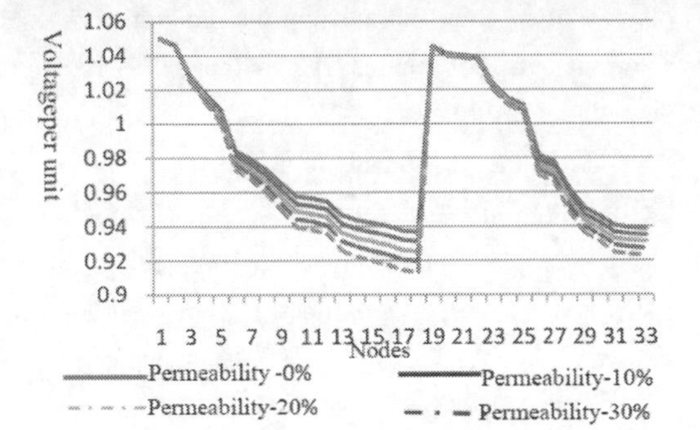

Figure 3. The node voltage curves with different penetration ratios.

Figure 4 shows the grid losses underdifferent penetration ratios. As the increasing of the electric vehicle penetration ratios, the grid losses also increase. The added peak-valley ratios essentially induce large grid losses, which could be derived by Cauchy-Schwarz inequality [11].

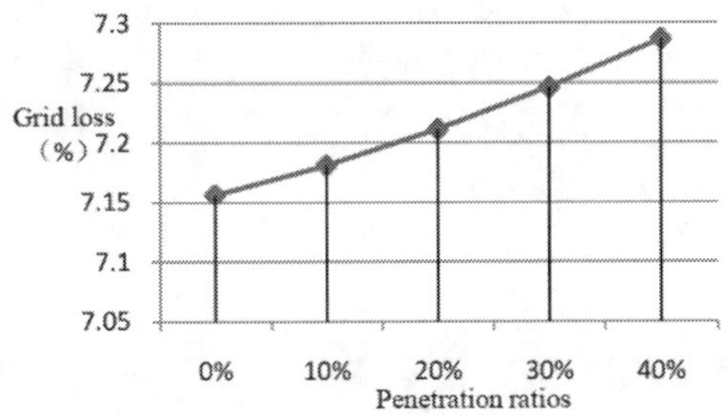

Figure 4. Grid losses with different penetration ratios.

3The vehicle responses analysis considering the time-of-use price

3.1 The response characteristic curve

According to consumer psychology, there is a lower threshold in the users consumer stimulus induced by prices, and in this dead zone, the user responses are zero. There is also a saturation value in the userconsumer stimulus, named as saturation zone. Between the dead zone and the saturation zone, there is a linear response zone, and in this area, the relationship between user responses and stimulus is linear. The peak-valley price ratio is defined as stimulus of the response curves, shown in (7), and the response curve is shown in Figure 5.

$$\beta = \frac{P_p - P_v}{P_0} \quad (7)$$

where$p_v$is the valley electricity price, and $P_p$is the peak electricity price,andP_0 is the initial electricity price.

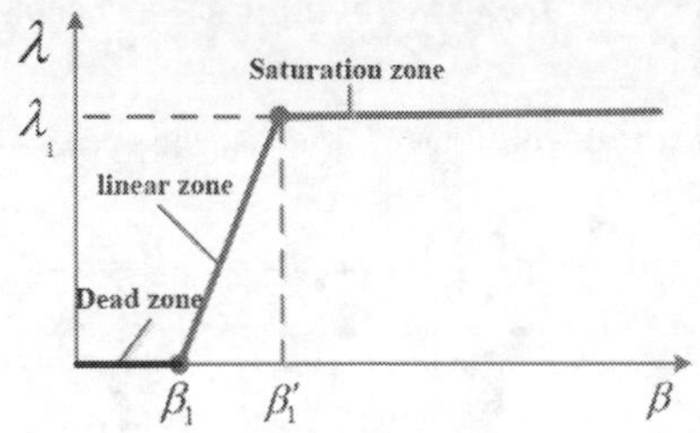

Figure 5. The response curve.

3.2 The charging time selection of electric vehicles under time-of-use price

Under time-of-use price, the charging behavior of no response users is not affected, while the response users would make changes to lower the costs. Assuming all the response users would choose a rational charging method, the start time of charging is shown as [10]

$$t_s = \begin{cases} t_1 + rand(t_1 + \Delta t - t_c) & 0 \le t_c < \Delta t \\ t_1 & \end{cases} \quad (8)$$

wheret_s is the start time of charging, and $t_c$is the charging duration time, and t_1 is the start time of valley electricity price, while $\Delta t$is the duration time of valley electricity price.

4 The optimal charging model of electric vehicles based on vehicles satisfaction index

The user satisfaction should be fully considered during setting time-of-use electricity prices, and it directly determines the success of the time-of-use prices. If the peak and valley price difference is too large, the charging method of users would change greatly, resulting in hypercorrection. And the great changes may lead to user resistances. On the other side, if the price difference is too little, fewer users would respond and change charging method. Consequently, when setting time-of-use electricity prices, both of users and the grid are supposed to be considered and the equilibrium point is achieved. Depending on the users and the grid, this paper constructs a multi-objective optimization model including the minimum peak-valley ratio, maximum consumption satisfaction index.

4.1The objective function

(1) The equation to minimize the load peak-valley ratio is shown as follows:

$$\min LPV(t_s)=1-\frac{\min\left[P_{l+ev}(t_s,t)\right]}{\max\left[P_{l+ev}(t_s,t)\right]} \quad (9)$$

where LPV is the load peak-valley ratio, and P_{l+ev}represents the power of all loads containing electric vehicles.

(2) Maximize consumption satisfaction index of electric vehicles

Consumption satisfaction indexmentioned here is related to the load curves coefficients before and after the implementation of time-of-use electricity price, and is shown as:

$$\max \varepsilon=\frac{1}{2}\frac{\sum\left|(P'_{evi}-\bar{P}'_{evi})(P_{evi}-\bar{P}_{evi})\right|}{\sqrt{(P'_{evi}-\bar{P}'_{evi})^2}\sqrt{(P_{evi}-\bar{P}_{evi})^2}} \quad (10)$$

where P_{evi}、$\bar{P}_{ev}$ are the powers of charging loads and average loads at the moment i before time-of-use electricity prices, and P'_{evi}、$\bar{P}'_{ev}$ are the powers of charging loads and average loads at the moment i after time-of-use electricity prices.

(3) Maximize charging costsatisfaction index

To achieve maximal charging costsatisfaction index, the following is satisfied:

$$\theta=\begin{cases}0 & C>C_p \\ \dfrac{C_p-C}{C_p-C_v} & C_v<C<C_p \\ 1 & C<C_v\end{cases} \quad (11)$$

where $C$is the total charging cost, andC_p、$C_v$are the charging costs when all users choose the peak-price and when they all choose valley-price, respectively.

4.2 Constraints

The constraints include the valley beginning time constraint, the valley time duration constraint, peak electricity price constraint, and valley electricity price constraint. These constraints are satisfied:

$$st\begin{cases}t_{s\min}<t_s<t_{s\max} \\ \Delta t_{\min}<\Delta t<\Delta t_{\max} \\ P_{p\min}<P_p<P_{p\max} \\ P_{v\min}<P_v<P_{v\max}\end{cases} \quad (12)$$

5 Solving models

5.1NSGA-Ⅱ algorithm and the procedure

The main parts of NSGA-Ⅱalgorithm are fast non-dominated sorting and calculating crowding level value [12].

In the fast non-dominated sorting method, each individual has two main parameters:$n_i$andS_i.$n_i$is the number of solutions dominating the individual$i$. S_i is the solutions set dominated by the individual$i$. The sorting procedure is as follows:

(a) Find the individuals satisfying $n_i=0$, and put them into the first level set F_1.

(b) Find the individual set S_j dominated by each individual j in F_1, and subtract 1 from the n_k of every individual k , if satisfy $n_k-1=0$, putthe individual k into F_2.

(c) Set F_1 asthe first non-dominated individuals set, and all the individuals in F_1 are the optimal individuals, and these individualsare not dominated. Each individual isgiven to a non-dominated sequence i_{rank} .

(d) Repeatstep (b) and step (c)in F_2 , until the individuals are all graded and are alsogiven to thenon-dominated sequences i_{rank} .

Crowding levelis used to evaluate the density of other individuals around anindividual, and it could keep the populationdiverse, and prevent partial accumulation problem. The schematic diagram of the crowding level is shown in Figure 6.

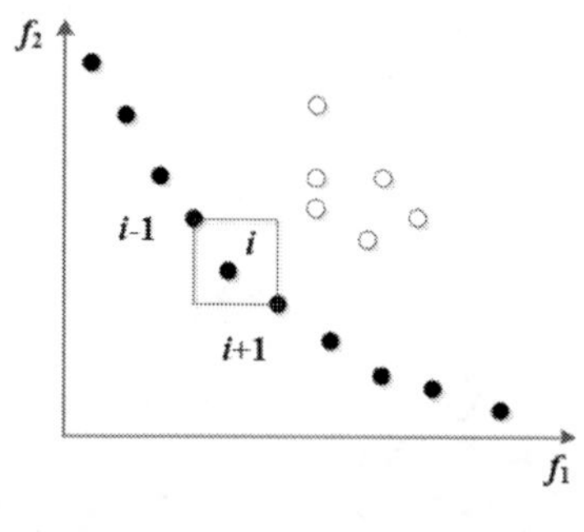

Figure 6. The schematic diagram of the crowding level.

The crowding level value of individual is calculated as

$$i_d=\frac{\sum_{j=1}^{m}\left(\left|f_j^{i+1}-f_j^{i-1}\right|\right)}{f_j^{\max}-f_j^{\min}} \quad (13)$$

where f_j^{i+1}、$f_j^{i-1}$are the j objective function values of individual $i+1$ and $i-1$, respectively, and m is the number of objective functions.The crowding level values of the boundary individualsare defined as infinity.

The main flowchart of NSGA-Ⅱ algorithm is shown in Figure 7.

The initial population P_t of size N is randomly generated, and a new offspring population Q_t is produced by genetic manipulation. The initial population and offspring population are combined together and achieve a new population $R_t = P_t \cup Q_t$. Thenon-dominated sorting is performed in R_t, and the crowding levels of individuals in F_i arecalculated. Then the best individual in R_t isselected.All individuals in F_1 are put into a new parent population P_{t+1} , if the number of individuals innew population P_{t+1} is less than N, then the individuals of next layer are placed into it. Otherwise, if the number of individuals inpopulation P_{t+1} is greater than N, the individuals choosing is performedaccording to the individual crowing level to ensure the number of individuals is N. Genetic manipulationproduces new offspring population Q_{t+1} from the new parent population P_{t+1} .The procedure is repeated until the algorithmsatisfies certain constraints.

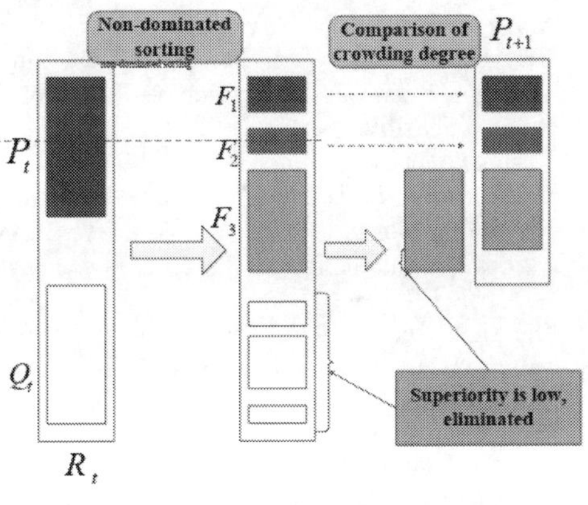

Figure 7. The main flowchart of NSGA-Ⅱ algorithm.

5.2 The optimal compromise solution

The multi-objective solution is not an optimal solution, but a non-dominated solution set. The optimal compromise solution is needed to be chosen from this set. Partial membership function is used to achieve the optimal compromise solution, and shown as [9]

$$u_i = \begin{cases} 1 & f_i < f_i^{\min} \\ \dfrac{f_i^{\max} - f_i}{f_i^{\max} - f_i^{\min}} & f_i^{\min} \leq f_i < f_i^{\max} \\ 0 & f_i > f_i^{\max} \end{cases} \quad (14)$$

where u_i is the satisfaction index of the individual objective function i , and $f_i^{\min}$ 、 $f_i^{\max}$ are the minimum value and maximum value of the objective function i , respectively.

The individual satisfactionindex in Pareto solution set is calculated as (15). The individual with maximum u is selectedas the optimal compromise solution.

$$u = \frac{1}{m}\sum_{i=1}^{m} u_i \quad (15)$$

5.3 Examples

Derived Pareto optimal solution set is shown in Figure 8.The content of Figure 8 indicates the Pareto optimal solution set has excellent diversity, and is evenly distributed. Consequently, the optimal model proposed by the algorithm has a strong global search capability.The derived optimal compromise solutionsare shown in Table 2.

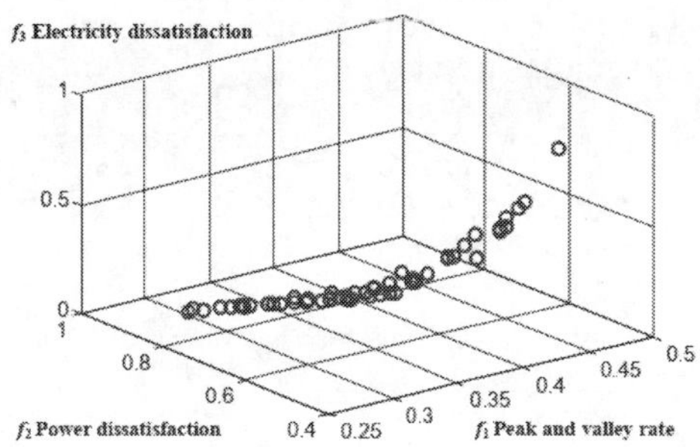

Figure 8. Paretooptimal solutionset.

Table 2. The optimal compromise solutions

t_s	Δt (hour)	v (Yuan)	P_v (Yuan)
	11	1.0449	0.5056

The derived objective function values of optimal compromise solutions are in Table 3. The optimal compromise solution tends to decreasethe peak-valley ratios of the grid and gets low consumption satisfaction index, but the costsatisfaction index is very high, and consequently, the user responseratiois also high.

Figure 9 shows the load power comparison curves under two charging modes, and it indicates the optimized charging loads play an important role in the peak clipping and valley filling of distribution grid.

Figure 10 gives the node voltage per unit valuesunder two charging modes. After optimized charging, the quality of voltage is improved, and the voltages of nodes 17 and 18 are both higher than 0.93, which satisfies the voltage requirement of 10kV grid.

Table 3. The objective function values of optimal compromise solutions

Response ratio (%)	Peak-valley ratio (%)	Consumption satisfaction index	Cost satisfaction index
76.88	33.76	0.6806	0.7689

6 Conclusion

Firstly, this paper analyzes the fluctuations to the grid during different electric vehicles charging. Then depending on the vehicle responses, the minimum peak-valley ratio and maximum satisfaction index optimization model is constructed considering both of the gird and vehicles. Finally, NSGA-II algorithm is utilized to solve the model equation. The analysis indicates:

1. In the optimized charging mode, the electric vehicles play an important rolein peak clipping and valley filling procedure, and also could improve efficiency of grid equipment.
2. In the optimized charging mode, the quality of the grid voltage is improved significantly, and losses of the grid are decreased.

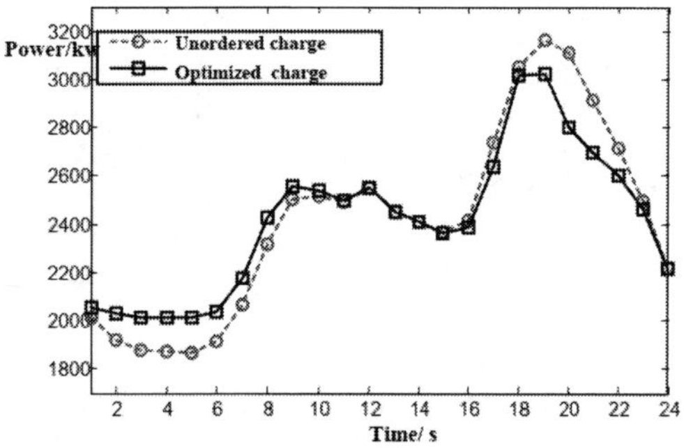

Figure 9. The load power comparison curves under two charging modes.

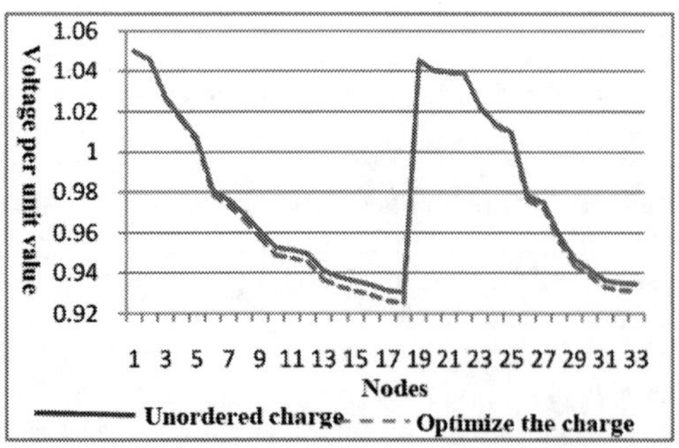

Figure 10. The node voltage per unit valueunder two charging modes.

The vehicle response ratio is only affected by time-of-use price in this paper, but unfortunately, in practical applications, many factors influence the vehicle responses, such as financial conditions and charging implementations. The accuracy of the response curves needs to be improved in the future.

Reference

1. HU Zechun, SONG Yonghua. Impacts and Utilization of Electric Vehicles Integration Into Power Systems[J]. Proceeding of the CSEE. 2012, 32(4); 1-10
2. Taylor M J, Alexander A. Evaluation of the impact of plug-in electric vehicle loading on distribution system operation [C] //IEEE power &Energy Society General Meeting, Calgary, Canada, 2009: 1-6.
3. Kejun Q, Chengke Z, Allan M, et al. Modeling of load demand due to EV battery charging in distribution systems[J] . IEEE Transactions on Power Systems, 26(2): 802-810.
4. Kristien CN ,EH, johan D. The Impact of Charging Plug-In Hybrid Electric Vehicles on a Residential Distribution Grid[J]. IEEE Trans on power systems, 2010（255）:371-380.
5. ZHAN Kaiqiao. SONG Yonghua, Coordination of Electric Vehicle Charging to Minimize Active Power Losses[J], Proceedings of the CSEE, 2012, 32(31): 11-18
6. NI Feng, ZHAO Ming-yu, Research on ordered charging of EV [J], Power Demand Side Management, 2015, 17(5): 16-19
7. LUO Min, ZHAO Wei, Research on Coordinated Charging Price of Electric Vehicle Based on Users' Price Sensitivity [J], Electrical & Energy Management Technology, 2015: 78-92.
8. GAO Yajing, LYU Mengkuo, Power Demand Price Elasticity Matrix Based on Discrete Attraction Model[J], Automation of Electric Power Systems, 2014, 38(13):103-107.
9. GE Shao-yun, HUANG Liu, Optimization of peak-valley TOU power price time-period in ordered charging mode of electric vehicle[J]. Power System Protection and Control , 2012, 40(10):1-5.
10. GONG Xin, LIN Tao. Impact of Plug-in Hybrid Electric Vehicle Charging on Power Distribution Network [J]. Power System Technology. 2012, 36(11): 30-35
11. CHEN Jia-sheng , ZHANG Jian-hua, Strategies for Electric Vehicle Charging with Aiming at Rwducing Network Losses[J]. Proceedings of the CSU-EPSA, 2012,24（3）:139-144.
12. GAO Yuan, Non-dominated Sorting Genetic Algorithm and Its Applications [D]. Zhejiang University, 2006

Designing and Modeling of Automated Anti-theft Electricity Distribution System

Awais Khan[1], Wei Xie[1], Suhail Aftab Qureshi[2], Muhammad Ilyas[3], JingChun Lin[4] and Guidong Liu[4]

[1] *College of Automation Science and Technology, South China University of Technology, Guangzhou 510641, China*
[2]*Electrical Engineering Department, UET Lahore, Pakistan*
[3]*Iqra International University Peshawar, Pakistan*
[4] *College of Electrical and Information Engineering, GuangDong Baiyun University, Guangzhou 51045, China*

Abstract. In this modern era electricity is the major component of human life. Its play a vital role to facilitate the human life by operating all modern electrical equipment. The consumption of electricity increases rapidly due to its uses in the domestic and industrial sector. In the distribution of electricity, there are few emerging issues leads to huge losses. The illegal connection of electricity in remote areas increases rapidly. Such theft of electricity from distribution line leads to huge losses in revenue of electricity. Pakistan facing huge energy crisis due to a shortage of electricity generation. This research project modeled and the designed the hardware prototype of automated anti-theft electricity distribution system. The proposed system detects the illegal load and burns it by sending high voltage signal from capacitor bank. The legal load is made safe and uninterrupted during execution of illegal load.

1 Introduction

Distribution of electricity is the major responsibility of the National Transmission and Dispatch (NTDC) company limited.As the length of the transmission line increases from generation plant causes an increase in line losses and another issue. Theft of electricity from distribution line in form illegal connection directly from the transmission line is the major issue in the collection of revenue. It is quite difficult for government institutions to check twenty-four the user connection to ensure 100% revenue. Such issue can be handled with the help technology development based on automation industry. It is a big challenge for NTDC to detect and handle such illegal connection from distribution line (DL). Each year, the electricity distribution companies face 20% to 25% line losses. According to WAPDA survey, these losses are round about 125 billion a year. These losses can destroy the power system of a country. Due to these losses, more than 20% consumers pay extra money.
There are two kinds of losses occurs in these systems.

1. Technical
2. Non-Technical

Losses due to structure and characteristics of the network are called Technical losses, while losses due to the stealing of electricity are called Non-technical losses. If someone is using electricity illegally, it is a silent crime [1]. Due to these losses government give a big relief to these companies [2], [3].

Generally, power system consists of three parts (1) Generation (2) Transmission (3) Distribution which is different from each other by their voltage values. Generally, the generation phase and transmission phase is far away from the city. So, the people cannot access them easily. The next phase that is the distribution phase is inside the city which is easily accessible to the people and within this phase, the main problem starts. The voltage level is also low that can be used directly.

To use electricity illegally there are many ways like (a) To decrease speed of meter by using hard metallic disks (b) Using large magnet disturbing electromagnetic field of electric meter (c) Taking connection directly from the distribution lines without using electric meter (d) By switching the electricity cables at meter connector box [4]. Our proposed research project is handling the direct contact from distribution line. It will detect and control the theft load by burning it. The controller portion can easily differentiate the legal and illegal or theft load. To handle the theft of electricity by tapping the meter in digital meter microprocessor will be used to record the value of their unit. Communication block is used to synchronize and analyze the values of power used at any instant. The stored values of power will be transmitted to the distributor. In detection block, if there will any mismatch between input power and output power microprocessor will detect it. If there is any illegal connection, then detection block will tell the reporting block to transmit the information to the respective authority through GSM module [4-5].

The energy billing system in Pakistan and all over the world can be changed using this system. This system of energy billing may help the energy distribution companies to minimize costs and increase

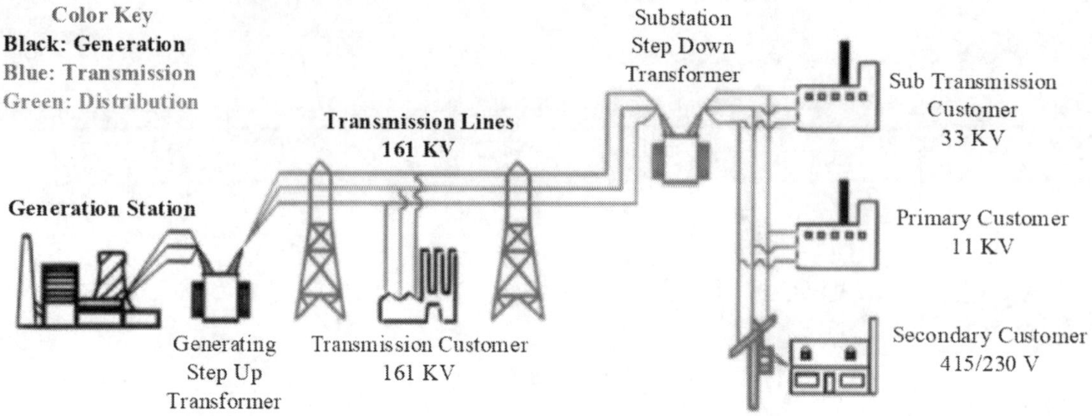

Figure 1. A general model of Electricity generation, transmission, and distribution system.

profits, to increase metering and billing accuracy and efficiency, and to contribute the energy in a sustainable way. Moreover, theft will also be reduced and theft load will burn as a penalty. The final advantage that makes this system superior then others is that: this system didn't use any communication module for theft detection so the system uses a common controller for comparing the power of consumer and distribution unit.

The proposed method is distributed in three stages.

1. Transmission phase
2. Distribution phase
3. User phase

At distribution phase, the power is stepped down by step down transformer and then supplied to the user. At this stage, adjustable rating circuit breakers will be used on connections of every user, due to the installation of these circuit breakers no user will exceed his required limit of load, if any user will cross the required limit the circuit breaker will be tripped and will be disconnected from supply. At user stage where conventional meters are used, will be replaced by smart meters having features of two-way communication system [6].

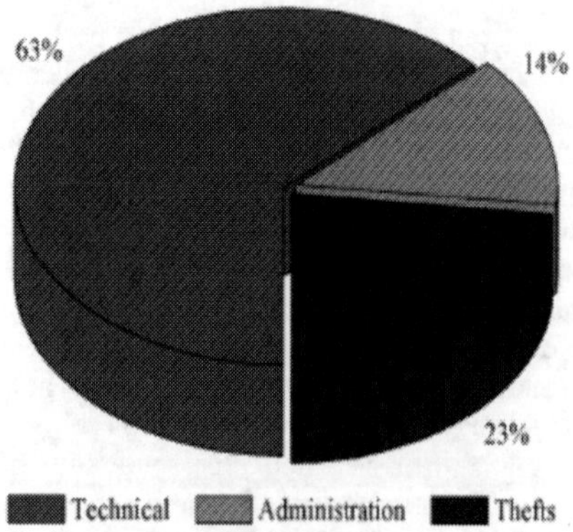

Figure 2. Estimated sources of grid losses.

The above figure represents that there are 23% theft losses amongst 100% lines losses. Technical losses are due to the design of the distribution network, the length of cables, transformers and loads. Technical losses are about 63% of total losses. Administration losses are due to by giving unknown connections or missing meters not illegal use of electricity. The administration losses are about 14% of total losses [6], [7], [8].

The goal of the project is to make a system which automatically detects and control illegal connection of electricity, the proposed project is about such automated system which will differentiate between actual and theft load and will control the theft of electricity by burning the theft load.

1. To make an algorithm that will give instruction to the microcontroller to monitor and control the theft load.
2. To design a microcontroller based system that will monitor and control actual and theft.
3. To implement a prototype of a smart WAPDA system.

This project includes building a hardware for the system, interfacing the hardware i.e. interfacing the microcontroller with relay driver, and interfacing different components like relays, energy meters, and capacitor bank.

This project will be very suitable for WAPDA because theft losses are one of the major problems of WAPDA due to this project WAPDA will overcome such losses and this project will also burn the theft loads as a penalty.

The proposed research project is organized as Section 3 defines the proposed scheme and model, Section 4 explains the software implementation. Section 5 is based on results and conclusion of the proposed scheme. Finally, Section presents the future recommendation.

2 Existing methods

In [9] G. L. Prashanthi, K. V. Prasad had proposed with GSM modules helps the company to monitor the amount of usage by this specified customer and generate bill periodically and send it to the customer via SMS, thus saving a lot of labor work, time and cost of reading. The proposed found to be a little bit complex as far as a distribution network is concerned, but it's an automated system of theft detection. It saves time as well as help to maximize profit margin for utility company working in electrical distribution network while in [10] S.Anusha, M.Madhavi, R.Hemalatha had done the project model to reduces the manual manipulation work and theft. Use of GSM in our system provides numerous advantages of the wireless network system. The government saves money by the control of theft in energy meter and also more beneficial for customer side and the government side. The metering IC ensures the accurate and reliable measurement of power consumed. Cost wise low when compared to another energy meter without automatic meter reading and theft control. In the paper [11] Kalaivani. R, Gowthami. M, Savitha.S, Mohanvel.S had done the system, in which service provider can collect the bill any time with a single message. The data collection and manipulation task become fast and easier. Any modification can be made to the code in less time. Changes in rate or unit calculation can be done very effectively. The project model reduces the manual manipulation work and theft. Use of GSM in our system provides the numerous advantages of wireless network systems. The metering IC ensure the accurate and reliable measurement of power consumed. Hence we are trying to manipulate cost wise low when compared to another energy meter without automatic meter reading and theft control. In some of the existing methods wireless communication system of energy meter used with Zigbee [12-13], relay control and GPRS [14-15]. The cryptographic method is used to secure the communication channel [16-17]and Zigbee for the transmission of data in a serial process [18-19].

3 Proposed scheme

The main goal of this paper is to design such smart system which will detect and prevent electricity theft. So for this purpose, some smart microprocessor based modifications will be used to locate the presence of electricity theft. The figure below shows the methodology of the proposed solution.

This goal can be obtained by comparing power supplied to the consumer and power used by the consumer. In measurement block flow of power will be calculated, in case of digital meter it will directly fetch the numeric value but in case of an analog meter, some extra infrared transmitter and receiver circuitry will be used to detect the revolution of the disk. So for continuous monitoring of power microprocessor will be used to record the values. Communication block is used to synchronize and analyze the values of power used at any instant. The stored values of power will be transmitted to the distributor. In detection block, if there will any mismatch between input power and output power microprocessor will detect it and there must be an illegal connection. If there is any illegal connection detection block will tell the reporting block which will transmit the information to the respective authority through GSM module. Now it is the responsibility of respective authorities to take action [5].

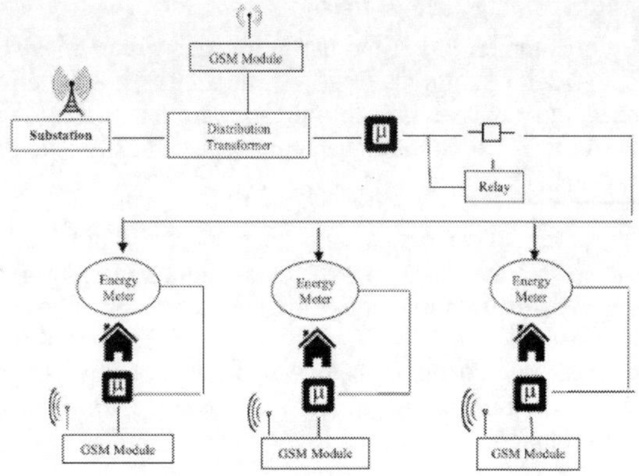

Figure 3. A birds eye view of methodology.

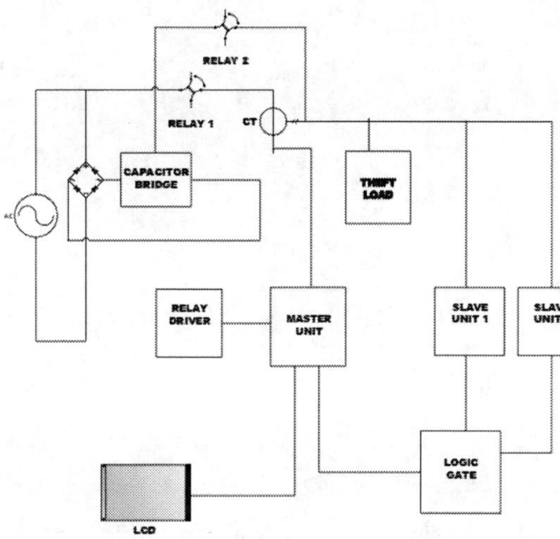

Figure 4. Block diagram of automated anti-theft electricity distribution system.

According to the block diagram shown below, the 220v AC supply is connected to relay 1 and bridge rectifier. Relay 1 is further connected to current transformer and slave units as well as bridge rectifier relates to capacitor bank having the ability to charge up to 600v DC. This circuitry is further connected with relay 2. Slave units relate to logic gates and it connected with the master unit. Master unit relates to LCD and relay driver. When there is AC supply through relay 1 slave units (meters) will be ON and bridge rectifier will convert AC supply to DC and charge capacitor bank. Capacitor bank relates to relay 2 that means relay 2 is always connected with high voltage DC, during normal operation when there is no theft load relay 2 is OFF and relay 1 is ON, whenever there is theft load relay 1 will be OFF and relay 2 will be turned ON and it will burn the theft load because it is high voltage DC. All these decisions are made by a master unit which collects data from current transformer and slave units whenever input of current transformer and output of slave units is equal circuits operates normally and whenever data is not equal it sends signals to relays with the help of relay driver and the circuit operates accordingly.

4 Software implementation

Before implementing the hardware, the circuit must be put under test, for the sake of simplicity, so that error checking and testing of the programming can be performed. Here Proteus version 8 is used to simulate the system.

Design of slave unit 1 and slave unit 2 that are energy meters are designed in Proteus version 8. Vsm signal generator is used for voltage and frequency source, as we increase the level of Vsm generator value of current changes as shown in below figure, the below figure shows the function of energy meter of calculating power consumption.

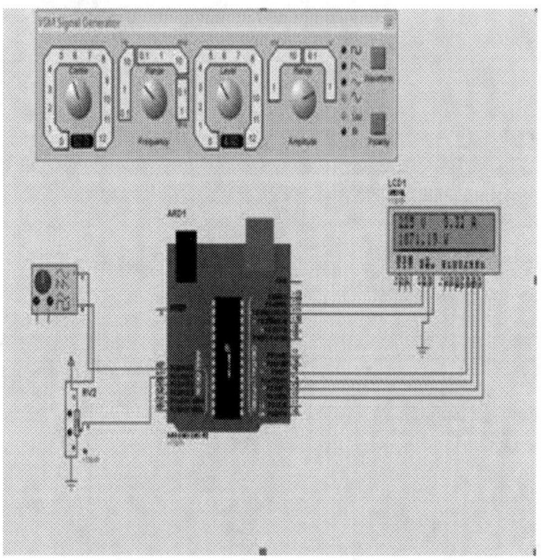

Figure 5(a). Simulation result of slave units.

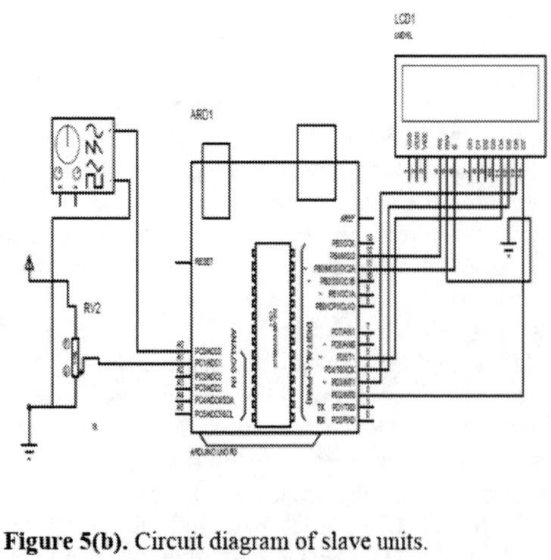

Figure 5(b). Circuit diagram of slave units.

For measuring the AC current, a current transformer (C.T) is used, where potential transformer (P.T) is used to measure the voltage of AC system .while C.T and P.T signals are sinusoidal so it will be converted to DC using bridge rectifier [8]. These signals will be transferred to the controller for displaying the information on LCD. All the information will be displayed on LCD. The figure 4 shown below indicates the LCD information. Here 3 potentiometers are used to check whether the system is running ok or not. Are the voltage and current value changes according to the change in resistance or not? After simulating the circuit, it was found that coding of the system and hardware is accurate according to demand. All the things are working ok, where the controller is sensing the theft and informing the controller which take necessary action i.e. turn off the official load and sending high voltage to the theft load, which basically damages the appliance.

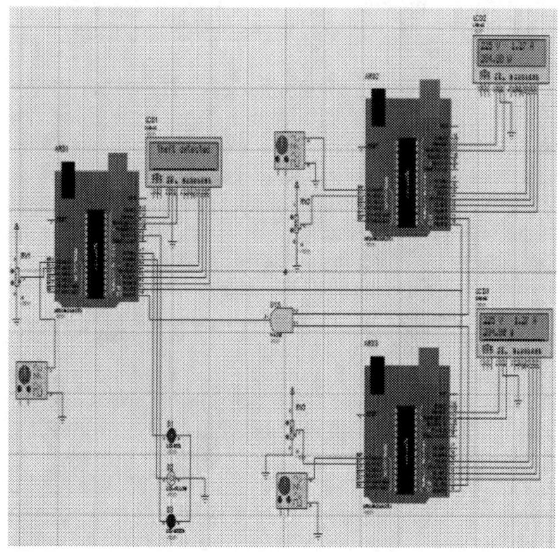

Figure 6. Simulation of the automated anti-theft electricity distribution system.

5 Results and conclusion

In this paper a verified idea for measuring the electrical power utilization of a load for two wire (phase and neutral) systems with the projected energy meter as a substitute to the conventional meters. This controller-based power meter has been applied (prototype) to deliver quantity up to 200W load from a 220 to 260 V line to neutral voltage. The planned energy meter can calculate power consumption for all loads situations. It doesn't own any rotating parts, which is a striking feature for the efficacies. The proposed meter includes a no-load threshold feature that will remove any tiptoeing effects in the energy meter. In addition, the process of reading the power consumption is smoothed by the LCD display that is very much closer than that for the analog power meters which decrease human errors while noting down the meter reading. Moreover, the system will be used to minimize the theft of electricity by sending HVDC on the theft load.

Automation in electricity distribution system can leads to smart distribution system with application of modern control strategy. The proposed project introduces a remarkable idea for controlling the theft load losses in electricity distribution system. The legal load is made safe and uninterrupted. The achieved results of hardware design can be further improved by determining the exact position of theft load with the application of advanced communication technologies.

6 Future recommendations

As our system have some limitations which can be overcome in future. This system is designed for two wireline system. A system that can be used for a four wire line system can be considered in future.

7 Acknowledge

This work was supported by Natural Science Foundation of Guangdong Province(2017A030313385), Science and Technology Project of Guangdong Province (2016B090927007,2017A040403064; 2017B090901040, 2017A040405023).

References

1. Depuru, S. S. S. R., Wang, L., Devabhaktuni, V., & Gudi, N. (2010, September). *Measures and setbacks for controlling electricity theft*, NAPS, *2010* (pp. 1-8). IEEE.
2. Mogilenko, A. V., and D. A. Pavlyuchenko. "*Expert System for Efficiency Comparison of Electric Network Companies, IJTPE, Transaction on Power Engineering* 1 (2009): 1-4.
3. Ramos, Caio CO, et al. *"Fast non-technical losses identification through optimum-path forest,* ISAP'09. 15th International Conference on. IEEE, 2009.
4. Aslam, Farhan, et al. *"Intelligent Modeling Scheme for Detection of Line Losses in Power Distribution System."* Computer Modelling and Simulation (UKSim), 2013 UKSim 15th International Conference on. IEEE, 2013.
5. Mehran University Research Journal of Engineering and Technology, Volume **32**, No. 3, July 2013 [SSN 0251-7821].
6. Kadurek, Petr, et al. *"Theft detection and smart metering practices and expectations in the Netherlands."* 2010 IEEE PES, ISGT Europe. IEEE, 2010.
7. Thomas B. Smith Department of Social and Behavioral Sciences, Zayed University, P.O. Box 19282, Dubai.
8. Dr. Mohd Yunus B Nayan1, Aryo Handoko Primicanta 2009. *Hybrid System Automatic Meter Reading.* ICCTD -09, pp : 264-267.
9. G. L. Prashanthi, K. V. Prasad *"Wireless Power Meter Monitoring with Power Theft Detection and* Intimation *system using GSM and Zigbee Networks"* IOSR Journal of Electronics and Communication Engineering (IOSRJECE), Volume **9**, Issue 6, Ver. I (NOV.-DEC. 2014).
10. S.Anusha, M.Madhavi, R.Hemalatha *"Detection of Power Theft using GSM"* IJARTET Volume **1**, Issue 3, Nov 2014.
11. Kalaivani. R, Gowthami. M, Savitha.S, Mohanvel.S *"GSM Based Electricity Theft Identification in Distribution System"* IJETT – Volume **8** Number 10 – Feb 2014.
12. Abhinandan Jain, Dilip Kumar, Jyoti Kedia, *''Design and* Development *of GSM based Energy Meter''* in IJERT, 2012.
13. S. Arun, Dr. Sidappa Naidu, *"Design and* Implementation *of Automatic Meter Reading System Using GSM, ZIGBEE through GPRS"* in IJARCSSE, 2012.
14. Young Hoon Lim, Moon Suk Chain, Jong Mock Baek, Sang-Yeom Lee *"An An Efficient Home Energy Management System Based on Automatic Meter Reading"* in IEEE International Symposium on Power Line Communication, 2011.
15. Xiujie Dong, Yan Yang, You Zhou *"The Design of Wireless Automatic Meter Reading System Based on SOPC"* in WASE ICIE, 2010 IEEE.
16. Syed Shahbaz Ali, Madiha Maroof, Sidrah Hanif *"Smart Energy Meter for Energy Conservation and Minimizing Errors"* in International Conference on Power Electronics, 2010.
17. Dr. Mohd Yunus B Nayan1, Aryo Handoko Primicanta *"Hybrid System Automatic Meter Reading"* in Proceeding of the ICCTD, 2009.
18. Vinu V Das, *"Wireless Communication System for Energy Meter Reading"* in International Conference on ARTCom, 2009.
19. Bharath, P. Ananth, N. Vijetha, S. Prakash, K.V.J. *"Wireless Automated Digital Energy Meter"* in Sustainable Energy Technologies, ICSET 2008.

Direct Torque Control with Adaptive PI Speed Controller based on Neural Network for PMSM Drives

Jia, Zhenyu[1],Kim, Byeongwoo[2]

[1]University of Ulsan, School of Electrical Engineering, 44610 Ulsan, South Korea

[2]University of Ulsan, School of Electrical Engineering, 44610 Ulsan, South Korea

Abstract. This paper presents an adaptive speed controller based on artificial intelligent technique to improvethe performance of classical Direct Torque Control (DTC) for Permanent Magnet Synchronous Motor (PMSM) drives. The proposed method applies back propagation (BP) based neural network (NN) to tune the parameters of classical proportional-integral (PI) speed controller. Comparisons between conventional PI speed controller and proposed method are carried out by Simulation.Simulation results demonstrate that conventional DTC system based on the proposed NN speed controller can achieve higher performance with fast speed response, small overshoot and robustness.

1 Introduction

Permanent magnet synchronous motors (PMSM) have become more popular in high-performance industrial drives. PMSMs have many advantages, such as high efficiency and power density, high torque-to-inertia ratio and simple structure.The main disadvantage of PMSMs is that flux and torque are inherent coupling, which makes them hard to control. Many advanced control theories have been proposed to make PMSMs more comparable.Filed-oriented control (FOC) [1] and direct torque control (DTC) [2] are the two main control techniques for PMSM drives. The FOC technique is widely used in PMSM drives. However, it performs not as well as predicted in practical engineering application due to the variations of motor parameters and inaccurate control mode [3]. In 1980s, DTC algorithm [4] was firstly introduced from ABB for induction motor drives.Decades past, it was already developed and used for variety of motor drives.

Compared with FOC, DTC is a significantly new concept.It possesses many characteristics of less dependence on machine parameters, fast dynamic response and less complexity without coordinate transformation or current regulation[5]. Classical DTC approach usually adopts a proportional-integral (PI) speed controller, torque and flux hysteresis controllers, and heuristic switching table for generating a variable switching frequency.Consequently, it can cause improper transient response with high overshoot, as well as high torque and flux ripples when applied to PMSM drives.

To solve the aforementioned problems of conventional DTC, varieties of optimization methods have been presented in the literature [6-10]. Many of them can obtain better results in ripple reduction and fixed switching frequency. However, most of them focus on replacing hysteresis controllers [6, 9], adopting space vector modulation (SVM) [7] and incorporating multilevel converter [8], thus adding more complexity, computational burden, or hardware cost to DTC system.

Speed controller, usually been neglected, also plays an important role in DTC technique for PMSM drives. A PI controller is widely applied to in the outer speed regulator loop due to its relative simple implementation and effectiveness [11]. However, the PI control approach cannot perform sufficiently well in nonlinear PMSM drives system with various uncertainties [12].Moreover, it's time consuming and hard to tune the proportional and integral gain parameters of PI controller. Recently, some new methods have been proposed to replace PI speed controller [10, 12]. In [10], artificial neural network (ANN) based on Kalman filter is used as speed controller. Although it can get better simulation results than conventional PI speed controller, it is relatively more complex by combining ANN and Kalman filter. Moreover, it lacks rapidity and stability compared with PI controller.

In this paper, a novel adaptive speed controller based on neural network PI (NN-PI) is presented. Neural network, which is capable of approaching to any nonlinear systems with uncertainties, is applied to tune PI parameters [13].Effective DTC control for PMSM drives can be obtained by optimizing PI controller parameters using NN with nonlinear self-adaptive ability. Simulation results demonstrate that proposed NN-PI speed controller can enhance the operating performance with faster transient response, smaller overshoot, better robustness, and even smaller torque and flux ripples compared with classical PI controller based DTC scheme.

2 Direct Torque Control

The basic principle of DTC is to select the optimal voltage vectors, which makes the stator flux linkage space vector rotates to required position while keeping magnitude within the hysteresis band and generate desired torque [10].
The speed controller uses the errors between reference speed and estimated speed as inputs to generate reference torque. In this paper, a NN-PI speed controller is applied to improve performance, which will be introduced in next part. The torque hysteresis and flux hysteresis controllers respectively correct errors between the set reference and estimated values to determine the outputs of proper operations in the closed loop [7]. These outputs, together with the information of the position of stator flux linkage space vector, will make the switching table generate corresponding switchingstates $\boldsymbol{S_{abc}}$of the voltage inverter [8]. An appropriate voltage vector will be decided based on the switching information.

The relationship between three phase stator voltages $\boldsymbol{V_{sabc}}$ and switching states can be described as (1):

$$\boldsymbol{V_{sabc}} = \begin{bmatrix} V_{sa} \\ V_{sb} \\ V_{sc} \end{bmatrix} = \frac{U_{dc}}{3}\begin{bmatrix} 2 & -1 & -1 \\ -1 & 2 & -1 \\ -1 & -1 & 2 \end{bmatrix}\begin{bmatrix} S_a \\ S_b \\ S_c \end{bmatrix} \quad (1)$$

Where U_{dc} is DC-bus voltage.

Applying Clark transformation, the α-β components of stator voltage space vector in stationary reference frame are obtained as (2):

$$\begin{bmatrix} V_\alpha \\ V_\beta \end{bmatrix} = \sqrt{\frac{2}{3}}\begin{bmatrix} 1 & -0.5 & -0.5 \\ 0 & \sqrt{3}/_2 & -\sqrt{3}/_2 \end{bmatrix}\boldsymbol{V_{sabc}}(2)$$

The DTC scheme works based on control of stator flux linkage space vector, which is estimated in α-β reference frame in (3) and (4):

$$\psi_\alpha = \int (V_\alpha - R_s i_\alpha)dt \quad (3)$$

$$\psi_\beta = \int (V_\beta - R_s i_\beta)dt \quad (4)$$

Where ψ_α and ψ_β are the α- and β- axis components of the estimated stator flux linkage, respectively.$i_\alpha$andi_β are the α-β components of stator current space vector. $R_s$is stator resistance.

The magnitude of stator flux linkage and electromagnetic torque are calculated by following equations (5), (6):

$$|\psi_s| = \sqrt{\psi_\alpha^2 + \psi_\beta^2}(5)$$

$$T_e = \frac{3}{2}P(\psi_\alpha i_\beta + \psi_\beta i_\alpha)(6)$$

Where P is the number of pole pairs.

3 Neural Network-PI Controller

Inspired by the marvellous function of neurons, the artificial neural network (ANN) has been developed very fast to solve big scale and complex problems recently. Numerous works [13-15] have demonstrated that ANN is suitable for dynamic and nonlinear control systems due to the capability of processing, analysing and learning. The speed controller of DTC structure is a nonlinear system, which hasdisturbances of load inertia and motor parametersvariation against temperature [12]. This makes it difficult for PI speed controller to perform sufficiently well for this DTC control system. This paper proposes an adaptive speed controller based on back propagation (BP) neural network, which is verified by simulation with higher performances.

3.1 Incremental PI Controller

The PI control theory has been one of the most widely used methods in industrial application. In continuous-time, the PI equation is expressed in (7). When the sampling period is very small, a discrete-time PI controller can be obtained as (8) by replacing integrate with summation and replacing derivative with difference quotient of (7).

$$\mathrm{u(t)} = k_p(e(t) + \frac{1}{T_i}\int_0^t e(\tau)d\tau)\ (7)$$

$$\mathrm{u(k)} = k_p(e(k) + \frac{T}{T_i}\sum_{j=0}^k e(j))\ (8)$$

Where u(t) and u(k) are controller outputs at continuous-time t and discrete-time k, respectively; $e(t)$ and $e(k)$ represent tracking errors at continuous-time t and discrete-time k, respectively;$k_p$is proportional gain; T_i is integral time constant;T is the sampling period. For convenience of program coding work, the equation (8) can be described as incremental form as (9) [15]:

$$\Delta\mathrm{u(k)} = k_p\big(e(k) - e(k-1)\big) + k_i e(k)(9)$$

Where k_i is the integral parameter gain.

3.2 Structure of NN-PI Speed Controller

The NN-PI speed controller using error back propagation (BP) algorithm consists of two parts. One is classical PI controller which can regulate motor speed in the closed loop. The other is BP neural network which can tune the parameters of k_p and k_i online to achieve the optimal performance. The schematic structure of NN-PI speed controller is shown as in figure 1.

https://doi.org/10.1051/matecconf/201816002011

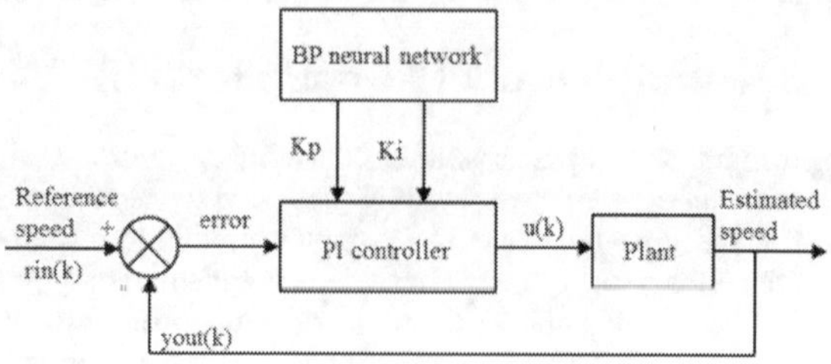

Figure 1. The structure of NN-PI speed controller.

The adopted BP neural network has three-layer, which is described as in figure 2. The numbers of neural nodes of input layer, hidden layer and output layer is J, I and 2, respectively. The inputs are the selected running states of system. The outputs are the parameters of PI controller.

As shown in figure 2, the neuron i of hidden layer at k can be expressed by following equations (10), (11):

$$O_i^{(2)}(k) = f(net_i^{(2)}(k)) \quad (10)$$

$$net_i^{(2)}(k) = \sum_{j=1}^{J} w_{ij}^{(2)} O_j^{(1)} \quad (11)$$

Where $net_i^{(2)}(k)$ and $O_i^{(2)}$ denote the input and output of neuron i, respectively; $w_{ij}^{(2)}$ denotes the connection weight parameter between j^{th} and i^{th} neurons; $O_j^{(1)}$ is the input the jth neuron; $f(x) = \tanh(x)$ istangent hyperbolic activation function.Similarly, the output $O_l^{(3)}(k)$ and input $net_l^{(3)}(k)$of output layer are given by equations (12), (13).

$$O_l^{(3)}(k) = g\left(net_l^{(3)}(k)\right) \quad l = 1,2 \quad (12)$$

$$net_l^{(3)}(k) = \sum_{i=1}^{I} w_{li}^{(3)} O_i^{(2)} \quad (13)$$

Where $g(x) = {}^{1}/_{2}\,(1 + \tanh(x))$ is the activation function of output layer.

The weights are updated using gradient descent method based on BP algorithm, which is given by (14):

$$\Delta w(k) = -\eta \frac{\partial E(k)}{\partial w} + \alpha \Delta w(k-1) \quad (14)$$

Where $E(k) = {}^{1}/_{2}\,(rin(k) - yout(k))^2$ is cost function, η is learning rate, α is momentum rate. The weights updating equations are given as follows:

$$\Delta w_{li}^{(3)}(k) = \alpha \Delta w_{li}^{(3)}(k-1) + \eta \delta_l^{(3)} O_i^{(2)}(k) \quad (15)$$

$$\Delta w_{ij}^{(2)}(k) = \alpha \Delta w_{ij}^{(2)}(k-1) + \eta \delta_i^{(2)} O_j^{(1)}(k) \quad (16)$$

Where $\delta_l^{(3)} = error(k) \frac{\partial y(k)}{\partial u(k)} \frac{\partial u(k)}{\partial O_l^{(3)}(k)} g'(net_l^{(3)}(k))$ and $\delta_i^{(2)} = f'(net_i^{(2)}(k)) \sum_{l=1}^{2} \delta_l^{(3)} w_{li}^{(3)}(k)$.

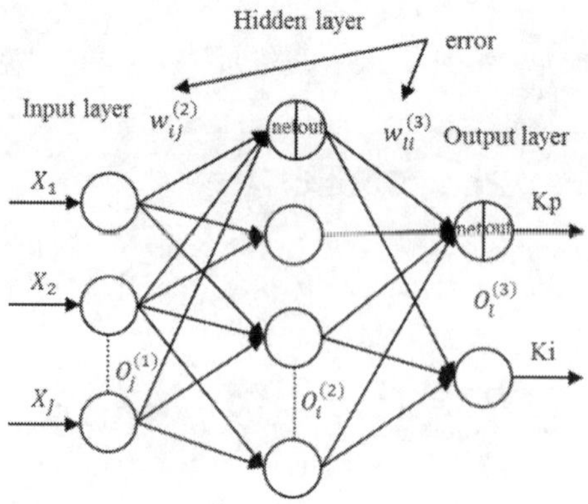

Figure 2. The structure of BP neural network.

4 Simulation and Results

The proposed method is evaluated by simulation on MATLAB.The parameters of PMSM drive system considered in this study are listed in Table 1.

Table1. Simulation parameters of PMSM.

Parameter	Values
Number of pole pairs P	4
Stator resistance R_s	1.2Ω
Magnet flux linkage	0.175Wb
d-axis inductance L_d	8mH
q-axis inductance L_q	8.5mH
Moment inertia J	$0.0008 kg \cdot m^2$
Viscous friction F	0.0001(N. m. s)

To demonstrate the performance of the presented NN-PI speed controller, the simulation results for a classical PI speed controller and the NN-PI speed controller are compared. The simulation model of proposed method is shown in Figure 3. Both of the methods are tested under same conditions with reference speed, load torque reference and simulation settings.

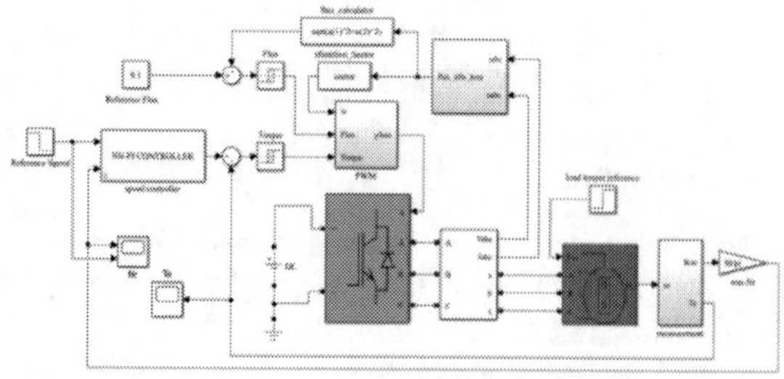

Figure 3. Simulation model of DTC system for PMSM drives with NN-PI speed controller.

The simulation is start at the speed on 600 r/min with a constant load torque 1.5N applied at 0.2s and sudden speed reduction to 550 r/min at 0.35s. The simulation results of speed respond are shown in Figure 4 and 5.

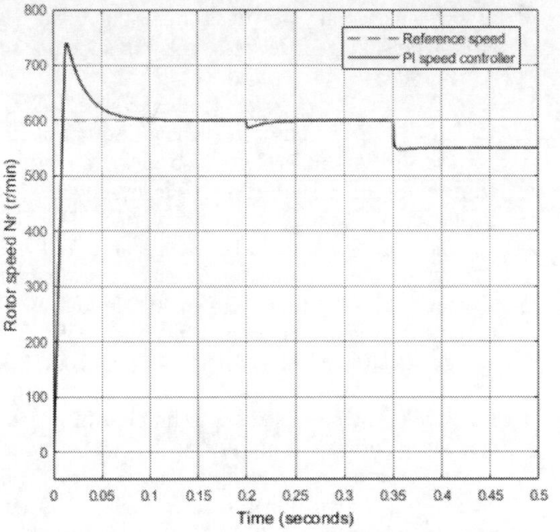

Figure 4. Speed response of conventional DTC system with PI speed controller.

It's clearly shown that the proposed method obtains great improvement in reduction of overshoot at the start-up period. Moreover, NN-PI speed controller has much better speed tracking performances than PI controller when facing load torque variations and sudden speed change.

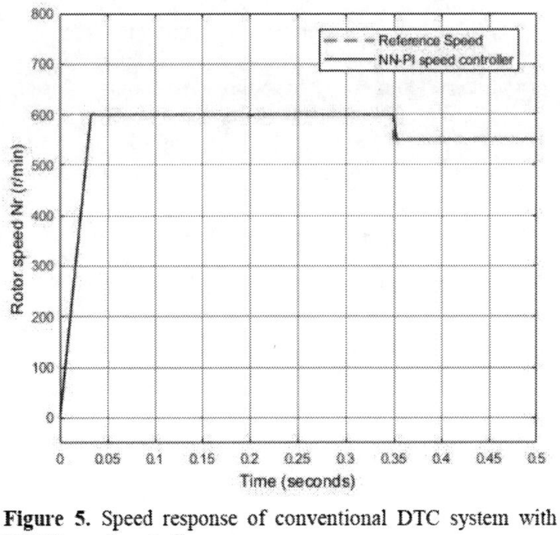

Figure 5. Speed response of conventional DTC system with NN-PI speed controller.

The comparison of Electromagnetic torque responses between conventional DTC system with PI controller and proposed DTC system with NN-PI controller is shown in Figure 6. As can be seen, the proposed method also get relatively better resultwith lower pulse and faster dynamic response in start-up period. Moreover, NN-PI also performs more stable under sudden load torque variationand speed reduction.

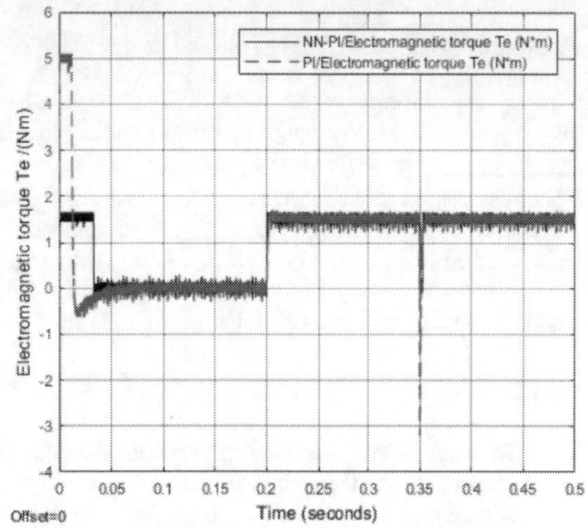

Figure 6. The response of electromagnetic torque of PI controller and NN-PI controller.

Figure 7 shows the flux response of both methods. It seems there is no big improvement. The proposed method can obtain stable flux response with small flux ripples under some system uncertainties.

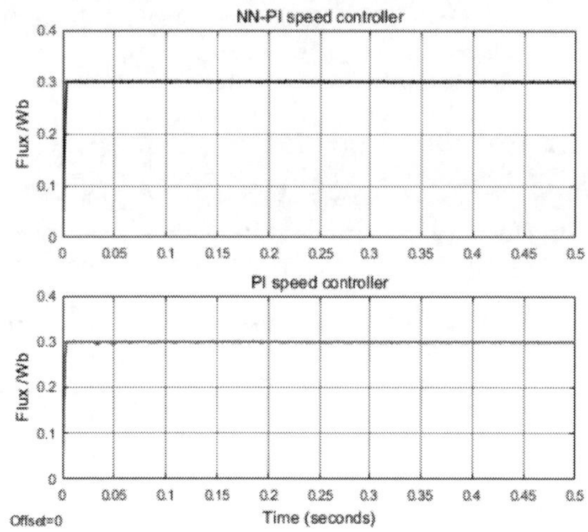

Figure7. The response of flux of NN-PI and PI controllers.

5 Conclusions

In this paper, an adaptive speed controller using BP neural network for DTC technique based PMSM drives is designed. BP neural network is applied to tune parameters of PI speed controller, which overcomes the disadvantages of conventional PI controller with high dependence on controlled model and time-consuming tuning work.

The simulation results demonstrate the effectiveness of the proposed NN-PI speed controller with faster response, much smaller overshoot and better robustness compared to conventional PI controller.

Further work will be concentrated on torque ripple reduction by more advanced theory.

ACKNOWLEDGEMENTS

I express my sincere gratitude towards my professor, Kim, Byeongwoo, for his supports; I am grateful to my colleagues for their helps.

The research was supported by the Industry Core Technology Development Project "Development of 2kW/kg, 100kW class IPMSM electric drive system with high efficiency cooling" through the Ministry of Trade, Industry and Energy (Grant Number : 10063006).

References

1. Ke Song, Wei guo Liu."Permanent Magnet Synchronous Motor Field Oriented Control and HIL Simulation," IEEE Vehicle Power and Propulsion Conference, Harbin, pp.4-6.(2008)
2. M. F. Rahman, L. Zhong, E. Haque, and M. A. Rahman, "A Direct Torque-Controlled Interior Permanent-Magnet Synchronous MotorDrive Without a Speed Sensor,"IEEE Trans. Energy conversion, vol. **18**, no. 1, (2003)
3. A. Haddoun, M.E.H Benbouzid, D. Diallo, R. Abdessemed, J. Ghouili, K.Srairi, "A loss-minimization DTC scheme for EVinduction motors," IEEE Conference Vehicle Power andPropulsion. pp. 7(2005)
4. I. Takahashi, T. Noguchi, "A new quick-response and high-efficiency control strategy of an induction motor," IEEE Trans. Ind. Appl, IA-22(5), 820-827 (1986)
5. W. Jiajia, G. Dawei, Z. Xin, L. Qingchun, "An efficiency optimization strategy of induction motors for electric vehicles," IEEE Vehicle Power and Propulsion Conference, pp, 1-5. (2008)
6. Y. Choi, H. Choi, and J. Jung, "Feedback linearization direct torque control with reduced torque and flux ripples for IPMSM drives,"IEEE Trans. Power Electron., vol.**31**, no. 5, pp. 3728-3737 (2016)
7. L. Tang, L. Zhong, M. F. Rahman, and Y. Hu, "A novel direct torque controlled interior permanent magnet synchronous machine drive with low ripple in flux and torque and fixed switching frequency," IEEE Trans. Power Electron., vol. **19**, no. 2, pp. 346–354, (2004)
8. C. A. Martins, X. Roboam, T. A. Meynard, and A. S. Carvalho, "Switching frequency imposition and ripple reduction in DTC drives by using a multilevel converter," IEEE Trans. Power Electron., vol. **17**, no. 2, pp. 286–297, (2002)
9. Y. Cho, K.-B. Lee, J.-H. Song, and Y.-I. Lee, "Torque-ripple minimization and fast dynamic scheme for torque predictive control of permanentmagnet synchronous motors," IEEE Trans. Power Electron., vol. **30**, no. 4, pp. 2182–2190, (2015)
10. F.Hamidia, MS.Boucherit, A.Larabi, M.Bouhedda, "Direct Torque Control with ANN Speed Controllerbased on Kalman filter for PMAM", IEEE Trans, 9th International Conference on Systems, Signals and Devices, (2012)
11. Y. Su, C. Zheng, and B. Duan, "Automatic disturbances rejection controller for precise motion control of permanent-magnet synchronous motors," IEEE Trans. Ind. Electron., vol. **52**, no. 3, pp. 814–823, (2005)
12. S. Li and Z. Liu, "Adaptive speed control for permanent magnet synchronous motor system with variations of load inertia," IEEE Trans. Ind. Electron., vol.**56**, no. 8, pp. 3050–3059, (2009)
13. H. Ji and Z. Li, Design of Neural Network PID Controller Based on Brushless DC Motor, Second International Conference on Intelligent Computation Technology and Automation, Vol.**3**, pp.46-49, (2009)
14. Sigeru Omatu, Michifumi Yoshioka, Toshihisa Kosaka and Hidekazu Yanagimoto, Neuro-PID Control of Speed and Torque of Electric Vehicle, International Journal on Advances in Systems and Measurements, vol. **3**, no. 1–2, (2010).
15. J. Kang, W.J. Meng, A. Abraham, H.B. Liu, An adaptive PID neural network for complex nonlinear system, Neurocomputing **135**, 79–85.(2014)

Preparation and Characterization of Nicke-iron Alloy Film as Freestanding Electrode for Oxygen Evolution Reaction

Mengqi Yao[1], Weilin Chen[2] and Wencheng Hu[1]

[1] *State Key Laboratory of Electronic Thin Films and Integrated Devices, University of Electronic Science & Technology of China, Chengdu 610054, China*

[2] *CNPC Beijing Richfit Information Technology Co. Ltd., Beijing 100007, China*

Abstract. This work reports the porous nicke-iron alloy film supported on stainless steel mesh as freestanding electrode for enhanced oxygen evolution reaction (OER) catalyst prepared from an one step electrodeposition method. Results indicated that the porous nickle-iron alloy film exhibits a low overpotential of 270 mV at 10 mA cm^{-2} and excellent electroconductibility. The superior OER properties can be attributed to its novel synthetic process, conductive substrate and porous structure. This work will provide a new strategy to fabricate alloy film for OER electrocatalyst.

1 Introduction

Hydrogen has been considered as a promising alternative to fossil fuels [1-3]. As our known, electrochemical water spliting is an efficient way to generate hydrogen [4,5]. Oxygen evolution reaction (OER) electrolysis, as an important reaction involved in water splitting, has attracted increasing attention for clean energy generation and efficient energy storage. The oxygen evolution reaction significantly influence the efficiency of overall water splitting due to its rough four electrons kinetic process [6,7]. Noble metal catalysts including RuO_2 and IrO_2 have been identified as efficient OER catalysts because of their low overpotential and small Tafel slope. However, the high cost, poor cycling stability and toxicity limit their large scale applications. First-row transition metal based materials have attracted enormous attentions for OER catalysts due to their low cost and high catalytic activity based on earth-abundant metals [8,9]. Among first-row transition metal materials, nickle and iron based materials possesses outstanding properties including relatively high electrical conductivity and highly catalytic performances for oxygen evolution reacion, making it potential candidates for OER catalysts [10-13]. However, NiFe based electrocatalysts mentioned in recent literatures, own low stability and poor conductivity [14-19]. Therefore, fabricating NiFe materials with enhanced OER properties attracts plenty of interests.

Recently, there were many discoveries and recent progresses being explored to improve OER catalytic activities. Scientists fabricated different microstructure and micromorophology by different methods to increase specific surface area and active sites to improve OER properties. Moreover, different elements were doped into NiFe based materials to increase defects in order to improve electrical conductivity of system. Meanwhile, combining conductive materials, such as graphene and Ag particles, with electroactive materias not only boost conductivity, but it also can expose more edge defects. For instance, Lu and his coworkers demonstrated that a three-dimensional (3D) architecture of NiFe layered double hydroxide (NiFe-LDH) significantly reduced the onset potential, yielded high current density at small overpotentials, and showed outstanding cycling stability in electrochemical oxygen evolution reaction [20]. Feng and his coworkers fabricated a of freestanding sandwiched NiFe nanoparticles encapsulated by graphene sheets with a low onste potential of 1.44 V (vs. RHE) [21]. Qi and his coworkers prepared highly porous nickle and iron oxide with improved crystallinity. The catalysts showed small overpotential, low Tafel slope and good stability for OER [22].

Most NiFe based OER catalysts are powders, which reported in the recent studies [23-26]. The binders, such as Nafion solution, were utilized to promote electroactive materials powders coating onto the surface of conductive substrates including Ni foam, carbon cloth and stainless steel mesh. However, the uses of binders significantly reduce the conductivity of the system. In other words, using binder will impede the transfer of ions and mass, giving rise to degraded OER performance. It is a common phenomenon that the glued OER powders always peel off easily during long-time electrochemical tests, since the influence of gas evolution on the surface of electrode [27]. Metal alloys, growing onto the surface of conductive substrates by electrodeposition method, have been wildly employed as high performance electrocatalysts. Although above mentioned metal alloy films own relatively high OER propertied compared with glued OER powder, the coatings usually show general electrocatalytic performance and stability, for the reason that the alloy film possesses little active sites and only the surface layers of electroactive materials are in contact with electrolyte. Therefore, exploring a new strategy to electrodeposit metal alloy filmwith abundant active sites for OER catalysts is of great interest.

In this study, porous nicke-iron alloy film supported on stainless steel mesh was synthesized by an one step electrodeposition method. We studied the OER properties of porous nickle-iron alloy films. We concluded that the one step electrodeposition method has advantages for oxygen evolution reaction as follows: (1) abundant active sites and large specific surface area; (2) relatively excellent crystallinity and improved cycling stability. The nicke-iron alloy film exhibits a low overpotential of 270 mV at a current density of 10 mA cm^{-2} and significantly improved electroconductibility, suggesting an enhanced OER property. The superior OER properties can be attributed to its novel synthetic process, conductive substrate and porous structure. This work will provide a new strategy to fabricate alloy film for OER electrocatalyst.

2 Experimental sections

2.1 Materials

Nickel sulfate hexahydrate ($NiSO_4 \cdot 6H_2O$), iron sulfate hexahydrate ($FeSO_4 \cdot 6H_2O$), boric acid (H_3BO_3), potassium hydroxide (KOH), ethyl alcohol (C_2H_5OH) and methanol solution (HCHO) were obtained from Chengdu Kelong Chemical Reagent Co. (Chengdu, China). Stainless steel mesh (304L, 300 mesh) and Ni foil were purchased from Chengdu Jinshan Chemical Reagent Co. (Chengdu, China). Ni foil (thickness: 1 mm) and deionized water with resistivity of 18.2 MΩ cm were used as counter electrode and solvent, respectively.

2.2 Preparation of nicke-iron alloy film

All chemicals were used as received without any further purification. The binary nickle-iron alloy film was prepared via a novel electrodeposition method. Briefly, 6.3g nickel sulfate hexahydrate ($NiSO_4 \cdot 6H_2O$), 3.3g iron sulfate hexahydrate ($FeSO_4 \cdot 6H_2O$) and 1.2 g boric acid (H_3BO_3) were dissolved in 65 ml distilled water. In addition, 5 ml methanol (HCHO) as reductant was added into the solution. The solution was stirred for 30min under magnetic condition. Then the transparent solution was transferred into a 100ml electroplating bath. A 2 cm×2 cm of stainless steel mesh (304L, 300 mesh) was used as cathode; 2 cm×2 cm nickel foil was used as anode. Prior to the electrodeposition process, the stainless steel mesh substrate was cleaned in 1 M HCl solution under ultrasonication condition for 10 min to remove oxide layer on the surface, and washed sequentially by deionized water and ethyl alcohol for several times. The electrodeposition systrn was put inside the oven at 60 °C for 10 minutes at the current density of 30 mA cm^{-2}. Then, the as-prepared nickle-iron alloy film was washed with distilled water, ethanol and acetone for six times and dried at 100 °C under vacuum condition. The obtained nickle-iron alloy film supported on stainless steel mesh was used for further materials characterizations and electrochemical tests.

3 Characterizations

The crystalline phase was characterized by X-ray diffractometer (Smartlab9, Pigaku, Japan.). The FESEM were used to evaluated micromorphologies and microstructure operated at 3.0 kV (Inspect F, FEI Co., U.S.).

The OER properties of electrode were characterized by the electrochemical station (PGSTAT302N, Netherlands) under a standard three-electrode system. 1.0 M KOH solution saturated by O_2 was used as electrolyte, 1cm×1cm nicke-iron alloy film supported on stainless steel mesh was used as a freestanding working electrode. Pt/C wire and Hg/HgO electrode was used as counter electrode and reference electrode, respectively. Linear sweep voltammetry (LSV) was performed in 1.0 M KOH solutions from 0 to 1.0 V at a scan rate of 2 mV s^{-1}. Electrochemical impedance spectroscopy (EIS) was measured at open circuit voltages from 10^6 to 10^{-2} Hz. All the potentials in this paper were versus the reversible hydrogen electrode (RHE) according the equation E(RHE) = E(Hg/HgO)+0.059 pH+0.098 V= E(Hg/HgO)+0.0591×14+0.098 V=E(Hg/HgO)+0.9254 V. The overpotential (η) corresponds to the equation η = E(RHE)-1.23 V. All the curves were reported with iR compensation.

4 Results and discussions

4.1 XRD analysis of nicke-iron alloy film

The crystalline structure of porous nicke-iron alloy film supported on stainless steel mesh and blank stainless steel mesh evaluated by X-ray diffractometer (XRD) are shown in Figure 1. The result indicates that the diffraction peaks located at 44.2°, 51.6° are well identified and can be successfully assigned to the (111) and (200) lattice planes of $FeNi_3$ (PDF card: 38-0419). Additionally, compared with the peaks of nicke-iron alloy film, the blank stainless steel mesh substrate shows nearly none obvious peaks, suggesting the phase purity of nicke-iron alloy film supported on stainless steel mesh. We can conclude that nickle-iron alloy is successfully fabricated via on step electrodeposition method. Besides, the particle size of nicke-iron alloy film supported on stainless steel mesh is 12.026 nm calculated by XRD curve using Scherrer equation, confirming a relatively excellent crystallinity.

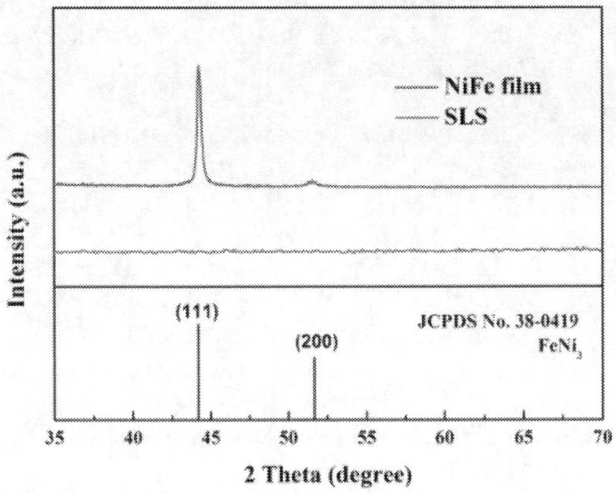

Figure 1. XRD pattern of nicke-iron alloy film and stainless steel mesh.

4.2 Micromorphology characterizations of nicke-iron alloy film

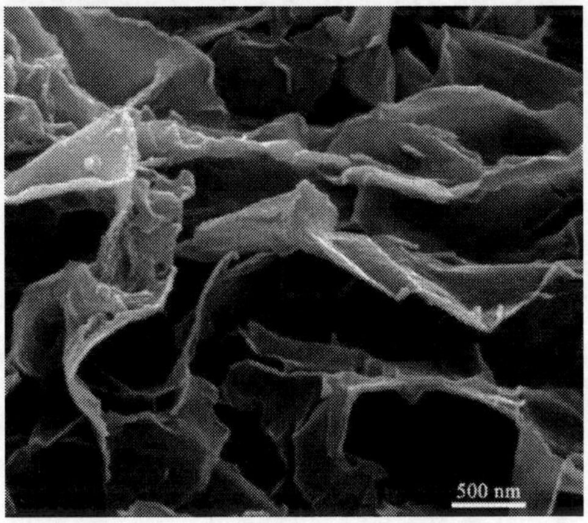

Figure 2. SEM image of nicke-iron alloy film.

Micromorphology and microstructure of porous nicke-iron alloy film supported on stainless steel mesh analyzed by scanning electron microscope (SEM) is shown in Figure 2. It can be observed that the nicke-iron alloy film was successfully electrodeposited on stainless steel mesh. As displayed in Figure 2, the nicke-iron alloy film was composed of relatively homogeneous nanosheets like microstructures. The thickness of each vertical nanosheet is approximately 80-120 nm and the extension of the nanosheets ranged from 200 nm to several thousand nanometers. The nanosheet like micromorphology can provide large specific surface area and abundant electrochemically active sites. The large specific surface area can boost release of O_2 from the surface of nicke-iron alloy film and prevent the film peeling from the surface of substrate under the impact of oxygen release during a long-term stability test, especially under high current density and vigorous gas evolution conditions. In other words, the nanosheet like structure enhance the cycling stability. Meanwhile, large specific surface area can provide more electrochemically active sites for the electrocatalysis leading to significantly enhanced electrocatalytic performance for oxygen evolution reaction [28, 29]. Besides, the nanosheet like micromorphologies also can facilitate diffusion of electron and OH^- and increase the contact area between electrolytes and catalytic active sites, leading to improved electrocatalytic performances.

4.3 OER properties of nicke-iron alloy film

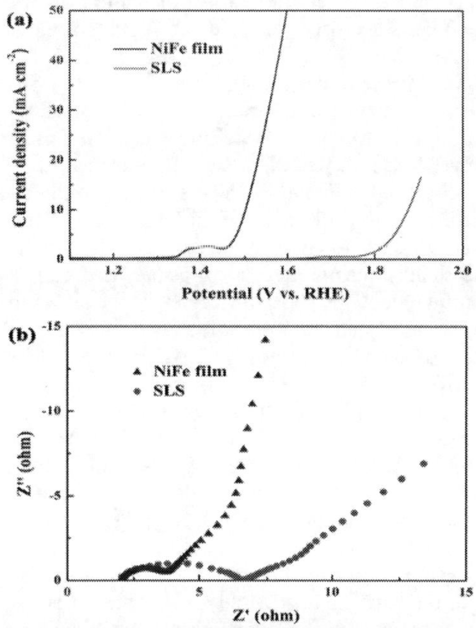

Figure 3. (a) The iR corrected polarization curves of nicke-iron alloy film and stainless steel mesh in 1.0 M KOH. (b) The EIS curves of nicke-iron alloy film and stainless steel mesh.

The electrochemical tests were analyzed under a standard three-electrode system using 1.0 M. Porous nickle-iron alloy film supported on stainless steel mesh was employed as the working electrode, a Hg/HgO was used as the reference electrode and a Pt plate was utilized as the counter electrode. Figure 3a reveals iR-corrected polarization curves of porous nicke-iron alloy film supported on stainless steel mesh and blank stainless steel mesh tested at a scan rate of 5 mV s^{-1}. As our known, the potential obtained at a current density of 10 mA cm^{-2} is a commen way to evaluate the OER property of catalysts. From Figure 3a, it can be demonstrated that nicke-iron alloy film shows a low overpotential of 1.50V (vs. RHE) at a current density of 10 mA cm^{-2}, whereas the stainless steel mesh presents potential of 1.87V (vs. RHE) at 10 mA cm^{-2}. The nicke-iron alloy film possess a low overpotential of 270 mV which is much lower than that of stainless steel mesh. The small oxidation process from LSV curve detected at 1.40 V before the onset of OER is corresponding to the formation of Ni and Fe species, which are the active sites to catalyse OER. The results indicate that the nicke-iron alloy film possesses higher electrocatalytic activity compared with stainless steel mesh due to its large specific surface area, abundant electrochemical active sites and synergistic effect of Ni and Fe atoms. The OER performance is superior to the recent studies. For example, Liu et al. fabricated mesoporous $NiFe_2O_4$ nanorods as efficient oxygen evolution catalyst, which shows overpotential of 342 mV at a current density of 10 mA cm^{-2} [29]; Zhang et al. synthesized NiFe LDH hollow microspheres with small onset overpotential of 239 mV at a current density of 10 mA cm^{-2} and a low overpotential of 316 mV at a current density of 10 mA cm^{-2} [30]. Besides, we also use the Tafel slope to evaluate the OER performance by Tafel equation:$\eta=a+b\log(j)$, where η (V) is the potential, b is the Tafel slope (mV dec^{-1})and j (mA cm^{-2}) is the current density. The results demonstrate that the Tafel slope of porous nickle-iron alloy fim is as low as 98 mV dec^{-1}, which is smaller compared with that of stainless steel mesh (152 mV dec^{-1}), suggesting nickle-iron alloy fim possess relatively excellent OER performance.

As displayed in Figure 3b, the electrochemical impedance spectroscopy (EIS) tests of nicke-iron alloy film and stainless steel mesh are investigated to further examine the sluggish charge transfer of as-prepared samples and illustrate the OER performances. It can be observed from the Nyquist plots that the solution transfer resistance (R_s) of two samples was nearly 2.2Ω. Besides, the charge transfer resistance (R_{ct}) of nicke-iron alloy film and stainless steel mesh is as low as 2.65Ω and 4.93Ω, respectively, suggesting that the nicke-iron alloy film possesses lower R_{ct} and easy sluggish charge transfer. The low R_{ct} can provide a faster OER kinetics and significantly improve the OER properties, such as lower overpotential and excellent stability. To our known, low R_{ct} can be attributed to combining electroactive material with the conductive substratre and abundant active sites which create charge transport pathways and improve electron transport [31-33].

5 Conclusions

In summary, we mention an one step electrodeposition method to prepare porous nicke-iron alloy film supported on stainless steel mesh freestanding electrode as oxygen evolution reaction catalyst. We concluded that the one step electrodeposition method has advantages for oxygen evolution reaction as follows: (1) abundant active sites and large specific surface area; (2) relatively excellent crystallinity and improved cycling stability.Nicke-iron alloy film shows a low overpotential 270 mV at a current density of 10 mA cm^{-2} and excellent electrical conductivity. We studied the OER properties of porous nickle-iron alloy films. In our paper, the nanosheets like structures provide highly exposed electrochemical active sites and large specific surface area which boost mass transport and facilitate oxygen overflow. The low cost and high catalytic activity of nicke-iron alloy film freestanding electrode provide a new strategy to fabricate alloy film for OER.

References

1. M. Dresselhaus, I. Thomas, Nature, 414, 332 (2001)
2. Y. Zheng, Y. Jiao, Y. Zhu, L.H. Li, Y. Han, Y. Chen, A. Du, M. Jaroniec, S. Qiao, Nat. Commun. 5, 3783 (2014)
3. X. Wang, W. Li, D. Xiong, D.Y. Petrovykh, L. Liu, Adv. Funct. Mater. 26, 4067 (2016)
4. Q. Lu, Y. Yu, Q. Ma, B. Chen, H. Zhang, Adv. Mater. 28, 1917 (2016)
5. F. Wang, P. He, Y. Li, T.A. Shifa, Y. Deng, K. Liu, Q, Wang, F. Wang, Y. Wen, Z, Wang, X. Zhan, L. Sun, J. He, Adv. Funct. Mater. 1602699 (2017)
6. Q. Lu, Y. Yu, Q. Ma, B. Chen, H. Zhang, Adv. Mater. 28, 1917 (2016)
7. R. Cao, W. Z. Lai, P. W. Du, Energy Environ. Sci. 5, 8134 (2012)
8. L. Han, S. Dong, E. Wang, Adv. Mater. 28, 9266 (2016)
9. H. Osgooda, S.V. Devaguptapua, H. Xub, J. Choc, G. Wu, Nano Today 11, 601 (2016)
10. E. Nurlaela, T. Shinagawa, M. Qureshi, D.S. Dhawale, K. Takanabe, ACS Catal. 6, 1713 (2016)
11. Z. Chen, J. Zhao, S. Yan, Y. Yuan, S. Liu, Mater. Lett. 157, 201 (2015)
12. X. Jia, Y. Zhao, G. Chen, L. Shang, R. Shi, X. Kang, G.I.N. Waterhouse, L. Wu, C. Tung, T. Zhang, Adv. Energy Mater. 6, 150258 (2016)
13. X. Xu, F. Song, X. Hu, Nat. Commun. 7, 12324 (2016)
14. S. Chen, J. Duan, M. Jaroniec, S. Qiao, Angew. Chem., Int. Ed. 52, 1356 (2013)
15. Y. Feng, H. Zhang, L. Fang, Y. Mu, Y. Wang, ACS Catal. 6, 4477 (2016)
16. C. Tang , H. Wang , H. Wang , Q. Zhang , G. Tian , J. Nie, F. Wei, Adv. Mater. 27, 4516 (2015)
17. C. Andronescua, S. Barwea, E. Ventosaa, J. Masaa, E. Vasiled, B.Konkenaa, S. Möllera, W. Schuhmann, Angew. Chem. Int. Ed. 56, 11258 (2017)
18. M Zhou, Q. Weng, X. Zhang, X. Wang, Y. Xue, X. Zeng, Y. Bando , D. Golberg, J. Mater. Chem. A 5, 4335 (2017)
19. X. Zhang, H. Xu, X. Li, Y. Li, T. Yang, Y. Liang, ACS Catal. 6, 580 (2016)
20. Z. Lu, W. Xu, W. Zhu, Q. Yang, X. Lei, J. Liu, Y. Li, S. Xiao, X. Duan, Chem. Sci. 50, 6479 (2014)
21. Y. Feng, H. Zhang, Y. Zhang, X. Li, Y. Wang, ACS Appl. Mater. Inter. 7, 9203 (2015)
22. J. Qi, W. Zhang, R. Xiang, K. Liu, H. Wang, M. Chen, Y. Han, R. Cao, Adv. Sci. 2, 199 (2015)
23. M. Gong, H. Dai, Nano Research 8, 23 (2015)
24. G. Wang, D. Zheng, D. Liu, J. Harris, J. Si, T. Ding, D. Qu, Electrochim. Acta 247, 722 (2017)
25. G. Liu, X. Gao, K. Wang, D. He, Jinping Li, Nano Research 10, 2096 (2017)
26. M. Yao, N. Wang, W. Hu, J. Electroanal. Chem. 782, 133 (2016)
27. C.C.L. McCrory, S. Jung, I.M. Ferrer, S.M. Chatman, J.C. Peter, T.F. Jaramillo, J. Am. Chem. Soc. 137, 4347 (2015)
28. P. He, X. Yu, X. Lou, Angew. Chem. Int. Ed. 129, 3955 (2017)

29. G. Liu, K. Wang, X. Gao, D. He, J. Li, Electrochim. Acta 211, 871 (2016)
30. C. Zhang, M. Shao, L. Zhou, Z. Li, K. Xiao, M. We, ACS Appl. Mater. Inter. 8, 33697 (2016)
31. M. Yao, N. Wang, W. Hu, J. Mater. Sci-Mater. El. 28, 11119 (2017)
32. M. Balogun, W. Qiu, H. Yang, W. Fan, Y. Huang, P. Fang, G. Li, H. Ji, Y. Tong, Energy Environ. Sci. 9, 3411 (2016)
33. M. Gong, Y. Li, H. Wang, Y. Liang, J.Z. Wu, J. Zhou, J.Wang, T. Regier, F. Wei, H. Dai, J. Am. Chem. Soc. 135, 8452 (2013)

Preparation and Characterization of Pure Organic Dielectric Composites for Capacitors

Xin Mao[1], Jinliang Lv[1], Fangfang Zhang[1], Bo Wu[1], Jie Yang[1], Lirong Du[1] and Xianzhong Tang[1]

[1] *State Key Laboratory of Electronic Thin Films and Integrated Devices, University of Electronic Science & Technology of China, Chengdu 610054, China*

Abstract. This work reports the excellent dielectric composites were prepared from polyimide (PI) and poly(vinylidene fluoride) (PVDF) via solution blending and thermal imidization or chemical imidization. The dielectric and thermal properties of the composites were studied. Results indicated that the dielectric properties of the composites synthesized by these two methods were enhanced through the introduction of PVDF, and the composites exhibited excellent thermal stability. Compared to the thermal imidization, the composites prepared by chemical imidization exhibited superior dielectric properties. This study demonstrated that the PI/PVDF composites were potential dielectric materials in the field of electronics.

1 Introduction

Polymer-based composites with high dielectric constant, low dielectric loss and easy processibility have attracted much attention in the fields of high-charge storage capacitors, energy, electrical engineering [1-8]. Polyimides (PIs) have extraordinary characteristics such as excellent mechanical and chemical properties, and outstanding thermal stability, so that they are being widely applied as matrix materials for the advanced applications [9-11]. However, pure PIs only have a low dielectric constant which could limit their applications in electronics [12, 13]. A series of related studies has been proposed to solve this problem [14-17]. A general method is to use high k materials as fillers to significantly improve the dielectric properties of polymers [18-23]. It is well known that polyvinylidene fluoride (PVDF) has a relatively high dielectric constant (10–14) [24, 25] which can be used as a high dielectric filler to improve the dielectric constant of PI.

In this study, polyimide-based dielectric composites were synthesized by two imidization methods. We studied the properties of such composites.

2 Experimental section

2.1 Materials

4, 4'-Diaminodiphenyl ether (ODA) and pyromellitic dianhydride (PMDA) were all analytical grade and purchased from Aladdin (Shanghai, China). PVDF (JHR-150) was supplied by Zhejiang Juhua Chemical Reagent Co. (Zhejiang, China). N, N-dimethylacetamide (DMAC) was obtained from Chengdu Kelong Chemical Reagent Co. (Chengdu, China). Acetic anhydride (Ac2O) and triethylamine (TEA) were purchased from Chengdu Jinshan Chemical Reagent Co. (Chengdu, China).

2.2 Preparation of PI/PVDF composites

The polyamic acid (PAA) precursor was polymerized from ODA and PMDA in freshly distilled DMAC under stirring at room temperature for 24 h. Then, different amounts of PVDF powders (10, 20, 30, 40, and 50 wt.%) were ultrasonically dispersed in DMAC for 1 h. In order to prepare homogeneous and different concentrations of PVDF/PAA solution, various weight fractions of PVDF/DMAC dispersion were added into the PAA solution with further vigorous stirring for 24 h. Then, the PAA was imidized using a standard thermal imidization method. Subsequently, the mixture was cast onto a piece of clean glass plate, and thermally imidized at 50, 100, 150, 200, and 250 °C for 1 h and 300 °C for 0.5 h to form PI/PVDF composite films. Consequently, composite films with 0, 10, 20, 30, 40, and 50 wt.% PVDF were obtained. Composites were also prepared by chemical imidization of PAA. Ac2O and TEA, acted as the dehydrating and catalyst agent respectively, were added simultaneously to the PVDF/PAA solution which was stirred for an additional 24 h to produce PI/PVDF composite. The whole synthesis process was carried out at ambient temperature under nitrogen protection. Afterwards, the PI/PVDF solution was precipitated in distilled water. Subsequently, filtering and washing with distilled water and methanol, and drying in a vacuum oven at 80 °C for

24 h to obtain PI/PVDF composite powders. After this, the composite powders were molded by hot pressing at about 150 °C under uniaxial pressure of 30 MPa for 1 h. For convenience, the sample codes were defined according to the amounts of PVDF, e.g. the thermally imidized PI/PVDF (70/30) blend was defined as T-30%, the chemical imidized PI/PVDF (70/30) blend was defined as C-30%.

3 Characterizations

Fourier Transform Infrared (FTIR) spectra were obtained using a Fourier transform infrared spectrometer (FTIR8400S) at wave numbers of 400–4000cm^{-1}, and samples were ground and mixed with potassium bromide and compressed into pellets for testing. The surface morphologies of the composites were investigated using a field-emission scanning electron microscope (FEI Co., USA) with an accelerating voltage of 20 kV. All the samples were coated with a homogeneous gold layer prior to analyses to avoid charge accumulation. The thermal behavior of the PI/PVDF composites was examined using a thermogravimetric analyzer (TA Q50, TA Instruments, USA) under nitrogen flow at a heating rate of 20 °C/min from room temperature to 800 °C. Dielectric behavior was measured using an Agilent 4294A precision impedance analyzer (Agilent Technologies, USA) within the frequency range of 100 Hz to 1 MHz. Before measurement, all the samples were sputter coated with aluminum on the two sides as electrodes. The relative permittivity of the samples is measured in a sandwich structure (aluminum /polymer/ aluminum) and calculated using the following equation (1):

$$\varepsilon_r = \frac{C \times d}{\varepsilon_0 \times A} \quad (1)$$

where ε_r is the relative permittivity of the hybrid films, C is the capacitance, d represents the thickness of the hybrid films, ε_0 is the dielectric constant in vacuum (ε_0 = 8.85 × 10−12 F/m), and A is the area of the electrodes. Reported data were the average value of at least three individual measurements.

4 Results and discussions

4.1 FTIR analysis

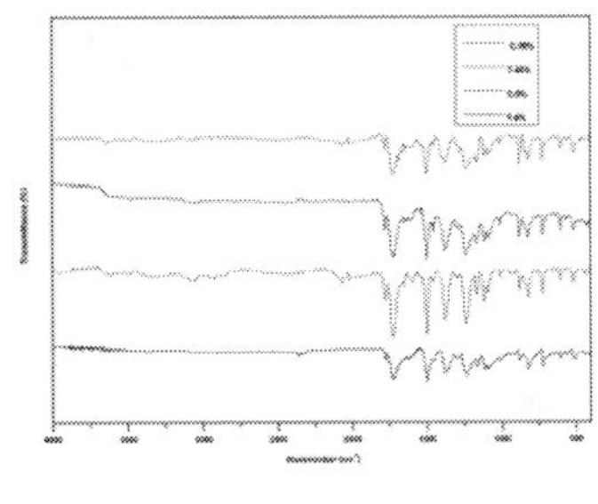

Figure 1. FT-IR spectra of composites.

The FTIR spectra of pure PI and the composites are shown in Fig. 1. Either after thermal imidization or after chemical imidization, the FTIR spectra show the typical absorption peaks of the imide groups of PI. The presence of absorption peaks at approximately 1780 cm^{-1} corresponds to the C=O asymmetric stretching vibration, 1720 cm^{-1} corresponds to the C=O symmetric stretching vibration, 1370 cm^{-1} corresponds to the C-N-C stretching vibration, and 720 cm^{-1} corresponds to the C=O bending vibration [26-28]. Furthermore, the absence of the characteristic absorption bands of PAA at approximately 1660 cm^{-1} corresponds to the amic acid C=O symmetric stretching vibration and 1535 cm^{-1} corresponds to the amide C-N asymmetric stretching vibration indicates complete imidization in the resulting composites [29, 30].

4.2 Microstructure of PI/PVDF composites

As shown in Fig. 2, in both two methods, the PVDF elliptic particles are uniformly distributed in the PI matrix without apparent aggregation. The surface morphology of composites synthesized by chemical imidization method was obviously better than that of composites synthesized by thermal imidization method at the same mass fraction. The uniform distribution of the composites may be propitious to enhance the dielectric constant. In addition, the phase separation could be attributed to the incompatibility between PI and PVDF [27].

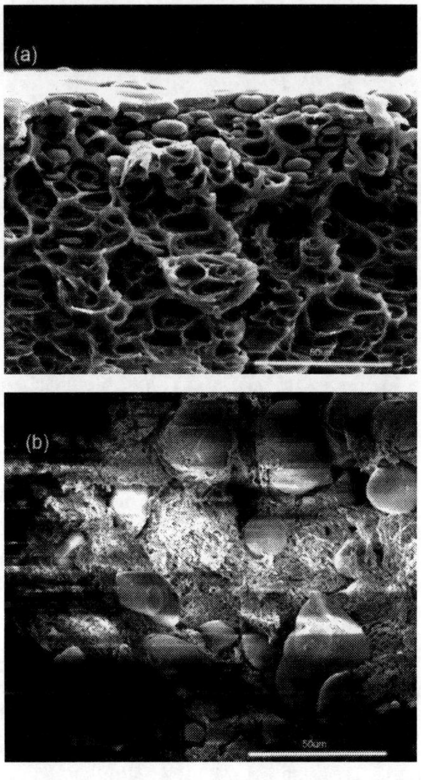

Figure 2. Surface morphology of composites (a): 30wt%PVDF synthesized by thermal imidization (b): 30wt%PVDF synthesized by chemical imidization.

4.3 Thermal analysis of composites

Thermal stability is one of the most important properties that should be investigated for potential high-performance engineering materials. As shown in Fig. 3, two stages of weight losses were observed from the pure PI. The first weight loss within 100 °C can be mainly attributed to the evaporation of residual solvent and the second stages were attributed to the degradation of PI. In both imidization methods, pure PI did not appear to suffer from a large weight loss when subjected to up to 500 °C. However, the decompositions of T-30% and C-30% both include three stages: the first weight loss stage starting below 100 °C was attributed to the removal of residual solvent; the second and the third weight loss at approximately 432 °C and 500 °C, which could be attributed to the difference in thermal stabilities between PVDF and PI [31]. Moreover, their characteristic yields after pyrolysis at 800 °C were over 40%. This indicates that the composites exhibited outstanding thermal stabilities.

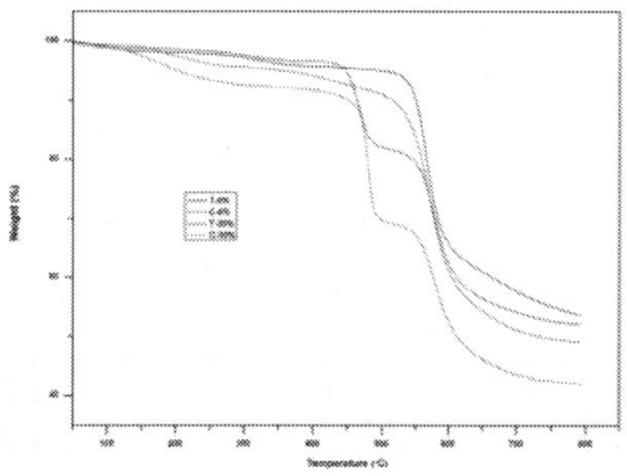

Figure 3. TGA curves of composites.

4.4 Dielectric properties of composites

Figs. 4(a) and 4(b) show the dependence of the dielectric constant and dielectric loss of the composites on the mass fraction of PVDF measured at room temperature and at 1 kHz. The dielectric constant of the composites synthesized by these two methods remarkably increased when PVDF was introduced into PI. This excellent enhancement in dielectric constant should be mainly attributed to the interfacial space charge polarization between PI and PVDF as well as the homogeneous dispersion of PVDF [32]. The dielectric constant of composites synthesized by chemical imidization method was much higher than that of composites synthesized by thermal imidization method at the same mass fraction. For instance, in chemical imidization, the dielectric constant was 7.745 when PVDF content was as high as 50 wt.%, which was about 1.5 times higher than that of composites synthesized by thermal imidization. The more uniform distribution of the composites synthesized by chemical imidization may be propitious to enhance the dielectric constant [33]. In addition, in thermal imidization method, the dielectric constant drastically increased to 5.7 when 30 wt.% PVDF was added, whereas when PVDF exceeded 30%, the dielectric constant decreased to 5.18. PVDF accumulation could hinder interfacial polarization and was responsible for decreasing relative permittivity. The dielectric loss of the composites synthesized by thermal imidization method was smaller than the dielectric loss of the composites synthesized by chemical imidization method. However, in both two methods, the dielectric loss of the composites was small, i.e., less than 0.1, which made it acceptable for practical applications. Furthermore, the frequency dependence of the dielectric constant and dielectric loss of the composites is presented in Figs. 4(c) and 4(d), respectively. The dielectric constant of the composites exhibited weak correlation with frequency and slightly decreased with the increase in frequency from 100 Hz to 1 MHz. The dielectric constant decreased at a higher frequency because the interfacial polarization and orientation polarization of dipoles could not rapidly follow the change in the external electric field [34, 35]. However, dielectric loss decreased at a low frequency range and then increased. The reason was that the space charge polarization contributed to the main part of the loss tangent while the conduction contributed less as the frequency increasing [33]. In addition, the maximum value of the dielectric loss was lower than 0.1 within the testing frequency range. This finding indicates that the PI/PVDF composites are promising dielectric materials for practical applications.

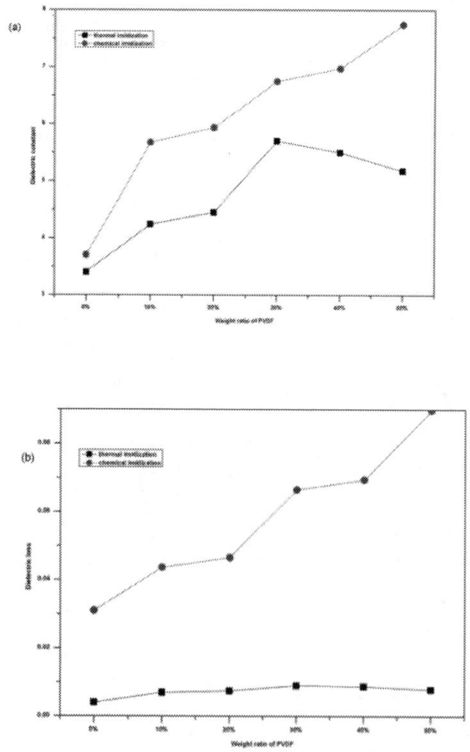

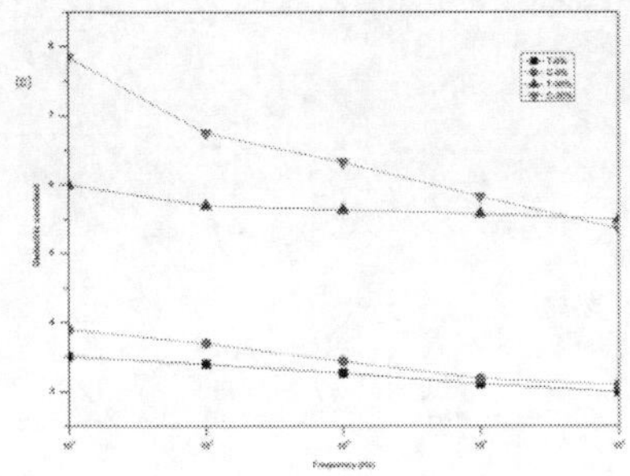

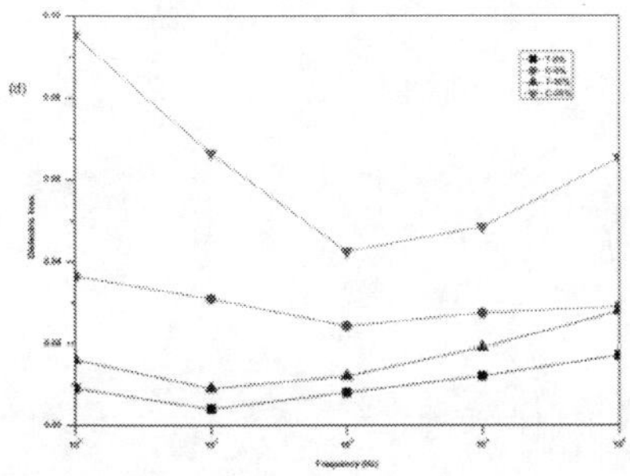

Figure 4. (a) Dielectric constant and (b) dielectric loss as a function of mass fraction of PVDF at 1 kHz; (c) dielectric constant and (d) dielectric loss at a function of frequency for PI/PVDF composites at room temperature.

5 Conclusions

All-organic PI/PVDF high dielectric composite materials were synthesized via thermal imidization or chemical imidization. The composites exhibited excellent thermal stability and dielectric properties. The characteristic yields after pyrolysis at 800 °C were over 40%. The dielectric constant of composites synthesized by thermal imidization rapidly increased to 5.7 at 1 kHz when PVDF content reached 30 wt.%, which was 1.7 times higher than that of pure PI, whereas dielectric loss was only 0.009. When PVDF content was as high as 50 wt.%, the dielectric constant of composites synthesized by chemical imidization was 7.745, which was about 2.1 times higher than that of pure PI. Furthermore, the dielectric constant of the composites was stable, and dielectric loss was lower than 0.1 within the testing frequency range (100 Hz–1 MHz). Thus, the PI/PVDF composites were demonstrated to be promising dielectric materials in the field of electronics. We will further improve the thermal and dielectric properties of composites in future research. In addition, we will also study issues such as material compatibility and thermal conductivity.

References

1. S. M. Yu, F. Qin, G. C. Wang, J. Mater. Chem. C 4, 1504 (2016)
2. Q. M. Zhang, H. Li, M. Poh, F. Xia, Z. Y. Cheng, H. Xu, C. Huang, Nature 419, 284 (2002)
3. J. M. P. Alaboson, Q. H. Wang, J. D. Emery, A. L. Lipson, M. J. Bedzyk, J. W. Elam, M. J. Pellin, M. C. Hersam, ACS Nano 5, 5223 (2011)
4. Y. Shen, Y. Lin, M. Li, C. W. Nan, Adv. Mater. 19, 1418 (2007)
5. K. Deshmukh, G. M. Joshi, RSC Adv. 4, 37954 (2014)
6. M. Arbatti, X. B. Shan, Z. Y. Cheng, Adv. Mater. 19, 1369 (2007)
7. X. Y. Huang, P. K. Jiang, Adv. Mater. 27, 546 (2015)
8. Y. Zhang, Y. Wang, M. Li, J. Bai, ACS Appl. Mater. Inter. 4, 65 (2012)
9. K. Vanherck, G. Koeckelberghs, IFJ. Vankelecom, Prog. Polym. Sci. 38, 874 (2013)
10. H. Y. Lu, C. Y. Chou, J. H. Wu, J. J. Lin, G. S. Liou, J. Mater. Chem. C 3, 3629 (2015)
11. X. F. Lei, Y. Chen, H. P. Zhang, X. J. Li, P. Yao, Q. Y. Zhang, ACS Appl. Mater. Inter. 5, 10207 (2013)
12. Z. M. Dang, J. B. Wu, L. Z. Fan, C. W. Nan, Chem. Phys. Lett. 376, 389 (2003)
13. Z. Chen, J. Zhao, S. Yan, Y. Yuan, S. Liu, Mater. Lett. 157, 201 (2015)
14. F. Xiang, H. Wang, K. C. Li, Y. H. Chen, M. H. Zhang, Z. Y. Shen, X. Yao, Appl. Phys. Lett. 91, 192907 (2007)
15. Z. M. Dang, Y. H. Lin, C. W. Nan, Adv. Mater. 15, 1625 (2003)
16. Y. J. Li, M. Xu, J. Q. Feng, Z. M. Dang, Appl. Phys. Lett. 89, 072902 (2006)
17. Z. M. Dang, H. Y. Wang, Y. H. Zhang, J. Q. Qi, Macromol. Rapid Comm. 26, 1185 (2005)
18. Y. Li, X. Huang, Z. Hu, P. Jiang, S. Li, T. Tanaka, ACS Appl. Mater. Inter. 3, 4396 (2011)
19. L. Dong, C. Xiong, H. Quan, G. Zhao, Scripta. Mater. 55, 835 (2006)
20. T. Hanemann, H. Gesswein, B. Schumacher, Microsyst. Technol. 17, 1529 (2011)
21. G. Wang, Y. Deng, Y. Xiang, L. Guo, Adv. Funct. Mater. 18, 2584 (2008)
22. H. Liu, Y. Shen, Y. Song, C. W. Nan, Y. Lin, X. Yang, Adv. Mater. 23, 5104 (2011)
23. Z. M. Dang, Y. Shen, L. Z. Fang, N. Cai, C. W. Nan, J. Appl. Phys. 93, 5543 (2003)
24. R. Gregorio, E. Ueno, J. Mater. Sci. 34, 4489 (1999)
25. A. DaSilva, C. Wisniewski, J. Esteves, R. Gregorio, J. Mater. Sci. 45, 4206 (2010)
26. L. Wang, J. Li, D. Wang, D. Wang, H. Li, Polym. Bull. 71, 2661 (2014)
27. M. Wahab, I. Kim, C. Ha, Polymer 44, 4705 (2003)
28. Y. Chen, D. Han, W. Ouyang, S. Chen, H. Hou, Y. Zhao, H. Fong, Composites: Part B 43, 2382 (2012)
29. H. Guo, M. A. B. Meador, L. McCorkle, D. J. Quade, J. Guo, B. Hamilton, M. Cakmak, G. Sprowl, ACS Appl. Mater. Inter. 3, 546 (2011)
30. H. Yeo, M. Goh, B. C. Ku, N. H. You, Polymer 76, 280 (2015)
31. J. Xia, J. Li, G. Zhang, X. Zeng, F. Niu, H. Yang, R. Sun, C. Wong, Composites: Part A 80, 21 (2016)

32. Z. M. Dang, J. K. Yuan, J. W. Zha, T. Zhou, S. T. Li, G. H. Hu, Prog. Mater. Sci. 57, 660 (2012)
33. J. Chon, S. Ye, K. J. Cha, S. C. Lee, Y. S. Koo, J. H. Jung, Y. K. Kwon, Chem. Mater. 22, 5445 (2010)
34. L. Zhang, X. Shan, P. Wu, Z. Cheng, Appl. Phys. A 107, 597 (2012)
35. Y. Zhang, Y. Wang, Y. Deng, M. Li, J. Bai, Mater. Lett. 72, 9 (2012)

Research on Frequency Control of Grid Connected Sodium-Sulfur Battery

Fenglin Zhang , Luan Chen, Haotian Chang, Jianxin Chen, Zhibin Deng, Yuncong Yang

School of Energy Science and Engineering, University of Electronic Science and Technology of China Chengdu, China

Abstract. Sodium sulfur battery is the only energy storage battery with large capacity and high energy density. It has a great application prospect in the peak load shifting of power grid, due to the lack of domestic research on it, it is urgent to evaluate the effect of grid-connection of sodium sulfur battery scientifically. According to the experimental data of the sodium sulfur battery project, the battery model is built. Compared with the real discharge curve, the error of the model simulation curve is small, so the battery model is effective. The AC / DC power grid model is built, and the rectifier and inverter control circuits are designed to simulate the scenario that the wind turbine and the battery are supplied to the passive load. The simulation results show that the grid-connected model of the sodium sulfur battery under the two control strategies can stabilize the larger frequency fluctuation.

1 Introduction

In recent years, the new type of energy storage battery has developed rapidly, and has been used more and more in the power grid. Energy storage battery not only can play the role of "peak load shifting", but also may be used as a large scale long-term energy storage device in the power grid in the long term.

As an important factor of power quality, the frequency of power grid is an important index to evaluate the effect of peak load shifting of sodium sulfur battery. So it is very important to develop a reasonable and effective integration strategy. Based on AC / DC power transmission model, this paper simulates the grid connected effect of sodium sulfur battery, because the energy storage battery is always combined with the new energy[1], the wind power model is used in the model. Document [2] proposed to use reactive power and voltage control to solve the problem of lithium battery grid connection, and the key to study the problem of peak load shifting is the power balance between power and load. Therefore, this paper is based on reactive power control, at the same time, the current data are collected to achieve the "power control", and then directly control the frequency[3] of the system after grid connected.

2 Sodium sulfur battery model

2.1 Model establishment

The most important thing to study the grid-connection problem of sodium sulfur battery is to establish the correct battery model[4], which is the basis of the follow-up work. According to the nature and working characteristics of sodium sulfur battery, the equivalent circuit model of sodium sulfur battery is established. The model is shown in figure 1.

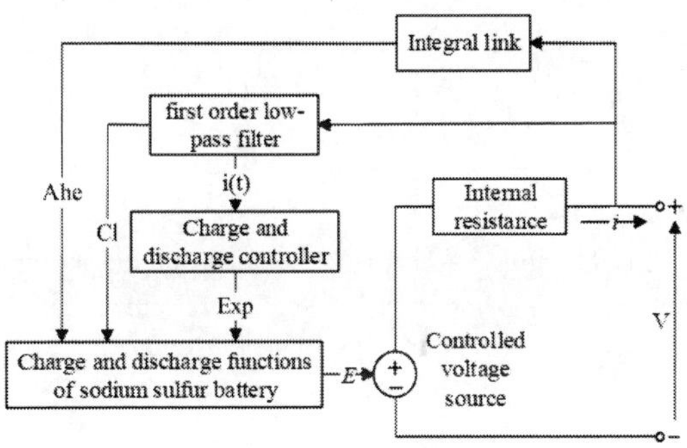

Figure 1. The circuit model of battery

In the above model, the charge and discharge functions of the source can best reflect the characteristics of the battery. According to the test data of sodium sulfur battery, the charging and discharging model[5] of the battery is established.

It is known that the electric potential of the battery during charging is E_c . The potential is related to a number of factors, which can be expressed by Formula (1).

$$E_c = E_0 - K\frac{Ah_m}{Ah_m - Ah_e}C_l - K\frac{Ah_m}{Ah_m - Ah_e}Ah_e + A\exp(-BAh_e) \quad (1)$$

In Formula (1), E_0 represents the constant voltage of the battery, K represents the polarization resistance, Ah_m and Ah_e represent the maximum capacity and extracted capacity of the battery respectively, and C_l represents low frequency current dynamics. In order to avoid the complexity of the formula, exponential voltage and exponential capacity are represented by A and B respectively.

Because of the nature of the battery, the charge and discharge curves of the battery are not coincident, and the potential functions of charging and discharging are also different. The expression of potential in the discharge process of sodium sulfur battery is shown as formula (2).

$$E_{dc} = E_0 - K\frac{Ah_m}{Ah_e + 0.1Q}C_l - K\frac{Ah_m}{Ah_m - Ah_e}Ah_e + A\exp(-BAh_e) \quad (2)$$

In formula (2), E_{dc} is used to represent the potential of the battery in the discharge process. Sodium sulfur batteries operate at very high temperatures, usually between 300 and 400 degrees Celsius, so the battery model should take into account temperature factors. The temperature dependent parameters in the model are constant voltage E_0, polarization resistance K and maximum capacity Ah_m. These parameters can be expressed with formula (3), (4) and (5).

$$E_0 = E_0(T) = E_0\big|_{T_N} + \frac{\partial E}{\partial T}(T - T_N) \quad (3)$$

$$K = K(T) = K\big|_{T_N} + \exp\left(\alpha\left(\frac{1}{T} - \frac{1}{T_N}\right)\right) \quad (4)$$

$$Ah_m = Ah_m(T_e) = Ah_m\big|_{T_e} + \frac{\Delta Ah_m}{\Delta T}(T_e - T_N) \quad (5)$$

In the above formula, T_N represents the nominal ambient temperature, T represents the internal temperature, T_e represents the ambient temperature, and α represents the Arrhenius rate constant for the polarization resistance.

Then the expression of the output voltage V_o can be obtained.

$$V_o(T) = E_c - R(T)i \quad (6)$$

$$V_o(T) = E_{dc} - R(T)i \quad (7)$$

The formula (6) is the expression of output voltage during the charging process, the formula (7) is the expression of output voltage during the discharging process, and $R(T)$ represents the internal resistance of the battery, and its expression is shown in formula (8).

$$R(T) = R\big|_{T_N}\exp\left(\beta\left(\frac{1}{T} - \frac{1}{T_N}\right)\right) \quad (8)$$

In formula (8), β represents the Arrhenius rate constant for the internal resistance.

2.2 Model parameters and verification

According to the established model of sodium sulfur battery, the relevant parameters of the battery are set up. The parameters are all the experimental data in the test of sodium sulfur battery, which are shown in Table 1.

Table 1. Model parameter of single sodium sulfur battery.

Parameter	Value
Discharge voltage	2.076-1.9V
Charging voltage	1.9-2.2V
Rated current	40A
Limiting current	120A
Limiting voltage	35V
Internal resistance	2.1-10mΩ
Working temperature	345-355℃
Rated capacity	0.595kWh

In order to evaluate the accuracy of the battery model, the data in the parameter list is the data of the single cell.

According to the established battery equivalent circuit and model, the simulation model of is established based on the battery model in MATLAB/Simulink. In order to evaluate the accuracy of the model, the discharge curve of the single cell was simulated, and the discharge curve of the battery under different discharge current was obtained. In order to reflect the difference of curves under different currents, the discharge current values are set to 20A, 40A and 60A, as shown in Figure 2.

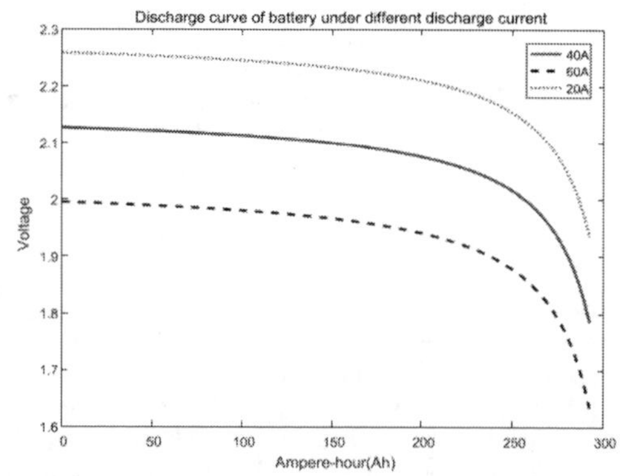

Figure 2. The discharge curves under different current.

The abscissa is the capacity of the cell, and the ordinate is the voltage of the battery. The three curves are obtained by the battery model built above, and verify the correctness of the model needs to be compared with the real discharge curve. The rated discharge current of the sodium sulfur battery in the actual project is 40A, so the real discharge curve of the battery with the discharge current of 40A is compared with the simulation curve to verify the correctness of the model, as shown in Figure 3.

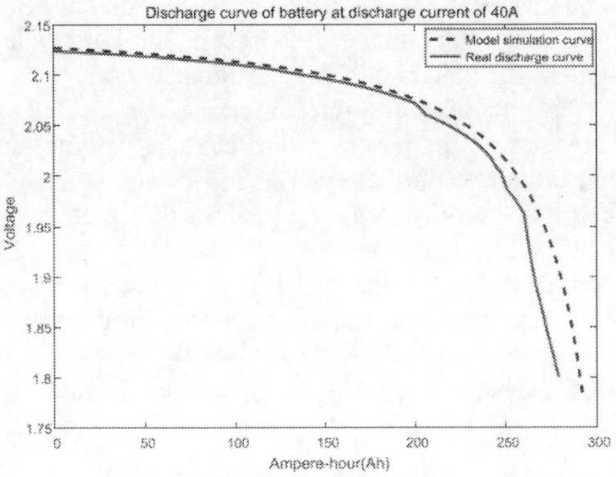

Figure 3. The discharge curves under 40A

It can be seen from the diagram that the discharge curve of the sodium sulfur battery model established in this paper has less error compared with the real curve, so it can be used for grid connected simulation experiment.

3 AC/DC power grid model

With the continuous development of new energy sources, sodium sulfur batteries are mainly used in the grid system composed of new energy sources. This paper selects the wind power and sodium sulfur battery connected to the grid, in order to test the effect of peak load shifting of batteries. In order to avoid excessive loss of electric energy in the transmission process causing the larger frequency fluctuation at the load side, the transmission process adopts the mode of direct current transmission.

AC/DC power grid model simulation based on MATLAB/Simulink. The transmission model is composed of a wind turbine model, a load model and a transmission and transformation model, the specific implementation methods of each model are described below.

3.1. Establishment of AC/DC power transmission model

The model simulates the situation of wind farm and battery supply power to passive load simultaneously. In order to test the effect of peak load shifting in sodium sulfur battery, step load is added to the model to investigate its transient[6] response.

The 3-phase source module is used to simulate a wind farm, and the Distributed Parameter Line function module is used to simulate the DC transmission line. The 3-phase parallel RLC load module is used to simulate the three-phase load.

The AC to DC rectifier consists of AC bus, transformer, filter circuit and bridge rectifier. The DC to AC part consists of filter, inverter, AC bus and transformer.

Because the simulation model is too complex, the basic components of the simulation model are described in block diagram, as shown in Figure 4.

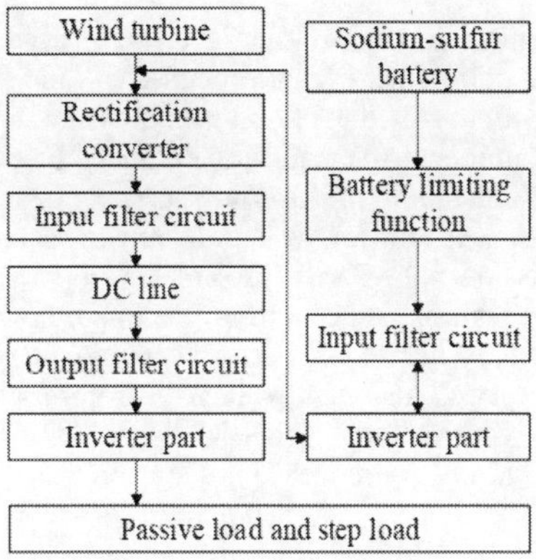

Figure 4. AC/DC power transmission model

In figure 4, the three-phase power generated by the wind turbine is transformed into a high voltage direct current through the rectifier bridge. After the filter circuit and the long distance transmission line, the power supply is supplied to the load through the inverter. When the battery is sufficient or the load is suddenly increased, the sodium sulfur battery supplies power to the load through the filter circuit and the inverter part, at the same time, the battery will absorb the electric energy from the grid when the battery capacity is insufficient or the load is suddenly reduced, so as to achieve the effect of peak load shifting.

3.2 Model of rectifier bridge control circuit

The control of rectifier is very important to DC transmission system. The power generated by the wind turbine needs to be converted into DC power by rectifier to be transmitted. In order to achieve the purpose of long distance transmission, it is necessary to boost the wind power. The DC voltage after transformation is easily disturbed by various disturbances, and the double closed-loop control strategy based on state feedback decoupling can solve this problem better. The simulation model is shown in Figure 5.

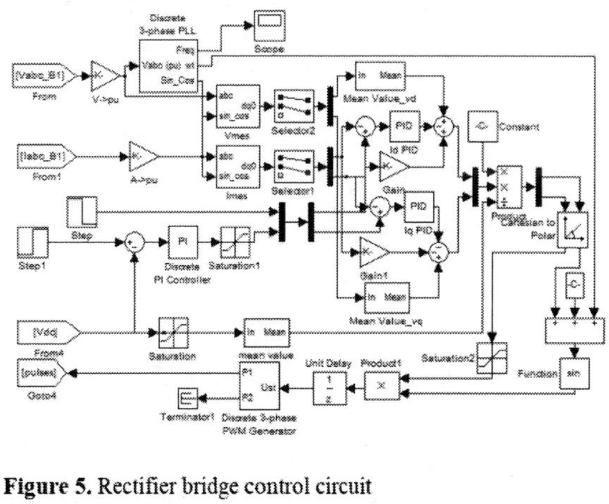

Figure 5. Rectifier bridge control circuit

In the above model, the reference value of DC voltage is 20Kv, and the reference value of reactive current is set to 0.

3.3 Model of inverter control circuit

The control circuit of the inverter is needed for the sodium sulfur battery grid-connected and the power supply circuit for the DC transmission line to load. The two inverter control circuits are described below.

The function of the inverter is converting high voltage direct current into low voltage alternating current in the transmission line. The control of the inverter is simpler than the control of the rectifier, the important role of DC transmission line to load power supply part is to ensure the stability of AC output voltage, therefore, only one AC voltage control loop is needed to achieve the control effect. The control model[7] is shown in figure 6.

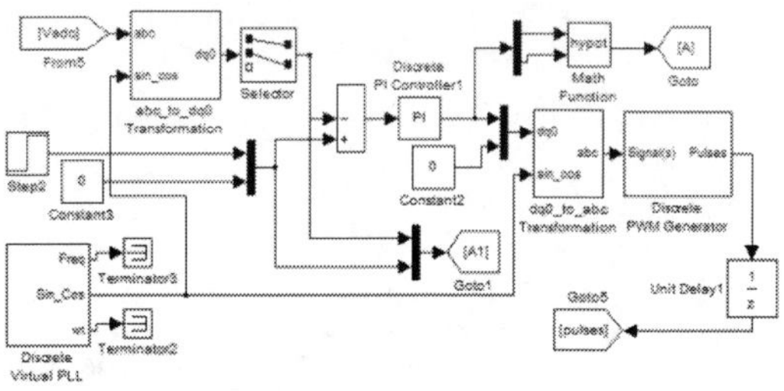

Figure 6. Inverter control circuit

The model can represent the inverter control model of sodium sulfur battery grid-connected. But the process of sodium sulfur battery grid-connected is more complex, and also need to consider some of the battery's own restrictions (battery voltage, capacity).

Fig. 7 is the control strategy with battery voltage as the standard in the control circuit of sodium sulfur battery.

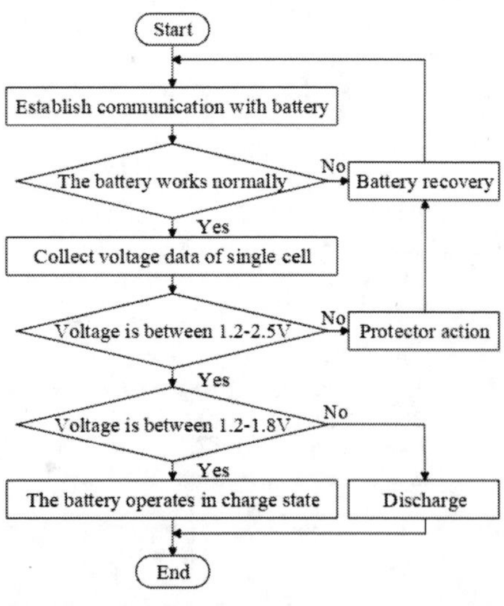

Figure 7. Control strategy with battery voltage as the standard

The diagram shows that the normal voltage range of single sodium sulfur battery is 1.2-2.5V, and the battery level is low between 1.2-1.8V, so the battery needs charging. In the 1.8-2.5V, the battery level is high, so the battery needs discharging. Similarly, the control strategy based on the battery capacity is also applicable to the block diagram shown in Figure 4, The normal capacity range of single sodium sulfur battery is 50-300Ah, and the battery level is low between 50-150Ah, so the battery needs charging. In the 150-300Ah, the battery level is high, so the battery needs discharging.

As one of the control indexes, the change of the frequency of the power grid affects the working state of the sodium sulfur battery. The experimental results show that the control strategy of sodium sulfur battery with its own limiting conditions can achieve the purpose of peak load shifting. So the application of the frequency is relatively simple, when the frequency drops, the battery should work in the discharge state. When the frequency rises, the battery should work in the charging state.

4 Grid connected simulation

Because the battery capacity and the battery voltage are two different control standards, the grid connected process of the sodium sulfur battery is simulated with these two control[8] standards respectively.

The capacity of the passive load in the simulation model is 5MW. In order to reflect the " peak load shifting " effect of sodium sulfur battery, the performance of sodium sulfur battery was simulated by increasing and reducing load. The simulation is divided into two stages. At the beginning of the simulation, the load capacity is reduced by 2MW, and then the load capacity is increased by 6MW at third seconds of simulation. Through the above settings to test the effect of two grid connected strategies on the frequency stabilization, the simulation results are shown in figure 8 and 9.

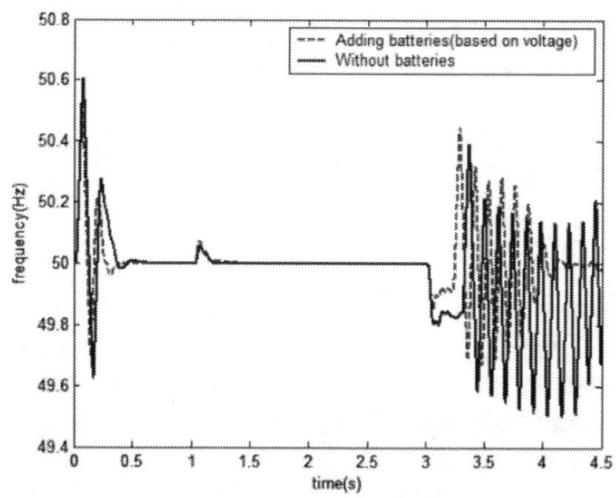

Figure 8. Frequency variation of power grid after adding sodium sulfur battery(based on voltage)

Figure 8 is the control strategy based on voltage as the standard, and the frequency fluctuation is obviously reduced in the diagram. Therefore, the peak load shifting effect of sodium sulfur battery is better.

The control strategy based on battery capacity as standard is shown in Figure 9.

 https://doi.org/10.1051/matecconf/201816003003

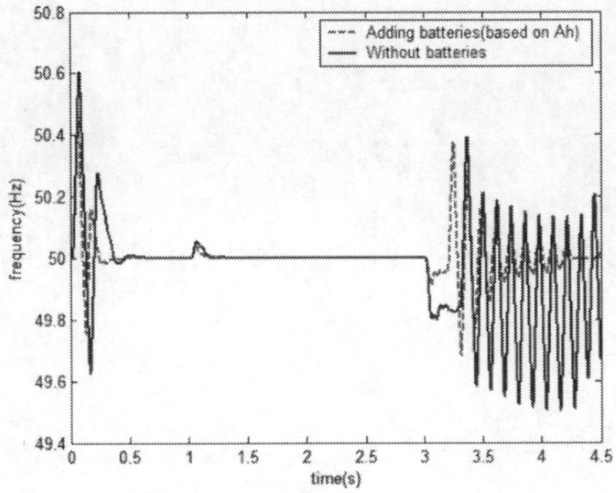

Figure 9. Frequency variation of power grid after adding sodium sulfur battery(based on Ah)

The capacity based control strategy has better control effect on frequency fluctuation, and the frequency fluctuation under the two control strategies is analyzed accurately. As shown in Table 2 and 3.

Table 2. Frequency fluctuation analysis(Load reduced by 2MW)

Index	Wind power	Adding batteries(V)	Adding batteries(Ah)
Maximum frequency	50.61	50.53	50.47
Minimum frequency	49.62	49.70	49.76
Fluctuation range	0.99	0.83	0.71
recovery time	0.3s	0.25s	0.16s

Table 3. Frequency fluctuation analysis (Load increased by 6MW)

Index	Wind power	Adding batteries(V)	Adding batteries(Ah)
Maximum frequency	50.41	50.43	50.39
Minimum frequency	49.53	49.71	49.73
Fluctuation range	0.88	0.72	0.66
recovery time	/	0.8s	0.62s

From the above analysis, it can be seen that when the sodium sulfur battery is connected to the grid, the battery capacity as the standard control strategy can better stabilize the frequency fluctuation. The frequency recovery time is short.

In order to verify the effect of sodium sulphide batteries, the follow-up work focused on the performance of the sodium sulfur battery in the actual operation.

5 Conclusion

The sodium sulfur battery model proposed in this paper can simulate the charging and discharging characteristics of the battery very well, and this model can provide a solid foundation for the follow-up study of sodium sulfur batteries. In the process of sodium sulfur battery grid connection, the control strategy with the battery capacity as the standard can better stabilize the grid frequency.

References

1. E.M.G. Rodrigues, G.J. Osório, R. Godina, A.W. Bizuayehu, J.M. Lujano-Rojas, J.C.O. Matias, J.P.S. Catalão, "Modelling and sizing of NaS (sodium sulfur) battery energy storage system for extending wind power performance in Crete Island", In Energy, vol. 90, part 2, pp 1606-1617, 2015.
2. ZY Hou, XH Hu, D Hui, "Design and Implementation of Grid-Connected Converter for Lithium Battery Energy Storage System", Power System Technology, vol. 36, issue 3, pp 246-251, 2012.
3. K.W. Gao, "Research on Coordinated Control Strategy of Wind Power and Energy Storage System for Improving Frequency Stability", Shenyang University of Technology, 2017.
4. Chung LW, Siam MFM, Ismail AB, Hussien ZF, "Modeling and simulation of sodium sulfur battery for battery energy storage system and custompower devices", National Power & Energy Conference: PECon 2004, Proceedings, pp.205-210, 2004.
5. Sarasua Antonio E , Molina Marcelo G, Mercado Pedro E ,"Dynamic model of sodium sulphur battery for application in microgrids Modelo dinámico de batería de sodio sulfuro para su aplicación en microredes", Revista de Ciencia y Tecnología, vol. 0, issue 20, pp. 5-10, 2013.
6. Mehdi A, Reama A, Benalla H, "Two Vector Based Direct Power Control of AC/DC Grid Connected Converters Using a Constant Switching Frequency", Journal of Power Electronics, vol.17, issue 5, pp.1363-1371, 2017.
7. Y Yu, K.P. Liu, J Yang, J Chen, "Small-signal Modeling and Decoupling Control of Hybrid Multi-terminal HVDC System", Proceedings of the CSEE, vol. 36, issue 1, pp 76-85, 2016.
8. Y.T. Li, "A Master's Thesis in Electrical Machines and Electrical Apparatus", School of Electrical Engineering Shenyang University of Technology, 2014.

A dynamic model of battery energy storage system based on the external characteristic equivalent

Xing Qu and Xinran Li

School of Electrical and Information Engineering, Hunan University, Changsha, Hunan Province, China 410082

Abstract. With the increasing application of battery energy storage in the power grid, there will be inevitably a large number of battery energy storage systems (BESS) in the future distribution network. It is very important to study the equivalent model of BESS which is suitable for comprehensive load modelling. In this paper, the battery group is established based on a type of general battery model, and then, a simulation platform of BESS based on MATLAB/Simulink is constructed, using a double closed loop control strategy with constant power as control object. Through the simulation analysis of grid-connected characteristics of BESS, a semi-mechanism dynamic model of BESS based on the external characteristic equivalent is proposed. This model represents the variation trend of the electrical variables of BESS during transient state, and considers the inertia characteristics of the BESS relative to power grid. By comparing the simulation results under different disturbance intensity, the proposed model can describe the transient characteristics of BESS, and can meet the performance requirements of the load modelling for the distribution network side.

1 Introduction

With the development of economy and society, on the one hand, the electricity demand is increasing continually and has obvious time characteristics, which make the peak-valley difference further expand. The application of energy storage technology in distribution network can effectively reduce peak-valley difference so as to realize peak load shifting, and then improve the comprehensive utilization ratio of distribution network assets [1]. On the other hand, small capacity and distributed intermittent renewable energy access will be widely distributed in the smart distribution network, and energy storage system (ESS) will be an essential device to smooth power fluctuation [2]. Battery energy storage is very suitable for smart distribution network because of its flexible configuration, fast response as well as no geographical environment restrictions [3]. Therefore, there will be a host of BESS in the future smart distribution network. Obviously, dynamic characteristics of BESS will affect the dynamic characteristics of the whole distribution network.

In the study of power system, load model has an important influence on the stability analysis of power system [4]. Since other models such as generator and transformer have been mature, the generalized load model considering the dynamic characteristics of BESS has become a key factor in the accuracy of power grid simulation results to the distribution network with BESS. Therefore, it is of engineering application value to study the dynamic model structure and parameter identification of BESS for generalized load model.

At present, many researches have been made on the modeling of energy storage batteries, and there are mainly two types of models: the electrochemical model [5], which has high precision but is extremely complex and is difficult to be applied to practical simulation [6]; the equivalent circuit model [7-9], which is widely used due to its simple modeling method, convenient parameter fitting as well as clear physical meaning.

The exiting researches on the modeling of energy storage system are mostly based on complex mathematical equations, or are simplified as a constant power source. Reference [10] establishes a detailed mathematical model of all parts of BESS for power system stability analysis, including battery group, inverter as well as controller. Reference [11] employs the general model of WECC that has complex model structure to verify the multiple application scenarios of BESS in power system. Reference [12] presents a generalized ESS model, using a set of linear differential algebraic equations describing (DAEs) the various types of ESS. Reference [13] proposes an electromechanical dynamic model based on wind-solar-storage hybrid simulation, but it is essentially a DC Thevenin equivalent model, without considering the influence of converter system. Reference [14] puts forward a general model of BESS taking account of grid-connected controller and converter that is equivalent to a first-order inertial link, but it is insufficient to describe the transient characteristics of BESS. Obviously, the above models are not applicable to describing the dynamic behavior of BESS in the generalized load model for the distribution network side, and now there is no dynamic model applicable to the generalized load modeling.

On the basis of the existing results, this paper studies the BESS dynamic model for the generalized load model for the distribution network side. Firstly, the simulation platform of BESS is built. Secondly, the dynamic behavior of grid-connected BESS is analyzed and its main features are extracted. On this basis, a dynamic model of BESS is proposed, which is suitable for generalized load modeling for the distribution network side. Then, the proposed dynamic model is verified and discussed. Finally, a summary of the whole paper is made.

2 Simulation platform of BESS

Battery monomer model generally include Thevenin model [7], PNGV model [8], and Run-time model [9] and so on. In this paper, the battery monomer model is based on the general model proposed in [15]. The battery group is usually composed of battery cells using series and parallel connection mode. This paper employs the equivalent circuit method to constitute the battery group for easily simulating [16].

In this paper, the BESS is incorporated into urban low voltage power grid, of which control strategy is double closed loop structure based on the d-q synchronous rotating coordinate system decoupling control method. The outer loop is the power loop, and the difference between the grid-connected power bus and the given power is used as reference current value of the d, q axis after passing through a PI controller. The inner loop is the current loop, and the difference between the grid-connected current and the reference current is used as reference voltage value of the d, q axis after passing through a PI controller as well as being decoupled; then, the PWM control signal of converter is obtained by anti-Parker transformation. The block diagram of the grid-connected control strategy is shown in Figure 1, where U_{dc} is the port voltage of battery group; i_d and i_q are the d and q axis components of the grid-connected current, respectively; u_{gd} and u_{gq} are the d and q axis components of the grid-connected voltage, respectively; u^*_d and u^*_q are the output voltage of the current inner loop, respectively; P^* and Q^* are the given active and reactive power, respectively; L is a filter inductance.

In the d-q coordinate system, the grid-connected active and reactive power can be expressed as

$$\begin{cases} P = 3/2(u_{gd} \times i_d + u_{gq} \times i_q) \\ Q = 3/2(u_{gq} \times i_d - u_{gd} \times i_q) \end{cases} \tag{1}$$

The BESS applied to distribution network generally absorbs or releases the active power, and the power factor is set as 1. When the direction of the d axis component of the voltage is in line with that of the resultant vector of the grid side voltage, namely u_{gq}=0, the formula (1) can be simplified as formula (2).

$$P = 3/2 \times u_{gd} \times i_d;\ Q = 0 \tag{2}$$

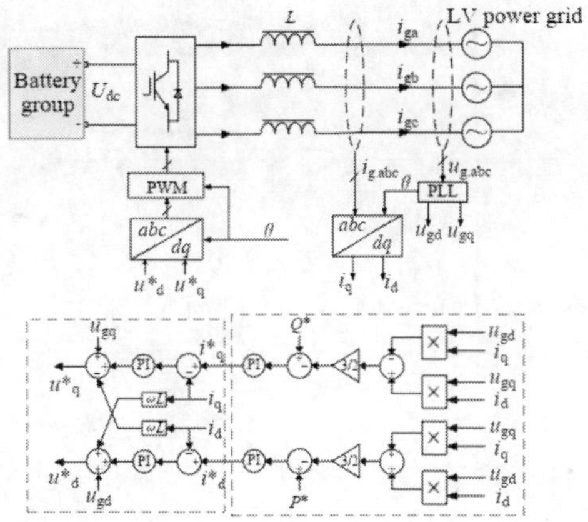

Figure 1. The control scheme of grid-connected converter.

3 Dynamic characteristics of BESS and its model

3.1 Dynamic characteristics of BESS

According to Figure 1 a simulation platform is constructed in MATLAB/Simulink to analyze the dynamic characteristics of BESS. When the voltage of power grid drops, through observing the change of the port voltage of battery group, grid-connected current, grid-connected power, the impact of the grid voltage on these variables is analyze, and thereby a mathematical model describing the relationship among them is established. Then the functional expression between grid voltage and grid connected power is got.

The power of battery group is set to 10kW, which is connected to the 0.4kV low-voltage power grid, and the power factor is 1. Supposing that a voltage drop with its duration being 0.1 seconds and its drop amplitude being 25% occurs, the curves of battery port voltage, grid-connected current and grid-connected power are recorded during 1 second before and after the voltage drop, which are shown in Figure 2.

Observing Figure 2, it can be found that BESS has the following characteristics during transient state: 1) In Figure 2 (a), the direct-current (DC) current of the battery group port increases about 5A, while the DC voltage of the battery group port decreases about 0.5V. 2) In Figure 2 (b), when the grid-voltage drops, the grid-connected current increases, but the increase magnitude is not proportional to the voltage drop amplitude. 3) In Figure 2 (c), similarly, when the grid voltage drops, the grid-connected power decreases, but the decrease magnitude is not proportional to the voltage drop amplitude, while the reactive power only fluctuates slightly at the moment when the voltage of the grid changes. 4) In Figure 2 (b) and 3 (c), during steady state and dynamic state, the grid-connected current and the grid voltage are in phase; when the grid voltage changes, the grid-connected power does not show step change, and there is a buffer process.

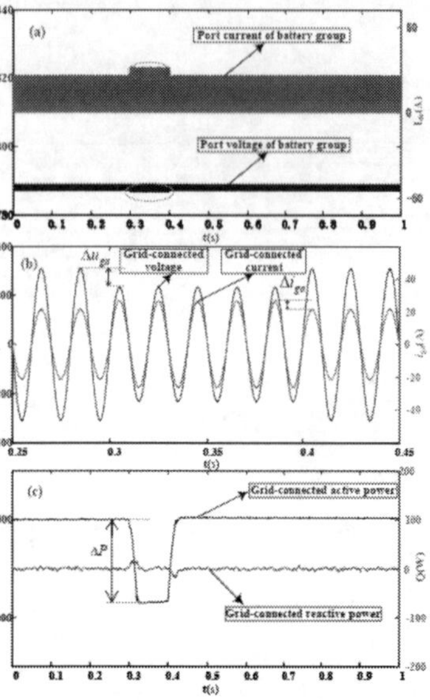

Figure 2. The transient characteristics of the BESS.

In term of the aforementioned analysis on dynamic characteristics of BESS, the following conclusions can be drawn: 1) During the grid voltage drop, although the port voltage of the battery group and the grid-connected power decrease, the port current of the battery increasing, while the total output power decreases. Therefore, it is not appropriate to take BESS as a constant power source. 2) During the dynamic period, the grid-connected voltage is determined by power grid, and the grid-connected current is determined by BESS and grid voltage, while the change of output power is determined by the combined effects of above all variables.

3.2 Dynamic model of BESS and its mathematical description

Concerning load modeling, the load model refers to the functional expression between the output power response and the input excitation voltage, which requires the model to be as simple as possible and the parameters as few as possible. Therefore, the dynamic model of BESS for distribution network generalized load modeling needs to meet the general requirements of load modeling and can describe the dynamic characteristics of BESS at grid-connected. In order to get the dynamic model of BESS, we will analyze and derive the model for load modeling based on the dynamic characteristics elaborated in Section 3.1.

The port voltage of the battery group is approximately constant (Figure 2 (a)) for the SOC can be considered unchanged in the microsecond and second time-scales, which determines the port voltage of the battery group almost unchanged. For converter system, its function is to convert DC to three-phase AC, so the port voltage can be equivalent to a three-phase voltage source from the perspective of functional equivalence. The equivalent three-phase voltage source is in phase with the grid voltage for the grid voltage and grid-connected current are the same phase (Figure 2 (b)). When the grid voltage drops, the grid-connected current increases (Figure 2 (b)). Thus, the equivalent three-phase voltage source can be connected with the power grid via a resistor. Furthermore, when the grid voltage has a step change, the power change has a buffer process, and from the point of view of the simplified model, a unity- gain inertia link can be used to represent.

Based on the above analysis, we can draw the dynamic model of BESS, of which equivalent structure is shown in Figure 3 and which is named as URT model in this paper. In Figure 3, $u_{b.abc}$ is the equivalent three-phase voltage source, which is in phase with the grid voltage $u_{g.abc}$; R is the resistance between the voltage source and the grid voltage; T is the inertial time constant; $i_{g.abc}$ is the gird-connected current.

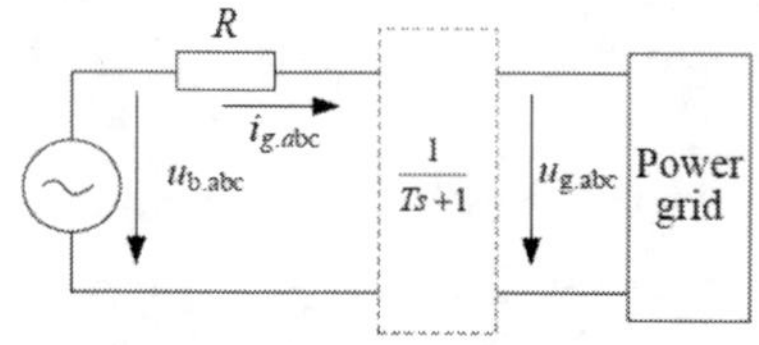

Figure 3. The dynamic model of the BESS.

It is important to point out that the inertial link in the red dashed frame is only used for the grid-connected power, but not for the voltage and current, and both sides of the inertial link can be understood as the grid voltage.

According to the equivalent model shown in Figure 3, the mathematical description such as formula (3)-(5) is given, where $i^*_{g.ab}$ is the conjugate of $i_{g.abc}$; P' is the intermediate value of grid-connected power.

$$i_{g.abc} = \frac{u_{b.abc} - u_{g.abc}}{R} \tag{3}$$

$$P' = u_{g.abc} \times i^*_{g.abc} \tag{4}$$

$$\frac{dP}{dt} = -\frac{1}{T}P + \frac{1}{T}P' \tag{5}$$

4 Model parameters and initialization

4.1 Model parameters

Known the structure of a model and its corresponding equations, the model parameters are obtained usually using system identification method. The first step of parameter identification is to determine the number of independent parameters and the identification initialization method.

The URT model has in total of 3 parameters: $u_{b.abc}$, R and T. When R or $u_{b.abc}$ is known, the $u_{b.abc}$ or R can be uniquely determined by combining the input voltage and the output power, and in other words, only one of R and $u_{b.abc}$ is an independent variable. In this paper, R is defined as independent variable, and then $u_{b.abc}$ is non-independent variable. Thus, the URT model has two independent variables, R and T.

4.2 Initialization

The initialization process of parameter identification is to determine the initial values of the non-independent and state variable. In URT model, the non-independent variable is $u_{b.abc}$, and the state variable is the model output response P_{mx}. Since the orders of magnitude corresponding to different variables may vary greatly, in order to improve the efficiency of identification, the per-unit system is generally adopted. We set U_b as per-unit value of $u_{b.abc}$, U_g as per-unit value of $u_{g.abc}$, and I as per-unit value of $i_{g.abc}$; the per-unit values of P_{mx} and R are still expressed with their actual value symbols.

Taking the grid voltage as the reference vector and passing through the Parker transformation, the d axis component of the grid voltage is equal to the resultant vector. Furthermore, the equivalent voltage source is in phase with the gird voltage. Therefore, U_b ,U_g and I can represent gird voltage, equivalent voltage source and grid-connected current, respectively, and then, the initialization process of model parameter identification is as follows.

1) Based on the steady-state grid voltage $U_g(0)$ as well as power $P(0)$, the initial value of grid-connected current is obtained by the formula (6).

$$I(0)=\frac{3}{2}\times\frac{P(0)}{U_g(0)} \quad (6)$$

2) The three-phase equivalent voltage source U_b is calculated by formula (7).

$$U_b=U_g(0)+I(0)\times R \quad (7)$$

3)The initial value $P_{mx}'(0)$ of the intermediate power P_{mx} is obtained by formula (8).

$$P_{mx}'(0)=\frac{3}{2}U_g(0)\times I(0) \quad (8)$$

4) Make the left end of the formula (5) be equal to zero, the initial value $P_{mx}(0)$ of the state variable (P_{mx}) is obtained.

$$P_{mx}(0)=P_{mx}'(0) \quad (9)$$

5 Simulation study

5.1 Description performance of URT model

In order to verify whether the URT model can describe the dynamic characteristics of BESS, taking the simulation platform of BESS introduced in Section 2 as the object and setting the voltage drops between 10% and 30% at grid-connected bus, a total of 3 sets of disturbance data as shown in Table 1, including the voltage and power, are collected to model BESS. Figure 4 is the tested power and model response power curves when the voltage drops are 15% and 25%, respectively.

Table 1. Sample instructions

Sample name	No .1	No .2	No .3
Voltage drop	15%	20%	25%

As shown in Figure 4, the model can well fit the sample data, which shows that the proposed URT model can effectively describe the dynamic characteristics of BESS and has good self-description ability. In addition, compared with Figure 4 (a) and (b), it can be found that the fitting effect is better than that of 20% when the voltage drop is 15%. The reason is that greater the voltage drop, the more obvious the transient power recovery process, and the power will go through a relatively long time to restore to the rated power state. Thus, the proposed URT model can well represent the dynamic characteristics of BESS.

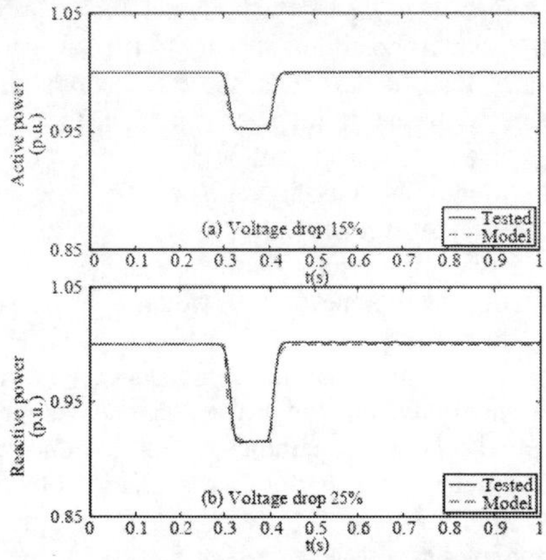

Figure 4. Power response curves of the model.

5.2 Generalization performance of URT model

The three samples in Table 1 correspond to that the grid voltage drops are 15%, 20% and 25% respectively. The No. 2 is chosen as the training sample to obtain the mode parameters via system identification. Then, the obtained parameters are used to fit the other two samples, and the fitting effect means the generalization performance of the URT model. To fit the No .1 and No. 3 using the No .2's parameters are called interpolation and extrapolation performance, respectively. The interpolation and extrapolation performance of URT model is shown in Figure 5.

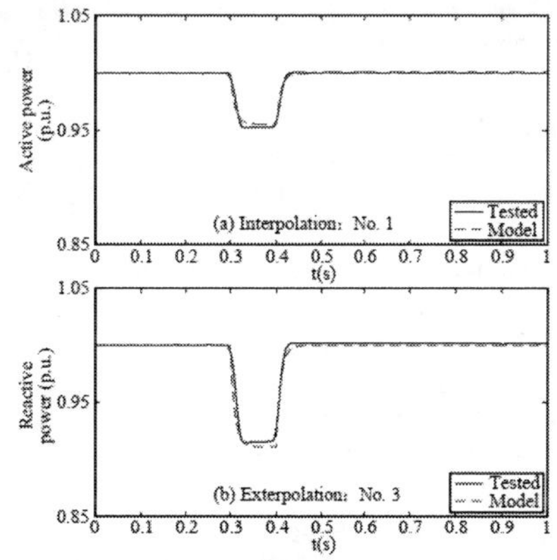

Figure 5. Generalization performance of the model.

As shown in Figure 5, the No.2' parameters can well fit the No. 1 and No. 2. The average fitting errors of interpolation and extrapolation performance are 0.000846 and 0.0016, respectively, which can be acceptable for load modeling. Therefore, the model has good generalization performance.

5.3 Discussion

1) Adaptability of model. At present, the application of BES focuses on the active power regulation of power grid or wind/PV station, such as energy storage frequency modulation, peak load shifting, and stabilizing the power fluctuation of distributed power supply. For power system load modeling, when considering the distribution network with BESS, BESS is generally regarded as a part of the generalized load, of which model is combined with the traditional load model to form a generalized composite load model. Thus, the BESS model studied in this paper is based on the scenario when the grid power factor equals 1. Besides, although the proposed mode is based on load modeling, it can be applied to other power system transient simulation.

2) Physical meaning of model parameters. This model contains three parameters, in which U represents the equivalent voltage source with the same phase as the grid voltage; R represents the impedance between the equivalent voltage and the grid voltage; T is not a general conceptual electrical parameter, but a time constant, which simulates the inertia characteristics of the whole BESS. Therefore, the UTR model is a semi-mechanism dynamic model.

3) Influence of control objects on model performance. In this paper, the modeling BESS is based on the given power, while the given variable is current, the BESS has different dynamic characteristics, of which external performance is a constant current source. Therefore, the URT model is not applicable to the BESS whose the given is current.

6 Conclusions

Energy storage battery plays an important role in smart grid. It is very important for accurate simulation of power system to study the dynamic model of BESS which is suitable for the generalized load modeling for the distribution network side. In view of the need of generalized load modeling, the simulation platform of BESS is constructed based on MATABLE/Simulink. From the perspective of generalized load, a dynamic model of BESS based on the external characteristics equivalent is proposed, and thereby the following three conclusions are drawn.

1) Under the constant power as the given condition, on the one hand, the dynamic characteristics of BESS do not conform to the constant power characteristic, and the output is nonlinear; on the other hand, it does not conform to the characteristics of the constant voltage source, and has inertia from the point of view of the grid.

2) When the grid voltage drop occurs at the grid-connected bus, the change of the port voltage of the battery group is very small, which can be considered basically unchanged, while the port current increases slightly, and the output power decreases but the internal loss of battery group increasing.

3) The URT model is simple in structure and less in parameters. The variation trend of grid-connected current caused by the change of the grid voltage is simulated from the power grid standpoint, and the inertia characteristic of BESS relative to power grid is also considered. Therefore, the model can well describe the dynamic behavior of BESS and can be applied to the load modeling for the distribution network side.

References

1. A. Oudalov, R. Cherkaoui, A. Beguin. Sizing and optimal operation of battery energy storage system for peak shaving application. IEEE Proceedings of Power Technology, Lausanne, Switzerland: IEEE, 5-10 (2007).
2. M. Beaudin, H. Zareipour, A. Schellenberglabe, et al. Energy storage for mitigating the variability of renewable electricity sources: An updated review. Energy For Sustainable Development , 14: 302–313 (2013).
3. Consulting Group of State Grid Corporation of China to Prospects of New Technologies in Power System. An analysis of prospects for application of large-scale energy storage technology in power system. Automation of Electric Power System, 37(1): 3-38 (2013).
4. K. H. Tseng, W. S. Kao, J. R. Lin. Load model effects on distance relay settings. IEEE, Transaction on Power Delivery, 18(4): 1140-1146 (2003).
5. W. D. Dennis, S. B. Vincent, B. Andre. Electrochemical modeling of lithium polymer batteries. Journal of Power Source, 110(2): 310-320 (2002).
6. R. J. Marijin, R. H. Boudewijn. Which battery model to use?. IET Software Let, 3(6): 445-457 (2009).
7. S. Barsali, M. Ceraolo. Dynamical models of lead-acid batteries: implementation issues. IEEE Transaction on Energy Conversion, 17(1): 16-23 (2002).
8. B. Zhang, L. Guo, H. Li, et al. PNGY model analysis of LiFePO4 li-ion battery for electric vehicle. Chinese Journal of Power Sources, 133(5): 417-421 (2009).
9. L. Benini, G. Castelli, A. Macci, et al. Discrete-time battery models for system-level low-power design. IEEE Trans. VLSI Systems, 9(5): 630-640 (2001).
10. C. Lu, C. Liu, C. Wu. Dynamic modeling of battery energy storage system and application to power system stability. IEE Proceeding on Generation Transmission Distribution, 142(4): 429-434 (1995).
11. X. Xu, M. Bishop, D. G. Oikarinen, et al. Application and modeling of battery energy storage in power systems. CSEE Journal of Power and Energy Systems, 2(3): 82-90 (2016).
12. á. Ortega, F. milano. Generalized model of VSC-based energy storage systems for transient stability analysis. IEEE Transactions on Power Systems, 31(5): 3369-3380 (2016).
13. H. Wang, Y. Tang, J. Hou, et al. Composition modeling and equivalence of an integrated power generation system of wind, photovoltaic and energy storage unit, Proceeding of the CSEE, 31(34): 1-9 (2011).

14. Y. Li, P. Jing, L.Wang, et al. A mathematical model of versatile energy storage system and its modeling by power system analysis software package. Power System Technology, 36(1):51-57 (2012).
15. M. Chen, G .A. Rincón-Mora. Accurate electrical battery model capable of predicting runtime and I-V performance. IEEE Transaction on Energy Conversion, 21(2): 504 - 511 (2006).
16. S. Peng, G. Shi, X. Cai, et al. Modeling and simulation of large capacity battery system based on the equivalent circuit method. Proceeding of the CSEE, 33(7): 11-18 (2013).

Substitution of Acetylene Black by Using Modified Flake Graphite Applied in Activated Carbon Supercapacitors

Peng Zhao[1], Chunmei He[2], and Wencheng Hu[1]

[1]State Key Laboratory of Electronic Thin Films and Integrated Devices, University of Electronic Science & Technology of China, Sichuan, Chengdu, 610054, P. R. China
[2]Experimental School of Shuangliu Middle School, Sichuan, Chengdu, P. R. China

Abstract. Flake graphite was mechanically modified at different times in N-methyl pyrrolidone under normal pressure. The results of the scanning electron microscopy, X-ray diffraction, and transmission electron microscopy suggested that the structure of the flake graphite was modified. The crystallinity of the flake graphite, and many defects were introduced into the material. The evaluation of capacitor performance by cyclic voltammetry, galvanostatic charge/discharge tests, and electrochemical impedance spectroscopy was also performed. Results showed that the electrochemical performance of flake graphite was strongly enhanced, particularly when it was exfoliated for 6 h. Moreover, the electrochemical capacitive properties of activated carbon were obviously enhanced through the substitution of acetylene black by flake graphite modified for 6 h.

1 Introduction

Supercapacitors (SCs), also known as electrochemical capacitors (ECs), hybrid capacitors, or ultracapacitors, are highly promising power devices that are superior to conventional physical capacitors and batteries in many aspects. These devices also play an important role in the energy-storage field [1, 2]. SCs show high power density, fast charging/discharging rate, sustainable cycling life, and excellent cycle stability [3, 4]. According to energy storage mechanism, SCs can be divided into two different types: double-layer capacitors (EDLCs) and faraday pseudocapacitor. EDLCs exhibit the capacitance that comes from pure electrostatic charges accumulated at the interface between an electrolyte and an electrode; while the pseudocapacitors typically show higher capacitances owing to the fast and reversible faradic redox reactions of electro-active materials of the electrodes [5-7].

Carbon-based electrode materials are most frequently used materrials for EDLCs include activated carbon, modified porous carbon, carbon nanotubes (CNTs), graphene, carbon aerogels, carbon nanofibers, etc[8]. Consequently, carbon materials are the ideal candidate electrode materials in SCs because of their easy accessibility, good processing ability, large surface area/porosity, low electrical resistivity, robust surface chemical environment, physicochemical stability, versatile existing forms, and low cost [9, 10]. Graphite is a carbon material that is widely used as the electrode material of SCs [11]. Graphite has a larger surface area and higher specific capacitance for an edge plane than other carbonaceous materials [12].

With regard to ESs, considering that the capacitance intensively relies on the electrolyte accessible surface of an electrode material, many researchers have focused on increasing the total surface area of graphite. For instance, mechanical grinding has been demonstrated as an effective approach [13-16]. After modification, the first cycle efficiency and capacity are improved. Nevertheless, some limitations exist in these methods [12, 17-20]. Thus, better methods are needed to overcome these defects.

We introduce a novel and simple method of modifying graphite by using a high-shear dispersing homogenizing emulsification machine for laboratory. The obtained graphite with high surface area has high specific capacitance and good Faradaic interfacial charge-transfer resistance (Rct). Moreover, this material can improve the electrochemical performance of activated carbon (AC) while doped into AC at an appropriate ratio. In this paper, flake graphite was chosen as the raw material in this dissertation. The High Shear Disperse Homogenizing Emulsification Machine for Laboratory was used to exfoliate it with different times. According to the results, its crystallinity was reduced, the specific surface area was increased and electrochemical performance was enhanced after flake graphite was exfoliated. Particularly, flake graphite exfoliated six hours (S6) owns the best performance as the specific capacitance of S6 is 99F g^{-1}, which is double of the raw material. Then S6 and acetylene black were added to the activated carbon using as conductive agent respectively. The result indicated that the electrochemical performance with S6was better, and the best content is 20%.

2 Experimental section

2.1 Sample preparation

All analytical-grade chemicals were supplied by Chengdu Changzheng Chemical Industrial, Ltd. (China), and used as received without further purification. Capacitor-grade AC with a specific surface area of approximately 2000 m^2 g^{-1} was purchased from Carbosino Material Co., Ltd. (China).

In a typical procedure, 0.5 g of flake graphite was dispersed in an appropriate volume of NMP, and the dispersion was heated in an oil bath at 60 °C overnight under intensive stirring. Thereafter, the dispersion was exfoliated with a high-shear-disperse homogenizing emulsification machine for laboratory for different times (2, 4, 6, 8, and 15 h) in a water bath at 70 °C under stirring. Thereafter, the resulting dispersion was centrifuged, repeatedly washed with ethanol, and dried in a vacuum at 80 °C. The obtained flake graphite modified for 2, 4, 6, 8, and 15 h were designated as S2, S4, S6, S8, and S15, respectively. The untreated flake graphite was referred to as S0.

2.2 Characterization

The X-ray diffraction (XRD) patterns of the five samples were obtained by using a Philips X'pert X-ray diffractometer operated at 1.5406 nm, 30 kV, and 20 mA. Data were collected at 2θ = 10° to 90°. A field-emission scanning electron microscopy system (Inspect F, FEI Co., USA) with an acceleration voltage of 20 kV was used to characterize the surface morphologies. Flake graphite exfoliated in NMP under normal pressure for 15 h was examined by transmission electron microscopy (TEM; JEM-100CX, JEOL, Japan).

2.3 Preparation of electrodes

At the first stage, the working electrodes were prepared by mixing 95 wt-% modified flake graphite with 5 wt-% SBR. Thereafter, the working electrodes were prepared by mixing 85 wt-% AC, 10 wt-% acetylene black (10 wt%, 20 wt-%, and 30 wt-% S6), and 5 wt-% SBR and were denoted as a, b, c, and d, respectively. On the basis of the weight of the electrode before and after coating, the mass of the deposited material can was calculated.

2.4 Electrochemical measurements

Electrochemical experiments were conducted by using Metrohm Autolab (PGSTAT 302N) with a classical three-electrode electrochemical cell at room temperature and normal pressure. The electrolytes were 7 M KOH and 1 M aqueous Na_2SO_4 solution at the two stages. A platinum tablet (1 cm × 2 cm) and a saturated calomel electrode were used as the counter electrode and reference electrode, respectively.

3 Results and discussion

XRD can be used to obtain the structural information of a material. Fig. 1 presents the XRD patterns of the five samples. Two diffraction peaks were observed near 25.5° and 54.6°, which corresponded to (002) and (004) reflections. The intensity of the two Bragg peaks gradually weakened, particularly the (002) Bragg peak, with increasing modification time, thus suggesting that many lattice defects were caused by the modification [19]. These defects led to the lower diffraction intensity of the samples than that of flake graphite.

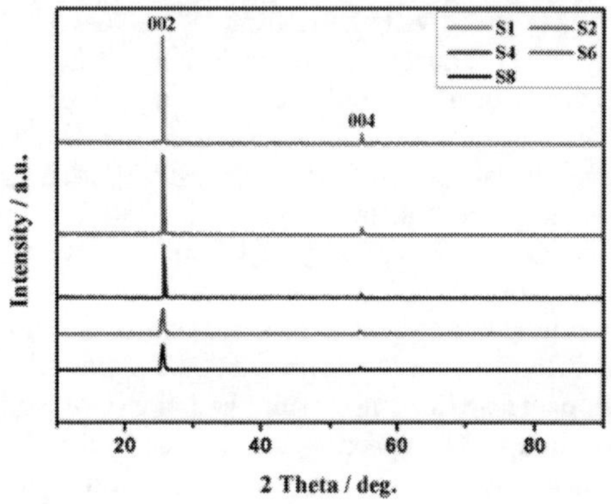

Figure 1. XRD patterns of the original and modified flake graphite.

SEM was used to show the morphologies of the materials (Figs. 2(a-c)). The flake graphite was composed of scaly grains measuring 10 μm to 20 μm. After treating flake graphite, the grain size was reduced and the large flakes were fragmented. With increased modification time, the specific surface area of flake graphite increased. The reason was that the high-shear disperse homogenizing emulsification machine for laboratory can provide high shear forces (22, 000 rpm), which undermine the Van der Waals forces of graphite layers. Given the greater transfer channels for ions, the charge/discharge performance at a large current may be strengthened.

Figs. 2(d) show the TEM images of the flake graphite exfoliated in NMP under normal pressure for 15 h. Flake graphite, which has a layer-to-layer film structure, has been exfoliated and cleaved into small pieces of sheet because of the high-energy shear force that breaks the covalent bond within the graphene planes [12]. This finding further proves the significant disruption of the ordered graphite structure throughout the long-term modification.

Figure 2. (a) SEM micrograph of S0; (b) SEM micrograph of S2; (c) SEM micrograph of S6; (d) TEM micrographs of S15.

Fig. 3(a) shows the CV curves of S0, S2, S4, S6, and S8 electrodes at a scan rate of 50 mV s^{-1} with potential windows ranging from -1 V to 0 V (vs. SCE) in a 7 M aqueous KOH solution. Flake graphite showed a nearly symmetrical rectangular curve, thus indicating good capacitive behavior. The distortion of the curve shape became progressively serious with increasing modification time. This distortion can be attributed to the deteriorated conductivity of flake graphite because of the intense destruction of the flake graphite structure with prolonged exfoliation time [8]. The area of the curve increased at the previous 6 h and then decreased because of the variation in the specific capacitance. The specific surface area of the electrode material was not the only factor that influenced the specific capacitance. The variation in the specific capacitance of S8 can be explained by its poorer pore structure than that of S6. Fig. 3(b) depicts the CVs of S6 at scan rates of 1, 5, 10, 50, and 100 mV s^{-1} with potential windows ranging from -1 V to 0 V (vs. SCE) in 7 M KOH aqueous solution.

The specific capacitance of the electrode measured by CV can be calculated from following equation:

$$C = \frac{1}{2 \cdot \Delta V \cdot m} \int_{V_{initial}}^{V_{final}} \frac{|I|}{(dV/dt)} dV \qquad (1)$$

where C is the specific capacitance (F g^{-1}), ΔV is the potential window (V), m is the mass of the carbon electrode (g), $V_{initial/final}$ is the starting/end potential in one cycle, $|I|$ is the instantaneous current at a given potential, and dV/dt is the potential scan rate. According to the equation, the specific capacitances of S0, S2, S4, S6, and S8 at a scan rate of 50 mV s^{-1} were 37, 45, 52, 65, and 63 F g^{-1}.

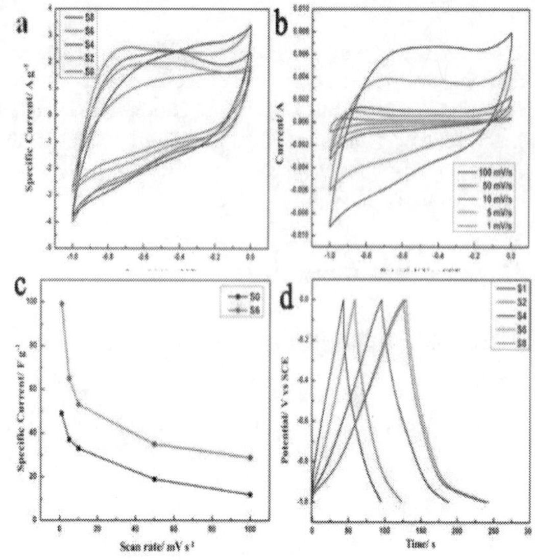

Figure 3. (a) CV curves of the five electrodes at the scan rate of 50 mV s^{-1} in 7 M aqueous KOH solution; (b) CV curves of the S6 at different scan rates in 7 M aqueous KOH solution; (c) The specific capacitance of the S0 and S6 at different scan rates derived from cyclic voltammetry curves; (d) Galvanostatic charge-discharge curves of S0, S2, S4, S6 and S8 electrodes at a current density of 1 A g^{-1}.

Fig. 3(c) presents the specific capacitance of S0 and S6 at different scan rates. At scan rates of 100, 50, 10, 5, and 1 mV s^{-1}, the specific capacitances of S0 were 49, 37, 33, 19, and 12 F g^{-1}, respectively; those of S6 were 99, 65, 53, 35, and 29 F g^{-1},respectively. The specific capacitance of S6 was nearly double that of S0. The galvanostatic charge/discharge curves of S0, S2, S4, S6, and S8 electrodes at a current density of 1 A g^{-1} were obtained within a stable window of -1 V to 0 V (Fig. 3(d)). S6 showed the highest specific capacitance corresponding to the CVs.

Fig. 4(a) shows Nyquist plots for S0, S2, S4, S6, and S8 electrodes in 7 M KOH aqueous solution with a frequency loop from 10^5 Hz to 10^{-2} Hz. All five Nyquist plots consisted of a semicircle in the high-frequency region and an arc in the low-frequency region. The two parts were ascribed to the charge-transfer process at the electrode/electrolyte interface and the diffusion process in solid, respectively [22]. The semicircle diameter showed the Faradaic interfacial Rct of the electrochemical system. The semicircle diameters of the exfoliated flake graphite electrodes were smaller than those of untreated flake graphite. This finding indicated that electrochemical reaction more easily occurred at the electrode/electrolyte interface of the modified flake graphite. The Rct values of S4, S6, and S8 were very close to one another. The experiment result further indicated that the modified flake graphite possessed excellent electrochemical capacitive properties.

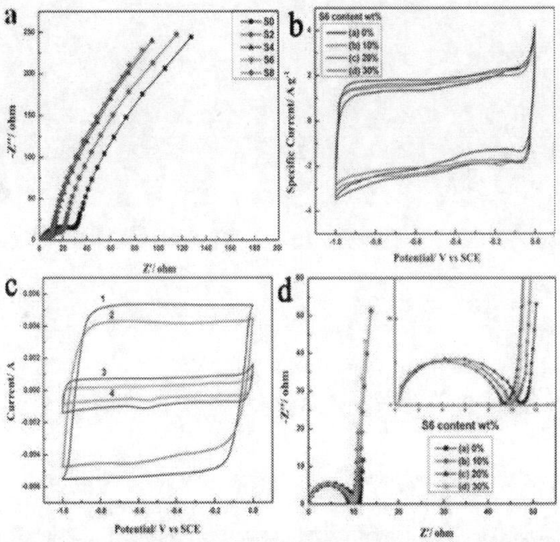

Figure 4. (a) Nyquist plots of S0, S2, S4, S6 and S8 electrodes. (b) CV curves of S6/AC electrodes with different S6 content at the scan rate of 10 mV s^{-1} in 1M Na_2SO_4 aqueous solution; (c) CV curves of the S6/AC electrode with 20 wt-% content of S6 at different scan rates in 1M Na_2SO_4 aqueous solution: (1) 100 mV s^{-1}; (2) 50 mV s^{-1}; (3) 10 mV s^{-1}; (4) 5 mV s^{-1}. (d) Nyquist plots of S6/AC electrodes with different S6 content. The inset shows the high frequency part of the Nyquist plots.

Fig. 4(b) shows the CV curves of S6/AC electrodes with different S6 contents at a scan rate of 10 mV s^{-1} with potential windows ranging from -1 V to 0 V (vs. SCE) in 1 M aqueous Na_2SO_4 solution. All curves showed a nearly rectangular shape and mirror-image characteristics, suggesting excellent electrochemical behavior. With increased S6 content, the area of the curve increased to different degrees. At a scan rate of 10 mV s^{-1}, the specific capacitances of the four electrodes were 168, 180, 208, and 195 F g^{-1}. The specific capacitance of S6/AC electrode with 20 wt-% S6 content was 23.8% higher than that of pure AC electrode. The enhancement in specific capacitance can be ascribed to the increase in conductivity and surface area because of the addition of the S6. Fig. 4(c) depicts the CVs of S6/AC electrode with 20 wt-% content of S6 at scan rates of 5, 10, 50, and 100 mV s^{-1} with potential windows ranging from -1 V to 0 V (vs. SCE) in 1 M aqueous Na_2SO_4 solution.

Fig. 4(d) shows the Nyquist plots of S6/AC electrodes with different S6 contents in 1 M aqueous Na_2SO_4 solution at a frequency loop of 10^6 Hz to 10^{-1} Hz. All Nyquist plots consisted of a semicircle in the high-frequency region and a line in the low-frequency region. The diameters of semicircles decreased from a to c and then enlarged from c to d, thus indicating that the Rct of S6/AC electrode with 20 wt-% S6 content was the smallest one and that S6 addition can enhance the electrochemical capacitive properties of AC.

4 Conclusions

In this study, a simple method was used to modify flake graphite. Flake graphite was exfoliated in NMP by using the high-shear disperse homogenizing emulsification machine for laboratory. The specific capacitance of flake graphite increased from 49 F g^{-1} to 99 F g^{-1} at a scan rate of 1 mV s^{-1} when it was modified for 6 h. The specific capacitance of S6/AC electrode with 20 wt-% S6 content was 23.8% higher than that of pure AC electrode, thus suggesting better electrochemical performance. The enhancement could be ascribed to the intrinsic specific capacitance of the modified flake graphite and the improvement in the hole utilization of AC resulting from the substitution of acetylene black by the modified flake graphite.

References

1. M. Winter and R. J. Brodd, Chem. Rev. **104,** 4245 (2004)
2. T.S. He, W.F. Mu, C.J. Zhang, Z.X. Jin and K.D. Cai, Adv. Mater. Res. **347**, 3645 (2012)
3. C. Liu, F. Li, L.P. Ma and H.M. Cheng, Adv. Mater. **22**, E28 (2010)
4. K. Liang, X.Z. Tang and W.C. Hu, J. Mater. Chem. **22**, 11062 (2012)
5. S. Yoda, K. Ishihara, J. Power Sources **81**, 162-169 (1999)
6. J. M. Ko, R. Y. Song, H. J. Yu, J. W. Yoon, B. G. Min and D. W. Kim, Electrochim. Acta **50**, 873 (2004)
7. A. Burke, J. Power Sources **91**, 37 (2000).
8. H.Q. Li, Y.G. Wang, C.X. Wang and Y.Y. Xia, J. Power Sources **185,** 1557 (2008)
9. E. Frackowiak and F. Beguin, Carbon **39**, 937 (2001)
10. A.G. Pandolfo and A.F. Hollenkamp, J. Power Sources **157**, 11 (2006).
11. H.Y. Wang and M. Yoshio, Electrochem. Commun. **8**, 1481 (2006)
12. H.Q. Li, Y.G. Wang, C.X. Wang and Y.Y. Xia, J. Power Sources **185,** 1557 (2008)
13. E. Gomibuchi, T. Ichikawa, K. Kimura, S. Isobe, K. Nabeta, and H. Fujii, Carbon **44**, 983 (2006)
14. M. Francke, H. Hermann, R. Wenzel, G. Seifert, and K. Wetzig, Carbon **43**, 1204 (2005)
15. F. Salver-Disma, A. Du Pasquier, J.M. Tarascon, J.C. Lassègues, and J.N. Rouzaud, J. of power sources, 291 (1999)
16. F. Salver-Disma, J.M. Tarascon, C. Clinard, and J.N. Rouzaud, Carbon 37, 1941 (1999)
17. R. Alcántara, P. Lavela, G. F. Ortiz, J. L. Tirado, R. Menéndez, R. Santamaría, and J. M. Jiménez-Mateos, Carbon **41**, 3003 (2003)
18. T.S. Ong and H, Yang, Carbon **38**, 2077 (2000)
19. F. Chevallier, L. Aymard, and J. Tarascon, J. Electrochem. Soc. **148**, A1216 (2001)
20. Y. J. Kim, Y. Horie, S. Ozaki, Y. Matsuzawa, H. Suezaki, C. Kim, N. Miyashita, and M. Endo, Carbon **42**, 1491 (2004)
21. C.S. Wang, G.T. Wu and W.Z. L, J. of Power Sources **76**, 1 (1998)
22. A.B. Yuan and Q.L. Zhang, Electrochem. Commun. 8, 1173 (2006)

Electrochemical Anti-corrosion System of Iron Tower Based on Solar Power Supply

Tian Tian and Shiwu Xiao

[1]*North China Electric Power University, School of Electrical and Electronic Engineering, 102206, Changping District, Beijing; China*

Abstract. Aiming at the serious problem of the corrosion of the transmission tower in the coastal area or in the harsh industrial area, a kind of electro-chemical anti-corrosion system based on solar power is designed. The system consists of a solar power module and an electrochemical anti-corrosion module: The solar power module consists of a solar panel, a photovoltaic controller, a accumulator and a constant potentiometer. The Electrochemical anti-corrosion modules include an anode block and an anode bed and reference electrode. The photovoltaic energy technology and forced current cathodic protection technology are used in the system, to achieve the effective protection of the tower anti-corrosion. Solar power supply to the nearest, solve the long-distance transmission loss and the high installation costs, form a simple structure, stable operation, low cost, clean and environmental protection, long service life of anti-corrosion system, with good economic efficiency and social benefits. It is of great significance to ensure the safe operation of the tower, maintain the normal operation of the power grid, and even promotes the optimization and upgrading of the industrial structure, save energy and reduces emissions, improve the safe and stable operation of the power system and the economic benefits, etc.

1 Introduction

With the rapid development of China's power industry, and the full construction of a strong smart grid, transmission tower as bearing structure of high voltage transmission lines, it is particularly important to guarantee the safe and reliable operation. In recent years, the coastal new nuclear power plants, thermal power plants, power plants, substations growing, a large number of power plants to send lines through the coastal tidal flat, long-term open-air transmission tower, by sea water, sea breeze impact and corrosion. Most of the existing power transmission tower adopts steel structure in our country, in the general atmospheric environment, the existing transmission tower anti-corrosion measures can make its normal service for many years, basically meet the transmission tower operation requirements. However, in coastal areas or in harsh industrial areas, environmental factors such as the wet and salty meteorological factors and severe air pollution will significantly reduce the durability of the transmission tower. This will not only increase the maintenance workload of the relevant departments, waste steel resources, improve the cost, and there will be serious when may affect the normal operation of the transmission line, and reduces the reliability of the transmission system operation [1].

In the references 1, the anti-corrosion of the transmission tower is analyzed in detail. At present, the anti-corrosion technology of the tower mainly includes hot-dip galvanizing, anticorrosive coating and external current cathodic protection. There are a lot of shortcomings in these technologies, mainly in: hot dip galvanized and anti-corrosion coating anti-iron tower corrosion method is not applicable to the coastal areas of humid and more salt environment, and in the construction, handling, welding and other processes easily lead to the damage, galvanized layer or paint layer loss of protection effect, and it will accelerate the corrosion of the tower, form pitting corrosion. In the references 2, an anti-corrosion device based on vibration-power generating of high-voltage transmission line is introduced. The anti-corrosion device based on vibration-power generating of high-voltage transmission line is complicated and the installation is inconvenient. It is not conducive to the construction and maintenance. At the same time, the insulation cost of high-voltage transmission line vibration power device is high, poor efficiency. Therefore, to seek more effective tower anti-corrosion technology has become a matter of urgency [2].

A kind of electrochemical corrosion tower system based on solar power is introduced in this paper. the use of solar power technology and forced current cathodic protection technology to fundamentally contain the metal corrosion tendency and corrosion rate, thus effectively extending the life of the tower, reducing the relevant departments Maintenance of the workload, and the system of low cost, long life, in the conservation of steel resources, reduce the cost of investment effect is significant. Therefore, the economic benefits of the system are obvious.

At the same time, the system structure is simple, the operation is stable, it has no influence on the existing equipment, the existing structure needs no change, the installation is simple and convenient. In addition, the use of solar power to provide additional current, solve the problem of using traditional energy pollution, clean and environmental protection. Solar energy nearby power supply, avoids long-distance transmission loss, maintenance simple, suitable for unattended situation. The polarization current size can be adjusted, the tower electric potential can be artificially set, and automatic tracking control. Therefore, the design result also has important economic and social significance.

2 Design principles and solutions

The system mainly uses the technology of compulsory current cathodic protection, which provides cathodic dc current to the corroded tower metal materials, and protects the tower against corrosion. Cathodic protection is a method of polarizing a metal cathode to prevent it from being corroded. It is an electrochemical protection method. Use the principle of corrosion cell, the tower metal structure which need to be protected is as cathode. Through the anode to the cathode to provide electrons uninterrupted, the structure is polarization first, and then the electrons are enriched on the surface of the structure, making it difficult to produce ions, thus greatly slows down the structure of corrosion speed [3].

The design system consists of solar power supply module and electrochemical anti-corrosion module. Solar power module consists of solar panel, photovoltaic controller, accumulator and constant potentiometer, electrochemical anti-corrosion modules include anode block and anode bed and reference electrode. The system structure diagram is shown in Figure 1.

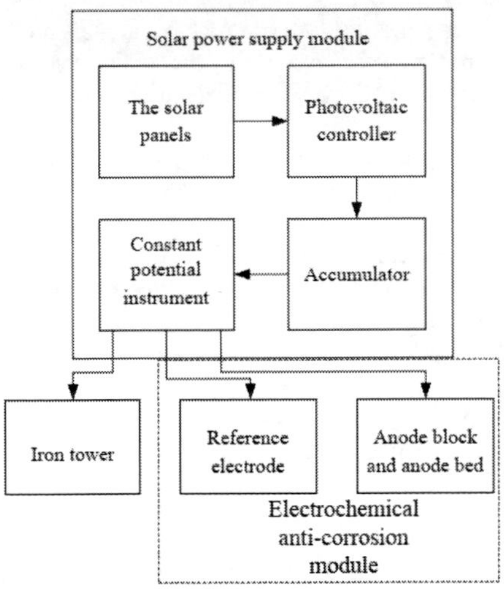

Figure 1. Block diagram of the anti-corrosion system.

The solar panel, photovoltaic controller, accumulators and potentiometer are connected in order. The output anode of the potentiostat is connected to the anode block and the anode bed with the wire, The output cathode of the potentiostat is connected to the lower part of the tower, The welding process is used to connect the wires directly to the internal steel so as to better protect the more easily corroded parts of the tower feet. The zero potential of the constant potentiometer is also connected to the cathode port of the tower lower part. The position of the connected power point is close to the output point of the output cathode of the potentiostat and the lower part of the tower, but it cannot be used to connect the zero to the output cathode with the short line. The sampling signal input terminal of the potentiometer is connected to the reference electrode.

The solar panel is installed on the tower, and the photovoltaic controller, battery and potentiometer are installed on the ground of the tower. The anode block and the anode bed are integrally formed and installed under the ground beneath the tower. The reference electrode is installed underground near the tower.

The protection process is as follows: the solar panel provides external power supply for the entire system, which converts solar energy into electricity in the light of sunlight. The role of PV controller is to control the working state of the solar power module, and to protect the battery through charging protection and over-discharge protection. The converted energy is stored in the battery by the PV controller and supplied to the potentiostat. Potentiostat give the tower a cathode current, and give the anode blocks and anode bed an anode current to constitute a corrosion cell, the reference electrode feedback is as constant standard, to control and adjust the polarization current output, and make the tower polarization within the scope of the protection potential, achieve the purpose of corrosion protection.

3 Key module analysis and design

3.1 Design of electrochemical anti-corrosion module and power analysis

The power analysis of electrochemical anti-corrosion was introduced in references 2, 3 and 4. Based on the analysis of the existing metal anti-corrosion methods, the system of electrochemical anti-corrosion of iron tower based on solar power generation is proposed in this paper. The electrochemical anti-corrosion module includes anode block and anode bed and reference electrode.

3.1.1 Anode block and anode bed

When choosing anodic material, it is necessary to take a comprehensive consideration of its characteristics, which generally requires good conductivity of anode materials, strong corrosion resistance, long life, large discharge, and sufficient mechanical strength.

The commonly used anode materials are graphite, stainless steel, high silicon iron, magnetic iron oxide, lead silver (2%) alloy and platinum-plated titanium tube (rods), etc.. These materials have good corrosion resistance characteristics. In this paper theoretical analysis, graphite is used as the anode block, but it is not limited to this in actual production, If necessary, it can also use waste steel, aluminum, zinc and other materials. Fill the anode block with coke as material and form an anode bed. The setting of coke anode bed can significantly reduce the actual consumption rate of anode materials, thus extending the service life of anode blocks [2,3,4].

3.1.2 Reference electrode

Commonly used reference electrode materials are Ag/AgCl electrode, Cu/CuSo4 electrode and Zn electrode. Here the reference electrode used Ag/AgCl electrode.

3.1.3 Power analysis of electrochemical anticorrosion of iron towers.

Access to the relevant provisions of the country, the bottom pillar of the tower can be simplified as a cylinder [2,3,4]: diameter $d = 0.1\text{m}$, length $L = 4\text{m}$, concrete diameter $D = 0.5\text{m}$. Surface area of tower feet: $S = \pi d L = 0.4\pi$, Select tower protection current density [2,3,4]: $i = 10\text{mA/m}^2$, Single tower pillar current: $I = iS = 12.5\text{mA}$, Total circuit current: $I_{\text{total}} = 4I = 50\text{mA}$, Required anode mass: $G = TgI / K = 0.4\text{Kg}$, Required anode block volume: $V = G / \rho = 95.8\text{cm}^3$.

Where T is the design life of the electrode 30a, g for the electrode consumption rate 0.158kg/A.a, K for the electrode effective utilization coefficient 0.5, I for the electrode operating current 0.1A.

In the design, the graphite body is taken as the anode block: diameter $d_{\text{graphite}} = 4\text{cm}$, length $L_{\text{graphite}} = 25\text{cm}$, take coke as anode bed: buried depth $t = 1.9\text{m}$, diameter $d_{\text{coke}} = 20\text{cm}$, length $L_{\text{coke}} = 60\text{cm}$. Access to information [5]:

Soil resistivity: $\rho_{\text{soil}} = 200\Omega\cdot\text{m}$, Coke resistivity: $\rho_{\text{coke}} = 50\Omega\cdot\text{m}$, Concrete resistivity, $\rho_{\text{concrete}} = 300\Omega\cdot\text{m}$, Contact resistance of graphite to coke:

$$R_1 = \frac{0.0171\rho_{\text{coke}}}{L_{\text{graphite}}} \times \ln\left(\frac{8L_{\text{graphite}}}{d_{\text{graphite}}} - 1\right)$$

$$= \frac{0.0171 \times 50}{0.25} \times \ln\left(\frac{8 \times 25}{4} - 1\right) = 13.31\Omega \quad (1)$$

Contact resistance of coke and soil [6] :

$$R_2 = \frac{\rho_{\text{soil}}}{2\pi L_{\text{coke}}} \times \ln\frac{2L_{\text{coke}}}{d_{\text{coke}}} \times \sqrt{\frac{4t + 3L_{\text{coke}}}{4t + L_{\text{coke}}}}$$

$$= \frac{200}{2\pi \times 0.6} \times \ln\frac{2 \times 0.6}{0.2} \times \sqrt{\frac{4 \times 1.9 + 3 \times 0.6}{4 \times 1.9 + 0.6}}$$

$$= 101.77\Omega \quad (2)$$

Anode potential

$$U_1 = I_{\text{total}}(R_1 + R_2)$$

$$= 0.05 \times (13.31 + 101.77) = 5.754V \quad (3)$$

Contact resistance of tower and concrete:

$$R_3 = \frac{\rho_{\text{concrete}}}{2\pi L_{\text{tower}}} \times \ln\left(\frac{4L_{\text{tower}}}{d_{\text{tower}}} - 1\right)$$

$$= \frac{300}{2\pi \times 4} \times \ln\left(\frac{4 \times 4}{0.1} - 1\right) = 60.506\Omega \quad (4)$$

Contact resistance of concrete and soil:

$$R_4 = \frac{\rho_{\text{soil}}}{2\pi L_{\text{tower}}} \times \ln\frac{4L_{\text{tower}}}{D_{\text{concrete}}} = 27.3268\Omega$$

$$= \frac{200}{2\pi \times 4} \times \ln\left(\frac{4 \times 4}{0.5} - 1\right) = 27.3268\Omega \quad (5)$$

Cathode potential:

$$U_2 = -I(R_3 + R_4)$$

$$= -0.0125 \times (60.506 + 27.3268) = -1.098\text{V} \quad (6)$$

Power of electrochemical anticorrosive consumption:

$$P = I_{\text{total}}(U_1 - U_2) = 0.05 \times (5.745 + 1.098) = 0.343 \quad (7)$$

3.2 Power supply module design and power analysis

The solar power module consists of a solar panel, a photovoltaic controller, a battery and a potentiostatic instrument.

3.2.1 Solar panels

The solar panel is the core of the module, providing an additional power source for the entire system, which converts solar energy into electricity in the light of sunlight.

Firstly, the power of solar panel is analyzed. The process is as follows:

According to the equation (7), the power of electrochemical anticorrosive consumption is 0.34W, and the system protects the tower continuously throughout the day. Therefore, the electricity consumption every day is:0.343×24=8.23WH. Considering the loss in the circuit and the storage process, the power supply efficiency is 0.7, and the effective sunshine time of the day is equal to 5h. The output power of the solar panel should be:

$$P = 8.23 / \eta / 5 = 2.35\text{W} \quad (8)$$

In design system, 12V/3W solar panels are selected.

3.2.2 Photovoltaic controller

PV controller main technical indicators:

1) Rated voltage: 12V,

2) Rated current: 30A,

3) Overload, short circuit protection: 1.25 times rated current 30 seconds, 1.5 times rated current 5 seconds, overload protection action, 3 times rated current short circuit protection.

3.2.3 Accumulator

Battery selection voltage is 12V, calculating battery capacity: 8.23WH/12V=0.68AH, That is, the daily power consumption is 12V/0.68 AH.

The design requires that the system can work normally during the continuous rain for 3 days. The battery daily power consumption should be controlled within 30%, thus requiring a 12V/2AH battery.

3.2.4 The potentiostat.

Main technical parameters of the potentiometer:

1) potential control range: ±2V, continuous adjustment,

2) Constant potential characteristics: ±0.1%(F, S),

3) Stability: ±0.5% (F,S)/24 hours

4 Corrosion process of the forced current cathodic protection

4.1 Theoretical analysis

In the case of forced current cathodic protection: there is an oxidation reaction on the boundary of the anode block (graphite, zinc block, scrap steel, etc.) and the hydrogen evolution reaction occurring on the protected cathode [7].

4.2 Model building

4.2.1 Model structure

To facilitate the modeling process and the ease of data selection, select zinc as the anode block, and the model structure is shown in Figure 2.

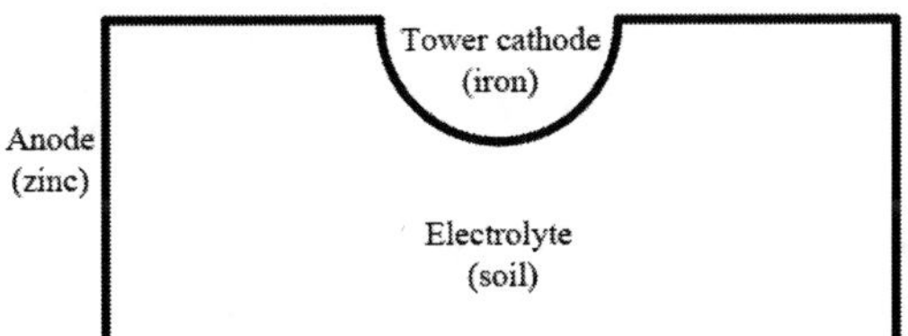

Figure 2. Model diagram of forced current cathodic protection.

4.2.2 Electrochemical current modeling

Use secondary current distribution interface for electrochemical current modeling, the diffusion of oxygen in the soil is described by the thin matter transfer interface. The oxidation and diffusion coefficient depends on the pore saturation of the pores.

4.2.3 Boundary conditions

The potential of the zinc anode is selected as the ground potential. Assuming that the reaction kinetics of the zinc anode is very fast, the polarization can be neglected, so the electrolyte potential is set as:

$$\phi_{l,Zn} = -E_{eq,Zn} + \phi_{s,Zn} = -E_{eq,Zn} \tag{9}$$

where $E_{eq,Zn}$ is the empirical measurement of the equilibrium potential of zinc/earth electrode. In the model, $E_{eq,Zn}$ is set to -0.68V. The oxygen concentration on the anode of Zn is set to the concentration in the atmosphere:

$$C_{O_2,Zn} = C_{O_2,ref} \tag{10}$$

Three different electrode reactions need to be considered on the boundary of the tower and soil, namely the oxidation of iron, the reduction of oxygen and the hydrogen dissociation reaction. The reaction equation is as follows:

$$\begin{cases} Fe \rightarrow Fe^{2+} + 2e^- \\ O_2 + 2H_2O + 4e^- \rightarrow 4OH^- \\ 2H_2O + 2e^- \rightarrow H_2 + 4OH^- \end{cases} \tag{11}$$

The kinetics of these reactions are modeled through the electrode surface nodes in the secondary current distribution interface. In this boundary condition, the external electrical potential of the rebar $\phi_{s,steel}$ is set as the applied battery potential -1V.

The electrode dynamics of rebar reaction are described according to the Tafel expression:

$$\begin{cases} i_{Fe} = i_{0,Fe} \times 10^{A_{Fe}\eta_{Fe}} \\ i_{O_2} = -\dfrac{c_{O_2}}{c_{O_2,ref}} i_{0,O_2} \times 10^{A_{O_2}\eta_{O_2}} \\ i_{H_2} = -i_{0,H_2} \times 10^{A_{H_2}\eta_{H_2}} \end{cases} \tag{12}$$

In modeling, some parameters of reference are shown in Table 1, and the potential of each reaction is calculated according to equation (13).

$$\eta = \phi_{s,steel} - \phi_l - E_{eq} \tag{13}$$

Table 1. Main parameters of simulation.

parameter	unit	Fe	O_2	H_2
Balance potential, *Eeq*	V	-0.76	0.189	-1.03
Exchange current density, *io*	A/m^2	7.1×10^{-5}	7.7×10^{-7}	1.1×10^{-2}
Tafel Slope, *A*	V/ decade	0.41	-0.18	-0.15

According to Faraday's law, the oxygen reduction reaction on the steel surface produces an oxygen flux, which can be modeled using the electrode electrolyte interface coupling node in the lean transfer interface. The other boundaries use symmetrical boundary conditions, and the initial values of the oxygen concentration are atmospheric.

5 Conclusion

This paper designs an electrochemical anti-corrosion system based on solar power. In view of the defects of the existing steel tower anti-corrosion technology, the power of electrochemical anti-corrosion protection of the tower was carried out by using the compulsory current cathodic protection technology, and solar power was used as the system energy. The anti-corrosion system based on solar power supply fundamentally curbs the corrosion tendency and corrosion rate of the metal, thus effectively prolong the life of the tower, reducing the maintenance workload of the related departments, and the system of low cost, long service life, to save steel resources, reduce the investment cost effect is remarkable. In addition, the system structure is simple, the operation is stable, it has no influence on the existing equipment, the existing structure needs no change, and installation is simple and convenient. The use of solar power to provide additional current has solved the problem of the use of traditional energy pollution. Solar energy nearby power supply, avoids long-distance transmission loss, maintenance simple, suitable for unattended situation. The polarization current size can be adjusted, the tower potential can be artificially set, and automatic tracking control. These features make the design of the tower of electrochemical corrosion protection system based on solar power has a good economic and social benefits, can be applied to the actual production, with a certain practical value.

References

1. Chengzhi Ma, Hongcang Yang, Qiyu Yu, Weizheng Liang, Yueling Lin. Comparison and Analysis on Anticorrosion Methods of Transmission Tower in Coastal Area[J]. Electromechanical engineering technology, 2014(12).141-144,218.
2. Ming Ge, Jinda Mao, Chen Zhang, Yiping Shang, Jiachen Bian, Yueming Liu,Wenbin Wu. A Design of Electrochemical Protection for the Tower Based on Vibration-Power Generating of Lines[J]. Transmission and Distribution Engineering and Technology, 2014(3).69-75
3. Xu Bai. The Application Study and the technology of Anti - corrosion of Substation Grounding Grid Based on Cathodic Protection Principle[D]. Master's thesis, North China Electric Power University, 2011.
4. Yunjin Yang. The simplified calculation of conductor shockproof[J]. Design of coal mine. 1980(3).18-20,40
5. Fengping Wang, Wanli Kang, Hemin Jing. Principles, methods and applications of corrosion electrochemistry [M]. 2008.
6. [6] Jianhong Wang, Quan Tang Anti-corrosion coating technology of concrete structure of Shantou Bay Bridge[J]. China New Technologies and New Products,2010(9).49
7. http:// cdn.comsol.com/ documentation/ 5.2.1.262 / COMSOL_Software License Agreement. zh_CN. Pdf

An Energy Loss Calculating Method for Wind Power System Based on the Shape Factor

Bo RUAN[1], Junjie QIAN[2], Dahai YOU[2], Tingting HOU[1], Xi CHEN[1], Zhen MEI[2]

[1] *Research Institute of Economics and Technology, Hubei Electric Power Company Limited, Wuhan, Hubei Province, China 430077*
[2] *State Key Laboratory of Advanced Electromagnetic Engineering and Technology, Huazhong University of Science and Technology, Wuhan, Hubei Province, China 430074*

Abstract. The large-scale access of the new energy makes various changes to the power system characteristics. For example, the acute volatility of the new energy such as wind power and photovoltaic energy makes the power-loss-calculation for the system more complex. The traditional typical daily method is inaccurate when used in new energy system because of its overlook of the generator output volatility. This paper proposes a new power-loss-calculation method for wind power system which based on the shape factor and gain a more accurate result. On the basis of this new calculating method, finds that the shape factor of the wind power plant for an hour period usually falls in a certain range. Therefore, proposes to directly use the expectation of shape factor in whole year to compute the annual energy loss with at least two values and once power flow calculation. And the acceptable relative error proves its large engineering practicability.

1 Introduction

With the development of global economy, human demand for energy is growing while the primary energy such as fossil fuel is drying up. As a result, the new energy like wind power gains a national encouragement for its renewability and cleanliness. However, the access of the new energy will have an influence on the energy-loss-calculation for the power system because wind power and photovoltaic energy have features of randomness, volatility and regionalism [1-3]. The traditional energy-loss-calculation method is the typical daily method, whose basic thinking is to calculate the electric energy loss of the representative day first, and then estimate the whole periodic energy loss [4]. This method entirely ignores the volatility of generation and load, considers the generator output and load remain unchanged within one hour and the power loss of everyday in the whole period is the same. Obviously, the wind power plant whose output changes sharply every minute is not suitable for this traditional method. However, if we adequately take the volatility into account, amount of measured data and repeated power flow calculation are required. It's too fussy and time-consuming to be applied in the practical projects. As the installed capacity of the new energy is growing, seeking an efficient and simple energy-loss-calculation method for wind power system is worth studying.

This paper proposes an energy loss calculating method for wind power system based on the shape factor, it uses hourly shape factors of the whole period to correct the wind power output curve and gain a more accurate loss result. In the further study, finds the hourly shape factors of wind power plant usually focus on a small area and considers to calculate the annual energy loss with the expectation of the whole year shape factors. In this way, it at least only needs to conduct once power flow calculation and has a small relative error.

2 The typical daily method for energy loss calculation

The traditional power loss calculating method is typical daily method. Its basic thinking is to calculate the energy loss of the typical day first, and then reckon the whole periodic energy loss through the typical-day's.

Using this method to calculate the monthly power loss:

(1). Consider the generator output and the load remain unchanged within every hour in the representative day. Use $I_{t,l}$ and $V_{t,m}$ to represent the current of line l and the voltage of node m in hour t, $R_{t,l}$ and $G_{t,m}$ to represent line resistance and node conductance and the energy loss of representative day □A_d is:

$$\Delta A_d = \left[\sum_{l=1}^{L} \left(3\sum_{t=1}^{24} I_{t,l}^2 R_{t,l} \right) + \sum_{m=1}^{M} \sum_{t=1}^{24} V_{t,m}^2 G_{t,m} \right] \quad (1)$$

(2). Consider the watt hour meter is more accurate than the power meter. When these two meters are inconsistent, correct the power graph by apportioning the electric quantity to per hour according to the load and generator output curve on the typical day.

(3). Use the corrected load and generator output to compute the line current and node voltage through power-flow calculation. And then take the result into the formula (1) to calculate the energy loss of the representative day.

(4). Consider the daily power loss is the same as the representative day within the whole month and as a result, multiply the power loss of typical day by the number of days in that month to get the power loss of the whole month.

The specific formulas and details of the typical daily method are not elaborated in this paper and they can be referred to in reference [5].

3 Energy loss calculation for wind power system based on the shape factor

3.1 Equivalent power method and the shape factor

Equivalent power method is a well-known electrical energy calculating method uses the average power because of its usage of the shape factor. For one line, the power loss within time T is:

$$\Delta A = 3\int_0^T I^2 R dt = 3I_{eq}^2 RT = \frac{P_{eq}^2 + Q_{eq}^2}{V^2} RT \tag{2}$$

where, I_{eq}, P_{eq} and Q_{eq} represent the equivalent value of the current, the active power and the reactive power respectively. The equivalent value is actually the root-mean-square value. The equivalent value of the current, the active power and the reactive power can be expressed by their average value as:

$$\begin{aligned} I_{eq} &= GI_{av} \\ P_{eq} &= KP_{av} \\ Q_{eq} &= LQ_{av} \end{aligned} \tag{3}$$

where, G, K and L are shape factors of the load curve $I(t)$, $P(t)$ and $Q(t)$ respectively.

The energy loss can be calculated by average power through the shape factor, and the average values are more available and accurate. The studies on shape factor show that it is decided by the form of the load curve. Actually, it is a reflection for the volatility of the load curve, the greater the volatility, the bigger the shape factor. It can be gain through mathematical derivation that the value range of the shape factor K is:

$$1 \le K \le \frac{1+\alpha}{2\sqrt{\alpha}} \tag{4}$$

where, α is the ratio of the minimum load to the maximum load called minimum load ratio. The concrete details can be referred to in reference [6].

3.2 An energy loss calculating method for wind power system based on the shape factor

The typical daily method has two assumptions that ignore the volatility of the generator output. First, considers the energy loss of everyday in the whole period is the same. The other is considering the generator output and the load remain unchanged within every hour.

This two assumptions can improve the calculation efficiency, but can also bring prodigious calculating error when applied in wind power system because of its acute volatility. As a result, it should be improved. First, for the representative day, this paper proposes the concept of wind power output expectation graph and it can be calculated as equation (5) below. The method collects the data of the wind power output per hour per day in the whole calculating period. Then the superposition of the data in per hour is divided by the number of days to acquire the expectation of wind power output per hour for that period. The 24 expectations constitute an expectation graph to replace the typical day curve.

$$P_{m,t} = \frac{\sum_{k=1}^{T_m} P_{k,t}}{T_m} \tag{5}$$

where, $P_{m,t}$ is the active power expectation of the wind power output in hour t, $P_{k,t}$ is the active power expectation of the wind power output in hour t on day k. T_m is the number of days in the whole period.

Secondly, for the output volatility within one hour, this paper brings in the shape factor. Concretely, uses the shape factor of every hour to correct the expression curve gained from the last step for a more accurate energy loss. The steps to calculate the shape factor are as follows:

Consider one hour is a period and the average of the active power is shown in equation (6):

$$P_{ave,t} = \frac{\sum_{k=1}^{T_m}\sum_{n=1}^{N} P_{k,n,t}}{T_m \times N} \tag{6}$$

The root-mean-square value of the active power is shown in equation (7):

$$P_{eq,t} = \sqrt{\frac{\sum_{k=1}^{T_m}\sum_{n=1}^{N} P_{k,n,t}^{\ 2}}{T_m \times N}} \tag{7}$$

So the shape factor of this period is shown in equation (8):

$$K_t = \frac{P_{eq,t}}{P_{ave,t}} = \sqrt{\frac{\sum_{k=1}^{T_m}\sum_{n=1}^{N} P_{k,n,t}^{\ 2}}{T_m \times N}} \Bigg/ \frac{\sum_{k=1}^{T_m}\sum_{n=1}^{N} P_{k,n,t}}{T_m \times N} \tag{8}$$

In the equation (6) to (8), $P_{ave,t}$ is the mean value of active power of wind farms during the period t, $P_{eq,t}$ is the RMS value of active power of wind farms at period t, $P_{k,n,t}$ is active power of wind farms at N sampling points during the period t, N is the number of sampling points per hour, and K_t is the shape factor of wind farms for the period t.

Then multiply the curve of the typical daily expected output by the corresponding shape factor to get the curve of the equivalent output of wind power, as shown in equation (9). Use the equivalent output curve to calculate the energy loss of one day and then estimate the total energy loss of the power system.

$$P_t = K_t \cdot P_{m,t} \tag{9}$$

where, P_t is the corrected active power of hour t, K_t is the shape factor of hour t.

4 Energy loss quick calculating method for wind power system based on the annual shape factor expectation

4.1 Annual shape factor expectation

Use the shape factor of each hour to corrects the expectation graph can gain an accurate result, but it needs lots of data and many times of power flow calculation. It's time-consuming and complex. When research the output characteristics of a wind power plant, it is found that although the output changes acutely in every hour, the most shape factors of one hour fall in a stable interval. As a result, considers to calculate the total energy loss using the expectation of the whole periodic shape factors. Take one year for example, it at least only needs two values that the annual electricity quantity and the annual shape factor expectation, and to conduct power flow calculation one time to gain the power loss of whole year.

Count the monthly generator output data of Xiangyang wind power plant somewhere and calculate its expectation and equivalent value to constitute the monthly output expectation curve and monthly output equivalence curve. As shown in figure 1. It is obvious that the expectation curve and equivalence curve have a totally the same variation tendency and it means that the shape factor of each hour changes little.

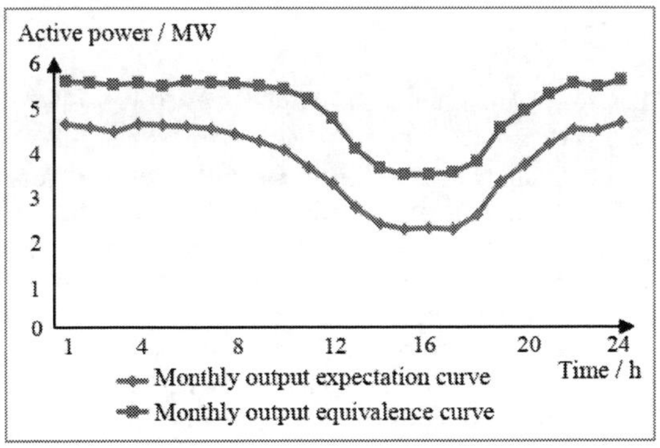

Figure 1. Monthly output expectation curve and monthly output equivalence curve (active power).

In order to further study the characteristics of wind power plant output for seeking an efficient energy loss calculating method, count the whole year output data of the Xiangyang wind power plant with the sampling frequency of 4 points per hour. Calculate the shape factor of each hour and make its frequency distribution histogram as figure 2.

It can be found from figure 2 that although the shape factors distribute in a wide range in the whole year and the maximum value reaches 2.6, most values concentrate within a small range. Take the range [1.2, 1.75] as a confidence interval, and its confidence coefficient reaches 96%. For the shape factor represents the volatility of the wind power system to some extent, it illustrates that most of the time, the volatility of the wind power system within an hour is close. Consider to use the expectation of whole year shape factor to replace the per hour shape factor for the correction of the expectation curve.

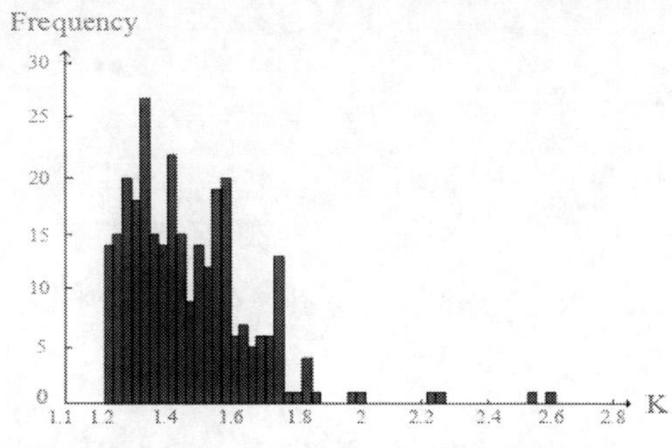

Figure 2. Frequency distribution histogram of shape factors K.

Use the Matlab to match the probability density distribution situation of whole year shape factors, as shown in figure 3.

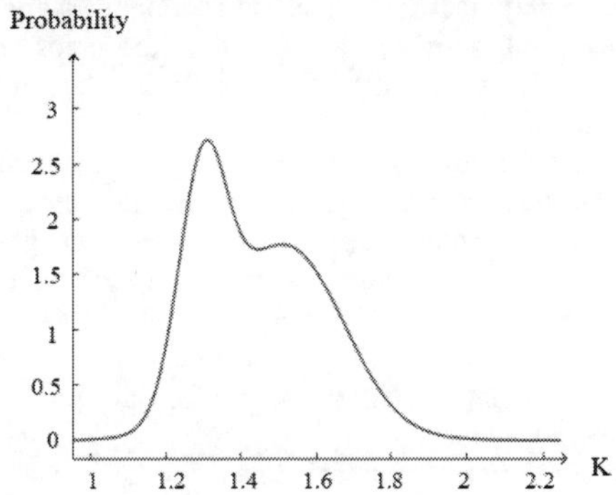

Figure 3. Matched probability density distribution of shape factors K.

As can be seen from the figure 3 that the probability distribution of shape factor presents a tendency of middle high and sides low. It reaches top when K is 1.33 and its form is similar to the normal distribution to some extent, although it is not symmetrical. Calculate the expectation of the whole year shape factors is $K_{av}=1.43$.

4.1 Use K_{av} to calculate annual energy loss

The structure of a 110kV main network is shown in figure 4.There are two power generators. One is named Longkou, which is a conventional power generation; the other is Xiangyang wind power plant which has been introduced in section 4.1 and its annual shape factor expectation has been calculated as 1.43. There are 5 loads and 4 lines in figure 4.

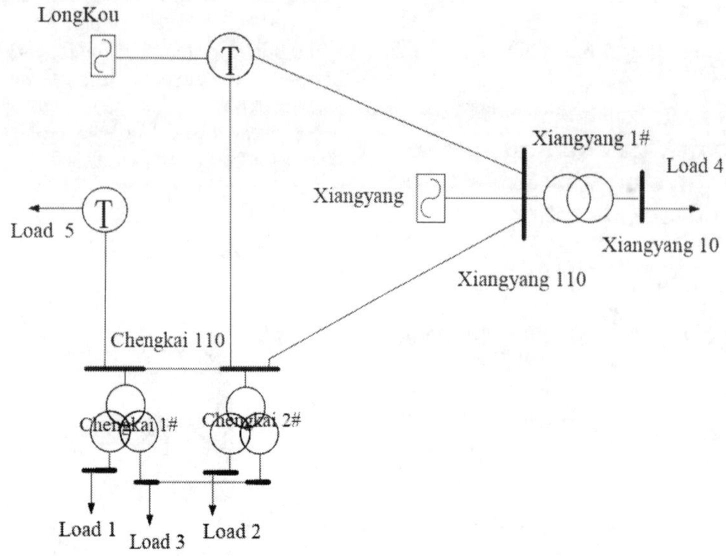

Figure 4. The structure of a 110kV main network.

Propose four kinds of solutions to calculate the annual energy loss of Xiangyang wind power plant. Use traditional typical daily method, per hour shape factor method and annual shape factor expectation method respectively to calculate the result and compare these methods.

As a reference, use the most complex way to calculate a reference value which is close to the real value. Collect wind power output data of Xiangyang wind farm 4 sampling points per hour. The power flow is calculated 4 times an hour, namely 96 times a day. The energy loss is accumulated day by day to acquire the annual energy loss. The accumulated energy loss value is set to be the reference value.

The reference value calculated by the power flow is 1953.8 MWh.

Solution one: use the traditional typical daily method to calculate the annual energy loss. Choose one day of the year as the representative day and calculate its power loss. Then multiply by the number of days in that year to gain the annual power loss. The load curves of the representative day are shown in figure 5 and figure 6.

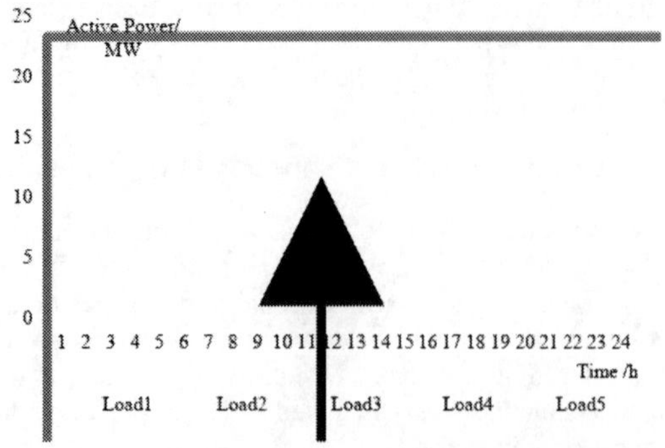

Figure 5. The active load curve of the representative day.

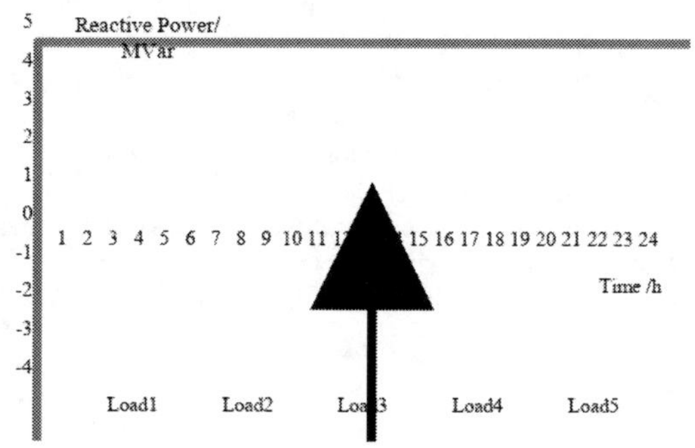

Figure 6. The reactive load curve of the representative day.

Solution two: use the shape factor of per hour to calculate the energy loss. Collect data to calculate the expectation curve of generator output and the shape factor of per hour, and then correct the output expectation curve with the average electricity. Get the curve of the typical daily equivalent output of wind farms for the year by multiplying curve of the typical daily expected output by the corresponding shape factor. And then use the equivalence curve to conduct power flow calculation for the energy loss of the typical day, and finally estimate the annual energy loss of the power system.

Solution three: use the shape factor expectation of whole year to calculate the annual energy loss. Collect data to calculate the expectation curve of generator output and the shape factor expectation of whole year. After correcting the output expectation curve with the average electricity, multiply the curve values by the shape factor expectation of whole year K_{av} to get an equivalent curve. Then calculate the energy loss of the representative day with the equivalent curve via the power flow calculation. Finally, multiplied with the number of days in the year to calculate annual energy loss.

Solution four: use annual electricity quantity and annual shape factor expectation to calculate the annual energy loss. Use the annual electricity quantity to calculate the average active power P_{ave} as equation (10):

$$P_{ave} = \frac{A_{y,P}}{T_y \times 24} \tag{10}$$

where, P_{ave} is the average active power of the wind power system this year, $A_{y,\,p}$ is the total electricity quantity this year, T_y is the number of days in that year.

Correct the P_{ave} with annual shape factor expectation K_{av} to gain the equivalent active power P_{eq} as equation (11):

$$P_{eq} = K_{av} \times P_{ave} \tag{11}$$

where, P_{eq} is the equivalent active power of the wind power system.

Assume that the active power output is the same in the whole year and it is equal to P_{eq}. Use the P_{eq} to conduct power flow calculation and gain the energy loss for an hour. As a result, the annual power loss can be calculated as equation (12):

$$\Delta A_{y4} = \Delta P \times 24 \times T_y \tag{12}$$

where, ΔP is the energy loss for an hour, ΔA_{y4} is the annual energy loss.

It should be illuminated that the process mode of reactive power is similar to the active power and not be detailed here.

It's obvious from above four solutions that the data and calculating steps needed have a simplified tendency from solution two to solution four. The solution four needs the least data and the simplest calculation steps and if its error is in the allowable range, this method has a strong engineering practicality.

Calculate the annual energy loss with these four solutions above respectively and compared with the reference value. The result is shown in table 1.

Table 1. Analysis of annual energy loss calculation result.

Calculation method of annual energy loss	Annual energy loss (MW)	Reference value (MW)	Relative error
Solution one	1547.5	1953.8	20.79%
Solution two	1846.8	1953.8	5.47%
Solution three	1833.4	1953.8	6.16%
Solution four	1802.7	1953.8	7.73%

As shown in table 1, the solution one uses traditional typical daily method and ignores the volatility of wind power system, which needs little data and simple calculating steps but has a large calculation error. The solution two uses the shape factor of 24 hours to calculate the energy loss and it needs the maximum data and the most complex calculating steps and gains a most accurate result among the four solutions. The solution three uses the shape factor expectation of whole year and its calculation error is little bigger than solution two. The solution four uses the shape factor expectation of the whole year to calculate the annual energy loss. Its calculation error is little bigger than solution two and three, but it only needs two values and once power flow calculation and the 7.73% is within the acceptable relative error range. If the environment around the wind power plant changes not particularly large, the shape factor expectation $K_{av}=1.43$ can be used in the calculation of later years and it only needs the annual electric quantity one value to calculate the annual energy loss.

5 Conclusion

This paper proposes a new energy loss calculating method based on the shape factor to solve the problem of the traditional method that can't be used in large volatility system. The new method replaces the traditional representative day curve with the expectation curve of the calculation period and makes use of the shape factor of per hour to correct the expectation curve for a more accurate result. This method changes the calculating thinking of the energy loss that use the shape factor to replace the traditional empirical value. The normal system without wind power plant can also take account of this method to improve the calculating accuracy. Additionally, for the shape factor per hour of the wind power plant is within a small range, this paper proposes to use the annual shape factor expectation to calculate the annual energy loss directly. It at least only needs two values and once power flow calculation and its calculation error is within an acceptable range. As a result, this method has a strong engineering practicability.

References

1. Pierre Pinson, Henrik Madsen, Henrik Aa. Nielsen, George Papaefthymiou, Bernd Kl¨ockl. From Probabilistic Forecasts to Statistical Scenarios of Short-term Wind Power Production[J]. Wind Energy (2008):1-13.
2. Pierre Pinson*, Henrik Aa. Nielsen, Jan K. Møller and Henrik Madsen, George N. Kariniotakis, Non-parametric probabilistic forecasts of wind power-required properties and evaluation[J]. Wind Energy. 2007; 10:497–516.
3. Feng Ni. The Study on Calculation Method of Theoretical Energy Loss in Low-voltage Distribution Network[M].Xi'an: Xi'an University of Technology, 2008 (IN CHINESE).
4. Weiji, Zhou.Li, Loss Analysis of Areas with Wind Power System[M].Xinjiang University.2012.
5. Gao Hui. Loss Calculation and Measures Analysis of Distribution Grid[J]. Hunan Electric Power (2004).
6. Liao Xueqi. Line Loss Calculation Analysis and Loss Reduction Measures of Rural Power Network[M] Water Power Press, 2006.

A Novel Unscheduled Islanding Detection Method for Microgrid

Hui Li[1], Zhaohan Li[1], Zhonghan Guo[1]and Junwei Lu[2]

[1]*College of Automation Engineering, Shanghai University of Electric Power, Shanghai, 200090, China*
[2]*School of Engineering Griffith University Brisbane, Qld 4111, Australia*

Abstract. Microgrid with its intelligent and flexible control characteristics conform to the trend of sustainable development of electricity, and when the microgrid in the unplanned island state, the successful detection of the island is a prerequisite, energy storage inverter as the key equipment in the microgrid system, island protection is one of the necessary functions. In this paper, an improved islanding detection method based on active frequency drift and q-axis reactive power perturbation is proposed. The method has the advantages of faster detection speed and minor influence on power quality, which makes the energy storage inverter with better output power quality when it works on grid-connected state, and can be detected the islanding state quickly from grid-connected mode to islanded mode. Finally, the validity and superiority of the improved island detection method are verified by simulation experiments.

1 Introduction

With the rapid development of distributed power generation, more and more renewable energy is converted into electricity sentto the grid through the inverter. When the grid-connected distributed power supply in the event of large power grid failure, the inverter is continue to supply the power to local load and run independently is called the island operation[1]. Taking into account the safety and quality of electricity, should quickly detect the island, take the appropriate control means, eliminating the system malfunction and run again.

At present in the islanding detection, according to different detection methods can be divided into passive and active method. Passive method is applicable in most cases and does not need to introduce hardware circuit, does not affect the power quality of the grid side, but it will fail when the microgrid power matches the load power, so it has a lot of detection blind. The active method includes active frequencydrift (AFD), sliding mode frequency offset method, output power perturbation method, etc. which is adding the disturbance signal actively in the inverter controller to achieve island detection, this method has the advantages of detection timely and small blind spots, but due to the introduction of interference will lead to power grid side of the power quality has declined. Therefore, to solve the contradiction between active detection method and passive detection method, one of the ideas is a combination of a variety of detection algorithms.

At present, the improvement of island detection method can be divided into two categories: faster detection and less harmonic content. Furong Liu analyzed the islanding detection blind zone and proposed a parameter optimization method of active frequency drift with linear positive feedback[2], but this method still causes interference to the power grid. Fenghuang Cai analyzed the harmonic problem after using active frequency drift method, and the harmonic content is reduced without affecting the detection efficiency[3], but it enlarges the detection blind zone while eliminating the harmonic wave. Fangrui Liu improve island detection efficiency or reduce the impact on the network respectively by optimizing the disturbance mode, controller and disturbance parameters[4-5], but only consider the detection effect and ignore the disturbance on the system. In this paper, an improved AFD islanding detection method is proposed, which combines the advantages of active frequency drift detection and the q-axis reactive power perturbation (IQF).

2 Island detection principle

As shown in Figure.1, the load power is supplied by the energy storage inverter and the grid when the energy storage inverter is connected to the grid. When the grid is disconnected, the energy storage inverter and the load form an island[6].

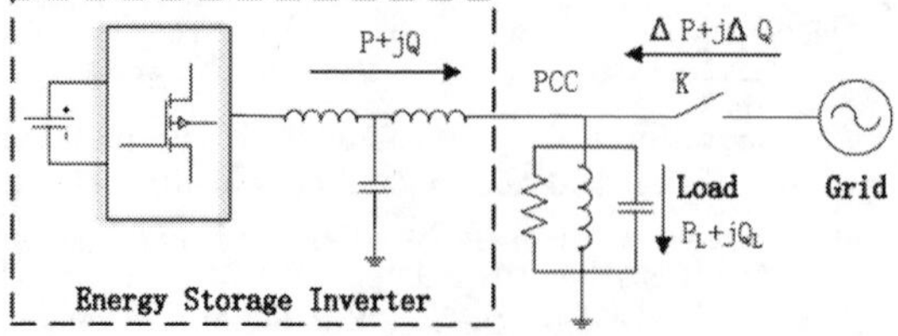

Figure 1. Grid power transmission system for Energy storage inverter.

The relationship between the branch powers in Figure. 1 can be expressed by equation (1)[7].

$$\begin{cases} P_L = P + \Delta P \\ Q_L = Q + \Delta Q \end{cases} \tag{1}$$

Where: P and Q are the active and reactive power output from the inverter; P_L and Q_L are the active and reactive power of the load respectively; and P and Q are the active and reactive power output from the grid.

By analyzing Figure 1 get the inverter side voltage, the relationship between the frequency as shown in equation(2)[8].

$$\begin{cases} U = \sqrt{PR} \\ f = f_0\left(1 - \frac{Q}{2PQ_f}\right) \end{cases} \tag{2}$$

Where: U is the inverter output voltage; f is the frequency of the inverter side; $Q_f = R\sqrt{C/L}$ is the quality factor of the load; $f_0 = 1/2\pi\sqrt{LC}$is the resonant frequency of the load; R, L, C are the values of load resistance, inductance and capacitance.

If the power of the inverter does not match the power of the load, the influence of the power variation on the voltage and frequency on the inverter side when the island is generated can be obtained as shown in equations (3) and (4). For example, if the power provided by the inverter is not enough to supply the load ($\Delta P>0$), the disconnection of the large power grid will cause the power of the load to decrease ($P_L\downarrow$), According to equation (3) determine the load terminal voltage, power down ($U\downarrow$, $f\downarrow$). In such load does not match the conditions, the application of passive island detection method to detect islands.

$$\begin{cases} \Delta P > 0 \rightarrow P_L \downarrow \rightarrow U \downarrow \rightarrow f \downarrow \\ \Delta P < 0 \rightarrow P_L \uparrow \rightarrow U \uparrow \rightarrow f \uparrow \end{cases} \tag{3}$$

$$\begin{cases} \Delta Q > 0 \rightarrow Q_L \downarrow \rightarrow f \uparrow \\ \Delta Q < 0 \rightarrow Q_L \uparrow \rightarrow f \downarrow \end{cases} \tag{4}$$

If the power of the inverter is matched with the load power, the magnitude and frequency of the voltage and current at the grid-connecting point will not change when the grid is disconnected. At this time, the traditional passive island detection method fails, so the active island detection method needs to be quoted.

3 Improved island detection technology

For active islanding detection method, AFD is a more classic and most studied active island detection method, which has the advantages of simple algorithm, easy implementation and fast detection speed. However, because AFD is the frequency of current to add disturbance, and active currency fluctuations will cause the output voltage fluctuations, thus affecting the power quality, so this paper presents an improved AFD method, by adding reactive current disturbance in the q axis, reduce the effect of inverter output power quality.

3.1 AFD islanding detection method

As shown in Figure 2, the AFD injects a constant current, slightly varying frequency current into the grid via an inverter. When the grid is disconnected, the frequency of the inverter output voltage is forced to drift until a limit value is exceeded and an isolated island is detected[9].

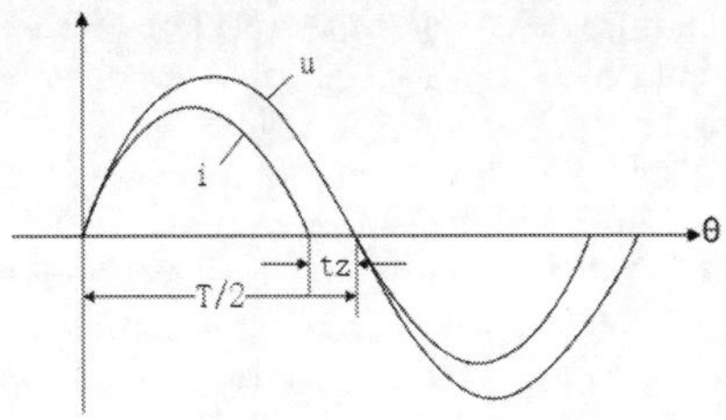

Figure 2. The value of 'u' and 'i' after the disturbance are added.

Here introduce the concept of the cut-off coefficient c_f, define c_f as the ratio of the time t_z of half-voltage period T/2 of the zero-crossing of the current zero-crossing lead (or lag) voltage.

$$c_f = \frac{2t_z}{T} \tag{5}$$

According to previous research experience[6], c_f can be expressed as shown in equation (6).

$$c_f = \begin{cases} c_f & \text{AFD} \\ c_f + k_f(f - f_0) & \text{AFDPF} \end{cases} \tag{6}$$

Where: c_f for the initial given truncation coefficient; k_f for the feedback gain; f_0 for the grid rated frequency.

Furong Liu and others have proved through research that the proper selection of the values of c_f and k_f can speed up the detection of islanding and reduce the detection blind zone. According to the research experience, $c_f = 0.02$ and $k_f = 0.1$ are selected here.

3.2 The principle of improved island detection method

Three-phase energy storage inverter control principle shown in Figure 3. The phase angle θ of the grid-side phase obtained after the phase-locking is obtained through the perturbation control to obtain the phase angle θ', and the grid-side voltage and current signals are transformed by using the phase angle θ' to obtain grid-side voltage and current values in the synchronous rotating coordinate system V_d, V_q, I_d, I_q; and then V_d, V_q and a given active power P_{ref}, reactive power Q_{ref} input power outer ring, thereby calculating the current inner reference i_{dref}, i_{qref}; obtained by the inner current loop voltage command u_d^*, u_q^* After anti-park, anti-clark transform command voltage into the PWM modulation module to generate six PWM signal control energy storage inverter work; finally obtained by the grid side voltage frequency f to determine whether the island. If it is determined that there is an island, then immediately change the control mode, the power of the outer ring into the voltage outer ring, through a given voltage and frequency reference control energy storage inverter to continue running, and thus from the grid PQ control mode to off-grid VF control mode.

Similar to the single-phase AFD algorithm, the three-phase AFD algorithm detects the output frequency of the energy storage inverter by changing the phase angle of the current and achieves the purpose of islanding detection. However, in the three-phase system, the output current is controlled based on the dq rotating coordinate system, and there is a coupling phenomenon between the three phase currents. The phase disturbance also affects the current of the other two phases at the same time, which further deteriorates the waveform. Therefore, this algorithm exists harmonic content of the larger problem.

As shown in Figure 3, unlike the AFD, the new method uses the locked phase angle θ' for the Clark transformation and the park transformation, and uses the frequency f obtained from the phase lock to generate the q-axis interference amount I_{qr}, Together with the I_q, I_{qref} as the current reference value I_q of the q-axis.

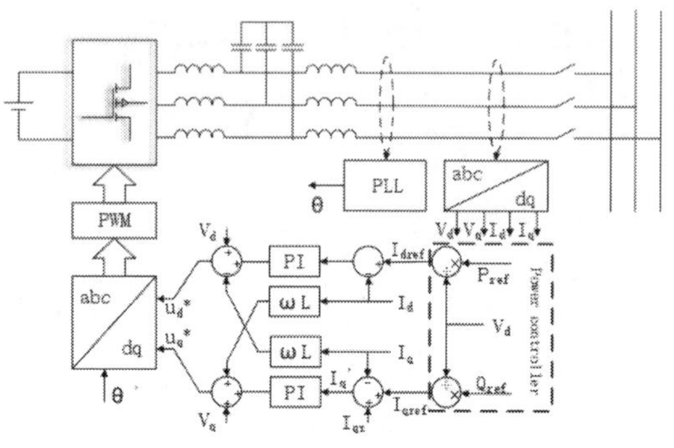

Figure 3. Control diagram of improved AFD island detection.

In [4], the relationship between the q-axis current component I_q and the grid-side frequency f is analyzed as shown in equation (7).

$$I_q = -\frac{2I_d Q_f}{f_0}(f - f_0) \tag{7}$$

Where: Q_f, f_0, I_d are constants. (7) can be seen as a function of I_q and f, which can be obtained in Figure 3 after the reactive current I_q' curve equation as shown in equation (8).

$$I_q' = -\frac{2I_d Q_f}{f_0}(f - f_0) + I_{qr} \tag{8}$$

Where: I_{qr} is the introduced perturbation.

For I_{qr}, can be used in different ways of disturbance, the project is more common positive feedback method [4].

$$I_{qr} = [C + K * (f - f_0)] * I_d \tag{9}$$

Where: C is a constant greater than zero; K is a positive feedback coefficient.

3.3Detection strategy of hierarchical perturbation method

Figure 3 to improve the island detection method by adding I_{qr}, which avoid the phenomenon of three-phase current coupling, active power affect the voltage and frequency, reactive power only affect the frequency, dq decomposition of the q axis after the impulse, Will not affect the d-axis active current, that will only affect the frequency, does not affect the voltage, thus ensuring the power quality.

For the island detection of multi-level perturbation algorithm in Figure 3, the application of θ' and I_{qr} is the key to this paper. The overall control flow chart shown in Figure 4. If the voltage side of the grid frequency exceeds the specified frequency range, determine the island; if not exceeded, to further determine the frequency deviation range, if 0.1Hz or less, adding q-axis current interference I_{qr}, this time for the first disturbance; More than 0.1Hz, increase the amount of q-axis current interference and the introduction of phase angle interference, until the island is generated, this time for the second disturbance; addition of disturbance, the need to determine the trend of frequency changes, making the direction and frequency of disturbance Change the direction of the same, thus speeding up the detection speed. This algorithm not only guarantees the power quality of the three-phase inverter when the network is connected, but also has a faster island detection speed.

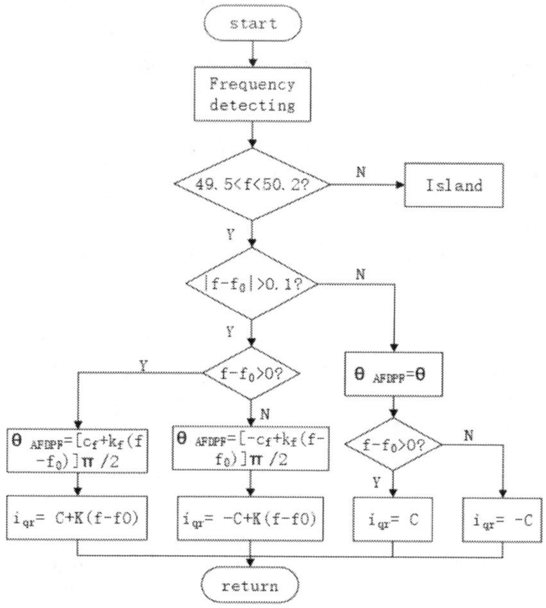

Figure 4. Flow chart of island detection.

4 Simulation verification and result analysis

In order to verify the correctness and feasibility of the above theory, simulation experiments are carried out by MATLAB/Simulink.

Simulation, the energy storage inverter began in the grid state, at t = 0.1s disconnect PCC point. Simulation, will be used passive method, AFD method, IQF method and improved AFD method for the island for island detection. According to GB/T33589-2017 "microgrid access power system technical requirements": the normal frequency of the power system to allow the value of ± 0.1Hz, from the country's major power system to run the actual operation, basically maintained at no more than ± 0.1Hz range. And when the common point frequency exceeds [49.5Hz ~ 50.2Hz] range, that the occurrence of the island.

In order to verify the output frequency change of the

inverter when the island is not matched, verify the correctness of equation (3) in the second section. In this paper, as shown in Figure 1, ΔP>0, ΔP<0, ΔQ>0, ΔQ<0 four conditions. The simulation results are shown in Figure 5.

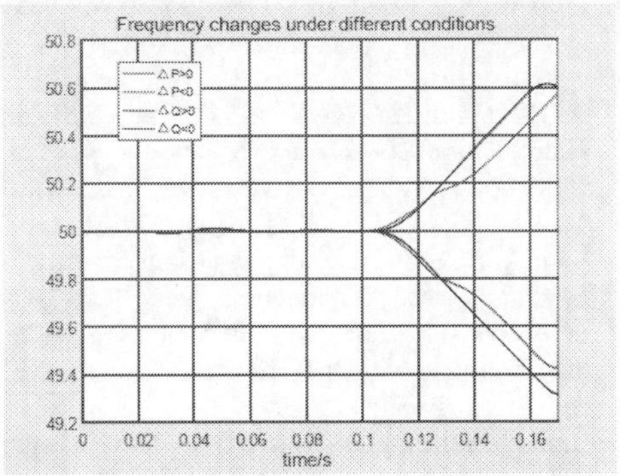

Figure 5. Inverter output frequency under different operating conditions

As can be seen in Figure 5, when ΔP>0, ΔQ<0, an island is formed and the frequency is decreased; when ΔP<0, ΔQ>0, an island is formed and the frequency is increased. Therefore, the experimental results in Figure5 coincide with those in equation (3), and verify the trend of inverter output frequency during islanding under power mismatch conditions. Island detection provides a theoretical basis. Meanwhile, the voltage is same as the frequency.

In order to verify the effectiveness of the algorithm, this paper simulates the isolated island detection conditions, so that the inverter power and load power match, that is, when using RLC load, then the circuit consumes the capacitive reactive and reactive power cancel each other , The load is equivalent to pure resistive load. The parameters of circuit and simulation shown in Table 1.

Table 1. Simulink parameters.

Parameter	Value
Load(R) / Active	14.52Ω/10kW
Load (L) / Reactive	18.48mH/25kW
Load (C) / Reactive	550uF/25kW
Load quality factor (Q_f)	2.5
Load resonant frequency (f_0)	50Hz
Rated active power (P)	10kW
Rated reactive power (Q)	0kvar
DC input voltage (U_{dc})	800V
c_f	0.02
k_f	0.1
C	0.02
K	0.1

Set the load resonance frequency f_0 = 50Hz to match the grid frequency. The test results are shown in Figure 6.

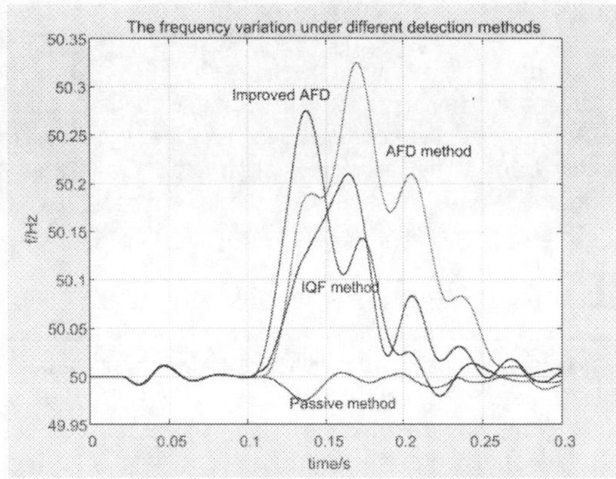

Figure 6. Inverter frequency of different island detection methods.

Before 0.1s, the energy storage inverter in the grid under the PQ control mode, due to the large grid voltage clamp, the frequency of this time can be stable at 50Hz. When the 0.1s PCC point is broken, the frequency and voltage fluctuate due to the loss of the support of the large power grid, but the change in the frequency detected by the different island detection algorithm is different. Among them, the passive detection method cannot detect the island state; IQF method to detect with about 60ms; AFD method can be detected within 55ms island state; and improved AFD method in 28ms to detect the island state. At the same time, the improved AFD method has little effect on the output current of the inverter when the grid is running, and the detection result is shown in Figure 7 Thus confirming the improvement of the fastness of the improved AFD method and the small effect on the power quality.

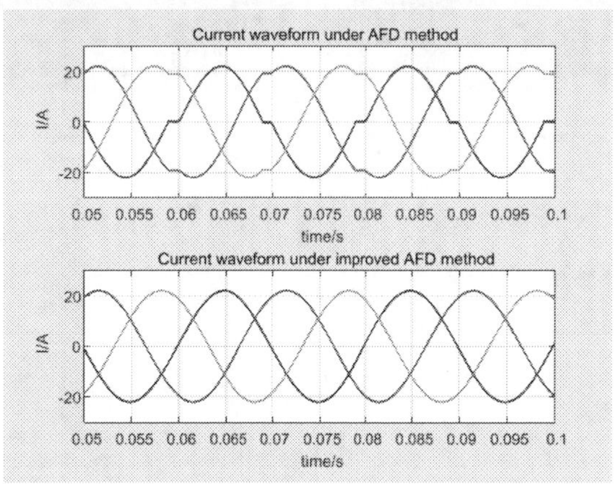

Figure 7. Inverter current of different island detection methods.

5 Conclusion

In this paper, an improved islanding detection method based on active frequency drift method and q-axis reactive power perturbation is proposed. The AFD method ensures the speed of detection. The introduction of q-axis reactive current disturbance reduces the influence of the output power quality of the inverter.

In this paper, the passive detection method, IQF method, AFD method and the improved AFD method were used to analyze the effect of the method, and the feasibility of the method was verified by simulation experiment. The experimental results show that the improved AFD method has the advantages of faster detection speed and less influence on the output power quality.

Islanding detection is the first step in the study of microgrid and off-grid conversion. How to realize the microgrid security after islanding detection is quickly shifted from the grid-connected mode to the off-grid mode, and how the microgrid bus voltage and frequency in the conversion process is fast stable, that is, to reduce the impact of micro-grid operation mode switching on the power grid and load, is an important issue that needs further study.

Acknowledgment

This work is supported by Shanghai International Science&Technology Cooperation Program (No. 15220710500) and Natural Science Foundation of Shanghai (No.15ZR1417600).

References

1. Guo-Kiang Hung, Chih-Chang Chang and Chern-Lin Chen. Automatic phase shift method for islanding detection of grid connected photovoltaic inverters[J]. IEEE Transaction on Energy Conversion, 2003, 18(1):169-173.
2. LIU Furong, KANG Yong, DUAN Shanxu, et al. Parameter Optimization of Active Frequency - Shift Algorithms [J]. Journal of China Electromechanical Engineering, 2008,28 (1): 95-99.
3. CAI Fenghuang, LIN Qiongbin, WANG Wu. Active Frequency Isolation Detection Method with Harmonic Compensation[J]. Power automation equipment,2014,34(9):54-58.
4. LIU Fangrui, KANG Yong, ZHANG Yu. Parameter Optimization of Active Frequency - Isolated Island Detection with Positive Feedback [J]. New Technology of Electrical Energy, 2008,27 (3): 22-25.
5. SmitaShrivastava, Shailendra Jain, Rajesh Kumar Nema, VijayshriChaurasia. Two level islanding detection method for distributed generators in distribution networks[J]. Electrical Power and Energy Systems,2017,87: 222-231.
6. ChuttchavalJeraputra, Prasad N. Development of a robust anti-islanding algorithm for utility interconnection of distributed fuel cell powered generation[J]. IEEE Transaction on Power Electronics, 2004, 19(5): 1163-1170.
7. ZHANG Liyan, WU Jun, CHEN Qihong, et al. Small blind fuel cell system and network islanding detection method [J]. Automation of Electric Power Systems,2017,41 (6): 98-102.
8. XIE Qiang. Research on Anti - islanding Detection of Energy Storage Inverter[D].University of Electronic Science and Technology, 2014.
9. CHEN Jiabin, CHEN Alian. An Improved Active Frequency Offset Island Detection Algorithm [J].Journal of Power Supply, 2016.

Design of Auxiliary Decision-Making System for Power Grid Fault Disposal

Hao Chen[1],Xinfan Jiang[1],Dijun Hu[1],Tao Wang[1] and Fan Zhou[1]

[1]State Grid Hunan Electric Power Company, China

Abstract. Accident treatment is the most important part in the daily monitoring, and the improvement of the intelligent level is of great significance to improve the efficiency of scheduling accident and shorten the time of the accident. This paper puts forward a set of intelligent power grid accident treatment decision system, introduces the background of the system, the composition module and the realization of functions, to facilitate the operation of the operator to automatically identify the grid failure, isolation and recovery, reduce the pressure of staff, guarantee the safe and stable operation of the power grid.

1 Introduction

The safety, stability and economy of electric power system are the main objectives of power grid operation. In the modern power system, computer system provides the function of alarm and fault information collection, this undoubtedly improve the systematic level of the power grid operation, and provides the technical conditions for fault information. But on the other hand, under the state of emergency and fault, due to the lack of a unified comprehensive failure data processing and analysis tools, the influx of information has made it difficult for operators to accurately filter information and identify faults, the dispatcher was unable to accurately judge the failure, leading to miss the best time to dealing with accidents and cause an accident.

Document 1 and Document 2 put forward the key technical points of implementing the intelligent decision support system for incident handling based on the dispatching automation system. Among them, Document 1 proposes such typical applications as fault isolation and recovery, intelligent solution generation and co-processing, thereby greatly improving the system's task handling capability and ensuring the safe and stable operation of the power grid.Typical applications as Document 2 refered,such as techniques of smart anti-error strategy, information flow control, clearly the basic architecture of the system and the functional requirements, and studied the accident disposal program generation, intelligent load shedding brownouts, power transfer intelligent decision-making system and so on, which can provide dispatchers with preconceived accident anticipation and simulated training platform for anti-accident drills during the normal operation of the power grid. Document 3 adapt a common fault location method based on the EMS platform for incident handling intelligent decision support system, According to the grid operation mode, power flow control, transmission line capacity constraints and accident management regulations, the document proposes a recovery control strategy suitable for the Guangzhou power grid, which makes the system provide ancillary decision-making solutions in real-time and research states, and making it convenient for dispatching operation personnel to carry out accident anticipation disposal and real-time recovery control, reduce personnel pressure, ensure safe and stable operation of power grid. In addition, document 5 presents a holistic framework for online incident-handling decision support for large power grids. The study also includes five aspects of comprehensive information support platform, prior risk perception, online fault diagnosis, post-event decision support and visual display.

The basis of the decision support system for power grid accident is the construction of an integrated information platform. The key is to establish a tracking and processing mechanism for all aspects of risk monitoring and analysis beforehand, during and after the event. Finally, visualization techniques are needed to create a panoramic visualization of incident handling and to improve accident handling emergency decision-making ability. In the past, the researches on the auxiliary decision-making of accident handling basically focused on one aspect, lacking the holistic solution. The typical representatives include the White Paper on "21st Century EMS Architecture" proposed by VLGO, focusing on the construction of basic information platform. The advanced control center (AC2) being developed by PJM, a regional power market operator in the United States, focuses on the electricity market and the platform architecture.

Therefore, there is an urgent need for an effective intelligent decision-making system for accident handling, control the operation of the power grid in real time by automated technology, timely detect, rapid diagnose and eliminate the problems in time, improve its automation level of scheduling and dealing with the accident, ensure the security of power system operation.

2 The basic framework of auxiliary decision-making system of power grid accident management

The intelligent auxiliary decision-making system of power grid accident management is located in the EMS platform, which shows the characteristics of the accident treatment of the regional power grid and provides intelligent technical support for fault diagnosis and method of power grid operation condition. Once in the event of an accident abnormality, the simple and complex faults in the power grid can be analyzed by capturing the real-time switching variable tidal current data and protecting the security information of the self-assembly device, so that the fault diagnosis and disposal scheme can be realized quickly and accurately, reduce the risk of misplaced and misuse, improve the efficiency and accuracy of accident handling. The system functional architecture of the auxiliary decision system is shown in Fig 1.

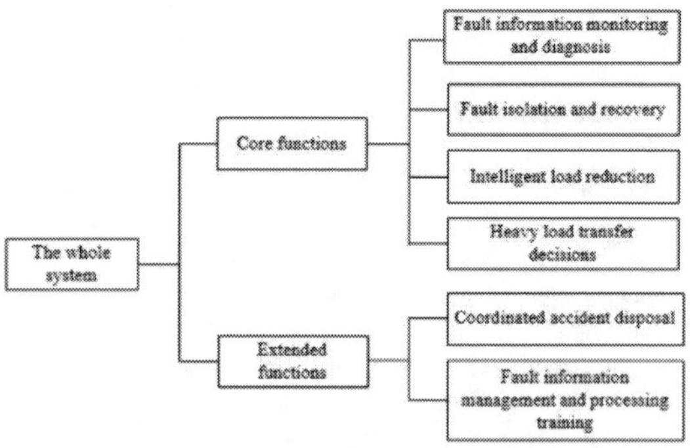

Figure 1. Schematic diagram of the functional architecture ofthe auxiliary decision-making system for power grid accident processing

The system requires the configuration of special accident handling assistant decision-making main server, through the configuration data interface, server and SCADA system can receive and monitor the power grid operation information in time. During the operation of the system, initiate different work which is responsible for the data flow control, fault diagnosis and fault disposal, from the application client we can view all kinds of information related to the fault.

Software design is based on the scheduling automation EMS platform, using independent application design and database design, we can fully share the SCADA various data, including real-time data and background parameters. The system can adapt to the SCADA system of different manufacturers to ensure the reliability of the data source. When the user switches the SCADA module of different manufacturers, the auxiliary decision system can still obtain the basic information reliably. The data flow of SCADA is carried out in a security configuration, so that the data can be read from SCADA module in one way, which will not affect the stability and security of the program and data of SCADA module.

3The core function of auxiliary decision-making system for power grid accident handling

3.1 Fault information monitoring and diagnosis

3.1.1Fault information detection

After obtaining three remote signals, protection signal and other information from the EMS system SCADA, security and other modules, use data mining technology for data filtering, removing error duplication information, and normal operation information not related to the accident, adding missing fault information, to ensure that troubleshooting has a comprehensive data base. The detection of auxiliary decision system is reflected in two aspects. In the normal operation of the grid, the auxiliary decision system continues to carry out data detection, analyze and accumulate the various real-time information of SCADA module, identify normal operation information, debugging information and redundant information. When capture to the information which conforms to the characteristics of fault, including accident alarm message, switch shift, relay protection action information, immediately use data mining module for signal filtering and completion, after pretreatment, sent to the fault analysis module.

3.1.2 Abnormal alarm of power grid

The auxiliary decision system can monitor the operation condition of the power grid in real time. Once the discovery of cross-section, line, the main transformer close to the operating limit or the more limited, by issuing an alarm message, automatically pop-up site map or the whole network trend map, abnormal equipment flashing means to remind the dispatcher processing restrictions on developments, ensure that the duty dispatcher does not ignore the monitoring of the running device because he is too focused on a particular detail during the event.

3.1.3Comprehensive analysis of power grid faults

After the grid failure actually happens, the fault analysis module automatically triggers. The fault analysis module uses the switch to change the information and protection action information, based on hierarchical intelligent technology to quickly locate the power outage range and lock the fault device, in a few seconds to complete the accident express, show to the dispatcher including fault equipment (when can not determine the only faulty equipment just provide the most likely suspicious equipment), power outage range, involving the power supply users, important users, reloading equipment list, and gives a preliminary evaluation of the switch to protect the action of the self-device.

Auxiliary decision system has two modes: real-time state and research state. Real-time state is based on SCADA information to trigger, all-weather automatic operation. The research state is mainly used when the basic data missing is too serious, dispatchers directly use the system to generate a fault handling program after setting up faulty device, or when the system is trouble-free, the dispatcher anticipate the accident.

3.2 Fault isolation and recovery

In the operation of power grid, the fault is more obvious, but there will be a small number of permanent failure, in order to prevent the permanent damage to the power equipment, classify the type of fault and make the appropriate fault isolation and recovery operations will be necessary.

The operation process of fault isolation firstly is operational environment preparation, enable the relevant data monitoring procedures, monitor data ready in place. The second is real-time access to fault data, the data acquisition process should be detailed scan each smaller sub-module will produce the corresponding fault information, and according to the corresponding data analysis algorithm to determine whether the failure to achieve the conditions of the higher, only when more than a certain threshold, immediately start the automatic reporting mechanism for data reporting. Finally, through the intelligent network topology analysis, automatically generate the corresponding accident handling and disposal program, and use pre-set isolation program to complete the fault isolation operation and complete the failure of the disposal process, so as to effectively complete the fault of the rapid isolation and rapid recovery of pressure load and other operations.

3.3 Intelligent load reduction

Intelligent load shedding power module is mainly to complete the main transformer, cross-section is more than expected threshold, when the phenomenon of low frequency, low voltage occur, the original automatic device will automatically eliminate in a short period of time, and thus lose the safety of power grid security possibilities, can not make stable and safe operation of the grid. If the grid failure is to N-2 it will likely lead to serious failures occur, in order to prevent the occurrence of such events, generally use the operation strategy of automatic processing, when the corresponding malfunction occurs, the third grid security line will be activated immediately to protect and run itself, thus increasing the security and stability of the power grid.

3.4 Heavy load transfer decisions

In order to improve the reliability of urban power grid, the research object of emergency control of power grid accident management will not only be 10KV distribution lines, but also extended to 220kV, 110kV system and substation 10kV bus, and direct access to the above voltage level of distribution power supply. The urban power grid system which considers the main network, distribution network and distribution power supply can ensure the continuous power supply of important users in the emergency of the power grid or in the event of natural disasters.

3.4.1 110kV line overload transfer

110kV line overload first according to the size of the overload to determine whether the load on the 110kV line may normal transfer, if can not transfer, then give the corresponding tips, if can transfer, give the transfer program, such as closing the corresponding mother joint switch and other operational programs, while the trend to check to determine whether the program will cause the equipment overload. 110kV line overload transfer based on its current operation mode may be normal operation mode, or may be a series operation mode.

For the normal operation mode, as shown in Figure 2 below, if L3 overload, through closing 7DL, disconnecting the 6DL to check the load transfer and the corresponding overload check.

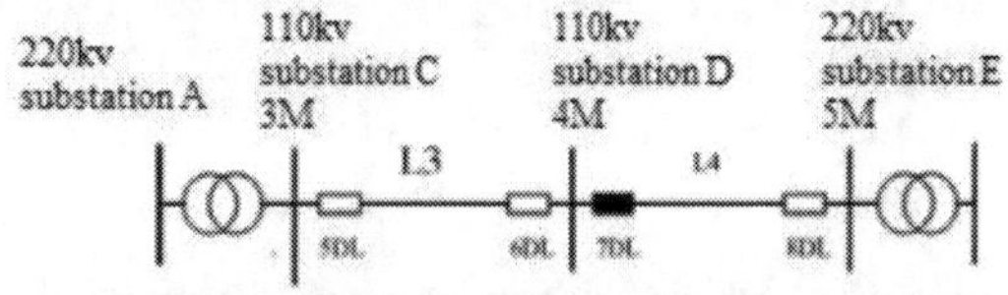

Figure 2. 110kV line common operation mode

For the series power supply mode, as shown in Figure 3, if the L1 overload, the program may be given, including closing 7DL and disconnecting 6DL, or closing 7DL and disconnecting 4DL and other operational programs.

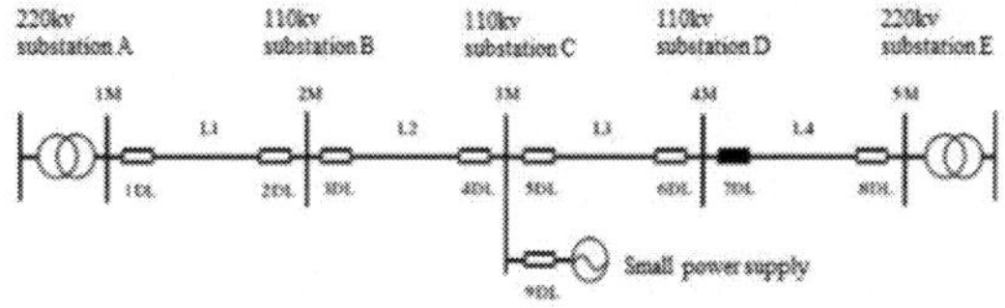

Figure 3. 110kV line multi-level serial supply operation mode

3.4.2 220kV line overload transfer

220kV line is usually running by ring network connection, the given principle of overload transfer program is based on the power flow to first determine the 220kV plant of the line, the searched 110kV line outlet load of 220kV carry out the transfer according to 220kV main transformer overload transfer mode, while the trend check.

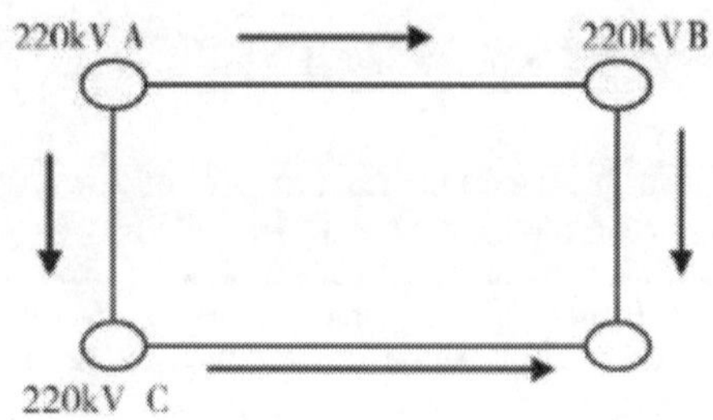

Figure 4. 220kV line ring network operation mode

If the line between 220kV A station and 220kV C station is overloaded, as shown in fig.4, the corresponding 110kV line transfer in 220kVC station will be preferred. If there is no plan, then the corresponding 110kV line load will be considered in the transfer of 220kVD station. The transfer load is calculated according to the injection sensitivity of node B and D station, and the load transfer is calculated accordingly.

3.4.3 220kV main transformer overload transfer

The line of the middle - voltage side of 220kV is combined to obtain the optimal combination line group to complete the transfer and to conduct the tidal current check. The main switching strategy is: 110kV contact line direct revolution, multi-level chain structure 110kV circuit switching, substation 110kV bus transmission, etc.

4 The expansion functions of power grid accident processing intelligent auxiliary decision system

4.1 Coordinated accident disposal

After the accident, decision-making system should be able to intelligently generate the list of key steps of accident treatment, and can intelligently divide work according to the number and post of duty dispatcher, each task can be completed in the step table, Other dispatchers can get a clear picture of the progress of the work done by the current dispatchers by looking at the real-time refresh of the task table, and can then be divided according to the schedule.

4.2 Fault information management and processing training

The auxiliary decision system provides a full range of fault case information management functions. Include event saving, retrieval, case library update, process report presentation and printout. You can retrieve the accident express, troubleshooting programs, processing information and so on.

The auxiliary decision system is an important tool for the training of dispatcher accident handling in the absence of an accident abnormality. Through the data interaction with the DTS module, the dispatcher in the auxiliary decision-making system to manually design the accident handling program is the process of DTS lesson plan training. On the other hand, demonstrating typical accident disposal cases in the system and giving a hint at the key processing node, can be more intuitive to the dispatcher who needs training, and can significantly improve the learning effect than simply reading the text system.

5 Conclusion

Based on the characteristic information of power grid production information, this paper designs the basic structure and basic functional requirements of the auxiliary decision system of power grid accident disposal. The system greatly improves the dispatcher's ability to deal with the power grid accident, reduces the decision-making time of the accident processing, makes the analysis of the grid accident gradually from the empirical type to the analysis, intelligent, improve the grid independent support and self-fast resilience after the accident. On the other hand, after the system is put into operation, the dispatcher enhances the understanding of the current operation mode of the power grid by simulating the operation of the power supply load and the overload load transfer, so that the power grid can operate more efficiently and reliably.

References

1. Liu Yang. Design and implementation of power grid incident management system[D]. University of Electronic Science and Technology of China, 2014.M. Ben Rabha, M.F. Boujmil, M. Saadoun, B. Bessaïs, Eur. Phys. J. Appl. Phys. (to be published)
2. Zhang Weihua. Research on intelligent auxiliary decision-making system of power grid accident handling [D]. South China University of Technology, 2011.
3. Yang Zeng, Liu Yuquan, Tan Xilin, Pan Kaiyan. An intelligent auxiliary system based on EMS platform
4. Yunnan Electric Power Technology, 2013,41 (05): 51-53.
5. Zhang Tao, Zhang Bin, Li Zhiqiang, Huang Xiangdong. Application of intelligent grid accident auxiliary decision system in scheduling [J]. Power Supply, 2009,26(05):26-28+36.
6. Shan Xin. Design and implementation of the auxiliary decision-making system for online accident management of power grid [A]. China electrical engineering institute. Proceedings of the annual meeting of the China institute of electrical engineering in 2013 [C]. China electrical engineering society: 2013:6.
7. Yanru Sun,Wei He. Online security risk prevention and control of power grid operation based on Smart Grid Dispatching Technology[J/OL]. Modern Industrial Economy and Informationization, 2017 (21) :55-56+61
8. Nanrong He. Design and implementation of power grid accident warning and disposal management system[D].Jilin University,2016.

9. Xiangyu Zhao, Chengliang Wang, Gang Yao,Xiang Fan,Guojiang Xiong,Xu Zhang,Luofang Zhu ,Xiaoxu Huang.Design and application of online accident risk early warning system for power grid based on South network accident regulation[J]. Electric Power,2016,49(S1):62-66.
10. Chengbin Gao. Research and system development of power grid operation risk assessment and accident warning[D].North China Electric Power University,2015.

The Strategy of Inverter Seamless Mode Switching in Master-Slave Independent Micro-grid

Hanhong Jiang[1], Yao Lu[1], You Wu[1] and Yi Wang[1]

[1]National Key Laboratory of Science and Technology on Vessel Integrated Power System, Naval University of Engineering, Wuhan, China

Abstract. In order to realize the uninterruptible power supply in the master-slave independent micro-grid system, the micro-grid inverter needs to realize the mode switching of the grid-connection/grid-disconnection. How to reduce the transient oscillation during switching, so as to effectively achieve seamless mode switching is a key issue to be solved. In this paper, a typical master-slave control independent micro-grid is used as an example, the strategy of mode switching is improved in two aspects. On the one hand, the state-following algorithm is adopted to improve the switching strategy of the outer loop. On the other hand, the current inner loop is taken by the H∞ robust controller to improve the robustness of the controller. Compared with the traditional PI control mode switching, this paper illustrates the feasibility of the proposed strategy through the simulation and experiment verification. The improvement strategy in this paper can effectively reduce the voltage and current oscillation during mode switching.

1 Introduction

With the problem of environmental pollution is more and more serious, new energy power generation has become a hot spot in various countries. Distributed generation (DG) technology and communication technology is the ever-changing development, which makes the application of micro-grid system more and more popular[1-3].

Master-slave control independent micro-grid system is an effective way to supply power to remote areas, which is difficult be covered by large power grids. The principle of independent micro-grids is to collect wind-photovoltaic energy and parallel in the synchronous generator through the inverter.

Due to the volatility of wind-photovoltaic energy, the independent micro-grid system generally needs to be equipped with a certain capacity battery energy storage system, which is used to stabilize the wind-photovoltaic volatility. Because the independent micro-grid system is independent of the large power grid, its stability is worse than the large power grid, so it is the hot topic of the current scholars[4]. How to ensure the critical loads in the independent micro-grid uninterrupted power supply, that is, to achieve the inverter grid-connection/grid-disconnection seamless smooth mode switch, which is a challenge to study the independent micro-grid technology. [5] proposes an indirect current control algorithm for seamless transfer of three-phase utility-interactive voltage source inverters. The propose method is able to provide critical loads with a stable and seamless voltage. [6] presents voltage magnitude and frequency control of a three-phase voltage source inverter to achieve a seamless transfer between grid-tied mode and intentional islanding mode. A virtual inductor with high-pass filter in synchronous d-q frame is proposed to improve the transient response. A novel seamless transfer of single-phase grid-interactive inverters between grid-connection and grid-disconnection modes is presented in[7].The voltage controller is used for compensating the filter capacitor current in grid-tied mode, and the voltage controller is used to regulate the output voltage in grid-disconnection mode. In [8], the scheme of the seamless transfer control strategy is investigated, and the proposed seamless transfer control strategy is not only capable of guaranteeing the uninterruptible load voltage, but also protecting fuel cell against the power demands beyond its allowed bandwidth. [9] proposes a phase-locked loop (PLL)-based seamless transfer control method between grid-connection and islanding modes in a three-phase grid-connection inverter.

In order to improve the stability during mode switching, this paper improves the strategy of mode switching in two aspects. On the one side, this paper improves the switching mode of the outer loop and proposes a kind of state-following switching strategy. On the other side, this paper adopts a H∞ robust controller instead of the current inner loop controller to ensure the robustness during switching and reduce the transient oscillation of voltage and current.

The structure of this paper is as follows: A typical topology of DC/AC hybrid isolated micro-grid system is used as an object in Section II Section III analyzes the characteristics of traditional mode switching and explains the existing problems. In Section IV, the design of H∞ robust controller is studied and an improved method of replacing the current inner loop with robust controller is given. In Section V, the feasibility and superiority of the proposed strategy are verified by simulation. The construction of the experiment platform and experiment verification is given in section VI Finally, Section VI gives the conclusions.

2 Master-slave control independent micro-grid system structure

In this paper, a typical topology of DC/AC hybrid isolated micro-grid system is used as an object of study, which includes photo-voltaic (PV), battery storage system (BSS), and synchronous generator (SG), as shown in Figure 1. SG is a master power supply, and renewable energy system (RES) inverter is a slave power supply[10-12]. The DC bus is constructed by a new energy power generation system. RES inverter and SG are connected in parallel through the isolation transformer to form a AC bus to bear the load in common. The inverter has two operating conditions with grid-connection and grid-disconnection power generation. Therefore, the control strategy has two modes: PQ or V/*f*, as shown in Figure 2.

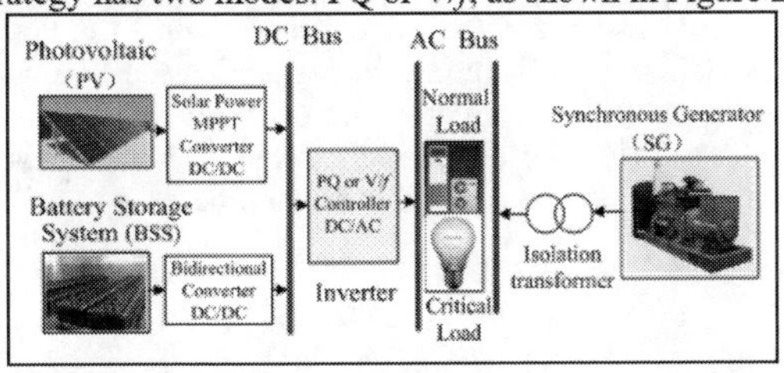

Figure 1. A typical master-slave control independent micro-grid system.

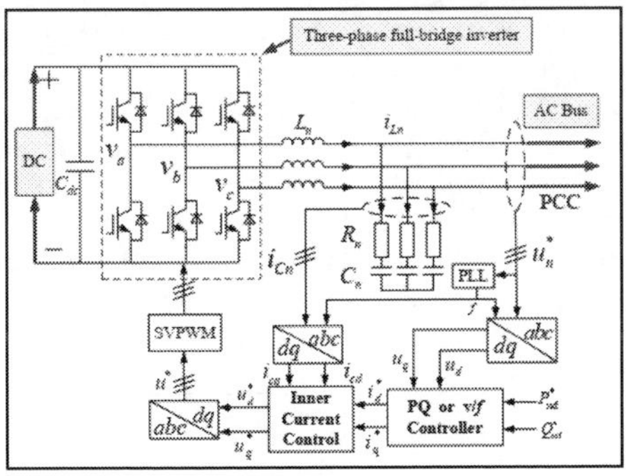

Figure 2. PQ or V/f control of three-phase inverter.

When the grid is connected, the synchronous generator system is the master power supply, which provides stable voltage and frequency support, the inverter is the slave power supply, using the PQ control mode. When the synchronous generator system is faulty or requires service, the inverter control mode is switched from PQ to V/*f*, and the critical loads are supplied by the inverter, thus ensuring uninterrupted power supply for critical loads.

3 Analysis of traditional mode switching

3.1 Grid-connection and grid-disconnection operation mode

The grid-connection mode of the inverter generally takes the typical PQ double-loop control structure as shown in Figure 3(a). The off-grid operation mode of the inverter is V/*f* double-loop controller as shown in Figure 3(b).

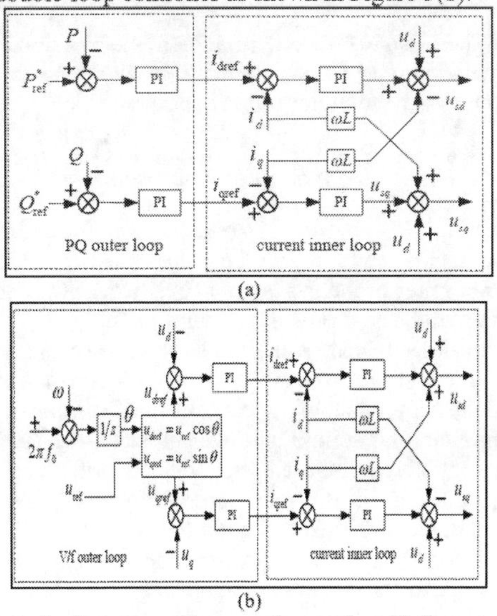

Figure 3. Inverter grid-connection and grid-disconnection operation mode: (a) PQ double loop control; (b) V/f double loop control.

In Figure 3, P and Q are the power feedback signals of the inverter, and P^*_{ref} and Q^*_{ref} are the power reference values. u_d, u_q, i_d, i_q are the inverter output voltage, current dq axis component. i_{dref} and i_{qref} are current inner loop reference values. u_d, u_q, i_d, i_q are the inverter terminal voltage, current dq axis component, respectively. u_{sd} and u_{sq} are the dq axis components of the output voltage control signal of the controller. ω is the frequency signal, u_{ref} is the voltage reference value, f_0 is the reference frequency 50Hz.

The dual-loop controller not only improves the quality of the three-phase output power, but also provides a buffering effect on both controller outputs during mode switching, reducing transient oscillations caused by controller switching.

3.2 The problem of traditional mode switching

As the inverter PQ and V/f controller is completely different, so to achieve the inverter grid-connection/grid-disconnection smooth switching, it is necessary to have some logic switches to realize the switching of different controllers. Switch controller structure is shown in Figure 4, where the grid controller is the PQ control mode, the independent controller is the V/f. The input of the controller is the difference between the feedback and the given reference value, and K1 and K2 are logic switches [13,14].

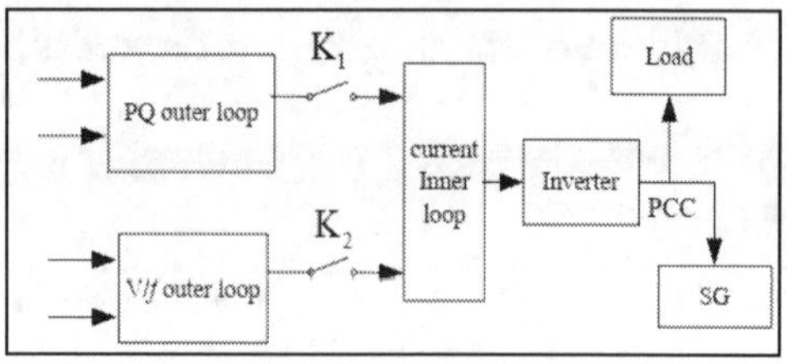

Figure 4. Traditional mode switching structure.

The reasons for the oscillation in the traditional mode switching process are the problem of outer loop controller and the problem of inner loop controller.

3.2.1 The problem of outer loop controller

When the PQ controller is running, the V/*f* controller is also running and the V/*f* output does not work. Since the output states of the two are not equal when switching, the output of the controller has a transition jump.

In order to reduce the transient shock of the switch, trying to set the V/*f* controller to zero before the switch, and the zero output of V/*f* controller and the output of PQ controller are still in a different state, which is not substantially eliminated the output with no match phenomenon of the two controllers, so the transient oscillation still exists. Figure 5 shows the simulation waveform that the controller does not initialize or initialize to zero when switching.

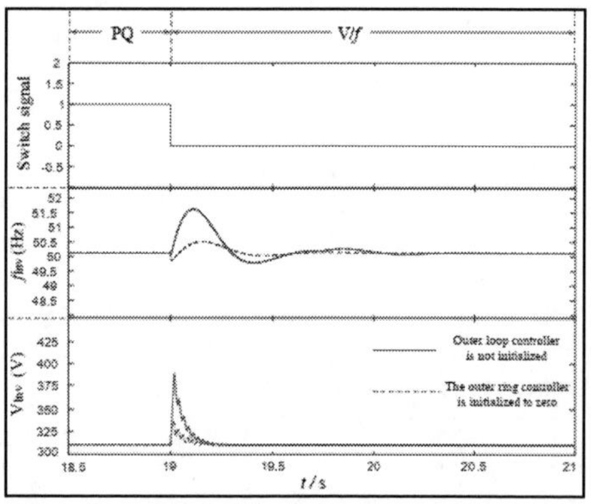

Figure 5. The effect of the initial output of the outer loop controller on the mode switching.

3.2.2 The problem of inner loop controller

The mode switching mainly completes the switching of the outer loop controller, and the current inner loop controller is the common part, that is, the current inner loop is not switched. If the parameters of the inner loop controller is not appropriate, it may cause the oscillation to be large during the switching process, and even the system is unstable. As shown in Figure 6, k_{ii}=0.24, k_{pi}=110, the micro-grid is switched at t=1. Simulation results show that, the system is stable in grid-connection cases (PQ), but the voltage and frequency of the system are unstable when switching to grid-disconnection cases (V/*f*).

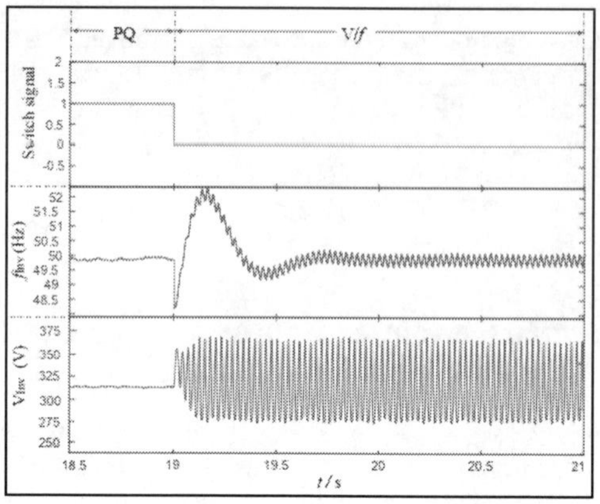

Figure 6. Influence of mismatch inner loop parameters on mode switching.

The simulation results in Figure 6 show that the control system parameters have different best match values for different control modes. However, in the micro-grid mode switching, only the outer loop controller switching, the inner loop control parameters are not switched, it is easy to cause the system instability after the mode switch. Therefore, a robust controller is required to replace the current inner loop to reduce the oscillation caused during the switching and to improve the controller's anti-interference capability. The following takes the H_∞ robust controller instead of the current inner loop and gives the controller design process.

4 The design of h∞ robust controller

4.1 The basic theory of H∞ robust control

The basic theory of H_∞ robust control is based on the development of liner quadratic regulation (LQR), which has become a mature optimal control theory, the optimal solution is the minimum value of the H_∞ norm. Because the parameters of H_∞ robust control system are easy to set, closed-loop feedback control can be realized effectively. Robust control has the characteristics of rapid response, strong anti-interference and dynamic performance, and it has been widely used in the field of control engineering. The basic principle of the H_∞ robust controller is shown in Figure 7 [15, 16].

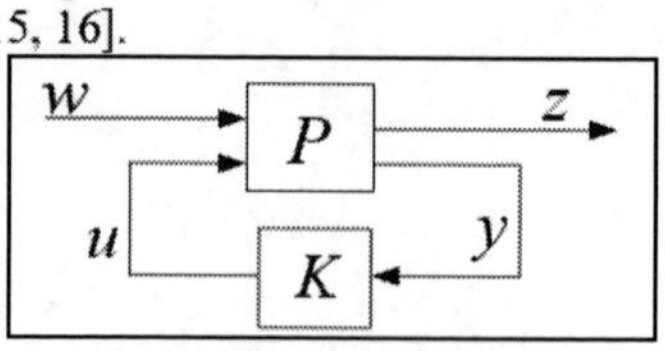

Figure 7. Standard H_∞ robust controller.

In Figure 7, *P* is a generalized controlled object, *K* is a controller, *w* is an external input, u is a control output, *y* is a measurement output, and *z* is a controllable output. The generalized state equation of the controlled object *P* is given as:

$$\begin{cases} \dot{x} = Ax + B_1 w + B_2 u \\ z = C_1 x + D_{11} w + D_{12} u \\ y = C_2 x + D_{21} w + D_{22} u \end{cases} \tag{1}$$

where, x is the state variable, then system open-loop transfer function in Figure 7 is:

$$\begin{bmatrix} z \\ y \end{bmatrix} = \begin{bmatrix} A & B_1 & B_2 \\ C_1 & D_{11} & D_{12} \\ C_2 & D_{12} & D_{22} \end{bmatrix} \begin{bmatrix} w \\ u \end{bmatrix} = \begin{bmatrix} P_{11} & P_{12} \\ P_{21} & P_{22} \end{bmatrix} \begin{bmatrix} w \\ u \end{bmatrix} = P \begin{bmatrix} w \\ u \end{bmatrix} \tag{2}$$

where,

$$\begin{bmatrix} A & B_1 & B_2 \\ C_1 & D_{11} & D_{12} \\ C_2 & D_{12} & D_{22} \end{bmatrix} = \begin{bmatrix} P_{11} & P_{12} \\ P_{21} & P_{22} \end{bmatrix} = P \tag{3}$$

The feedback branch function in Figure 7 is set to:

$$u=Kx \tag{4}$$

Then the closed-loop transfer function of w to z is:

$$T_{zw}=P_{11}+P_{12}K(I-P_{22})^{-1}P_{21} \tag{5}$$

In engineering applications, the H_∞ robust control problem is:

① It is needed to find K to make the H_∞ norm of T_{zw} minimum, that is, $\min\|T_{zw}\|\leq\gamma_0$ (γ_0 is a very small positive number).

② When the condition ① is satisfied, the controlled object $P(s)$ is stabilized in the closed loop system.

4.2 Design of H∞ robust controller for current inner loop

dq coordinate system of the current loop control equation is:

$$\begin{cases} L\dfrac{di_d}{dt} = -Ri_d + \omega L i_q + u_{sd} - u_d \\ L\dfrac{di_q}{dt} = -Ri_q - \omega L i_d + u_{sq} - u_q \end{cases} \tag{6}$$

Considering the parameter uncertainty, it is assumed that the deviation of the parameter R and L is ΔR and ΔL, respectively. That is:

$$\begin{cases} L = L_0 + \Delta L \\ R = R_0 + \Delta R \end{cases} \tag{7}$$

where L_0 and R_0 are the measured values of the inductance and the resistance, ΔR and ΔL are the disturbance values of the inductance and the resistance. Therefore, (6) can be written as:

$$\begin{cases} (L_0 + \Delta L)\dfrac{di_d}{dt} = -(R_0 + \Delta R)i_d + \omega(L_0 + \Delta L)i_q + u_{sd} - u_d \\ (L_0 + \Delta L)\dfrac{di_q}{dt} = -(R_0 + \Delta R)i_q - \omega(L_0 + \Delta L)i_d + u_{sq} - u_q \end{cases} \tag{8}$$

According to the equation of state equation (1), the controlled object P with parameter perturbation can be expressed as:

$$\begin{cases} \dot{x} = (A + \Delta A)x + B_1 w + (B_2 + \Delta B)u \\ z = C_1 x + Du \\ y = C_2 x \end{cases} \tag{9}$$

where x is the state variable, y is the measured output, z is the evaluation signal, w is the external input of the corresponding dimension, u is the control output of the corresponding dimension. A, B_1, B_2, C, D are constant matrices. ΔA, ΔB is the perturbation matrix.

According to the H_∞ robust controller design requirements, ΔA, ΔB satisfy the equation: $[\Delta A\ \Delta B]=E\Sigma(t)[F_a, F_b]$, where E, F_a, F_b is a constant matrix. $\Sigma(t)$ is an uncertain matrix that satisfies $\Sigma^T(t)\Sigma(t)\leq I$. The specific definitions of the matrices are as follows.

$$x = \begin{bmatrix} x_1 \\ x_2 \\ x_3 \\ x_4 \end{bmatrix} = \begin{bmatrix} i_d - i_{dref} \\ \int_0^t (i_d - i_{dref})dt \\ i_q - i_{qref} \\ \int_0^t (i_q - i_{qref})dt \end{bmatrix}, \quad y = \begin{bmatrix} i_d - i_{dref} \\ i_q - i_{qref} \end{bmatrix},$$

$$A = \begin{bmatrix} -\dfrac{R_0}{L_0} & 0 & w & 0 \\ -1 & 0 & 0 & 0 \\ -w & 0 & -\dfrac{R_0}{L_0} & 0 \\ 0 & 0 & -1 & 0 \end{bmatrix}, \quad B_1 = \begin{bmatrix} 0 & 0 \\ 1 & 0 \\ 0 & 0 \\ 0 & 1 \end{bmatrix},$$

$$B_2 = \begin{bmatrix} \dfrac{1}{L} & 0 \\ 0 & 0 \\ 0 & \dfrac{1}{L} \\ 0 & 0 \end{bmatrix}, \quad C_2 = \begin{bmatrix} 1 & 0 & 0 & 0 \\ 0 & 0 & 0 & 0 \\ 0 & 0 & 1 & 0 \\ 0 & 0 & 0 & 0 \end{bmatrix},$$

$$\Delta A = \begin{bmatrix} \delta_1 & 0 & w & 0 \\ 0 & 0 & 0 & 0 \\ 0 & 0 & \delta_1 & 0 \\ 0 & 0 & 0 & 0 \end{bmatrix}, \quad \Delta B = \begin{bmatrix} \delta_2 & 0 & 0 & 0 \\ 0 & 0 & \delta_2 & 0 \end{bmatrix}^T$$

where, $\delta_1 = \dfrac{L_0\Delta R - R_0\Delta L}{(L_0 + \Delta L)L_0}$, $\delta_2 = \dfrac{\Delta L}{(L_0 + \Delta L)L_0}$, the maximum values of the δ_1 and δ_2 parameter deviations are σ_1 and σ_2, that is, $|\delta_1|\leq\sigma_1$, $|\delta_2|\leq\sigma_2$. From the above can be obtained as follows:

$$\Sigma(t) = \begin{bmatrix} \dfrac{\delta_1}{\sigma_1} & 0 & 0 & 0 \\ 0 & \dfrac{\delta_1}{\sigma_1} & 0 & 0 \\ 0 & 0 & \dfrac{\delta_2}{\sigma_2} & 0 \\ 0 & 0 & 0 & \dfrac{\delta_2}{\sigma_2} \end{bmatrix}, \quad E = \begin{bmatrix} 1 & 0 & 1 & 0 \\ 0 & 0 & 0 & 0 \\ 0 & 1 & 0 & 1 \\ 0 & 0 & 0 & 0 \end{bmatrix}$$

$$F_a = \begin{bmatrix} \sigma_1 & 0 & 0 & 0 \\ 0 & 0 & \sigma_1 & 0 \\ 0 & 0 & 0 & 0 \\ 0 & 0 & 0 & 0 \end{bmatrix}, \quad F_b = \begin{bmatrix} 0 & 0 \\ 0 & 0 \\ \sigma_2 & 0 \\ 0 & \sigma_2 \end{bmatrix}$$

The defined evaluation signal z is:

$$z=\begin{bmatrix}\sqrt{q_1}&0&0&0\\0&\sqrt{q_2}&0&0\\0&0&\sqrt{q_3}&0\\0&0&0&\sqrt{q_4}\\0&0&0&0\\0&0&0&0\end{bmatrix}x+\begin{bmatrix}0&0\\0&0\\0&0\\0&0\\\sqrt{\rho_1}&0\\0&\sqrt{\rho_2}\end{bmatrix} \quad (10)$$

q_1, q_2, q_3, q_4, ρ_1, ρ_2 are the parameters that need to be optimized to determine the H_∞ robust controller design process. Generally, the optimal value is found by repeated iteration method.

The optimal H_∞ robust controller design is solved by the existence of a scalar $\lambda > 0$ such that the Riccati inequality satisfies:

$$A^T P+PA+P(B_1 B_1^T+\lambda^2 EE^T)P+C^T C+\frac{1}{\lambda^2}F_a^T F_a - (PB_2+\frac{1}{\lambda^2}F_a^T F_b)R^{-2}(B_2^T P+\frac{1}{\lambda^2}F_b^T F_a)<0 \quad (11)$$

There are positive definite solution P>0, where the H_∞ robust controller of the quadratic stability of the closed-loop system is given as follows:

$$K=-R^{-2}(B_2^T P+\frac{1}{\lambda^2}F_b^T F_a) \quad (12)$$

When the H_∞ robust control law is used to instead of the current inner loop, the current inner loop H_∞ robust control law is as follows:

$$\begin{bmatrix}u_{sd}\\u_{sq}\end{bmatrix}=Kx+\begin{bmatrix}R_0 i_{dref}-\omega L_0 i_{qref}+L_0\dot{i}_{dref}+u_d\\R_0 i_{qref}+\omega L_0 i_{dref}+L_0\dot{i}_{qref}+u_q\end{bmatrix} \quad (13)$$

The current inner loop control block diagram taken in this paper is shown in Figure 8 below:

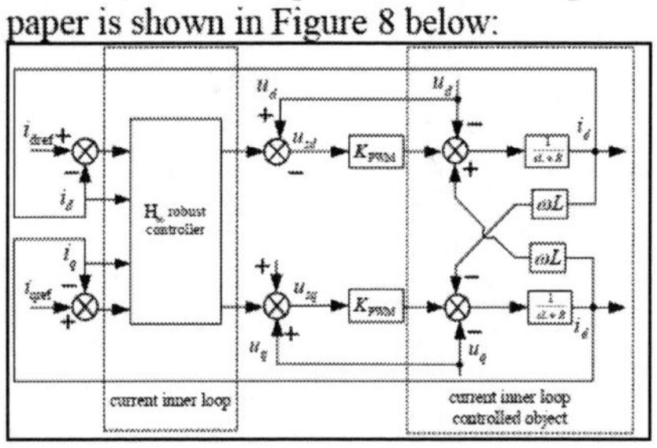

Figure 8. Block diagram of current inner loop H_∞ robust control system.

The parameters of the controller in this paper are shown in Table 1.

Table 1. The parameters of micro-grid.

parameters	parameter-values
Filter inductance L_0	47(mH)
Inductance deviation ΔL	±30% 47 L_0
Equivalent resistance R_0	0.2(Ω)
Resistance deviation ΔR	±20% R_0
Grid frequency f	50 (Hz)
AC voltage	380 (V)

The robust controller box is used to solve the H_∞ robust controller parameters, the final H_∞ controller parameters are:

$$K=\begin{bmatrix}-23.3343&-120.4234&-0.0221&0.6533\\-0.0221&0.6533&-23.3343&-120.4234\end{bmatrix} \quad (14)$$

4.3 The improvement of the outer loop controller switching mode

In order to achieve seamless smooth mode switching and reduce the transient oscillation during switching, this paper not only takes the H_∞ robust controller instead of the current inner loop PI controller to improve the robustness of the system.

In this paper, the state-following algorithm is adopted to improve the switching mode of the outer loop controller The method designs the output status of the V/*f* controller and the PQ controller as a negative feedback, which is the input of the PQ controller and the V/*f* controller respectively, to ensure the output status of both controllers always keep the same.

As shown in Figure 9, when the control mode is PQ, K_1, K_4, K_6 are closed, K_3, K_5, K_2 are open. The output status of the V/*f* controller remains consistent with the output status of the PQ controller at any time. Similarly , when the control mode is V/*f*, K_3, K_5, K_2 are closed and K_1, K_4, K_6 are open. The output status of the PQ controller remains consistent with the output status of the V/*f* controller at any time.

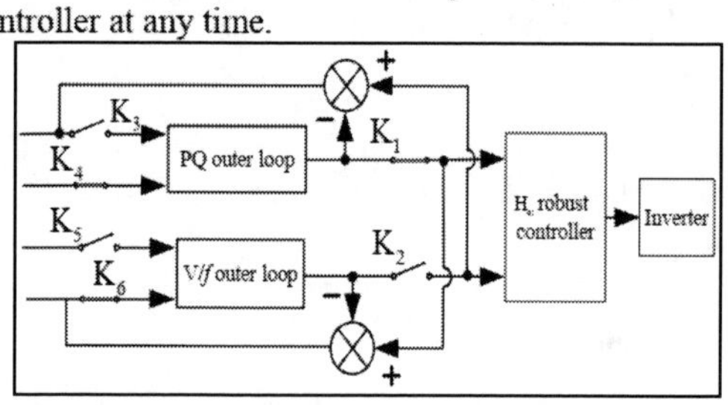

Figure 9. State-following of outer loop controller mode switch.

5 Simulation verification

In order to verify the feasibility of the proposed method, the simulation model is built in Matlab/Simulink. The parameters of the simulation model are shown in Table 2.

Table 2. The parameters of micro-grid.

parameters	parameter-values
Inverter rated power P_{Inv0}	15kVA
Generator rated power P_{SG0}	30kVA
Filter capacitor C	47uF
Filter inductance L	2.5mH
Equivalent series resistance R	0.2Ω
DC side voltage	710V
Rated frequency	50Hz
Current inner loop k_{pi} , k_{ii}	2.4, 110
V/*f* outer loop k_{pu} , k_{iu}	4, 8
PQ outer loop k_p , k_i	0.81, 6

5.1 PI control mode switch

Scenario setting: At t=0s, the inverter independently assumes a sensitive load of 5kW, and the synchronous generator independently bears a load of 10kW. At t=7s, the inverter is connected to the network, and the control mode is switched from V/*f* to PQ. At t=15s, the inverter output power increased to 9kW. At the time of t=19s, the inverter runs grid-disconnection, the inverter and the synchronous generator return to the initial state of t=0s.

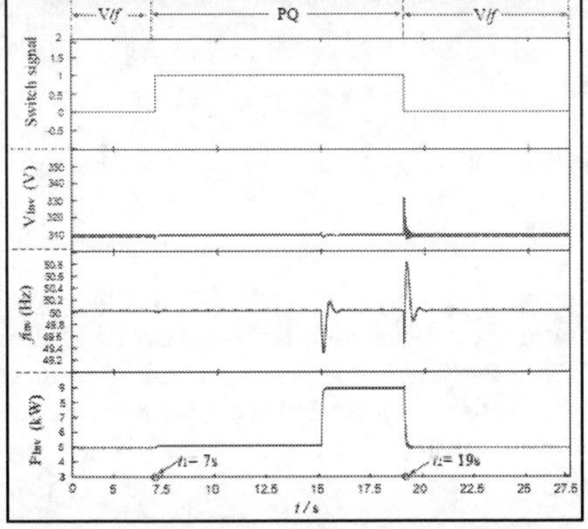

Figure 10. Mode switching simulation waveform with PI controller.

It can be seen from Figure 10, the current loop takes PI controller. V/*f* switches to PQ at t = 7s. Due to take the state-following algorithm, the inverter voltage and frequency waveform does not appear obvious transient fluctuation. At t = 19s, during the mode PQ switching to V/*f*, the inverter output power suddenly becomes smaller, and the voltage and frequency have a significant transient oscillation during the switching. Figure 11 shows the voltage and current waveforms during mode switching.

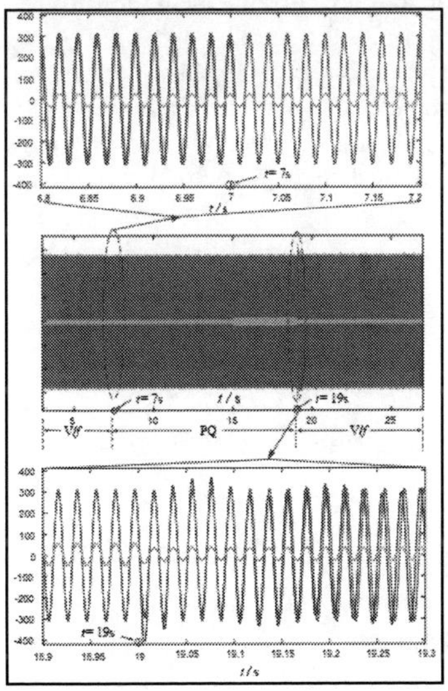

Figure 11. PI control the voltage and current simulation waveforms.

As shown in Figure 11, after taking the state-following, the voltage and current are transition smooth during V/*f* switching to PQ. But in the PQ switch to V/*f* moment, the inverter output voltage appears to be overshoot and oscillation phenomenon.

5.2 H$_\infty$ robust control mode switching

Scenario settings: the same as scenario setting in 5.1.

As shown in Figure 12, it is a simulation experiment waveform when the current inner loop takes H$_\infty$ robust control and the grid/grid-disconnection mode is switched.

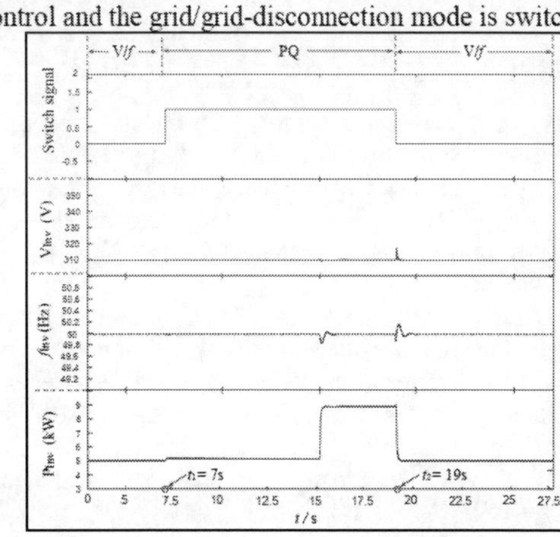

Figure 12. H$_\infty$ robust control mode switching simulation waveforms

It can be seen from Figure 12 that smooth and seamless mode switching can be achieved when H$_\infty$ robust control is adopted in the inner loop. Compared with the waveform of Figure 10, the voltage amplitude oscillation of the inverter becomes noticeably smaller and the frequency does not fluctuate during the period when PQ is switched to V/*f*. Figure 13 shows the voltage and current waveforms during mode switching

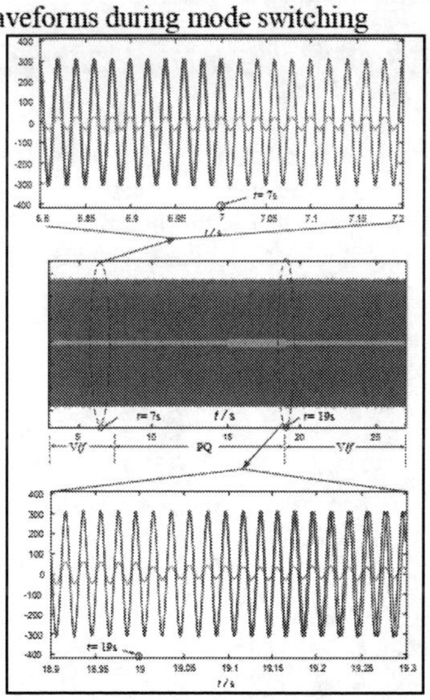

Figure 13. H$_\infty$ robust control the voltage and current simulation waveforms.

Compared with Figure 11 and Figure 13, it can be seen that the voltage and current of PI control and H$_\infty$ robust control are smoothed when V/*f* is switched to PQ mode at t=7s.

However, when the current loop is H_∞ robust controller, the PQ mode is switched to V/*f* mode at t=19s, voltage and frequency does not appear overshoot and oscillation phenomenon as shown in Figure 11. The simulation results show that the H_∞ robust control can realize the seamless switching of the inverter.

6 Experiment verification

6.1 The construction of the experimental platform

This paper sets up a set of experimental system, the experimental system structure is shown in Figure 14. DC bus side is replaced by a battery, the battery DC output voltage is 710V. Inverter rated power is 15kVA, the inverter with the sensitive load is P_{Load2} = 3kW. K_{PCC} is a grid breaker.

Using the inductance L_{Line}=3mH to simulate the line length, the simulate line length is 3kM. Taking the "back-to-back" converter to simulate the synchronous generator, its rated power is 30kVA,and its input is 380V urban electricity. The fixed load of synchronous generator is P_{load0}=11kW. The load that can be switched is P_{Load1}=4kW.

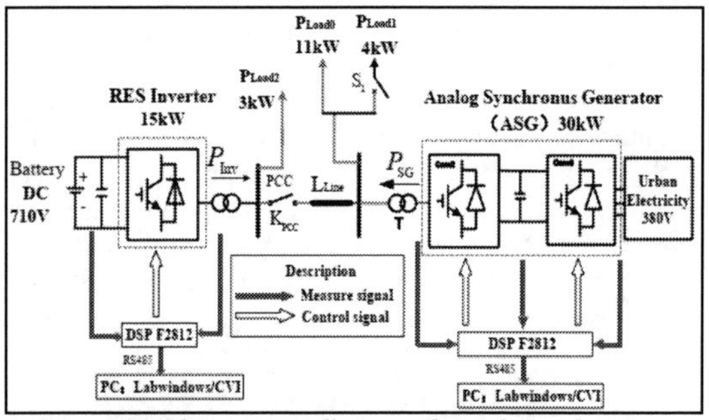

Figure 14. The structure diagram of experimental system.

The main control chip for the converter is the DSP F2812. System of the host computer monitoring is the use of LabWindows/CVI, it is through the serial port to real-time access to device parameters. The experimental system is shown in Figure 15.

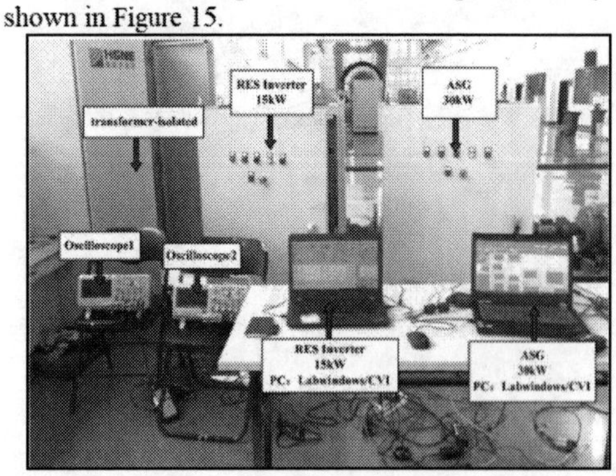

Figure 15. The devices of experimental system.

6.2 The current loop is taken by the PI controller, the inverter is switched from PQ to V/f mode

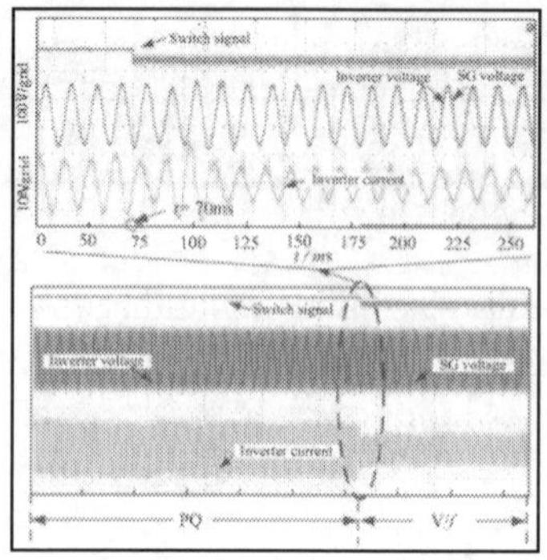

Figure 16. Mode switching voltage and current waveform with PI controller.

At t=0ms, the inverter and the analog synchronous generator are connected to the grid, and the shared load is 14kW. At t=75ms, the inverter is grid-disconnection and bear the sensitive load 3kW independently, the control mode is switched from PQ to V/*f*. The voltage and frequency curves are shown in Figure 16 when the current inner loop is taken by the PI controller.

As can be seen from Figure 16, the amplitude of the voltage fluctuates slightly and the current waveform has a significant oscillation during PQ switching to V/*f*. When the current inner loop is taken by PI controller there is a significant transient shock.

6.3 The current loop is taken by H∞ robust controller, the inverter is switched from PQ to V/f mode

As shown in Figure 17, the voltage is almost no transient oscillation and the current fluctuation is also significantly reduced during the mode switching since the current inner loop is taken by H_∞ robust control. Compared with Figure 16, the transient oscillation during switching is smaller in Figure 17 since H_∞ robust control is adopted, and the seamless mode switching can be achieved better.

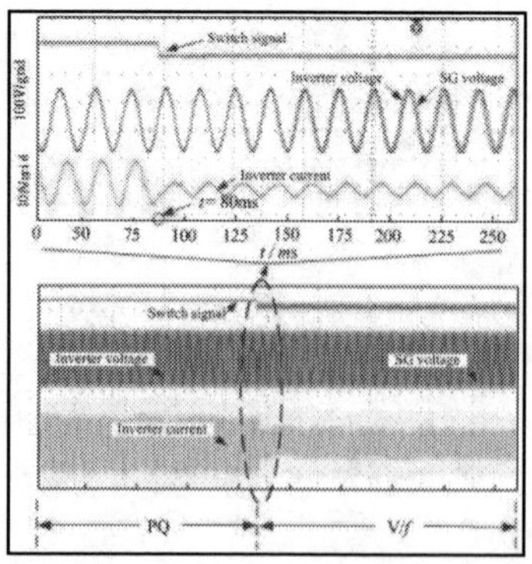

Figure 17. Mode switching voltage and current waveform with H_∞ robust controller.

7 Conclusion

In this paper, the seamless mode switching of independent micro-grid inverter is studied, and the reason of transient oscillation of the traditional mode switching is analyzed. In order to effectively realize the seamless mode switching of the inverter, this paper has improved in two aspects. The simulation and experiment are given to illustrate the effectiveness of the proposed method. This paper mainly completed the following three aspect works.

1) In this paper, the dual-loop control structure of micro-network inverter is given, the traditional grid-connection/grid-disconnection mode switching is analyzed, and the influence of control parameters on mode switching is explained. Outer loop controller inappropriate switching timing and poor robustness of current inner loop controller are the main cause of transient oscillation.

2) The state-following algorithm is used to improve the switching timing to ensure that the output of the two outer loop controllers are seamlessly connected at the time of switching.

3) The H_∞ robust controller is used to instead of the current inner loop to improve the adaptability and robustness of the current loop control, which can further improve the power quality during switching.

References

1. R. Teodorescu, F. Blaabjerg, "Flexible control of small wind turbines with grid failure detection operating in grid-disconnection and grid-connection mode," *IEEE Trans. Power Electron*, **19**, 1323–1332 (2004)
2. S. Adhikari, F. Li, "Coordinated V-f and P-Q control of solar ptovoltaic generators with MPPT and battery storage in microgrid" *IEEE Trans. Smart Grid,* **5**, 1270–1281(2014)
3. J. Kwon, S. Yoon, S. Choi, "Indirect Current Control for Seamless Transfer of Three-Phase Utility Interactive Inverters," *IEEE Trans.Power Electron*, **27**, 773-781(2012)
4. Y. Abdel-Rady, I. Mohamed, Amr A. Radwan, "Hierarchical Control System for Robust Microgrid Operation and Seamless Mode Transfer in Active Distribution Systems," *IEEE Trans. Smart Grid*, **2**, 352–362(2011)
5. J. Kwon, S. Yoon, S. Choi, "Indirect Current Control for Seamless Transfer of Three-Phase Utility Interactive Inverters," *IEEE Trans. Power Electron*, **27,** 773-781(2012)
6. Z.Q. Guo, D.S. Sha, X.Z. Liao, "Voltage magnitude and frequency control of three-phase voltage source inverter for seamless transfer," *IET Power Electron*, **7**, 200–208(2014)
7. Z.l. Yao, L. Xiao, Y.G. Yan, "Seamless Transfer of Single-Phase Grid-Interactive Inverters Between Grid-connection and Grid-disconnection Modes," *IEEE Trans. Power Electron*, **25**, 1597-1603(2010)
8. W.P. Zhang, D.H. Xu, X. Li, "Seamless Transfer Control Strategy for Fuel Cell Uninterruptible Power Supply System," *IEEE Trans. Power Electron*, **28**, 717-729(2013)
9. T.V. Tran, T.W. Chun, H.H. Lee, "PLL-Based Seamless Transfer Control Between Grid-connection and Islanding Modes in Grid-connection Inverters," *IEEE Trans. Power Electron*, **29**, 5218-5228(2014)
10. Z. Liu, J.J. Liu, "Indirect Current Control Based Seamless Transfer of Three-phase Inverter in Distributed Generation," *IEEE Trans. Power Electron*, **29**, 3368-3383(2014)
11. I.J. Gabe, V.F. Montagner, H. Pinheiro, "Design and implementation of a robust current controller for VSI connected to the grid through an LCL filter," *IEEE Trans. Power Electron*, **24**, 1444–1452(2009)
12. R. Tirumala, N. Mohan, C. Henze, "Seamless transfer of grid-connection PWM inverters between utility-interactive and grid-disconnection modes," *IEEE Appl. Power Electron. Conf*, 1081–1086(2002)
13. T.S. Hwang, S.Y. Park, "A seamless control strategy of a distributed generation inverter for the critical load safety under strict grid disturbances," *IEEE Trans. Power Electron*, **28**, 4780–4790(2013)
14. Q. Lei, S. Yang, F.Z. Peng, "Multi-loop control algorithms for seamless transition of grid-connection inverter," *IEEE-Appl. Power Electron. Conf. Expo*, 844–848(2010)
15. L.A. Maccari, J.R. Massing, L. Schuch, C. Rech, H. Pinheiro, V.F. Montagner, R.C.L.F. Oliveira, "Robust H_∞ control for grid connected PWM inverters with LCL filters," *10th IEEE/IAS International Conference*, 1–6(2012)
16. G.D. Chen, M.Y. Yang, L. Yu, "Mixed H2/H_∞ optimal guar-anteed cost control of uncertain linear systems," *International Conference on Machine Learning and Cybernetics*, **2**, 784-788(2003)

Study on a New Operational Mode of Economic Operation of Islanded Microgrids Using Electric Springs

Zhiyu Zhao[1], Keyou Wang[1], Guojie Li[1], Xiuchen Jiang[1], Yin Zhang[2]

[1]*Department of Electrical Engineering, Shanghai Jiao Tong University, 200240, Shanghai, China*
[2]*Shanghai Electric Power Technical Research Institute of State Grid, 200247, Shanghai, China*

Abstract. With the increasing penetration of intermittent renewable energy sources (RESs) into microgrids, the original operation mode of power generation determined by load demand faces severe challenges due to the uncertainties of the RESs power output. The electric springs(ESs), as an emerging technology has been verified to be effective in enabling load demand to follow power generation and stabilizing fluctuation of RESs output. This paper presents a new mode of economic operation for island microgrids including non-critical loads with embedded electric springs. Its connotation includes that i) the capacity of energy storage can be reduced through the interaction of the energy storage system (ESS) and the electric springs, ii) the electric springs reduce the stress of peak load regulation and operational cost and iii) the demand of microgrids system for ramping ability of generation units is reduced with the buffer of the electric springs. Numerical results show that the coordinated operation between electric springs and energy storage system of microgrids can bring down the investment cost for the ESS and short-term operational cost in the aspect of economic dispatch, reducing requirements for the capacity and ramp ability of the energy storage system in microgrids. Energy buffering can be achieved with lower cost and the load demand can follow power generation in the new operational mode of islanded microgrids using electric springs.

1 Introduction

Traditionally, load demand determines power generation in operational mode of power system [1]. With the increasing penetration of intermittent renewable energy sources (RESs), just like photovoltaic and wind power into microgrids year by year, it is difficult to accurately predict the total amount of power generation and achieve real-time power balance between load demand and power generation. The following problems, including voltage fluctuations and frequency flicker [2], are more and more obvious. Large grids have some self-regulation capability for voltage fluctuations, while small and islanded microgrids have poor regulation capabilities with the risk of damage to electrical equipment. It has been proposed that using battery, superconducting or high-speed flywheel energy storage device to offset the difference between power generation and load demand is one of the effective ways to solve this problem. However, the energy storage system (ESS) such as batteries are expensive with higher cost and the used batteries are the main source of contaminants. The operational mode of microgrids is moving to "power generation determine load demand". Demand energy management research has arisen, including load scheduling [3], differentiated electricity prices and so on. However, most of those methods are suitable for hour-interval load requirements management, not for real-time energy balance. The new technology of the electric springs (ESs) has recently been proposed for load-side energy management with a large absorption of intermittent renewable power, instantaneously transferring energy fluctuations to non-critical loads and automatically adjusting the power consumption of non-critical loads, with automatic balance between power generation and consumption.

This paper proposed the new mode of economic operation for island microgrids including non-critical loads with embedded electric springs. The capacity of energy storage can be reduced with the electric springs [4]. Due to the interaction of the ESS and the ESs, the timely adjustment of the power consumption of non-critical loads can reduce the charge and discharge power. The double layer optimal dispatch algorithm for islanded microgrid with ESs has been proven to be effective to determine the minimum ESS capacity meeting a certain scenario. When the peak load comes, the electric springs can reduce the power system's demand for the ability of peak load regulation, so as to enhance short-term economic benefits. The demand of microgrids for ramping ability of the ESS or other generation units is reduced with the buffer of the electric springs with lower cost for design and manufacture of generation units. The coordinated operation between the ES and ESS of microgrid can bring down the investment cost of the ESS and short-term operational cost in the aspect of economic dispatch. Energy buffering can be achieved for lower cost and the load demand can track power generation in the new operational mode of islanded microgrids using electric springs. The excellent function of the energy buffer of the ESs will provide a novel access for the construction of the future microgrids. Three advantages of ESs mentioned above have been demonstrated in the model of optimal dispatch in a hypothetical microgrid example.

2 Basic principles of electric springs in islanded microgrids

This hypothetical case is performed by an islanded microgrid containing wind power, ESS, diesel generators and two kinds of load. Critical loads require higher power quality supplies and do not allow for relatively large voltage fluctuations. Non-critical loads, such as electric heaters, have a wide voltage range of operation. The voltage of the ESs will changes according to the real-time power balance. Normally, power generation will follow the load demand without the ES.

Typical, a three-phase electric spring connect in series with a non-critical load. The series combination connects in parallel with critical load. The ES itself does not participate in the regulation of the active power of the microgrid system, but changes the power consumed by changing the voltage of the non-critical load.

There are two kinds of operational mode of the ES, including capacitive mode and inductive mode. The capacitive mode matches with different loads, shown in Fig.1. The magnitude of the voltage of the ES determines the angle of the voltage vector between the critical load and the non-critical load, and the impedance factor angle of the non-critical load also affects it [5].

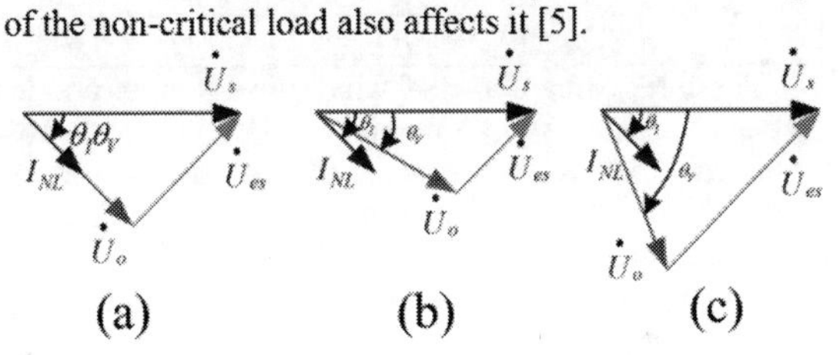

Figure 1. The Capacitive Modes of ES connected with (a) Resistive load. (b) Resistive-inductive load. (c) Resistive-capacitive load.

Where I_{NL} is current of non-critical load. The vector formulations about voltage and power between the power spring and the load are as follows:

$$\dot{U}_s = \dot{U}_o + \dot{U}_{es} \tag{1}$$

$$P_{cl} = 3\times re(\dot{U}_s \times (\frac{\dot{U}_s}{Z_{cl}})^*) \tag{2}$$

$$P_{nl} = 3\times re(\dot{U}_o \times (\frac{\dot{U}_o}{Z_{nl}})^*) \tag{3}$$

$$P = P_{cl} + P_{nl} \tag{4}$$

Where $\dot{U}_s$ and $\dot{U}_o$ denotes the voltage of critical and non-critical load of single phase. And $\dot{U}_{es}$ denotes the voltage of the ES of single phase. The voltage of non-critical load moves between 0.8 and 1.0p.u. with the rated voltage 0.9p.u., for peak and valley load regulation. The voltage of critical load will be maintained at 1.0p.u. if without a sudden emergency.

3 Double layers optimal dispatch algorithm for islanded microgrid with electric springs

The economic dispatch of microgrid is divided into day-ahead and intraday dispatch. Intraday dispatch requires for short-term or ultra-short-term load forecasting data. The energy buffer of the ES can be applied to day-ahead dispatch and real-time adjustment of intraday dispatch.

Double layers optimal dispatch algorithm for islanded microgrid with ESs can be more suitable for real-time dispatch. The 1st layer optimal dispatch is to minimize the operational cost within a day and meanwhile constrained by definite operating constraints of islanded microgrid. Based on the solution of the 1st layer, the 2nd layer optimal dispatch can be used to obtain the minimum capacity requirement of ESS that can meet the current micro-grid, the least ramping ability requirement of the generator units and the obvious effect of lowering the pressure of peak load regulation in microgrid with the buffering of the electric springs.

3.1. The 1st Layer- Optimal Dispatch Formulation of the Islanded Microgrid

The 1st layer optimal dispatch algorithm is similar to the traditional economical dispatch of microgrid [6]. This optimization objective can be formulated as generation costs function of diesel generator:

$$\min\{\sum_{t=1}^{N}[\lambda D_t]\} \tag{5}$$

Where N denotes the number of dispatch intervals within a day. λ and D_t are the generation unit price and output power of diesel generator in the tth interval.

Considering power output, capacity and actual scenario of this microgrid, multiple definite operating constraints are as follows:

$$0 \le W_t \le W_{\max} \tag{6}$$

$$|W_{t+1} - W_t| \le R_{w\max} \tag{7}$$

$$-S_{c\max} \le S_t \le S_{d\max} \tag{8}$$

$$|S_{t+1} - S_t| \le R_{s\max} \tag{9}$$

$$0 \le D_t \le D_{d\max} - D_r \tag{10}$$

$$|D_{t+1} - D_t| \le R_{d\max} \tag{11}$$

$$W_t + S_t + D_t = P_{cl-t} + 0.81P_{nl-t} \tag{12}$$

Where Equation (6)-(7) demonstrate the power output upper limits constraints and ramping constraints of wind power. Equation (8)-(11) are similar constraints of ESS and diesel generators. D_r is spare capacity of diesel generator groups to prevent the extreme scenarios of islanded microgrid. In general, $D_r = 0.1D_{d\max}$ Equation (12) represents the real-time power balance in the microgrid considering critical and non-critical loads.

When the ESS is discharged, $S(t) \geq 0$ and the state of charge (SOC) can be expressed as:

$$SOC(t) = (1-\eta)SOC(t-1) - \frac{S(t)\Delta t}{\eta_d E_b} \quad (13)$$

When the ESS is charged, $S(t) \leq 0$ and

$$SOC(t) = (1-\eta)SOC(t-1) - \frac{S(t)\eta_c \Delta t}{E_b} \quad (14)$$

The SOC of ESS changes within safe operating range. And in order to ensure the ESS's ability to adjust of the next day, the change in SOC within a day shall not exceed a proportion of the maximum capacity:

$$SOC_{\min} \leq SOC_t \leq SOC_{\max} \quad (16)$$

$$| SOC_N - SOC_1 | \leq \delta \quad (17)$$

The solution of the 1st layer optimal dispatch can be used to determine the smaller operational cost, regardless of the electric springs.

3.2 The 2nd Layer- Collaborative Optimization based on Interaction of ESS and ES

The 2nd layer optimal dispatch focuses on the interaction of ESS and ES. The ES can adjust the non-critical load demand, thus reducing current and power of charge and discharge of the ESS based on the same voltage. This optimization objective can be formulated as the minimum of total power of charge and discharge of ESS:

$$\min\{\sum_{t=1}^{N} | S_t |\} \quad (18)$$

If the interaction of ESS and ESs is permitted, the constraints of power balance can be amended and strengthened as:

$$S_{1-t} + 0.81P_{nl-t} = S_{2-t} + (0.81 + k_t)P_{nl-t} \quad (19)$$

Where S_{1-t} and S_{2-t} are the solutions of ESS from the 1st and 2nd layer optimal dispatch. The constraint of non-critical load power regulation coefficients k_t is as follows:

$$0.64 \leq 0.81 + k_t \leq 1 \quad (20)$$

If the 2nd layer optimal dispatch focuses on pressure relief of peak load regulation, the interaction of diesel generators and ES will lower the cost of diesel generators or other dispatchable units during peak load period. And Equation (19) can be modified as:

$$D_{1-t} + 0.81P_{nl-t} = D_{2-t} + (0.81 + k_t)P_{nl-t} \quad t \in \Omega \quad (21)$$

Where Ω is the periods set that requires the peak load regulation, determined from the 1st layer optimal dispatch. D_{1-t} is also the solution of diesel generator from the 1st layer.

The 2nd layer optimal dispatch demonstrates the benefits of the energy buffering of the ES, with the applications of reducing the cost of investment of the ESS and pressure peak load regulation. Double layers optimal dispatch algorithm with ESs can be suitably used in economical dispatch in the future islanded microgrid on the new operational mode.

The mathematical nature of double layers optimal dispatch algorithm for islanded microgrid with the ES can be attributed to the second-order cone programming (SOCP) problem and can be effectively solved.

4 Numerical results

This paper takes the microgrid of region of Beijing Olympic Winter Games as the hypothetical case. The formulations of double layers optimal dispatch algorithm for islanded microgrid with the ES are programmed in MATLAB and the summarized SOCP model is solved using Gurobi [7]. The simulations are carried out on an Intel Core i3-4130 CPU running at 3.40 GHz with 8 GB of RAM.

The microgrid is expected to be configured with wind power capacity of 2MW, energy storage capacity of 17.3 $MW \cdot h$ and diesel generation capacity of 600kW. Part of parameters of the output units in the microgrid are shown in Table 1.

Table 1 Parameters of the Islanded Microgrid

Name	*Value*	*Name*	*Value*
$S_{c\max}$	800kW	$S_{d\max}$	600kW
$R_{s\max}$	400kW	$R_{w\max}$	700kW
η	0.00001	$R_{d\max}$	560kW
η_d	0.9	E_b	17.3 $MW \cdot h$
$SOC_{\min}$	0.2	$SOC_{\max}$	0.9
δ	0.05	η_c	0.85
λ	3.2yuan/($kW \cdot h$)	N	48

The forecasting curve of wind power output and load of the region is shown in Fig 2. The total load in this microgrid consists of critical and non-critical load. The ratio of the critical load to the total load is different at different dispatch interval of one day. The ESS will firstly charge, then discharge and finally charge because of the constraint of $| SOC_N - SOC_1 | \leq \delta$ with the benefits of sustainability in the normal operational mode of the islanded microgrid.

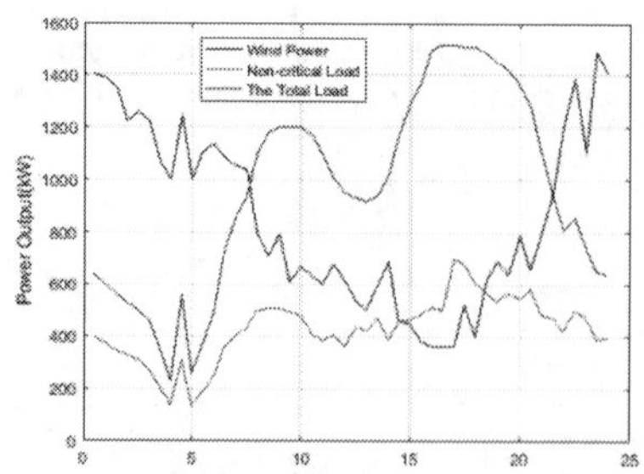

Figure 2. Forecasting Curve of Wind Power Output, Non-critical Load and Total Load

The solution of traditional optimization dispatch is shown in Fig.3 with the lowest cost based on the forecasting RES and constraints of generation units regardless of the ES. The ESS still has the capacity margin when non-peak load. During the peak load from 15:00 to 20:00, the ESS reaches the maximum output and diesel generators are put into operation from Fig.6. The RESs output the maximum power as soon as possible within the operational constraints.

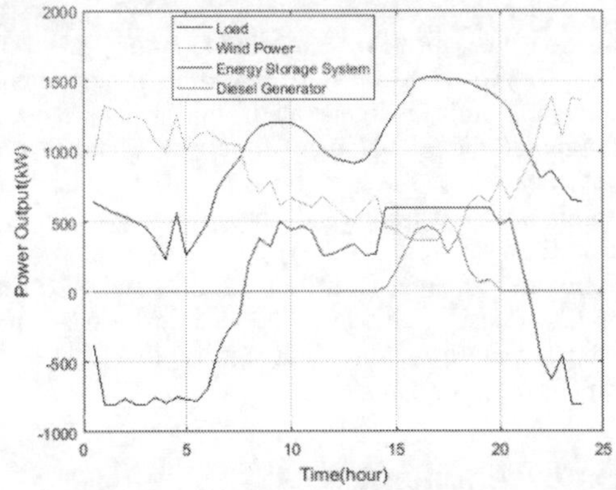

Figure 3. Solution Curve for Optimization Dispatch without the ES

4.1. Reduction and Determination of the Capacity of ESS

The SOC curve of ESS can be obtained when the capacity of the ESS changes in the range of $[0.6E_b, E_b]$ as Fig.4 and it can be concluded that one day's operational cost of islanded microgrid will be maintained at 4423.22 yuan if the capacity of ESS is above $0.6E_b$ after many tests. But cost will rise and efficiency will decline (It is unnecessary to discharge firstly, then charge when the wind power is rich in the red cycle) if the capacity of the ESS below the threshold of $0.6E_b$ ($0.599E_b$ more precisely).

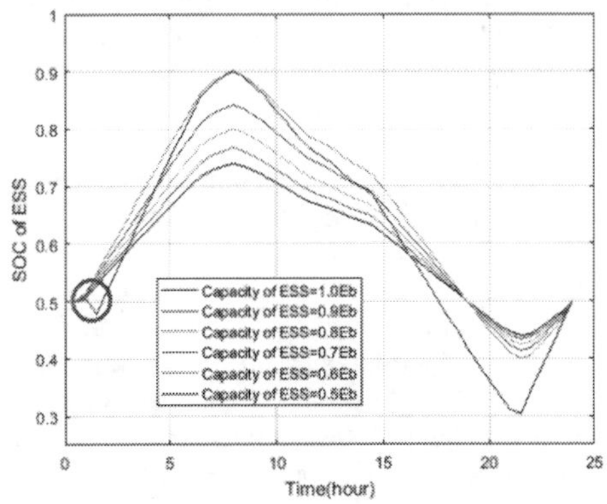

Figure 4. SOC Curve when the Capacity when the ESS Changes

If the 2nd layer collaborative optimization based on interaction of ESS and ES is enabled with the $0.6E_b$ capacity of the ESS, the charge and discharge power of the ESS is reduced, and the variation range of the SOC is smaller (that is to say, a part of the energy storage capacity is surplus if not considering the operational margin of microgrid.), shown in Fig.5-6.

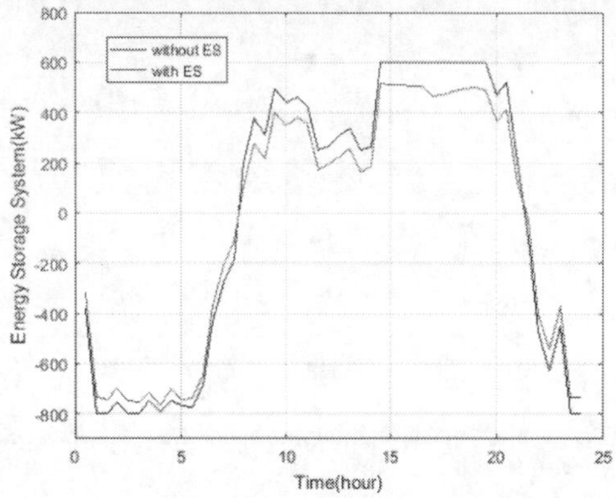

Figure 5. Charge and Discharge Power Curve of the ESS with and without the ES

The further numerical results demonstrate that the operational cost of one day can still be maintained at 4423.22 yuan when the capacity of the ESS drops to with the help of ES. That is to say, the capacity of the ESS decreased by 7.8%. However, the operational cost of one day has soared to 5488.89yuan without the ES when the capacity of the ESS is adopted.

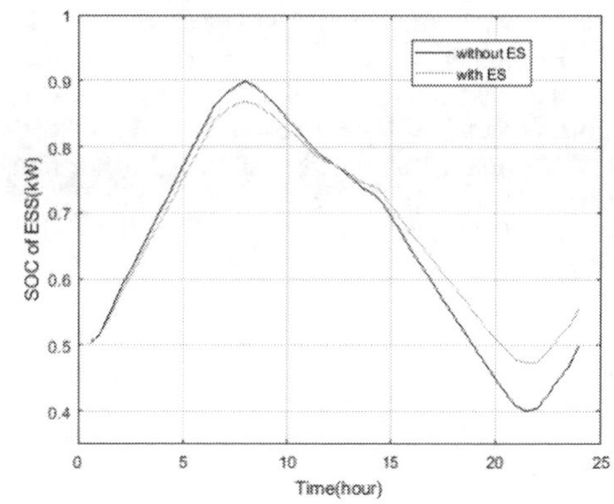

Figure 6. SOC Curve of the ESS with and without the ES

4.2. Pressure Mitigation of Peak Load Regulation

When the electric springs are used for peak load regulation of microgrids, the mere participation of collaborative optimization with generation units that generate short-term costs, like diesel generators, directly reduces operational costs of one day, mitigating pressure of peak load regulation simultaneously. The economical dispatch results with ES only applied in peak load regulation are shown in Fig.7.

The function of the electric spring under the peak load can be summarized as the electric spring for passive load regulation only for a short time.

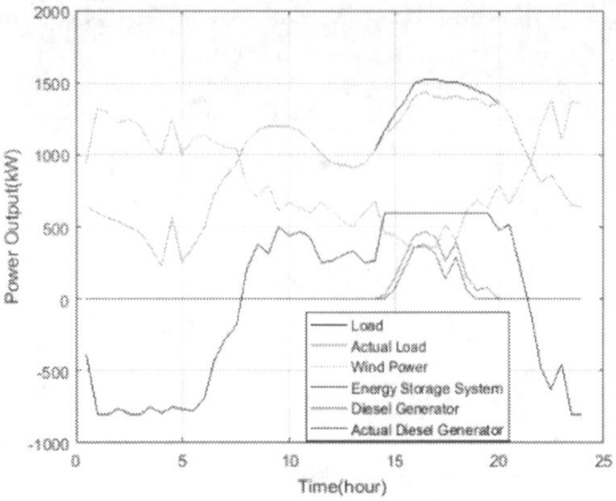

Figure 7. Solution Curve for Optimization Dispatch with ES for Peak Load Regulation

4.3. Reduction and Determination of Requirements for Ramping Ability

The ES can be used to reduce the requirements of ramping ability of generator in islanded microgrid. When the ramping capacity $R_{s\max}$ of the ESS drops from 400kW to 352kW without ES, the operational cost in one day will be maintained at 4423.22 yuan without the ES. When the ramping capacity of the ESS is below 352kW, the operational cost in one day of microgrid will rise. The optimal dispatch result of microgrid with the ES is shown in Fig.8. Numerical simulation results show that the operational cost in one day will not yet increase when the ramping capacity of ESS is reduced to 265kW with the help of ES, that is to say, electric spring makes ramping capacity decreased by 25% maximum.

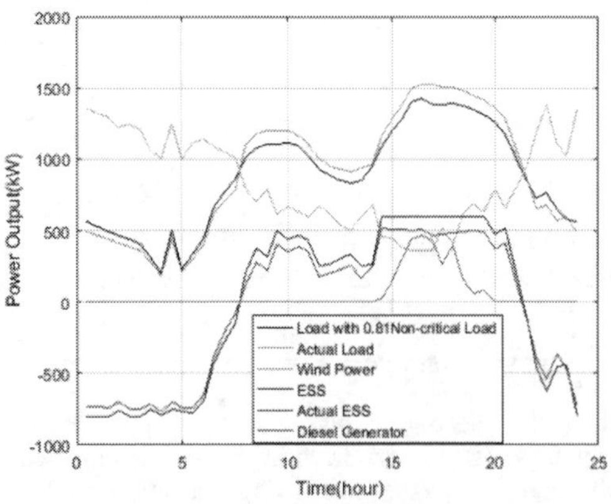

Figure 8. Optimal Dispatch Result of Microgrid with ES

If the power springs can be widely installed in the microgrid, the need for ramping ability of the generator units will be greatly reduced, thereby reducing the manufacturing costs. The function of the electric spring cooperated with the ESS can be summarized as the electric spring for interactive operation for a long time.

5 Conclusion

The study on a new mode of economical operation of islanded microgrids using electric springs has been pushed forward. The double layers optimal dispatch algorithm for islanded microgrids using ES is proposed in this paper. The mode of "load demand following power generation" has been effectively verified into operation for island microgrids including non-critical loads with embedded electric springs with three important benefits including that i) the reduction of capacity of ESS through the interaction of the ESS and the ES, ii) pressure mitigation when peak load regulation using the electric springs with lower cost and iii) reduction of requirements for ramping ability of generation units with the energy buffering of the ES. Numerical results, obtained on a hypothetical case, show that the coordinated operation between electric springs and energy storage system of microgrids can bring down the investment cost in capacity and ramping ability of the ESS and short-term operational cost in one day. The electric springs can be reliably used in future microgrids to change the traditional operating modes.

Acknowledgment

This work was supported by State Grid Corporation of China Science & Technology Project on Principle, Design and Control Technology of New Operation Mode of Microgrid Based on Electric Spring.

References

1. D. Westermann and A. John, "Demand matching wind power generation with wide-area measurement and demand-side management,"IEEE Trans. Energy Convers., vol. 22, no. 1 , pp. 145–149, 2007.
2. Cheng M, Zhu Y. The state of the art of wind energy conversion systems and technologies:A review[J]. Energy Conversion and Management, 2014(88):332-347.
3. A. Mohsenian-Rad, V. W. S. Wong, J. Jatskevich, R. Schober, and A.Leon-Garcia, "Autonomous demand-side management based on gametheoretic energy consumption scheduling for the future smart grid,"IEEE Trans. Smart Grid, vol. 1 , no. 3 , pp. 320–331, 2011.
4. C. K. Lee, and S. Y. Hui, "Reduction of Energy Storage Requirements in Future Smart Grid Using Electric Springs," IEEE Trans. Smart Grid, vol. 4, no. 3, pp. 1282-1288, 2013.
5. N. R. Chaudhuri, C. K. Lee, B. Chaudhuri et al., "Dynamic Modeling of Electric Springs," IEEE Trans. Smart Grid, vol. 5, no. 5, pp. 2450-2458, 2014.
6. S. Conti et al, "Optimal dispatching of distributed generators and storage systems for MV islanded microgrids," IEEE Trans. Power Del., vol. 27, no. 3, pp. 1243-1251, Jul. 2012
7. Gurobi Optimizer 7.5, Jul. 2017 [Online]. Available: http://www.gurobi.com

A New Security Warning Model about Power Grid

HuiJing Bi[1], HongYan Zhang[1] ,YueMei Jiang[2] and XiLan Zhao[1]

[1]*State Grid Hebei Power Company, Training Centre, Shijiazhuang, China*
[2]*Beijing Kedong Electric Control System Co., Ltd, Beijing, China*
[3]*State Grid Gansu Electric Power Company, Dispatch department, Tianshui, China*

Abstract. In recent years, lots of power grid blackouts at home and abroad indicates that the meteorological disasters caused by external meteorological conditions has gradually rose to major contradiction of power grid security. Basing on release information of meteorology, lightning monitoring information and power transmission equipment monitoring information, it established a weather warning model about power grid security, combined with real-time security analysis. Firstly, mathematical models of various types of weather conditions and weather risk assessment model grid were built, then actual operating conditions of a certain regional power grid and model results were compared, the comparison result prove the accuracy of the warning model, and provides a strong recommendation for decision-making.

1 Introduction

Meteorological natural disasters are the main factors threatening the safety of modern power grids. According to the statistics, meteorological disasters caused by the grid accident accounted for about 70% of the total number of accidents. In 2002, the United States, Missouri suffered raging snow disaster, resulting in nearly 40% of the 500kV transmission line outage. In 2008, China's southern rain and snow continued extreme weather, resulting in more than 10kV line parking up to 7500, 35kV substation outage load of 40 million kW, resulting in more than 33 million users out of power, 110 million people affected by electricity, direct economy Loss of more than 1500 million yuan. Therefore, extreme weather disasters on China's current and future safe operation of the power grid constitutes a serious threat[1-2].

In recent years, all levels of power companies to increase investment, efforts to enhance the weather forecast real-time, professional and fine degree[3-6]. At present, although some of our power dispatching center or power grid emergency command center to build the grid weather warning system[7-8], but these systems and technical solutions focus on meteorological information collection, monitoring, forecasting and early warning, in the security warning did not power grid real-time Safety analysis and the probability of meteorological disasters combined with the grid, did not achieve quantitative risk assessment. At the same time, these systems are not deep enough to study the outage model of power grid equipment for different meteorological disasters, and do not realize the accurate quantitative analysis, and the grid dispatching center still lacks effective early warning means for grid weather hazard monitoring. Therefore, this paper studies and builds the mathematical model of meteorological conditions and the meteorological risk assessment model of the grid, and uses the meteorological information of the meteorological department, the lightning monitoring information and the meteorological information of the power transmission equipment to establish the relationship between the meteorological model and the risk assessment model, For all types of meteorological conditions under the grid dispatch department defense and decision-making to find new ways.

2 System structure and working principle

The early warning system model consists of several modules: weather forecast sub-module, lightning detection information sub-module, transmission equipment detection system sub-module, grid state estimation sub-module, meteorological risk on-line analysis and early warning sub-module and meteorological risk result and alarm sub-module The As shown in Figure 1.

The early warning system first collects, integrates and processes various meteorological information, geographic information, power transmission and transformation equipment model, and realizes the above information and grid model, as well as the integration with the online running state, based on real-time steady-state power system data, Probability Analysis of Equipment Discontinuance for Weather Safety. Then, static safety risk analysis is carried out based on the static safety assessment index system of grid meteorological safety. Finally, the static safety assessment and early warning of power grid are carried out according to the grid risk assessment index.

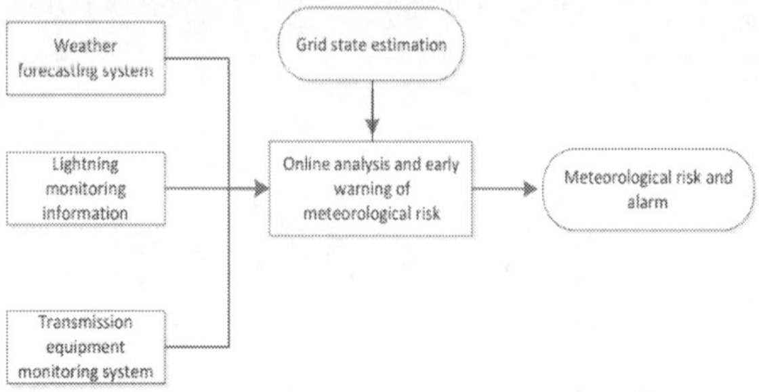

Figure 1. New grid weather weather warning model structure diagram

3 Probabilistic model of power plant equipment outage

3.1 Probabilistic model of power plant equipment outage under lightning conditions

In order to evaluate the accuracy of the thunder zone forecast, use R_{POD} (probability of detection, POD), R_{FAR} (false alarm ratio, FAR)and R_{LDR} (lighting detection probability, LDP)， As shown in (1) - (3):

$$R_{\text{POD}} = \frac{E \cap A}{A} \times 100\% \quad (1)$$

$$R_{\text{FAR}} = \frac{E \cap \bar{A}}{E} \times 100\% \quad (2)$$

$$R_{\text{LDP}} = \frac{(E \cap A) \times Max\{p', p\}}{A \times p} \times 100\% \quad (3)$$

In the formula, E is the area of the predicted lightning area; A is the actual lightning area; $\bar{A}$ is the actual area of non-lightning area; $E \cap A$ is the area of lightning area that is accurate; $E \cap \bar{A}$ is the area of lightning area predicted by mistake; p' is the predicted lightning density; p is the actual lightning density; $E \cap A \times \max\{p', p\}$ is the number of successful mines; $A \times p$ is the actual lightning strikes.

3.2 Probability model of power equipment outage under wind condition

In the air pressure, rainfall, temperature, relative humidity and many other micro-meteorological factors, the greatest threat to the power grid is caused by the wind caused by the wind flashover. The wind speed at the time of wind flashover is random and can be used to model the probability of airborne outage.

The transmission line design specification specifies the minimum gap d_c between the live part and the tower component under the corresponding wind conditions. When the wind speed v is much smaller than the critical wind speed v_c, the wind angle is smaller and the minimum air gap will be much larger than the minimum gap d_c, so the line will not occur wind partial flashover, that is, the probability of wind bias trip P_M is 0; when the wind speed v is much larger than the critical wind speed v_c, the wind angle is large, the minimum air gap will be far less than the minimum gap d_c, so the line will occur wind flash, that is, the probability of wind bias trip P_M is 1; when the wind speed v is near the critical wind speed v_c, the probability of wind deviation P_M is between $(0, 1)$, and the same trend as wind speed. In order to make the outage probability model more simple and practical, we use the piecewise linear model approximation as shown in Figure 2.

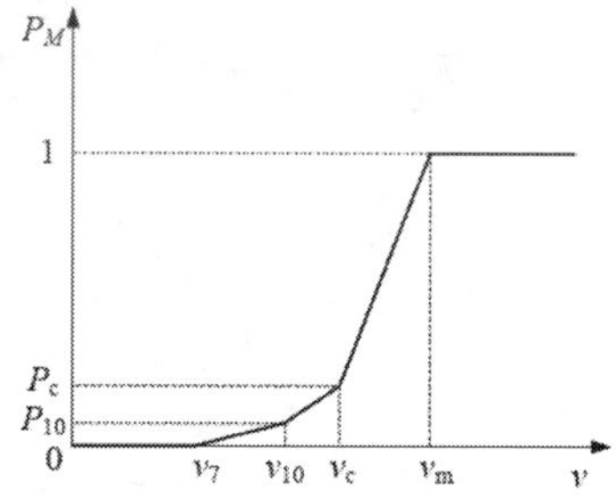

Figure 2. transmission line wind offset device model

Figure 2 in the wind speed turning points v_7、v_{10} were 7 and 10 wind speed, v_c is critical wind speed of the line, P_7 , P_{10} and P_c are the probability of outage at the corresponding wind speed, v_m is the minimum wind speed that must occur for tripping, the probability of outage is 1. P_{10}, P_c, v_c and v_m can be obtained by linear fitting of the wind speed data of the grid micro-meteorological monitoring equipment and the wind-skew trip accident record.

3.3 Probability model of power plant equipment outage under typhoon condition

Typhoon is one of the major catastrophic weather that affects our coastal areas and some inland areas. The probability model of typhoon and wind deviation is based on wind speed as the main influencing factor, and the same model can be used.

According to the latitude and longitude coordinates of the tower, the tower towers within the influence range of the typhoon are selected, and the typhoon forecast wind speed is the main influencing factors. According to the probability model of the wind line deviation of the transmission line, the typhoon outage probability of the tower is calculated. Finally, the typhoon outage probability of each tower is calculated and the typhoon outage probability of the transmission line is calculated.

$$P_Y = 1 - \prod_{i=1}^{n} (1 - P_{Yi}) \quad (4)$$

Where P_Y is the typhoon outage probability of the line; n is the number of towers for the line; P_{Yi} is the typhoon outage probability of the i-th pole tower.

3.4 The Probability Model of Equipment Outage in Integrated Meteorological Condition

Let the transmission line i lightning outage probability P_{Li}, micro-meteorological outage probability P_{Mi}, typhoon outage probability P_{Yi}. The micro-meteorological and typhoon outage probabilistic models take the wind speed as the main influencing factor, so the two take the maximum value as the wind outage probability P_{Wi}, ie

$$P_{Wi} = \max\left(P_{Mi}, P_{Yi}\right) \quad (5)$$

The probability of lightning outage and the probability of wind outage are the logical series relationship, the probability of the total meteorological outage of the transmission line i is

$$P_{Ti} = 1-(1-P_{Li})(1-P_{Wi}) \quad (6)$$

4 Meteorological risk assessment model for power grid

The meteorological conditions of the components (transmission lines, transformers) of the grid are usually not the same during meteorological disasters (such as lightning, wind), so the probability of failure (outage) is different and the probability of the state of the grid is different , The corresponding operational risk is not the same. The purpose of the operational risk assessment is to find out the risk of the system operating state, that is, its probability and consequences of a larger system state, and the system state according to the size of the risk indicators in descending order for the dispatcher to provide the current risk of large systems status information.

Assuming that P_{0j} and $P_{1j}(t)$ are the probability that the component j is in the run and outage state at time t, ie, $P_{0j}(t)$+ $P_{1j}(t)$ = 1, the outage probability $P_{1j}(t)$ can be calculated from the meteorological outage probability.

Let the state of the system state S, in the running state of the composition of the composition C_0, in the outage state of the components constitute the set C_1, then t time system state S probability can be calculated using the following formula

$$P_S = \prod_{j\in C_0} P_{0j}(t)\cdot\prod_{j\in C_1} P_{1j}(t) \quad (7)$$

The probability of all system states can be calculated by equation (5). If the number of elements in the grid is N, the total number of system states $N_T = 2^N$, if N is large, then N_T is very large and the risk assessment of all system states is In this paper, we consider only the system state of the number of outage components $N_f \leq 2$, and select the top probability of the top N_S system states from the N_T system states for risk assessment.

The operating risk index of any system state S in the top probability of the top N_S system states from the maximum system state is:

$$Risk(S) = P_S \cdot Sev(S) \quad (8)$$

Where $Risk(S)$ is the operational risk index of the system state S, P_S is the probability of the system state S, obtained from the meteorological outage probability calculation model, $Sev(S)$ is the consequence of the system state S, and the consequences reflect the severity of the outage failure in the system state , "Consequences" is very broad, can be the degree of line overload, the degree of bus voltage limit, the degree of cross-section of the trend and so on. Define the consequences according to the scope of the assessment. According to the definition of the scope of the assessment of the consequences of the risk assessment indicators can be divided into two categories, one is the system risk assessment indicators, and the other is the cross-sectional risk assessment indicators.

4.1 System risk assessment index and system risk alarm

The early warning model uses the MSFS (margin of system flow security) as the evaluation index, the specific expression is as follows:

$$\mathrm{MSFS}_S = P_S \cdot \min_{a\in A}\left(1-\frac{L_a}{L_{a,\max}}\right) \quad (9)$$

Where: MSFS_S is the total safety margin of all line currents in the system if there is no line overload in the system state S; P_S is the probability of the system state S; L_a is the trend of line a, $L_{a,\max}$ is the maximum allowed for line a transmit active power; A is the line set in the system.

Because all the selected system state must be a power flow calculation, if you need to assess the system state is very large, the calculation is quite large. Because these system states are based on the ground state on the basis of one or more line outage (breaking fault), in order to improve the calculation speed, it can be broken through the DC method to calculate the power flow L_a. DC method to break the flow of the calculation of the use of ground state power flow of the DC node charge matrix B_0 factor table, do not re-form the factor table, can greatly speed up the breaking calculation.

$$\mathrm{MSVS}_S = \min\{\mathrm{MSHVS}_S, \mathrm{MSLVS}_S\} \quad (10)$$

Where MSVS_S is the total safety margin of all bus voltages in the system if there is no bus voltage in the system state S; MSHVS_S is the upper limit safety margin of the system voltage, MSLVS_S is the lower limit of the system voltage margin, and its definition (9), (10), so MSVS_S should take the minimum value of MSHVS_S and MSLVS_S.

$$\mathrm{MSHVS}_S = P_S \cdot \min_{b\in B}\left(1-\frac{V_b}{V_{b,\max}}\right) \quad (11)$$

$$\mathrm{MSLVS}_S = P_S \cdot \min_{b\in B}\left(\frac{V_b}{V_{b,\min}}-1\right) \quad (12)$$

In the above formula: V_b for the bus b voltage; $V_{b,\max}$ for the bus b voltage upper limit; $V_{b,\min}$ for the bus b voltage lower limit; B for the system bus set.

It is not possible to give the information of the amplitude of the bus voltage, but the fast decomposition method based on the compensation principle is used to calculate the current flow. Matrix $\mathbf{B}'$ and $\mathbf{B}''$ factor table, do not re-form the factor table, you can quickly calculate the voltage V_b.

$$\mathrm{DSO}_S = P_S \cdot \max_{a \in A} \left(\frac{L_a}{L_{a,\max}} - 1 \right) \quad (13)$$

Where DSO_S is the total overload risk of all line currents in the system if there is a line overload in the system state S, and the consequence is the overload level of the most overloaded lines in all overload lines; A is the set of overload lines in the system The If there is no line overload, then DSO_S is 0.

4.2 Section risk assessment index and section risk warning

Margin of flow-gate flow security (MFFS) calculation formula is as follows:

$$\mathrm{MFFS}_S = P_S \cdot \left(1 - \frac{\sum_{a \in FA} L_a}{L_{F,\max}} \right) \quad (14)$$

Where MFFS_S is the total safety margin for all line currents in section F in the cross section of system state S; $L_{F,\max}$ is the maximum transmission power limit for section F ; FA is the line set in section F.
Degree of flow-gate overload (DFO) calculation formula is as follows:

$$\mathrm{DFO}_S = P_S \cdot \left(\frac{\sum_{a \in FA'} L_a}{L_{F,\max}} - 1 \right) \quad (15)$$

Where DFO_S is the total overloaded power of the active power transmitted in section F, i.e., the overload of the transmission power of the section F, if the active power transmitted by section F exceeds the transmission power limit in system state S; FA' is the set of lines that are overloaded in the section F.

4.3 Operational risk assessment index calculation

Based on the above operational risk assessment indicators, the calculation of the operational risk assessment indicators, calculation and calculation of the system alarm can be shown in Figure 3.

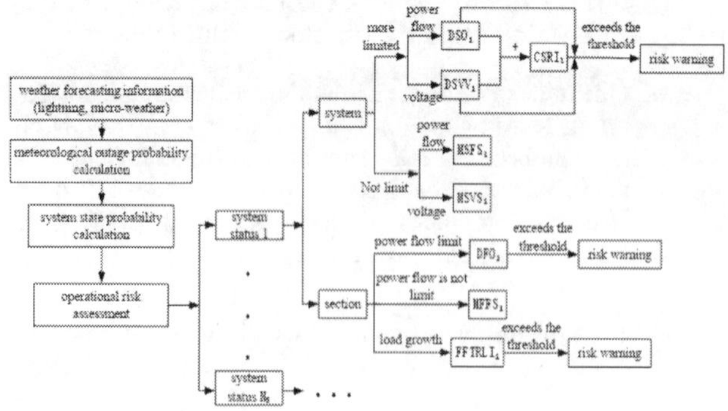

Figure 3. Operational risk assessment index calculation

5 System operation example

In order to verify the accuracy and practicability of the model, this paper takes the relevant data of a certain power grid as an example to detect. The weather data is provided by the grid weather station and transmitted through the network to the meteorological information server database of the dispatching automation system area Ⅲ. Through the development of the interface program, you can directly access the meteorological information database, and regularly extract the relevant weather information.

Part of the data for the December 2013 period of the grid was predicted using the method of 1 minute ahead of the line lightning strikeout (the results are shown in Table 1.). The actual occurrence of 12 lightning accident occurred in the grid, of which three times due to a large number of data missing, cannot be calculated for the probability of outage can be calculated 9 outage accident occurred in the 220kV or 500kV line, based on this method can In this paper, the probability of stopping the outage line relative to the other lines in the minefield is indeed high (basically all in the forefront), and the validity and correctness of the method are proved again.

Table 1. Probability prediction of power outage

Fault time	Fault line voltage level(kV)	Fault line name	Forecast results	Predict the probability of lightning strikeout	Line outage probability sort / mine area line number
14:43:00	220	Line YiZao 4884	success	0.567176	1/4
17:52:00	220	Line NingQian 2D91	success	0.452720	3/5
17:29:00	110	Line LiFu 5941	unpredictable	-	-
15:15:00	500	Line ShanLi 5366	success	0.2202	5/6
16.18:00	110	Line QingNing 481 /line QingNing 482	unpredictable	-	-
18:15:00	500	Line FangZhu 2873	success	0.913716	1/5
14:48:00	500	Line DuShan 5392	success	0.769957	1/7
16:21:00	220	Line DuTian 4868	success	0.376876	2/2
19:52:00	±500	Line NingQian 2D91	unpredictable	-	-
00:59:00	500	Line FanHui 5904	success	0.988842	1/6
21:09:00	220	Line HanMiao 4846	success	0.698580	2/4
13:26:00	220	Line GeTian 2830	success	0.619922	2/10

The power grid 220kV transformer equivalent to the high side, and 500kV connection line and 1000kV UHV line outside the end as a balance node, a total of 389 nodes, 470 branches; power generation total output power of 21467MW, The load of the province is 13635MW, the provincial delivery of about 7800MW.

The power flow in the ground state can be obtained by calculating the power flow in the state of the system running normally. The trend of all branches in the ground state is not limited by the power flow calculation. 47 dynamic sections were identified by the dynamic section identification program. It is shown that there are two sections of the cross-section of the trend, respectively, by the 220kV ZhenGang 2D55 line and the ZhenGang 2D56 line composed of two lines 1; by the 220kV ZhuZhen 4841 line and ZhuZhen 4842 line composed of two lines The limit of the cross section trend is shown in Table 2.

Table 2. Cross section trend limit

No.	Section actual power	Transmission power limit	Cross section trend limit	Trend limit percentage
1	505MW	318MW	187MW	58.9%
2	376MW	341MW	35MW	10.3%

As can be seen from Table 2, the trend of the cross-section of the maximum percentage of the limit, the most serious limit.

The probability of occurrence of meteorological factors and their outages is shown in Table 3. The probability of outage in the table indicates the total probability of meteorological outage for each return line in the double-circuit line, the probability of meteorological outage for each return line in the return line is equal.

Table 3. The probability of outage of each branch of the system

Branch	Affect meteorological factors	Outage probability
YuChen 4723/4727 line	Lightning	0.16
ShuangDeng 4C85 line	typhoon	0.067
ShuangAn 4C81/4C82 line	typhoon	0.11
LuoGu 2768 line	Micro meteorology	0.4

When calculating the risk value of the grid, the consequence *C* uses the cumulative value of the limit of the branch flow and the limit of the cross-section trend to characterize the total trend of the system. For the sake of space, only the first five branches of the system with the largest risk of failure are listed, as shown in Table 4.

Table 4. Fault risk value

Fault number	Fault probability	Stop the branch	Risk value
1	2.78×10^{-1}	LuoGu 2768 line	61.62MW
2	7.94×10^{-2}	YuChen 4723 line	17.57MW
3	5.29×10^{-2}	LuoGu 2768/YuChen 4723 line	11.67 MW
4	5.16×10^{-2}	ShuangAn 4C81 line	11.47MW
5	3.44×10^{-2}	LuoGu 2768/ShuangAn 4C81 line	7.62MW

It can be seen from Table 4. that the risk value of the "LuoGu 2768 line" outage failure is the largest and the probability of failure is the largest. The risk value exceeds the yellow warning threshold (30MW), and the yellow risk alarm should be given; the risk value of other faults Smaller than the blue warning threshold (0), should be given a blue risk alarm.

6 Analysis conclusion

Based on the meteorological information of the Bureau of Meteorology, the monitoring information of lightning and the monitoring information of transmission equipment, this paper establishes the model of grid safety weather warning based on real-time safety analysis of power grid. And the actual operation status of a provincial power grid is compared with the model calculation result. The comparison result shows the accuracy of the early warning model and provides a strong suggestion for the decision. In addition, the grid risk and safety evaluation index system is the technical problem of power supply control and operation decision support. The static risk safety evaluation index of grid in this project still needs to be tested, adjusted and perfected in practice, and based on long-term operation Experience to further develop other risk factors outside the meteorological factors, and ultimately establish a comprehensive grid safety and safety evaluation index system.

References

1. Yin Yonghua, Guo Jianbo, Zhao Jianjun, et al. "Preliminary analysis of large scale blackout in interconnected North America power on August 14 and lessons to be drawn," *Power System Technology*, vol. 27, pp. 8-11 (2003).
2. Chen Yonjin, Ren Xia, Huang Wenying. "Model and analysis of power system reliability evaluation considering weather change," *Automation of Electric Power Systems*, vol. 28, pp. 17-21 (2004).
3. LIU Wengying, YANG Nan, ZHANG Jianli. "Complex Grid Failure Propagating Chain Model in Consideration of Adverse Weather," *Proceedings of the CSEE*, vol. 32, pp. 53-59(2012).
4. LuoYi, WangYingying, WanWei, et al. "Fault chains model for cascading failure of gfid," *Automation of Electric Power Systems*, vol.33, pp. 1-5(2009).
5. Yu Qun, Guo Jianbo. "Statisics and self-organized criticality characters of blackouts in China electric power systems," *Automation of Electric Power Systems*, vol. 30, pp.16-21(2006).
6. Cao Yijia, Jiang Quanyuan, Ding Lijie. "Self-organized criticality phenomenon for power system blackouts," *Power System Technology*, vol. 29, pp.1-5(2005).

7. Ding Ming, Han Pingping. "Research on power system cascading failure with complex system theory," Journal of Hefei University of Technology: *Natural Science*, vol. 28, pp.1047-1052(2005).
8. Wang Jiaming, Liu Wenying, Zhang Jianli. "Online Early Warning Model for Cascading Failure in Complex Power Grid Under Severe Weather", *Power System Technology*, vol.36, pp.239-244(2012).

Application of Component Technology in Power Simulation Training

Xi-Lan ZHAO[1] , Geng Wang [1], Jian-mei Yang[1] and Feng GAO[2]

[1]State Grid Gansu Power Company., Tianshui, China
[2]Beijing Kedong Electric Control System Co. Ltd., Beijing, China

Abstract. With the rapid development of power grid and the implementation of integrated management and control mode, the knowledge, technology and skills of power supply production management personnel have put forward higher requirements, and the establishment of simulation system for the integration of power grid regulation and control is to improve the relevant professionals Quality and skill level of the important means, component-based modeling method can quickly establish the entire power grid of the secondary model, is gradually becoming a new trend in the field of power simulation. The research results of this paper have been successfully applied in the power grid of Gansu Province, with cost-effective, more complete and more realistic simulation, and got good training effect.

1 Introduction

According to the guiding ideology of national energy development, the State Grid Gansu Electric Power Company fully implement the scientific concept of development, clearly to steadily promote the integration of power grid regulation. In accordance with the requirements of the new management system, power grid regulation and control personnel and substation operation and maintenance personnel to assume the power grid security, stability, economic operation of the responsibility is growing, and with the power grid a large number of new technology and new equipment applications, need to greatly improve the regulation of professionals The quality and skill level in order to ensure its ability to master the operation characteristics of new equipment and control the operation of the power grid to establish the integration of power grid regulation and control system [1] is to improve the quality of relevant professionals and skills level of the important means, The focus of the integrated simulation is the secondary system model of the grid. The secondary system model of the power grid refers to all the plant stations of the whole power grid, and establishes a detailed mechanism model for the secondary control, action time, action logic and action information of the secondary equipment such as its protection, communication, monitoring and control, station AC and DC , Complete and true reproduction of the secondary system of light plate signal, protection of information, messages and other secondary system of logical behavior and its interactive relationship [2].

Power system simulation A large number of research contents mainly focus on the research of algorithms. However, due to the long-term concern of the algorithm research, making the actual application of the project, there are the following situations: the development of the secondary system model is often carried out in a project, the coupling between the various functional modules is high, making the simulation system design, There are many problems with maintenance and upgrades. Different simulation engineering secondary system model has many functions and applications similar, but because of its lack of standardization, the interface specification is not uniform and other issues, it is difficult to directly use it to reuse. Making the secondary system model of the flexibility and reusability of the key advantages of difficult to play, engineering development most of the time are used in the reconstruction of these secondary models, which makes the power training simulation system has become a costly system engineering, In the regulation of integrated management model this problem is more prominent. Therefore, it is an urgent requirement and key to promote the development of power simulation technology by encapsulating the secondary system model of the power grid into a simulation component with standardized interface, reusable, versatile, scalable and low coupling.

2 Development of Power Simulation Modeling Technology

Before the 1990s, the realization of the electric power training simulation system basically adopted the modular modeling method. The modular modeling method adopts the process-oriented design. The simulation system can only be used to describe the model by the process function. Scale simulation of complex systems. With the development of modular design, such as attributes, methods, examples and other concepts, the mid-90s gave birth to the development of object-oriented modeling technology to promote the power training simulation system to object-oriented modeling direction. Object-oriented simulation technology breaks through the shackles of traditional concepts, makes the modeling process close to people's normal logical thinking mode. The established model has inherent extensibility and reusability, which is conducive to the establishment of graphical modeling and simulation environment, Thus providing support for simulation analysis of large complex systems. However, object-oriented design methods are also flawed, the degree of coupling between simulation applications is high, scalability and reusability are not strong, the lack of object semantic information, the model is not easy to support multi-application, is not conducive to the development of joint simulation system. As a result of object-oriented modeling methods, component-based software modeling methods are gradually becoming a new trend in power simulation. Compared with the traditional power simulation system modeling technology, the componentized power simulation system modeling technology embodies the idea of software reuse, improves the reusable and extensible features of the simulation system, and makes the simulation system easy to upgrade and maintain, Thus improving the development efficiency.

3 Application of Component Modeling Technology

Process-oriented programming reuse functions, object-oriented programming reuse classes, and component programming reuse specific functional integrity modules. The basic idea of component software technology is to build a software framework that divides large and complex software applications into a series of software unit components that can be implemented, easy to develop, understand and adjust. Each component is functionally defined, individually designed, separately encoded, and each component provides some standard and simple application interfaces that allow the user to set and adjust parameters and attributes. Users can organically combine multiple components from different sources to quickly form a complex application that meets the actual needs. Therefore, component-based software system solutions, the development of high efficiency, low investment, low maintenance costs, reuse ability, software upgrades simple.

The software framework is an important way to achieve large granularity reuse. Is a reusable "semi-finished" software for a field (including business areas such as ERP, and computing, such as a GUI) that implements the common parts of the field and provides a series of well-defined variable points To ensure flexibility and scalability.

In order to solve the above problem, it is proposed that the secondary model of the power grid is difficult to reuse, manage the confusion and build difficult. This paper draws up the idea of "software reuse", and proposes a construction scheme of power system secondary system simulation modeling platform based on component technology. We first set up the power simulation software framework. Second, the need for the secondary model of the power grid to be component-based design, packaging; again, for the simulation components to establish a standardized external data interface; Finally, the establishment of user-friendly graphical modeling tools to complete the simulation component management and simulation system structures Function, so as to quickly achieve the entire power grid secondary system modeling. The following is an example of how to establish a modularization model for the secondary system of the grid by taking the RCS-931 device as a line protection:

(1) to establish a suitable software for the integration of network control simulation application framework

Compared with the ordinary components, the framework of a larger particle size, reflecting the common domain. The integrated simulation software framework includes at least the following components: the relationship between the components and the interaction mechanism, a series of variable points and variable point behavior adjustment mechanism.

Component replacement is achieved through a bus mechanism in the software framework, adding the already identifiable components with relatively independent functions to the system via the bus. Developers through the variable point of the behavior adjustment mechanism, the simulation component bound to the software frame of the variable point, resulting in the final realization of power grid integrated simulation system, this process is called the software framework of the instantiation process. Through the use of software framework, developers can focus on the specific part of the system, which greatly improved the integration of simulation system development efficiency and quality.

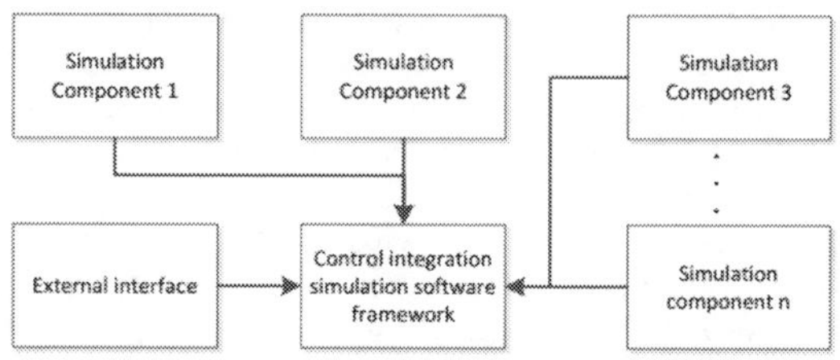

Figure 1.control integration simulation software structure diagram

Software framework includes data management, event management, simulation component management, simulation interface management.

Data management: Due to the real-time requirements of the simulation system, a large amount of data exchange is required during the simulation run. Our software framework provides the use of data bus to achieve rapid exchange of high-volume data, providing data distribution management services.

Event Management: You can broadcast events, or you can dynamically update the execution list by simulating the sequence of events triggered by each component.

Simulation component management: Provides component registration, configuration component parameters, component initialization, component addition, delete and other functions. Provide service management, time management and other services for each simulation component on the platform.

Simulation interface management: for the system simulation process to provide interface operations, including human-machine interface start, operation, pause, recovery, fault and other external trigger events.

(2) The line protection RCS-931 unit is packaged in a package

1) package part model. The components of the RCS-931 unit are divided into signals, platens, switch handles, buttons and other types. All components are encapsulated in the part model file according to their type, as shown in Table 1:

Table 1. Example of RCS-931 operating part model

Control type	Name	Status
<Signal>	Protection action	OFF
<Signal>	Reclosing action	OFF
< Plate >	DC power air switch	ON
< Plate >	AC power air switch	ON
< Plate >	Main protection	ON
<Button>	Signal return	OFF
<Switch handle>	Reclosing method	Single-phase

2)encapsulation logic model. Line protection RCS-931 device is composed of multiple protection function plug-in, respectively, the implementation of different protection functions, each protection plug-in internal logic is defined as an expression, all the logical expression saved to the internal logic model file, Constitute the internal logic model, as shown in Table 2.

Table 2. RCS-931 internal logic model example

Function protection plugin	Action conditions	Action information
<Current differential protection>	<Differential current = 1 && Plates: Main protection=ON>	<Signal: protection action ON>
<Split Differential protection>	<Split differential current = 1 && Main protection=ON >	<Signal: protection action ON>
< Reclose >	<Breaker trip = 1 && switch handle: reclosing mode = Single-phase >	<Signal: reclosing action ON>

The operating conditions in the table<Differential current = 1 && Plates: Main protection=ON> Meaning of the logical expression: the line is differential and the master protection plate is in the input state condition, the current differential protection will act. Action information <signal: protection action : ON> Meaning of logical expression: current differential protection action to send protection action information.

(3) establish a standardized data interface for the RCS-931 component

The component interface adopts the XML design criterion, the component model is separated from the component interface, and the component model and the component interface hierarchy can be expanded independently. XML-based interface is simple, easy to expand, easy to achieve standardization of interface data, as shown in Table 3.

Table 3. RCS-931 component interface example

Data type	Interface data
Grid fault data （Tianshui station TianPu line A 20% of the ground at the permanent fault, grounding 100%）	<Station name> Tianshui</station name> <Line name> Tianpu </ line name> <Fault type> line fault </ fault type> <Fault name> single phase ground </ fault name> <Fault difference> A phase </ fault difference> <Fault location> 20% </ fault location> <Degree of grounding> 100% </ degree of grounding> <Duration> 9999s </ duration>
Operating displacement data(Exit the Tianshui Station TianPu Line RCS-931 Main Protection Plate)	<Station name> Tianshui </station name> <Line name> Tempe line </ line name> <Device name> RCS-931 </ device name> <Device number> 1 </ device number> <Control type> plate </ control type> <Control name> main protection </ control name> <Target status>OFF </ target status>

(4) through the graphical component modeling tools to complete the above components of the management and simulation system to build the work

Component library provides users with a wealth of simulation components, and the establishment of a unified simulation component information description, in order to better and faster to build a power simulation system provides a strong support. Component library structure shown in Tabel 4:

Table 4. Component library structure

Protection name	Protection type	Manufacturer
PRS-723A	Breaker	NARI TECHNOLOGY CO., LTD
PRS-723AH	Breaker	NARI TECHNOLOGY CO., LTD
PRS-723CF	Breaker	NARI TECHNOLOGY CO., LTD
PSL-631A	Breaker	Guodian Nanjing Automation Co.,LTD
PSL-631C	Breaker	Guodian Nanjing Automation Co.,LTD
PSL-631U	Breaker	Guodian Nanjing Automation Co.,LTD

Graphical component modeling tool is a component management, component assembly interface, the role is to ensure that users in accordance with the correct method of implementation, allowing users to more simply complete the simulation system to build the work. As shown in Tabel 5 below, users only need to select the component model from the device component library to complete the construction of the secondary model of the grid.

Table 5. Graphical component modeling platform

Station	Device type	Device name	Protection name	Protection number
Tianshui	Bus	500kV bus I	BP-2B	1
Tianshui	Bus	500kV bus I	BP-2B	2
Tianshui	Bus	500kV bus II	BP-2B	1
Tianshui	Bus	500kV bus II	BP-2B	2

4 Application

The research results have been successfully applied to the State Grid Gansu Electric Power Company, and the relevant production personnel have been trained in the normal operation skills, abnormal and accident handling skills of the grid, and achieved good results. System structure shown in Figure 2:

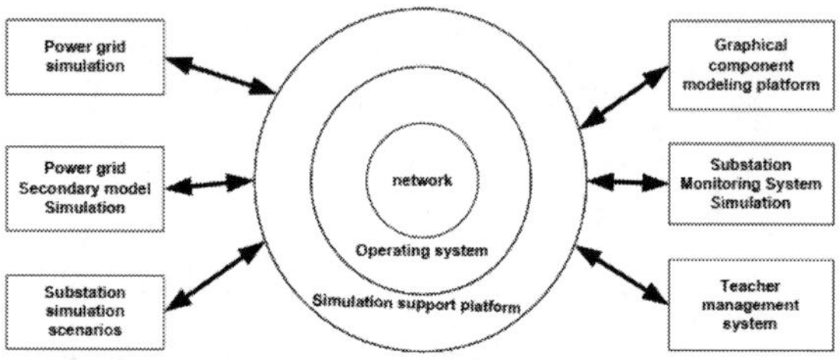

Figure 2. Power simulation system structure diagram

Power grid simulation, power network secondary model simulation, substation three-dimensional scene simulation, substation monitoring system simulation, teacher management system and graphical component modeling platform constitute the entire software system, the module through the simulation run support platform [3] to achieve the simulation process time Synchronization, information interaction and coordination control, Figure 3 gives the overall structure of the software system.

(1) grid simulation to achieve the power grid transient, medium and long-term process of integrated simulation [2], power grid simulation of a device active, reactive, current, voltage and other remote measurement and circuit breaker and the state of the remote signal sent to the Other emulation applications, and receive operational events sent by other emulation applications and emulate

(2) The simulation of the secondary simulation module of the power grid is based on the simulation of the power grid. The logic behavior of the secondary control system, the measurement system, the error prevention system, the AC system, the DC system, the relay protection and the automatic device are simulated. Sub-equipment and secondary circuit simulation results sent to the grid simulation, substation scene and substation monitoring simulation module, and receive the three modules of the operating instructions.

(3) substation monitoring system simulation: simulation of substation monitoring system functions, man-machine interface and operation process, [4] have a detailed discussion.

(4) substation scene simulation is the use of OPENGL [5] [6] three-dimensional engine development. It uses the model editor, the scene editor to establish the substation equipment and scene model, to carry out the scene roaming, virtual operation and other training, the image of the reproduction of the entire production process of substation.

(5) graphical component modeling platform to complete the management of the secondary network simulation components and power grid secondary simulation system to build the work.

(6) The teacher management module is responsible for the whole process of management, monitoring, control, training before the preparation, training, analysis, monitoring, control and management, as well as post-training evaluation and other functions and man-machine composition.

5 Summary

With the rapid development of power grid, process-oriented or object-oriented modeling method has been unable to meet such a large and complex power grid secondary system. This paper draws on the idea of software reuse in software engineering, and proposes to build a power simulation modeling platform based on the component software framework. First, the establishment of a modular software framework. Secondly, the framework of the core components of the standardized thinking of the design package logic model. Again, establish a standardized data interface for the component. Finally, the establishment of graphical modeling tools to complete the management of simulation components, modeling and simulation system to build and other functions. To help users quickly configure the simulation system to achieve the reproduction of the simulation process, thereby enhancing simulation modeling efficiency.the system has been successfully applied to the State Grid Gansu Electric Power Company, and achieved good results.

References

1. Jiang Yefeng, Wei Wenhui, Gao Feng, et al. "Design and implementation of integrated simulation and simulation application based on SG – OSS". *China Electric Power,* vol.47, pp. 80-85 (2014).
2. Huang Wentao, Wei Wenhui, Wu Jihao. "Multi - level joint simulation system based on digital physics hybrid simulation technology of substation". *Power Grid Technology,* vol. 36, pp. 95-98 (2008).
3. Yang Xuanhuai, Lin Changnian, Wang Guoping, et al. "A comprehensive simulation platform of power system based on high - level architecture". *Power Grid Technology,* vol. 33, pp. 98-103 (2009).
4. Hai Xiaotao, Lin Wei, Gao Feng, Ge Xianjun. "Northwest 750kV AC and DC hybrid power network integrated simulation system". *Power Grid Technology,* vol. 36, pp. 254-259 (2012).

5. Li Weiqing, Wu Huizhong, Lin Changian. "Research and Implementation of Virtual Environment for Substation Simulation System". *Journal of System Simulation* , vol. 18 pp. 123-126 (2006).
6. Hou Jun, Li Weiqing, Lin Changian. "Research on Key Technology of Three - dimensional Interactive Scene Simulation in Substation". *Power Grid Technology*, vol. 29 pp. 70-75 (2005).

 https://doi.org/10.1051/matecconf/201816004007

A simulation training evaluation method for distribution network fault based on radar chart

Xu Yuhang[1], Li Tongning[1], Wang Yili[1], Lin Yibin[1], Wei Kuntu[1] and Zhang Wei [2]

[1] State Grid Xiamen Electric Power Supply Company, 361000 Xiamen, FuJian, P. R. China
[2] Power Grid & Distribution Department, Jicheng Electronics Co., Ltd,250100 Jinan, Shandong, P. R. China

Abstract. In order to solve the problem of automatic evaluation of dispatcher fault simulation training in distribution network, a simulation training evaluation method based on radar chart for distribution network fault is proposed. The fault handling information matrix is established to record the dispatcher fault handling operation sequence and operation information. The four situations of the dispatcher fault isolation operation are analyzed. The fault handling anti-misoperation rule set is established to describe the rules prohibiting dispatcher operation. Based on the idea of artificial intelligence reasoning, the feasibility of dispatcher fault handling is described by the feasibility index. The relevant factors and evaluation methods are discussed from the three aspects of the fault handling result feasibility, the anti-misoperation correctness and the operation process conciseness. The detailed calculation formula is given. Combining the independence and correlation between the three evaluation angles, a comprehensive evaluation method of distribution network fault simulation training based on radar chart is proposed. The method can comprehensively reflect the fault handling process of dispatchers, and comprehensively evaluate the fault handling process from various angles, which has good practical value.

1 Introduction

With economic development, people's demand for reliability of power supply continues to increase. In order to ensure the safety, economy and reliability of the power supply quality, higher requirements are set for the professional level of the distribution network dispatching operators [1-3].

Dispatcher simulation training is an important means of daily training and accident inversion for dispatchers [4-7]. However, it is difficult for the dispatcher simulation training to reasonably evaluate the training process and the business level of the dispatcher. At present, the dispatcher simulation training system mainly includes the substation simulation training system and the power network simulation training system [8-14]. The scoring methods mainly include traditional addition and subtraction, manual intervention, artificial intelligence matching, etc..

The simulation training system of distribution network dispatching includes two cases: normal simulation training and fault simulation training. The fault simulation training is to simulate the real-time operation status of the distribution network when the fault occurs, including the protection of alarm matters, switching action items, terminal fault alarm signals, remote signal variation, telemetry mutation and power flow distribution etc.. According to the complex and changeable information, the dispatcher can accurately determine the interval of the fault, and quickly operate the corresponding equipment isolation fault, at the same time, the relevant load that is affected by the fault will be restored to power supply. Although the frequency of occurrence of fault in distribution network is not high, but the effect is significant, although the feeder automation system can automatically isolate the fault, but the fault isolation and recovery as the basic scheduling dispatcher, has been the foundation and key distribution network simulation training.

There are few researches on the evaluation of distribution network fault simulation training. The reference [9] proposed a man-machine combined intelligent evaluation method of substation simulation. This method has great improvement in fault-tolerant case, but requires human intervention to correct the result. The reference [10] proposed a substation simulation training method based on the combination of expert system and weight function. This method introduced expert mechanism and weighting function, but did not describe fault tolerance and fuzziness of dispatcher operation. The reference [11-12] proposed a three-dimensional scene-based distribution network training simulation system, but none of the dispatcher training evaluation method was described. The reference [13] proposed a method of transformer substation operation training evaluation based on the principle of maximum matching. This method evaluated dispatchers strictly in accordance with operating procedures but did not describe the dispatcher's incomplete error operation. The reference [14] proposed a two-level evaluation method which used fuzzy comprehensive evaluation method to evaluate dispatcher's training situation synthetically. This method has more considerations on the dispatcher's rating factors, but the model is more complex and the evaluation result is affected by the weight coefficient, and it does not describe the dispatcher's incomplete error operation.

In summary, in order to make dispatcher daily training and accident inversion work effectively guide and ensure the distribution network safe, economical, efficient and reliable operation. It is of great significance to study a simulation training evaluation method for distribution network fault, which can tolerate dispatcher isolation and transfer to non-standard operation, and can effectively reflect dispatcher training operation and business level.

Based on the characteristics of the simulation training of distribution network fault, this paper presents a comprehensive evaluation method of distribution network fault simulation based on radar chart from the three dimensions of the fault handling result feasibility, the anti-misoperation correctness and the operation process conciseness.

2 Establishment of distribution network fault simulation training model

2.1 Fault handling information matrix model

The distribution network dispatching simulation training system includes two kinds of situations: normal simulation training and fault simulation training, in which fault simulation training has been the focus of dispatcher simulation training.

The key factors to evaluate the fault handling level of dispatchers can be generally summarized into three aspects. The first aspect is the feasibility of fault handling, that is, the smaller the fault isolation range is, the smaller the outage range will be affected by the fault, and the less the loss of electricity, the higher the level of fault handling. The second aspect is the correctness of anti-misoperation, that is, the dispatcher fault handling process must meet the anti-misuse rules to prevent further expansion of the fault range due to misoperation. The third aspect is the conciseness of the operation process, that is, the more concise the dispatcher's fault handling process is, the shorter the fault handling time is, and the higher the fault handling level is.

The fault handling information matrix (*FHIM*) is set up to describe the operation sequence and operation information of the equipment in the process of the dispatcher's fault treatment. The details are as follows:

$$FHIM = \begin{bmatrix} ft_{11} & ft_{12} & ft_{13} & ft_{14} & ft_{15} & ft_{16} & ft_{17} \\ ft_{21} & ft_{22} & ft_{23} & ft_{24} & ft_{25} & ft_{26} & ft_{27} \\ \vdots & \vdots & \vdots & \vdots & \vdots & \vdots & \vdots \\ ft_{m1} & ft_{m2} & ft_{m3} & ft_{m4} & ft_{m5} & ft_{m6} & ft_{m6} \end{bmatrix} \quad (1)$$

Wherein, the first column is the name of the operating device. The second column is the device operating status, 1 for closing, 0 for opening. The third column is the device operating time interval, which is determined by the time difference between this operation and the next operation, and the unit is second. The fourth column is the common load number of the power outages and transmission, including the common load number of the closing power supply and the ordinary load of the switching power supply, in which the ordinary load of the transmission is expressed as positive number, and the common load of the power outages is expressed as negative numbers. The fifth column is the number of important loads of power outages and transmission, in which the number of power transmission important loads is expressed as positive numbers, and the number of power outages important loads is expressed as negative numbers. The sixth column is the load of power outages and transmission, in which the load of power transmission is expressed as positive number, and the load of power outages is expressed as negative number. The seventh column is the feeder ID of the equipment, and the loop-switch belongs to the two feeder lines associated with it. *m* is the dispatcher fault handling steps.

2.2 Fault isolation range analysis

When the fault occurs, the fault area must be determined first, and then the fault area is isolated. However, during actual training, the actual fault isolation range of the dispatcher is not necessarily the same as the standard fault isolation range set by the trainer. As shown in Fig.1, there are four cases of fault isolation, in which S1 and S2 are substation outlet switches, A~J is sectionalizer switch, and K is loop-switch. Area CD is the fault point. There are four cases in the process of isolating fault: inclusion isolation, upstream isolation, downstream isolation and branch isolation.

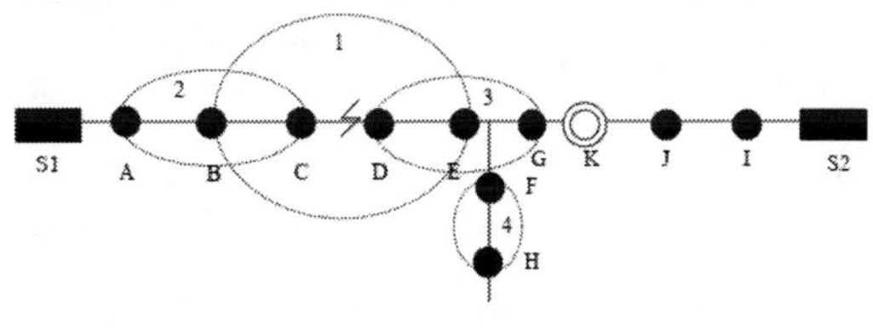

Figure 1. Four cases of fault isolation.

The so-called inclusion isolation means that the dispatcher's actual fault isolation range includes the standard fault isolation range, as shown by the ring 1 in Fig.1. It can isolate the fault correctly, and will not cause the line to lose power again when the fault is isolated and transferred to the load. This dispatcher isolation and load transfer operations are feasible.

The so-called upstream isolation means that the actual fault isolation range of the dispatcher is in the upstream of the standard fault isolation range, but there is no intersection relationship between them, as shown by the ring 2 in Fig.1. It can correctly isolate the upstream fault range and fail to isolate the downstream fault region. In the process of load transfer, it leads to the feeder voltage loss on the other side of the loop-switch. This isolation operation of the dispatcher is partially feasible without causing the side line loss of power.

The so-called downstream isolation means that the actual fault isolation range of the dispatcher is in the downstream of the standard fault isolation range, but there is no intersection relationship between them, as shown by the ring 3 in Fig.1. It can correctly isolate the downstream fault area and fails to isolate the upstream fault region, and it will cause the substation outlet switch to trip again during load transfer. The isolation operation of the dispatcher is partially feasible without causing the trip of the outlet switch again.

The so-called branch isolation means that the actual fault isolation range of the dispatcher is on the branch path, and has nothing to do with the standard fault isolation range, as shown by the ring 4 in Fig.1. It can not isolate fault area completely. In the process of load transfer, it will cause the substation outlet switch to trip again, and lead to feeder power loss on the other side of the loop-switch. This isolation operation is completely infeasible for dispatchers.

2.3 Anti-misoperation rules set

Dispatchers in the process of equipment operation should strictly abide by the anti-misoperation rules. The wrong operation sequence and operation method will lead to heavy equipment overload and reduce the service life of equipment. Seriously, they will lead to arc short circuit and line trip seriously endanger personal safety.

The anti-misoperation lock operation rules table (*ALRT*) is established to record the rules prohibiting dispatcher operation. The details are as follows:

$$ALRT = [al_1, al_2, ar_3, al_4, al_5, al_6, al_7, al_8] \quad (2)$$

Wherein, the al_1 indicates that the disconnector is closed or opened when the feeder has power. The al_2 indicates that the grounding switch is closed when the feeder is energized. The al_3 indicates that the operation causes the feeder current to cross the load limit. The al_4 indicates that the operation causes the feeder voltage to cross the limit. The al_5 indicates that the feeder is powered on while the inspection card is being hung up. The al_6 indicates that the feeder is powered off while the security card is being hung up. The al_7 indicates that the feeder is powered on while the grounding card is being hung up. The al_8 indicates that the operation causes the feeder to loop for a long time.

3 A fault simulation training evaluation method based on radar chart

The simulation training evaluation of distribution network is mainly to realize the evaluation of the results of the fault handling and the feasibility of the treatment process. It is a problem that has a close connection with people's subjective consciousness. Based on artificial intelligence, simulating human subjective consciousness is one of the methods to solve the problem.

3.1 Fault handling result feasibility evaluation

According to the fault handling information matrix and fault isolation range described in Section 2, the factors related to the fault handling result are the range of isolation, the number of critical users losing power, the number of general users losing power, the power loss load and so on. The latter three factors are the concrete manifestation of the former factor. Therefore, the fault handling result feasibility evaluation index (*FHRF*) is set up to evaluate the fault handling results from the three aspects of the number of critical users losing power, the number of general users losing power, and the power loss load. The details are as follows:

$$\begin{cases} FHRF = \alpha_1 \cdot CLLE + \alpha_2 \cdot GLLE + \alpha_3 \cdot LLE \\ \alpha_1 + \alpha_2 + \alpha_3 = 1, \alpha_1 \ge \alpha_2 \ge \alpha_3 > 0 \end{cases} \quad (3)$$

Wherein, the *CLLE* is a critical load power loss evaluation index, described as:

$$CLLE = e^{-\left|AC - \sum_{i=1}^{n} ft_{i5} - cl\right|} \quad (4)$$

The *AC* is the number of critical loads contained in the faulty feeder, and the *cl* is the standard value of the number of critical users losing power set by the trainer.

The *GLLE* is a general load power loss evaluation index, described as:

$$GLLE = \begin{cases} \dfrac{\min(AG - \sum_{i=1}^{n} ft_{i4}, gl)}{\max(AG - \sum_{i=1}^{n} ft_{i4}, gl)} & AG - \sum_{i=1}^{n} ft_{i4} \ne 0 and gl \ne 0 \\ e^{-\max(AG - ft_{i4}, gl)} & AG - \sum_{i=1}^{n} ft_{i4} = 0 or gl = 0 \end{cases} \quad (5)$$

Where, the *AG* is the total number of general loads contained in the feeder and the *gl* is the standard value of the number of general load power loss set by the trainer. The *i* (*i*= 1, 2, …, *n*) is the number of operation rows for the fault feeder in the *FHIM*.

The *LLE* is a power loss load evaluation index, described as:

$$LLE = \begin{cases} \dfrac{\min(AL - \sum_{i=1}^{n} ft_{i6}, ll)}{1/2 \times (AL - \sum_{i=1}^{n} ft_{i6} + ll)} & AL - \sum_{i=1}^{n} ft_{i6} \ne 0 and \ ll \ne 0 \\ e^{-\max(AL - \sum_{i=1}^{n} ft_{i6}, ll)} & AL - \sum_{i=1}^{n} ft_{i6} = 0 or \ ll = 0 \end{cases} \quad (6)$$

Where, the *AL* is the load of the fault feeder, and the *ll* is the standard value of the fault handling power loss load set by the trainer. The α_1 is a critical load weighting factor for power loss, the α_2 is a general load weighting factor for power loss, and the α_3 is the weight factor of power loss load. According to the distribution network dispatching operation experience and dispatching expert evaluation, the range of α_1 is [0.45, 0.55], the general value is 0.5; the range of α_2 is [0.25, 0.35], the general value is 0.3; the range of α_3 is [0.15, 0.25], the general value is 0.2.

3.2 Anti-misoperation correctness evaluation

Based on the *ALRT*, anti-misoperation checking is performed for each step in the *FHIM*. If the operation steps violate any one of the *ALRT*, this step operation violates the anti-misoperation lock criterion, and the accumulated value is *Anum*. The anti-misoperation correctness evaluation index (*AMCE*) is established to prevent the misoperation of dispatcher fault handling. The details are as follows:

$$\begin{cases} AMCE = \alpha_4 \cdot \dfrac{n - Anum}{n} + \alpha_5 \cdot e^{-Anum} \\ \alpha_4 + \alpha_5 = 1, \alpha_4 \geq \alpha_5 \geq 0 \end{cases} \quad (7)$$

Wherein, the α_4 is the correct step weight factor for anti-misoperation, and the α_5 is the error step weight factor for anti-misoperation. According to the distribution network dispatching operation experience and dispatching expert evaluation, the range of α_4 is [0.5, 1], the general value is 0.6; the range of α_5 is [0, 0.5], the general value is 0.4.

3.3 Operation process conciseness evaluation

The fault handling time evaluation index (*FHTE*) is established to evaluate the simplicity of the dispatcher fault operation, which is specifically as follows:

$$\begin{cases} FHTE = \alpha_6 \cdot \dfrac{n - Bnum}{n} + \alpha_7 \cdot e^{-Bnum} + \alpha_8 \cdot e^{\lambda} \\ \alpha_6 + \alpha_7 + \alpha_8 = 1, \alpha_6 \geq \alpha_7 \geq \alpha_8 > 0 \end{cases} \quad (8)$$

$$\begin{cases} \lambda = \dfrac{-\sum_{i=1}^{n} ft_{i3}}{LTime}, (\sum_{i=1}^{n} ft_{i3} > LTime) \\ \lambda = 0, (\sum_{i=1}^{n} ft_{i3} \leq LTime) \end{cases} \quad (9)$$

Wherein, the *Bnum* is the number of steps in violation of the single step operation limit (*STime*). The *LTime* is the fault handling total time set by the trainer, and the unit is seconds. The α_6 is a help-increase weight factor, the α_7 is an attenuation weight factor, and the α_8 is a total time weight factor. According to the distribution network dispatching operation experience and dispatching expert evaluation, the range of α_6 is [0.45, 0.65], the general value is 0.5; the range of α_7 is [0.25, 0.45], the general value is 0.3; the range of α_8 is [0.15, 0.35], the general value is 0.2.

3.4 Comprehensive evaluation based on radar chart

As shown in Fig.2, the concentric circles A1~A4 represent the perfect boundary (evaluation as 1), the excellent boundary (evaluation as 0.9), the good boundary (evaluation as 0.8), and the passing boundary (evaluation as 0.6) respectively. The X-axis, Y-axis and Z-axis respectively show the result evaluation, anti-misoperation evaluation and process evaluation, and x0, y0, z0 is the full score, which is 1. The difference between the axes is 120 degrees, the O is concentric circles, and the concentric circle A1 radius is R. The x, y, z respectively represent the mapping coordinate positions of the real-time evaluation situations of *FHRF*, *AMCE* and *FHTE* in the X-axis, Y-axis and Z-axis coordinates. The details are as follows:

$$\begin{cases} x = FHRF \cdot R \\ y = AMCE \cdot R \\ z = FHTE \cdot R \end{cases} \quad (10)$$

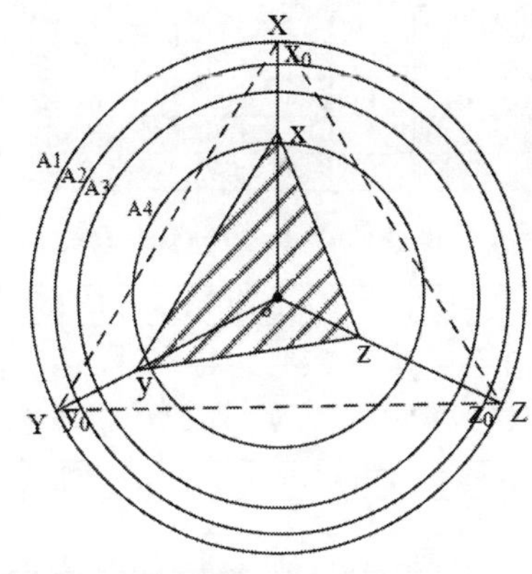

Figure 2. Comprehensive evaluation radar chart.

The comprehensive evaluation index of distribution network fault simulation training is described by the ratio of triangle xyz area and triangle $x_o y_o z_o$ area:

$$FHCE = (xy + xz + yz)/3 \quad (11)$$

4 Case Analysis

As shown in Fig. 3, the distribution network, F1-F6 represents 6 feeders, and S1-S6 is the power outlet switch of six feeders. The 11-66 solid point is the closing sectionalizer switch and the 151-562 hollow point is the split loop-switch. The x1/x2/x3 respectively represent the general load / critical load / total area load (A) contained in the minimum distribution area. If the Fault1, Fault2 and Fault3 faults occur, the operation process of the dispatcher is shown in Table 1, and the weight parameter adopts the default value described in Section 3. The other parameters, such as Table 2, are described in Section 3, and the evaluation results of each fault are shown in Table 3.

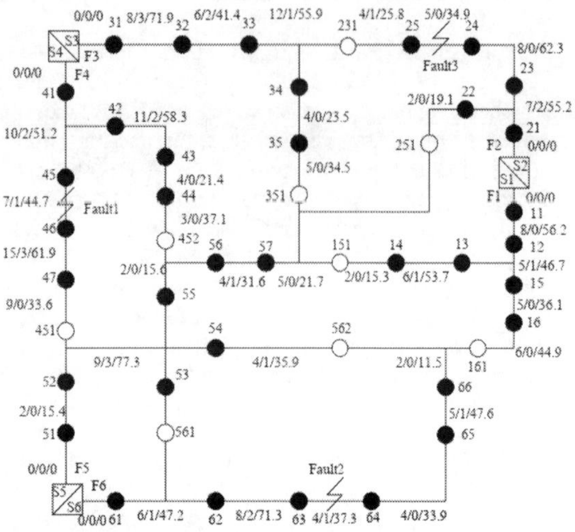

Figure 3. Typical distribution network diagram.

Table. 1 Dispatcher fault handling procedures table

Step	**1**	**2**	**3**	**4**	**5**	**6**	**7**	**8**	**9**	**10**	**11**	**12**	**13**	**14**
Equ	41	42	46	S4	452	451	63	S6	64	161	23	24	S2	231
State	0	0	0	1	1	1	0	1	0	1	0	0	1	1
Time	12	18	20	10	15	12	25	14	18	19	20	16	15	16
Feeder	F4	F4	F4	F4	F4,F5	F4,F5	F6	F6	F6	F6,F1	F2	F2	F2	F2,F3

Table. 2 Fault handling important parameter table

Para	gl_1	gl_2	gl_3	cl_1	cl_2	cl_3	ll_1	ll_2	ll_3	*STime*	*LTime*
Val	7	4	5	1	1	0	44.7	37.3	39.4	20	80

Table. 3 Dispatcher fault handling evaluation table

Fault	***CLLE***	***GLLE***	***LLE***	***FHRF***	***AMCE***	***FHTE***	***FHCE***
Fault1	0.135	0.412	0.636	0.318	1	1	0.546
Fault2	1	1	1	1	1	0.685	0.79
Fault3	0.001	0.104	0.1	0.032	1	1	0.367

The average of the single fault evaluation results in Table 3 can be obtained as 0.568 for the three fault treatments. If using the percentile evaluation, the dispatcher fault handling result is 56.8 points.

5 Conclusion

The fault handling information matrix is established to record the dispatcher fault handling operation sequence and operation information. The four situations of the dispatcher fault isolation operation are analyzed. The fault handling anti-misoperation rule set is established to describe the rules prohibiting dispatcher operation.

Based on the idea of artificial intelligence reasoning, the feasibility of dispatcher fault handling is described by the feasibility index. The relevant factors and evaluation methods are discussed from the three aspects of the fault handling result feasibility, the anti-misoperation correctness and the operation process conciseness. A comprehensive evaluation method of distribution network fault simulation training based on radar chart is proposed. The method can comprehensively reflect the fault handling process of dispatchers, and comprehensively evaluate the fault handling process from various angles, which has good practical value.

References

1. Wang Guanghui, Li Baowei, Hu Zechun, etal. *Challenges and future evolution of control center under smart grid environment.* Power system technology.**35**(2011).
2. Liu Keyan, Sheng Wanxing, Zhang Dongxia,etal. *Big Data Application Requirements and Scenario Analysis in Smart Distribution Network.*Proceedings of the CSEE,**35**(2015).
3. Wu Guopei,Liu Yuquan.*Research and application of technology support system for smart distribute grid.*Power system protection and control.**38**(2010).
4. Li Min, He Hua, Gu Quan, etal . *An integral simulatoraiming for the training both dispatcher and maintenance man .* Automation of Electric Power Systems, **32**(2008).
5. Jiang Yefeng, Wei Wenhui, Gao Feng, etal. *Design and implementation of dispatcher and operator training simulator based on SG-OSS.* Electric power. **47**(2014).
6. Wei Wenhui, Shi Junjie, Gao Ping. *Dispatcher training simlllator of China National Power Dispatching & communication center.*Power system technology，**32**(2008).
7. Yang Yongbo, Gong Qingwu, Zha Xiaoming. *Virtual substationoperator training system based on high level architecture.* Journal of System Simulation, **21**(2009).
8. Zhao Xingtao. *Improvement of automatic scoring of the powerplant simulator training.* Automation of Electric Power Systems. **26**(2002).
9. Li Xuejian,Liu Feng. *Design of Intelligent Evaluation System in Substation Simulation System.* Computer technology and development. **24**(2014).
10. Zhang Fan,Jin Honghe,Yuan Cheng, etal. *Intelligent evaluation system for substation simulation training.* Journal of electric power science and technology, **24**(2009).
11. Qiao Hui，Gong Qingwu，Jiang Chuanwen.*Design Of Experiential Gaming Model-Based Simulation Training System for Distribution Network Operation.* Power System Technology. **35**(2011).
12. Li Gang,Tian Dongmei.*Research on Three-dimensional Training Simulator of Distribution System.* Guangdong electric power. **21**(2008).
13. He Hua,Gu Quan,Xia Keqing.*A New Evaluation Method for Substation Training Simulator Based on Maximal Matching Principle.*Automation of Electric Power Systems.**32**(2008).
14. Chang Xianrong , Xing Jinfeng , Zhang Jie. *Dispateher tranining evaluation based on multi-grade fuzzy comprehensive evaluation method.* Power System Technology. **29**(2005).

A Survey of Cooperative Driving between Auxiliary Autonomous System and Human Driver

Qijie Zou[1,2], Haoyu Li[1,2], Rubo Zhang[3] and Tengda Pei [1,2]

[1]*Information Engineering College, Dalian University,116622 DaLian, China*
[2]*BeiDou High Precision Positioning Service Technology Engineering Laboratory, 116622 DaLian, China*
[3]*Mechatronic Engineering and Automation Institute, Dalian Minzu University, 116600 DaLian, China*

Abstract. The cooperative driving is a main direction of intelligent vehicle development since safety-critical tasks must be executed by human and autonomous system. An intelligent vehicle is equipped with a full range of sensors, which are larger and faster than the perception and computational ability of a human driver, while the latter has a comprehensive ability to adapt to unexpected events. According to their respective advantages, the cooperative driving between human and autonomous system can have new synergies. The emphasis of this paper is to survey the current state of the art of cooperative driving, with a specific focus on decision-making and motion planning levels and correlative algorithms. Such researches enable the autonomous system to compensate the human driver in dangerous or uncomfortable circumstance. This paper provides insights into the scope of decision-making and motion planning for cooperative driving, as well as the shortcomings and tendencies.

1 Introduction

During the past decade, the traffic accidents caused by the improper driving of human driver have increased, and become the main cause of traffic accidents with the increase of the density of vehicles. The improper driving is due to the state of driver, such as tired, distracted, and negligent. Several ambitious projects of autonomous vehicles (or semi-autonomous vehicles), e.g., Google's self-driving car and Tesla's Autopilot, have realized rapid progresses towards commercialization. But from the 2016 Disengagement of Autonomous Mode Report reports [1], self-driving vehicles still need a long time to develop. Meanwhile, intelligent Driver Assistance Systems (DASs) have the potential to improve safety by observing and interacting with the driver. Therefore, there is an increasing interest to keep the human driver in the control loop that needs cooperative driving with the autonomous system to deal with complicated situations. In the field of intelligent vehicle, the study of cooperative control has just started, which is completely different from the existing DASs, since it requires more autonomous driving ability.

The ultimate goal of the cooperative driving system (CDS) focuses on combining the autonomous and the semi-autonomous system with the mature DASs to achieve synergy. Drawing upon fundamental research in understanding of the driver, prediction of vehicle trajectory and the control of collaboration of between human and system, there has been some research progress in this field. To limit the scope of this survey, we focus on the aspects of decision making, motion planning, driver modelling and collaboration controlling for auxiliary autonomous system. When the system finds the deviate trajectory of vehicle, it can predict the driver intention, make decision about the behaviour of vehicle and steer the vehicle safely and smoothly given the driver input.

The remainder of the paper is structure as follows: in section 2, a high level overview of the conceptual architecture and some of key components are presented. Section 3 reviews the decision making and motion planning of autonomous systems, which solve the problem of vehicle trajectory prediction and robust control. Section 4 surveys the understanding of the driver, in order to predict driver input and infer driver intention. The decision methods of collaboration timing are discussed in section 5. Section 6 concludes with remarks on the state of the art and potential areas for future research. Although the intelligent vehicle has no full-autonomous system, we refer to the intelligent control system of vehicle as autonomous system in this paper.

2 Overview of the framework in CDS

2.1 The problem description

In the CDS, the autonomous driving system collaborates with the driver to steer the vehicle, which can combine human's intention with the high-performance computing and full range of sensing capabilities of the intelligent vehicle to achieve the safe, comfort, and low workload driving. Therefore, cooperative driving problem can be described as a synergic effect optimization of two agents, a human driver is natural and an autonomous system is artificial. The graphical representation of this relationship is shown in Figure 1.

The vehicle dynamics is described as $\dot{x} = f(x,u)$. r_D and r_A are the reference roads of human driver and autonomous system. u is the input into the vehicle dynamics, which is the result of conditional integration of r_D and r_A. If the trajectory prediction is no exception, the driver instruction solely controls the vehicle. Otherwise, the control commands will be based on the vehicle state, the reference path, and the action of the driver. $u = g(x,u_D,r_A)$, whereas u_D is the input of driver [2-3].

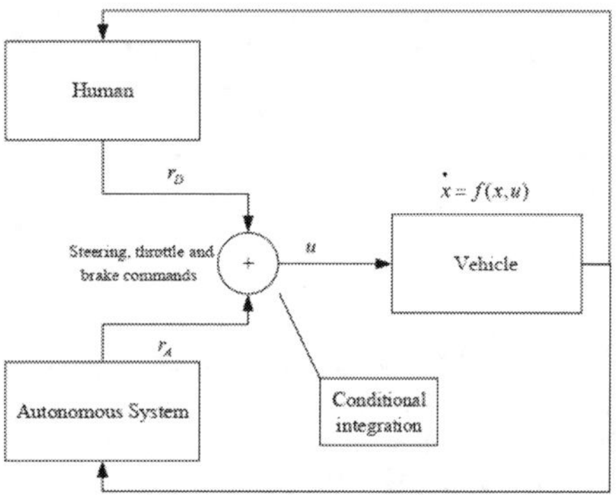

Figure1. The problem description of cooperative driving.

2.2 General framework

The main components used in CDS are *the hierarchy of decision-making process of autonomous driving system, the cooperative interface and the driver model*, cf. Figure 2.

The autonomous driving system is a vehicle self-management agent that generates reference paths and the control command. The cooperative interface controls the integration of human instructions and the reference paths of vehicle when it is aware of unusual driving behaviors. A human driver is the cooperative partner of CDS, which need to understand and predict the intention of human driver.

Therefore, researches in this field depend on the decision-making, motion planning and trajectory prediction of autonomous system, the share control, driver modelling, etc. The subsequent sections review the progresses of related researches.

2.2.1 Decision making

As shown in the blue dotted rectangle of figure 1, the hierarchical decision-making process which is a general term for decision-making and planning in autonomous system can be divided into four levels: *Route Planning, Behavioral Decision-making, Motion planning, Local Feedback Control*. Through the collaboration between the components, system receive real-time environmental information which include the traffic condition of the existing road and driver's own intention to generate global optimal path, obtain the global driving tasks. And then based on some algorithms [4-5], the autonomous system transforms the driving task into the rational driving behavior of the decision, based on the related factors, it is planned to be the driving trajectory which include expected velocity and expected path. Finally, complete the driving.

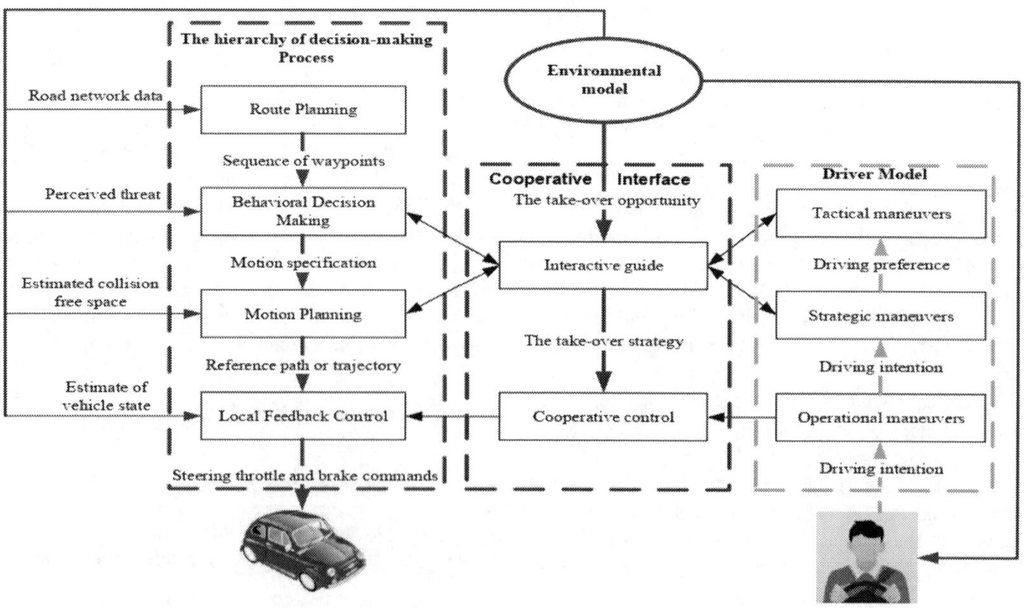

Figure 2. The overall conceptual architecture of cooperative driving.

Table 1. The introduction of rout planning algorithm.

Algorithm	Summary	Development and application	Disadvantage	Solution
Artificial Potential Field	According to the principle of the target point, it attracts robot and obstacles repels robot. It builds a stepped scalar field and then finds a feasible path according to the decline gradient of the scalar field [12]	Planning aspects of mobile robots [13-16]	The kinematics or dynamics of intelligent vehicles are not considered in the planning process, so that the path of planning is not smooth, which is not conducive to the following path tracking.	Through geometrical method to smooth the path planning, construct basic curve set in advance, and then use these basic get planning path curve are interconnected, making the whole path is smooth,
Methods Based on Roadmap	The two visible points are connected in a straight line to form a roadmap, and then the search algorithm is used to find a feasible path.	Visibility Graph [17-18] and Voronoi graph [19]		
Cell Decomposition	To divide the whole environment into convex and unobstructing regions which are called units, [20-22] Thus, the environment is discretized and the path planning problem is transformed into a search problem.	After the map which is based on the graph representation is constructed, the search algorithm is used to find the no-collision path between the starting point and the target point.		By introducing the vehicle's kinematic and dynamic constraints to the process of path planning, with the aid of vehicle dynamics, to obtain a more reasonable path planning, the common algorithms including the kinematics model with curvature polynomial representation and RRT, etc

2.2.2 Cooperative interface

We summarize the research on the switching and integration of control between drivers and autonomous driving systems, and call it *Cooperative Interface.* This part includes: Identify the timing of cooperative driving [6-9], and implement a cooperative strategy [10-11]. The former gets the time to identify drivers and autonomous system is identified by means of trajectory prediction. the latter is used to determine who controls or partial control the driver and the autonomous driving system.

2.2.3 Driver model

Most autonomous driving systems, whether it's advanced driver assistance systems or the cooperative autonomous driving system that we're concerned about, all they need to be focused on the driver model. In essence, the driver model is the system's description, understanding, evaluation and prediction of the driver, which directly influences the performance of cooperative control between the autonomous driving system and the driver.

Currently, the following research objectives of driver model have been considering by the most scholars and researchers. Firstly, it is the intentional prediction, where the driver model could predict the driving intention of the driver base on its driving prior, and further enhance the cooperation efficiency between the system and driver. The second object is dissociation evaluation, since the model will prevent the improper driving behavior that caused by the driver dissociation. In addition, the evaluation has other functions such as the early warning of driving risk. At last, the driving behavior modelling will perform a analysing of the characteristics of the action of driver’s wheel steering, usage of brake and gas pedal, so as to realize personalized driving assistance.

3 The hierarchy decision making process

3.1 Rout planning

In the CDS, the vehicle is required to have strong autonomous planning and independent decision-making ability to cooperate with the driver at the right time. The rout planning of intelligent vehicle draws on the related research of wheeled robot and develops very rapid.

There are three main types of rout planning methods: artificial potential field [12-16], methods based on roadmap [17-19], and cell decomposition [20-22]. The introduction to these algorithms is shown in Table 1. And also introduce the development and improvement of some of these methods in the field of intelligent vehicle.

In recent years, RRT (rapidly exploring random tree) algorithm and its many variants are also gradually applied in intelligent vehicle path planning [23-25]. As a typical random sampling method, RRT can combine the dynamic constraints of the vehicle during the path planning. However, the sampling efficiency of the algorithm is the main factor affecting the real-time performance of planning, and how to improve the sampling efficiency is a major problem in the future.

3.2 Behaviour decision making

Driving behavior decision-making is an important function module of intelligent car, but also the focus and difficulty of autonomous driving technology research. Driving behavior decision-making is based on the perception of the system to obtain environmental information and the current state of the vehicle to make safe and reasonable driving operation,such as vehicles to follow, lane, overtak, import traffic, etc., its capability is an important indicator of intelligent vehicle intelligence level indicators.

Rule-based decision-making method is one of the main methods of intelligent vehicle driving behavior decision-making [26], In the 2007 DARPA City Challenge, the Boss Smart Car of CMU (Carnegie Mellon University) used a rule-based approach to automate vehicle driving decisions [27-28]. But the method is robustness for high dynamic traffic environment.

In order to make the driving decision system more robust, Wei et al. [29-30] proposed a method based on PCB (prediction and cost function-based algorithm) to realize the high speed of intelligent vehicle, Naranjo et al. [31] proposed an intelligent vehicle lane changing method based on fuzzy control, and Perez et al. [32] proposed a fuzzy logic-based decision-making system, but fuzzy rules dependent on the relevant empirical knowledge are often difficult to obtain.

There are also some probabilistic methods for driving behavior decisions is proposed [33], such as based on Dynamic Probabilistic Networks (DPN)[34], Bayesian network[35-37], and based on stochastic automates [38]

3.3 Motion planning

The movement of the intelligent vehicle can be divided into longitudinal movement and lateral movement [39], and the corresponding control is speed tracking control and path tracking control.

3.3.1 Speed tracking control

For speed tracking control, commonly used algorithms fall into two categories: control based on dynamic model and control method based on dynamic model. The former includes sliding mode control, PID feedback control based on kinetic model. The latter includes linear PID, fuzzy PID, and adaptive neural network- fuzzy control PID.

3.3.2 Path tracking control

Path control, ie, lateral control, common control methods include feedback control based on geometric or kinematic models, PID control, Robust control, synovial control, pole placement, optimal control, model predictive control, The fuzzy control.

The evaluation and description of these algorithms are shown in Table 2. In this table, we introduce the application of various algorithms in speed tracking control and path tracking control. The control methods included in the table are verified by the real vehicle.

The information in the table includes: whether the algorithm needs the vehicle dynamic model or not, Whether the algorithm needs artificial setting parameters or artificial setting rules when applying the algorithm. The accuracy of the proposed algorithm is related to the efficiency of control. And also introduce whether the algorithm apply in the speed planning control or path tracking control and can get the information of the improvement about this algorithm.

We use N to represent NO, Y to YES, U to Unrelated, R to Related, P to Path tracking control, and S to Speed tracking control.

Table 2. The evaluation and description of motion planning algorithms.

Algorithm	Use the vehicle dynamics model	Need artificial parameters and rules	Relationship between model precision and control effect	Speed tracking control	Path tracking control	Improvement
sliding mode control	Y	N	U	Y [40-43]	Y [54]	P: Using frequency shaping [55], loop forming [56-57], H∞ control [58] and other methods to improve
Feedback Control Based on Dynamic Model	Y	N	R	Y [44-47]	N	S: Consider modeling the whole dynamic system, such as longitudinal motion dynamics [44-45]
Linear PID	Y	Y	R	Y [48-52]	Y [57]	S: combine with reinforcement learning algorithm [48] P: Introduce proportional gain scheduling [60] and nested PID [61] to improve the control effect.
Fuzzy PID	N	Y	U	Y [53]	Y	S: Using neural networks to express fuzzy rules [53] P: genetic algorithm is introduced in fuzzy logic to optimize the membership function and rules [62,63]

4 Driver Model

Driver modelling is widely studied and used in the field of intelligent driving. Researchers set up models of the different research priorities with different approaches. Driving tasks are accomplished with the help for collaboration between the driver and the system in CDS, so the mutual perception evaluation and prediction between those two parts are the prerequisite and foundation for ensuring the safety and comfort of driving.[64]

With the help of human-machine interaction technology, the autonomous system can obtain the real-time status information such as: driver's concerns, status, behavior and posture and so on. Based on the information, the system can predict the behavior and intention of the driver and offer an evaluation to the driver. Generally, the work that the driver needs to do in the driving of the vehicle can be divided into two categories: workload and situation awareness, which are the driver factor that most researchers generally agree is used to predict driving efficiency and safety the definition as follows:

1) Situation Awareness: Predict what will happen and react accordingly [65]

2) Workload: Questionnaire or test results which to assess the cost and difficulty of the driver [66]

Based on this, we divide the purpose of driver model into three categories: Prediction of the driver's behaviour, inference of the driver's intention, identification of the driver's cognition.

4.1. Behavior model

The driver's behavior modeling which is used to reproduce or predict the behavior of the driver is the basis of the latter two models. Therefore, it can help the autonomous system to better determine the driver's information, and finish the driving work perfectly. The indicators which represent the physical attributes of the driver and reproduces the mathematical expression of the driving process include: steering wheel Angle, speed and other parameters [67]. The behavior of the driver to operate the vehicle, which has strong adaptability, needed to be considered by multiple variables, such as Limb model, the distribution of eye focus area, the use of foot pedals, and so on. We set up a pilot model to analyze and predict driver behavior by HMM (Hidden Markov Model), Bayesian network, machine learning and other methods.[68-69]

4.2. Inference of driver intention

After determining the driver's behavior, status and other information, we then model the driver's intention. The same driver's choice of driving behavior with the different intentions in different time may be particularly aggressive or conservative, when dealing with the same driving scenarios, such as overtaking and lane-changing. And the choice of driver behavior in some state, such as drunkenness and fatigue, becomes an unestimable parameter. Therefore, we need to model the driver's intention to solve the problems in these states.

A common driver's intention determination method includes based on driver characteristic and based on vehicles dynamic state. Both of those methods obtain the required data by using vehicle sensor, such as in-vehicle cameras velocity sensor and steering wheel angle sensor, to perception driver intension. But the method based on driver characteristic determine the intension and degree of activity of the driver by using vision sensor to detect relevant characteristics, such as the pupil diameter of the driver. The frequency and time of blinking, hand motion [70-71], head motion [72], foot motion, etc. another method is based on vision sensors, speed sensors, steering wheel angle sensor, and other unit to monitor the vehicle driving speed, traveling path, and leading vehicle distance, and analyses and judge the state of the driver. the measurement reflects the actual driving behavior of the driver, and then deduce the driver's intentions.

4.3. Cognitive model

Although a qualified driver's operation to vehicle is within a certain range, and the driver behavior modeling is based on the driver universal model which get rid of the driver's personality. Then with different drivers' physical and psychological differences, there are differences in the number of drivers, such as reaction time, driving skill, age, gender, etc. in other words, different driver have different Driving Style. Based on the previous driver behavior and intention, researchers judge the ability of the driver through a series of subtasks to obtain relevant features, such as reaction time, the driver on the environment target recognition rate [73]. In addition, the current domestic use statistical principles to analyze the correlation between the drivers' response indicators and traffic accidents, and then analyze the differences between different drivers of these indicators. And other country's researchers mainly studied the learning and adaptation process of the driver's adaptive cruise system to analyze driver's dependence on system [74], and evaluation of vehicle track tracking and vehicle driving capability by comparing the actual driving path [75].

In order to establish efficient and accurate driver model, there is no doubt that a great deal of data is necessary to build a model. Thanks to the current sensor and data storage capabilities, the acquisition of large amounts of data has become a breeze. How to use these massive amounts of data is a top priority. In the method to solve the problem of the driver modelling, machine learning method for large data processing has obvious advantages, especially the application of deep learning approach.

5 The timing of collaboration between driver and autonomous system

With the help of cooperative driving technology, we dynamically eliminate the cognitive differences between the driver and the autonomous system, achieving consultation, learning and enhancing the ability of autonomous systems to deal with uncertainty, diversity and complexity issues. However, it's still a core research how to switch reasonably the right to drive, and efficient to achieve man-machine integration. With the improvement of autonomous system capabilities, this problem will not disappear, but with the enhancing ability to the autonomy of the system, the cooperative driving technology cooperative driving technology will become a part of the field of autonomous interaction.

In Disengagement Report [1], by statistics we can find that due to improper interaction between the two sides caused by the driver involved in improper participation, lack of participation caused by the situation, accounting for nearly half of the total proportion of detachment .Similarly, if the autonomous system find abnormal phenomena ,it should predict the result of abnormal driving behavior and summary up the reason, rather than rigidly intervent in the driver control competence to trigger a strong distrust of driver. Therefore, both sides of the CDS in the completion of mutual perception evaluation and prediction under the premise, it need to be further completed, such as the determination of the suitability of the autonomous system, the degree of influence on driving safety after the changed driving right [76], the possibility and cause of manual intervention by the driver on the operation of the autonomous system [77] and so on. And determine the difference between the operation of the driving autonomous system and the independent operation of the driver [78-79] and so on, then determine the timing of the interaction so as to real

5.1. Based on threat assessment

It's necessary for cooperative driving technology to determine the time of Man-machine integration. Now the common timing for the determination can be divided into two: (1) Intelligent vehicle system to travel safety and arrive destination for the constraints and objectives; (2) In addition to safety, autonomous intelligent vehicle systems also meet multiple goals, energy consumption, smoothness, stability, etc. The former for this autonomous system cannot complete the driving task of the situation. The latter means that when the performance of the autonomous system is not good for the existing technology, the research work is still focused on the completion of driving tasks. Most of the existing researches belong to the first case.

The threat-based assessment also falls into the first case, it is based on the starting point of the study to predict the collision situation; according to the research perspective, respectively predict the vehicle driving track and other cars road environment in driving track to determine the next period of time (usually 0.5 ~ 5s) the possibility of collision occurred. Meanwhile the car degree of self-confidence is also man-machine rights changing and integration of the main basis [80-81] In other words, if factors that threaten driving safety factors make the system of confidence less than the threshold or the degree of danger is higher than the threshold, the autonomous system will choose to switch to please driver for help or switch the driving right during the driving process

5.2. Based on trajectory prediction

Trajectory prediction is also the first case--driving safety requirements. Lane offset as a cause of road accidents is a major reason, is a variety of factors (road conditions, fatigue driving, drunk driving) the combined effect of the results. In general, the autonomous system obtains the road condition environment through the vehicle sensor, based on the prediction of the future state of the vehicle to determine the danger, and then in the planning layer through the visual / auditory / tactile way to pass the dangerous information to the driver, change the driver for dangerous perception and judgment, to assist in driving to avoid or reduce the occurrence of lane departure accidents. Or a system that corrects the vehicle's travel trajectory when the danger is determined, by controlling the steering / braking / driving control and obtaining the driving right when the vehicle is about to be out of the lane.

The common approach to the timing of man-machine integration, which includes threat assessment or trajectory prediction, can still be seen as a solution to the path planning and decision-making problems in the driving process. In these solutions, reinforcement learning method of multi-objective has just need through the interaction with the environment can get optimal or approximate optimal learning strategies, the advantages of application in the future research is worth looking forward to.

6 Future researches

From the above survey, the conclusion can be reached that CDS is a fusion technology of different agents, which can achieve synergy and its research will promote the development of driverless vehicle technology.

The motion planning and decision-making researches have been mature, but the understanding and description of drivers are still in the initial stage of research. Furthermore, the dynamic integration of natural intelligence and artificial intelligence is the key challenge problem.

Thus, first extension of the CDS could be learning ability. CDS will learn driving experience through daily driver behaviors, and establish vehicle decision-making and planning model, which can make the system supplement the driver's actions when detecting dangerous and abnormal. Secondly, a new pattern humanoid intelligent vehicle will be reach Level 3 and Level 4 which can switch the control smoothly between the human and the autonomous system.

References

1. URL:https://www.dmv.ca.gov/portal/dmv/detail/vr/autonomous/disengagement_report_2016
2. R Li, et al.arXiv preprint arXiv: 1704.00866 (2017).

3. V.A Shia, et al. IEEE Transactions on Intelligent Transportation Systems 15(6), 2696-2709(2014).
4. D.C.K Ngai, et al, Intelligent Transportation Systems Conference, (2007).
5. D.C.K Ngai, N.H.C Yung, IEEE Transactions on Intelligent Transportation Systems, 12(2): 509-522(2011).
6. M.M Minderhoud. Towards The New Horizon Together. Proceedings Of The 5TH World Congress On Intelligent Transport Systems, Paper NO. 2145, (Korea, 1998).
7. P. Fancher, *Research on desirable adaptive cruise control behavior in traffic streams.* (UMTRI-2003-14, 2003)
8. T.P Alkim, G Bootsma, S.P Hoogendoorn. Intelligent Vehicles Symposium, IEEE 1198-1203 (2007).
9. F. Viti, S.P. Hoogendoorn, T.P. Alkim, et al, Intelligent Vehicles Symposium, 745-750 (IEEE,2008)
10. G.A. Mann, N. Small, Human System Interactions (HSI), 2012 5th International Conference on. IEEE, 35-42, (2012)
11. B.D. Argall, T.D. Murphey, Proceedings of the AAAI Fall Symposium on Artificial Intelligence for Human-Robot Interactions (2014)
12. O. Khatib, International Journal of Robotics Research5(1) ,90-98(1986)
13. J. Barraquand, B. Langlosi, J.C. Latombe, IEEE Transactions on Systems, Man, and Cybernetics 22(2), 224-241(1992)
14. E. Rimon, D.E. Koditschek, IEEE Transactions on Robotics and Automation,8(5), 501-518(1992)
15. J. Chuang, N. Ahuja, IEEE Trans. Syst., Man, Cybern. B, Cybern28(5), 729-736(1998)
16. S.S. Ge, Y.J. Cui, IEEE Transactions on Robotics and Automation16(5), 615-620(2000)
17. T. Lozano-Perez, Communications of the ACM22, 560-570(1979)
18. H. Rohnert, Information Processing Letters23, 71-76(1986)
19. O. Takahashi, R.J. Schilling, IEEE Transactions on Robotics and Automation 5(2), 143-150(1989)
20. J.T. Schwartz, M. Sharir, IEEE Trans. Robot. Autom RA-36(3), 345-398(1983)
21. R.A. Brooks, IEEE Transactions on Systems, Man, and Cybernectics SMC-13(3), 190-197(1983)
22. R.A. Brooks, T. Lozano-Perez, IEEE Transactions on Systems, Man, and Cybernetics SMC-15(2), 224-233(1985)
23. W. Xu, J. Wei, J.M. Dolan, et al, Robotics and Automation (ICRA), 2012 IEEE International Conference on. IEEE, 2061-2067(2012).
24. L. Ma, J. Xue, K. Kawabata, et al, IEEE Transactions on Intelligent Transportation Systems16(4), 1961-1976(2015)
25. M. Du, T. Mei, H. Liang, et al, Sensors16(1), 102(2016)
26. A, Niehaus, R.F. Stengel, IEEE Transactions on Vehicular Technology43(3), 626-634(1994)
27. C. Urmson, J. Anhalt, D. Bagnell, et al, Journal of Field Robotics25(8), 425-466(2008)
28. C.R. Baker, J.M. Dolan, International Cofference on Intelligent Robots and Systems, 1752-1758(2008).
29. J. Wei, J.M. Dolan, 2009 IEEE Intelligent Vehicle Symposium, 1015-1020(2009)
30. J. Wei, J.M. Dolan, B. Litkouhi, 2010 IEEE Intelligent Vehicles Symposium, 512-517, (University of California, San Diego, CA, USA, 2010)
31. J.E. Naranjo, C. Gonzalez, R. Garcia, et al, IEEE Transactions on Intelligent Transportation Systems9(3), 438-450(2008)
32. J. Perez, V. Milanes, E. Onieva, et al, Proceedings of the 2011 IEEE International Conference on Mechatronics, 188-193, (Istanbul, Turkey, 2011)
33. M. Ardelt, C. Coester, N. Kaempchen, IEEE Transactions on Intelligent Transportation Systems13(4), 1576-1585(2012)
34. J. Forbes, T. Huang, K. Kanazawa, In Proceedings of the International Joint Conference on Artificial Intelligence, 1878-1885(San Francisco, CA, USA, 1995)
35. R. Schubert, K. Schulze, G. Wanielik, IEEE Transactions on Intelligent Transportation Systems11(3), 607-616(2010)
36. R. Schubert, G. Wanielik, IEEE Intelligent Transportation Systems Magazine, 6-19(2011)
37. R. Schubert, IEEE Transactions on Intelligent Transportation Systems13(1), 354-364(2012)
38. C. Unsal, *Intelligent navigation of autonomous vehicles in an automated highway system: Learning methods and interacting vehicles approach*, (1997)
39. M. Mitschke, H. Wallentowitz. *Dynamik der* kraftfahrzeuge, (Berlin, Springer, 1972)
40. S.B. Choi, J.K. Hedrick, Proceedings of the 1995 Vol5, 3106-3110, (American Control Conference, 1995)
41. J.C. Gerdes, J.K. Hedrick, Control Engineering Practice5(11), 1607-1614(1997)
42. R. Rajamani, H.S. Tan, B.K. Law, et al, IEEE Transactions on Control Systems Technology8(4), 695-708(2000)
43. X.Y. Lu, J.K. Hedrick, 2003. Proceedings of the 2003Vol 1, 36-41(American Control Conference, 2003)
44. D. Anderson, *Splined speed control using SpAM (Speed-based Acceleration Maps) for an autonomous ground vehicle*, (2008)
45. P.N. Currier, *Development of an automotive ground vehicle platform for autonomous urban operations*, 2008.
46. F.W. Rauskolb, K. Berger, C. Lipski, et al, Journal of Field Robotics25(9), 674-724(2008)
47. J. Bohren, T. Foote, J. Keller, et al, Journal of Field Robotics 25(9), 598-614(2008)
48. I. Miller, M. Campbell, D. Huttenlocher, et al, Journal of Field Robotics25(8), 493-527(2008)
49. S. Moon, I. Moon, K. Yi, Control Engineering Practice17(4), 442-455(2009)
50. S. Thrun, M. Montemerlo, H. Dahlkamp, et al, Journal of field Robotics23(9), 661-692(2006)
51. J. Leonard, J. How, S. Teller, et al, Journal of Field Robotics25(10), 727-774(2008)

52. B.J. Patz, Y. Papelis, R. Pillat, et al, Journal of Field Robotics25(8), 528-566(2008)
53. H. Khayyam, S. Nahavandi, S. Davis, Expert systems with applications39(3), 3874-3885(2012)
54. C. Hatipoglu, U. Ozguner, K.A. Redmill, IEEE transactions on intelligent transportation systems4(1), 13-22(2003)
55. H.S. Tan, J. Guldner, S. Patwardhan, et a, IEEE/ASME transactions on mechatronics4(3), 258-272(1999)
56. J.Y. Wang, M. Tomizuka, Asian Journal of Control2(3), 140-154(2000)
57. M. Tai, P. Hingwe, M. Tomizuka, IEEE/ASME Transactions on Mechatronics9(4), 609-618 (2004)
58. C. Urmson, J. Anhalt, D. Bagnell, et al, Journal of Field Robotics25(8), 425-466(2008)
59. G.V. Raffo, G.K. Gomes, J.E. Normey-Rico, et al, IEEE transactions on intelligent transportation systems10(1), 92-102(2009)
60. A. Broggi, M. Bertozzi, A. Fascioli, et al, International Journal of Intelligent Control and Systems3(4), 409-441(1999)
61. R. Marino, S. Scalzi, M. Netto, Control Engineering Practice19(12), 1459-1467(2011)
62. E. Onieva, J.E. Naranjo, V. Milanés, et al, Applied Soft Computing11(1), 1303-1309(2011)
63. J. Guo, P. Hu, L. Li, et al, IEEE Transactions on Vehicular Technology61(7), 2913-2924(2012)
64. S.A. Mostafa, M.S. Ahmad, A. Ahmad, et al, 2016 2nd International Symposium on. IEEE, 106-111(ISAMSR, 2016).
65. E.C. Adam. *Fighter cockpits of the future*, 12th DASC, 318-323(, 1993)
66. R, Fuller, Accident Analysis & Prevention37(3), 461-472(2005)
67. J.L. Coyte, B. Li, H. Du, et alNeural Networks (IJCNN), 2014 International Joint Conference on. IEEE, 940-946(2014)
68. N. Kuge, T. Yamamura, O. Shimoyama, et al, *A driver behavior recognition method based on a driver model framework* (SAE Technical Paper, 2000)
69. V.A. Shia, Y. Gao, R. Vasudevan, et al, IEEE Transactions on Intelligent Transportation Systems15(6), 2696-2709(2014)
70. C. Tran, M.M. Trivedi, IEEE Transactions on Industrial Informatics8(1), 178-187(2012)
71. M.B. Holte, C. Tran, M.M. Trivedi, et al, IEEE Journal of selected topics in signal processing 6(5), 538-552(2012)
72. E. Murphy-Chutorian, M.M. Trivedi, IEEE Transactions on intelligent transportation systems11(2), 300-311(2010)
73. J.C.F. De. Winter, R. Happee, M.H. Martens, et al, Transportation research part F: traffic psychology and behaviour, 27, 196-217(2014)
74. C.M. Rudin-Brown, H.A. Parker, Transportation Research Part F: Traffic Psychology and Behaviour, 7(2), 59-76(2004)
75. F. Biral, M. Da. Lio, E. Bertolazzi, Intelligent Vehicles Symposium, 2005. Proceedings, 36-41 (2005)
76. C. Mobus, M. Eilers, Digital Human Modeling, 413-422(2009)
77. G.S. Aoude, B.D. Luders, K.K.H Lee, et al, Intelligent Transportation Systems (ITSC), 2010 13th International IEEE Conference on. IEEE, 1855-1862(2010)
78. T.M. Roehr, Y. Shi, Proceedings of the 10th International Symposium on Artificial Intelligence, Robotics and Automation in Space, 507-514(Sapporo, Japan. 2010)
79. Jiajia Chen, *Research on Decision Making* System *of Autonomous Vehicle in Urban Environments* (University of Science and Technology of China, 2014)
80. D.I.F. Ferguson, D.A. Dolgov, *Modifying behavior of autonomous vehicle based on predicted behavior of other vehicles*, (U.S. Patent No. 8,457,827. 4 Jun. 2013)
81. L.O. Stenneth, V. Boroditsky, *Mode transition for an autonomous vehicle* (U.S. Patent No. 9,365,213. 14 Jun. 2016)

Causal Analysis to a Subway Accident: A Comparison of STAMP and RAIB

Yao Zhou[1] and Fei Yan[2]

[1]*School of electronic and information engineering, Beijing Jiaotong University, Beijing 100044, China*
[2]*National Engineering Research Center of Rail Transportation Operation and Control System, Beijing Jiaotong University, Beijing 100044, China*

Abstract. Accident investigation and analysis after the accident, vital to prevent the occurrence of similar accident and improve the safety of the system. Different methods led to a different understanding of the accident. In this paper, a subway accident was analysed with a systemic accident analysis model – STAMP (System-Theoretic Accident Modelling and Processes). The hierarchical safety control structure was obtained, and the system-level safety constraints were obtained, controllers of the physical layer were analysed one by one, and put forward the relevant safety requirements and constraints, the dynamic analysis of the structure of the safety control is carried out, and the targeted recommendations are pointed out. In comparison with the analysis results obtained by the Rail Accident Investigation Branch (RAIB). Some useful findings have been concluded. STAMP treats safety as a control problem and reduces or eliminates causes of the accident from the controlling perspective. Whereas RAIB obtains causes of the accident by analysing the sequence of events related to the accident and reasons of these events, then chooses one(or more)event(s) as the immediate cause and some of the key events as causal factors. RAIB analysis is based on the sequential event models, but STAMP analysis provides us with a holistic, dynamic way to control system to maintain safety.

1 Introduction

Accident investigation and analysis theory is the way that we find out causes of the accident, helping us to understand why the accident happened and how it happened. It determines whether we can prevent similar incidents from happening. Most accident reports are written inadequately about causes of the accident, and the analysis frequently stops after finding someone to blame--usually a human operator--and the opportunity to learn important lessons is lost [1]. It is also acknowledged that our understanding of accidents remains incomplete and that accidents will continue to occur within complex socio-technical systems [2].

Traditional cause--effect accident models suggest that complex systems accidents are caused by events such as catastrophic equipment failure or an unsafe human action. However, as system complexity has increased over time, many accidents do not simply resulted from such trigger events. These accidents are contributed by different factors, because there are complex phenomenon within the normal operational variability of a system [3]. As a typical complex socio-technical system, subway system involves many stakeholders, so the safety of subway cannot be ignored. With the increase of the speed of subway, global subway accidents continue to increase, resulting in a large number of casualties, which has seriously threatened the people's life. Therefore, accident analysis method that are suitable for the subway accident analysis is urgently needed.

There are currently three dominant accident causation models: Rasmussen's (1997) risk management framework (e.g. [4]); Reason's (1990) omnipresent Swiss Cheese model (e.g. [5]); and Leveson's (2004) Systems Theoretic Accident Modelling and Processes model (STAMP e.g. [6]). STAMP can be used to identify the problems that need to be answered to fully understand why the accident occurred. It provides the basis for maximizing learning from the events [1]. In STAMP, an accident is regarded as involving a complex process, not just individual events. STAMP can present a dynamic process which led to the accident, we can know what safety constraints were violated through the safety control structure, and what systemic control failures happened in the accident. Thereby obtaining targeted measures to make up for system safety vulnerabilities.

The aim of this paper is to compare the analysis results of RAIB with that of STAMP. STAMP was used to analyse an accident and the analysis results were compared with the analysis results of RAIB. Some difference between these methods were found. In order to achieve this aim, we should:

1. Analyse the accident (the passenger trapped in train doors and dragged at Clapham south station) using STAMP, from the analysis process to show how this method is used to find out causes of the accident.
2. In this case, compare causes of the accident obtained by two methods and find the difference among methods and which method can provide a more thorough and complete analysis.

2 The passenger trapped in train doors and dragged at Clapham South station

At about 08:00 hrs on the morning of Thursday 12, March 2015, a passenger's coat became trapped in the doors of a northbound Northern line train at Clapham South Station on the London Underground. The train departed and the passenger was dragged along the platform, fell to the ground, became separated from her coat and then fell into the gap between the train and the platform [7].

The passenger sustained a broken arm, and injured to her shoulder and face. There was no damage to the train, or to the railway infrastructure, and London Underground staff authorised reopening of the station at 09:11 hrs.

3 Methods

3.1 The analysis methods

3.1.1 RAIB

The Rail Accident Investigation Branch (RAIB) is a British government agency that investigates accidents and incidents which occur on the UK main line networks (Network Rail and Northern Ireland Railways), London Underground, other metro systems, tramways, heritage railways and the UK part of the Channel Tunnel in order to find a cause, not to lay blame. Created in 2005, it is required by law to investigate accidents causing death, serious injuries or extensive damage. It also has authority to investigate incidents that could have resulted in accidents. It currently has two bases - Derby and Farnborough - to allow it to respond quickly to accidents.

3.1.2 STAMP

The accident causation model called STAMP was developed by Prof. Nancy Leveson from MIT. Accidents are seen as resulting from inadequate control. The STAMP accident causation model is built on three basic concepts - safety constraints, a hierarchical safety control structure, and process models [8]. STAMP is a constraints-based model which focuses on the interactions between system components and the control mechanisms used throughout the work system [9]. Unlike conventional accident causation models, STAMP is not based on chain of events. It is based on system theory where each level or the organization plays a major role in contributing to an accident or attaining successful system safety controls. Thus STAMP prevails conventional accident models by accounting for organizational factors, human error, and adaptation to change over time. In STAMP, system safety is not achieved by preventing component failure measures; in fact, it is achieved by enforcing safety constraints continuously [10]. Therefore, accidents do not occur because of failure of components, they occur because of ineffective safety constraint where main focus is not on how to prevent failure, but on how to design better safety controls [6].

3.2 Accident analysis process

3.2.1 RAIB analysis process

The purpose of a RAIB investigation is to improve railway safety by preventing future railway accidents or by mitigating their consequences. It is not the purpose of such an investigation to establish blame or liability [7]. RAIB analysis process is as follows:

Firstly, a summary of the accident tells us what happened and some basic information related to the accident. Then the sequence of events are extracted. Key facts and analysis are composed of four steps: Identification of the immediate cause, identification of causal factors and the factors affecting the severity of consequences and previous occurrences of a similar character. Finally, according to this investigation, comprehensive recommendations for this investigation are made.

3.2.2 STAMP analysis process

STAMP analyses each component and controller of the system according to the safety control structure. Then the safety constraints that were violated at each level of this control structure were obtained, and the reasons that safety constraints was violated were also found. The main steps of STAMP analysis are as follows:

1. Identify the system(s) and hazard(s) involved in the loss.
2. Identify the system safety constraints and system requirements associated with that hazard.
3. Document the safety control structure in place to control the hazard and enforce the safety constraints.
4. Determine the proximate events leading to the loss.
5. Analyse the loss at the physical system level.
6. Moving up the levels of the safety control structure, determine how and why each successive higher level allowed or contributed to the inadequate control at the current level.
7. Examine overall coordination and communication contributors to the loss.
8. Determine the dynamics and changes in the system and the safety control structure relating to the loss and any weakening of the safety control structure over time. Finally, generate recommendations.

4 Applying the analysis models to the accident

4.1 RAIB analysis output

The RAIB analysis output see Rail Accident Investigation report: Passenger trapped in train doors and dragged at Clapham South station 12 March 2015 [7].

4.2 STAMP model analysis

4.2.1 Establish chain of events

While the event chain does not provide the most important causality information, the basic events related to the loss do need to be identified so that the physical process involved in the loss can be understood [1]. The event chain has been described in table 1.

Table 1. Proximity of events.

Time	Events
19:59:25	Train stops, and the passenger entered the fourth car using the rearmost double door
19:59:26	Doors open, the passenger step back out of the car and step into the gap of about 300mm between the yellow line and the train
19:59:39	CSA raises baton
19:59:41	Train operator presses door close button
19:59:45	Doors start to close (but do not fully close)
19:59:48	Train operator presses door open buttons because pilot light does not illuminate
19:59:50	Train operator presses door close button
19:59:51	Doors open with normal delayed response from button operation
19:59:56	Doors start to close, trapping the passenger's coat
19:59:57	Doors closed, the passenger noticed that the coat was caught
19:59:50	Train operator presses start buttons
19:59:54	Train starts to move
19:59:58	Passenger dragged past CSA
19:59:58	CSA starts to run and lowers his baton
20:00:00	Train operator applies emergency brake at train speed of 31km/h(He saw this unusual passenger movement on the in-cab screen)
20:00:07	Train stops

4.2.2 Identify the system(s) and hazard(s) involved in the loss

The systems or controllers involved in the accident mainly are: the interlock system, the train operator and CSA (the customer services assistant). The physical process is: when all passengers on board, interlock system detects whether all doors are locked in the fully closed position, and if so, the pilot light in the driving cab will light up. The train operator will press the start button after CSA raise the baton, the system hierarchical control structure diagram is shown in Figure 1, which from bottom to top is the physical layer, the operating layer and management layer, and the hierarchical control structure shows the main components of the system and the interaction between them.

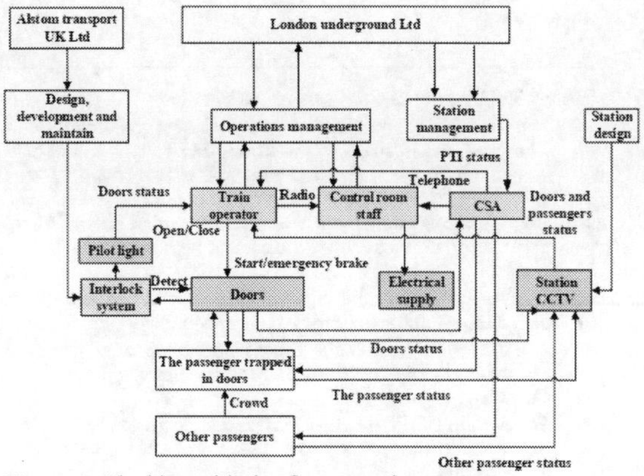

Figure 1. The hierarchical safety control structure.

An accident or loss event for the train moving processes can be defined as the death, illness or injury due to the passengers (or objects carried by passengers) trapped in train doors and dragged. The hazard is that someone trapped in doors while the train is moving. System-level safety constraints as follows:

(1) The train control system must effectively detect objects between the doors before starting.
(2) Before starting, the safety of the platform and the platform train interface (PTI) must be ensured.
(3) Must ensure passengers stay in a safe position on the platform when train enter/leave the station.
(4) Available and effective measures must be taken to deal with the injured person in case of danger.
(5) Emergency treatment measures must be used to protect the passengers and the relevant personnel, minimizing casualties and losses.

4.2.3 Analysing the physical process

Identifying the physical and operational controls and any potential physical failures, dysfunctional interactions and communication, or unhandled external disturbances that contributed to the events [8]. The goal is to determine why the physical controls in place were ineffective in preventing the hazard [11]. Physical processes are shown in figure 2.

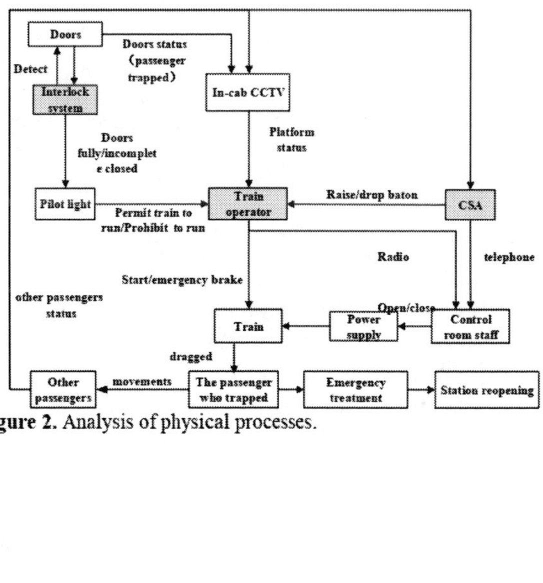

Figure 2. Analysis of physical processes.

The analysis results of physical layer are shown as follow:

Safety Requirements and Constraints Violated
- The train door control system must be able to detect whether there is an object was trapped between the door leaves during the process of closing the door, not only to detect the distance between the door leaves.
- Door control system must ensure that nothing between the doors, then provide the train operator with the doors is fully closed.
- The position where the detector located must ensure the distance will not affect the detection capability.
- The design of the detector must take into account the effect of the door seals flexibility on the detection capability.
- The station CCTV must be installed in accordance with the relevant standards, and have a record of installation and maintenance.
- The in-cab CCTV system must always ensure the train operator can see the entire length of the train.
- The in-cab CCTV system must be qualified products, to ensure the data can be saved in real time.
- The design of the pushback system must consider the train operation direction and the direction of the possible force, ensue the door can be opened in case of emergency.
- The design of the edge of the platform must meet relevant standards

Failures and Inadequate Controls
- The door control system provide the door is fully closed to train operator, but there is an object was trapped between the door leaves.
- The train's door control system did not detect that an object was trapped between the door leaves.
- The detector does not detect the passenger's clothes, because clothes are far away from the detector.
- The in-cab CCTV had no record of accidents related video.
- The pushback system does not work when the passengers trapped, causing the passenger could not remove the coat from the door.

Physical Contextual Factors
- The design of train doors is intended to ensure that they will not be detected closed if there is a 5 mm visible gap between the door seals when these are not compressed. But the coat material compressed and the gap between the outer faces of the door seals was only 4 mm.
- The detector at the top of the door, and the coat was trapped about 0.5 meters above floor level, so the detector is less sensitive to objects that are positioned lower down.
- Only one leaf of the door with the pushback system, and the passenger apply the force on the fixed leaf.

4.2.4 Analysing the higher levels of the safety control structure

This part will analyse accident involved operators, from the following aspects: safety-related responsibilities, context, unsafe decision and control actions, and mental model flaws [12].

The train operator:
The train operator should check the entire platform-train interface, it is necessary to check again before opening the door, before closing the door and before leaving the station, so that the train operator can promptly eliminate the unsafe state on the platform. Analysis results as follows:

Safety-related responsibilities
- Monitor the train operation.
- Focus on the CCTV, ensure that there is no exceptions at PTI,
- To communication with CSA, grasp the platform status in real time.
- When the view is hindered, should actively look for other ways to check the PTI, and pay attention to other signs may cause exceptions.
- After several times to confirm the safety of platform train interface, then press the start buttons.
- Must always pay close attention to CCTV during the leave station process, ensure the PTI safety.
- Must take emergency measures immediately when necessary.

Context
- The train operator had been driving for 14 years, very skilled and experienced in their own work, and also clear all kinds of emergency response measures.
- The station was crowed, blocking the train operator's vision and can't see the complete doors.

Unsafe decision and control actions
- Press the start buttons while the passenger was trapped in the door.
- Press the close buttons after see that the CSA raises the baton without other confirmation.
- Didn't carefully observe the CCTV and didn't find that the passenger who was trapped in the door.

Mental model flaws
- After the Pilot light indicates that doors fully closed, thought that there is no exceptions.
- Did not know the passenger was trapped in the door.
- Did not know CSA did not find the trapped passenger.
- In order to reduce the station crowd, want to minimize the train stop time

The CSA:
The CSA was performing station assistant duties, including: managing the platform train interface and involve managing passenger behaviour and movements on the platform. Analysis results as follows:

Safety-related responsibilities
- Must ensure that the platform-train interface is safe during the train stop and depart process.
- Effectively manage and organize the passengers on the platform to ensure that the passengers are in a safe position.
- Must ensure that the door is safe then raise the baton.
- Must alerts the passengers to be careful and inform the train operation of station status.
- Must take emergency measures immediately when necessary.

Context
- The CSA about 10 meters from the door where the passenger was trapped.
- CSA needs to organize a large number of passengers during the busy morning peak.
- Must ensure safety and efficiency during door closure and train departure.
- CSA do the same work at other stations.

Unsafe decision and control actions
- Raise the baton while the passenger was trapped in the door.
- No further confirmation of the platform-train interface safety.
- The CSA heard shouting on the platform but did not confirm.
- The CSA did not organize the passengers to stand in a safe position

Mental model flaws
- Did not know the passenger was trapped in the door.
- Thought sounded more like an argument than someone in difficulty.
- Thought that there is no exceptions before raise the baton.

4.2.5 Coordination and communication

To establish an effective safety control structure, effective coordination and communication between parties not in direct hierarchical control levels is important [8]. The development and management team must provide complete operation manuals to the communication between the train operator, the CSA and electricity control operator. In this accident, the CSA uses a baton to transmit signals to the train operator, there is no specific provision about what conditions need to be met before CSA raise the baton, the responsibility distribution of the CSA is not clear.

4.2.6 Dynamic analysis of safety control structure

Most major accidents result from a migration of the system toward reduced safety margins over time. Pressure from commercial competition is one cause of this degradation in safety. It is, of course, a very common one [13]. Usually there are precursors signalling the increasing risks associated with these changes in the form of minor incidents and accidents, but in this case, as in so many others, these precursors were not recognized [13]. For example, in this accident, similar incident didn't attract enough attention and appropriate adjustment measures were also didn't carried out.

In many cases, such as the busy morning peak, in order to maintain the train throughput and expedite prompt boarding of trains, delay of the train will lead to the station congestion and speed up the production of potential risks, this means stop time need to be reduced as much as possible and PTI risk need to be controlled at the same time. Therefore, train operator and CSA will reduce the safety margin, such as the CSA need to minimize the station congestion and to ensure the PTI safety at the same time, the workload is large. In addition, with the extension of working hours, the operator may encounter some common abnormal activities on weekdays, may ignore the rules and regulations. In this accident, the CSA heard shouting on the platform, did not confirm the PTI was safe and give the confirmation signal.

Like the operators, the device will also reduce the safety margin over time. The equipment should always be checked regularly and the equipment breakdown which appeared frequently should be paid enough attention [10]. In this accident, the detection ability of door control system would decrease over time or lose detection capability, that didn't attracted enough attention, if the attention had been paid to the detected faults, a good solution would be taken.

4.2.7 Issue recommendations

Some suggestions are put forward from the aspects of equipment safety constraints, hardware and software design, inappropriate control behaviour and so on.

Physical device:

1. Improving the door control system, so that it can detect objects trapped in train doors, such as installing the sensing element on the door leaf.
2. Only when the distance between the doors meet the requirements, and there is no objects in train doors, the door is fully closed. Considering the installation position of the detector, to ensure the detection accuracy is not affected.
3. Because the congestion will make the train operator's vision is blocked, suggest an appropriate increase in the number of station CCTV or change the location of the CCTV.
4. Installing pushback mechanism on each door leaf, so the force can be applied in any direction and the door will be slightly opened, so that passengers can save themselves if they are trapped.

Operators:

1. The train operator shall always clear the status of the PTI and on-board equipment status in real time and need to confirm again before start, when the operation condition is satisfied, then press the start button. Take emergency measures immediately in case of emergency situations.
2. The CSA should always clear the status of the PTI and use the way of patrol back and forth to check the PTI, and check any abnormal situation may result in danger to ensure the safety of passengers.
3. During the busy morning peak, the CSA is responsible for too many duties, suggest to increase the staff members and reduce the work time, besides, to improve the organization efficiency and reduce the work mistake.
4. Regularly carry out operation and emergency treatment training to operators, pay more attention to the problems that often appeared in daily work. In case of emergency, the proper measures should be taken quickly to minimize the casualties and losses.

5 Discussion

5.1 Comparison of causal analysis process

From the results of the two causal analysis, they are clearly described the accident. According to the analysis of RAIB, we know the details of the accident and sequence of events, but there is no complete and adequate answer to the question of why the accident happened and what is the relationship between them, such as why the train operator presses the start button without being able to see the entire door and why the CSA raises the baton without confirming that the PTI is safe and so on. The accident analysis method used by RAIB can be summarized as shown in Figure 3.

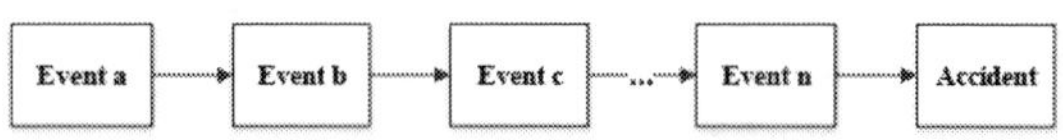

Figure 3. RAIB analysis process.

RAIB provides sufficient data involved in a particular accident, such as the environment in which the accident occurred, equipment failure and operator failures at the time of the accident, and the similar accident that occurred before.

A more complete accident analysis should find out why the accident happened and how to prevent it from reoccurring in the future [6]. Causal analysis of STAMP start from the lower layer of the system control structure, check whether the control command in the structure is properly executed, if not, why? Then from the physical layer and operate layer to analyse the various components and controllers involved in the accident one by one, determine the safety requirements and constraints that they violated, and unsafe control action, have a holistic, systematic understanding of the accident [14]. The use of STAMP does not lead to identifying single causal factors or variables. Instead it provides the ability to examine the entire sociotechnical system design to identify the weaknesses in the existing safety control structure and to identify changes that will not simply eliminate symptoms but potentially all the causal factors, including the systemic ones [1]. STAMP provides an overview of how the system should have been controlled and how the accident in question should have been prevented from happening as well as the relations between actors, and the context in which the accident occurred [9]. Not only all of the failure were identified, but the context and mental model flows were also analysed.

5.2 Comparison of accident causes

RAIB analyses the cause of the accident by accident-related events, selects the event which closest to the accident as the immediate cause and other related events as the causal factors. STAMP believes that the accident was caused by a violation of the safety constraints, in this accident, including: the CSA did not fully see the door and raise the baton, during the busy morning peak, only one CSA to check the entire platform is difficult to find out the abnormal situation and may send the wrong message to the train operator. Besides, the train door control system cannot detect the objects between the doors and the train operator cannot see the entire doors in the CCTV in-cab and press the start button. For the reason that CSA did not find the trapped passenger, we cannot simply say that CSA did not fulfil his duties, London Underground's work configuration is not reasonable, did not take into account the peak hours, a CSA cannot check the entire platform. Some adjustments should be made, such as adjusting the number of CSA or adding additional ancillary measures according to the situation at different times to ensure the safety of the platform train interface.

6 Conclusions

RAIB analysis is more of an objective analysis and explanation as the third party in terms of the accident, RAIB can provide the analyst with all the events and details associated with the accident, however, we do not know the relationship between these events and what role it has played in the accident. Therefore, RAIB analysis can be used as an information source for us to understand details of the accident. STAMP analysis provides an overall understanding of the organization, management, and operational framework of the system by the hierarchical control structure. This is what the other methods do not have. Besides, STAMP analyses the relationship between various causal factors and causes for the human error at that time, and provides a comprehensive perspective to ensure the safety of system.

References

1. N. Leveson, Engineering a Safer World: Systems Thinking Applied to Safety, MIT Press, 2011.
2. E. Hollnagel, Ergonomics, **50**(6), 961-962 (2004).
3. P.V.R.D. Carvalho, Reliab Eng Syst Safe, **96**(11), 1482-1498 (2011).
4. I. Svedung, J. Rasmussen, Safety Sci, **40**(5), 397-417 (2002).
5. W. Ward, J. Brewbaker, English Journal, **94**(6), 111 (2005).
6. H. Altabbakh, M. Alkazimi, S. Murray, G. Katie, J. Loss Prevent Proc, **32**, 109-119 (2014).
7. Branch, R.A.I., Rail accident report: Passenger trapped in train doors and dragged at clapham south station - 12 march 2015 .
8. A. Dong, Application of CAST and STPA to railroad safety in China, 2012.
9. P.M. Salmon, M. Cornelissen, M.J. Trotter, Safety Sci, **50**(4), 1158-1170 (2012).
10. N. Leveson, Safety Sci, **42**(4), 237-270 (2002).
11. M. Ouyang, L. Hong, M.H. Yu, Q. Fei, Safety Sci, **48**(5), 544-555 (2010).
12. P. Underwood, P. Waterson, G. Braithwaite, Safety Sci, **82**(42), 129-143 (2016).
13. N.G. Leveson, Safety Sci, **49**(1), 55-64 (2011).
14. K. Hardy, F. Guarnieri, Using a Systemic Model of Accident for Improving Innovative Technologies: Application and Limitations of the STAMP model to a Process for Treatment of Contaminated Substances, 2011.

Hybrid Force and Position Control Strategy of Robonaut Performing Object Transfer Task

Gang Chen, Yu-Qi Wang , Qing-Xuan Jia and Pei-Lin Cai

School of Automation, Beijing University of Posts and Telecommunications, Beijing 100876, China

Abstract. This paper proposes a coordinated hybrid force/position control strategy of robonaut performing object transfer operation. Firstly, the constraint relationships between robonaut and object are presented. Base on them, the unified dynamic model of the robonaut and object is established to design the hybrid force/position control method. The movement, the internal force and the external constraint force of the object are considered as the control targets of the control system. Finally, a MATLAB simulation of the robonaut performing object transfer task verifies the correctness and effectiveness of the proposed method. The results show that all the targets can be control accurately by using the method proposed in this paper. The presented control method can control both internal and external forces while maintaining control accuracy, which is a common control strategy.

1 Introduction

The complexity and difficulty of space missions increase alongside the deepening of human exploration about space. Therefore, a new type of space robot called robonaut is designed to deal with this situation, and it can be used to assist or replace the human astronauts to carry out the on-orbit operation tasks in order to reduce the astronauts' working pressure and the cost of space operations. The robonaut consist of three branches, in which one is generally called the coupled torso while the other two are robot arms. Due to the existence of redundant arms, its flexibility, operation ability, operating range and other aspects of performance are far better than single arm robot, and its coordination ability and control accuracy are significantly improved compared with the multi robot because of the existence of coupled torso. The robonaut, with its strong collaboration and wide operating range, can efficiently and reliably assist astronauts in performing many on-orbit operations, including mating, twisting, handling, complex parts assembling and space station maintenance etc. As a typical dual arm coordination task, object transfer is representative in orbit operation, so it is of great significance to study the coordinated operation method in the process of object transfer.

The position control strategy is difficult to meet the requirements of the interaction between the robot and the environment when the robonaut performs coordinated manipulation tasks. To finely perform the given task, compliance control method is necessary which means it is vital to control the necessary operation force in the coordinated manipulation tasks, as well as limit the undesirable force to the system meanwhile. There are several main manners of dual-arm coordinate operate control technique, such as impedance control[1-2], hybrid force/position control [3-4], intelligent control[5-6]. The hybrid force/position control method is relatively simple and does not rely on environmental information. Besides, the accuracy of this method is relatively higher than other methods. Given this, the hybrid force/position control method is adopted in this paper.

A number of relevant researches have been carried out which concerning the hybrid force/position control during the processes of object transfer. C. V. Albrichsfeld [7] analyzed the relationship in two robots holding a single object by introducing a concept of 'virtual stick', and then they presented a symmetric non-master/slave hybrid position/force coordinated control scheme using the equations which derive from virtual stick. A. I. Tuneski [8] extended an active compliance control scheme which is capable of adjusting its parameters to the unknown system compliance. Furthermore, a general model was derived by discussing different contact cases of the object. However, all the methods mentioned above existing the following deficiencies: firstly, these articles could not achieve the external force compliance control only considering the internal force when the object is subjected to external force, which may cause the failure of compliance control. Secondly, some of the methods directly divide the object into two parts and then fixed on the end of each manipulator. In this way, a rather better simulation results can be obtained, but these methods are limited to theoretical analysis and are not suitable for practical operation.

For the deficiencies above, more specific study will be carried out in subsequent chapters. Specifically, in chapter II, the constraint relationships between robonaut and object will be presented. Then, in chapter III, the unified dynamic model of the robonaut and object is going to be established to design the hybrid force/position control method. Moreover, a simulation study will be implemented to verify the algorithm control in chapter IV. Finally, the comprehensive review and summary will be done in the final chapter.

2 Dynamic modeling

The structure of robonaut studied in this paper is shown in Figure 1. It consists of three branches: the coupled torso branch (branch 1) and two arm branches (branch 2 and branch 3). Generally, one end of the coupled torso branch is fixed at the base, the other end is connected with the root of the two arm branches. The ends of the arms are free to move to complete the operational tasks. Assume that branch 1 has $n_1-\text{DOF}$, branch 2 and branch 3 have $n_2-\text{DOF}$ and $n_3-\text{DOF}$ respectively.

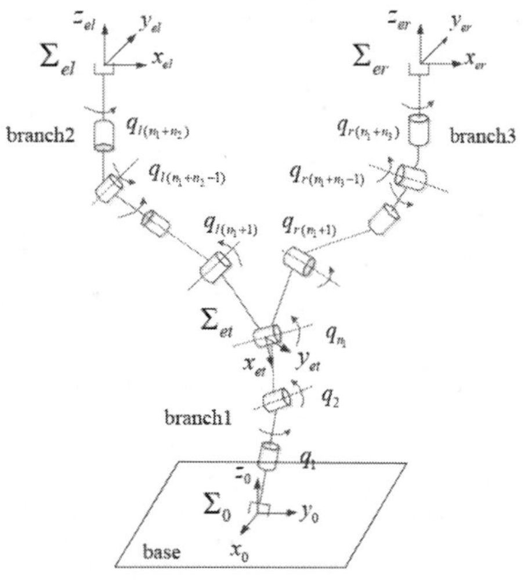

Figure 1. The structure of robonaut

2.1 Constraint relationships

Figure 2 shows the coordinated model of object transfer operation. Point O_c is the mass of the object and the object frame is attached at it. For simplicity, let us ignore the size of object, and translate the origins of the left and right arms end frames to the point O_c, named as $O_{cw}(w=l,r)$.

When we assume that the stiffness of the object is large enough to ignore the deformation, the pose constraint between robonaut and object can be written as

$$\begin{cases} \boldsymbol{P}_c = \boldsymbol{P}_l = \boldsymbol{P}_r \\ {}^0\boldsymbol{R}_c = {}^0\boldsymbol{R}_l\,{}^l\boldsymbol{R}_c = {}^0\boldsymbol{R}_r\,{}^r\boldsymbol{R}_c \end{cases} \tag{1}$$

The velocity constraint can be written as

$$\dot{\boldsymbol{X}}_c = \dot{\boldsymbol{X}}_l = \dot{\boldsymbol{X}}_r \tag{2}$$

The acceleration constraint can be written as

$$\ddot{\boldsymbol{X}}_c = \ddot{\boldsymbol{X}}_l = \ddot{\boldsymbol{X}}_r \tag{3}$$

The force constraint relationship is shown in Figure 3. The output forces of left and right arms, the external constraint force of robonaut system are $\boldsymbol{F}_w = \left[\boldsymbol{f}_w^{\mathrm{T}} \;\; \boldsymbol{n}_w^{\mathrm{T}}\right]^{\mathrm{T}} (w=l,r,f)$, and translate them to the point O_c, named as $\boldsymbol{F}_{cw} = \left[\boldsymbol{f}_{cw}^{\mathrm{T}} \;\; \boldsymbol{n}_{cw}^{\mathrm{T}}\right]^{\mathrm{T}} (w=l,r,f)$.

The resultant force of the object can be expressed as

$$\boldsymbol{F}_c = \boldsymbol{W}\boldsymbol{F}_e + \boldsymbol{F}_{cf} \tag{4}$$

where $\boldsymbol{W} = \left[\boldsymbol{E}_6 \;\; \boldsymbol{E}_6\right]$, $\boldsymbol{F}_e = [\boldsymbol{F}_{cl}^{\mathrm{T}},\; \boldsymbol{F}_{cr}^{\mathrm{T}}]^{\mathrm{T}}$.

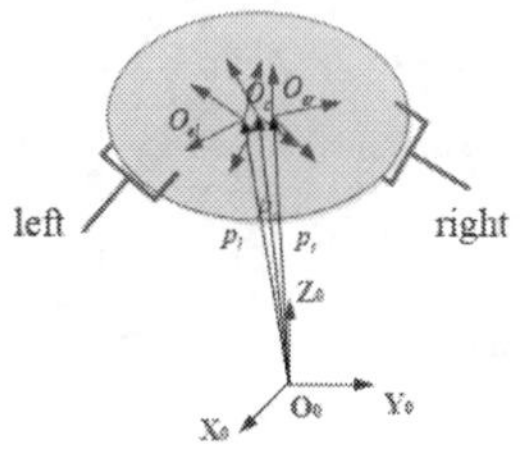

Figure 2. The coordinated model of object transfer operation.

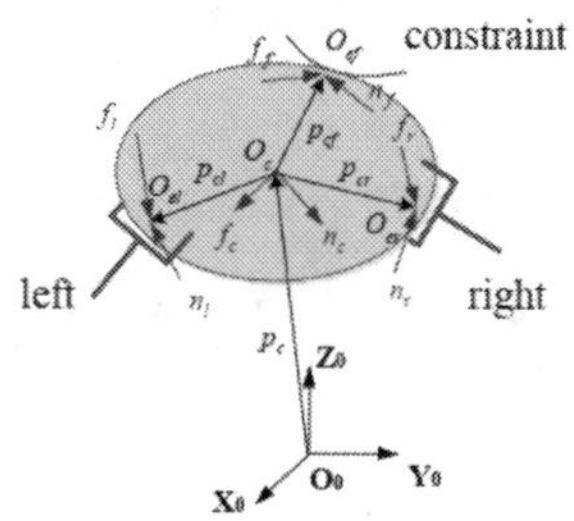

Figure 3. The force model of object transfer operation.

2.2 Dynamic models

In this paper, we intend to use the movement, the internal force and the external constraint force of the object as the control targets of the control system. Firstly, the dynamic models of the object and the robonaut are established. Base on them, the unified dynamic model is constructed, which means the object and the robonaut as a whole.

2.2.1 Object dynamic model

The dynamic equation of object is given by using the Newton-Euler equation

$$\begin{cases} m_c\dot{\boldsymbol{v}}_c = \boldsymbol{f}_c \\ \boldsymbol{I}_c\dot{\boldsymbol{\omega}}_c + \boldsymbol{\omega}_c \times (\boldsymbol{I}_c\boldsymbol{\omega}_c) = \boldsymbol{n}_c \end{cases} \tag{5}$$

where m_c, $\boldsymbol{I}_c$ represent the object's mass and inertia tensor, respectively.

Equation (5) can be rewritten as

$$\boldsymbol{F}_c = \boldsymbol{M}_c\ddot{\boldsymbol{X}}_c + \boldsymbol{C}_c \tag{6}$$

where $\boldsymbol{M}_c = \begin{bmatrix} m_c\boldsymbol{E}_3 & \boldsymbol{0}_{3\times3} \\ \boldsymbol{0}_{3\times3} & \boldsymbol{I}_c \end{bmatrix}$, $\boldsymbol{C}_c = \begin{bmatrix} \boldsymbol{0}_{3\times1} \\ \boldsymbol{\omega}_c \times (\boldsymbol{I}_c\boldsymbol{\omega}_c) \end{bmatrix}$.

2.2.2 Robonaut dynamic model

The velocity of the ends of the arms can be written as $\dot{X}_w=\left[v_w^{\mathrm{T}},\omega_w^{\mathrm{T}}\right]^{\mathrm{T}}(w=l,r)$, where v_w and ω_w are the linear and angular velocity respectively. By differential kinematics, it can be obtained that

$$\begin{cases}\dot{X}_l=[v_l^{\mathrm{T}},\omega_l^{\mathrm{T}}]^{\mathrm{T}}=J_l\dot{q}_l=\left[J_{l1},J_{l2}\right]\dot{q}_l\\ \dot{X}_r=[v_r^{\mathrm{T}},\omega_r^{\mathrm{T}}]^{\mathrm{T}}=J_r\dot{q}_r=\left[J_{r1},J_{r3}\right]\dot{q}_r\end{cases} \quad (7)$$

where $J_w(w=l,r)$ represent the Jacobian matrixes of the left and right manipulators respectively.

According to equation (7), the kinematics equation can be thus written as

$$\dot{X}_e=\begin{bmatrix}\dot{X}_l\\ \dot{X}_r\end{bmatrix}=\begin{bmatrix}J_{l1} & J_{l2} & \mathbf{0}_{6\times n_3}\\ J_{r1} & \mathbf{0}_{6\times n_2} & J_{r3}\end{bmatrix}\dot{q}=J\dot{q} \quad (8)$$

where $J\in\mathbf{R}^{12\times(n_1+n_2+n_3)}$ represents the Jacobian matrix of the robonaut.

From equation (8), we can get

$$\begin{cases}\ddot{X}_e=\dot{J}\dot{q}+J\ddot{q}\\ \ddot{q}=J^{+}\left(\ddot{X}_e-\dot{J}\dot{q}\right)\end{cases} \quad (9)$$

where J^{+} represents pseudo-inverse of J.

By using Lagrange-Euler method, the dynamics of the robonaut in joint space coordinates suggested in this paper can be described as follows

$$A(q)\ddot{q}+2B(q)\dot{q}_m\dot{q}_n+C(q)\dot{q}^2+D(q)+J^{\mathrm{T}}F_e=\tau \quad (10)$$

where $A(q)$ is an $n\times n$ symmetric positive definite inertia matrix. $A(q)\ddot{q}$ represents the inertia forces vector. $B(q)$ is an $n\times\left[n(n-1)/2\right]$ Coriolis matrix. $2B(q)\dot{q}_m\dot{q}_n$ represents the Coriolis forces vector. $C(q)$ is an $n\times n$ centrifugal matrix. $C(q)\dot{q}^2$ represents the centrifugal forces vector. $D(q)$ is an $n\times 1$ gravitational matrix. In addition, $n=n_1+n_2+n_3$.

Substituting (9) to(10), the dynamics of the robonaut in Cartesian space coordinates can be described as

$$M\ddot{X}_e+C+F_e=U \quad (11)$$

Where

$$\begin{cases}M=J^{+\mathrm{T}}A(q)J^{+}\\ C=J^{+\mathrm{T}}\left[2B(q)\dot{q}_m\dot{q}_n+C(q)\dot{q}^2+D(q)\right]-M\dot{J}\dot{q}\\ U=J^{+\mathrm{T}}\tau\end{cases}$$

2.2.3 Robonaut system unified dynamic model

From(4), the output force of robonaut can be obtained as

$$\begin{aligned}F_e&=W^{+}\left(F_c-F_{cf}\right)+\left(E_{12}-W^{+}W\right)\begin{bmatrix}\eta_l\\ \eta_r\end{bmatrix}\\ &=W^{+}\left(M_c\ddot{X}_c+C_c\right)-W^{+}F_{cf}+W_iF_i\end{aligned} \quad (12)$$

① The first term on the right side of the equation, represents the driving force of robonaut on the object, where $W^{+}=[E_6,\ E_6]^{\mathrm{T}}/2$.

② The second term on the right side of the equation, represents the external constraint force of robonaut system.

③ The third term on the right side of the equation, $\left(E_{12}-W^{+}W\right)$ represents the null space of W, $\eta_w(w=l,r)$ represents any 6×1 vector. Due to the characteristics of null space, we can get that $W\left(E_{12}-W^{+}W\right)\left[\eta_l^{\mathrm{T}},\ \eta_r^{\mathrm{T}}\right]^{\mathrm{T}}=\mathbf{0}_{6\times 1}$. So, this part of force produces the internal force of the object. Also, $W_i=\left[E_6,\ -E_6\right]^{\mathrm{T}}$, $F_i=(\eta_l-\eta_r)/2$.

Substituting (3), (12) to (11), the unified dynamic model is constructed as

$$U=M_s\ddot{X}_c+C_s-W^{+}F_{cf}+W_iF_i \quad (13)$$

where:

$$\begin{cases}M_s=M+W^{+}M_c\\ C_s=C+W^{+}C_c\end{cases} \quad (14)$$

3 Control algorithm

In this paper, the internal force, the movement and the external constraint force of the object are considered as the control targets of the hybrid force/position control system. To simplify the problem, the internal force at the mass center of the object can be chosen in object transfer operation task.

The position control law of object can be designed as

$$\ddot{X}_c=\bar{S}\left[\ddot{X}_{cd}+K_d\left(\dot{X}_{cd}-\dot{X}_c\right)+K_p\left(X_{cd}-X_c\right)\right] \quad (15)$$

where X_{cd}, $\dot{X}_{cd}$ and $\ddot{X}_{cd}$ represent the desired pose, velocity and acceleration respectively. K_p, K_d are derivative and proportional feedback gains respectively. $\bar{S}$ is an 6×6 position-selecting matrix.

The external constraint force control law of the robonaut system can be designed as

$$F_{cf}=S\left[F_{cfd}+K_{pf}\left(F_{cfd}-F_{cf}\right)+K_{df}\left(\dot{F}_{cfd}-\dot{F}_{cf}\right)\right] \quad (16)$$

where F_{cfd} represent the desired external constraint force. K_{pf}, K_{df} are derivative and proportional feedback gains respectively. S is an 6×6 force-selecting matrix.

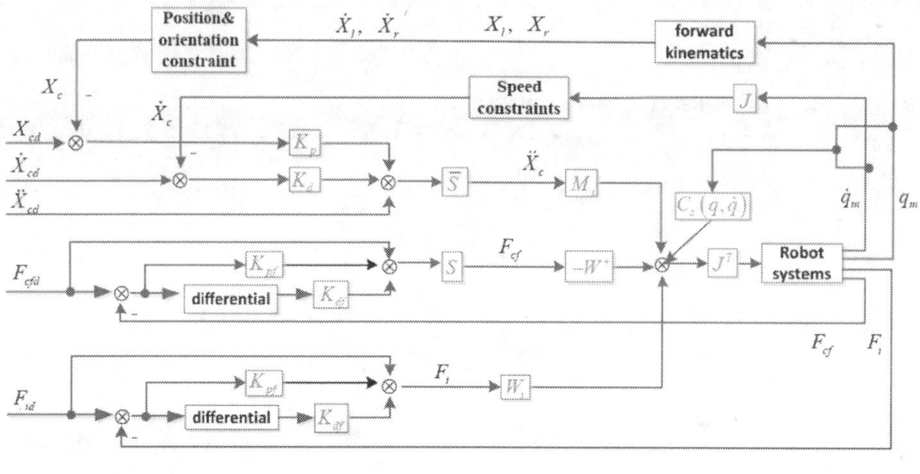

Figure 4. Hybrid force/position control structure of robonaut.

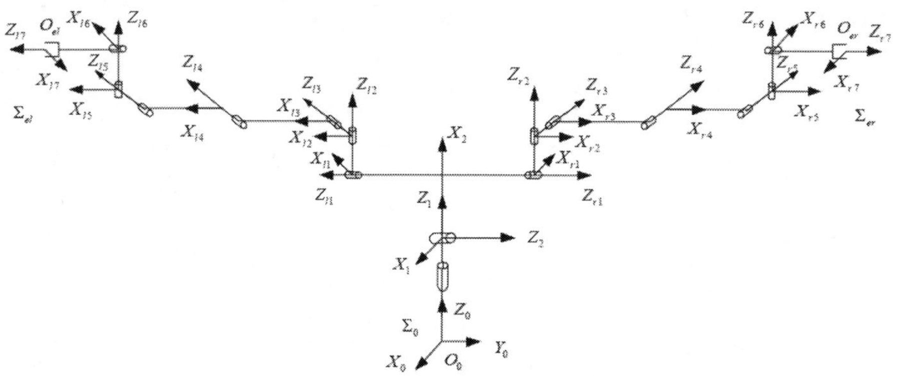

Figure 5. D-H frames of robonaut in simulation.

The internal force control law of the object can be designed as

$$F_{ic} = F_{id} + K_{pi}\left(F_{id} - F_i\right) + K_{di}\left(\dot{F}_{id} - \dot{F}_i\right) \tag{17}$$

where F_{id} represent the desired internal force.

Substituting(15), (16) (17) to (13) it can be obtained that the control joint torque

$$\begin{aligned}\tau = J^{\mathrm{T}}\Big[& M_s\bar{S}\left(\ddot{X}_{cd} + K_d\left(\dot{X}_{cd} - \dot{X}_c\right) + K_p\left(X_{cd} - X_c\right)\right) \\ & + C_s - W^{+}S\left(F_{cfd} + K_{pf}\left(F_{cfd} - F_{cf}\right) + K_{df}\left(\dot{F}_{cfd} - \dot{F}_{cf}\right)\right) \\ & + W_i\left(F_{id} + K_{pi}\left(F_{id} - F_i\right) + K_{di}\left(\dot{F}_{id} - \dot{F}_i\right)\right)\Big]\end{aligned} \tag{18}$$

The proposed control structure is shown in Figure 4.

4 Simulation Study

The robonaut involved in this simulation is shown in Figure 5. The coupled torso has 2 rotation joints and the two arms both have 7 rotation joints. The D-H parameters of the robonaut are listed in Table 1.

Table 1. D-H parameters of the robonaut in simulation.

	α_{i-1} / rad	a_{i-1} / m	θ_i / rad	d_i / m
t1	0	0	0	0.45
t2	-π/2	0	-π/2	0
al1	π	0.3	-π	0.2
al2	π/2	0	π/2	0.15
al3	π/2	0	0	0.15
al4	0	0.5	0	0.15
al5	0	0.5	0	0.15
al6	-π/2	0	-π/2	0.15
al7	-π/2	0	-π	0.2
ar1	0	0.3	π	0.2
ar2	-π/2	0	-π/2	0.15
ar3	-π/2	0	0	0.15
ar4	0	0.5	0	0.15
ar5	0	0.5	0	0.15
ar6	π/2	0	π/2	0.15
ar7	π/2	0	π	0.2

The task involved in this simulation is that the robonaut holds an object to terminal point along a desired arc trajectory on the constraint surface. Assuming that the radius of the arc is 0.1m, and the object must remain in contact with the constraint surface during the movement. Set the total control time to be 10s.

Assuming that the constraint surface is parallel to the inertial frame yOz plane. The origin of the constraint frame is set at the center of the arc trajectory. Then the homogeneous transformation matrixes representing the poses of constraint frame with respect to inertial frame is $\boldsymbol{R}_{0c}$ =[0, 0, -1; 1, 0, 0; 0, -1, 0].

The initial position of object is $\boldsymbol{P}_{c\_int}$ =[-0.5m, 0.1m, 1.5m], the corresponding joint angle is q_{t0} =[-1.34°, -15.88°], q_{l0} =[47.09°, -131.02°, 186.84°, -94.48°, 157.38°, 161.55°, 38.06°], q_{r0} =[-53.89°, 142.63°, -194.46°, 90.40°, -152.17°, -164.75°, -50.64°]. The desired internal force is set as $\mathbf{0}_{6\times1}$, the desired external constraint force is set as $\mathbf{0}_{6\times1}$.

Apply the algorithm mentioned above to this case, then the trajectory , the internal force and the external constraint force of the object result to be Figure 6, Figure 7 and Figure 8 respectively, which verifies the effectiveness of the algorithm discussed in this paper.

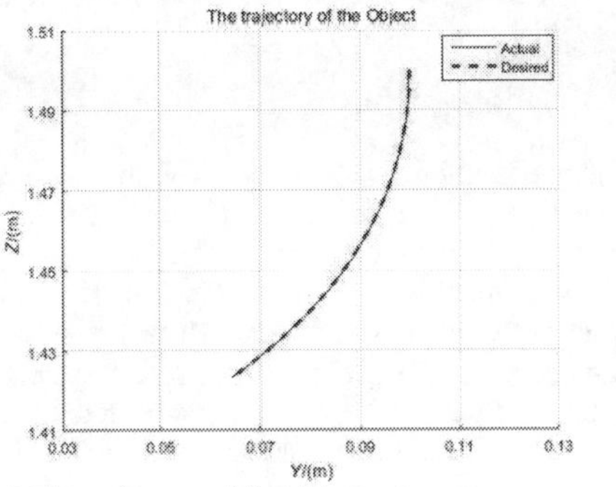

Figure 6. The trajectory of the object.

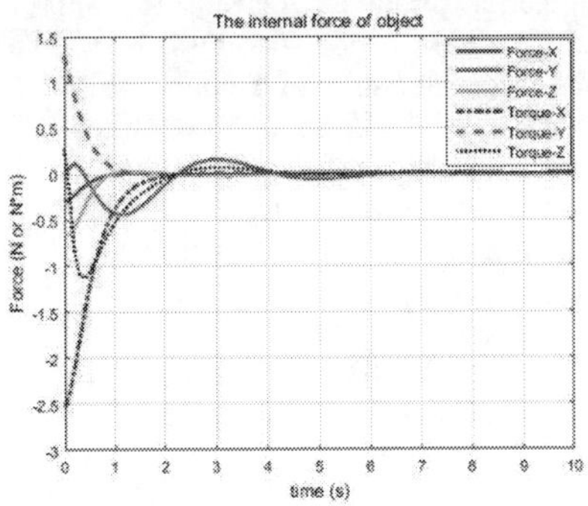

Figure 7. The internal force of the object.

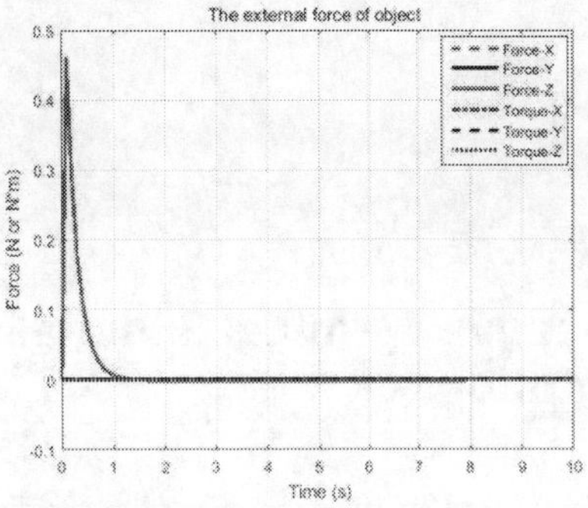

Figure 8. The external force of the object.

5 Conclusion

This paper proposes the coordinated hybrid force/position control strategy of robonaut performing object transfer operation. The constraint relationship between two manipulators and object is presented at first. Afterward the unified dynamic model of the robonaut and object is established to design the hybrid force/position control method of the robonaut. The movement, the internal force and the external constraint force of the object are considered as the control targets of the control system, thereby achieving simultaneous control of the internal force and the external force of the object. Finally a simulation study verifies the effectiveness of the algorithm.

References

1. Y L. Kim, B S. Kim, J B Song. *IEEE CASE.*, 1074(2012)
2. J. Lu, J P. Yan, J J. Chen. Con Theo & Appl. **20**(1), 85(2003).
3. W F. Xu, R X. Zhou, D S. Meng. Journal of Astronautics, **34**(10), 1353(2013).
4. K. Zhang, Q. Wei, W S.Chang. ROBOT, **24**(1), 44(2002).
5. S T. Lin, A K. Huang. J Intell Rob Syst Theor Appl, **19**(4), 393(1997).
6. A. A. G. Siqueira, M. H. Terra. *ACC*, **17**(3), 418(2009).
7. C V. Albrichsfeld, H. Tolle. Control Eng. Pract., **10**(2), 165(2002).
8. A I. Tuneski, M K. Vukobratovic, G M. Dim. Acou Spe & Sig Proce News. **215**(4), 385(2001).

Adaptive Terminal Sliding-Mode Control for Servo Systems with Inertia Variations

Xiang Wang, Yifei Wu , Fei Yan, Yang Gao and Zhen Xu

School of Automation, Nanjing University of Science and Technology, 210094 Nanjing, China

Abstract. Inertia variations in servo systems greatly affect the control performance. This paper presents an adaptive terminal sliding-mode controller to deal with the problem. Instead of using traditional mathematics model, a characteristic model, which has more advantages in describing time-varying dynamics, is adopted to describe the servo system with inertia variations. The parameters of characteristic model are identified by the recursive least squares algorithm. Then, an adaptive terminal sliding-mode controller is designed based on the characteristic model. Theoretical analysis proves that the quasi-sliding mode is reached in finite steps. Simulation results demonstrate the improvement of tracking performance of the proposed controller.

1 Introduction

Servo systems have been widely applied to various modern industries, including machine tools, robots, satellite antennas, radars and manipulators. In recent years, modern industries have higher requirements for fast response, high precision and robustness under uncertainties than ever. However, in practical production, the load inertia of servo systems varies frequently due to different types of products. Load inertia variations have great influence on the flexibility of the transmitted torque between motor and load and cause inaccuracies and oscillations, which makes it challenging for controller design.

In the literature, lots of methods have been reported to deal with inertia variations. Firstly, inertia identification is a straightforward technique to compensate for inertia variations. A model reference adaptive method was adopted in [1] for inertia identification and the feedforward compensation gain was tuned accordingly. The work in [2] combined the extended state observer with the inertia identification method and proposed an adaptive controller for speed regulation servo systems. Several kinds of inertia identification algorithms and tuning schemes are presented in [3, 4]. In addition to inertia identification, other control schemes like sliding-mode control [5, 6] and the active disturbance rejection control [7] were also employed to restrain inertia variations. The parametric uncertainties can also be encapsulated into a lumped nonlinearity function which is identified by fuzzy approximator and neural network.

The studies mentioned above have their own advantages. However, most of these control schemes are complicated and unsuitable to practical application due to the high-order and nonlinearity of the traditional mathematics model. Motivated by this, we adopt an engineering discretization modeling method called characteristic modeling [8]. The key idea of this modeling method is that the input and output characteristics of a system can be described by a slow time-varying difference equation, if the system satisfies certain conditions. The characteristic model reduces the complexity of traditional mathematics model and is good at describing time-varying dynamics. So the characteristic model of the servo system is established to adapt to inertia variations. Focusing on control algorithm, the terminal sliding-mode strategy is chosen to combine with the characteristic modeling method. Then an adaptive terminal sliding-mode controller (ATSMC) is proposed. The characteristic model is mainly used to describe parametric variations and the terminal sliding-mode control is used to restrain disturbances. The improvement of tracking performance of the proposed ATSMC controller is verified by simulations.

The rest of this paper is organized as follows. The dynamics model and control problem of the servo system is formulated in Section 2. The characteristic modeling process is presented in Section 3. Section 4 develops the ATSMC controller and presents the stability analysis. Section 5 conduct the simulation studies and present the results. Finally, some conclusions are given in Section 6.

2 Dynamics model and problem formulation

The dynamic model of the servo system is the mostly widely used model for controller design. Before introducing the characteristic model, we first give the dynamic model of the servo system, which can be described as

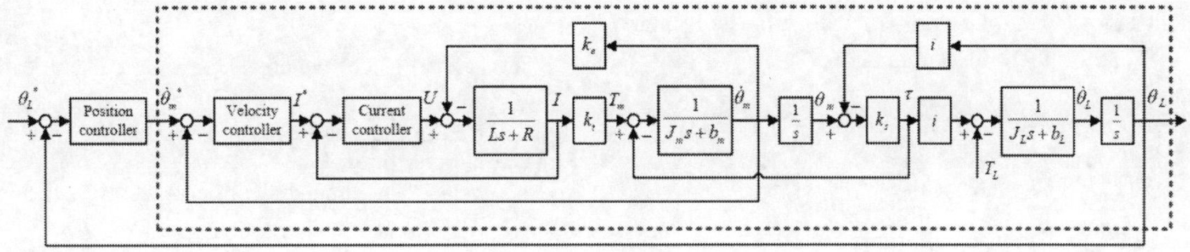

Figure 1. Block diagram of the servo control system

$$\begin{cases} U(t) = k_e \dot{\theta}_m(t) + RI(t) + L\dot{I}(t) \\ J_m \ddot{\theta}_m(t) + b_m \dot{\theta}_m(t) = T_m - \tau \\ J_L \ddot{\theta}_L(t) + b_L \dot{\theta}_L(t) = i\tau - T_L \\ T_m = k_t I(t) \\ \tau = k_s(\theta_m - i\theta_L) \end{cases} \quad (1)$$

where $U(t)$ and $I(t)$ are the stator current and voltage, respectively. R and L are the stator resistance and inductance, respectively. θ_m and θ_L are the angular displacement of the motor and the load, respectively. J_m, J_L, b_m and b_L are the inertia and the viscous friction coefficient of the motor and the load, respectively. T_m, T_L and τ are the motor torque, the load torque and the elastic torque between the motor and the load, respectively. k_e, k_t and k_s are the back electromotive force coefficient, the torque coefficient and stiffness coefficient respectively. i is the gear ratio.

In order to achieve good position tracking performance, a structure of three loops is employed in thecontrol system, including a current loop, a velocity loop and a position loop, as shown in Figure 1. PI controllers are adopted as two inner loop controllers (velocity controller and current controller).

In the following sections, we will introduce the design process of position controller based on the characteristic modeling method.

3 Characteristic model

Characteristic modeling is an effective engineering discretization modeling method. Unlike traditional modeling methods which require the accurate dynamic analysis of the plant, characteristic modeling is mainly based on the dynamic characteristics and performance requirements of the system. Characteristic modeling has been successfully applied to engineering practice (Di, 2014; Meng, 2009; Zhou 2012). First, we present the theory of characteristic modelling.

Consider a nonlinear system

$$\dot{x}(t) = f\left(x, \dot{x}, \cdots, x^{(n)}, u, \dot{u}, \cdots, u^{(m)}\right) \quad (2)$$

where x and u denote the state and input of the system, respectively. Define

$$\begin{aligned} & x = x_1, \dot{x} = x_2, \cdots, x^{(n)} = x_{n+1}, \\ & u = u_1, \dot{u} = u_2, \cdots, u^{(m)} = u_{m+1} \end{aligned} \quad (3)$$

then (3) can be rewritten as

$$\dot{x}_1(t) = f\left(x_1, \cdots, x_{n+1}, u_1, \cdots, u_{m+1}\right) \quad (4)$$

Assumption 1 [8]. The nonlinear system (4) has the following properties

1) Single input single output.

2) The order of input is 1.

3) If $x_i = 0$ and $u_j = 0$, then $f(\cdot) = 0$.

4) $f(\cdot)$ is continuous and differentiable to all the variables x_i and u_j, and all partial derivatives are bounded.

5) $\left|f\left(x(t+\Delta t), u(t+\Delta t)\right)\right| - \left|f\left(x(t), u(t)\right)\right| < M\Delta t$, where $M > 0$ and Δt is the sampling time.

6) All the variables x_i and u_j are bounded.

Lemma 1 [8]. For the system (3), if assumptions 1) - 4) in Assumption 1 are satisfied, then the characteristic model of the system can be established in the form of a second-order difference equation as

$$\begin{aligned} x(k+1) = & f_1(k)x(k) + f_2(k)x(k-1) \\ & + g_0(k)u(k) + g_1(k)u(k-1) \end{aligned} \quad (5)$$

If the system is stable and satisfies assumptions 5) and 6) in Assumption 1, then

a) The parameters $f_1(k)$, $f_2(k)$, $g_0(k)$ and $g_1(k)$ are slow time-varying.

b) The ranges of the parameters can be determined beforehand.

c) In dynamic process, under the same input, selecting a suitable sampling time can guarantee that the error between characteristic model output and practical plant output is maintained within a permitted small range. In steady state, the two outputs are equal.

If the controlled plant is a minimum-phase system, for the simplicity in engineering, only one item $g_0(k)u(k)$ is chosen as the input item.

Now we apply this method to establish the characteristic model of the servo system. The control object of position controller includes the dynamic model of the servo system, the velocity controller and the current controller. Then we consider the three parts as a whole. The system control input is $\dot{\theta}_m^*(t)$ and the output is $\theta_L(t)$. Based on the dynamic model (1), Assumption 1 is satisfied. According to Lemma 1, the characteristic model can be established as

$$x(k+1) = f_1(k)x(k) + f_2(k)x(k-1) + g_0(k)u(k) \quad (6)$$

where $x(k) = \theta_L(t)$ and $u(k) = \dot{\theta}_m^*(t)$. The characteristic parameters $f_1(k)$, $f_2(k)$ and $g_0(k)$ are estimated online by the following RLS algorithm.

$$\begin{cases} \hat{\theta}(k) = \hat{\theta}(k-1) + K(k)\left[x(k) - \phi(k-1)^T \hat{\theta}(k-1)\right] \\ P(k) = \frac{1}{f}\left[I - K(k)\phi(k-1)^T\right]P(k-1) \\ K(k) = \frac{P(k-1)\phi(k-1)}{f + \phi(k-1)^T P(k-1)\phi(k-1)} \end{cases} \quad (7)$$

where $\phi^T(k-1) = [x(k-1)\ x(k-2)\ u(k-1)]$, $\hat{\theta}^T(k) = \left[\hat{f}_1(k)\ \hat{f}_2(k)\ \hat{g}_0(k)\right]$ and f is the forgetting factor. The Initial parameters are selected as $\hat{\theta}^T(k) = [1.6\ -0.6\ 0.00001]$ and $P(0) = 10^6 I$.

4 Controller design and stability analysis

Terminal sliding-mode control is generally used to achieve a finite-time convergence and speed up the convergence rate. The design process of ATSMC controller is introduced as follows.

The sliding function is defined as

$$s(k) = x(k) - r(k) \quad (8)$$

where $r(k)$ is the desired position of the servo system.

The ATSMC scheme is composed of equivalent control and switching control. The former forces the system to evolve on the sliding-mode surface, and the latter ensures the robustness to disturbances.

The total control action of ATMC is given as

$$u(k) = u_{eq}(k) + u_{sw}(k) \quad (9)$$

where

$$u_{eq}(k) = \frac{1}{\hat{g}_0(k)}\begin{bmatrix} -\hat{f}_1(k)x(k) - \hat{f}_2(k)x(k-1) \\ +r(k+1) \end{bmatrix} \quad (10)$$

$$u_{sw}(k) = \frac{1}{\hat{g}_0(k)}\left\{s(k) - \beta T s(k)^{q/p} - \varepsilon T \operatorname{sgn}[s(k)]\right\} \quad (11)$$

where β and ε are positive constants, T is the sampling time, $0 < q/p < 1$ is a ratio of odd integers, which guarantees that the sign of sliding function remains intact.

Lemma 2 [9]. Consider the scalar dynamical system

$$z(k+1) = z(k) - lz(k)^{\alpha} + g(k) \quad (12)$$

where $l > 0$ and $0 < \alpha < 1$ is a ratio of odd integers. If $|g(k)| \le \gamma$, $\gamma > 0$, there is a finite number $k^* > 0$ such that

$$|z| \le \psi(\alpha) \cdot \max\left\{\left(\frac{\gamma}{l}\right)^{1/\alpha}, l^{1/(1-\alpha)}\right\} \quad (13)$$

where function $\psi(\alpha)$ is defined as

$$\psi(\alpha) = 1 + \alpha^{\frac{\alpha}{1-\alpha}} - \alpha^{\frac{1}{1-\alpha}} \quad (14)$$

Define the lumped the identification error as

$$d(k) = \tilde{f}_1(k)x(k) + \tilde{f}_2(k)x(k-1) + \tilde{g}_0(k)u(k) \quad (15)$$

where $\tilde{f}_1(k) = f_1(k) - \hat{f}_1(k)$, $\tilde{f}_2(k) = f_2(k) - \hat{f}_2(k)$ and $\tilde{g}_0(k) = g_0(k) - \hat{g}_0(k)$ denote the parameter identification error.

Based on the realistic servo system, we make the following assumption.

Assumption 2. The lumped identification error is bounded as

$$|d(k)| < \mu, \forall k \ge 0 \quad (16)$$

where μ is a positive constant.

Theorem 1. Consider the system (6) along with the sliding function (8), if $d(k)$ satisfies Assumption 2 and the control parameters satisfy that $\beta > 0$, $0 < q/p < 1$ is a ratio of odd integers and $\varepsilon > \mu / T$, then the control law (9) guarantees that the quasi-sliding-mode is reached in finite steps.

Proof. Rewrite the system equation (6) as

$$\begin{aligned} x(k+1) &= f_1(k)x(k) + f_2(k)x(k-1) + \hat{g}_0(k)u(k) \\ &\quad + \tilde{g}_0(k)u(k) \end{aligned} \quad (17)$$

Substituting the controller (9) into (17), we obtain

$$\begin{aligned} x(k+1) &= r(k+1) + f_1(k)x(k) - \hat{f}_1(k)x(k) \\ &\quad + f_2(k)x(k-1) - \hat{f}_2(k)x(k-1) + s(k) \\ &\quad - \beta T s(k)^{q/p} - \varepsilon T \operatorname{sgn}[s(k)] + \tilde{g}_0(k)u(k) \end{aligned} \quad (18)$$

That is,

$$\begin{aligned} s(k+1) &= s(k) - \beta T s(k)^{q/p} - \varepsilon T \operatorname{sgn}[s(k)] \\ &\quad + \tilde{f}_1(k)x(k) + \tilde{f}_2(k)x(k-1) + \tilde{g}_0(k)u(k) \end{aligned} \quad (19)$$

Then (19) can be written as

$$s(k+1) = s(k) - \beta T s(k)^{q/p} - \varepsilon T \operatorname{sgn}[s(k)] + d(k) \quad (20)$$

Taking (16) into consideration, we have

$$|d(k) - \varepsilon T \operatorname{sgn}[s(k)]| < \mu + \varepsilon T \quad (21)$$

Then, based on Lemma 2, there exsits a finite number $k_s^* > 0$ such that $\forall k > k_s^*$

$$|s(k)| \le \psi\left(\frac{q}{p}\right) \cdot \max\left\{\left[\frac{\mu + \varepsilon T}{\beta T}\right]^{p/q}, (\beta T)^{1/(1-q/p)}\right\} \quad (22)$$

which demonstrates that the sliding function converges into the bounded set in finite steps.

5 Simulation studies and results

Simulations are carried out in Matlab to investigate the performance of the proposed ATSMC controller. A block diagram of ATSMC control scheme is shown in Figure 2. Based on the sampled input-output data of the servo system, the characteristic parameters are identified online by the RLS algorithm, which updating the parameters of the ATSMC controller. The PI controller is used for comparison. The system parameters used in simulations are listed in Table 1. To verify the adaptability to inertia variations, simulations are conducted in two cases, as shown in Table 2. To verify the robustness, a sinusoidal disturbance torque is added at 1.5s. The command position is a 60° step signal.

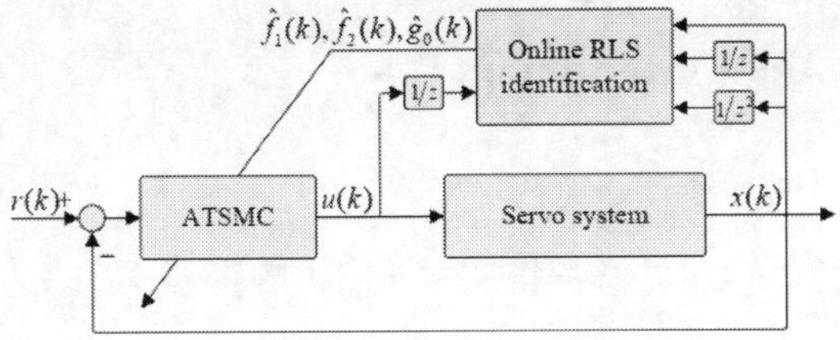

Figure 2. Block diagram of ATSMC control scheme.

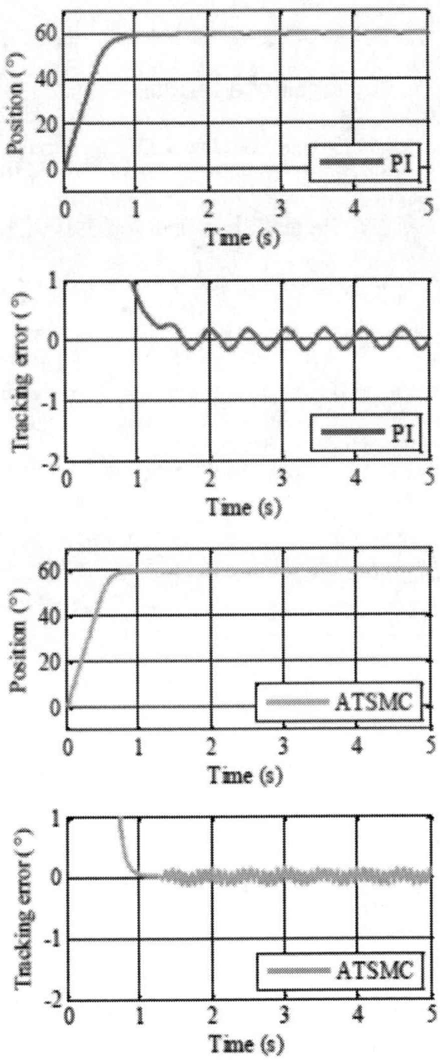

Figure 3. Position trajectories and tracking errors in case 1.

After simulation adjustment, the parameters of PI controller are selected as $k_p = 5.8$ and $k_i = 0.15$. The parameters of ATSMC controller are $\beta = 0.5$, $q = 9$, $p = 13$, $\varepsilon = 10$. The sampling time is $T = 0.005$. The RLS forgetting factor is $f = 0.995$. All of these parameters are fixed in case 1 and case 2 in order to investigate the adaptability and robustness. The position trajectories and tracking errors of the two controllers in case 1 and case 2 are shown in Figure 3 and Figure 4, respectively.

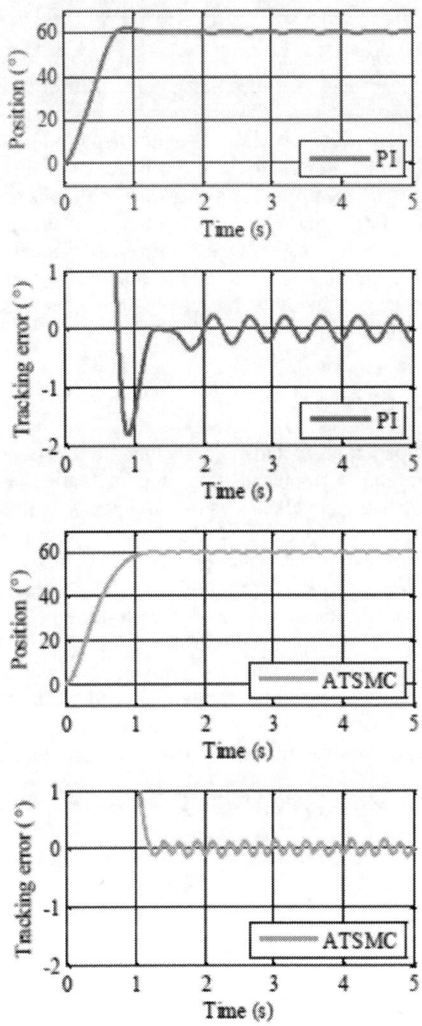

Figure 4. Position trajectories and tracking errors in case 2.

Table 1. System parameters.

Parameter	Value	Unit
R	1.3	$kg \cdot m^2$
L	37.5	$N \cdot m/krpm$
k_e	67.2	$kg \cdot m^2$
k_t	1.11	$N \cdot m/krpm$
k_s	3×10^6	$N \cdot m/rad$
J_m	0.000323	$kg \cdot m^2$
b_m	0.015	$N \cdot m/krpm$
J_L	20.5/204.6	$kg \cdot m^2$

b_L	0.024	N·m/krpm
T_L	10	N·m
i	178	

Table 2. Two simulation cases.

Case	Load inertia	Inertia ratio
1	20.5	2:1
2	204.6	20:1

It can be observed from the simulation results that when load inertia varies from case 1 to case 2, the ATSMC controller keeps a good tracking performance with small overshot while PI controller produce a large overshoot in case 2. On the other hand, PI controller has larger steady-state error due to the disturbance torque. The ATSMC controller restrains the disturbance torque effect and the steady-state errors are smaller. This demonstrate that the ATSMC controller has better adaptability to inertia variations and stronger robustness to disturbances.

6 Conclusion

In this paper, characteristic modeling method is extended to the servo systems with inertia variations. The characteristic model reduces the complexity of traditional mathematics model, and has more advantages in describing time-varying dynamics. Then, based on the characteristic model, a ATSMC controller is designed to improve robustness and achieve finite-time boundedness property. The adaptability and robustness of the proposed controller have been verified by simulations.

The characteristic modeling method provides a feasible method of low-order controller design for complex systems. It is more practical in digital control systems for its discrete-time form and easy implementation. For further improvement of tracking performance, the online identification method under circumstances of external disturbances are the future investigations.

Acknowledgment

This research was supported by National Natural Science Foundation of China under Grant No. 61333008, 61673217, 61673214, 61673219 and 61773211.

References

1. Z. Li, C. Liu, F. Meng, K. Zhou, Proc. Inst. Mech. Eng. C 227, 1481 (2013)
2. S. Li, Z. Liu, IEEE Trans. Ind. Electron. 56, 3050 (2009)
3. S. Jee, J. Lee, Int. J. Precis. Eng. Manuf. 13, 1655 (2012)
4. J. Jin, S. Wang, S. Huang, *IEEE International Conference on Mechatronics and Automation* (IEEE, Tianjin, 2014)
5. J. Geng, Y. Sheng, X. Liu, Tans. Inst. Meas. Ctrl. 36, 604 (2014)
6. L. Qi, S. Bao, H. Shi, Tans. Inst. Meas. Ctrl. 37, 875 (2015)
7. Y. Xia, L. Dai, M. Fu, J. Frankl. Inst. 351, 2299 (2014)
8. H. Wu, J. Hu, Y. Xie, *Characteristic model-based intelligent adaptive control* (China Science and Technology Press, Beijing, 2009)
9. S. Li, H. Du, X. Yu, IEEE Trans. Auto. Ctrl. 59, 546 (2014)

Integral Command Filtered Backstepping Control of a Flexible UAV

Ding Han[1], Lin Yan[2], Guozheng Yan[1] , Xiaoliang Wang[3], Dengping Duan[3]

[1]School of Electronic Information and Electrical Engineering, Shanghai Jiao Tong University, China
[2] Computer Science, University of Utah, America
[3] School of Aeronautics and Astronautics, Shanghai Jiao Tong University, China

Abstract. Airships, as the significant UAV, have a need for greater autonomy in their new missions. Therefore, airship flight control systems require precise dynamic modeling, taking into account the effect of flexibility and the interaction with aerodynamic forces. This research effort develops an efficient modeling of the autonomous flexible airship. The formalisation used is based on the Lagrange method. The resulting model includes the rigid body motion, the elastic deformation, and the coupling between them. Based on the precise flexible dynamic model, a novel backstepping nonlinear controller with integral action is proposed for motion control systems. The resulting feedback controller is able to adapt to actuator performance limitations, such as limitations in magnitude and rate of change of rudder, than conventional backstepping controller without integral action. With the deformation considered, the presented controller could resist the flexible uncertainty effect, and the system's trajectory tracking ability is significantly improved. The approach guarantees exponential stability of a compensated tracking error in the sense of Lyapunov.

1 Introduction

Airships, as autonomous aerial vehicles, have recently gained importance. The range of tasks they are designed to support is growing. For example, climate research, data collecting, surveillance, and more closer to us, we always saw them as means of advertising [1]. The resurgence of airships in the past few years has created a need for accurate dynamics models and simulation tools to study their flight behavior and design their control systems. Several researchers have presented airship dynamic models. Tischler et al. proposed the nonlinear equations of motion and developed the simulation program HLASIM for the design of heavy lift airships [2]. Jex and Gelhausen and Amann used different aerodynamic computational methods to predict the control responses of the Skyship-500 airship [3, 4]. Analytical linear dynamics models using aerodynamic derivatives were also applied to study the flight stability and control design of airships [5, 6]. Li and Nahon presented a modeling procedure for airship dynamics simulation and validated their model with published experimental data [7]. All of these works were researched by using the assumption of rigid body behavior. Sometimes the flexibility effects are modeled as a perturbation. However, real airships experience elastic deformations and their structural flexibility has been an important consideration in the history of airship development. In other vehicles, such as light aircraft, the introduction of the flexibility in the dynamic model becomes essential [8, 9,10]. Thus we try to contribute to the study of the elastic deformation of the airships by considering the effect of flexibility. To design the control system, the deformation of the airship is considered acting on the motion rather than as a perturbation.

Recently, with increasing interest in airships and a drive to make them more autonomous, there has been increased research effort on airship modelling and control. Airships have different nonlinearities from heavier-than-air aircraft, such as added mass terms that are negligible for other aircraft. The airship has been less extensively studied than that for other aircraft. Due to the flexible structures and the time-varying characters, obtaining the precise mathematical models valid for relevant flight conditions is more difficult. Traditionally, to overcome this problem, airship models are usually linearized, and the flight controllers are designed for a selected flight condition. Several control techniques are then applied, usually with excellent results [10]. However, if the state of the airship is too different from the flight condition used for design, the performance of controller is not as optimal as before, and the airship model might fail to accurately capture the system behavior. In this situation, linear control laws can be prone to misbehavior or even failure. Gain scheduling methods have been successfully developed to deal with different operating points [11]. However, the disadvantages of these methods are that the controllers are not global used and a wide range of flight conditions need to be estimated，which is a complicated task.

Nonlinear control methods have been developed to overcome the shortcomings of linear design approaches.

For example, feedback linearization has been used to handle the nonlinear equations of motion, generating controllers suitable for the entire flight envelope, if a precise knowledge of the aircraft model exists [12]. For security reasons, as well as simplicity and flexibility, a global nonlinear control is more interesting than a linearized and decoupled one. Lyapunov function based on backstepping control is described in [13], and have been proposed in [14], [15], and [16]. In reference [14], Godhavn derives a full nonlinear model and allows large variation in velocities. He used splines to make references smoothly and backstepping control to track them [14]. In reference [15], the dynamical extension is proposed. Fierro and Lewis consider the specific vehicle dynamics to convert a steering system (kinematic model) command into control inputs for the actual vehicle [15]. Encarnacao and Pascoal provide a control method that combines both the trajectory tracking and path following problems and shows a good tracking and following performance [16]. In the presence of parametric uncertainties, an alternative solution is to design the controller using the adaptive backstepping design method. For example, Härkegård has presented an adaptive backstepping design in [17], complemented with a control allocation method to compute actuator values. However, these designs do not take into account input saturation and rate limitation of the actuators, which will constrain the efficiency of the designed control law, or even cause an accident.

In recent control systems, the overall performance of the motion control systems is restricted by the limitation of magnitude and rate of change of the actuators, such as rudder angles. What's more, the elastic deformation of airship during flight motion process is also an complicating factor for control systems. So it is of great importance to design advanced controllers to achieve the bandwidth improvement.

In this work, a command filtered backstepping control law with integral action is designed for airship flexible dynamics. The resulting control law is able to track the given yaw reference considering the limitations in magnitude and rate of change of actuators. The proposed control law shows the good performance when considering actuator position and rate limits.

The paper is organized as follows: Section 2 establishes the mathematical flexible model of the airship. Section 3 presents a detailed derivation of the command filtered backstepping control law with integral action and deformation considered to deal with airship dynamics. Some simulations are carried out to show the performance and stability of the proposed controller in section 4. Finally, in section 5, the paper is concluded with some observations.

2 Dynamics modeling of flexible airships

Generally, the hull of an airship is a flexible inflating structure. The shape of an airship hull will change under the influence of aerodynamics. So the changes in shape, aerodynamics and attribute parameters need to be considered in the dynamics modeling. The main differences of flexible dynamic modeling and the rigid dynamic modeling are the hull shape change model, the aerodynamic model, the added masses and attribute parameters model.

The aerodynamic force will be changed with the flow condition and the shape of flexible airship. For obtaining the aerodynamics under the conditions of different flow condition and flexible shape, we use the semi-experiential estimate method [18] in this paper. But the change of local relative velocity and shape along the x axis of airship hull was contained in the aerodynamic model.
The coordinate of airship is shown in Fig.1.

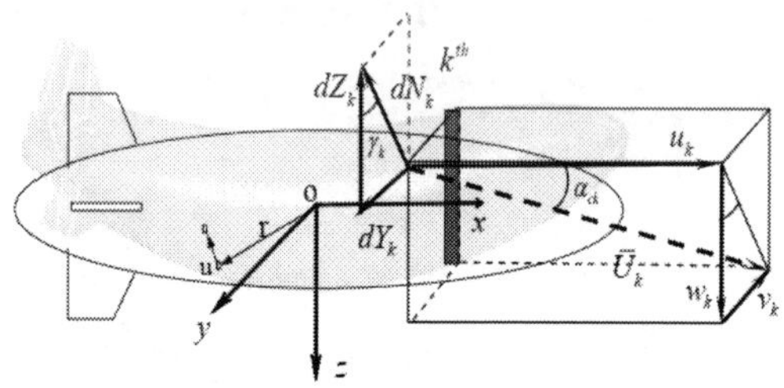

Figure 1. The body coordinate and parameters definition.

2.1 The attribute parameters model

The attribute parameters of an airship include the mass and inertia. Based on the structure and shape characteristics of airship, the airship was divided into different parts. The mass distribution of an airship includes the structure and interior gas. The hull was divided into N_g panels. The interior gas was divided into N_{xg} cylindrical segments. The fins and cabin were seen as concentrated mass parts. This method can take into account the mass distribution and change of shape of the airship in detail.

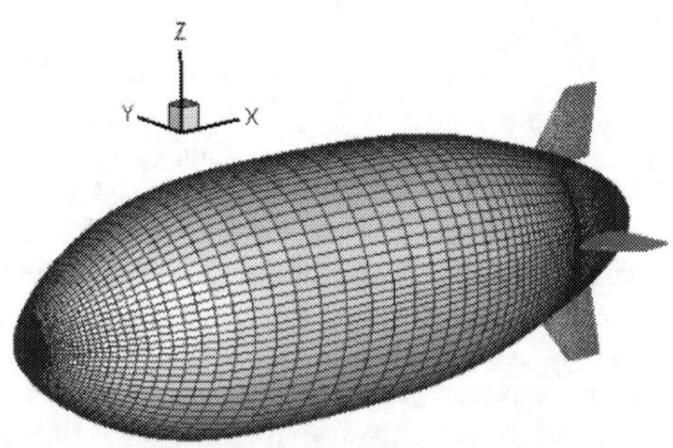

Figure 2. The discrete airship surface.

Thus, the coordinate of every panel is

$$\mathbf{r} = \begin{bmatrix} x(g) \\ y(g) \\ z(g) \end{bmatrix}, \qquad g = 1,2,\cdots,N_g$$

The mass of every panel is $\Delta m(g), \quad g = 1,2,\cdots N_g$

In addition, every natural mode shape can be discretised,

$$\mathbf{\Phi}_i(\mathbf{r}) = \begin{bmatrix} \Phi_{xi}(\mathbf{r}) \\ \Phi_{yi}(\mathbf{r}) \\ \Phi_{zi}(\mathbf{r}) \end{bmatrix} = \begin{bmatrix} \Delta x_i(g) \\ \Delta y_i(g) \\ \Delta z_i(g) \end{bmatrix}, \qquad g = 1, 2, \cdots, N_{xg}$$

2.2 The flexible dynamic model of airship

Through the dynamic equations of flexible body, the dynamic model of flexible airship can be obtained [19],

$$\begin{bmatrix} \mathbf{M}_{11} & \mathbf{M}_{12} & \mathbf{M}_{13} \\ \mathbf{M}_{21} & \mathbf{M}_{22} & \mathbf{M}_{23} \\ \mathbf{M}_{31} & \mathbf{M}_{32} & \mathbf{M}_{33} \end{bmatrix} \begin{bmatrix} \dot{\mathbf{v}}_0 \\ \dot{\mathbf{\Omega}} \\ \ddot{\mathbf{q}} \end{bmatrix} = -\begin{bmatrix} \mathbf{0} \\ \mathbf{0} \\ \mathbf{S_E} \end{bmatrix} + \begin{bmatrix} \mathbf{F_I} \\ \mathbf{M_I} \\ \mathbf{Q_I} \end{bmatrix} + \begin{bmatrix} \mathbf{F_G} \\ \mathbf{M_G} \\ \mathbf{Q_G} \end{bmatrix} + \begin{bmatrix} \mathbf{F_{AS}} \\ \mathbf{M_{AS}} \\ \mathbf{Q_{AS}} \end{bmatrix} + \begin{bmatrix} \mathbf{F_{AD}} \\ \mathbf{M_{AD}} \\ \mathbf{Q_{AD}} \end{bmatrix} + \begin{bmatrix} \mathbf{F_C} \\ \mathbf{M_C} \\ \mathbf{Q_C} \end{bmatrix} \quad (1)$$

where, $\mathbf{M}_{sys}$ is the symmetry mass matrix,

$$\mathbf{M}_{sys} = \begin{bmatrix} \mathbf{M}_{11} & \mathbf{M}_{12} & \mathbf{M}_{13} \\ \mathbf{M}_{21} & \mathbf{M}_{22} & \mathbf{M}_{23} \\ \mathbf{M}_{31} & \mathbf{M}_{32} & \mathbf{M}_{33} \end{bmatrix} \quad (2)$$

$\mathbf{S_E}$ is the interior elastic force vector, $\mathbf{S_E} = \mathbf{Kq}$, $\mathbf{K}$ is the stiffness matrix,

$$K_{i,j} = K_{N+i,N+j} = \int_L EI\Psi_i''\Psi_j''dx, i, j = 1, 2\cdots, N \quad (3)$$

where EI is the bend stiffness, $\Psi_i'' = \dfrac{d^2\Psi_i}{dx^2}$.

The right term of the dynamic model of flexible airship is the exterior forces, $\mathbf{F_I}, \mathbf{M_I}, \mathbf{Q_I}$ are the inertia force and moment; $\mathbf{F_G}, \mathbf{M_G}, \mathbf{Q_G}$ are the gravity and gravity moment; $\mathbf{F_{AS}}, \mathbf{M_{AS}}, \mathbf{Q_{AS}}$ are the buoyancy and buoyancy moment; $\mathbf{F_{AD}}, \mathbf{M_{AD}}, \mathbf{Q_{AD}}$ are the aerodynamic force and moment; $\mathbf{F_C}, \mathbf{M_C}, \mathbf{Q_C}$ are the control force and moment. u, v, w are velocity components, $\mathbf{v}_0 = (u, v, w)$, p, q, r are roll, pitch and yaw, $\mathbf{\Omega} = (p, q, r)$. ψ is velosity potential[19].

2.3 Kinematics of airship

The kinematics equations of airship are:

$$\begin{bmatrix} \dot{\varphi} \\ \dot{\vartheta} \\ \dot{\psi} \end{bmatrix} = \begin{bmatrix} 1 & \sin\varphi\tan\vartheta & \cos\varphi\tan\vartheta \\ 0 & \cos\varphi & -\sin\varphi \\ 0 & \dfrac{\sin\varphi}{\cos\vartheta} & \dfrac{\cos\varphi}{\sin\vartheta} \end{bmatrix} \begin{bmatrix} p \\ q \\ r \end{bmatrix} \quad (4)$$

$$\begin{bmatrix} \dot{x} \\ \dot{y} \\ \dot{z} \end{bmatrix} = \begin{bmatrix} \cos\psi\cos\vartheta & \sin\psi\cos\vartheta & -\sin\vartheta \\ \cos\psi\sin\vartheta\sin\varphi - \sin\psi\cos\varphi & \sin\psi\sin\vartheta\sin\varphi + \cos\psi\cos\varphi & \cos\vartheta\sin\varphi \\ \cos\psi\sin\vartheta\cos\varphi + \sin\psi\sin\varphi & \sin\psi\sin\vartheta\cos\varphi - \cos\psi\sin\varphi & \cos\vartheta\cos\varphi \end{bmatrix}^T \begin{bmatrix} u \\ v \\ w \end{bmatrix} \quad (5)$$

where, ϑ, ψ, φ are the pitch, yaw and roll angle respectively; x, y, z are the global position coordinates respectively.

3 Backstepping control design and error analysis

Several places in this paper refer to filtering of a signal x_c^0 to produce a signal x_c and its derivative $\dot{x}_c$. This is referred to as command filtering. The state space implementation of such a filter is[20]

$$\begin{aligned} \dot{x}_1 &= x_2 \\ \dot{x}_2 &= -2\zeta w_n x_2 - w_n^2(x_1 - x_c^0) \end{aligned} \quad (6)$$

where $x_c = x_1$ and $\dot{x}_c = x_2$. Note that if x_c^0 is bounded, then x_c and $\dot{x}_c$ are bounded and continuous. The transfer function from x_c^0 to x_c is

$$\frac{X_c(s)}{X_c^0(s)} = H(s) = \frac{w_n^2}{s^2 + 2\zeta w_n s + w_n^2} \quad (7)$$

which has a unity gain at low frequencies, damping ratio ζ and undamped nature frequency w_n. The error $|x_c^0(t) - x_c(t)|$ is small if the bandwidth of $X_c^0(s)$ is less than the bandwidth of $H(s)$. If the bandwidth of x_c^0 is known and the goal of the filter is to generate x_c and its derivative with $|x_c^0 - x_c|$ small, then the designer simply chooses w_n sufficiently large.

It is to be noted that, the yaw control system of the flexible airship can be obtained as the second-order form from equations (1) and (4),

$$\begin{cases} \dot{\psi} = \dfrac{1}{\cos\vartheta} r \\ \dot{r} = \dfrac{1}{\mathbf{M}_{22}(3,3)}[-0 + \mathbf{M_I}(1,3) + \mathbf{M_G}(1,3) + M_{AS}(1,3) + \mathbf{M_{AD}}(1,3) + \mathbf{M_C}(1,3) \\ \qquad - \mathbf{M}_{21}(3,1:3)\dot{\mathbf{v}}_0 - \mathbf{M}_{23}\ddot{\mathbf{q}}(1:4,1)] \end{cases} \quad (8)$$

where, $\mathbf{M_C}(1,3) = k\delta_r$, k is rudder coefficient.

Let $x_1 = \psi$, $x_2 = r$, $u_{input} = \delta_r$. Then we analyze the second-order state-space equations as follow

$$\begin{cases} \dot{x}_1 = G_1 x_2 \\ \dot{x}_2 = f_2 + G_2 u_{input} \end{cases} \quad (9)$$

where the

coefficients $G_i(i = 1, 2)$ and f_2 are $G_1 = \dfrac{1}{\cos\vartheta}$,

$$G_2 = \frac{1}{\mathbf{M}_{22}(3,3)} k,$$

$$f_1 = \frac{1}{\mathbf{M}_{22}(3,3)}[-0 + \mathbf{M_I}(1,3) + \mathbf{M_G}(1,3) + M_{AS}(1,3) + \mathbf{M_{AD}}(1,3) - \mathbf{M}_{21}(3,1:3)\dot{\mathbf{v}}_0 - \mathbf{M}_{23}\ddot{\mathbf{q}}(1:4,1)]$$

Considering the virtual system

$$\dot{x}_1 = G_1 \upsilon \tag{10}$$

υ is the virtual term, and defining the position tracking error

$$z_1 = x_{1d} - x_1 \tag{11}$$

where x_{1d} is the reference state.

And the Lyapunov function

$$V_1 = \frac{1}{2} z_1^T z_1 \tag{12}$$

whose dynamics can be derived as

$$\dot{V}_1 = z_1^T \dot{z}_1 = z_1^T (\dot{x}_{1d} - \dot{x}_1) = z_1^T (\dot{x}_{1d} - G_1 \upsilon) \tag{13}$$

To drive the tracking error to zero asymptotically, the traditional choice is to pick

$$\upsilon = G_1^{-1} (\dot{x}_{1d} + k_1 z_1) \tag{14}$$

In order to introduce the integral action, we choose the behavior of the virtual control as

$$\upsilon = G_1^{-1} (\dot{x}_{1d} + k_1 z_1 + \lambda \xi) \tag{15}$$

where k_1 and λ are positive constants at the disposal of the designer and

$\xi = \int_0^t z_1(\tau)\, d\tau$ is the integral action of the tracking error.

By introducing the integral action into the stabilizing function, we can make sure that the tracking error is converged to zero at the steady state despite the presence of the disturbance and inaccuracy, just like the influence of flexible model.

Then consider the tracking error

$$z_2 = \upsilon - x_2 = G_1^{-1} (\dot{x}_{1d} + k_1 z_1 + \lambda \xi - x_2) \tag{16}$$

Hence $\dot{z}_1 = G_1 z_2 - k_1 z_1 - \lambda \xi$ (17)

$$\dot{z}_2 = G_1^{-1} (\ddot{x}_{1d} + k_1 \dot{z}_1 + \lambda \xi) - (f_2 + G_2 u_{input}) \tag{18}$$

The control Lyapunov function

$$V_2 = \frac{\lambda}{2} \xi^T \xi + \frac{1}{2} z_1^T z_1 + \frac{1}{2} z_2^T z_2 \tag{19}$$

The derivative with respect to time of equation (19) reads

$$\begin{aligned} \dot{V}_2 &= \lambda \xi^T \dot{\xi} + z_1^T \dot{z}_1 + z_2^T \dot{z}_2 \\ &= -k_1 z_1^T z_1 + z_2^T \left((G_1^T + \lambda G_1^{-1} - k_1^2 G_1^{-1}) z_1 + k_1 z_2 - k_1 \lambda G_1^{-1} \xi + G_1^{-1} \ddot{x}_{1d} - f_2 - G_2 u_{input} \right) \end{aligned} \tag{20}$$

In order to render the non-positivity of the control Lyapunov derivative function, we choose the input as

$$u_{input} = G_2^{-1} \left((G_1^T + \lambda G_1^{-1} - k_1^2 G_1^{-1}) z_1 + 2k_1 z_2 - k_1 \lambda G_1^{-1} \xi + G_1^{-1} \ddot{x}_{1d} - f_2 \right) \tag{21}$$

Then $\dot{V}_2 = -k_1 z_1^T z_1 - k_1 z_2^T z_2 \le 0$.

4 Simulation results

Table 1. Parameters of model.

$m(kg)$	2682.1	$I_x(kg \cdot m^2)$	108340
$I_y(kg \cdot m^2)$	288190	$I_z(kg \cdot m^2)$	217220
$m_{11}(kg)$	505.7455	$m_{22}(kg)$	5073.7
$m_{33}(kg)$	5073.7	$m_{44}(kg)$	68879
$m_{55}(kg)$	585400	$m_{66}(kg)$	585400
$L(m)$	50	(x_G, y_G, z_G) (m)	(0,0,3.605)

The following controller parameters are used: $k_1 = 0.1$, $k_2 = 0.1$, and $\lambda = 0.01$; the initial conditions are: $\psi = 0$ an $\delta_r = 0$. The reference trajectory chosen for ψ_d is:

$$\psi_d = \begin{cases} (1 - e^{-0.05t}) \cdot \frac{\pi}{2} & 0 \le t < 100 \\ \frac{\pi}{2} + (1 - e^{-0.05(t-100)}) \cdot \frac{\pi}{6} & 100 \le t < 200 \\ \frac{\pi}{2} + \frac{\pi}{6} - (1 - e^{-0.05(t-200)}) \cdot \frac{\pi}{3} & 200 \le t \le 400 \end{cases}.$$

In order to close to practical applications, the following limitations in magnitude and rate of change of actuator (rudder angle) have also been considered in the simulation model.

$$F_T = 270\,\text{N}, |\delta_r| \le 30\,\text{deg}, |\dot{\delta}_r| \le 5\,\text{deg/s}.$$

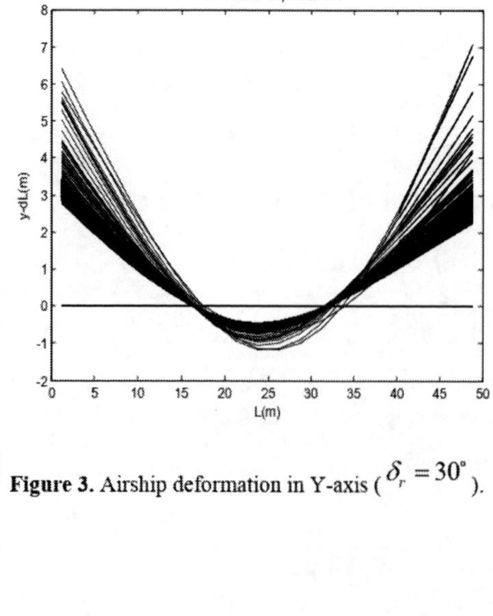

Figure 3. Airship deformation in Y-axis ($\delta_r = 30^\circ$).

Figure 3 shows the results of the flexible airship in the open-loop conditions. Rudder was set $\delta_r = 30^\circ$ and airship kept turning circle with the speed $u = 8.7m/s$. From the results, we can clearly see the deformation of the airship in Y-axis due to the force and moment of the rudder. Compared to the middle of the airship, the nose and the rear of the airship have larger deformation, about 6m and 7m respectively in max. When the airship turned circle with constant radius, the deformation of the nose and rear kept at about 3m.

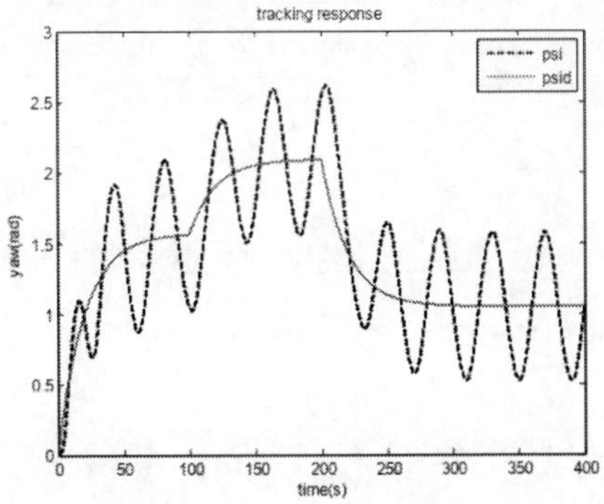

Figure 4. Time evolution of yaw performance without the integral action (solid is reference and dotted is tracking response).

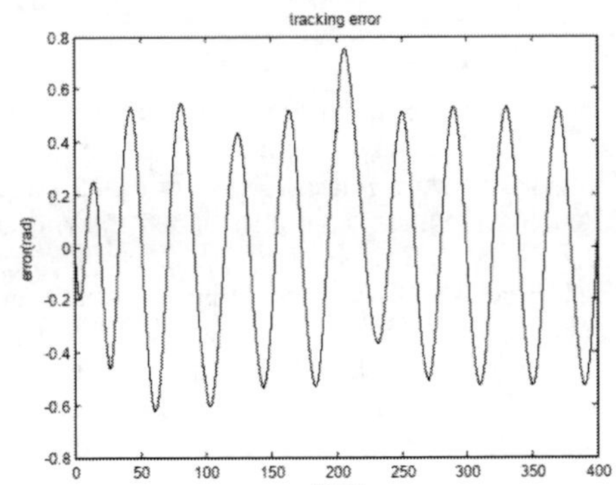

Figure 5. Tracking error without integral action.

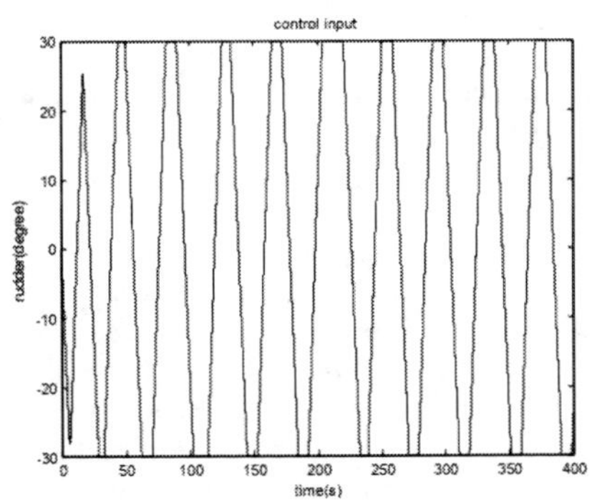

Figure 6. Control input without integral action.

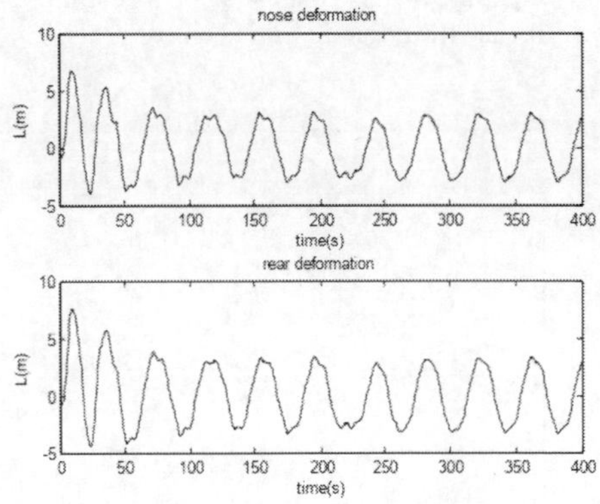

Figure 7. Deformation in Y-axis (no integral action).

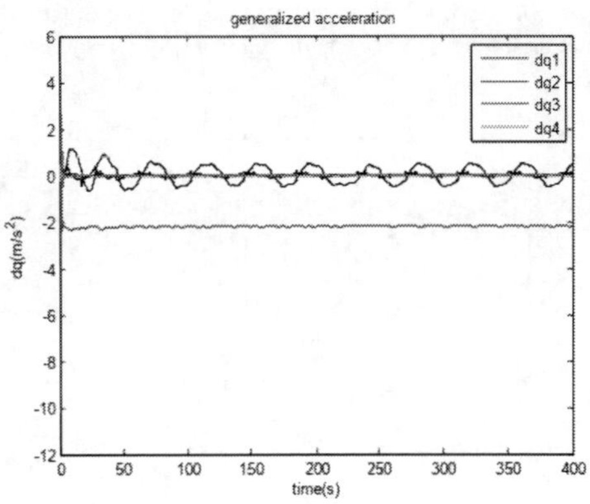

Figure 8. Time evolution of generalized acceleration (without integral action).

Our proposed integral command filtered backstepping nonlinear controller for flexible system is based on the Lyapunov function design methodology and essentially has nonlinear control properties. Because the system can be proven stable by the Lyapunov function through energy consumption, the backstepping controller with integral section has much higher stability range than the conventional controller (without integral section)

The reference signal is trapezoidal input in order to close to the practical applications and slightly smoothed to avoid discontinuities that would demand instantaneous changes in the control signals.

From our control design, we can see that the integral action can be easily removed from the backstepping controller by setting λ =0. Figs 4 to 8 are the simulation results for the backstepping without the integral action. At time t=100, reference is commanded from pi/2 rad to 2pi/3 rad; then at time t=200s, reference is decreased with a magnitude pi/3 rad. From the tracking error plot in Fig.5, we can see that there is a dynamic error of about 0.6 radians. This is because of the bandwidth limitation of the conventional backstepping controller, which can be seen from control input (Fig.6). The rudder keeps saturating due to the rate limitation of the actuator. So the tracking error vibrates near the equilibrium point. During the tracking process, the deformation of airship is fluctuating due to the control input (see Fig.7), which may influence the accurate flight control.

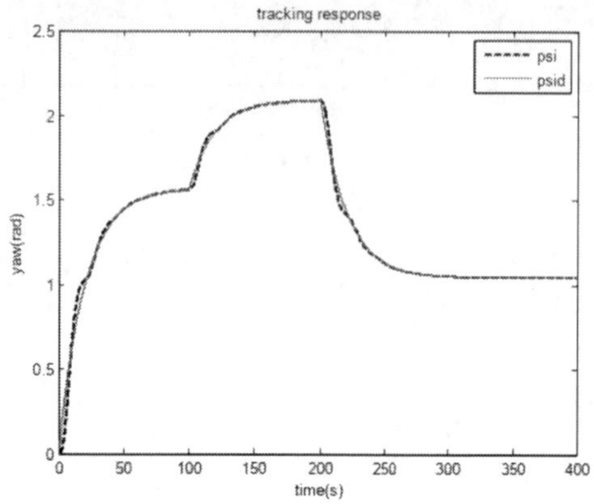

Figure 9. Time evolution of yaw performance with integral action (solid is reference and dotted is tracking response).

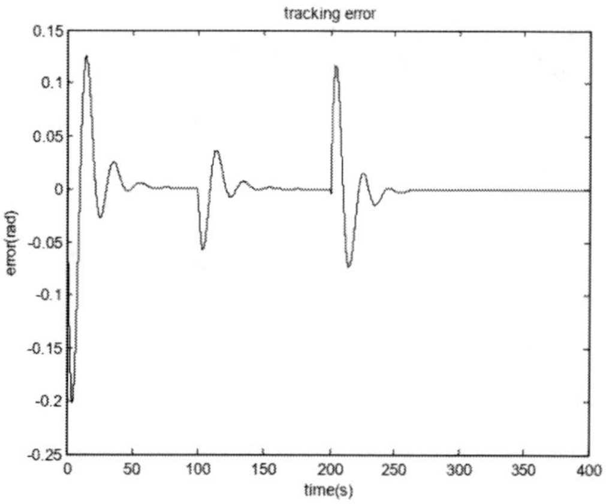

Figure 10. Tracking error of the backstepping controller with the integral action.

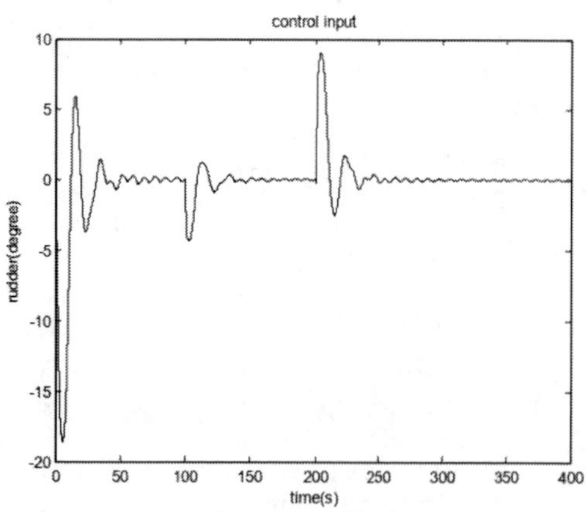

Figure 11. Control input with integral action.

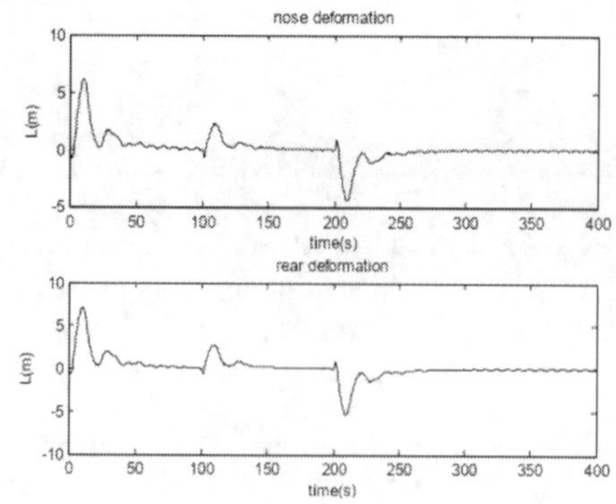

Figure 12. Airship deformation in Y-axis (with integral action).

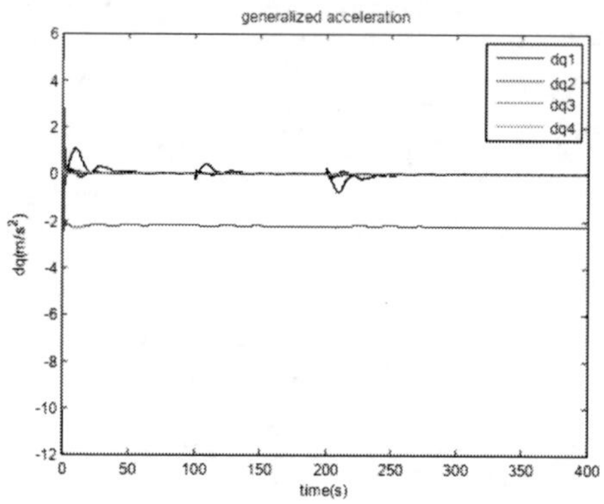

Figure 13. Time evolution of generalized acceleration (with integral action).

The same command position reference is used for the integral backstepping nonlinear controller. The performance is shown from Fig.9 to fig.13. With the backstepping nonlinear controller added integral action, we can obtain that the tracking error has been greatly reduced to below 0.01 radians during the steady process as can been seen in Fig.10. Even at the steep changes, the tracking error is only about 0.13 radians, which is much smaller than that of the conventional controller. The fast tracking performance is preserved. Fig. 9 shows a better match to the reference, in comparison with Fig. 4. Comparing Fig. 11 with Fig.6, the rudder response is rather smooth when moderate changes in yaw angle are demanded. What is more, with the integral action, the backstepping controller is adapted to the bandwidth limitation of the rate of actuators and makes the simulation more close to reality. From Fig.12 we can see that the deformation tendency of airship just opposite to the rudder (control input), and rapidly converge to zero, which is further improvement than that of Fig.7. Fig.13 shows that the integral backstepping controller considering flexible factors could resist the flexible uncertainty and converge the generalized acceleration to steady values. All of these above prove that integral backstepping control with flexible factors considered is more adapted to limitation of actuators and shows better performance to control the system against the flexible uncertainty presented in the motion control systems.

5 Conclusion

In this paper, the dynamic modeling of the flexible airship has been presented, which offers the precise model for control near the truth. And then it has shown the process for the design of a nonlinear controller employing Lyapunov and command filtered Backstepping techniques to control the yaw angle of the flexible system. In order to make up the flexible uncertainty factor and the limitation of the rate of actuators, integral action is considered in the design controller. Numerical simulations are performed to verify the performance of the proposed control law. The results are shown that the proposed integral command filtered backstepping control algorithm could stabilize the flexible system and drive the flexible airship to the desired trajectory of yaw angle.

References

1. Oh S, Kang S, Lee K, et al. Flying display: autonomous blimp with real-time visual tracking and image projection[C]//Intelligent Robots and Systems, 2006 IEEE/RSJ International Conference on. IEEE, 2006: 131-136.
2. Tischler, M.B., Ringland, R.R., and Jex, H.R., "Heavy Airship Dynamics," Journal of Aircraft, Vol.20, No.5, 1983, pp.425–433.doi:10.2514/3.44888
3. Jex, H.R., and Gelhausen, P., "Pre- and Post-Flight-Test Models Versus Measured Skyship-500 Control Responses," 7th AIAA Lighter-Than-Air Technology Conference, AIAA, NewYork, 1987, pp.87–97.
4. Amann, J.H., " A Comparison of a Nonlinear Flight Dynamic Simulation of an Airship with Flight Test Results," 7th AIAA Lighter-Than-Air Technology Conference, AIAA, NewYork, 1987, pp.78–86.
5. Cook, M.V., Lipscombe, J.M., and Goineau, F., "Analysis of the Stability Modes of the Non-Rigid Airship," The Aeronautical Journal, Vol.104, No.1036, 2000, pp.279–290.
6. Schmidt, D.K., "Modeling and Near-Space Stationkeeping Control of a Large High-Altitude Airship," Journal of Guidance, Control, and Dynamics,Vol.30,No.2,2007,pp.540–547.
7. Li, Y., and Nahon, M., "Modeling and Simulation of Airship Dynamics," Journal of Guidance, Control, and Dynamics, Vol.30, No.6,2007, pp.1691–1700.
8. Calise A., Nakwan K., 2002 " Adaptive compensation for flexible dynamics". AIAA Guidance, navigation and control conference. Montery, Canada.
9. Bianchin M., Quaranta G., Mantegazza P., 2003 "State space reduced order models for static aero elasticity and flight mechanics of flexible aircrafts". 17 national conference AIDAA, Italy,.
10. McLean, D.. Automatic Flight Control Systems. Prentice Hall. 1990
11. Nichols, R.A., Reichert, R.T., and Rugh, W.J.. Gain scheduling for controllers: A flight control example. IEEE Trans. Contr. Syst. Tech., 1993,1(2): 69-78.
12. Ochi, Y. and Kanai, K.. Design of restructurable flight control systems using feedback linearization. Journal of Guidance, Control, and Dynamics, 1991,14(5),903-911.
13. M. Krstic, I. Kanellakopoulos, and P. V. Kokotovic, Nonlinear and Adaptive Control Design, John Wiley Sons, Inc, New York, 1995.
14. E. Freund and R. May, "Nonlinear Path Control in Automated Vehicle Guidance," IEEE Transactions on Robotics and Automation, 1997,13(1):pp.4940.
15. R. Fierro and F. L. Lewis, "Control of a nonholonomic mobile robot: Backstepping kinematics into dynamics," in Proceedings of CDC'95, 34th IEEE Conference on Decision and Control, New Orleans, LA, USA, Dec. 1995, pp. 3805-3810.
16. P. Encarnacao and A. Pascoal, "Combined trajectory tracking and path following for marine craft," 40th IEEE Conference on Decision and Control, Mar. 2001.
17. Härkegård, O., "Backstepping and Control Allocation with Applications to Flight Control," Ph.D. Thesis, Linköping Univ., Linköping, Sweden,2003.
18. S.P.Jones, J.D.Delaurier, Aerodynamic estimation techniques for aerostats and airships, Journal of aircraft, VoL.20 No.2,1983.
19. Yuwen Li, Dynamics Modeling and Simulation of Flexible Airships, Doctor thesis, 2008.
20. Han D, Wang X, Chen L, et al. Command-filtered backstepping control for a multi-vectored thrust stratospheric airship [J]. Transactions of the Institute of Measurement and Control, 2015

Control and Simulation of Torque Converter Lock-up Clutch

Li-mei Zhao[1, 2], Huai-chao Wu[2], Zheng-hu Peng[2], Cheng Zhang[2], Shi-hao Tang[1],Yun-xiang Long [3] and Guo-qiao Li[3]

[1] *Key Laboratory of Advanced Manufacturing Technology of Ministry of Education, Guizhou University, China*

[2]*College of Mechanical Engineering, Guizhou University, Guiyang , China*

[3]*Guizhou Winstar Hydraulic Transmission Machinery Co., Ltd, Zunyi, China*

Abstract. Considering the requirements of dynamic performance and quality during the torque converter locking process, a two parameters latch-up control strategy was designed, and a simulation method to study the control performance by use of processor in loop (PIL) simulation technology was proposed. Through the study of the practically applied requirements of a loader, a two parameter lock-up control strategy with turbine speed signal and gear signal as main control parameters was developed. According to the control strategy, the logic diagram of real-time locking and unlocking control of the latch-up clutch was established by using Matlab/Simulink. Then, PIL technology was used to realize the co-simulation of CCS and Simulink. Simulation results show that the latch-up control model is correct and the lock-up clutch can effectively lock and unlock in accordance with the lock-uprule.

1Introduction

The application of torque converter in vehicles greatly simplifies the vehicle handling, improves ride comfort, safety, crossing country and average vehicle speed, but the biggest defect of hydraulic transmission is inefficiency. In order to reduce the fuel consumption of torque converter, a controllable latch-up clutch is installed between the pump wheel and the turbine of the torque converter. When the driving condition of the car reaches the setting target, the lock-up clutch is controlled to lock, the pump wheel is integral to the turbine, and the torque converter then turns into mechanical transmission. To a certain extent, it relieves the problem of low transmission efficiency.

Under the condition that the converter structure type and parameters have been determined, control of lock-up clutch is mainly focus on the improvement of lock-up control quality and the rationality of lock-up rule. The main purpose of lock-up control is to reduce the lock-up impact and improve vehicle dynamic performance. The quality of lock-up process mainly depends on the control during engagement process. In recent years, many domestic scholars have done a great deal of research work in optimizing the lock-up points, realizing multi parameter lock-up control, and control methods such as buffer control by manipulating oil pressure[1-4].In this study, the lock-uprule of the lock-up clutch of a loader was studied, and the simulation model of the lock-up control was established, and the co-simulation of CCS and Simulink analysis was realized by using PIL detection technology.

2 Lock-up control law of the torque converter

2.1The rule of latch-up control

The rule of latch-up control is generally divided into single parameter control and two parameters control. And two-parameter control is a control strategy with two signals as the main control signals and the rest for the auxiliary control signals. In this study, loaders generally work on sites where the environments are severe, so two-parameter control was chosen to reduce the impact of the environment on the equipment. In the two-parameter control system, the turbine speed signal and the gear signal were chosen as the main control parameters and the steering signal, the throttle signal and the braking signal were chosen as the auxiliary control parameters.

The clutch control can be divided into dynamic locking, economic locking, steering state locking, braking state locking and integrated locking. And the rule of integrated locking has good power performance and fuel economy as well as the operability of the vehicle, and it also ensures great probability for the integrated locking performance of the vehicle for different applications. Therefore, the integrated locking rule was chosen to research on the control system. The integrated locking rule designed of the loader was as follows:

1)Whether the clutch would lock or not is determined by the turbine speed in the braking state;

2)In the steering condition, the unlocking of the latch-up clutch is determined by throttle angle, and the latch-up clutch is not locked when the throttle angle is less than 20%;

3)Locking would prevent the shift from running smoothly when shifting, so the lock-up clutch needs to be unlocked;

4)Whether the latch-up clutch would lock or not can be judged by the rotation speed of the turbine when not to change gear;

5)According to different gears and using purposes, at the first gear, the clutch must be unlocked in order to obtain greater torque for starting, climbing, and high power; at the second gear, the locking is determined by the throttle angle. And when the throttle angle is less than 20%, the clutch is unlocked in order to obtain greater torque; when the throttle angle is more than 20%, whether the clutch would lock or not is judged by the rotation speed of the turbine and the throttle angle in order to make the vehicle more efficient; At the third gear, the clutch needs to be locked in order to make the vehicle get higher speed and larger utilization of power.

2.2 Determination of lock-up point of the lock-up clutch

In order to ensure the dynamic performance of the vehicle, double parameters (turbine speed-throttle angle) lock-up rule is usually adopted to select the lock-up points. Under a certain throttle angle, the intersection of two turbine shaft output torque curves under hydraulic conditions (unlocked) and mechanical conditions (locked) respectively is selected, the turbine speed of the intersection point is the lock-up point at the throttle angle. The torque of the torque converter is the output torque of the pump wheel. It can be expressed by equation (1)[5-6].

$$M_B = \lambda_B \gamma n_B^2 D^5 \tag{1}$$

Where D is the effective diameter of the circulating circle of torque converter, D=0.397m; λB is the torque factor; γ is the working liquid density (kg/m3), nBis the pump wheel speed (r/min).

The relationship between the torques and rotational speeds of turbine and pump wheel are shown as equation (2), (3).

$$M_{Ti}=K_i M_{Bi} \tag{2}$$

$$n_{Ti}=i n_{Bi} \tag{3}$$

Where M_T is the turbine torque (N·m); K is the torque coefficient; M_B is the pump torque (N·m);$n_{Ti}$is the turbine output speed (r/min); n_{Bi} is the pump wheel speed (r/min); *i* is the speed ratio.

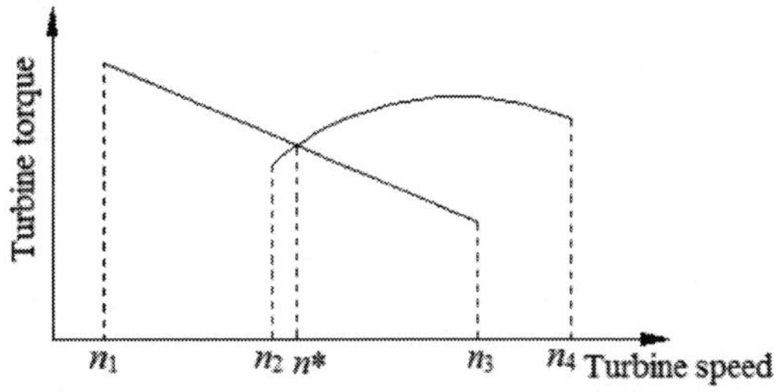

Figure 1. Schematic diagram of double-parameter power lock-up point design

At a certain throttle angle, according to the original data of the engine and formula equation (1), (2), (3), the output curve of the turbine shaft under hydraulic conditions (unlocked) and mechanical conditions (locked)can be calculated under a certain throttle angle. The design of the lock-up point diagram is shown as Figure 1, where $n*$ is the lock-up point.

3 Modeling and Simulation of torque converter lock-up control

According to the lock-up control rule established, the DSP control system of lock-up clutch was simulated and analyzed by adopting the concept of model design. By using CCS and Simulink combined simulation method, the control flow chart was built according to the lock-up rule in Simulink/Stateflow. Then, TI compiler (Mainly CCSIDE integrated development environment) oriented code was generated by Matlab Real-Time-Workshio (RTW). The code compile, link and download were completed, then run the program on the TMS320F28335 development board to realize the simulation system[7-13].

3.1Establishment of latch-up control model

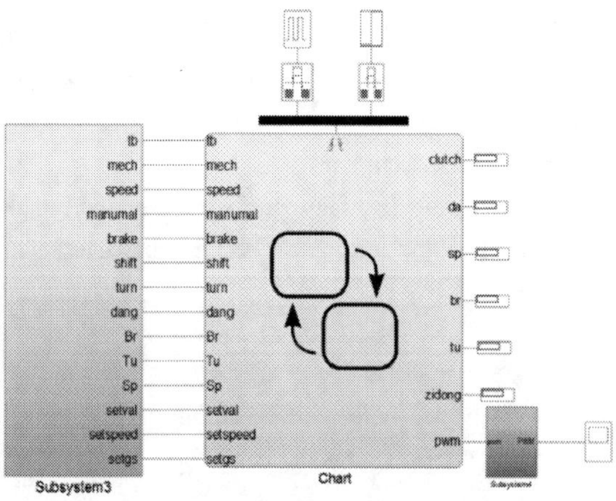

Figure 2. Simulation model of lock-up clutch control

The model of the latch-up control system established in Simulink is shown in Figure 2. Signals such as automatic / manual switch signal, turbine speed signal, shift signal, brake signal, speed signal and turn signal were simulated in Simulink and fed into the lock-up control system in Stateflow. In these signals, the shift signal and the turbine speed signal are the main control signals, while the rest are auxiliary signals. These signals can be set were encapsulated into subsystem Subsystem3. The signals were fed into the logical control in Stateflow/chart, then, according to the signals, the right PWM wave is output by logical judgment. In the output signals, the Clutch of 1means the latching state and the Clutch of 0 means the unlock status. The Zidong of 1 represents the braking state; the Zidong of 0 represents the manual state. The Da, SP, Br, and Tu respectively indicate the initial gear position, speed, braking and turning signals of the vehicle. PWM is the final output duty cycle signal. It acts directly on the latch-up clutch signal valve in the hydraulic system and realizes the control of the combination cavity and the release cavity oil pressure, thus realizes the locking and unlocking control of the clutch.

The Stateflow logic control flow chart inside the Chart is shown in Figure 3.

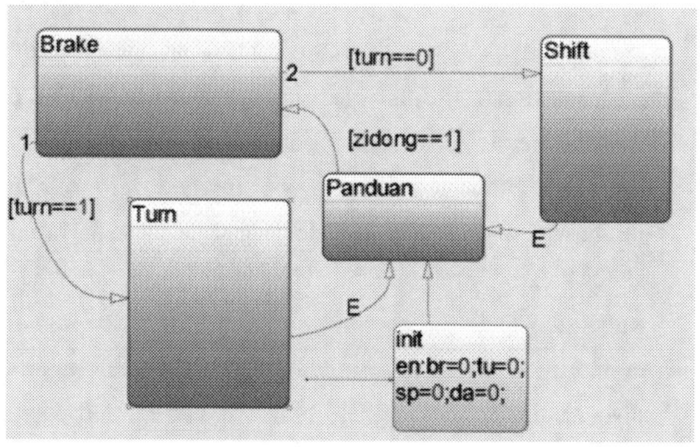

Figure 3. Stateflow logic control flow chart

At the beginning of the operation, the program was initialized in the init state. Then, shift, speed, brake and turn signals were cleared. After the initialization processing, the signal get into the judgment module (Panduan). When the signal Zidong= 1the state shifted, whether the program entered into the next cycle was determined by other signals. If Zidong= 0, it was always in the Panduan state, executed the Panduan module repeatedly. If the condition[zidong==1] was satisfied, the program was transferred to the braking state and the brake logic control was performed. When signal brake= 0, the turn signal was judged. In chart, when the signal turn=1, program would execute steering logic control. If the signal turn=0, the shift signal was judged and the program would perform the program under the condition of a shift signal, or a program without a shift signal.

3.2The hardware of DSP control simulation system

Hardware is mainly ZQ28335 board and XDS100USB emulator developed by Hefei 00 electronic as shown in Figure 4. In the development board, digital tubes were used to display changes in gears, speed, and other signals. The blocking value and gear changes, etc. were set by the matrix keyboard in the development board. The emulator was used to connect the development board and CCS software to realize the co-simulation of hardware and software. First, the environment variables that Simulink and CCS match were set up in the Windows system.Then, the models those consistent with the development board and emulator were selected, and the simulation model was built in Simulink. The interfaces of the chip used in the simulation model were set. Finally, the model in the Simulink was compiled into the development board by emulator, and the co-simulation of the lock-up control system with Simulink and CCS was realized.

Figure 4. Development board and emulator

3.3 Analysis of simulation results

The signal brake= 0 and signal turn= 0were set, and the transmission was placed in first gear, these signals detected by the PIL processor in the loop. The simulation output of the PWM wave is as shown in Figure 5.By changing the duty cycle of PWM wave in real time, the high speed on-off valve in hydraulic system can be controlled, so that the lock-up clutch can be controlled indirectly. As shown in Figure 5, the duty cycle of the output PWM wave is very small. So that the high-speed on-off valve in the hydraulic control system controlled by the PWM wave was off, lock - up clutch was controlled lock. The simulation result satisfied the lock control law.In the former 30s,the amplitude of the wave is zero because the flow chart designed leaves sufficient time for each signal to occur, and also because each trigger signal has a time delay during its production

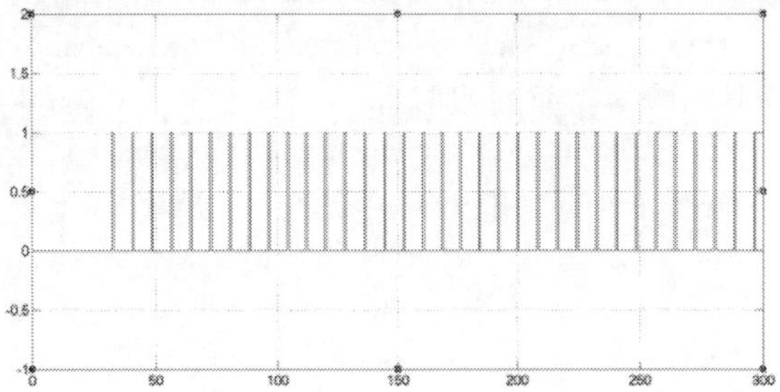

Figure5. The simulation output PWM wave in first gear

In the simulation process, the transmission was set to second gear, throttle angle and turbine speed were changed. In order to facilitate observation of changes of output PWM wave when the locking condition was satisfied and the locking condition was not satisfied in second gear, in the former 150s the throttle angle was set to 35%, the turbine speed was set to 2000rpm, which satisfied the lock condition, and 150 seconds later set throttle angle was set to 15%, the turbine speed was set to 1000rpm, which did not satisfy the locking conditions. These signals detected by the PIL processor in the loop. The simulation output of the PWM wave is as shown in Figure 6. As shown in Figure 6, the former 30sis the transition time for program execution. From 30 s to 150 s, the throttle angle and the turbine speed meet the locking conditions, the duty cycle of the output PWM is 70%.And the PWM would control the clutch locked. 150s later the throttle angle and the turbine speed do not meet the locking conditions, the duty cycle of the output PWM is 30%. And the PWM would control the clutch unlocked. From the above analysis, it can be seen that the simulation results meet the control requirements of the lockup clutch.

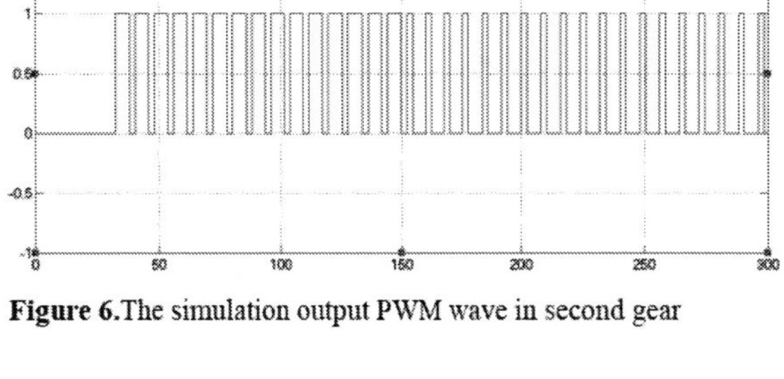

Figure 6. The simulation output PWM wave in second gear

4 Conclusions

According to the established control rule of lock-up clutch of torque converter, the simulation system of lock-up process of lock-up clutch was built by the method of co-simulation of MATLAB/Simulink and CCS, the lock-up processes under the conditions of different gears were simulated, and the simulation results were analyzed. The results show that the established control ruleis reasonable and meet the engineering requirements. It is proved that the PIL simulation method presented in this paper is feasible and effective.This study presents a feasible scheme for the ECU development and the design of other similar control systems.

Acknowledgements

This project is supported by training plan for high-level innovative talent in Guizhou Province (Grant No. Q. K. H. P.T. R. C[2016]5659)，the science and technology major project of Guizhou Province in China (Grant No. Q. K. H. Z. D. Z. X. Z [2013]6015) and the introduction of talent fund of Guizhou University in China (Grant No.G. D. R. J. H. Z[2014]41).

References

1. B. Long, SCI. & TEC. INFO. 34, 37(2007)
2. J.Hu, D.Qin, X.Jing, J. of Chongqing Univ.27,1(2004)
3. S.Yang, H. Yuand J.Liu, J. OF MECH. ENG.50, 152(2014)
4. X.Jiang and D. Qin, Chongqing Univ.22(2014)
5. Z. Hang, S. Zhao and Q. Li, ShanghaiAUTO. 4, 13(2000)
6. L.Yan and H.Huang, Jiling Univ. 15(2005)
7. C.Liu and W. Zhang, Shandong Univ. of Tech. 20(2009)
8. Z.Song and B.Zhang, Hefei Univ. of Tech. 12(2010)
9. D.Zhang and A.Ge, Jilin Univ. 30(2005)
10. Z.Li and Y.Lei, Jilin Univ. 23(2014)
11. E.Hans, Thies Dr, Arthur Wetzel. Drive System Tech.4,1(1998)
12. Kato Yoshiaki, Akasaka Ryokoh and Yamazaki Tsutomu.J.of Japan. Soc. of Trib.42,485 (1997)
13. Q.Chen, D.Qin and X.Ye, J. of Sys. SL.22,669(2010)

Research on Ship Trajectory Tracking with High Precision Based on LOS

Liu Hengzhi[1], Geng Tao[2], Zhang Wei[3]

[1]College of Physics and Electronics, Henan University, Henan Kaifeng 475004, China
[2]College of Physics and Electronics, Henan University, Henan Kaifeng 475004, China
[3]Henan Academy of Science Applied Physics Institute Co.,Ltd,Henan Zhengzhou450000, China

Abstract. Aiming at how precise to track by LOS, a method is proposed. The method combines the advantages of LOS simplicity and intuition, easy parameter setting and good convergence, with the features of GPC softening, multi-step prediction, rolling optimization and excellent controllability and robustness. In order to verify the effectiveness of the method, the method is simulated by Matlab. The simulation's results show that it makes ship tracking highly precise.

1 Introduction

The role of oceans is very important for humans. On the one hand, the ocean is related to the security of a country and safeguarding the inviolability of the sea is an important manifestation of the sovereignty and dignity of a country. On the other hand, the ocean attracts mankind with its enormous resources such as marine living resources, marine energy, marine mineral resources, marine tourism resources, sea water resources , etc. The further development and utilization of marine resources have an increasingly important strategic significance for the long-term sustainable development of one country.

In order to develop marine high and new technology, develop marine resources, promote national economic construction and national defense construction, all kinds of marine vehicles must be vigorously developed. The ship is a very important kind of marine carrier, and the tracking of ship track is an important performance of the ship.

In the field of ship motion control, high precision of ship trajectory tracking control has been a key research. Heading control is the core of ship tracking. As a classical target tracking method, Line Of Sight (LOS) navigation algorithm uses the simple geometric relationship between point and point to judge visibility. It shows the advantages of simplicity and intuition, easy parameter setting and good convergence. The algorithm is independent of the dynamics controller and needs less design parameters. The expected heading's acquisition is only related to the position of ship and the given track, and the expected heading angle can be efficiently obtained.

Previous researchers use PID to control the heading[1-2]. PID control[3]can solve the trajectorytracking problem to some extent. But with actuator's amplitude limited, the integrator is prone to saturation and the heading appears oscillation and overshoot, which leads to low precision tracking and even instability of the system. Generalized Predictive Control (GPC) is a kind of predictive control method and developed with the study of adaptive control[4]. It is with three basic characteristics of model prediction, rolling optimization and feedback correction, also with an excellent control performance and robustness. Besides, on account ofthe use of the traditional parameter model (CARIMA model), the number of parameters is small and estimating parameters online becomes easy.

This paper proposes a method using GPC controller to control the expected heading angle obtained by LOS. The method is that the GPC controller is used to track the expected heading angle of LOS. By predicting the next output of few moments and optimizing the parameters of the controlled object in real time, the steering angle of the ship can be calculated. Finally, the ship can track with high precision.

2 LOS navigation algorithm

LOS navigation is reflected in the intuitive understanding of helmsman's steering and ship motion. The ship's heading command can be calculated by the target position. As long as following the heading command, the ship can keep tracking effectively[5]. InLOS navigation algorithm, ship navigates by calculating LOS angle[6-7].

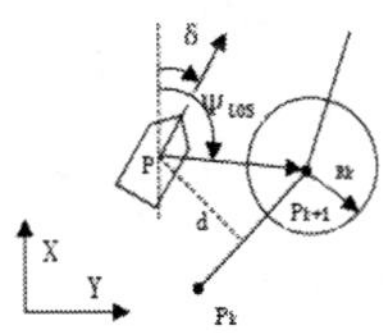

Figure 1. Schematic diagram of LOS navigation algorithm

The angle is taken from positive direction of x axis to the ray on the coordinate plane ,which points from the origin of coordinates to the point (x, y) . The four quadrant tangent calculation is defined, so $\arctan(y, x) \in (-\pi, \pi)$ is the expected heading angle. As shown in Figure 1, the position of LOS is $P_{los} = [x_{los}, y_{los}]$ and the current position of the ship is $P = [x, y]$ in the earth coordinate system. The expected heading is obtained by using the ship's position and expected track, thus the ship's track is controlled indirectly. Then the LOS angle can be calculated by the following formula:

$$\psi_{los} = \frac{\pi}{2} - \arctan\left(\frac{y_{los} - y}{x_{los} - x}\right) \tag{1}$$

It is shown in Figure 2 that the position of the LOS is located between the preceding point (x_k, y_k) and the last road point (x_{k+1}, y_{k+1}) , which represents two adjacent waypoint points. (x, y) represents the ship's current position. Take (x, y) as the center of the circle and R as the radius ($R = nL_{pp}$, the length of ship is L_{pp} , $n = 2 \sim 5$) to draw an arc .When the ship is close to the track section, the arc will intersect the track at two points $A_B(t_i)$, $A_F(t_i)$. The line of LOS is between the ship position (x, y) and the point $A_F(t_i)$. And the expected heading is between the LOS and the horizontal coordinate X . According to the expected heading, tracking the virtual point can make the ship's track converge gradually to the given track. Under the action of the heading controller, the ship gradually converges to the expected track along the expected heading at a given speed.

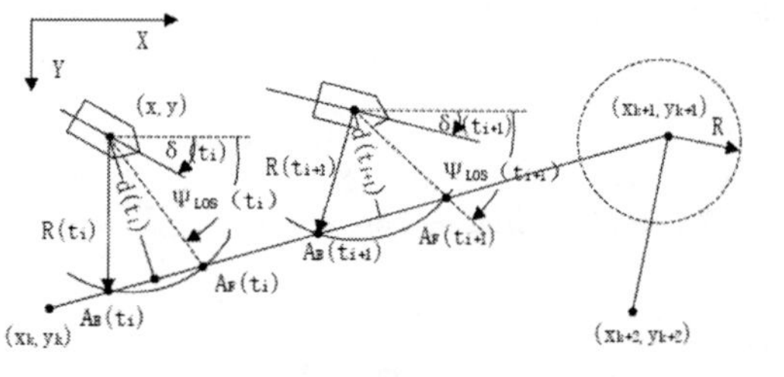

Figure 2. Schematic diagram of LOS navigation algorithm with fixed radius

When the ship has tracked along the current track, it needs a certain update principle to switch the expected track to the next track, so as to achieve the purpose of continuous tracking along the given track. When the controlled ship's position (x, y) satisfies the following relations:

$$(x_{k+1} - x)^2 + (y_{k+1} - y)^2 \le R^2 \tag{2}$$

it is necessary to convert the current expected heading point into the next expected heading point. A new LOS angle is calculated by using the position of the ship and the data of the next expected heading point.

3 High precision heading control

3.1Establishment of ship heading model

As formost of the ship motion and its control problems, the main variable is the movement speed u , traverse speed v , yaw rate r . And the effects of heaving, pitching and rolling motion can be ignored. Therefore, it is considered that ship motion is in plane. It simplifies the problem of ship's motion to only three degrees of freedom[8]. Considering that the ship mass distribution is symmetric in xz plane, $I_{XY} = I_{YZ} = 0$ and $y_G = 0$, the ship motion equation is simplified as follows [9]:

$$\begin{cases} m\left(\dot{u} - rv - x_G r^2\right) = X \\ m\left(\dot{v} + ru - x_G \dot{r}^2\right) = Y \\ I_z \dot{r} + m x_G\left(\dot{v} + ru\right) = N \end{cases} \tag{3}$$

From the literature [10], it can be seen that the upper form can be expressed as the following simplified form:

$$\begin{cases} \dot{\eta} = J(\eta)v \\ M\dot{v} = -C(v)v - D(v)v + \tau + \tau_E \end{cases} \tag{4}$$

Where, $\eta = [x \quad y \quad \psi]^T$, $v = [u \quad v \quad r]^T$. M is inertial parameter matrix , $C(v)$ is Coriolis centripetal force matrix, $D(v)$ Damping parameter matrix. Where $\tau = [\tau_u \quad 0 \quad \tau_r]^T$, τ_u is propulsion, τ_r is yaw torque. Where $\tau_E = [\tau_{uE} \quad \tau_{vE} \quad \tau_{rE}]^T$, τ_{uE} 、 τ_{vE} 、 τ_{rE} are the disturbances of sea acting on the advancing, transversal and yawing respectively.

Assuming that the initial state of a ship is uniform linear motion and the initial values of all variables is zero, the transfer function [10] of rudder angle to heading is as follows:

$$G(s) = \frac{\psi(s)}{\delta(s)} = \frac{K(T_3 s + 1)}{s(T_1 s + 1)(T_2 s + 1)} \tag{5}$$

As a ship is with large inertia, its dynamic characteristics are important in the low frequency range . Therefore, the Nomoto model[11]can be used to describe the upper form, which is to simplify three order model

$$G(s) = \frac{\psi(s)}{\delta(s)} = \frac{K}{s(Ts + 1)} \tag{6}$$

In this paper, the CARIMA model (also called ARIMAX model) is used to describe the Nomo to model:

$$A(z^{-1})\psi(k) = B(z^{-1})\delta(k) \tag{7}$$

where z^{-1} is unit delay operator, $A(z^{-1})$ and $B(z^{-1})$ are polynomial coefficients
where

$$A(z^{-1}) = 1 - (1 + \frac{T}{T + T_s})z^{-1} + \frac{T}{T + T_s}z^{-2},$$

$$B(z^{-1}) = -\frac{KT_s^2}{T + T_s}$$

and T_s is the sampling period.

3.2 GPC control strategy

As shown in Figure 3, the position and rudder angle of the ship are fed back to the LOS algorithm and the GPC controller respectively. The expected headingis calculated by the LOS, then the GPC controller tracks the heading angle to obtain a new command rudder angle. Ships can track continuously by repeating the above operation.

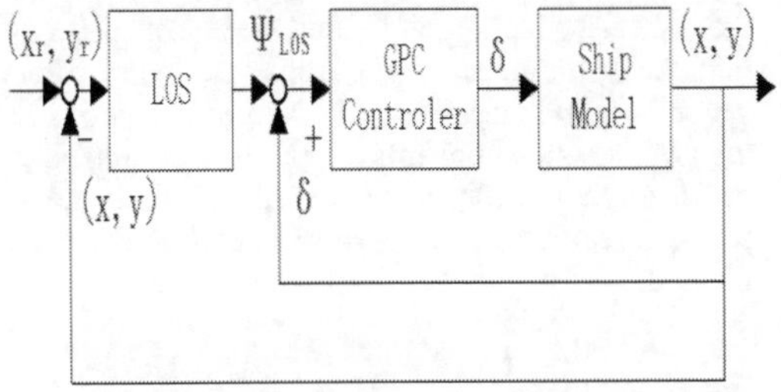

Figure 3. The schematic diagram of the control process

GPC is described by CARIMA model. The GPC algorithm proposed by [12] is used as the control step:

(1)According to the polynomial coefficients $A(z^{-1})$, $B(z^{-1})$ in the ship heading model, the predicted output is obtained as following:

$$\psi(t+1)=\sum_{i=1}^{n} a_{1,i}\psi(t+1-i)+\sum_{i=0}^{m} b_{1,i}\Delta\delta(t-d-i) + \sum_{i=0}^{r} c_{1,i}e(t+1-i) \tag{8}$$

here ψ is the heading, δ is the rudder angle, e is the external white noise, and $d+1$ is the pure delay of the system.

(2)The prediction output of p step's heading is calculated by lower recursion.Where ψ_m is form vector:

$$\psi_m(t+k)=\sum_{i=1}^{n} a_{1,i}\psi_m(t+k-i) + \sum_{i=1}^{m} b_{1,i}\Delta\delta(t+k-d-1-i) + \sum_{i=1}^{r} c_{1,i}e(t+k-i) \tag{9}$$

where $k=1,\cdots,p$,and p is prediction step size

$$\Delta\delta(t+i)=\begin{cases}0, & i\ge 0\\ \Delta\delta(t+i), i<0\end{cases},$$

$$e(t+i)=\begin{cases}0, & i>0\\ e(t+i), i\le 0\end{cases}$$

(3)The actual operation requires that the heading approximates to the first-order inertial system and gradually approaches the set point ψ_r . Considering the delay of the system, the $p-d$ step reference trajectory is computed recursively by the lower form, ψ_r is form vector, α is the diffusion coefficient,

$$\begin{cases}\psi_r(t+d)=\psi_m(t+d)\\ \psi_r(t+d+j)=\alpha\psi_m(t+d+j-1)+(1-\alpha)\psi_s\end{cases} \tag{10}$$

(4) The deviation of the rudder angle should be as small as possible in the heading keeping, and the rudder angle should not be changed too frequently.Therefore, objective function[13-14]is as following:

$$J=\min\{(\psi_r-\psi)^T(\psi_r-\psi)+\lambda\Delta\delta^T\Delta\delta\} \tag{11}$$

here λ is weighted coefficient. The command angle of the t moment is obtained by minimizing the objective function,

$$\delta(k)=\delta(k-1)+\Delta\delta(k) \tag{12}$$

where

$$\Delta\delta(k)=(1,0,\cdots,0)(G^TG+\lambda I)^{-1}G^T(\psi_r-\psi_m)$$

Where

$$G=\begin{bmatrix} b_{1,0} & & & 0\\ b_{2,0} & b_{1,0} & & \\ \cdots & \cdots & \ddots & \\ b_{p-d,0} & b_{p-d-1,0} & \cdots & b_{1,0}\end{bmatrix}$$

and

$$b_{k,0}=b_{1,k-1}+\sum_{j=1}^{k_1} a_{1,j}b_{k-j,0},\quad k=2,3,\cdots;\quad k_1=\min\{k-1,n\}$$

4 Simulation and verification

In simulation, the mariner model in GNC is used [15]. The length of ship is L=160. The initial parameters are set. n=5, the speed of ship is 20 kn , the speed of rudder angle is $5°/s$ .The track is set from starting point 1 (0,0), to the turning point 2 (3000,3000), to the turning point 3 (6000,2000), to the turning point 4 (8000,8000), to the turning point 5 (4000,6000), to the arrival point 6 (2000,8000). Prediction step size is $p=10$, and the sampling period is $T_s=2$.

The simulation results are shown in Figure 4-6. Figure 4 is the expected track and the simulation track, in which the dashed line represents the expected track, and the solid part represents the simulation track. It can be seen from the image that the expected track is consistent with the simulated track. As is shown in Figure 5, there are four corners in the simulation process. The first, second, fourth turns is $\le 90°$ and the third turns is $>90°$. According to Figure 4, it can be seen that the ship can track well at the first, second and fourth turn; the third expected heading angle changes greatly, which makes the heading's change of simulation delay greatly, but the change of expected track and simulation track is consistent. All angles on the trajectory can converge to the expected heading angle.

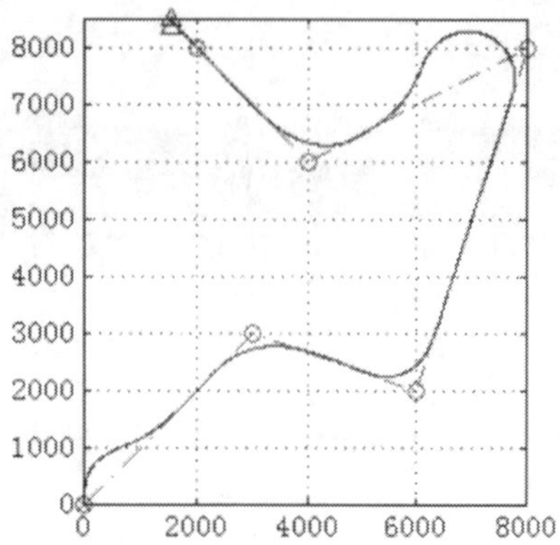

Figure 4. The expected track and the simulation track

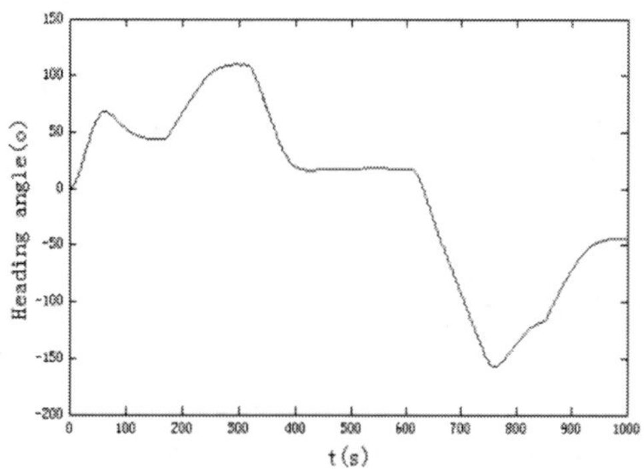

Figure 5. Heading angle curve

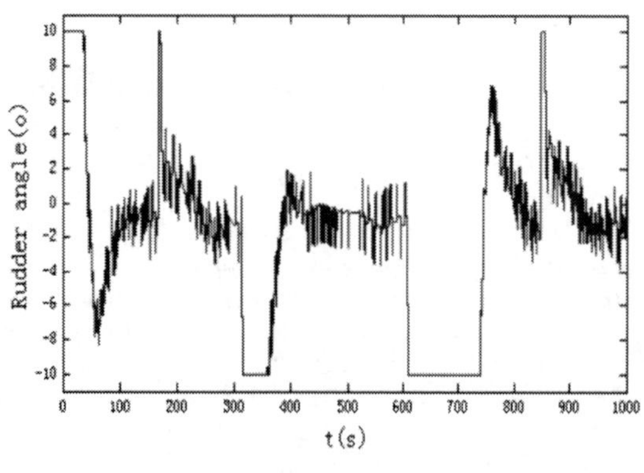

Figure 6. Rudder angle curve

In the straight path, two tracks almost coincide. In the case of small turning, the overshoot is very small, because the GPC has the characteristics of multi-step prediction, rolling optimization and feedback correction. In the case of large turning, the overshoot appears when switching is large. But it is in line with the actual situation when turning. The actual track is convergence to the expected track. Therefore, this method can make ship trajectory tracking highly precise.

5 Conclusion

This paper studies the ship's track tracking, according to this method, can achieve high-precision track tracking.

In this paper, LOS navigation algorithm and high precision heading tracking algorithm based on GPC are studied. LOS navigation algorithm can imitate the human visualization according to the setting. By means of judging the distance , the expected point is tracked, thus the expected heading angle can be obtained. The heading tracking algorithm designed in this paper can track well, and has good dynamic and static characteristics. Through theoretical deduction and simulation results, it is proved that the method can make ship track very effectively. Therefore, this research will make contributions to some insights and implication both theoretically and practically.Some physical experiments will be done in the future research to further verify the effectiveness of this method.

References

1. TIAN Yong,WANG Dan,PENG Zhou-hua,LIU Lu, Journal of Dalian Maritime University, 41,(4):14-18,(2015)
2. HAN Peng,ZHOU Ze-cai,TANG Hao, Applied Science and Technology,44,(4):1-5, (2017)
3. ZENG Zhi-gang,YANG hai,HUANG Wang-jun, Research of Mobile Robot Navigation with Self-Adaptive Filtering and Fuzzy PID,22,(05):953-957,(2015)
4. PENG Yan,WU Wei-qing,LIU Mei,USV Traching Control Based on Cascade GPC-PID,21,(5):245-248,(2015)
5. Fossen T I, Breivik M, Skjetne R, University of Science and Technology(NTNU), NO.-7491,(2003)
6. Khaled N, Chalhoub N G, Nonlinear Dynamics, 73,(1-2):897-906, (2013)
7. Oh S R, Sun J, Ocean Engineering, 37,(2-3):289-295(2010)
8. LIU Yang, LIU Mei-jie, Leader/Follower Formation Control of Underactuated Surface Ships, 20, (5):980-983,(2013)
9. Jia Xinle,Yang Yansheng, Chuan Bo Yun Dong Shu Xue Mo Xing-Ji LiJian Mo Yu Bian Shi Jian Mo(in chinese) ,(1999)
10. Zhang Y, Hearn G E, Sen P, Oceanic Engineering IEEE Journal of, 21,(4):513-527 (1996)
11. Nomoto K, Taguchi K, Honda K, et al.International Shipbuilding Progress, 4,(35):354-370 (2009)
12. Jin Yuanyu,Gu Xingyuan.Gai Jin De Guang Yi Yu Ce Kong Zhi Suan Fa(in chinese),19, （3） :8-13 (1990)
13. HU Yao-hua, XU Su-wu, Control design for ship tracking based on prediction of desired headings,27,(12):1641-1646,(2010)
14. HE Pei,QU Xiangju,Wu Zhe,Study of optimization method for flight path following,29,(7):598-601,(2003)
15. Perez T, Fossen T I,Modeling Identification & Control, 28,(1):19-30(12) (2007)

Modeling and Analysis of a Closed-Loop System for High-Q MEMS Accelerometer Sensor

Yalin Wang and Yongjun Yang

Hebei Semiconductor Research Institute, 050057 Shijiazhuang, China

Abstract. High-Q sensing element is desirable for high performance while makes the loop control a great challenge. This paper presents a closed-loop system for high-Q capacitive MEMS accelerometer which has achieved loop control effectively. The proportional-derivative(PD)control is developed in the system to improve the system stability. In addition, pulse width modulation (PWM) electrostatic force feedback is designed in the loop to overcome the nonlinearity. Furthermore, a sigma-delta (ΣΔ) modulator with noise shaping is built to realize digital output. System model is built in Matlab/Simulink. The simulation results indicate that equivalent Q value is reduced to 1.5 to ensure stability and responsiveness of the system. The effective number of bits of system output is 14.7 bits. The system nonlinearity is less than 5‰. The equivalent linear model including main noise factors is built, and then a complete theory of noise and linearity analysis is established to contribute to common MEMS accelerometer research.

1 Introduction

The capacitive microaccelerometer with high measurement precision, stable output, low temperature drift, is one of the most widely used MEMS sensors. High performance MEMS accelerometer with high-Q, low Brownian noise, are increasingly needed in oil exploration, inertial navigation and guidance, micro-gravity measurement in space, earthquake prediction and other fields.

Traditional MEMS accelerometer system research is concentrated in open-loop with analog output[1]. Recently more research on high-Q MEMS accelerometer is focused on the closed loop system, mainly divided into analog and digital closed-loop[2-4]. Closed-loop with analog output makes quality block movement near the equilibrium position improving the linearity[5]. But it is easily affected by the precision and noise of circuit meanwhile it is not conducive to the subsequent signal processing because of analog output. Another high-order ΣΔ loop system realizes low power consumption, low circuit complexity and digital output. But high order system lead to instability and coupling problem between mechanical and circuit[6]. This work will present a system which effectively reduces these unfavorable factors mentioned above. Moreover the theory on noise and linearity will contribute to the MEMS system research.

In this paper, we study a digital closed-loop system for high-Q MEMS accelerometer sensor which combines PD controller with ΣΔ modulator. The structure is organized as follows: Section 2 presents an electrical model of MEMS sensitive structure, and designs a PD control loop. Section 3 puts forward a new structure of ΣΔ modulator, and simulates the system model. Section 4 establishes the equivalent linear model, puts forward a theory of noise and linearity analysis. Finally, Section 5 concludes the paper and points out further work.

2 MEMS model and PD control loop

2.1 The system-level design

The system-level block diagram is shown in Fig.1.

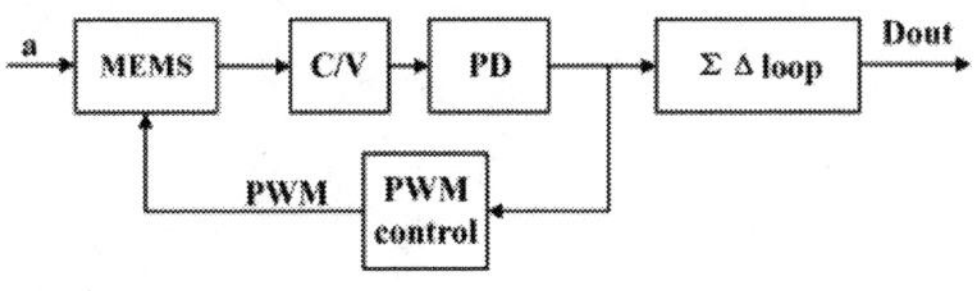

Figure 1. System-level block diagram

The system consists of MEMS mechanical structure, capacitor voltage conversion circuit, PD control loop with PWM feedback and ΣΔ modulator. The external acceleration signal causes a displacement of the MEMS sensitive structure's mass. The displacement causes the change of differential capacitance and converts to analog voltage. PD control loop ensures the stability and linearity of the system. Analog voltage is converted to digital signal by ΣΔ modulator and quantization noise is shaped to decrease the system noise level.

2.2 Electrical model and working principle of the MEMS sensitive structure

Different from traditional sandwich structure, the MEMS sensitive structure studied in this paper is the comb type, as showed in Fig.2.

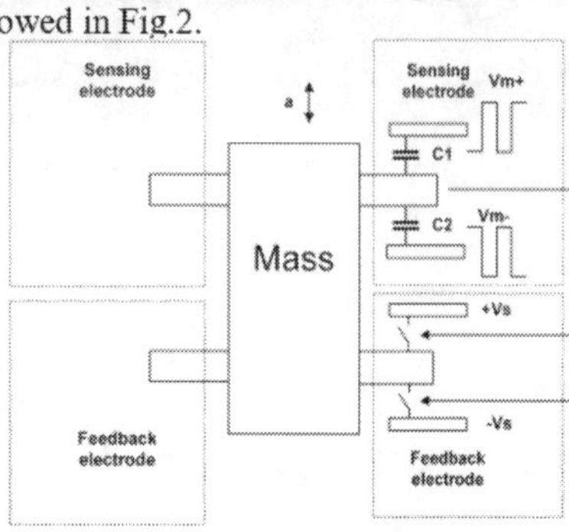

Figure 2. The simplified model of MEMS sensor

It consists of movable mass and fixed electrode with bilateral symmetrical structure. Sensing electrode is connected to front-end circuit realizing capacitor-voltage conversion (C/V). Feedback electrode is connected to feedback voltage. The device of MEMS accelerometer can be modeled as a 2nd order spring-damping system with external acceleration input. The equation of motion can be written as:

$$m\frac{d^2x}{dt^2}+b\frac{dx}{dt}+kx=ma(t) \quad (1)$$

Where m is the proof mass, x is the displacement of the mass, b is the damping coefficient, k is the spring coefficient. The acceleration-to-displacement transfer function is written as:

$$M(s)=\frac{X(s)}{A(s)}=\frac{1}{s^2+s\frac{b}{m}+\frac{k}{m}}=\frac{1}{s^2+s\frac{\omega_0}{Q}+\omega_0{}^2} \quad (2)$$

Where ω0 is the resonant frequency, and Q is the quality factor.

The mass moves near the balance position to ensure the system work in a linear state[7]. The PD closed-loop as follows will realize this condition. When PD control is added,the system transfer function is given as:

$$M'(s)=\frac{M(s)K_{C/V}}{1+M(s)K_{C/V}Kfb(Kp+K_d s)} \quad (3)$$

Where $K_{C/V}$ is the capacitance to voltage conversion factor, K_{fb} is the voltage to force conversion factor, K_p is proportional coefficient and K_d is differential coefficient.

The MEMS electrical model with PD control is shown in Fig.3. The equivalent Q(Q_E) and resonant frequency ω of system is given as:

$$Q_E=\frac{\sqrt{m(k+KpKfbK_{C/V}}}{b+KdKfbK_{C/V}} \quad (4)$$

$$\omega=\sqrt{\frac{k+KpKfbK_{C/V}}{m}} \quad (5)$$

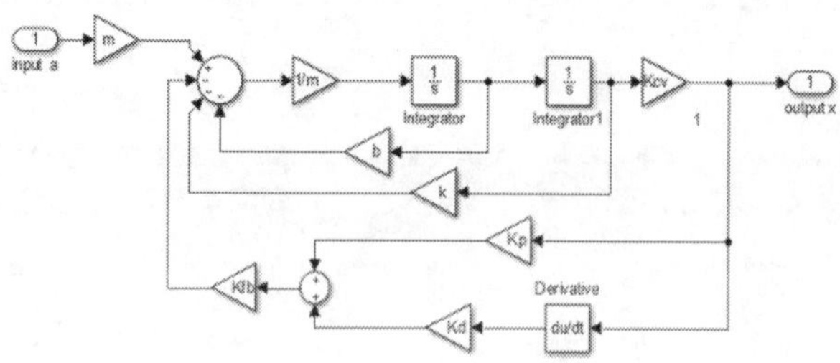

Figure 3. The Simulink model of closed-loop MEMS with PD control

The optimal system is presented by setting appropriate parameters. According to the bode diagram (Fig.4) and square input response(Fig.5), the Q_E is 1.5 while open loop's Q is 50.2, and ω is up to 2200Hz while ω_0 is 1050Hz. The PD control improved the stability by reducing the quality factor and broadened bandwidth of system.

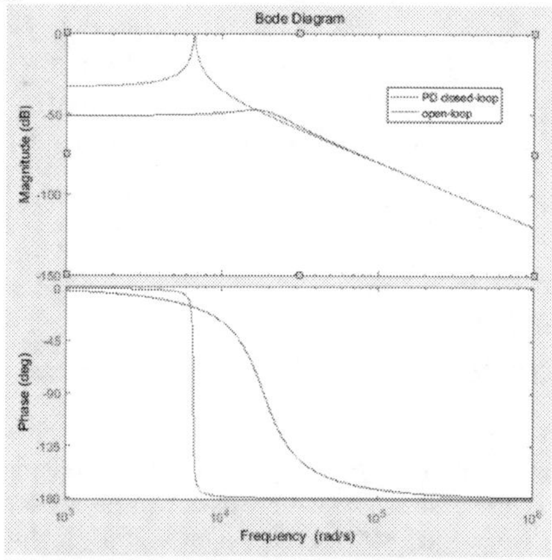

Figure 4. The system bode diagram

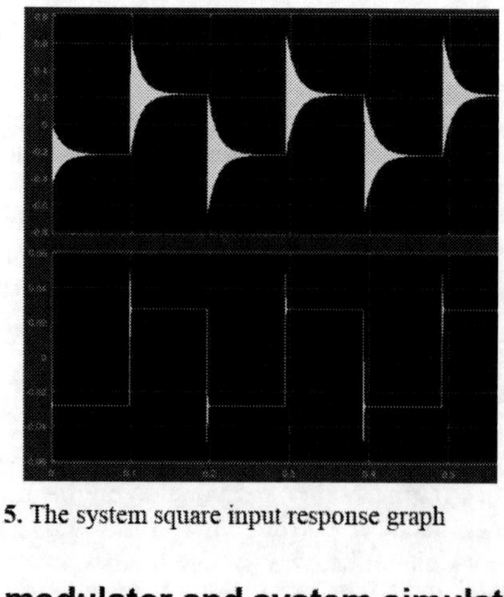

Figure 5. The system square input response graph

3 ΣΔ modulator and system simulation

The single loop ΣΔ modulators are sorted into two typical architectures called feed-forward (FF) and multiple-feedback (MF) architecture. The FF architecture has definite zeros and higher low-frequency gain than MF to achieve high linearity. But it has worse stability than MF.

 https://doi.org/10.1051/matecconf/201816005008

Meanwhile there are adjustable poles in the MF architecture to ensure stability, but it is poor in linearity[8]. Based on common ΣΔ modulators [9], the architecture in this paper as shown in Fig. 6, combined the FF with the MF with both comprehensive advantage: stability and linearity. There is phase delay in the system(Fig.5), the front-end phase compensator for ΣΔ modulator is necessary.

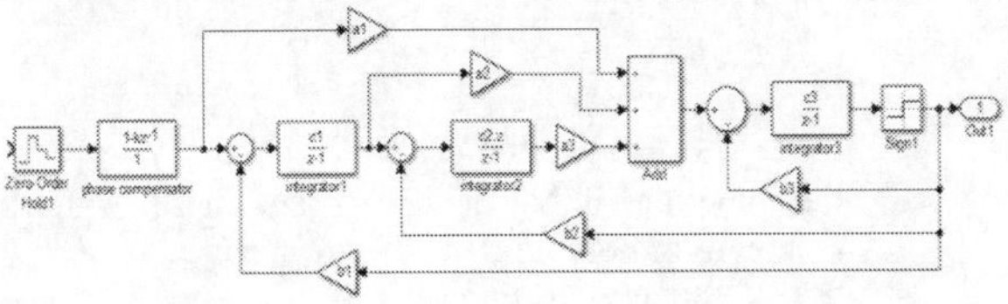

Figure 6. Architecture of the ΣΔ modulator

Where k=0.25 to provide appropriate phase compensation. The optimal value of other parameters are determined by the scanning parameter method: a1=0.4, a2=0.3, a3=0.3, b1=0.3, b2=0.3, b3=0.5, c1=0.5,c2=0.3,c3=0.4. The transfer function is given as:

$$H=\frac{0.16z^4-0.402z^3+0.3673z^2-0.1387z+0.016}{z^5-3.589z^4+4.844z^3-2.908z^2+0.6544z} \quad (6)$$

When input signal is a sine wave with a magnitude of 10 and a frequency of 80 Hz, oversampling ratio is 256, the simulation results are showed in Fig. 7.

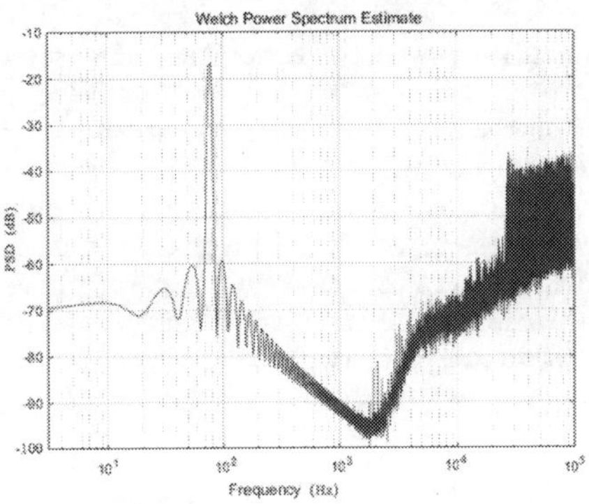

Figure 7. Output signal power spectral density

The simulation results show that the signal-to-noise ratio(SNR) is 90dB, the effective number of bits is 14.7 bits.

4 Linearity and noise analysis

4.1 Linearity analysis

The total nonlinearity of a closed-loop system is affected by the both forward signal path and feedback path[10]. Because the displacement is near the balance place by using PD control, the forward path nonlinearity is significantly depressed. The feedback path nonlinearity will dominate. Nonlinearity in feedback path is mainly composed of two parts: the mismatch between the actuation capacitors; and the non-zero proof mass displacement due to the finite loop gain. In [10],feedback force is given as:

$$F_{fb}(V_{fb})=\frac{T_{fb}}{T_s}\left[\frac{\varepsilon A(V_p-V_{fb})^2}{2(d_2+x)^2}-\frac{\varepsilon A(V_{fb}-V_n)^2}{2(d_1-x)^2}\right] \quad (7)$$

The square law relationship between the electrostatic feedback force and the applied voltage leads to relatively poor nonlinearity up to several percent. In this work, time-control(pulse width modulation force feedback) rather than traditional voltage-control actuation is developed. The PWM control principle is shown in Fig.8.

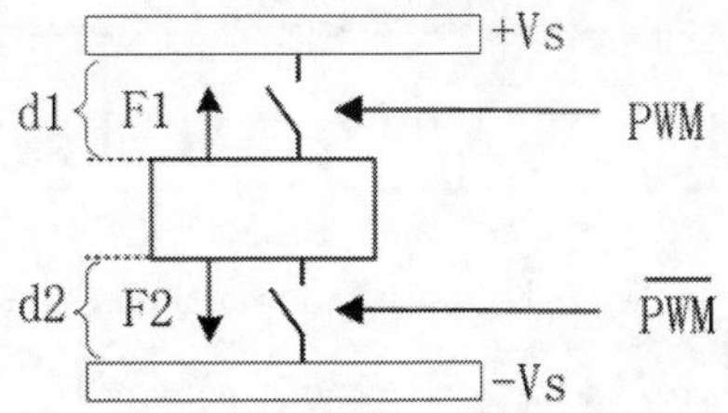

Figure 8. PWM control principle diagram

PWM signal controls the switch between movable mass and electrode with the voltage(+Vs&-Vs). The feedback force F_{fb} is proportional to the pulse width of the PWM signal(Φ). The feedback force is given as:

$$F_{fb}=F1-F2=\frac{T_s-\Phi}{T_s}F1-\frac{\Phi}{T_s}F2 \quad (8)$$

Where T_s is the sampling time, $F1$, $F2$ is the electrostatic force as follows:

$$F1=\frac{\varepsilon\varepsilon_0 A(-V_s-V_s)^2}{2d_1^2} \quad (9)$$

$$F2=\frac{\varepsilon\varepsilon_0 A(V_s+V_s)^2}{2d_2^2} \quad (10)$$

Where ε_0 is the dielectric constant in a vacuum, ε is the dielectric constant between mass and plate electrode, d_1 is the distance between upper plate electrode and d_2 is the distance between under plate electrode.

Equation (8)~(10) shows that mismatch between the actuation capacitors ($F1 \neq F2$) in balance state is no longer the system nonlinearity factor, while displacement will still lead to the nonlinearity , thus an applicable loop gain is needed for linearity. A verilog-A model of the MEMS accelerometer was developed in Cadence. The nonlinearity (NL) VS mismatch and the loop gain is given as figure 9.

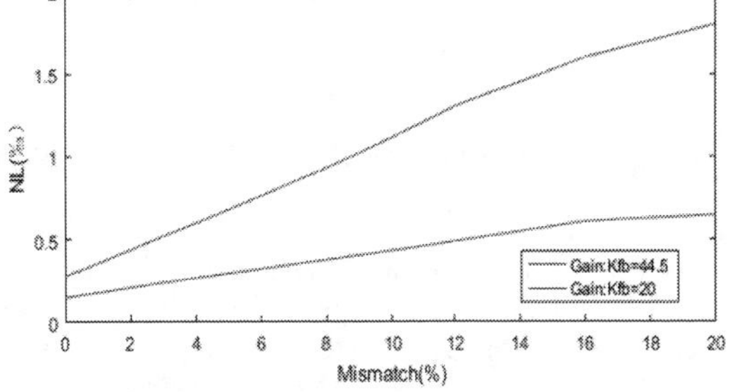

Figure 9. Nonlinearity (NL) VS mismatch and loop gain

The simulation result indicates that mismatch increases to 20, NL is changed about 0.4‰ when the loop gain is 44.5.

4.2 Noise analysis

The system noise consists of mechanical noise(Brownian noise), charge amplifier noise, quantization noise, integrator noise. Based on noise factors, the equivalent linear model is given as Fig.10

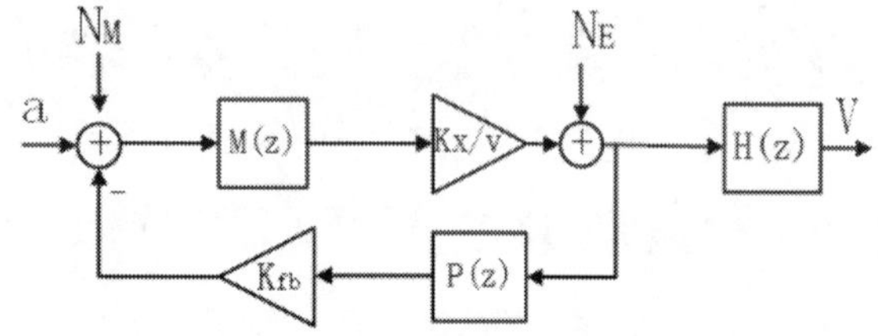

Figure 10. System equivalent linear model with noise factors

Where NM is mechanical noise, NA is charge amplifier noise, P(z) is PD controller transfer function. H(z) is ΣΔ modulator, the noise model is showed in Fig. 11.

It is helpful to get the noise transfer function to evaluate the noise impact on the system noise

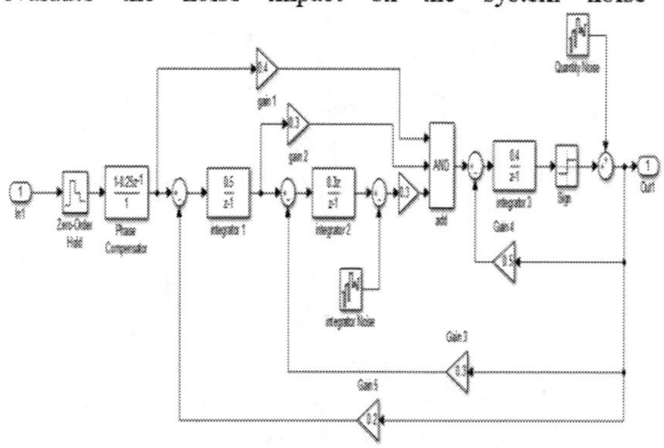

Figure 11. ΣΔ modulator structure with noise factors

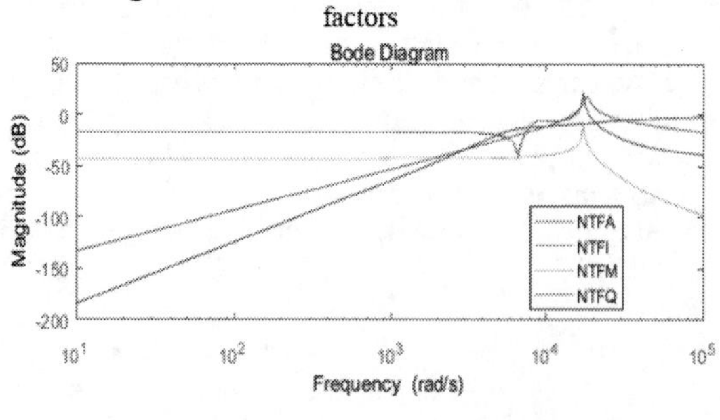

Figure 12. Simulated frequency responses of the noise transfer functions

Where K_b is Boltzmann constant($1.38\times10^{-23}J/K$), T is Kelvin temperature, m is the proof mass, b is the damping factor. Because of vacuum packaging, b is small, leading to small mechanical noise. The mechanical noise of performance. The noise transfer function (NTF) can be written as:

$$NTFM=\frac{K_{X/V}M(z)H(z)}{1+K_{fb}K_{X/V}M(z)P(z)} \tag{11}$$

$$NTFA=\frac{H(z)}{1+K_{X/V}K_{fb}M(z)P(z)} \tag{12}$$

$$NTFI=\frac{0.12z^3-0.336z^2+0.312z-0.096}{z^4-3.589z^3+4.489z^2-2.918z+0.6587} \tag{13}$$

$$NTFQ=\frac{z^3-3z^2+3z-1}{z^3-3.4z^2+3.782z-1.384} \tag{14}$$

where $NTFM, NTFA, NTFI$, $NTFQ$ respectively are the transfer function of mechanical noise, charge amplifier noise, integrator noise and quantization noise.

The frequency responses of the noise transfer functions are plotted in Fig .12.

At low frequency (f<10000rad/s) as shown in Fig.12, $NTFA > NTFM > NTFI > NTFQ$, thus charge amplifier noise and mechanical noise are the main noise source. Mechanical noise is caused by the interaction of the air molecules surrounding called Brownian movement.It is a white noise, and it's power spectrum is defined as follows :

$$a_n{}^2=\frac{4K_bTb}{m^2} \tag{15}$$

MEMS researched in this paper is nearly 20 $ng/\sqrt{Hz}$. Compared with other noise, it is almost negligible. The charge amplifier noise becomes dominant noise source.

Figure 13 describes the front-end interface circuit including charge amplifier and MEMS sensor. Vm is the reference voltage of MEMS sensor, Vcom is the common mode voltage of the amplifier, and Vx is the output of amplifier. When an acceleration comes, the mass will change its displacement, and the charge amplifier will convert the capacitance variation to voltage. When mass is balanced without extra acceleration:

$$C0=\frac{\varepsilon\varepsilon_0 A}{d_0} \tag{16}$$

When input acceleration lead to a displacement x,

$$C1=\frac{\varepsilon\varepsilon_0 A}{d_0-x} \tag{17}$$

$$C1=\frac{\varepsilon\varepsilon_0 A}{d_0+x} \tag{18}$$

$$\begin{aligned}\Delta C=C1-C2&=C0\left(\frac{d_0}{d_0-x}-\frac{d_0}{d_0+x}\right)\\&=C0\left[\sum_{n=0}^{\infty}\left(\frac{x}{d_0}\right)^n-\sum_{n=0}^{\infty}\left(-\frac{x}{d_0}\right)^n\right]\\&=2C0\left[\frac{x}{d_0}+\left(\frac{x}{d_0}\right)^3+\left(\frac{x}{d_0}\right)^5+\dots\right]\\&\approx 2C0\frac{x}{d_0},(x<<d_0)\end{aligned} \tag{19}$$

In $\phi 1$ and $\phi 2$, the charges stored in different capacitors are written as:

$$Q_{C1}(\phi 1)=(V_m-V_{com})*C1 \tag{20}$$

$$Q_{C2}(\phi1)=(V_m+V_{com})*C2 \quad (21)$$

$$Q_{Cf}(\phi1)=V_{com}*Cf \quad (22)$$

$$Q_{C1}(\varphi2)=(-V_m-V_{com})*C1 \quad (23)$$

$$Q_{C2}(\varphi2)=(V_{com}-V_m)*C2 \quad (24)$$

$$Q_{Cf}(\phi2)=(V_{com}-V_x)*Cf \quad (25)$$

When ph1 changes to ph2,s charges change of the feedback capacitor Cf is given as:

$$\Delta Q_f=\Delta Q_f(\phi1)-\Delta Q_f(\phi2)=2V_m*(C1-C2)=V_xCf \quad (26)$$

Thus

$$V_x=2V_m\frac{C1-C2}{Cf}=4V_mC0\frac{x}{d_0*Cf} \quad (27)$$

The charge amplifier noise mainly consists of operation amplifier noise, flicker noise, reference voltage noise and switch noise[11]. The flicker noise which be reduced by correlated double sampling technique will no longer be the main noise in this paper. To calculate the noise, the charge amplifier schematic can be simplified as Fig. 14.

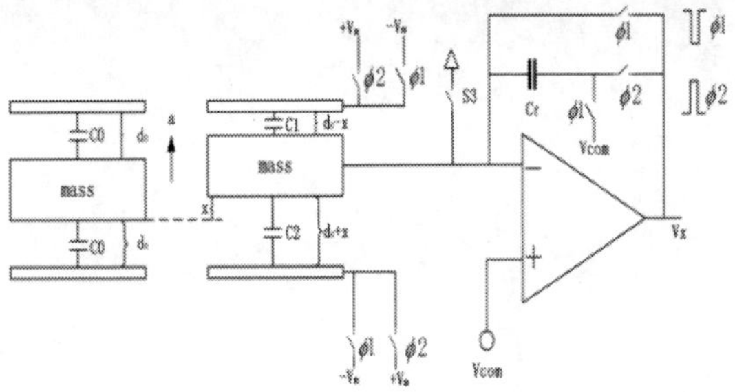

Figure 13. Front-end interface circuit

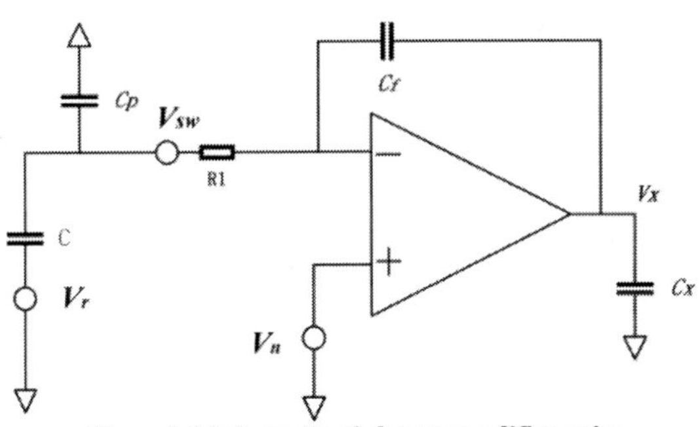

Figure 14. Schematic of charge amplifier noise

The operation amplifier noise (V_n),the switch noise (V_{sw}) and the reference voltage noise (V_r) are written as:

$$V_n=\sqrt{\frac{16K_bT(C+Cp)}{3f_sCxCf}} \quad (28)$$

$$V_{sw}=\frac{1}{Cf}\sqrt{\frac{2K_bTR1g_mC}{f_s}} \quad (29)$$

$$V_r=\frac{V_{rf}}{Cf}\sqrt{\frac{g_mC}{f_s}} \quad (30)$$

Where K_b is Boltzmann constant($1.38\times10^{-23}J/K$), f_s is the sampling frequency, Cx is the load capacitance, Cf is the feedback capacitance, C is the sensing capacitance, and Cp is the parasitic capacitance, V_{rf} is the reference voltage noise at low-frequency (assumed to be white noise with an in-band spectral density of 8 $nV/\sqrt{Hz}$), $R1$ is the equivalent resistance of the switch, and g_m is the trans-conductance of the input transistor of the charge amplifier.

According to the design parameters, three main types of noise are calculated. In this paper, $f_s=256KHz$, $Cx=10PF$, $Cf=12PF$, $C=280PF$, $Cp=15PF$, $g_m=3ms$, $R1=180\Omega$. Thus, $V_n=0.4253\mu g/\sqrt{Hz}$, $V_{sw}=0.1702\mu g/\sqrt{Hz}$, $V_r=1.2076\mu g/\sqrt{Hz}$. The switch noise has less impact on system than operation amplifier noise and reference voltage noise. The reference voltage noise should be reduced firstly in the design of system-level low noise optimization.

5 Conclusions

In this paper, we establish a closed-loop system for high-Q MEMS accelerometer sensor, clarify the equivalent electrical model of MEMS accelerometer sensor and design a PD closed-loop with proper coefficient to improve the stability of the system. Then we design a new structure of sigma-delta modulator with good stability and high linearity to realize digital output. Finally, we put forward a linearity analysis theory, use PWM technology in q feedback loop to improve linearity and build a compete noise analysis theory which will provide the theoretical basis for the latter circuit design and MEMS accelerometer sensor system research. The future research work includes the low-power interface circuit design and function test after fabrication.

References

1. B Borovicl, et al, "Open-loop versus closed-loop control of MEMS devices choices and issues", Journal of Micromechanics and micorengineering, pp1917－1924,2005
2. Zhaohui Song, Teng Sun, and Jian Wu, "System-level simulation and implementation for a high Q capacitive accelerometer with PD feedback compensation", Microsyst Technol, vol. pp. 2233–2240, 2015.
3. Yin Liang, et al, "High resolution interface circuit for closed-loop accelerometer", Journal of Semiconductors, vol. 32, pp.(045005)1-8, 2011
4. Yuntao Liu et al, "Stability Analysis in High-Order Electromechanical Sigma-Delta Accelerometer", Key Engineering Materials, vol 483, pp 716-720, 2011
5. Bin Wang, et al, "Model of Closed-loop Detection System for Capacitive MEMS Accelerometer and its IC Realization", Advanced Materials Research, vol 468, pp 2170-2175, 2012
6. Shaodong Chen, Honglin Xu, "Design analysis of a high-Q micromechanical capacitive accelerometer

system", IEICE Electronics Express, vol.14, pp.1-6, 2017

7. Mingjun MA, Zhong He, and Hui Jie, "A combined modulated feedback and temperature compensation approach to improve bias drift of a closed-loop MEMS capacitive accelerometer", Frontiers of Information Technology & Electronic Engineering, vol.16, pp.497-510, 2015
8. Zhang Meimei, Quan Haiyang and Zhang Fuqiang, "Comparisons of Feed-Forward and Multiple-Feedback Sigma-Delta Modulators for MEMS Accelerometers", MATEC Web of Conferences, 08003,2016
9. Zhimin Li, Terri S. Fiez, "A 14 Bit Continuous-Time Delta-Sigma A/D", IEEE Journal of Solid-State Circuits, vol. 42, pp.1873-1883, 2007.
10. Y.F.Dong, Kraft.M, "Force feedback linearization for higher-order electromechanical sigma-delta modulators, Journal of Micromechanics and micorengineering", vol.16, pp.54-60, 2006
11. H. Kulah, J. Chae, and K. Najafi, "Noise analysis and characterization of a sigma-delta capacitive microaccelerometer", IEEE Journal of Solid-State Circuits, vol.41, pp.352－361, 2006

Reconfigurable Fault Tolerant Control for Spacecraft Based on Modified IMM Algorithm

Li Xiaoyun[1], Du Wei[2]

[1]*Beijing Institute of Spacecraft System Engineering, Beijing, China*
[2]*Beijing Institute of Control Engineering, Beijing, China*

Abstract. This paper proposes an integrated fault detection, diagnosis, and reconfigurable control method for attitude tracking of a spacecraft. A novel IMM algorithm, based on the unscented Kalman filter and an index related to the closed-loop system performance, is presented to detect and diagnose the faults. To achieve steady attitude tracking, the sliding mode variable structure controller is designed. When a fault is detected and isolated, the controller structure is reconfigured to compensate the faulty system to maintain the system performance. A simulation example evaluating the attitude tracking process is employed, which demonstrates the efficiency of the proposed approach.

1 Introduction

Modern engineering systems are becoming more and more sophisticated. The requirements for their reliability, availability, and security also grow significantly. In the spacecraft attitude control system, failures of its actuator or sensor may cause serious problems and need to be detected and isolated as soon and as accurately as possible. However, due to the limit of mass, cost and other factors, spacecraft cannot be configured enough measurement devices to detect faults. So the question of how to maximize the use of analytical redundancy to diagnose and isolate faults is urged to be solved in practical projects.

A fault tolerant control system is a control system that possesses the ability to accommodate system failures automatically, and to maintain overall system stability and acceptable performance. Generally speaking, an active fault tolerant control system compensates for faults either by selecting a precomputed controller or by synthesizing a new control strategy on-line. Reconfigurable fault tolerant control as an active fault tolerant control method, fault detection and diagnosis (FDD) is its prerequisite.

In this paper, FDD is carried out using interacting multiple-model (IMM) algorithm in that the system model sequence is assumed to be a Markov chain with transition probabilities, which has superiority in fault diagnosis and state estimation for system involving structural as well as parametric changes. IMM algorithm has been used in maneuvering target tracking [1], FDD of dynamic systems [2], and so on.

In the traditional IMM algorithm, the extended Kalman filter (EKF) is the estimation unit and based on linearization technique can only achieve first-order accuracy. While the unscented Kalman filter (UKF) is based on the unscented transform technique. The state variable is represented by a minimal set of carefully chosen sample points, which are able to approximate the posterior mean and covariance of the output variable with second-order accuracy [3, 4]. Further, it is not necessary to compute explicitly the Jacobians in the UKF.

In the meanwhile, in the traditional IMM algorithm, model probabilities are used to provide an indication of the model in effect at a given time. However, when a control action is introduced and combined with a multiple-model estimator to form a closed-loop system, the effect of the feedback will force the residuals from different models to be similar and this affects the discrimination property of the filter. Such a situation is even more serious in reconfigurable control design because the objective of a reconfigurable control is to make the performance of the reconfigured system as close to that of the prefault system as possible [5]. Inspired by the multiple-model switching and tuning approach, in this paper, a performance index is used to provide additional information for reliable fault diagnosis.

Furthermore, the attitude tracking controller of a sliding mode variable structure controller is designed. Once a fault is detected, the reconfigurable fault tolerant controller is adopted to accommodate the faulty system.

2 Kinematics and dynamics models

Taking into account spacecraft need large angle maneuver and bring singularity, the attitude kinematics equation is described by unit quaternion

$$\begin{cases} \dot{q}_0 = -\frac{1}{2}\boldsymbol{q}_{13}^{\mathrm{T}}\boldsymbol{\omega} \\ \dot{\boldsymbol{q}}_{13} = \frac{1}{2}(q_0\boldsymbol{I}_3 + \boldsymbol{q}_{13}^{\times})\boldsymbol{\omega} \end{cases} \tag{1}$$

where $\boldsymbol{\omega} = [\omega_x, \omega_y, \omega_z]^{\mathrm{T}}$ is the angular velocity; $(q_0, \boldsymbol{q}_{13}) \in R \times R^3$ is the attitude quaternion satisfying $\boldsymbol{q}_{13} = [q_1, q_2, q_3]^{\mathrm{T}}$ and $q_0^2 + \boldsymbol{q}_{13}^{\mathrm{T}}\boldsymbol{q}_{13} = 1$; $\boldsymbol{I}_3$ denotes an 3×3 identity matrix; the notation $q_{13}^{\times}$ denote a skew-symmetric matrix.

Ignoring the flexibility and liquid sloshing, the attitude dynamics equitation can be described by a rigid body model as follows

$$\boldsymbol{J}\dot{\boldsymbol{\omega}} + \boldsymbol{\omega}^{\times}\boldsymbol{J}\boldsymbol{\omega} = \boldsymbol{T}_c + \boldsymbol{T}_d \tag{2}$$

where $\boldsymbol{J}$ is the inertia matrix of a spacecraft. $\boldsymbol{T}_c$ and $\boldsymbol{T}_d$ denote the vectors of control torques and disturbance torques respectively. $|\boldsymbol{T}_d(t)| \le T_{d\max}, T_{d\max} > 0$.

3 FDD based on IMM algorithm

Since faults system cannot be modelled well by a single set of equations of state that varies continuously. A more appropriate mathematical model for such a system is the so-called stochastic hybrid system. It differs from the conventional stochastic systems in that its state may jump as well as vary continuously.

Euler discretization from equations (1) and (2), the following model can be available

$$\begin{cases} \boldsymbol{\omega}(k+1) = T\boldsymbol{J}^{-1}(m(k+1))[\boldsymbol{T}_c(k) + \boldsymbol{T}_d(k) - \\ \qquad \boldsymbol{\omega}(k)^{\times}\boldsymbol{J}(m(k+1))\boldsymbol{\omega}(k)] + \\ \qquad \boldsymbol{\omega}(k) + \boldsymbol{\zeta}_1(k, m(k+1)) \\ q_0(k+1) = -\frac{T}{2}\boldsymbol{q}_{13}^{\mathrm{T}}(k)\boldsymbol{\omega}(k) + q_0(k) \\ \boldsymbol{q}_{13}(k+1) = \frac{T}{2}[q_0(k)\boldsymbol{I}_3 + \boldsymbol{q}_{13}(k)^{\times}]\boldsymbol{\omega}(k) + \boldsymbol{q}_{13}(k) \end{cases} \tag{3}$$

where T is the sampling interval; $m(k)$ is the discrete-valued modal state at time k ; ζ_1 is the discrete time process noises.

Supposed that state vector

$$\boldsymbol{x}(k) = [\boldsymbol{\omega}^{\mathrm{T}}(k)\ q_0(k)\ \boldsymbol{q}_{13}^{\mathrm{T}}(k)]^{\mathrm{T}} \tag{4}$$

measurement vector

$$\boldsymbol{z}(k) = \boldsymbol{x}(k) + \boldsymbol{\eta}(k) \tag{5}$$

where $\boldsymbol{\eta}(k)$ is the discrete time measurement noises.

Based on the above, the stochastic hybrid system of attitude control system is can be described as follows

$$\begin{cases} \boldsymbol{x}(k+1) = \boldsymbol{A}(k, m(k+1))\boldsymbol{x}(k) + \\ \qquad \boldsymbol{F}(k, m(k+1))\boldsymbol{M}(k) + \\ \qquad \boldsymbol{H}(k, m(k+1))\boldsymbol{\zeta}(k, m(k+1)) \\ \boldsymbol{z}(k) = \boldsymbol{C}(k, m(k))\boldsymbol{x}(k) + \boldsymbol{\eta}(k, m(k)) \end{cases} \tag{6}$$

where $\boldsymbol{A}, \boldsymbol{F}, \boldsymbol{H}, \boldsymbol{C}$ are the corresponding dimension matrices; $\boldsymbol{M}$ is the sum of $\boldsymbol{T}_c$ and $\boldsymbol{T}_d$. $\boldsymbol{\zeta}$ and $\boldsymbol{\eta}$ are independent, representing system noises and measurement noises.

The system model sequence assumed to be a Markov chain with transition probabilities

$$P\{m_j(k+1) \mid m_i(k)\} = \pi_{ij}, \forall m_i, m_j \in \boldsymbol{S} \tag{7}$$

with

$$\sum_j \pi_{ij} = 1, \quad i, j = 1, 2, \cdots, N \tag{8}$$

where $P\{\cdot\}$ denotes probability; π_{ij} is the transition probability from the model i to the model j ; the event that m_j is in effect at time k is denoted as $m_j(k) \square \{m(k) = m_j\}$; $\boldsymbol{S} = \{m_1, m_2, \cdots, m_N\}$ is the set of all possible system models.

A complete cycle of the IMM-based FDD scheme with UKF as its model-matched filter is described as follows

(1) interaction of the estimates

model probability predicting [5]:

$$\mu_j(k+1 \mid k) = \sum_i \pi_{ij}\mu_i(k) \tag{9}$$

probability mixing:

$$\mu_{i|j}(k) = \pi_{ij}\mu_i(k) / \mu_j(k+1 \mid k) \tag{10}$$

estimate mixing:

$$\hat{x}_j^0(k \mid k) = \sum_i \hat{x}_i(k \mid k)\mu_{i|j}(k) \tag{11}$$

covariance mixing:

$$\begin{aligned} \boldsymbol{P}_j^0(k \mid k) = \sum_i \{&\boldsymbol{P}_i(k \mid k) + [\hat{x}_j^0(k \mid k) - \hat{x}_i(k \mid k)]\cdot \\ &[\hat{x}_j^0(k \mid k) - \hat{x}_i(k \mid k)]^{\mathrm{T}}\}\mu_{i|j}(k) \end{aligned} \tag{12}$$

(2) unscented Kalman filter

initialization:

$$\begin{aligned} \boldsymbol{x}_0 &= \mathrm{E}[\boldsymbol{x}(0)], \\ \boldsymbol{P}_0 &= \mathrm{E}[(\boldsymbol{x}(0) - \boldsymbol{x}_0)(\boldsymbol{x}(0) - \boldsymbol{x}_0)^{\mathrm{T}}] \end{aligned} \tag{13}$$

sample points and their weights:

$$\chi_0 = \hat{x}(k \mid k), \quad \omega_0 = \kappa / (n_x + \kappa) \tag{14}$$

$$\begin{aligned} \chi_i &= \hat{x}(k \mid k) + (\sqrt{(n_x + \kappa)\boldsymbol{P}_j(k \mid k)})_i, \\ \omega_i &= 1/[2(n_x + \kappa)], \quad i = 1, 2, \cdots, n_x \end{aligned} \tag{15}$$

$$\begin{aligned} \chi_i &= \hat{x}(k \mid k) - (\sqrt{(n_x + \kappa)\boldsymbol{P}_j(k \mid k)})_i, \\ \omega_i &= 1/[2(n_x + \kappa)], \quad i = n_x + 1, \cdots, 2n_x \end{aligned} \tag{16}$$

where n_x is the dimension of state variable; $\kappa \in R$ satisfies $n_x + \kappa \neq 0$; $(\sqrt{(n_x+\kappa)\boldsymbol{P}_j(k\mid k)})_i$ is the ith row of the matrix square root.

sample points through the nonlinear function:

$$\xi_i(k+1\mid k) = f_j(\chi_i) \tag{17}$$

state variable and covariance predicting:

$$\hat{x}_j(k+1\mid k) = \sum_{i=0}^{2n_x} \omega_i \xi_i(k+1\mid k) \tag{18}$$

$$\begin{aligned}\boldsymbol{P}_j(k+1\mid k) = \sum_{i=0}^{2n_x} \omega_i[(\xi_i(k+1\mid k) - \hat{x}_j(k+1\mid k))\cdot \\ (\xi_i(k+1\mid k) - \hat{x}_j(k+1\mid k))^{\mathrm{T}}] + \boldsymbol{Q}_j(k+1)\end{aligned} \tag{19}$$

measurement derived from sample points predicting:

$$\varsigma_i(k+1\mid k) = \boldsymbol{C}_j(k)\xi_i(k+1\mid k) \tag{20}$$

measurement and covariance predicting:

$$\hat{z}_j(k+1\mid k) = \sum_{i=0}^{2n_x} \omega_i \varsigma_i(k+1\mid k) \tag{21}$$

$$\begin{aligned}\boldsymbol{S}_j(k+1) = \sum_{i=0}^{2n_x} \omega_i[(\varsigma_i(k+1\mid k) - \hat{z}_j(k+1\mid k))\cdot \\ (\varsigma_i(k+1\mid k) - \hat{z}_j(k+1\mid k))^{\mathrm{T}}] + \boldsymbol{R}_j(k+1)\end{aligned} \tag{22}$$

cross-covariance of measurement and state variable predicting:

$$\begin{aligned}\boldsymbol{P}_{xz} = \sum_{i=0}^{2n_x} \omega_i[(\xi_i(k+1\mid k) - \hat{x}_j(k+1\mid k))\cdot \\ (\varsigma_i(k+1\mid k) - \hat{z}_j(k+1\mid k))^{\mathrm{T}}]\end{aligned} \tag{23}$$

UKF gain:

$$\boldsymbol{W}_j(k+1) = \boldsymbol{P}_{xz}\boldsymbol{S}_j^{-1}(k+1) \tag{24}$$

state and covariance updating:

$$v_j(k+1) = z_j(k+1) - \hat{z}_j(k+1\mid k) \tag{25}$$

$$\begin{aligned}\hat{x}_j(k+1\mid k+1) = \hat{x}_j(k+1\mid k) + \\ \boldsymbol{W}_j(k+1)v_j(k+1)\end{aligned} \tag{26}$$

$$\begin{aligned}\boldsymbol{P}_j(k+1\mid k+1) = \boldsymbol{P}_j(k+1\mid k) - \\ \boldsymbol{W}_j(k+1)\boldsymbol{S}_j(k+1)\boldsymbol{W}_j(k+1)^{\mathrm{T}}\end{aligned} \tag{27}$$

(3) model probability updating

likelihood function:

$$L_j(k+1) = \frac{\exp[-\frac{1}{2}v_j(k+1)^{\mathrm{T}}\boldsymbol{S}_j^{-1}(k+1)v_j(k+1)]}{\sqrt{|(2\pi)\boldsymbol{S}_j(k+1)|}} \tag{28}$$

model probability:

$$\mu_j(k+1) = \frac{\mu_j(k+1\mid k)L_j(k+1)}{\sum_j \mu_j(k+1\mid k)L_j(k+1)} \tag{29}$$

(4) FDD logic

residual vector:

$$\varepsilon_j(k+1) = \left\| z(k+1) - \boldsymbol{C}_j(k+1)\hat{x}_j(k+1\mid k+1)\right\|_2 \tag{30}$$

index:

$$J_j(k+1) = c_1\varepsilon_j^2(k+1) + c_2\sum_{i=k-D+2}^{k}\varepsilon_j^2(i) \tag{31}$$

where $c_1 \geq 0$ and $c_2 \geq 0$ are constants, and D is a moving window.

FDD logic:

$$h(t) = \arg\min_j J_j(k+1) \tag{32}$$

where $h(t)$ denotes the model in effect at time t and $t \in [kT+T, kT+2T)$.

Moreover, the aggregated state estimate, which is to be used in the following attitude tracking controller, is obtained as a probability-weighted sum of the updated state estimate from all the filters in the IMM algorithm

$$\hat{x}(k+1\mid k+1) = \sum_j \hat{x}_j(k+1\mid k+1)\mu_j(k+1) \tag{33}$$

4 Reconfigurable fault tolerant control scheme

4.1 Attitude tracking controller design

It is assumed that $(\psi_i, \vartheta_i, \gamma_i)$ and $(\psi_f, \vartheta_f, \gamma_f)$ are the initial attitude angles and final attitude angles, and the corresponding unit quaternion are $(q_{i0}, \boldsymbol{q}_{i13})$ and $(q_{f0}, \boldsymbol{q}_{f13})$, $\boldsymbol{q}_{i13} = [q_{i1}, q_{i2}, q_{i3}]^{\mathrm{T}}$, $\boldsymbol{q}_{f13} = [q_{f1}, q_{f2}, q_{f3}]^{\mathrm{T}}$.

To achieve the desired attitude tracking, the Lyapunov function is chose as follows [6, 7].

$$V = \sum_{i=0}^{3}(q_i - q_{fi})^2 \tag{34}$$

from(1) and (34),

$$\begin{aligned}\dot{V} = \omega_x(q_{f0}q_1 - q_{f1}q_0 - q_{f2}q_3 + q_{f3}q_2) + \\ \omega_y(q_{f0}q_2 - q_{f1}q_3 - q_{f2}q_0 + q_{f3}q_1) + \\ \omega_z(q_{f0}q_3 - q_{f1}q_2 - q_{f2}q_1 + q_{f3}q_0)\end{aligned} \tag{35}$$

to satisfy $\dot{V} \leq 0$, assuming the following switching function $\boldsymbol{s} = [s_1, s_2, s_3]^T$ and

$$\begin{cases} s_1 = \omega_x + k_1(q_{f0}q_1 - q_{f1}q_0 - q_{f2}q_3 + q_{f3}q_2) \\ s_2 = \omega_y + k_2(q_{f0}q_2 - q_{f1}q_3 - q_{f2}q_0 + q_{f3}q_1) \\ s_3 = \omega_z + k_3(q_{f0}q_3 - q_{f1}q_2 - q_{f2}q_1 + q_{f3}q_0) \end{cases} \tag{36}$$

where $k_i > 0, i = 1,2,3$. So $\boldsymbol{s} = 0$ when the system come into the sliding mode area and $\dot{V} \leq 0$.

the constant rate reaching law

$$\dot{\boldsymbol{s}} = -s_0\,\mathrm{sgn}(\boldsymbol{s}) - \lambda\boldsymbol{s} \tag{37}$$

where $s_0 > 0, \lambda > 0$.

To avoid chattering, sgn function in (37) is replaced by tanh function as follows

$$\tanh(\frac{\boldsymbol{s}}{\varepsilon}) = \frac{\sinh(\boldsymbol{s}/\varepsilon)}{\cosh(\boldsymbol{s}/\varepsilon)} = \frac{e^{s/\varepsilon} - e^{-s/\varepsilon}}{e^{s/\varepsilon} + e^{-s/\varepsilon}} \tag{38}$$

where $\varepsilon > 0$ and

$$\dot{s} = -s_0 \tanh(\frac{s}{\varepsilon}) - \lambda s \tag{39}$$

To achieve attitude tracking, the inequation $s^{\mathrm{T}}\dot{s} < 0$ should come into existence, so the following controller can be obtained

$$\boldsymbol{M}_c = \boldsymbol{\omega}^{\times}\boldsymbol{J}\boldsymbol{\omega} - \boldsymbol{M}_d - \frac{1}{2}\boldsymbol{K}(J)\boldsymbol{P}(q_f)\boldsymbol{G}(\omega)\boldsymbol{q} - s_0 \tanh(\frac{\boldsymbol{s}}{\varepsilon}) - \lambda \boldsymbol{s}] \tag{40}$$

4.2 Fault tolerant controller design

In this paper, three modes are considered: normal mode, actuator fault and sensor fault mode. Actuator fault would be modeled by annihilating the appropriate column of the control input matrix $\boldsymbol{F}$, that is, the elements of the appropriate column of matrix $\boldsymbol{F}$ become zeros. Similarly, sensor fault would be modeled by annihilating the appropriate row of the output matrix $\boldsymbol{C}$.

In the normal mode, controller (40) is used. In the fault mode, the fault tolerant controller is described by the following mathematical expression based on the on-line FDD information.

$$\boldsymbol{M}_c(t) = \boldsymbol{M}_{j^*}(t) \tag{41}$$

where $j^* = h(t)$.

5 Numerical simulation

In this section, a numerical simulation example evaluating the attitude tracking process is employed to demonstrate the efficiency of the proposed approach. $J_x = 1600kg \cdot m^2$, $J_y = 2900kg \cdot m^2$, $J_z = 2350kg \cdot m^2$, $(\psi_i, \vartheta_i, \gamma_i) = (18^\circ, -85.8^\circ, 0^\circ)$, $(\psi_f, \vartheta_f, \gamma_f) = (85.8^\circ, 0^\circ, 0^\circ)$, $T = 0.01\text{s}$, $M_d = 0.005\sin(\omega_0 t)\text{Nm}$, $k_i = 0.1$, $\omega_0 = 1\, rad/s$, $\varepsilon = 0.5$, $\lambda = 1000$; initial parameters for UKF: $\hat{x}_j(0|0) = x(0)$, $\boldsymbol{P}_j(0|0) = 100I_7$, $c_i = 0.5$; process noises and measurement noises are mutually independent with zeros mean, and their covariances are $\boldsymbol{R}_j = 10^{-6} I_7$ and $\boldsymbol{Q}_j = \mathrm{diag}(10^{-8}, 10^{-8}, 10^{-8}, 0,0,0,0)$ respectively. Transition probability matrix is used as follows. Based on the parameters, 50 Monte Carlo numerical simulations are employed respectively for three cases.

$$\Pi = \begin{bmatrix} 0.9983 & 0.0017 \\ 0.0011 & 0.9989 \end{bmatrix}$$

(1) Normal mode

In this case, the system of spacecraft is normal and not any actuator or sensor fault. As can be seen from Fig. 1, the sliding mode variable structure controller can meet the attitude tracking requirement, which has demonstrated the effectiveness of the designed controller.

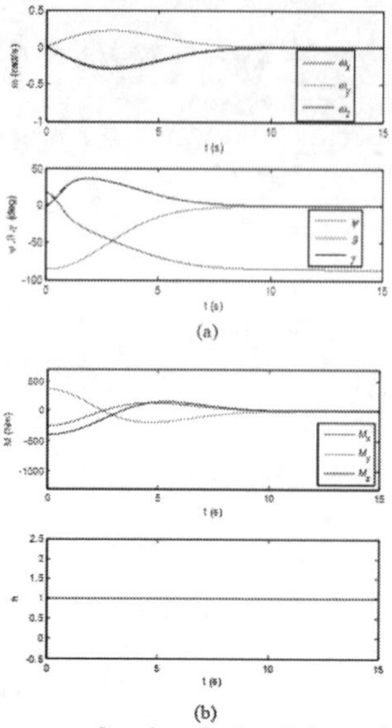

Figure 1. The curve of angle velocity, Euler angles, control moments and h function (normal mode)

(2) Actuator fault mode

The process is divided into 2 models: from 0 second to 6 seconds, the system of spacecraft is normal; from 6 seconds to 15 seconds, the first component of the corresponding actuator is failure but sensors are not failure.

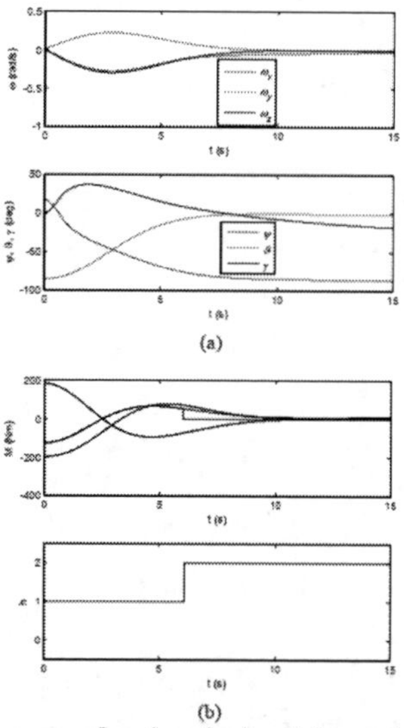

Figure 2. The curve of angle velocity, Euler angles, control moments and h function (actuator faulty system without controller reconfiguration)

Table 1 The performance of proposed scheme (actuator fault)

	CDID	FA	IFID	MFD	NMD
Value	0.9965333	0.00088	0	0.0025867	0

In this table, CDID is correct detection and identification, FA is false alarm, IFID is incorrect fault identification, MFD is missed fault detection, NMD is no mode detection.

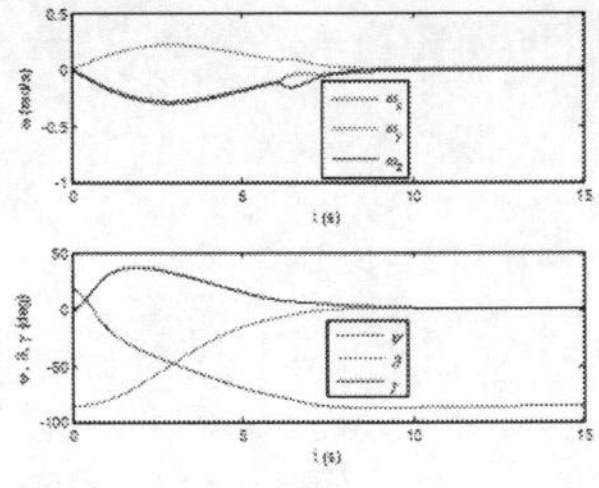

(a)

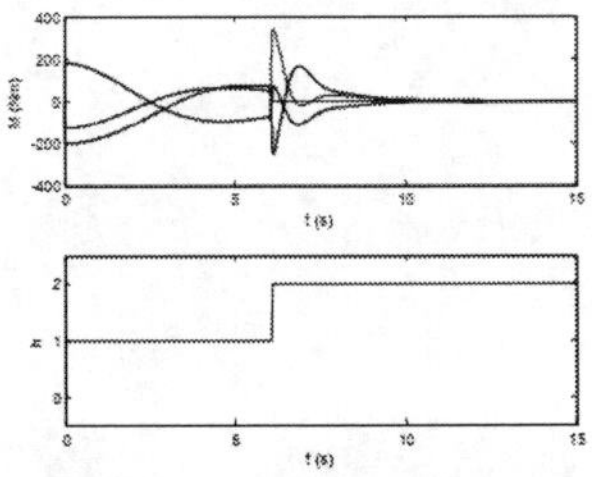

(b)

Figure 3. The curve of angle velocity, Euler angles, control moments and h function (actuator faulty system with controller reconfiguration)

(3) Sensor fault mode

The process is divided into 2 models: from 0 second to 6 seconds, the system of spacecraft is normal; from 6 seconds to 15 seconds, the first component of the corresponding sensor is failure but actuators are not failure.

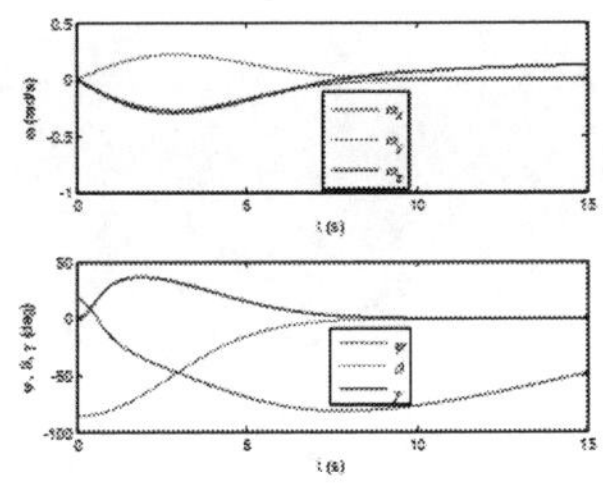

(a)

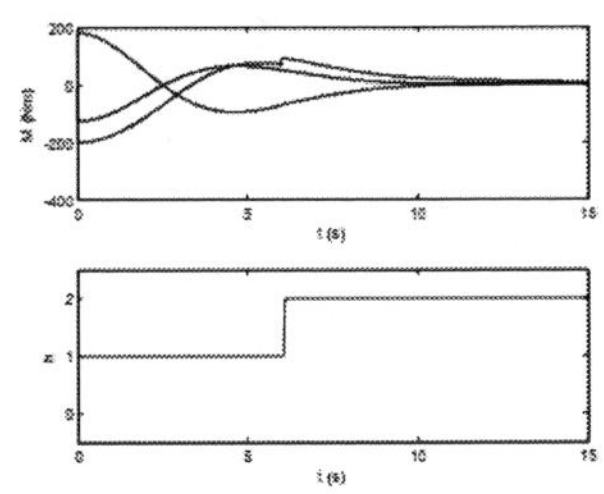

(b)

Figure 4. The curve of angle velocity, Euler angles, control moments and h function (sensor faulty system without controller reconfiguration)

Table 2 The performance of proposed scheme (sensor fault)

	CDID	FA	IFID	MFD	NMD
Value	0.9976667	0.001	0	0.0013333	0

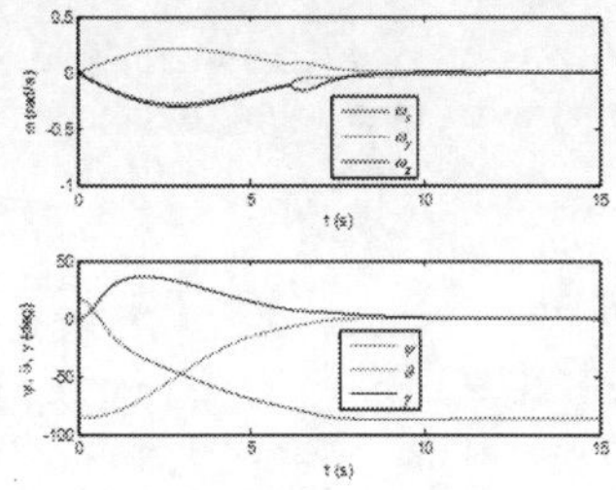

(a)

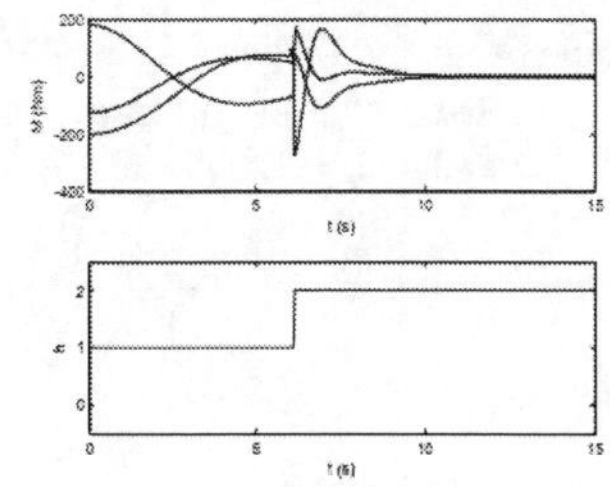

(b)

Figure 5. The curve of angle velocity, Euler angles, control moments and h function (sensor faulty system with controller reconfiguration)

From the h in the Fig. 2 to Fig. 5, and the table 1 and 2, the IMM algorithm with the UKF as the basic estimation unit and the output residual as the fault diagnosis index can detect and diagnose the fault condition in real time, and has high accuracy, low false alarm rate and false negative rate. Probability error is mainly caused by time delay.

In the case (2) and (3), the actuator or sensor is fault, but the faulty system is not reconstructed. From Fig. 2 and Fig. 4, there is a large attitude bias because the original designed controller for faulty system loses efficiency. However, based on the on-line FDD information, the real-time reconfiguration can be carried out. In Fig.3 and Fig.5, the sliding mode variable structure fault tolerant controller is adopted. Compared Fig.3, Fig.5 with Fig.2, Fig.4, the performance of the reconfigured system is very close to that of the normal system, which has tested the feasibility of reconfigurable fault tolerant method.

6 Conclusion

An integrated fault detection, diagnosis, and reconfigurable control scheme for a spacecraft with attitude tracking. FDD has been carried out using the modified IMM algorithm, based on the unscented Kalman filter and an index related to the closed-loop system performance. The sliding mode variable structure fault tolerance controller is designed based on this IMM algorithm. The Monte Carlo simulation results show that the IMM algorithm with the output residual as the fault diagnosis index can detect and diagnose the fault condition in real time. Steady attitude tracking has been achieved using fault tolerant controller is adopted to accommodate the faulty system to make the performance of reconfigured system to be close to that of the normal system. The simulation results have indicated that the proposed FDD and reconfigurable control scheme can deal with faults effectively. The disadvantage is that there is a certain delay in the diagnosis, how to design parameters more rationally and to improve the algorithm to reduce the delay time is the further content need to study.

References

1. Y.F. Han, Application of IMM Fusion Algorithm in Super-Low Altitude Target Detection [J], Systems Engineering and Electronics, 2003, 25(1): 15-17.
2. Y.M. Zhang, X.R. Li, Detection and Diagnosis of Sensor and Actuator Failures Using IMM Estimator [J], IEEE Transactions on Aerospace and Electronic Systems, 1998, 34(4): 1293-1313.
3. S. J. Julier, J. K. Uhlmann, H. F. Durrant-Whyte. A New Approach for the Nonlinear Transformation of Means and Covariances in Filters and Estimators. IEEE Transactions on Automatic Control. 2000, 45(3): 477-482
4. K. Xiong, C.W. Chan, H.Y. Zhang, Detection of Satellite Attitude Sensor Faults Using the UKF[J], IEEE Transactions on Aerospace and Electronic System, 2007, 43(2): 480-491.
5. Y.M. Zhang, J. Jiang, Integrated Active Fault-Tolerant Control Using IMM Approach [J], IEEE Transactions on Aerospace and Electronic Systems, 2001, 37(4): 1221-1235.
6. K.M. MA, Z.Y. Dong, Variable Structure Attitude Control of Spacecraft Large Angle Maneuver Flight[J], Flight Dynamics, 2004, 22(2): 45-48.
7. L.J. Hu, F. C. Liu, Research on Attitude Maneuver Control of Large-ScaIe Satellite Based on Variable Structure Control System with Sliding Mode [J], Aerospace Shanghai, 2011, (1): 23-27.

Fault-Tolerant Path Planning of the Space Manipulator with Single Joint Failure Based on the Dexterity Space

Jia-Yi Tan[1], Gang Chen[2], and Yu-Qi Wang[2]

[1]Beihang University, Beijing 100191, China
[2]Beijing University of Posts and Telecommunications, Beijing 100876, China

Abstract. To enable the space manipulator to complete the original task efficiently after any single joint fails, a fault-tolerant path planning method for the manipulator with single joint failure is proposed based on dexterity space in this paper. On the base of solving the degraded workspace, the dexterity space of the manipulator with single joint failure is established by constructing the dexterity index, and then the traditional A* algorithm is improved to complete fault-tolerant path planning in the dexterity space. The correctness and validity of fault-tolerant path planning based on improved A* algorithm are verified by simulating experiments with 7R manipulator.

1 Introduction

Due to the harshness of the space environment and the complexity of its own structure, joint locked failure occurs easily during the orbit operation for the space manipulator[1]. The degree of freedom of the space manipulator with the joint failure is reduced, which reduces the reachability of its terminal attitude, thereby causing the space operation task to fail to complete normally. Therefore, it is necessary to carry out fault-tolerant path planning with taking into account the reachability of the terminal attitude to make the space manipulator continue the task after joint locked failure occurs.

Aiming at fault-tolerant path planning for faulty manipulator, Khatib [2] et al. proposed an artificial potential field method, which make the proposed path smoother, but with local optimal problems still existing. Du[3] et al. combined the genetic algorithm with simulated annealing algorithm to propose a new algorithm. At present, many scholars utilize A* algorithm widely to search the shortest path in the workspace by selecting the appropriate valuation function[4]. Therefore, the A* algorithm is going to be utilized to carry out fault-tolerant path planning research of the manipulator in this paper.

Much path planning often requires perceive and analyze the environment and according to the performance of manipulator itself. Roberts[5] et al. carried out joint movement planning under the premise that there are high manipulability at the time of any joint failure based on the degraded manipulability. Zhao[6] et al. completed the algorithm design of fault-tolerant path planning with the assessment of fault tolerance performance based on the central degree. When the space manipulator performs certain tasks, we expect to obtain a set of all possible gestures of the spatial position to ensure completing the task with high quality. Therefore, the dexterity index that represents the reachability of the terminal attitude[7] is going to be constructed in this paper. The traditional A* algorithm is improved by increasing the flexibility index in the cost function, and then fault-tolerant path planning is completed utilizing the improved A* algorithm.

In summary, aiming at the space manipulator with any joint locked failure, the dexterity index that represents the reachability of the terminal attitude is constructed by considering three posture parameters of the end of the space manipulator, and then the dexterity space is established on the base of solving the degraded workspace of the manipulator. The traditional A* algorithm is improved by increasing the dexterity index in the cost function, and then fault-tolerant path planning of the space manipulator with single joint failure is completed utilizing the improved A* algorithm. The dexterity index constructed in this paper realizes the characterization of the terminal attitude of the space manipulator with single joint failure. The trajectory searched based on the improved A* algorithm has meet the higher requirements of the terminal attitude for the manipulator and ensured that the manipulator can still complete the task.

2 The establishment of the dexterity space for the space manipulator

In this section, the dexterity space is constructed by considering three attitude parameters of the end of manipulator, and then the degraded workspace of the manipulator is solved based on the Monte Carlo method.

At the same time, the dexterity of each space point is calculated and the dexterity space of the space manipulator with single joint failure is established later.

2.1 Kinematic modeling of the manipulator with joint failure

The kinematic model of n DOF manipulator is established first before the establishment of the dexterity space, as shown in Figure 1.

Under normal condition, the symbols $L_i(i=1,2,\dots n)$ and $J_i(i=1,2,\dots n)$ represent the link i and joint i of the manipulator respectively and $\Sigma_i(i=0,2,\dots,n)$ represents the body coordinate system of L_i . As to joint angle sequence $\boldsymbol{\theta}=\{\theta_1,\theta_2,\cdots,\theta_{n-1},\theta_n\}$, the positive kinematics equation of the manipulator is:

$$ {}^0_n\boldsymbol{T} = {}^0_1\boldsymbol{T}(\theta_1)\,{}^1_2\boldsymbol{T}(\theta_2)\cdots{}^{n-2}_{n-1}\boldsymbol{T}\,{}^{n-1}_{n}\boldsymbol{T}(\theta_n) \tag{1} $$

When the joint J_i locked failure occurs, the link L_{i-1} and L_i are going to be connected to be a new link $\tilde{L}_i$. If the joints, links and coordinate systems of the manipulator are located before the locked joint, their symbols are unchanged; on the contrary, they only need to add the "~" above their symbols, and the body coordinate systems fixed to the connecting link $\tilde{L}_i$ is Σ_{i-1} and $\tilde{\Sigma}_{i+1}$, as is shown in Figure 2.

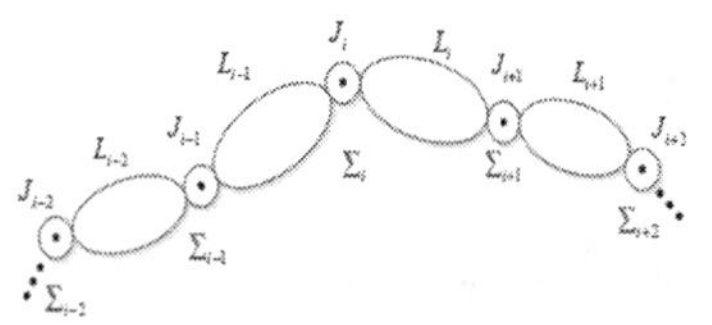

Figure 1. The symbols of links, joints and coordinate systems under normal condition.

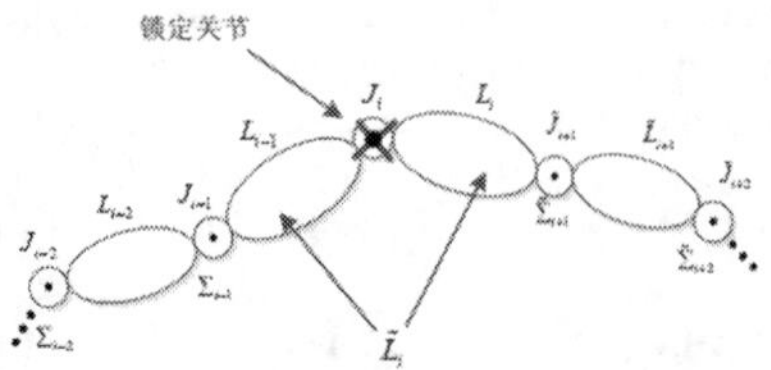

Figure 2. The symbols of links, joints and coordinate systems under fault condition.

2.2 The construction of the dexterity index for the faulty manipulator

The dexterity index which represents the reachability of the terminal attitude is going to be constructed by three attitude parameters of the manipulator.

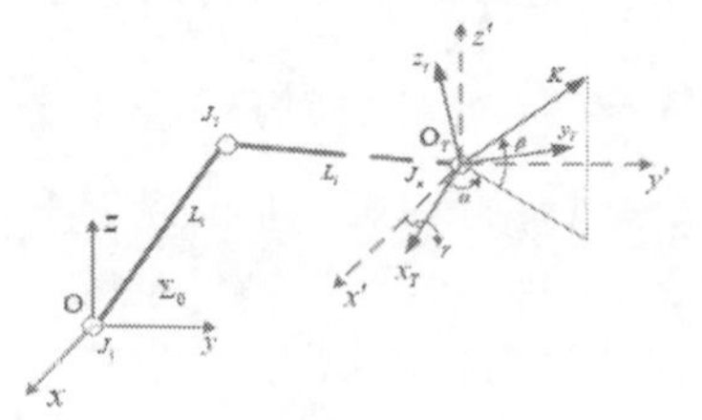

Figure 3. The diagram of solving dexterity.

Figure 3 is a configuration of the manipulator in the base coordinate system Σ_0 . Combined with the coordinates of the end of the manipulator in the certain space point $\boldsymbol{P}=\left[p_x,p_y,p_z\right]$, The terminal attitude of the manipulator is:

$$ \boldsymbol{T}=\begin{bmatrix} k_xk_x\mathrm{vers}\gamma+c\gamma & k_yk_x\mathrm{vers}\gamma-k_zs\gamma & k_zk_x\mathrm{vers}\gamma+k_ys\gamma & p_x \\ k_xk_y\mathrm{vers}\gamma+k_zs\gamma & k_yk_y\mathrm{vers}\gamma+c\gamma & k_zk_y\mathrm{vers}\gamma-k_xs\gamma & p_y \\ k_xk_z\mathrm{vers}\gamma-k_ys\gamma & k_yk_z\mathrm{vers}\gamma+k_xs\gamma & k_zk_z\mathrm{vers}\gamma+c\gamma & p_z \\ 0 & 0 & 0 & 1 \end{bmatrix} \tag{2} $$

The dexterity can be defined as follows:

$$ \left.\begin{aligned} F(\alpha,\beta,\gamma)&=\begin{cases}0 & \text{unreachable attitude}\\ 1 & \text{reachable attitude}\end{cases}\\ D&=\frac{\sum_{m=1}^{a}\sum_{n=1}^{b}\sum_{k=1}^{c}F(\alpha,\beta,\gamma)}{\sum_{m=1}^{a}\sum_{n=1}^{b}\sum_{k=1}^{c}1} \end{aligned}\right\} \tag{3} $$

where $F(\alpha,\beta,\gamma)$ indicates whether the attitude of the certain point is reachable and D represents the dexterity at the point.

2.3 The establishment of the dexterity space for the faulty manipulator

The dexterity space is going to be established by introducing the dexterity index in every point in the degraded workspace, to represent the reachability of the terminal attitude. Before the dexterity space, the degraded workspace corresponding to joint locked failure needs constructing first.

2.3.1 The establishment of the degraded workspace

The process of the establishment of the degraded workspace, which is constructed by the Mont Carlo method [8], is as follows:

STEP1. Locking the fault joint at θ_{i0} ; Generating N random numbers between 0 and 1, and obtaining the random angle values of the healthy joints: $\theta_i=\theta_{i\min}+(\theta_{i\max}-\theta_{i\min})N_i$. $\theta_{i\max}$ and $\theta_{i\min}$ are the upper bound and the lower bound of the ith joint angle respectively;

STEP2. Substituting the random healthy joint angles and the locked fault joint angle from STEP1 into Eq.(1), and then obtaining corresponding terminal positions of manipulator;

STEP3. The degraded workspace consists of all terminal positions generated in STEP2.

The construction of dexterity space is based on the established degraded workspace.

2.3.2 The establishment of the dexterity workspace

According to Eq.(2) and Eq.(3), the dexterity index of all points in the degraded workspace can be calculated, and then the dexterity space is obtained.

STEP1. Selecting a certain point in the degraded workspace, and recording its coordinate value;

STEP2. Assigning the values of α, β, γ in $[0,2\pi]$ at the regular intervals of $\Delta\theta$, and the total number of attitudes is:

$$N_{\text{post}} = \left(\frac{2\pi}{\Delta\theta}\right)^3$$

STEP3. Combining the position of the selected point with every assigned attitude, getting the corresponding configurations based on Eq.(2), and then carrying out the inverse kinematic[9];

STEP4. If there exists solution of inverse kinematic, the corresponding attitude is reachable, otherwise it is unreachable. Then the dexterity in the current point D is the ratio of the number of reachable attitudes and the total number of attitudes;

STEP5. Traversing all points in the degraded workspace, and repeating the process in STEP1~STEP4, the dexterity space is obtained.

Thus, the dexterity space that is the significant reference to the fault-tolerant path planning is established.

3 Fault-tolerant path planning of the space manipulator

Considering that the A* algorithm has the characteristics of strong autonomy and avoid the unreachable region flexibly, fault-tolerant path planning is going to be carried out utilizing A* algorithm, which is going to be improved based on the dexterity. The dexterity space is going to be rasterized after being established first and then the traditional A* algorithm is improved, completing fault-tolerant path planning in the dexterity space to obtain a path that represents the reachability of the terminal attitude for the manipulator.

3.1 The rasterization of the dexterity space

Set a cuboid with its length, width and height in parallel with x, y, z of the inertial frame respectively and then set the side lengths a, b, c of the rectangular so that the cuboid can envelop the dexterity space fully. The number of small cube is:

$$N_{\text{cube}} = \lceil a/\Delta l \rceil \lceil b/\Delta l \rceil \lceil c/\Delta l \rceil \quad (4)$$

Thus, the dexterity space can be seen as composed by a number of small cube units, and the coordinate and dexterity of each cube unit can be represented by the center point of the small cube unit.

3.2 Fault-tolerant path planning based on improved A* algorithm

As a heuristic search algorithm, the A* algorithm is widely utilized in solving the optimal path and its cost function is:

$$f(n) = g(n) + h(n) \quad (5)$$

To obtain the trajectory meeting the terminal attitude of the manipulator, the cost function is improved according to the dexterity index, getting the following expression:

$$f(n) = g(n) + h(n) + Q(D) \quad (6)$$

where $Q(D)$ represents the function evaluating the reachability of the terminal attitude for the manipulator at a certain point and is just concerned with the dexterity D. Set the threshold $k \in [0,1]$ and $Q(D)$ is:

$$Q(D) = \begin{cases} \infty, & D < k \\ \delta/D, & D \geq k \end{cases} \quad (7)$$

where δ represents the search coefficient of the dexterity. δ and k can be determined according to the actual needs of the task, changing the weight of the dexterity in the path search.

After adding the dexterity index, $Q(D)$ becomes the priority determination item in the path search.

According to the improved A* algorithm and the actual operational requirements, the process searching for a reachable path is as follows:

STEP1. Regarding the starting point as the current node, calculating to the costs $f(n)$ of all the small cube value adjacent to the current node, and taking the center point of the small cube whose $f(n)$ is the smallest as the next node of the current node;

STEP2. Regarding the next node obtained at STEP1 as the current node, and repeating STPE1;

STEP3. Repeating the above step, until the next node is the target node. And then by connecting all the obtained node, a reachable path is formed from the starting point to the target point.

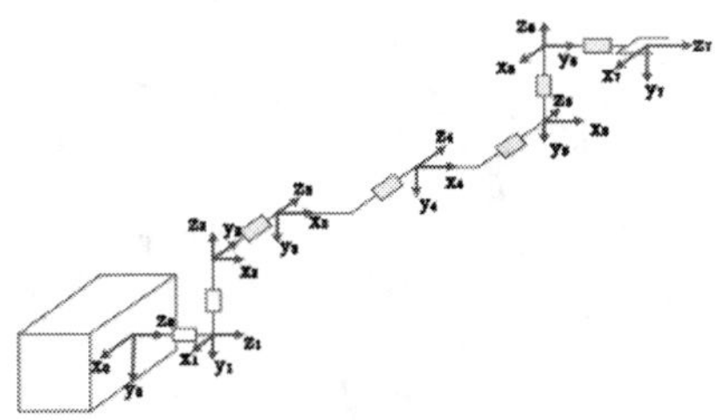

Figure 4. The initial configuration of 7R redundant manipulator.

The planning path based on the improved A^* algorithm may not be the shortest path. So we select the appropriate search coefficient of the dexterity δ and threshold k according to the actual task requirements, to meet both the expected attitude and the shortest distance.

4 Simulation

4.1 The establishment of kinematic model for the space manipulator

As shown in Figure 4 7R redundant manipulator is chosen as the object of study, whose DH parameters are shown inTable 1.

Table 1. The DH parameter table of the initial configuration for 7R manipulator.

Link	a_{i-1}/m	α_{i-1}/°	d_i/m	α_{i-1}/°
1	0	0	0.6	0
2	0	90	0.5	90
3	0	–90	0.5	0
4	5	0	0.5	0
5	5	0	0.5	0
6	0	90	0.5	–90
7	0	–90	0.6	0

4.2 The establishment of the dexterity space for the space manipulator

Assume that the joint 4 has failed and the locked angle is 75° , the degraded workspace is obtained via MATLAB simulation, as is shown in Figure 5. The number of traversal is selected as 1000 In this paper.

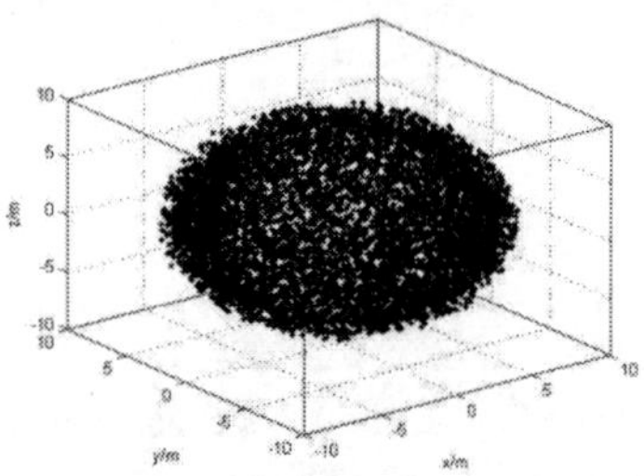

Figure 5. The degraded workspace when $\theta_4 = 75°$.

On the base of solving the coordinates of each point for the degraded workspace, considering that the step size would affect the calculation, take $\Delta\theta = 72°$, and then the dexterity space can be obtained according to the solve steps of the dexterity space , as is shown in Figure 6. It can be seen from Figure 6, that there is a certain rule of distribution for the degraded workspace and it is a gradual process. To reveal the internal dexterity, select some of the interface map, for instance, the planes x=0 and y=0 where the dexterity distributes with the shape of "X"; in the plane z=0, the dexterity is higher and more uniform, as is shown from Figure 7 to Figure 9.

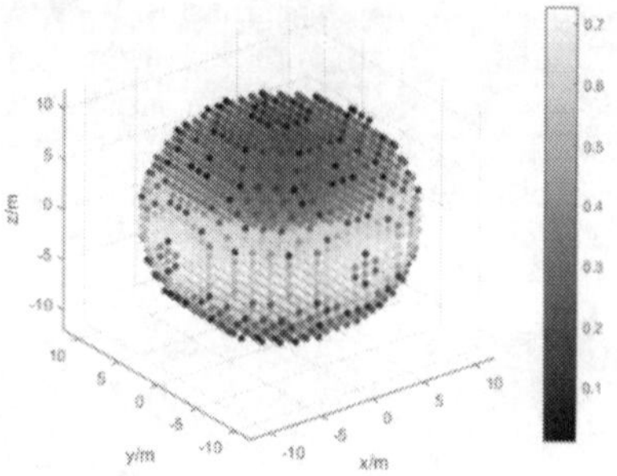

Figure 6. The degraded workspace of the manipulator with single joint failure.

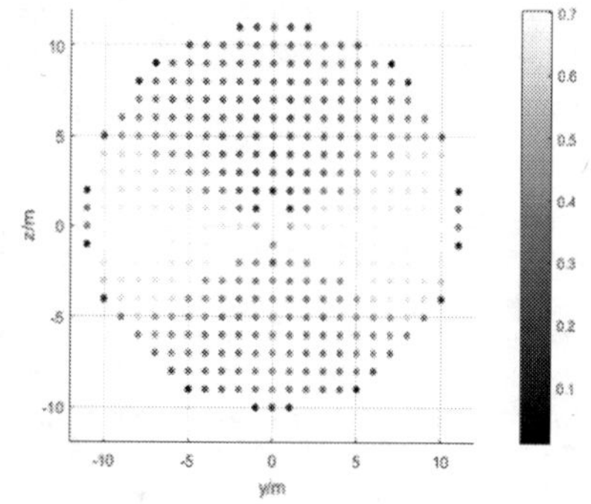

Figure 7. Section view of the degraded workspace when x=0.

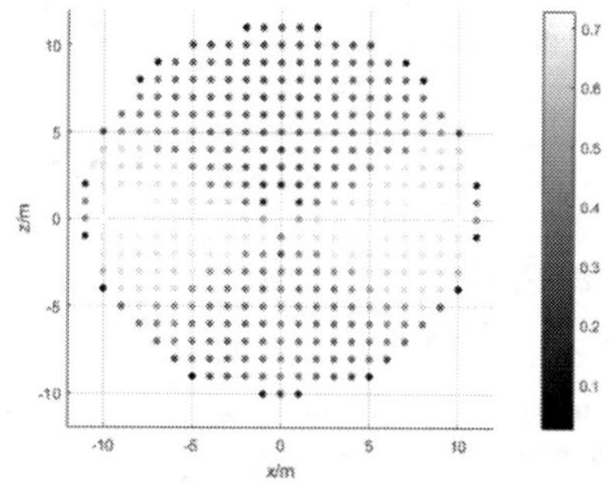

Figure 8. Section view of the degraded workspace when y=0.

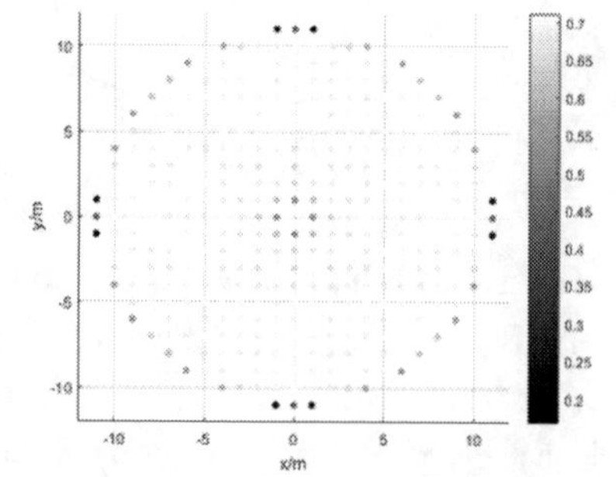

Figure 9. Section view of the degraded workspace when z=0.

4.3 The fault-tolerant path planning of the space manipulator

Assume that the joint 6 is faulty and the locked angle is 149 °, the coordinates of the starting point and the ending point for the task are as follows:

$$x_{\text{int}} = 0, y_{\text{int}} = -6, z_{\text{int}} = 3$$

$$x_{\text{des}} = 0, y_{\text{des}} = 3, z_{\text{des}} = 1$$

The simulation results of the path planning are shown in Figure 10. Comparing the difference between the two paths, it can be found that the path using improved A* algorithm keeps higher dexterity. The Figure 11shows the trend of dexterity where the path node passed by. It can be seen that the dexterity of the improved path increased a lot and the dexterity improved overall:

$$\varepsilon=(0.624-0.3992)/0.3992=84.37\%$$

As to the distance comparison between the two paths, the total distances of the paths are 10.66m and 11.32m respectively. It explains that although the distance in the path searched by improved A* algorithm has increased, while the dexterity improved, it is not obvious.

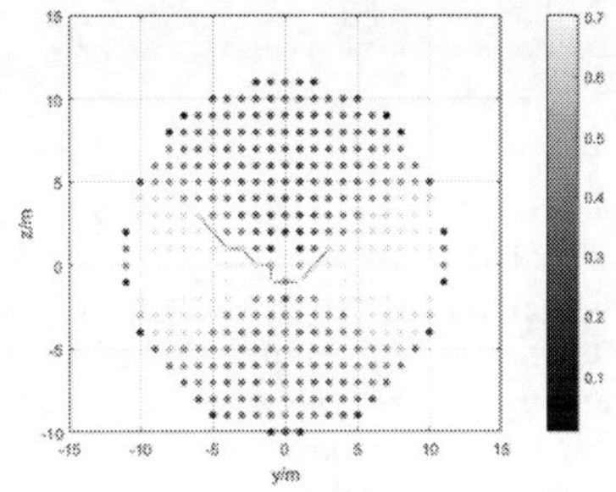

(a) Path search based on improved A* algorithm

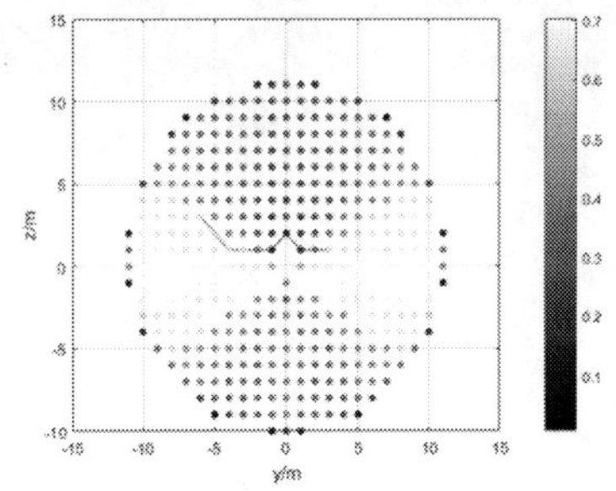

(b) Path search based on unmodified A* algorithm

Figure 10. The comparison of the experiment for fault-tolerant path planning.

The dexterity obtained by utilizing the improved A* algorithm is overall 84.37% higher than the previous, while the distance is 7% longer than the previous. In the actual missions, it tend to pay an additional 7% of the distance to get the 80% increase in dexterity. Therefore, it is feasible to utilize the improved A* algorithm with dexterity to carry out path planning.

The simulation results show that the improved A* algorithm based on the dexterity can solve the problem of path planning for the manipulator with single joint failure. In summary, the correctness and validity of fault-tolerant path planning method for the manipulator based on the dexterity space can be verified by the simulation results.

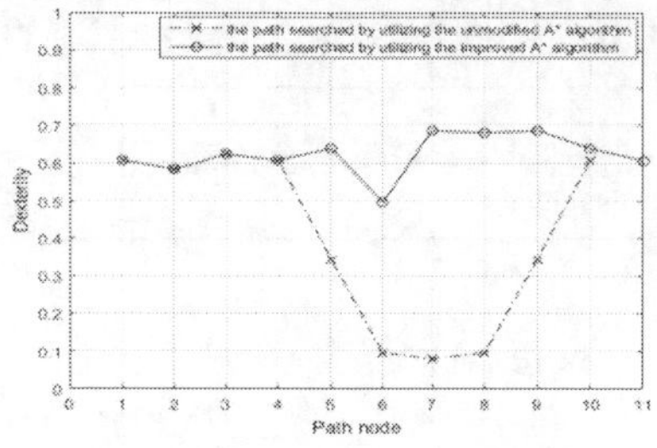

Figure 11. The comparison of the dexterity for the two paths.

5 Conclusion

A fault-tolerant path planning method for the manipulator with single joint failure is proposed based on dexterity space in this paper. The dexterity index is constructed by considering three posture parameters, and then the dexterity space is established on the base of solving the degraded workspace. In the meantime, the traditional A* algorithm is improved based on the dexterity, and then fault-tolerant path planning comes true utilizing the improved A* algorithm. The simulation results show that the planning path based on the improved A* algorithm in the dexterity space can meet the requirements of the higher attitude of the end of the manipulator so that the manipulator can continue to complete the original task after the joint failure occurs.

References

1. Q X. Jia, T. Li, G. Chen, et al. *MFI. IEEE*, 1(2014).
2. O. Khatib. IEEE. *ICRA*, 500(1985).
3. Z Z. Du, G D.Liu. Computer Simulation. **26**(12), 118(2009).
4. M Z. Zhao, Z D. Meng. Jour of HuaZhong Unive of Sci and Tech. 205(2008).
5. R G. Roberts, A A. Maciejewski. IEEE Trans on Robotics & Auto. **12**(4), 543(1996).
6. J. Zhao, K L. Zhang, Y B. Yao. Mech Sci and Tech for Aero Engi. **25**(3), 261(2006).
7. S C. Liu, J Q. Xu, X F. Ma. Robot. **11**(6), 12(1989).
8. J. Rastegar, B. Fardanesh. Mech & Mach Theory. **25**(2), 233(2014).
9. W. Xu, Y. She, Y. Xu. Acta Astro , **105**(1), 201(2014)

Simulation on Spacecraft Formation Flight and Formation Reconfiguration

Chao Li Ye Yan and Yue-neng Yang

National University of Defense Technology,College of Aerospace Science and Technology, 410073, Changsha,China

Abstract. Spacecraft formation flight refers to two or more spacecraft according to a certain formation or arrangement of flight, it has important application value for deep spatial exploration, spatial science experiment, ground investigation and military and so. This paper studies the spacecraft formation flight design and formation reconfiguration based on STK. Firstly, the Clohessy-Wiltshire (CW) equation is used to describe the relative motion of the near-circular orbit and deduce the relative orbital dynamics model. Then, based on the dynamic method of the CW equation, the spatial circular formation is designed and the STK is applied to simulate it. Finally, based on the above formation, a simple multi-impulse formation reconfiguration is performed and the simulation test is verified by STK. The simulation results show that the absolute error of orbital elements of the spacecraft is calculated by the relative orbital dynamics model is less than 10^{-5}, and the expected formation can be completed under the condition of two-body environment, and the feasibility of simple multi-pulse formation reconfiguration is proved successfully.

1 Introduction

Spacecraft formation flight technology refers to the multi-spacecraft constructed cluster, the system is closely coordinated with each other, so as to form a unified organic whole, and the performance goes beyond the traditional single spacecraft system [1]. Relative motion dynamics, control and formation design optimization are essential for formation flight technology, so considering the design and reconfiguration attracted the attention of many scholars [2].

Pan[2]emphatically expounds the development of spacecraft formation flight and its key technology and introduced that spacecraft formation flight play an important role for GPS and relative navigation technology, it laid a solid foundation for the research work of many scholars. Abbott[3] present a method for finding a minimum-power dipole solution for a given set of desired forces, which does not rely on arbitrary parameterizations as have prior approaches, and the structure enables further analysis of numerical conditioning and convergence. Wu [4]designed a linear quadratic control algorithm based on HILL equation dynamics model to verify the feasibility of that control algorithm and its simulation realization method, and laid a certain foundation for formation reconfiguration. Seo[5] investigates Collision avoidance strategies for multiple UAVs (Unmanned Aerial Vehicles) based on geometry and Numerical simulations are performed to demonstrate the performance of the proposed strategies. Jiang[6]studied two-point boundary value problems of relative motion, the algebraic method was used to study the geometric properties of the relative orbit and provided a reliable mathematical basis for the formation design. Meng [1] further studied the theory and method involved in the two-body orbit formation of relative motion dynamics and formation design, for the relative navigation, control and formation optimization design provides a reliable theoretical basis.

Aiming at the multi-purpose and multi-tasking of spacecraft formation, design the formation becomes the key and foundation of the task. Therefore, the spatial circular formation is designed based on the C-W equation dynamic method and by applying simple multi-impulse to achieve the formation reconfiguration, and the simulation test is verified and analyzed based on STK [6].

2 Design Spatial Circular Formation Based on Dynamic Equation

2.1 Spatial circular formation

The spatial circular formation refers to the around spacecraft with reference spacecraft as the center and the relative motion trace is circle orbit. In addition, the phase difference between the three spacecraft is 120° and uniform distribution on the spatial circle.

The orbital elements of the reference spacecraft as follows:

Table 1. The orbital elements of the reference spacecraft.

	a /km	e	i /°	Ω /°	ω /°	M /°
Reference Spacecraft	7000	0	60	10	20	30

2.2 Formation design based on C-W equation

Model design flow chart as follows:

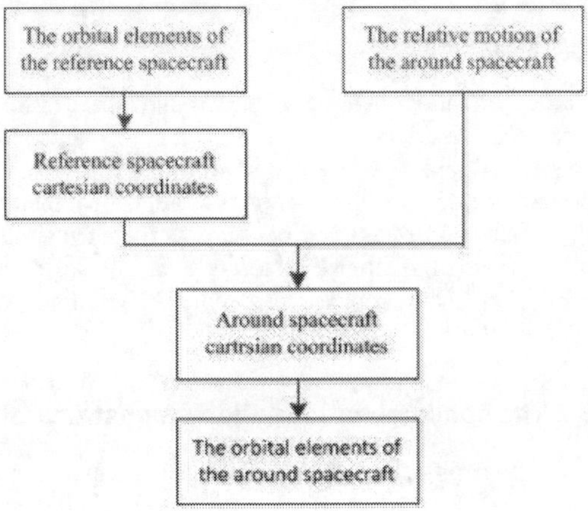

Figure 1. Model design flow.

Assuming that the earth is a homogeneous sphere and regardless of any perturbation factor, $[x,y,z]^T$ is the relative position vector of the spacecraft relative to the reference spacecraft, by solving the CW equation[1]:

$$\begin{cases} \ddot{x} = 2n\dot{y} + 3n^2 x \\ \ddot{y} = -2n\dot{x} \\ \ddot{z} = -n^2 z \end{cases} \tag{1}$$

The analytical form of equation solution can be obtained:

$$\begin{cases} x(t) = \frac{\dot{x}_0}{n}\sin(nt) + \left(-3x_0 - \frac{2\dot{y}_0}{n}\right)\cos(nt) + 2\left(2x_0 + \frac{\dot{y}_0}{n}\right) \\ y(t) = 2\left(3x_0 + \frac{2\dot{y}_0}{n}\right)\sin(nt) + \frac{2\dot{x}_0}{n}\cos(nt) - 3(2nx_0 + \dot{y}_0)t \\ \quad + (-\frac{2\dot{x}_0}{n} + y_0) \\ z(t) = \frac{\dot{z}_0}{n}\sin(nt) + z_0\cos(nt) \end{cases} \tag{2}$$

x_0, y_0, z_0 is the initial state of the relative motion of a given reference frame.

If the formula $\dot{y}_0 = -2nx_0$ can be satisfied, the drift will be eliminated so that relative motion formation is stable [7-8]. $n = \sqrt{\mu / a_{ref}}$ is the average orbital angular velocity of reference spacecraft and μ is geocentric gravitational constant. Further processing the stable relative motion solution, get the simpler form:

$$\begin{cases} x = -A\cos(nt+\theta) \\ y = 2A\sin(nt+\theta) + L \\ z = -B\cos(nt+\varphi) \end{cases} \tag{3}$$

among them:

$$\begin{gathered} A = -\sqrt{x_0^2 + (\dot{x}_0/n)^2}, \begin{cases} \cos\theta = -x_0/A \\ \sin\theta = \dot{x}_0/(nA) \end{cases} \\ B = -\sqrt{z_0^2 + (\dot{z}_0/n)^2} \\ \begin{cases} \cos\varphi = -z_0/A \\ \sin\varphi = \dot{z}_0/(nA) \end{cases}; L = -\frac{2\dot{x}_0}{n} + y_0 \end{gathered} \tag{4}$$

A, θ, B, φ, L are the amplitude and phase of the elliptical orbital plane, the amplitude and phase along the normal direction of the orbital surface, and the position of the ellipse center in the reference frame. On this basis, considering that the center is located at the origin of reference frame, it is necessary to meet the constraints $L = -\frac{2\dot{x}_0}{n} + y_0 = 0$.

From the analysis, it can be:

$$\begin{cases} \cos\theta = -x_0/A \\ \sin\theta = \dot{x}_0/(nA) \\ y_0 = \frac{2\dot{x}_0}{n} \\ \dot{y}_0 = -2nx_0 \\ \cos\varphi = -z_0/B \\ \sin\varphi = \dot{z}_0/(nB) \end{cases} \tag{5}$$

The relative motion state of around spacecraft relative to reference spacecraft $x_0, y_0, z_0, \dot{x}_0, \dot{y}_0, \dot{z}_0$ is obtained. The formation design ultimately get the initial motion state of each spacecraft in the formation, the Cartesian coordinate state of the geocentric inertia system, or the orbital elements. Therefore, the formation design of the spacecraft formation based on the C-W equation is as follows:

(1). According to the orbital elements of the reference spacecraft calculate the position vector R_c and velocity vector V_c of the reference spacecraft in the geocentric inertial system.

(2).According to the expected radius, the around spacecraft in the orbital coordinate system relative to the reference spacecraft position r, speed v is calculated.

(3).Can be converted by the conversion matrix $Q = Q_z[-\Omega]Q_x[-i]Q_z[-u]$ ($Q_j(\theta)$ is the elementary transformation matrix that turns θ around the j axis),the around spacecraft in the orbit coordinate system relative to the reference spacecraft position r , the speed v through the coordinate conversion to the inertial coordinate system;

(4).The position R_h and velocity V_h of the around

spacecraft in the geocentric inertial system can be obtained by adding the sum of the position r, vector v in step (3) and the position R_c, vector V_c of the reference spacecraft.

(5).According to the position R_h and velocity V_h of the around spacecraft, the orbital elements of the spacecraft can be calculated.

2.3 Simulation results and analysis

Based on the above mathematical model, can calculate the orbital elements of the three around spacecraft, and then according to the following six graphs can be analyzed: The semi-major axis a of ellipse of the around spacecraft is less than 50 meters. Eccentricity e difference is less than 10^{-10}; Inclination i difference is less than 0.1°. The right ascension of ascending node Ω is less than 0.1°. True anomaly ω are 0°, 120° and 240° respectively. The Argument of latitude M are 50°, 290° and 170° respectively.

The orbital elements of the around spacecraft are shown in Table 2:

Combined with orbital elements of the above four spacecraft, STK is used to make the simulation and the following simulation results are obtained.

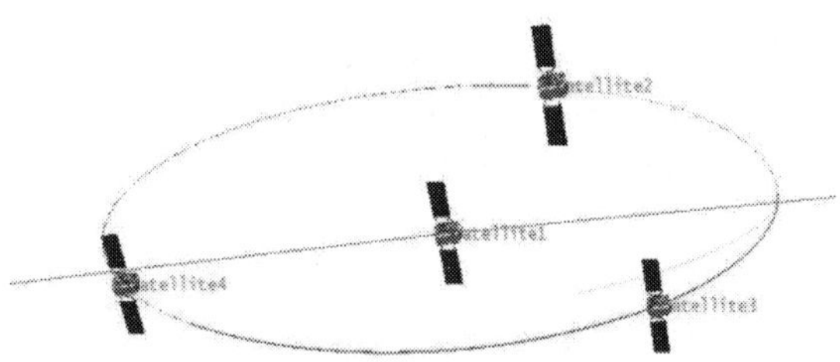

Figure 2. The spatial circular formation obtained by STK.

When the results of the mathematical model are calculated and visual simulation is made by STK, four spacecraft formation is the spatial circular formation, which that three around spacecraft was 120° phase difference. It can be seen that the C-W equation is used to design formation, and the correctness of the mathematical model and simulation experiment is proved.

3 Formation reconfiguration based on simple multi-impulse

3.1 Carpe formation

Carpe formation refers to the around spacecraft take reference spacecraft as center, one of the around spacecraft and reference stars are coplanar, and the around one do an elliptical motion. The relative motion of the other two around spacecraft is perpendicular to the orbit plane of reference spacecraft, and make a linear oscillatory motion, phase difference between the three spacecraft is 90°.

3.2 The condition of formation reconstruction

The initial phases of the three spacecraft in the spatial circle are: 0° , 120° , -120° . It is known that the motion equations of the three around spacecraft relative to the reference spacecraft are:

$$\begin{cases} x = -500\cos(nt+\theta) \\ y = 1000\sin(nt+\theta) \\ z = -500\sqrt{3}\cos(nt+\theta) \end{cases} \tag{6}$$

According to the orbital elements of the reference spacecraft in the spatial circle, the orbital elements of Sat_1', Sat_2' and Sat_3'in the CarPe formation can be calculated. The calculation results are shown in the table 3.

In order to reconstruct the CarPe formation, the three around spacecraft are maneuvered by adding a simple multi-impulse so that it can be realized.

Table 2. The orbital elements of the reference spacecraft and three around spacecraft.

	a /km	e	i /°	Ω /°	ω /°	M /°
ref	7000	0	60	10	20	30
Sat_1	7000.000072	7.14286×10^{-5}	59.9946	10.0053	49.9974	0
Sat_2	7000.000232	7.14286×10^{-5}	60.0067	10.0028	289.999	120
Sat_3	7000.000232	7.14286×10^{-5}	59.9988	9.992	170.004	240

Table 3. The orbital elements of the reference spacecraft and three around spacecraft.

	a/km	e	i/°	Ω/°	ω/°	M/°
ref	7000	0	60	10	20	30

Sat_1'	7000	7.14286×10^{-5}	60	10	20	7.105×10^{-15}
Sat_2'	7000	0	60.0026	10.0036	20.0006	30
Sat_3'	7000	0	59.9974	9.99638	19.9994	30

3.3 Design elliptical formation on orbit plane

An impulse is applied to the Sat_1, which is maneuvered to perform elliptical flight relative to the reference spacecraft on the spacecraft's orbital plane. Since this maneuver is a coplanar orbit maneuver, it is necessary to apply impulse to the normal direction to eliminate the relative motion of the vertical direction. ΔV_1 is impulse which applied to the spacecraft, the relative motion equation as follows:

$$\begin{cases} x = -500\cos f \\ y = 1000\sin f \\ z = -500\sqrt{3}\cos f + \dfrac{\Delta V_1}{n}\sin(f - f_{o1}) \end{cases} \tag{7}$$

To eliminate the relative motion of the vertical direction of orbital plane (that is the Z direction), according to the above equation:

$$\begin{cases} \dfrac{\Delta V_1}{n}\sin f_{o1} + 500\sqrt{3} = 0 \\ \dfrac{\Delta V_1}{n}\cos f_{o1} = 0 \end{cases} \tag{8}$$

The available f_{o1} solutions are $f_{o1} = k\pi + \frac{1}{2}\pi, k = 0,1,2,\ldots n$, then $\Delta V_1 = 500\sqrt{3}n \approx 0.93358 m/s$.

From the above parameters:

$$\begin{cases} \Delta V_1 = nB_c = n\left[B_f^2 + B_0^2 - 2B_f B_0 \cdot \cos(\eta_f - \eta_0)\right]^{1/2} \\ M_1 = -\eta_0 - \eta_c + \dfrac{\pi}{2} \end{cases} \tag{9}$$

$$\begin{cases} \text{n}\eta_c = \dfrac{B_f \sin(\eta_f - \eta_0)}{B_c} \\ \cos\eta_c = \dfrac{B_f\cos(\eta_f - \eta_0) - B_0}{B_c} \end{cases} \tag{10}$$

$$n = \sqrt{\mu / a^3} \quad t_1 = M_1 / 57.3n \tag{11}$$

The time of the impulse t_1 can be obtained.

3.4 Design oscillation on the vertical direction of spacecraft orbital plane

First, it is necessary to eliminate the shifting ΔL between the formation center of the spatial circle and the formation center of CarPe. By applying a radial control impulse M_s once:

$$\begin{cases} \Delta V_s = -\dfrac{1}{2}n\Delta L \\ M_s = -\eta_0 + \dfrac{\pi}{2} \\ t_s = M_s / 57.3n \end{cases} \tag{12}$$

Substituting the formation parameters, the amount of the impulse along the normal direction and the time of the application can be applied.

And then three times along the trace control impulse would be applied:

$$\begin{cases} \Delta V_{T1} = \dfrac{n}{8}A_c = \dfrac{n}{8}\left[A_f^2 + A_0^2 - 2A_f A_0 \cdot \cos(\theta_f - \theta_0)\right]^{1/2} \\ M_{01T} = -\theta_0 - \theta_c \end{cases}$$

$$\sin\theta_c = \frac{A_f \sin(\theta_f - \theta_0)}{A_c} \qquad \cos\theta_c = \frac{A_f\cos(\theta_f - \theta_0) - A_0}{A_c} \tag{13}$$

$$\Delta V_{T2} = -2\Delta V_{T1} = -2\Delta V_{T3} \quad n = \sqrt{\mu / a^3} \quad t = M_T / 57.3n$$

$$M_{02T} = M_{01T} + \pi$$

$$M_{03T} = M_{02T} + \pi$$

The initial formation parameters of Sat_2'and Sat_2 are obtained from the previous contents:

$$\begin{cases} A_0 = -\dfrac{1}{2} \quad B_0 = \dfrac{\sqrt{3}}{2} \quad \theta_0 = 120^\circ \quad \eta_0 = 120^\circ \\ A_f = 0 \quad B_f = 1 \quad \theta_f = 0^\circ \quad \eta_f = 0 \end{cases} \tag{14}$$

Substituting the formation parameters to obtain the amount of impulse along the trace and the time of the application.

Finally, a control impulse M_W is applied once along normal direction of the orbital plane:

$$\begin{cases} \Delta V_W = nB_c = n\left[B_f^2 + B_0^2 - 2B_f B_0 \cdot \cos(\eta_f - \eta_0)\right]^{1/2} \\ M_W = -\eta_0 - \eta_c + \dfrac{\pi}{2} \end{cases}$$

$$\sin\eta_c = \frac{B_f \sin(\eta_f - \eta_0)}{B_c} \qquad \cos\eta_c = \frac{B_f\cos(\eta_f - \eta_0) - B_0}{B_c} \tag{15}$$

$$n = \sqrt{\mu / a^3} \quad t = M_W / 57.3n$$

Substituting the formation parameters, the amount of the impulse along the normal direction and the time of the application are obtained. Calculating the amount of the applied impulse and the time to get the following two tables:

Table 4. Impulse amount during reconfiguration (m / s).

	Frist trace direction	Second trace direction	Third trace direction	Normal direction	Radial direction
Sat_1	0	0	0	0.9336	0
Sat_2	0.04716	-0.09433	0.047163	1.426070	-0.1617
Sat_3	0.0876	-0.1752	0.0876	1.4261	0.1617

Table 5. Apply the time interval during reconfiguration (s).

	Frist trace direction	Second trace direction	Third trace direction	Normal direction	Radial direction
Sat_1	0	0	0	4371.3872	0
Sat_2	1457.1291	2914.2581	2914.2581	1633.4966	5342.8065
Sat_3	1457.1291	2914.2581	2914.2581	1280.7615	3399.9678

3.5 Simulation results and analysis

According to the above results that the spatial circular formation can complete the formation reconfiguration successfully, and the simulation process of the reconfiguration as follows:

(1) The state before reconfiguration:

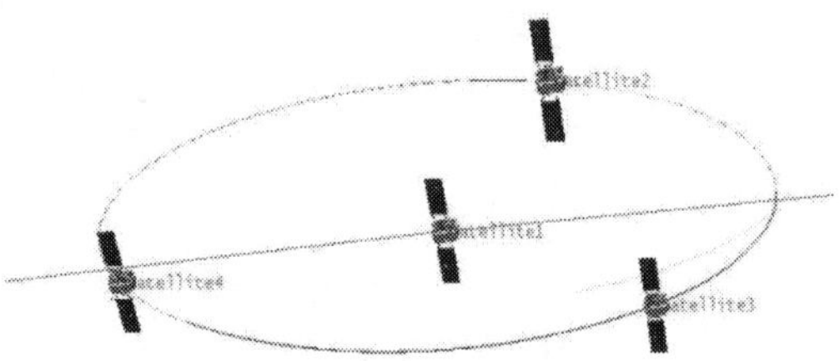

Figure 3. The state before reconfiguration.

(2) During the process of orbital maneuver, the relative motion between the spacecraft is:

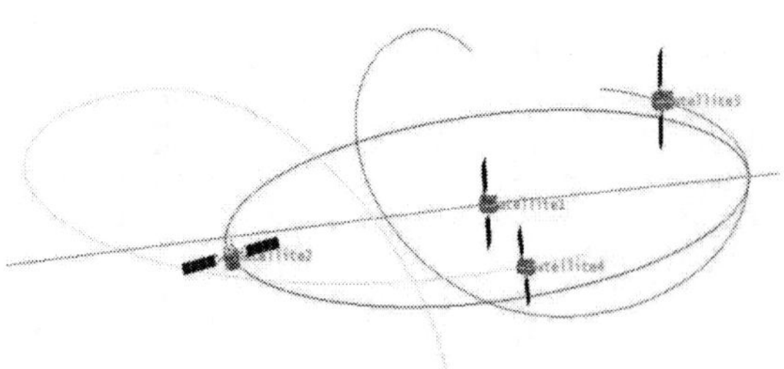

Figure 4. State of motion during reconfiguration(1).

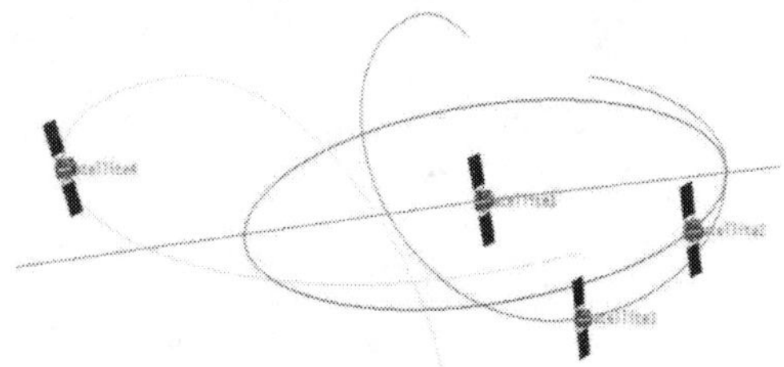

Figure 5. State of motion during reconfiguration(2).

(3) Maneuver is completed, Carpe formation as shown:

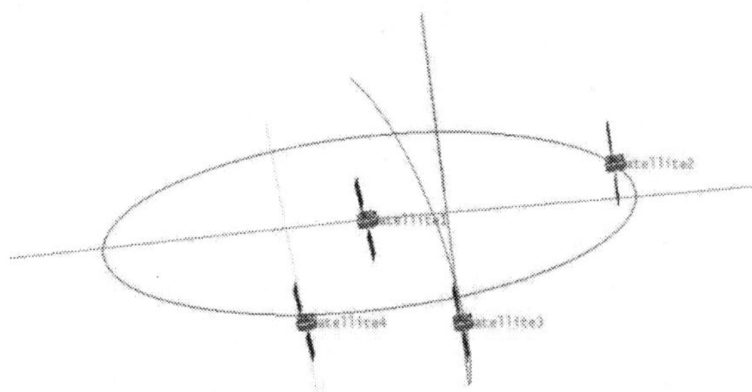

Figure 6. The state after reconfiguration.

According to the goal of the formation reconfiguration, taking advantage of the results which are calculated by the mathematical model to carrying out visualization simulation by STK, the formation of the four spacecraft is changed from the spatial circle formation by applying multiple impulses (Table 4, Table 5),gradually become the CarPe formation, the above analysis process can successfully verify the correctness of the mathematical model and the simple multi-impulse formation reconfiguration.

4 Conclusion

Based on the characteristics of spacecraft formation flight, this paper designs the spatial circular formation, and takes advantage of the simple multi-impulse to complete the formation reconfiguration task, which provides a more effective solution for the complicated and changeable tasks in spatial. The whole formation system cooperates with each other to realize the multi-purpose and multi-tasking of the spacecraft formation network [9-10], which can be applied to the formation design of the specific task.

References

1. Meng Yunhe. Spacecraft formation flight introduction [M], Beijing: National defense industry press, 2014.
2. Pan Keyan. The development of spacecraft formation flight and its key technology [J], Telemetry remote control, 24 (5): 9-15,2003.
3. Jake J.Abbott, Joseph B. Brink, and Braxton Osting. Computing Minimum-Power Dipole Solutions for Interdipole Forces Using Nonlinear Constrained Optimization With Application to Electromagnetic Formation Flight[J], IEEE Robotics and automation letters,2(2) :1008-1014,2017.
4. Wu Xia, Zheng Jianhua. Maintain Control and simulation of formation flight based on STK [J], Computer simulation, 24 (3): 66-68,2007.
5. Joongbo Seo, Youdan Kim.Collision Avoidance Strategies for Unmanned Aerial Vehicles in Formation Flight[J],IEEE Transactions on aerospace and electronic systems,0018-9251(c),2016.
6. Jiang Fanghua. Research on the relative motion of spacecraft formation flight[D], Beijing: Tsinghua University, 2009.
7. Zhang Jinwu. Research on unmanned aerial vehicle formation flight technology[J], Ship electronic engineering,,35 (8): 9-12,2015.
8. Wu Yun-hua, Cao Xi-bin. Research on autonomous relative navigation algorithm of formation flight spacecraft [J]; Journal of Harbin institute of technology, 39 (3): 354-358,2007.
9. Gao Youtao. Research on the modeling and control of flight dynamics of spacecraft formation[D], Nanjing: Nanjing university of aeronautics and astronautics, 2010.
10. Yu Ping, Zhang Honghua. Typical modal and formation control method for elliptical orbit formation flight[J], Journal of astronautics,,26 (1): 7-12,2005.

Automobile safety technology and its improvement

Xiao-Yang Wang[1],Yi-Hang Jiang[1],Ren-Bo Li[2]and Ming-Zhang Chen[3]

[1a]Wuhan University of Technology ,Wuhan, China
1Wuhan University of Technology ,Wuhan, China
[2]Wuhan University of Technology ,Wuhan, China
[3]Wuhan University of Technology ,Wuhan, China

Abstract. With the development of society, and the sharp increase in car sales, traffic accidents have become increasingly frequent and serious .At the same time, with the rapid development of science and technology, automotive technology has also been rapidly improved, automotive safety technology gradually involved in all aspects of the car. This paper introduces the main automotive safety technology at present, proposes a kind of airbag system, and describes its specific working methods.

1 Introduction

With the continuous improvement of the living standards of the society and the people, the automobile industry as an economic pillar is also developing rapidly. In recent years, the car ownership in China has been growing rapidly. Cars are no longer luxuries. According to scientific statistics, China has exceeded 140 million vehicles in 2014, accounting for 2% of the world's total. As early as 2013, the number of vehicles had broken 137 million mark, which is about 5.7 times the size ten years ago, as shown in Figure 1. When the automobile has brought efficient development to our society and rapid growth to our economy, the widespread use of the automobile has also brought us some negative effects, such as traffic congestion, environmental pollution, traffic accidents. Among them, the impact caused by traffic accidents is the most direct, causing the biggest loss, making the most widely damage.

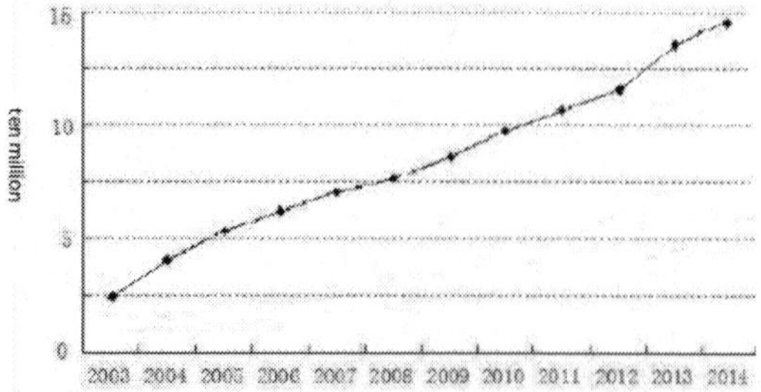

Figure 1. The number of Chinese cars

With the continuous popularization of cars, the number of cars in China is constantly increasing, the speed of cars is increasing, and traffic accidents are becoming more frequent. According to the data on the cause of death of sick residents by the health department, 166,906 people died of motor vehicle accidents in 2012, more than double that of a decade ago, as shown in Figure 2.If appropriate measures are not taken, traffic accidents will be the third leading cause of global diseases and injuries by 2020[1]. In addition, according to the relevant reports, the economic losses of automobile traffic accidents accounted for 1% of GDP in low-income countries and 2% in high-income countries. Global car traffic accident damage is estimated at $518 billion per year[2]. Therefore, the automotive safety technology has become one of the themes of the development of international automotive high and new technology.

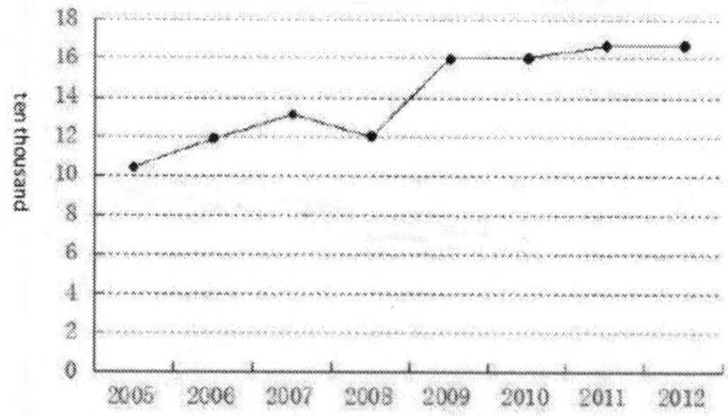

Figure 2. Death toll of motor vehicle accidents

Since 1980s, with the rapid development of computer technology and the development of multi rigid body and finite element software, computer simulation technology has been widely used in the study of vehicle passive safety, pedestrian and occupant protection and related issues. Such as the deduction of mechanical analytical formula of typical body structure components by Wierzbick. T et al. The calculated results are in good agreement with the experimental data, providing sufficient basis for the conceptual design of automobiles. Through the collision test and computer simulation analysis, the vehicle passive safety problems, such as frontal and side crashworthiness, occupant restraint system matching, pedestrian protection and so on, are deeply studied by Huang.M[3]. et al. As shown in Figure 3.Car manufacturers in developed countries have mastered the positive development design technology for the passive safety of vehicles. In addition, the simulation software MADYMO self-developed by TNO company in the Netherlands after joining the AUTO DOE module, successfully achieved the parameter optimization and performance control of the occupant restraint system, and is widely used in automotive passive safety development.

Figure 3. Computer simulation

The energy absorbing space in front of automobile is often limited, so the research and development of materials with high energy absorption efficiency has become the focus of research. Some of the energy absorbing efficiency of higher materials have developed PC/PET materials such as GE, Dow STANDFOAM materials, Tsinghua University and other units are also committed to the development of some new energy absorbing materials, such as solid-liquid composite based on nano porous materials[4].

A parametric human finite element model is developed by University of Michigan and other units, which can significantly improve the development efficiency of a variety of human models[5]. The main idea is based on different characteristics (age, weight and height) of the parametric shape and size of the skeleton, derived from human body model of non standard finite element model standard by radial basis function grid changes, to obtain the characteristic parameters (age, weight, height) finite element model of the corresponding the crowd. It has laid a good foundation for the development of automotive safety technology, and also developed adaptive occupant protection system for various populations and working conditions in the future, so as to cope with various situations of real traffic accidents.

In addition, the study of the driver's attention has also made progress. Scholars have divided the driver's distraction into cognitive distraction and visual distraction, the former refers to the driver's cognitive thinking is occupied by other things (such as voice dialogue), the latter refers to the driver did not focus on the road environment (when using a cell phone or navigation system). The research shows that distraction will significantly increase the driver's workload, make it change the vehicle operating behavior, response time becomes longer, the accident rate increased. For example, Volvo Trucks proposed a solution to monitor the driver's eye movement through the camera to determine whether the driver is distracted, and use the sound and LED lights to provide different levels of reminders. Toyota also uses the camera to observe the driver's head movements in its Pre-crash Safety System, and measure the distance between the upper and lower eyelids to determine whether the driver is dozing.

Some achievements have been made in the development of active safety technology in China. The THMR-V intelligent vehicle developed by Tsinghua University is equipped with CCD camera and navigation system to realize the intelligent driving function. Geely Automobile put forward the concept of GTMS safety development, in order to reduce the casualties of traffic accidents as the development goal, combined with electronic stability control system, adaptive cruise system, front / rear collision warning system and G-Netlink system in one. The safety protection of occupants before collision, collision and collision is realized.In addition, some domestic universities have studied car collisions. The research team of Tsinghua University, Huang Shilin, Zhou Qing and Zhang Jinhuan, carried out a structural transformation design for a van, and tested the crashworthiness of vehicle structure through vehicle crash simulation calculation. Jilin University Lin Yi's research team conducted a structural improvement design and restraint system optimization of the FAW CA6350 van and evaluated the vehicle crashworthiness. Zhang Junyuan independently studied the integrated child restraint system and developed a one-piece child seat to protect the safety of children occupants. Chongqing University Li Xiaohong's team conducted a technical transformation of the Chang'an SC6370 Van and improved the passive safety performance of the vehicle. Hunan University President Zhong Zhihua's research team made technical improvements to the Liuqi N1 Car, set up an airbag research center and developed a variety of car crash occupant protection technology and equipment.

2 Automotive Safety Technology

Karl Benz, a famous German engineer, invented the first car in history in 1886 and so far the reform and development of the automobile industry has lasted for nearly 130 years. In the meantime, automobiles have spread to all parts of the world and have rapidly penetrated into the social economy and people's daily life, and have become the most important and most common means of transportation for people. However, in recent years, due to the rapid popularization of automobiles, traffic safety has become more and more serious. The recent "World Disaster Report" pointed out that about 1.3 million people died in traffic accidents each year, and the number of injuries is up to 30 million. The world lost as much as $3,000 billion due to traffic accidents, of which up to two-thirds were lost in developing countries. Therefore, car safety performance is of great significance to occupant safety.

The R & D of automotive safety technology mainly starting from the whole, the first is the most important thing is to ensure the vehicle safety and comfortable situation as far as possible to prevent the occurrence of traffic accidents, the second is when the vehicle traffic accidents, to minimize the damage caused to the people in the car. At present, automobile safety is divided into two categories: active safety and passive safety.

2.1 Active safety technology

Active safety technology means the reasonable design of automobile and scientific structure in the automotive R & D and production process,using the very advanced technology in automobile industry, as far as possible to avoid or reduce the vehicle in normal driving sudden collision accidents, to improve the safety of their performance, to reduce the risk of a traffic accident then, to reduce the losses caused by the occupant. In recent years, active safety technology has become an important research object of R & D institutions of major automotive companies.

2.1.1 Chassis active control technology

Vehicle active safety control is usually realized by active safety control of chassis. Electronic brake control is one of the most widely used technologies of chassis active control technology. Electronic stability control system (ESC) is considered to be one of the signs of vehicle active safety technology development, by comparing the running state of driver's intention and the actual vehicle system, reasonable distribution of longitudinal and lateral tire force, dynamic behavior and precise control of the vehicle, the driver will try to follow the road, reducing the probability of traffic accident the. According to the U.S NHTSA statistics, ESC can reduce 34% of bicycle accidents and 71% of rollover accidents; for SUV, it can reduce 59% of bicycle accidents and 84% of rollover accidents[6].

2.1.2Security early warning technology

The security warning technology detects the road environment information through the ultrasonic, vision and radar sensors, and analyzes the various information to remind and warn the driver, thereby reducing the occurrence of accidents. At present, the technology mainly through the use of visual sensors in the car, to detect the surrounding environment, auxiliary driver operation. With the development of vision technology, the security warning system based on monocular vision sensor will be more widely used in vehicles, and data fusion with other sensors will further improve the security of security warning technology.

2.2 Passive safety technology

Passive vehicle safety technology refers to the automobile after the collision, using the protection devices to protect the people inside and outside the automobile as much as possible, to minimize damage. Today, most of the car mounted occupant protection devices include anti-collision body, pre tightening seat belts and airbags and so on.

2.2.1 Safety belt

The seat belt was invented by Chaire L. Strath in 1935. It has been used as a must for automobiles. Seat belts are the most representative and major vehicle protection devices in the history of automotive passive safety system's research and development. In the process of automobile traffic accidents, it mainly uses the occupant restraint to absorb and buffer most of the energy generated by collision, so as to avoid the occupants hit into the cockpit or the happening of second collision, as much as possible to reduce occupant injuries and economic losses. According to NHTSA estimates, the use of seat belts reduced 45% of fatal injuries and 50% moderate to severe injuries to car front occupants, while 60% and 65% for light truck occupants. It saved the lives of 12802 people in the United States in 2014[7-8]. The use of seat belts is shown in Figure 4.

Figure 4. Proper use of seat belts

2.2.2 Airbag

Airbags were first invented by the Swedes. The first patent for American airbags was made in the 1950s, and some factories started to develop airbags in the 1970s. In 1972, General Motors took the lead in large-scale airbag field test. After the 1980s, more cars equipped with airbags, airbag technology tended to improve. In the 1990s, the number of airbag installations increased rapidly and there were many new technology products. The first airbag to prevent side impact was first developed by General Motors in 1996. In the 21st century the airbag has become a very common and vital safety device in automobiles.

The airbag is an auxiliary occupant restraint system, which works with the seat belt to prevent occupants from being hurt by the car's interior trim. It mainly includes the controller, sensor, gas generator, airbag bag, warning lamp and other components. After the controller receives

 https://doi.org/10.1051/matecconf/201816005012

the signal, it immediately begins to analyze and judge whether to open the bag. It will immediately to send the ignition instructions if it is necessary. After receiving the ignition command, generator quickly generate enough gas to air filled gas until the airbags fully open now, and then contact with the soft passenger airbag, preventing the occurrence of two collision and car parts, reducing the traffic accidents caused by the occupant injuries and economic losses. The working principle is shown in Figure5.The startig airbag is shown in Figure 6.

Sensor: One of the most important components of a car's airbag, used to capture signals such as speed and acceleration when a car collides, and send the signal to the controller.

Controller: Used to receive and analyze the speed or acceleration signal of the vehicle, and determine whether to send the ignition signal according to the analysis. In the case of collision, the more accurate judgment of the controller, and can accurately control the ignition time, the more the airbag can be opened correctly, to maximize the protection of the occupant.

Gas generator: When the car collides, the gas generator quickly generates enough gas and immediately inflates to the airbag. The gas can not be harmful to the human body, no pollution to the air, and the gas temperature not too high to avoid burning the human body. At the same time, it requires high stability and high reliability.

Airbag bag: when the airbag expands, the explosive force is large and the temperature is high, which requires the airbag bag of high strength, good elasticity, thermal performance is strong, at the same time with high stability, aging resistance, small volume after folding, and can quickly expand, and the air bag can not scratch the human face.

Among them, the gas generator is the main component of the airbag system, its role is to provide sufficient gas with explosive speed in a very short period of time. When the occupant is in contact with the airbag, the airbag hold up the occupant's head, to avoid touching other parts and reduce the injury.

At present, the typical products are active pop-up hood and pedestrian protection airbags. The active pop-up cover is triggered by a warning sensor installed on the bumper and other locations. When the vehicle collides, the hood is popped up by gunpowder or spring to increase the cushioning space to reduce the head and upper limb injury. The technology has been applied in the Jaguar XJ, Citroen C6 and other models. The installation of outer airbag in the car's head also appeared in the Ford, Fiat Stilo, Toyota and other models. The hood airbag developed by Ford in 2007 was mounted above the bumper and up to the hood surface. After that, Toyota launched a 360 degree airbag protection system, between the bumper and the engine cover, between the engine cover and the front windshield, the fender, the rearview mirror were installed five airbags to the front all wrapped, thereby enhancing the pedestrian protection effect.

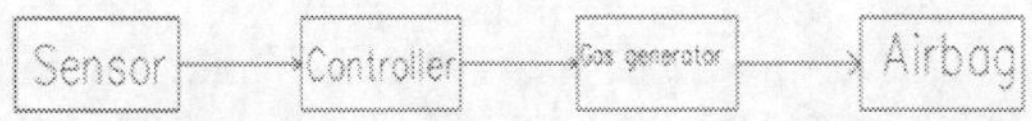

Figure 5. Working principle of airbag

Figure 6. The startig airbag

3 New airbag system

At present, the traffic accidents in bad weather have attracted people's attention, and also exposed the problems in the safety of the car. From the data of deaths from various types of car accidents in China in 2004, turnover accounted for about 60%, while the car fall accounted for about 20%, the collision object fell to 10%~20%. Among them, the death rate of car crashes increased year by year, and the mortality rate was the highest among the bicycle accidents. The US NHTSA statistics show that rollover accidents are a major form of fatal accidents. In 2014, the number of passengers killed by rollover in the United States accounted for 33% of the total number of fatalities[9]. On the other hand, the city flood disaster in recent years in sudden extreme rainstorms caused by frequent, which caused a large number of cars were flooded, vehicle floating or sliding instability, and caused some people drowned. In many traffic accidents, collision accidents account for a large proportion (accounting for about 70%). Collision accidents often lead to economic losses and accident disputes, and even lead to casualties. But at present, when most vehicles collide, only the airbag inside the vehicle is opened, the side protection device is very few, and it can not reduce the damage caused by collision.

Therefore, this paper improves the traditional airbag to reduce the economic losses and casualties caused by traffic accidents and bad rainstorm environment.

3.1 Basic constitution

The new airbag system mainly includes the controller, sensor, gas generator, airbag bag, warning lamp and other components. In addition,the airbag system including cab safety gasbag, set in front of the car rear airbags and set up in the floor and roof of the car body airbags, each equipped with airbag airbag triggering device the pressure sensor; and a start switch are respectively connected with the input end of the controller, the output terminal of the controller and personal safety airbag, airbag and airbag connecting body collision airbag triggering device. In the storm, the airbag is located in the bottom can timely use buoyancy to make the car floating in the water, convenient personnel escape, to avoid drowning personnel. The airbag around the roof and the roof can be started when the car is hit or overturned, reducing the impact and protecting the safety of the people in the car. Install 4 parachute packs on the four corners of the roof to deal with dangerous situations like falling. Installing 4 parachute packs on the four corners of the roof, to deal with dangerous situations like falling.The canopy should be for textile materials, light weight, good elasticity, good chemical stability, the umbrella rope woven rope or hollow core requirements, compact structure, soft, good elasticity, unevenness of elongation is small. The umbrella line is the suture zone, the parachute, rope connecting material to all parts of the requirements high strength, good smoothness and uniform twist stability. In addition, device should also be parachute bag is fixed in the roof, and trigger to open it when necessary.

3.2 Working mode

The working principle of the new airbag system is shown in figure 7.

It is found that the greater the water level is, the greater the force required to escape the door when escaping. Therefore, the shorter the escape time is, the greater the probability of success is. When water depth is 70-80cm away from the ground, it is the critical depth of vehicle escape. Therefore, in the rainstorm environment, the water level rises rapidly, and the judgment of water level is the key. The airbag system installs pressure sensors at the bottom of the car to detect the water depth.

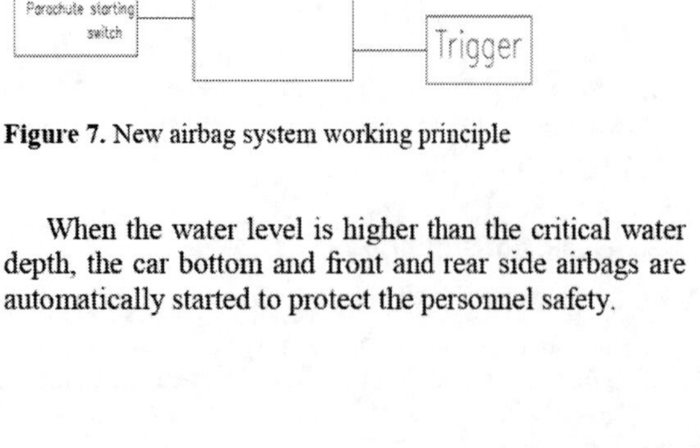

Figure 7. New airbag system working principle

When the water level is higher than the critical water depth, the car bottom and front and rear side airbags are automatically started to protect the personnel safety.

When the car is in danger of collision, that is, when the impact force (deceleration) exceeds the set value, the signal is produced. After starting the system, the four positions airbag is inflated, and the crew airbag system also works.

In addition, when the car has a rollover or falling danger, the system senses the inclination parameters of the vehicle through the vehicle interior attitude sensor. When the angle reaches the critical point, the starting signal is generated. All the airbags inflate and release the four angles parachute, which reduces the impact force of the car landing, and protects the personnel safety as much as possible.

To sum up, the airbag device has a variety of starting methods, including water pressure passive start mode, the active start mode of rainstorm, traffic accident initiation mode, high-altitude falling mode, etc., to adapt to the dangerous situation of the car in different environmental conditions

Water pressure passive start mode: when the vehicle is immersed in water for a certain depth, water pressure reaches a critical value, the water pressure sensor transmits a signal to the MCU, MCU control car airbag, car using buoyancy to float in the water.

The active starting mode of the rainstorm: when the rain enters the car and the car has people or someone finds that the car has been immersed in danger, it starts. The starting system is the radio remote start system. After starting the system, the airbag system of the car bottom and the system of the four position airbag begin to work

The traffic accident start mode: when the car driving start system in a relatively flat area of traffic accident. The collision analysis control is connected to the gate element to control the system. Analysis of the essence of control when the collision impact force (deceleration) signal exceeds the set value, the system. After the start of a four airbag occupant work the airbag system is working.

Falling mode: when the automobile from the start when the parachute into high altitude, with the sensor control and gate components. Essence of sensor control by attitude sensor vehicle status, exceeds the critical value generated signal. After the system starts in four directions and start the air bag and the vehicle airbag, passenger airbag system work.

4 conclusion

Safety is the eternal theme of automobile technology development. With the development of society and economy, Car ownership keeps increasing; the urbanization promotes the change of urban and rural traffic environment; and the problem of automobile safety becomes more and more serious. In this paper, the current automotive safety technology is summarized; in addition, in view of the bad weather vehicle safety problems, a new airbag system is proposed, and its working mode is described. The effect of the new air bag needs to be tested by experiments. In the future, its' safety will be analyzed by the way of physical simulation. It means how this air bag copes with the environment of rainstorm and the roller accidents which lead to high fall accidents ,will be analyzed. Then, further improvements will be made according to the analysis results.

References

1. World Health Organization. World report on road traffic injury prevention [R]. Geneva: WHO, 2004. (in Chinese)
2. WANG Xiaojing. The review of ITS for traffic safety [C]// 2008 Fourth China Intelligent Transport Annual Meeting Thesis, 2008-09. (in Chinese)
3. Wiezbicki,T.,Recke.L.,Abramowicz.W.,Gholami.T.,J .Huang.Stress Profiles in Thin-Walled Prismatic Columns Subjected to Crush Loading-11.Bending,Computers&Structures Vol.51No.6,P625-64 1.
4. SUN Yueting, XU Xiaoqing, XU Chengliang, et al. Elastomeric cellular structure enhanced by compressible liquid filler [J]. Sci Rep, 2016, 6: 26694.
5. SHI Xiangnan, CAO Libo, Reed M P, et al. Effects of obesity on occupant responses in frontal crashes: a simulation analysis using human body models [J]. Comput Methods Biomech Biomed Eng, 2015, 18: 1280-1292
6. NHTSA-200727662, Federal motor vehicle safety standards; electronic stability control systems; controls and displays [S]. Washington DC: National Highway Traffic Safety Administration, 2007
7. National Center for Statistics and Analysis. Lives saved in 2014 by restraint use and minimum-drinking-age laws, DOT HS 812 218 [R]. Washington, DC: National Highway Traffic Safety Administration, 2015
8. Kahane C J. Lives saved by vehicle safety technologieand associated Federal Motor Vehicle Safety Standard1960 to 2012, DOT HS 812 069 [R]. Washington, DC:
9. National Center for Statistics and Analysis. Passenger vehicles: 2014 data (Traffic safety facts), DOT HS 812 302 [R]. Washington, DC: National Highway Traffic Safety Administration, 2016.

Adaptive GridMap Building Algorithm for Mobile Robot Based on Multiway Tree

Jing Zhou[1,2], Xiaogang Ruan[1,2], Pengfei Dong[1,2] and Jingjing Zhang[1,2]

[1]*Faculty of Information Technology, Beijing University of Technology, Beijing 100124, China*
[2]*Beijing Key Laboratory of Computational Intelligence and Intelligent System, Beijing 100124, China*

Abstract. In order to solve the problem that uniform grid map occupies large storage space in the process of large-scale map creation, a new adaptive grid map building algorithm based on multiway tree is proposed. The environment map is divided into nine grids. The algorithm determines whether the grid is completely occupied, partially occupied or vacant. Then the part occupied gridsare further subdivided into smaller grids. The algorithm continues to determine whether the smaller grid is completely occupied, partially occupied or vacant.Repeat the above segmentation process until the entire map search is completed and the accuracy requirements are met. Finally, the scale of grid map does not depend on human experience but presenting adaptive characteristics. The algorithm is compared with the uniform grid method.The simulation results show that the algorithm has convergence.Andcompared with the uniform scale raster map, itgreatly saves the storage space.

1 Introduction

Three key issues of mobile robot navigation are positioning, map building and path planning[1]. The autonomous map building in the field of robotics research is a fundamental and important problem, which is the precondition to achieve autonomous navigation.

At present, the most widely used methods of map creation are geometric feature method, topological method and grid method.

A geometric feature map is represented by related geometric features (points, lines, planes). This method can find the shortest path, but the disadvantage is that the more obstacles.The longer the search time, especially for the search of circular obstacles may be invalid. In 2011, Zhu Jianguo et al. [2] extracted linear features from lidar data to describe the environment. The established environmental feature maps were accurate and effective, and the data were small compared with raster maps. In 2013, GuoYupeng [3] used lidar data to extract the geometric features of the environment for the semantic classification of indoor environment.

Topological map is a topological structure graph showing indoor environment as node and related connection line. The composition is fast and simple, but it is difficult to create large scale maps, and has no discrimination ability for similar or symmetrical scenes. In 1994, Gaussier P, ZrehenS[4] and other online learning topological graph were used to realize obstacle avoidance of mobile robot. In 2012, W Yan and others [5] built Growing Neural Gas (GNG) indoor environment cognitive map, and the robot's position was obtained through the global image machine on the ceiling. Wang Dongshu et al. [6] constructed a model of nodes with growing neural gas network as a topological node, which makes the robot easy to understand the environment. In 2013, RuanXiaogang et al.[7] proposed a dynamically increasing and decreasing self-organizing feature map (DGPSOM) algorithm, which uses a small number of neurons to represent a large amount of environmental information. In 2017, Alitappeh, RJ, et al. [8] proposed an efficient distributed deployment strategy, which can optimize the allocation of robot teams in the environment of topological graph representation.

Grid method is suitable for a variety of complex environments, and it is better than geometric feature method and topology method in modeling, updating and processing. Therefore, the method is favored by scholars and researchers in recent years. Grid map was first proposed by Elfes and Moraves[9] in 1985. The idea is to divide the whole work environment into homogeneous grids, and point out the possibility of obstacles in each grid. In 1990s, Konolige[10] and Thrun[11] established and optimized the grid map method, and applied it to the problems such as Simultaneous Localization and Mapping (SLAM). The geometric topological graph and the grid topological graph are established respectively for the sacred [12] and the ThrunS[13]. In 2010, Li Tiancheng[14] and so on based on fan grid map path planning for robots, increase the diversity of grid morphology. GuoLijin [15] and Duan Yong [16] section proposed the creation algorithm of adaptive grid map of variable precision, instead of the traditional uniform scale map with variable precision raster map. That saves the storage space. Although the grid method in robot environment cognition and map building showed many advantages, in the construction of large scale environment map.It still exists some shortcomings with the map to expand the scope and improves the grid resolution.The algorithm will consume huge storage space with great computational complexity.

Therefore, in order to solve the problem of occupying large amount of memory space in the process of large scale map creation, a new adaptive raster map building algorithm based on multiway tree is proposed.

2 Related works

2.1 Method for creating uniform raster map

The grid map usually refers to the traditional uniform scale raster map. The method first divides the external environment into many small units called grids. Then uses probability values to represent the possibility that each grid is occupied. A uniform scale raster map divides the whole environment into a homogeneous unit gridand each grid giving a value of 0-1 intervals to represent the state of the grid. 0 means completely empty, and 1 means full occupancy. In this way, the whole environment space can be divided into occupied space and vacant space.

An environment map represented by a grid. The resolution of the environment space is related to the size of the grid. The traditional grid scale is determined by human experience.When the environment changes, you need to redetermine the scale by human experience. Increasing resolution means increasing computing time and computer memory consumption. With the expanding of human detection environment and unstructured environment, adding high dimensional entity, these not only lead to environmental information problem caused the massive increase of information storage, but fusion of massive data become more application of grid method for obstacle.

2.2 Grid method based on quadtree

Document [9] adopts the principle of four fork tree to construct adaptive raster map, which belongs to a kind of variable precision raster map. The quadtree of each tree is divided into 4 sub nodes. Each node also has three states, according to the different state of continued segmentation. The composition will not stop until the entire map exists only completely occupied and vacant two states. The composition method optimizes the traditional method to a certain extent and saves the storage space, but the real-time composition of the composition needs to be improved.

In this paper, the existing grid method is improved from two aspects of saving data storage space and improving the real-time composition.

3 Adaptive raster map creation algorithm

3.1 The basic principle of multiway tree

The basic idea of adaptive raster map building algorithm stems from multiway tree. Different from the general multiway tree, this tree must follow two basic principles.

(1)The gray node is uncertain state nodes. Each graynode should continue to be divided into nine sub nodes.

(2)Black and white nodes determine the statement ofthenodes.Black represents to be completely occupied and white say vacant.They are leaf nodes, indivisible. The structure of the multiway tree is shown in Figure 1.

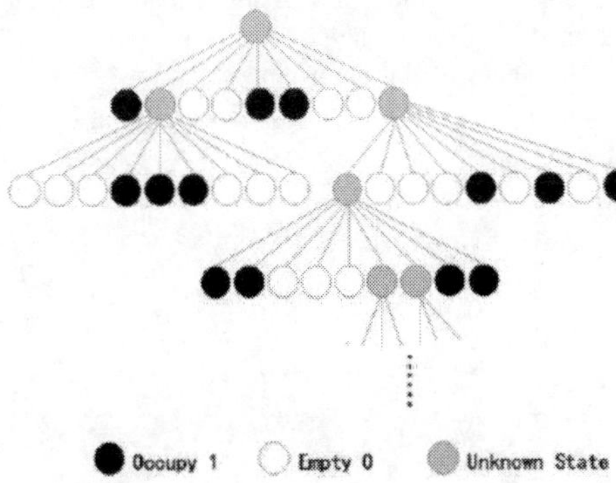

Figure 1. Multiway tree structure diagram.

3.2 Adaptive raster map creation process

According to the principle of the multiway tree, an adaptive raster mapis created, which called variable precision raster map. The segmentation process of the raster map is shown in Figure 2.

(1) The robot roughly circles around the whole environment, and divides the whole unknown environment into nine square structures.

(2) Firstly, judge the nine sub grid state grid which is empty or occupied and which is in an unknown state. Secondly, the unknown grid environment for detailed exploration are further subdivided into small squares structure.

(3) Update the map.

(4) Repeat steps (2) and (3) until the entire map search is completed and the accuracy requirements are met. Then the composition is completed.

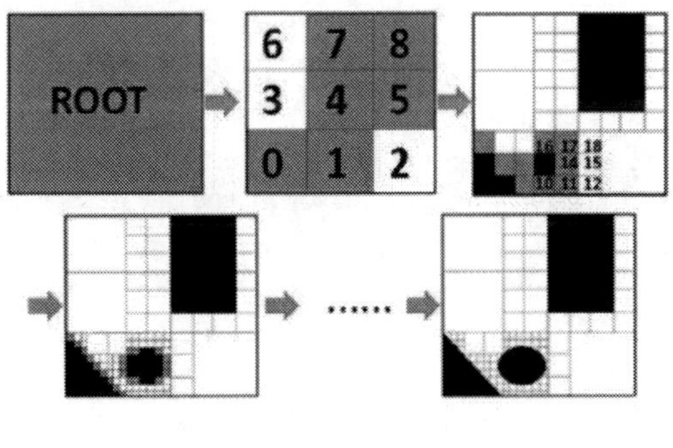

Figure 2. Adaptive raster map creation process.

3.3 Adaptive raster map representation

The father grid is in an unknown state grid segmentation, nine sub grid in the position number of the layer as shown in Figure 2. The N subdivision is obtained after the grid serial number is $X_1X_2X_3...X_n$ (n bits). X_i, which according to the position of grid at the i layer, selecting from 0,1,2,3,4,5,6,7 and 8. For grids that continue to be down, taking the left to right order to add the location number of the sub grid to the back of the parent grid. If the grid number after three times splits is 152, a subdivision is continued.The nine sub grid is numbered1520, 1521, 1522, 1523, 1524, 1525, 1526, 1527, 1528. The location of the grid in the map can be obtained by grid numbers.

The scale of the environment map is $a \times b$. The grid is numbered $X_1X_2X_3...X_n$, and the central coordinate of the grid is (x, y).Then (x, y) can be expressed by the following formula.

$$\begin{aligned} x &= a\cdot\left(\frac{1}{2}+\frac{1}{3}\cdot\lambda_1+\frac{1}{3^2}\cdot\lambda_2+\frac{1}{3^3}\cdot\lambda_3+...+\frac{1}{3^n}\cdot\lambda_n\right) \\ y &= b\cdot\left(\frac{1}{2}+\frac{1}{3}\cdot\varphi_1+\frac{1}{3^2}\cdot\varphi_2+\frac{1}{3^3}\cdot\varphi_3+...+\frac{1}{3^n}\cdot\varphi_n\right) \end{aligned} \quad (1)$$

When $X_1, X_2, X_3,..., X_n$ take 0, 3, 6, $\lambda_1, \lambda_2, \lambda_3,...., \lambda_n$ =-1.
When $X_1, X_2, X_3,..., X_n$ take 1, 4, 7, $\lambda_1, \lambda_2, \lambda_3,...., \lambda_n$ =0.
When $X_1, X_2, X_3,..., X_n$ take 2, 5, 8, $\lambda_1, \lambda_2, \lambda_3,...., \lambda_n$ =1.
When $X_1, X_2, X_3,..., X_n$ take 0, 1, 2, $\varphi_1, \varphi_2, \varphi_3,...., \varphi_n$ =-1.
When $X_1, X_2, X_3,..., X_n$ take 3, 4, 5, $\varphi_1, \varphi_2, \varphi_3,...., \varphi_n$ =0.
When $X_1, X_2, X_3,..., X_n$ take 6, 7, 8, $\varphi_1, \varphi_2, \varphi_3,...., \varphi_n$ =1.

3.4Probability for occupying grids

Suppose the robot sensor detects the distance range $[d_{\min}, d_{\max}]$, the angle range is $[\phi_{\min}, \phi_{\max}]$. The final state of each gridis x_i, namely one of two states, each state $\{OCC, EMP\}$ .Correspond to a probability value according to the probability value of the map update.The initial state probability of x_i is set $p\{x_i = OCC\} = p\{x_i = EMP\} = 0.5$, namely grid undetected. When the robot moves to steps, the sensor sees an obstacle in front of it and the sensor reads $D(s) = [d(s), \phi(s)]^T$. As is shown in Figure 3.

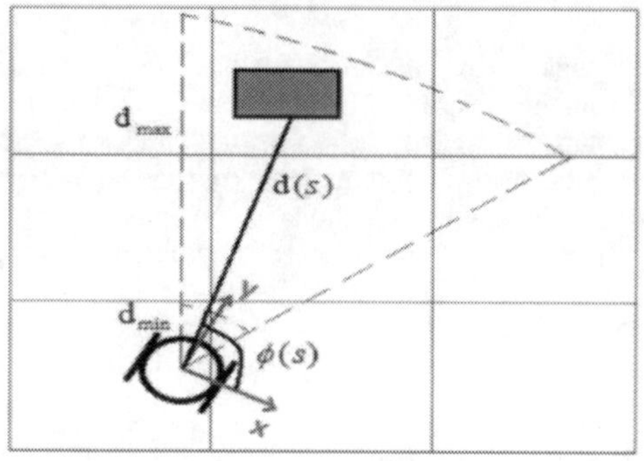

Figure 3. Laser sensor ranging principle.

Estimate which grids are covered according to the position and pose of the robot $X_z(s)$. Calculate the probability value of the covered grid and update it. The return value $d(s)$ and return value $\phi(s)$ based sensor angle and distance are used to calculate the probability of obstacles in grid i. The algorithm is as follows.

$$p\{x_i(s) = OCC \mid X_Z(s), d(s)\} = \frac{(d_i - d_{\min})^2}{(d(s) - d_{\min})^2} \quad (2)$$

$$p\{x_i(s) = OCC \mid X_Z(s), \phi(s)\} = 1 - \frac{(\phi_i - d_{\min})^2}{\phi_{\max}{}^2} \quad (3)$$

Among them, d_i is the distance between the robot and the center of the grid i. ϕ_i is the angle between the x axis of the robot coordinate system and the center of the grid i. Assuming that the two events of $d(s)$ and $\phi(s)$ are independent of each other, the probability of obstacles in the grid i can be obtained according to the robot's current pose and sensor readings.

$$\begin{aligned} & p\{x_i(s) = OCC \mid X_Z(s), D(s)\} = \\ & p\{x_i(s) = OCC \mid X_Z(s), d(s), \phi(s)\} = \\ & p\{d(s) \mid X_Z(s), \phi(s), x_i(s) = OCC\} \times \\ & p\{x_i(s) = OCC \mid X_Z(s), \phi(s)\} / \\ & p\{d(s) \mid X_Z(s), \phi(s)\} = \\ & p\{d(s) \mid X_Z(s), x_i(s) = OCC\} \times \\ & p\{x_i(s) = OCC \mid X_Z(s), \phi(s)\} / \\ & p\{d(s) \mid X_Z(s)\} \end{aligned} \quad (4)$$

Also as we have

$$\begin{aligned} & p\{d(s) \mid X_Z(s), x_i(s) = OCC\} = \\ & p\{x_i(s) = OCC \mid X_Z(s), d(s)\} \times \\ & p\{d(s) \mid X_Z(s)\} \end{aligned} \quad (5)$$

$$\begin{aligned} & p\{d(s) \mid X_Z(s)\} = \\ & p\{x_i(s) = OCC \mid X_Z(s)\} \times \\ & p\{d(s) \mid X_Z(s)\} \end{aligned} \quad (6)$$

Therefore, the formula (4) is used

$$\begin{aligned}&\mathrm{p}\{x_i(s)=OCC\mid X_Z(s),D(s)\}=\\&\mathrm{p}\{x_i(s)=OCC\mid X_Z(s),\mathrm{d}(s)\}\times\\&\mathrm{p}\{x_i(s)=OCC\mid X_Z(s),\phi(s)\}/\\&\mathrm{p}\{x_i(s)=OCC\mid X_Z(s)\}\end{aligned}\tag{7}$$

Similarly, according to the robot's current pose and sensor readings, the probability that there is no obstacle in the grid i can be obtained.

$$\begin{aligned}&\mathrm{p}\{x_i(s)=EMP\mid X_Z(s),D(s)\}=\\&\mathrm{p}\{x_i(s)=EMP\mid X_Z(s),\mathrm{d}(s)\}\times\\&\mathrm{p}\{x_i(s)=EMP\mid X_Z(s),\phi(s)\}/\\&\mathrm{p}\{x_i(s)=EMP\mid X_Z(s)\}\end{aligned}\tag{8}$$

Also as we have

$$\begin{aligned}&\mathrm{p}\{x_i(s)=EMP\mid X_Z(s),D(s)\}=\\&1-\mathrm{p}\{x_i(s)=OCC\mid X_Z(s),D(s)\}\end{aligned}\tag{9}$$

$$\begin{aligned}&\mathrm{p}\{x_i(s)=EMP\mid X_Z(s),\mathrm{d}(s)\}=\\&1-\mathrm{p}\{x_i(s)=OCC\mid X_Z(s),\mathrm{d}(s)\}\end{aligned}\tag{10}$$

$$\begin{aligned}&\mathrm{p}\{x_i(s)=EMP\mid X_Z(s),\phi(s)\}=\\&1-\mathrm{p}\{x_i(s)=OCC\mid X_Z(s),\phi(s)\}\end{aligned}\tag{11}$$

And at the statement of without having the sensor observation value *D*(s)

$$\begin{aligned}&\mathrm{p}\{x_i(s)=EMP\mid X_Z(s)\}=\\&\mathrm{p}\{x_i(s)=OCC\mid X_Z(s)\}\end{aligned}\tag{12}$$

Substitute (8) ~ (11) for (7)

$$\begin{aligned}&\mathrm{p}\{x_i(s)=OCC\mid X_Z(s)\}=\\&1-[1-\mathrm{p}\{x_i(s)=OCC\mid X_Z(s),\mathrm{d}(s)\}]\times\\&[1-\mathrm{p}\{x_i(s)=OCC\mid X_Z(s),\phi(s)\}]/\\&[1-\mathrm{p}\{x_i(s)=OCC\mid X_Z(s),D(s)\}]\end{aligned}\tag{13}$$

To substitute (13) for (7)

$$\begin{aligned}&\mathrm{p}\{x_i(s)=OCC\mid X_Z(s),D(s)\}=\\&\mathrm{p}\{x_i(s)=OCC\mid X_Z(s),\mathrm{d}(s)\}\times\\&\mathrm{p}\{x_i(s)=OCC\mid X_Z(s),\phi(s)\}/\\&[1+2\times\mathrm{p}\{x_i(s)=OCC\mid X_Z(s),d(s)\}\times\\&\mathrm{p}\{x_i(s)=OCC\mid X_Z(s),\phi(s)\}-\\&\mathrm{p}\{x_i(s)=OCC\mid X_Z(s),d(s)\}-\\&\mathrm{p}\{x_i(s)=OCC\mid X_Z(s),\phi(s)\}]\end{aligned}\tag{14}$$

The probability that there is no obstacle in grid i is

$$\begin{aligned}&\mathrm{p}\{x_i(s)=EMP\mid X_Z(s),D(s)\}=\\&1-\mathrm{p}\{x_i(s)=OCC\mid X_Z(s),D(s)\}\end{aligned}\tag{15}$$

Next, you need to update the grid map. The convergence of the algorithm is proved that when the probability of the grid occupancy is greater than or equal to $\mathrm{p}_{\max}$, the grid is occupied and represented by 1. However, when the probability of the occupancy of the grid is less than or equal to $\mathrm{p}_{\min}$, the grid is considered empty and represented by 0. Besides, when the grid scale is less than half of the length and width of the robot, it is no longer subdivided. Under this condition, if the minimum level of the grid is occupied by a probability greater than 0. Otherwise, the grid is assumed to be occupied and represented by 1. If the smallest grid is occupied, the probability is equal to 0 and the grid is empty, expressed in 0. According to the above analysis, it can be seen that the environment map constructed by the algorithm has the condition of ending the segmentation explicitly. So the convergence of the adaptive grid algorithm is proved.

4 Simulation experiment

Suppose that the simulation environment is a $50m\times50m$ static indoor environment, as isshown in Figure 4.

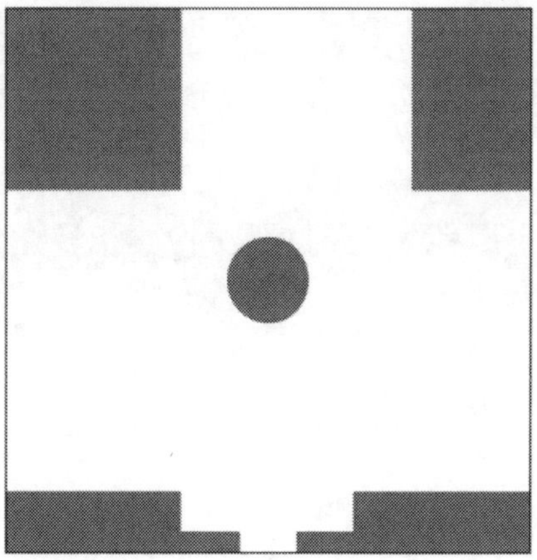

Figure 4. Static simulation environment.

The simulation experiment is carried out on the Turtlebot2 robot simulation system and the robot diameter is 42cm. The data collection of motion model is completed by odometer and gyroscope. The data acquisition of measurement model is completed by RPLIDAR-A2 360 degree laser radar scanner. The laser radar's height is 40.5mm, and the width is 72.5mm. The range of distance is the best of 0.15m~8m. The scanning angle is from 0 degree to 360 degree. The range resolution and the angular resolution is respectively less than 0.5mm and 0.9 degrees. A single ranging time is 0.25ms. The measuring frequency of laser radar is more than 4000Hz and the scanning frequency is 10Hz. Besides, the typical frequency of scanning one week is the situation of 400 sampling points in one scan. And in this simulation environment can achieve fast and accurate positioning. The covariance matrix of the motion noise is $Q(s)=diag\{0.1^2,0.1^2,0.02^2,(1.0\times\pi/360)^2,...,(1.0\times\pi/360)^2\}$, and the covariance matrix of the observed noise is $R(s)=diag\{0.02^2,(0.2\times\pi/360)^2\}$.

The traditional uniform scale grid method and the adaptive grid algorithm based on multiway tree are used to compose the simulation environment of Figure 4. The simulation results are compared with those shown in Figure 5 (a) and Figure 5 (b).

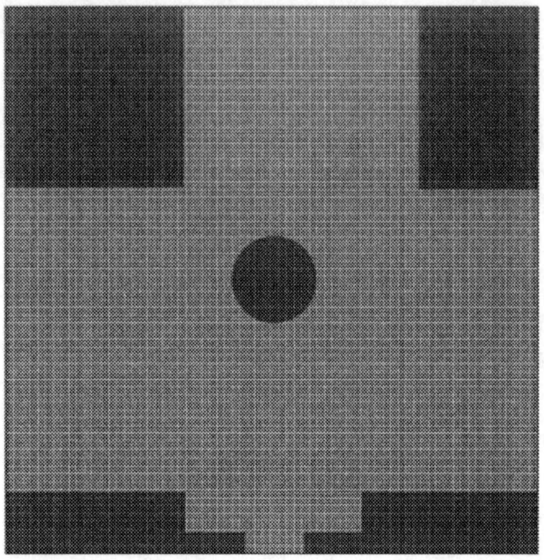

Figure 5(a). Uniform scale grid method composition.

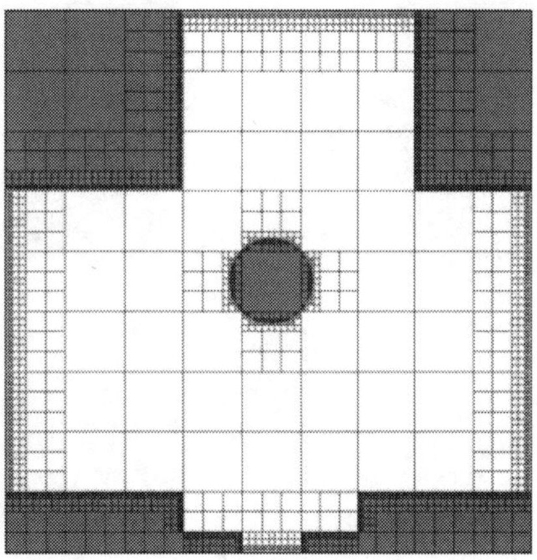

Figure 5(b). Adaptive grid composition.

The traditional uniform grid method and the adaptive grid algorithm based on the multiway tree simulation results data statistics are shown in Table 1.

Table 1. Data statistics of simulation results.

	Occupy storage space	Running time
Uniform scale grid map	3.6M	0.86s
Adaptive raster map	0.25M	2.08s

Figure 5 (a), (b) and Table 1 simulation results show that the adaptive grid map construction process consumes more time than traditional uniform scale map. But the storage space is saved by more than 10 times, reaching the purpose of designing the algorithm.

5 Conclusion and prospect

An adaptive grid map building algorithm for mobile robots based on multiway tree is proposed in this paper. Grid partition based on the idea of multiple tree. Each time, dividing the grids into full occupied, partially occupied or vacant according to the state of them. And then gradually in-depth subdivision until the entire map is completely in two states about occupied and vacant. The algorithm solves the problem that the traditional uniform scale raster map occupies massive storage space in the process of large-scale map creation and avoids the data explosion caused by the expansion of search range. This paper compares the adaptive raster map creation algorithm with the uniform scale raster map creation algorithm. Simulation results show that the algorithm not only has the convergence, but also saves the storage space greatly.

The adaptive grid map creation algorithm based on multiway tree is applied to the large scale environment with uneven distribution of obstacles, which can obtain obvious advantages. But for distributed small scale environment, the advantage of the algorithm is not outstanding. Because most of the grids are constantly segmented in a uniform environment. And this process will consume more time, and save less storage space. How to reduce the time consumption and reduce the waste of storage space is a subject worthy of further study.

References

1. Z.W. Wang,G. Guo, *Status and Prospect of mobile robot navigation technology*. J. Robot. **25**(5):470-474 (2003)
2. J.G. Zhu, Y. Gao, K.J. Li, X.S. Gao. *Indoor mobile robot in unknown environment to create the feature map of.* J. Computer Measurement and Control. **19** (12): 3044-3046(2011)
3. Y.M. Guo, S.R. Liu, B.T. Zhang. [J]. *Journal of Ningbo University*. J. **26**(4):35-39(2013).
4. P. Gaussier, S. Zrehen, *Topological Neural Map For Online Learning - Emergence Of Obstacle Avoidance In A Mobile Robot. Z.* BRIGHTON, ENGLAND:M I T PRESS, 55 HAYWARD ST, CAMBRIDGE, MA 02142 (1994)
5. W. Yan, C. Weber, S. Wermter, *A neural approach for robot navigation based on cognitive map learning. C.* International Joint Conference on Neural Networks. IEEE. 1-8 (2012)
6. D.S. Wang, H.T. Wang. University. *Environment exploration and map building of autonomous robot in unknown environment*. J. Journal of Zhengzhou. **45** (4): 52-57(2013)

7. X.G. Ruan, J.X. Gao. *Research on robot path planning in dynamic changes based on SOM diagram. J. Control Engineering*. **20**:61-68(2013)
8. R.J. Alitappeh, GAS Pereira, A.R. LCA Araujo, Pimenta, *Multi-robot Deployment using Topological Maps*.J.JOURNAL OF INTELLIGENT & ROBOTIC SYSTEMS. **86**(3-4):641-661 (2017)
9. H.P. Moravec, A. Elfes, *High resolution maps from wide-angle sonar*. Proceedings of the 1985 IEEE International Conference on Robotics and Automation . St. Louis USA:IEEE Press. 116-121 (1985)
10. K. Konolige, *Improved Occupancy Grids for Map Building*. J. Autonomous Robots. **4**(4):351-367 (1997)
11. S. Thrun, D. Fox, F. Burgard, *Probabilistic Mapping Of An Environment By A Mobile Robot*.IEEE International Conference on Robotics and Automation, 1998. Proceedings. 1546-1551 (1998)
12. Y. Zhuang, X.D. Xu, W. Wang. *Construction and self localization of mobile robot geometry topological Hybrid Map*. J. Control and Decision Making, **20** (7): 815-822(2005)
13. S. Thrun, *Learning metric-topological maps for indoor mobile robot navigation*. J. Artificial Intelligence. **99**(1):21-71 (1998)
14. T.C. Li, S.D. Sun, Y. Gao, *Global path planning of mobile robot based on fan grid map*.J.Robot.**32**(4):547-552 (2010)
15. L.J. Guo, W.X. Shi, Y. Li, F.X. Li, *Adaptive raster map creation algorithm based on four fork tree*. J. Control and Decision Making. **26**(11):1690-1694 (2011)
16. Y. Duan, X.H. Xin, *Navigation of Mobile Robot Based on Uncertain Grid Map*. J. Control Theory& Applications .**23**(6):1009-1013 (2006)

Simultaneous Localization and Mapping of Mobile Robot Based on Improved RBPF

Jinging Zhang[1,2], Xiaogang Ruan[1,2], Pengfei Dong[1,2] and Jing Zhou[1,2]

[1]*Faculty of Information Technology, Beijing University of Technology, Beijing 100124 , China*
[2]*Beijing Key Laboratory of Computational Intelligence and Intelligent System, Beijing 100124, China*

Abstract. The traditional SLAM based on RBPF has the problem of constructing high-precision map which requires large amounts of particles to make the calculation complexity and the phenomenon of particle depletion caused by particle degradation. Aiming at these problems, an improved RBPF particle filter based on adaptive bacterial foraging optimization algorithm and adaptive resampling is proposed for mobile robot SLAM problem. Firstly, the introduction of adaptive bacterial foraging algorithm to RBPF making the distribution of particles before resampling closer to the real situation. Then use the adaptive resampling method makes the newly generated particles closer to the real movement, thereby increasing the robot position estimation accuracy and map creation accuracy. The experimental results show that this method can improve the practicability of the system, reduce the computational complexity, improve the operation speed and get more effective particles while guaranteeing the accuracy of the grid map.

1 Introduction

Robot positioning is to determine the location of the robot in the environment, and in order to build an unknown environment map and must understand the location of the robots' position.Early studies have independently studied the two models, ignoring their relevance and making the study fall into a temporary pause. In 1986, Smith and Cheesemanet proposed the SLAM problem at first [1].The mobile robot starting from an unknown location, uses the sensor to observe the environment incrementally during the movement while synchronizing the established map based on the established map. The problem has always been a hotspot and difficult point in the field of mobile robot research [2], which is considered to be the key to the real autonomy of robot navigation.

At present, the solution of SLAM problem is mainly based on the Kalman filter (KF) algorithm and the algorithm based on particle filter (PF). Kalman filter based on the various methods often require a relatively high degree of linear system, and requires a larger amount of computing and storage space. The particle filter is a basic statistical tool, the core is based on the Bayesian sampling estimation of the order of important sampling (sequential importance sampling, SIS). It becomes the more commonly used SLAM solution because it has relatively short storage space, the system of linear and high Gaussian requirements are not high, when the number of particles enough time to approximate any probability distribution and so on.

The Rao-Blackwellized Particle filters proposed by Doucet et al. are effective methods to solve SLAM problems [3]. Each particle represents a possible motion trajectory of the robot and has its own global map [4].

They correspond to the trajectories of the particles. The algorithm can well approximate the joint probability density of the robot pose and the environment map. However, when the number of particles and the size of the environment map increase, the algorithm will take up a lot of memory space, and the global map copy need to take up a lot of memory space in resampling process [5].While the resampling process ignores a large number of useful particles, resulting in reduced particle diversity. Therefore, while reducing the computational complexity and reducing the number of particles required to build an accurate map, it is necessary to ensure that the particle diversity becomes a problem that needs to be addressed. The application of the algorithm has been improved in [6-9], but there are still some shortcomings.

This paper applies the bacteria foraging optimization algorithm and system resampling to the mobile robot SLAM problem based on the existing research. Compared with the traditional method, this method introduces the bacterial foraging algorithm into RBPF and the latest robot observation information into the sampling distribution, and makes the distribution of the particles close to the real motion before resampling, and accelerates the convergence of the particles. So the number of particles used in the traditional SLAM method can be reduced while the accuracy of the state prediction is guaranteed, and the calculation amount of the algorithm can be effectively reduced. In addition, considering that the resampling process is susceptible to sample depletion, the reuse of the system can be used to improve the particle degradation phenomenon while maintaining the diversity of particles and makes the newly generated particles closer to the real movement.

2 General description of SLAM based on RBPF

2.1 Rao-Blackwellized transform

The key idea of SLAM based on RBPF is to estimate its possible pose information $x_{1:t} = x_1, x_2, \cdots, x_t$ that is, posterior probability $p(x_{1:t}|z_{1:t}, u_{1:t-1})$, through the robot's observation information $z_{1:t} = z_1, z_2, \cdots, z_t$ and the odometer control information $u_{1:t} = u_1, u_2, \cdots, u_t$, and use it to compute the union based on the joint posterior probability distribution map m and pose $x_{1:t}$,

$$p(x_{1:t}, m \mid z_{1:t}, u_{1:t-1}) = p(m \mid x_{1:t}, z_{1:t}) \bullet p(x_{1:t} \mid z_{1:t}, u_{1:t-1}) \quad (1)$$

From the formula (1) we can see that the posterior probability $p(m|x_{1:t}, z_{1:t})$ of the map is calculated based on the known $x_{1:t}$and $z_{1:t}$. The algorithm uses the particle filter to estimate the posterior probability $p(x_{1:t}|z_{1:t}, u_{1:t-1})$ of the robot pose, and each particle represents a possible path. For $p(m|x_{1:t}, z_{1:t})$, we can use kalman filter algorithm for analytical calculation after obtaining $p(x_{1:t}|z_{1:t}, u_{1:t-1})$.

2.2 Use RBPF for map building

The general procedure for RBPF-SLAM is as follows:

Step1.Initialization:When t = 0s, according to the robot motion model prior probability $p(x_0)$ selected N particles, recorded as $x_0^{(i)}(i = 1,2, \cdots N)$, each particle corresponding to the weight $\omega_0^{(i)} = 1/N$, according to the target state a priori probability$(\tilde{x}_0^{(i)}, p_0^{(i)})$.

Step2.Sampling: Generates the next generation particle set $\{x_t^{(i)}\}_{i=1,2,\cdots,N}$ from the particle set $\{x_{t-1}^{(i)}\}_{i=1,2,\cdots,N}$ according to the proposed distribution of π samples, usually using the odometer motion model$p(x_t|x_{t-1}^{(i)}, u_{t-1})$ as the proposed distribution π .

Step3.Importance weight calculation: According to the principle of importance sampling, the importance weight of each particle is calculated by:

$$w_t^{(i)} = \frac{p(x_{1:t}^{(i)}|z_{1:t}, u_{1:t-1})}{\pi(x_{1:t}^{(i)}|z_{1:t}, u_{1:t-1})}$$
$$\propto w_{t-1}^{(i)} \cdot \frac{p(z_t|m_{t-1}^{(i)}, x_t^{(i)})p(x_t^{(i)}|x_{t-1}^{(i)}, u_{t-1})}{\pi(x_t|x_{1:t-1}^{(i)}, z_{1:t}, u_{1:t-1})} \quad (2)$$
$$\propto w_{t-1}^{(i)} p(z_t|m_{t-1}^{(i)}, x_t^{(i)})$$

Step4.Resampling: resampling when the number of valid particles is less than the preset threshold value Nth. After resampling, all particles have the same weight.

Step5.Update map: update the corresponding map $p(m^{(i)}|x_{1:t}^{(i)}, z_{1:t})$ according to the pose $x_{1:t}^{(i)}$ and historical observation information $z_{1:t}$ of the particle.

It is easy to calculate when using the odometer motion model as a proposed distribution, but when the accuracy of the sensor information is significantly higher than that of the odometer movement model, only a small fraction of the particles is located in the observed high likelihood region so that they increase in the updated weight. In addition, resampling leads to particle degradation, loss of important particles, and therefore requires a large number of particles to fully cover the high observed likelihood of the region, thereby increasing the computational complexity.

3 Improved RBPF-SLAM algorithm

3.1 Brief introduction of bacteria foraging optimization algorithm(BFOA)

Passino was inspired by the foraging behavior of escherichia coli in 2002 and proposed a BFOA (bacteria foraging optimization algorithm) [10]. This is a simple and effective random global optimization algorithm, which points out a new direction for practical engineering problems that many traditional optimization methods can not satisfy [11]. BFO algorithm is not yet perfect because the proposed time later, and the country is also from the beginning of 2007 to study it[12].

The BFO algorithm implements global search through collaboration and competition among individuals. The system is initialized into a set of random solutions. Each bacterium corresponds to a random solution, which is updated by chemotaxis [13]. The replication process is used to optimize the survival process of the fittest and the local search stagnation problem is improved by migration. These three steps are nested to allow the bacteria to move toward nutrient-rich areas to achieve group intelligence [14]. The position of the bacteria after each chemotaxis is updated as follows:

$$P^i(i, j+1, k, l) = P^i(i, j, k, l) + C(i)\frac{\Delta(i)}{\sqrt{\Delta^T(i)\Delta(i)}} \quad (3)$$

$P^i(i, j+1, k, l)$ is the location of the i-th trend, the k-th copy operation and the z-th migration; C (i) is the chemotaxis of the bacteria step ;$\frac{\Delta(i)}{\sqrt{\Delta^T(i)\Delta(i)}}$ is the random vector of the direction at which the bacterial i is turned over; $\Delta(i)$ is the random vector of arbitrary generation of the bacterial i, which is in the range of [-1,1].

3.2Adaptive Bacterial Foraging Optimization Algorithm

The ABFO algorithm is an improvement of the BFO algorithm. In order to achieve the survival of the fittest and the size of the group to maintain the same purpose, the classic BFO algorithm replication operations mainly through the ability to split the foraging bacteria and poor bacteria will be eliminated. The migration operation of the improved algorithm is combined with the random migration of the flora to produce the replication operation of the better flora. The classical algorithm framework is adjusted: the copy operation of the classical BFO algorithm is integrated into the migration operation of the ABFO algorithm and the replication loop is removed. This allows the algorithm to reduce the complexity of time at the same time more conducive to achieving rapid convergence. The accuracy of the optimal value will also be improved due to the introduction of dimension adaptive learning algorithms.

3.2.1 Improved chemotaxis steps

Bacteria swimming search directly affects the convergence rate of the algorithm, which is the key in the operation of chemotaxis. The improved algorithm adds the variation law of the cosine function to the swimming step, C_{max} and C_{min} are the maximum and minimum values of the swimming step, N_s is the total number of chemotaxis, and C (i, k) is the chemotaxis step of the bacteria i at the kth swimming. Its improvement is expressed as:

$$C(i,k) = \frac{C_{max} - C_{min}}{2}\cos\left(\frac{\pi}{N_s}k\right) + \frac{C_{max} + C_{min}}{2} \quad (4)$$

where$\omega = 0,1,\cdots,N_s$.

3.2.2 Dimensional adaptive learning algorithm

In some cases, the fitness of the bacteria and the overall quality of the solution are improved, but some dimensions are far from the optimal solution. The algorithm takes longer to search and increase the optimization time. In order to avoid the above problems, we introduce the dimension adaptive learning algorithm to improve the BFO algorithm.

Suppose there are d dimensions, The position of a bacterium in high dimensional space before the start of adaptive learning is expressed by $p_{k=0} = [x_0^1, x_0^2, \cdots, x_0^d]$, $y = [y^1, y^2, \cdots, y^d]$ for each dimension of the y^i value is used to adjust x^i, that is, y is the initial value of 0 for the measurement of learning changes.$\frac{\alpha}{\beta(k)}$is the function of the number of learning k, where α is the initial maximum learning step, β(k) variable k non-linear increasing function, where $\beta(k) = k^{1.2}$; c is an adaptive learning factor [15]. The basic flow of the algorithm is as follows:

Step1.Initialization. Initialize the dimension d, the number of iterations per dimension is T and the initial learning speed $V_{k=0} = [0,0,\cdots,0]$, $p_{k=0} = [x_0^1, x_0^2, \cdots, x_0^d]$, $y_{k=0} = [y_0^1, y_0^2, \cdots, y_0^d] = [0,0,\cdots,0]$.

Step2.Set the cycle parameters. Where the dimension factor $i = 1,2,\ldots,d$, the number of adaptive learning $k = 1,2,\ldots,T$.

Step3.The ithdimension of $p_{k-1} = [x_{k-1}^1, x_{k-1}^2, \cdots, x_{k-1}^d]$ is subjected to the kth learning according to the following formula:

$$V_k^i = \frac{\alpha}{\beta(k)} \cdot rand + c \cdot V_{k-1}^i (5)$$

$$y_k^i = x_{k-1}^i + V_{k-1}^i (6)$$

Get the new position vector $p_k=[x_k^1, \cdots, x_k^{i-1}, y_k^i, x_k^{i+1}, \cdots, x_k^d]$.

Step4.After the i-dimensional update, get the new vector p_k, calculate the corresponding fitness function value $f(p_k)$.If $f(p_k) \geq f(p_{k-1})$, then the fitness function value is not improved after dimension learning, which is not conducive to the optimization of the whole solution and continue to follow Step3 to learn. Let $V_k^i = V_{k-1}^i$, $c = 0.5$, turn Step6.

Step5.Update the location. Increase the learning factor c and continue to learn in this dimension, that is let $V_k^i = y_k^i$,$c = 2$.

Step6: Judgment. Update k value and turn Step3 if k <T; otherwise, update the value of i. When i = d, the dimension adaptive learning process ends and outputs the learning result $p_k = x_k^1, x_k^2, \cdots, x_k^d$.

The speed learning factor c is changed according to the previous different speed update effect. When the new speed leads to the improvement of the fitness function value, it is shown that the speed is beneficial to the improvement of the overall quality of the solution. Then increase the intensity of learning, take c> 1 (such as Step5, so c = 2). On the contrary, it means that the speed is too large should reduce the next learning speed, take c<1 (such as Step4, so c = 0.5).

3.2.3 Improved migration operations

The BFO algorithm with stochastic migration according to the fixed probability does not preserve the favorable position information searched by the algorithm in advance, which leads to the large difference between the random new position and the optimal solution, and is not conducive to the convergence of the algorithm. In this paper, the migration operation is improved and the elite bacteria are initialized as the initial point of the Tent chaotic map. The mathematical expression of Tent mapping is:

$$x_{k+1} = \begin{cases} \frac{x_k}{a} & x \in [0,a] \\ \frac{(1-x_k)}{(1-a)} & x \in (a,1] \end{cases} \quad (7)$$

The system is in chaotic state and the mean value of chaotic sequence is 0.5 When $a \in (0,1), x_0 \in [0,1]$. Newly generated bacteria can jump out of local optimum and prevent search stagnation due to randomness of chaotic models. At the same time, the local optimal point is the initial point of the chaotic motion and the domain point that produces the local optimal point after chaotic search is beneficial to the convergence of the algorithm.

ABFO algorithm migration operation integrates the classic BFO algorithm replication operation and dimension adaptive learning algorithm, the specific steps are as follows.

3.2.4 ABFO algorithm operation steps

The operation of the ABFO algorithm is as follows:

Step1. initializes the parameters. $i = j = k = 0$ .The size of bacteria is S, the step size is C, the search space dimension is d, the number of chemotaxis cycles is N_c, the number of migration is N_{ed} ,the probability of migration is P, and the maximum number of chemotaxis in one direction is N_s.The number of iterations of each dimension in the dimension adaptive learning algorithm is t, the maximum swimming step is C_{max} ,and the minimum swimming step is C_{min} .

Step2. Initializes the flora position. Randomize the location of the bacteria and calculate the initial fitness value f for each bacterium.

Step3. Sets the loop variable. Where the number of migration cycles $l = 1,2,\dots,N_{ed}$, the number of chemotaxis cycles $j = 1,2,\dots,N_c$.

Step4. Entering chemotactic cycle and carrying out chemotaxis operation.

(1) Flip: Flip according to equation (3) and calculate the fitness function value f on the new position;

(2) Swimming: according to equation (4) to calculate the swimming step if the bacteria in the new position on the fitness function value has improved.The bacteria swim until the fitness function value no longer improves or reaches a predetermined maximum chemotaxis step.

Step5. Dimensional Adaptive Learning. The adaptive fitness of the bacteria with the best fitness function value (elite bacteria) was evaluated after the completion of the first chemotaxis and the position of the elite bacteria was updated.

Step6. Entering migration cycle and carrying out migration operation. Identify elite bacteria and adapt them to dimensional learning and update accuracy. For other bacteria, the random probability is generated and compared with the fixed migration probability P_{ed}.All the number of bacteria less than P_{ed} (recorded as m) is accumulated. S bacteria are initialized by Tent chaotic mapping. Replace the m bacteria that meet the migration conditions before.

Step7.Judgment.If the cycle is finished, the current optimal solution is dimensionally adaptive to obtain a higher precision elite bacteria, and the optimal solution is updated to the higher resolution and output it; Otherwise, go to Step4 and continue to search until the end of the algorithm to meet the conditions.

3.2.5 Improved resampling algorithm

The resampling method was first proposed by Gordon in [16] in 1993, and its basic idea is based on the weight of the sample. Small particles of particle size are removed and the weight of the particles is retained, so that the performance of the particle filter can be improved to a certain extent. There are many literatures on the resampling algorithm, such as polynomial resampling [16], hierarchical resampling [17] and system resampling [18] and so on. The resampling algorithm in the traditional RBPF-SLAM resets the importance of the sequence, and as the resampling continues to reduce the diversity of the particles, the filter diverges [19].So the key to improving filter performance is when and how to resample. The normalized weights are expressed by $\tilde{w}^{(i)}$.The number of effective particles N_{eff} is introduced to assess whether the current number of particles can present a true posteriori probability, defined as:

$$N_{eff} = \frac{1}{\sum_{i=1}^{N}(\tilde{w}^{(i)})^2} \tag{8}$$

Perform resampling when the number of valid particles is below the pre-set threshold N_{th}.

The traditional resampling method of particle degradation is almost powerless because it is a random way. The weights of the particles will be replicated many times, and the low-weight particles will be removed. So that the path and map information represented by these low-weight particles will be deleted, resulting in some knowledge of past information. On the other hand, the weight particles are replicated many times, and many of the offspring particles inherit the same path and map estimates, which greatly reduces the particle diversity. Which may lead to "particle depletion" problem, so that the algorithm can only meet the consistency requirements in a short time.

The migration operation of ABFO algorithm integrates the copy operation and dimension adaptive learning algorithm of classical BFO algorithm. A new resampling method is proposed based on the idea of migration in the ABFO algorithm, that is, the particles in the RBPF are simulated in the ABFO algorithm. The particles in RBPF were simulated in ABFO algorithm, and the specific steps were as follows:

Step1. Selecting the current optimal particle as the elite particle and the accuracy of the particle is improved by adaptively learning it.

Step2. The probability of migration is generated for each non-elite particle and is compared with the fixed migration probability. If the probability of migration is less than the fixed migration probability, the particle is eliminated and the number of bacteria less than the fixed migration probability is accumulated.

Step3. The chaotic maps are generated by using the elite particles as the initial solutions of the Tent mappings, resulting in S feasible solutions and extracting the m-bacteria from the S feasible solutions to replace the depleted particles.

3.3 Improved RBPF-SLAM algorithm

The improved RBPF-SLAM algorithm steps are as follows:

Step1. Particle initialization. Generating the pose vector $\left\{x_0^{(i)}\right\}$ of the initial particle at k=0 based on the environmental information, and the average distribution of the particle weights $w_0^i = 1/N, i = 1,2,\cdots,N$. N is the number of particles

Step2.First step prediction of pose. Based on the motion model of the robot, the initial pose $x_t^{(i)}$ of the robot is estimated according to the pose $x_{t-1}^{(i)}$ and the odometer control information u_{t-1}. And the estimated pose $x_{t|t-1}^{(i)}$ of each particle at time t is obtained.

Step3.Perform a scan match based on the map $m_{t-1}^{(i)}$.First get the real observation at time t through the sensor;Then we use the nearest neighbor method to correlate the data, and use the formula $w_t^i = w_{t-1}^i p\left(z_t^i | x_{t|t-1}^{(i)}\right)$ to update the particle weights.Down the weight of the normalization of the use of the formula:

$$w_t^i = w_{t-1}^i / \sum_{i=1}^{N} w_t^i \tag{9}$$

Step4.Second step prediction of the pose of the robot. Based on ABFO algorithm: the particle set $\left\{x_{t|t-1}^{(i)}\right\}$ is taken as the initial bacterial population using the fitness function of formula (1).The individuals in the flora were chemically transformed, replicated and migrated respectively, and then the second updated particle set $\left\{x_t^{(i)}\right\}$ was obtained.

Step5.The map update is based on the posture information of the updated particle set and the weights are updated and normalized according to the formula in Step 3;

Step6.Calculating the number of effective particles N_{eff} and settting a threshold N_{th}. The particles are resampled using the improved algorithm if $N_{eff} < N_{th}$.

Step7.Using the weighting formula to calculate m^i according to the robot pose x_t^i and the observation information z_t ,and update the map. Go to Step2 if you still need to calculate the next moment of the posture, otherwise the end.

4 Experimental results and analysis

Simulation of the laboratory environment to build the initial environment map, where the white part that can pass the part of the black part of the non-accessible part (obstacle part) as shown in Figure 1:

Figure1. Initial environment map.

In the case of the same number of particles, the two algorithms are applied as follows:

By contrast, it is found that the RBPF-SLAM algorithm has a serious inconsistency when the number of particles is the same in the same environment and the accuracy of the improved algorithm is obviously higher than that of the traditional RBPF-SLAM algorithm.

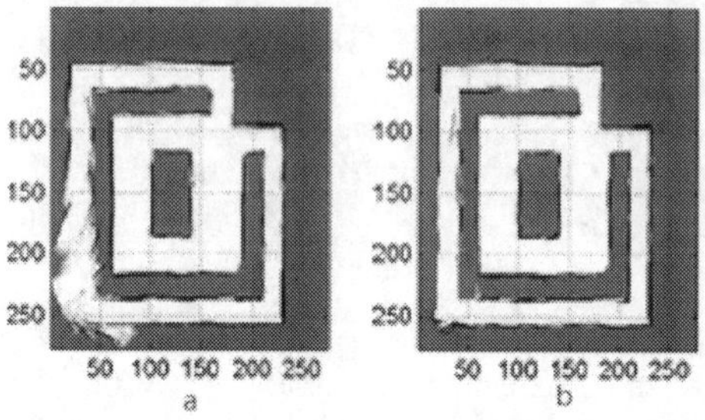

Figure 2. Construction of maps with the same number of particles(a.Grid map for traditional algorithms;b.Grid map for improved algorithms).

In order to verify the accuracy of the algorithm, the simulation experiment is carried out in MATLAB environment. The algorithm is analyzed by comparing the matching degree of the traditional RBPF-SLAM algorithm with the improved RBPF-SLAM algorithm to the particle trajectory. The results are shown in Figure3.

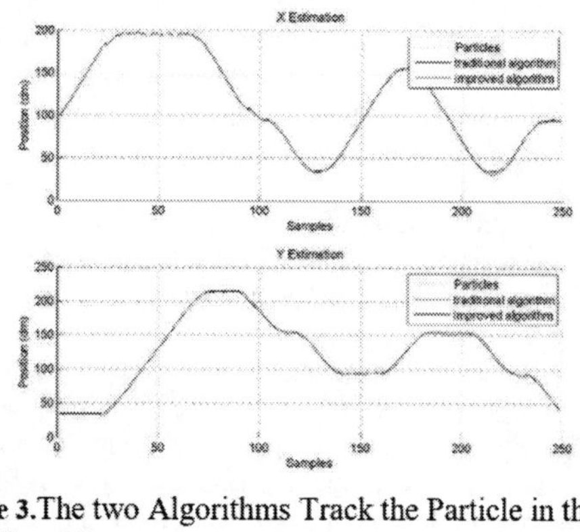

Figure 3. The two Algorithms Track the Particle in the x, y Coordinate System.

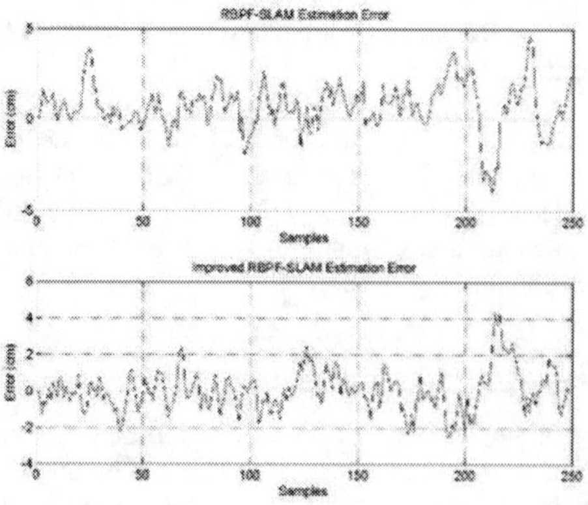

Figure 4. Estimationerror.

The estimation error of the improved RBPF-SLAM algorithm is obviously improved. The estimated error in the composition process is shown in the following Figure4:

In order to investigate the improvement of the phenomenon of particle resynchronization by the improved resampling method, the traditional resampling method and the method studied in this paper are run 9 times respectively, and the change of the mean value of the effective particles of the two algorithms is calculated and the simulation.

In order to investigate the improvement of the particle degradation phenomenon by the resampling method based on the adaptive bacterial foraging algorithm migration, the particles diversity of the two methods are represented by the traditional resampling method and the average Hamming distance of the particles produced by the improved method studied in this paper.The simulation results are shown in Table 1. The resampling method proposed in this paper can greatly improve the diversity of particles, and the average value is increased from 0.348 to 0.992.

Table 1. Diversity comparison results.

	1	2	3	4	5	6
Tradition	0.39	0.30	0.32	0.35	0.39	0.34
Improve	1.03	0.98	0.99	0.95	0.98	1.02

5Conclusion

In this paper, an adaptive bacterial foraging algorithm is proposed to introduce RBPF-based mobile robot synchronization location and map construction method.Firstly, the adaptive bacterial foraging algorithm is introduced into RBPF, which makes the particles close to the real motion before resampling.Then the improved migration operation in the ABFO algorithm is applied to the particle resampling part, which improves the diversity of the particles and increases the number of effective particles.The experimental results show that the improved RBPF algorithm and the traditional RBPF algorithm are used in the SLAM to improve the tracking accuracy and improve the accuracy of the estimation error, and more effectively improve the particle degradation phenomenon.In future work, the map building algorithm will be further optimized to achieve a precise map of a slightly larger environment.

References

1. Smith R, Self M, Cheeseman P. Estimating uncertainspatial relationships in robotics[M]//COX I J, WILFONG G T.*Autonomous Robot Vehicles*. New York: Springer, 167-193 (1990)
2. Dissanayake G, Huang S, Wang Z,et al. A review of recent developments in simultaneous localization and mapping[C].6th International Conference on Industrial and Information Systems(ICIIS), Kandy, Sri Lanka,477-482 (2011)
3. Murphy K P.Bayesian map learning in dynamic environments [C] //Proc of Neural Info Proc Systems (NIPS). Denver: MIT Press ,1015-1021 (2000)
4. Doucet A, De Freitas N, Murphy K, et al. Rao-Blackwellised particle filtering for dynamic Bayesian networks[C]//Proceedingsof the Sixteenth Conference on Uncertainty in Artificial Intelligence San Francisco, USA,176-183 (2000)
5. LipingQu,Hongjian Wang. An overview of robot SLAM problem[C]// International Conference on ConsumerElectronics,Communicationsand Net-works (CECNet),Xian-ning,China,1953-1956 (2011)
6. Grisetti G, Stachniss C, Burgardw W. Improved techniques for grid mapping with Rao-Blackwellized particle filters[J]. Robotics,**23**: 34-46 (2007)
7. Dong J F, Wijesoma W S, Shacklock A P. An efficient rao-blackwellized genetic algorithmic filter for SLAM [C]//Proceedings of 2007 IEEE International Conference on Robotics and Automation,April10-14,2007,Roma,Italy.Piscataway:IEEEPress,2427-2432 (2007)
8. Won D, Chun S, Sung S, et al. INS/vSLAM system using distributed particle filter [J]. International Journal of Control, Automation, and Systems,1232-1240 (2010)
9. Mengyin Fu, Hao Zhu, Yi Yang, Meiling Wang, Zhihong Deng. A navigation map building algorithm using refined RBPF-SLAM. 2016 IEEE Chinese Guidance, Navigation and Control Conference (CGNCC), 2483 – 2487 (2016)
10. Jin Li. Research and Application of Hybrid Frog Leap Algorithm in Small Habitats[D].Xian: Journal of Xidian University, (2012)
11. Passino K M. Biomimicryof Bacterial Foraging for Distributed Optimization and Control [J]. Control System Magazine, 22(3):52-67 (2002)
12. Xiaolong Liu. Improvement and Application of Bacteria Feeding Optimization Algorithm [D]. Guangzhou: South China University of Technology, (2011)
13. Nandita S, Amitava C, SugataM.Fuzzy VQ based imagecompression by bacterial foraging optimization algorithm with varying population. Proceedings of the 2015 Third International Conference on Computer, Communication,Control and Information Technology (C3IT),1-6 (2015)
14. Li M S,Ji T Y,Tang WJ, et al. Bacterial Foraging Algorithmwith Varying Population[J]. BioSystems, 100(3):185-197 (2010)
15. Gordon N J, Salmond D J, Smith A F M. Novel approach to nonlinear/non-Gaussian Bayesian state estimation [J].IEE Proceedings F Reader and Signal Processing,140(2):107-113 (1993)
16. FengC, Wang M, Qing-Bo J I. Analysis and comparison of resampling algorithms in particle filter [J]. Journal of System Simulation ,21(4):1101-1105 (2009)

17. Jinxai Yu, Cai Z X, Duan Z H. Survey on some key technologies of mobile robot localization based on particle filter [J]. Application Research of Computers,24(11):9-14 (2007)
18. MerweR V D, Doucet, Freitas N D, et al. The unscented particle filter [J]. Adavances in Neural Information Processing Systems, 13:584-590 (2001)
19. Havangi R, Taghirad H D, Nekoui M A, et al. Asquare root unscented FastSLAM with improved proposal distribution and resampling [J]. IEEE Transactions on Industrial Electonics,61(5):2334-2345 (2014).

A Dynamics Controller Design Method for Car-like Mobile Robot Formation Control

Baofang Wang , Chen Qian and Qingwei Chen

School of Automation, Nanjing University of Science and Technology, 210094 Nanjing, China

Abstract. A dynamics controller design method based on characteristic model is proposed for the formation control problem of car-like mobile robots. Only kinematics controller is not enough for some cases such as the environment is rugged, and the dynamics parameters of the robot are time-varying. Simulation results show that the proposed method can improve the responding speed of the mobile robots and maintain high formation accuracy. First, we obtain the kinematic error state equations according to the leader-follower method. A kinematics controller is designed and the stability is proved by Lyapunov theory. Then the characteristic model of the dynamics inner loop is established. A sliding mode controller is designed based on the second order discrete model, and the stability of the closed-loop system is analyzed. Finally, simulations are designed in MATLAB and Microsoft Robotics Developer Studio 4 (MRDS) to verify the effectiveness of the proposed method.

1 Introduction

Formation control problem of multi robots has become an interesting and challenging subject for researchers. In the wake of developments in sensors, servo motors and control technology, the performance of robot core components has been improved accordingly. Robots have been widely used in various situations.

In trajectory tracking and formation control problems for car-like mobile robots, numerous researchers have focused on the design of kinematics controllers. This is reasonable for many cases because of two factors: Firstly, most car-like robots have their own motor drive and control modules which contain the dynamics inner control loop capable of tracing the velocity command signal; Secondly, traditional robot workspaces are flat ground, and the mass and velocity of the robot are always low. In these situations, kinematics controllers are enough for robot trajectory tracking and formation control. Dynamics analysis is not necessary. However, in recent years, car-like robots have faced increasingly complicated situations due to the requirements of different tasks [1-3], such as geographical exploration, rescue after disaster and agriculture cultivation. Robots may have higher velocity and larger mass of load, and the workspaces are always rough terrain with obstacles. In these tasks, only kinematics controllers cannot satisfy the performance requirement. And researches on dynamics controllers become more and more necessary.

Characteristic model based adaptive controller design method is proposed in [4]. A second order discrete time-varying model is acquired by parameter identification. The characteristic model is established according to the input and output signals of the system without accurate mathematical system model. And the controller based on this model has good adaptive ability to the change of system parameters. So, when the dynamics model of the robot is unknown and the mass and velocity of the robot are time-varying, the characteristic model based dynamics controller design method may have some advantages.

Numerous researchers have focused on the kinematics controller design for car-like mobile robot formation control problem. In [5], a new vision based formation system was proposed. An adaptive kinematics controller and an adaptive camera controller were designed. The authors of [6] proposed a PD-type controller for car-like mobile robots to compensate the delay effect of data transmission, and the stability of the system was proved. Back-stepping method was used to design car-like robots kinematics controller in [7]. Besides kinematics controllers, some researchers have considered the dynamics controller design problem. A nonlinear control law was derived using Lyapunov analysis for formation control of car-like mobile robots using robot dynamics in [8]. The authors of [9] presented a sliding mode tracking control scheme using the simplified kinematics model of car-like robots.

In this paper, we propose a dynamics controller design method for car-like robots considering the time-varying dynamics parameters and the rough terrain workspace.

According to the leader-follower method, the controller of outer kinematics loop is designed based on the error state equations. In the inner dynamics loop, a dynamics controller or compensation controller is proposed based on the characteristic model. The proposed method is tested using MRDS and MATLAB to verify the effectiveness of the proposed controller.

2 Kinematics model and controller

Using the typical triangle formation as an example, the leader-follower method and the model of the car-like mobile robot are shown in Figure 1. The rear wheels of the robot are fixed parallel to the car body and allowed to roll or spin but not slip. The front wheels can turn to the left or right, but the two wheels must be parallel or satisfy Ackermann steering geometry. For the convenience of explanation, we only consider the situation of parallel front wheels. All the corresponding kinematics parameters are also presented in Figure 1. l is the wheelbase. $p=[x,y,\theta]^{\mathrm{T}}$ represents the state of the car-like mobile robot, where $(x,y)^{\mathrm{T}}$ is the position of the rear wheel centre and θ is the angle between the vehicle frame orientation and axis X.

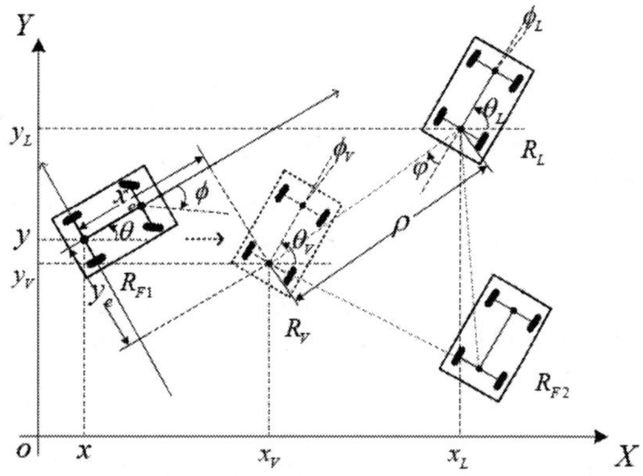

Figure 1. Leader-follower method.

R_{L} is the leader robot. R_{F1} and R_{F2} are the followers which need to keep triangle formation with the leader robot. Taking R_{F1} as an example, R_{V} is the virtual robot which R_{F1} should follow. If the position error $\mathbf{e}^*=[x_{\mathrm{V}}-x,y_{\mathrm{V}}-y,\theta_{\mathrm{V}}-\theta]^{\mathrm{T}}$ converge to zero, the formation is accomplished. It is the same with R_{F2}. The position of R_{V} is $[x_{\mathrm{V}},y_{\mathrm{V}},\theta_{\mathrm{V}}]^{\mathrm{T}}$ which can be acquired by the position of R_{L} and the formation parameters $[\rho,\varphi]^{\mathrm{T}}$ as follows,

$$\begin{cases} x_{\mathrm{V}}=x_{\mathrm{L}}-\rho\cos(\theta_{\mathrm{L}}+\varphi) \\ y_{\mathrm{V}}=y_{\mathrm{L}}-\rho\sin(\theta_{\mathrm{L}}+\varphi) \\ \theta_{\mathrm{V}}=\theta_{\mathrm{L}} \end{cases}, \quad \begin{cases} v_{\mathrm{V}}=v_{\mathrm{L}} \\ \phi_{\mathrm{V}}=\phi_{\mathrm{L}} \end{cases}. \tag{1}$$

The control input of the car-like robot is $[v,\phi]^{\mathrm{T}}$, where v is the velocity of the rear wheels and ϕ is the steering angle of the front wheels. The kinematics model of the car-like mobile robot is as follows [8],

$$\begin{cases} \dot{x}=v\cos\theta \\ \dot{y}=v\sin\theta \\ \dot{\theta}=v\tan\phi/l \end{cases}. \tag{2}$$

Transform the coordinate from the global coordinate to the robot coordinate, the position error $\mathbf{e}^*$ is transformed into

$$\mathbf{e}=\begin{bmatrix} x_e \\ y_e \\ \theta_e \end{bmatrix}=\mathbf{Te}^*,\quad \mathbf{T}=\begin{bmatrix} \cos\theta & \sin\theta & 0 \\ -\sin\theta & \cos\theta & 0 \\ 0 & 0 & 1 \end{bmatrix}. \tag{3}$$

Take the derivative of (3) and combine with the kinematics model (1), the error state functions can be obtained as

$$\dot{\mathbf{e}}=\begin{bmatrix} \dot{x}_e \\ \dot{y}_e \\ \dot{\theta}_e \end{bmatrix}=\begin{bmatrix} v_{\mathrm{V}}\cos\theta_e+vy_e\tan\phi/l-v \\ v_{\mathrm{V}}\sin\theta_e-vx_e\tan\phi/l \\ v_{\mathrm{V}}\tan\phi_{\mathrm{V}}/l-v\tan\phi/l \end{bmatrix}. \tag{4}$$

The kinematics controller is designed as

$$\begin{cases} v=v_{\mathrm{V}}\cos\theta_e+k_1x_e \\ \phi=\arctan(\dfrac{v_{\mathrm{V}}}{v}\tan\phi_{\mathrm{V}}+\dfrac{k_2lv_{\mathrm{V}}}{\theta_ev}y_e\sin\theta_e+\dfrac{k_3l}{v}\theta_e) \end{cases}, \tag{5}$$

where $k_1,k_2,k_3>0$. Consider the following Lyapunov function

$$V=\frac{1}{2}x_e^2+\frac{1}{2}y_e^2+\frac{1}{2k_2}\theta_e^2. \tag{6}$$

Combining (4), the derivative of (6) is given by

$$\begin{aligned} \dot{V}&=x_e\dot{x}_e+y_e\dot{y}_e+\frac{1}{k_2}\theta_e\dot{\theta}_e \\ &=x_ev_{\mathrm{V}}\cos\theta_e-x_ev+y_ev_{\mathrm{V}}\sin\theta_e \\ &\quad+\frac{1}{k_2}\theta_e(v_{\mathrm{V}}\tan\phi_{\mathrm{V}}/l-v\tan\phi/l) \\ &=-x_e(v-v_{\mathrm{V}}\cos\theta_e)-\frac{\theta_ev}{k_2l}(\tan\phi \\ &\quad-\frac{v_{\mathrm{V}}}{v}\tan\phi_{\mathrm{V}}-\frac{k_2lv_{\mathrm{V}}}{\theta_ev}y_e\sin\theta_e) \end{aligned} \tag{7}$$

Taking the controller (5) into (7), we have

$$\dot{V}=-k_1x_e^2-\frac{k_3}{k_2}\theta_e^2\le 0. \tag{8}$$

Therefore, when the kinematics controller is selected as (5), the Lyapunov function $\dot{V}\le 0$ and $\dot{V}=0$ only if $x_e=0,\theta_e=0$. So the kinematics outer loop is asymptotically stable.

3 Dynamics model and controller

As car-like mobile robots are widely used in different fields, there appear more and more situations when the dynamics parameters of the robot are time-varying, such as sprinklers, road sweeping robots and other transport robots with time-varying load. In these cases, the load of the robot is time-varying which influences the rotary inertia of the robot wheels and the whole mass of the robot. Therefore, an adaptive dynamics controller is designed to improve the accuracy of formation control.

In some cases, we should directly design the kinematics controller based on the characteristic model acquired by system identification. The system structure of dynamics controller design is shown in Figure 2.

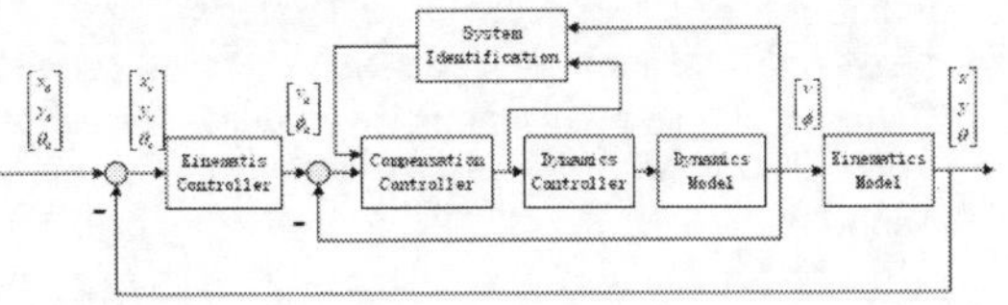

Figure 2. System structure of dynamics controller design.

When the robots have their own drive and control modules, we could not change their original dynamics controller. In this case, a dynamics compensation controller is designed to improve the control performance. The system structure of compensation controller design is shown in Figure 3.

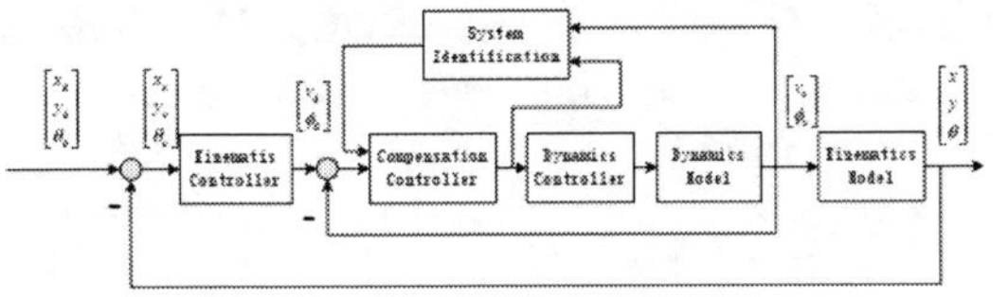

Figure 3. System structure of compensation controller design.

In Figure 2 and 3, $[x_d, y_d, \theta_d]^T$ represents the expected position. $[x_e, y_e, \theta_e]^T$ is the position error. $[v_d, \phi_d]^T$ is the expected input signal of kinematics outer control loop. $[v, \phi]^T$ is the actual kinematics input signal, and $[x, y, \theta]^T$ is the actual position of the robot.

Characteristic modeling method is based on the dynamic characteristics and control performance requirements of the plants. The characteristic model of the robot dynamics loop is acquired through system identification. It compresses the information of high-order model into several characteristic parameters which is different from the reduced-order model of a high-order system. Characteristic modeling method offers a way to design a feasible low-order controller for various complicated plants, and has been successfully applied to many situations. When the system satisfies the following conditions, it can be expressed as a second-order discrete time-varying system.

Consider a nonlinear system

$$\dot{x}(t) = f(x, \dot{x}, \cdots, x^{(n)}, u, \dot{u}, \cdots, u^{(m)}) . \quad (9)$$

Define

$$\begin{aligned} x = x_1 , \dot{x} = x_2 , \cdots , x^{(n)} = x_{n+1} \\ u = u_1 , \dot{u} = u_2 , \cdots , u^{(m)} = u_{m+1} \end{aligned} . \quad (10)$$

Then (9) can be written as

$$\dot{x}_1(t) = f(x_1, \cdots, x_{n+1}, u_1, \cdots, u_{m+1}) . \quad (11)$$

Assuming the system satisfies the following conditions:

(a) Single input single output;
(b) The order of the control input u is 1;
(c) $f(\square) = 0$ when all the variables $x_i = u_i = 0$;
(d) $f(\square)$ is continuous and differentiable to all the variables x_i and u_i, and all partial derivatives are bounded;
(e) $|f(x(t+T), u(t+T))| - |f(x(t), u(t))| < MT$, where T is the sampling period and M is a positive constant.

Lemma 1 [4]. If a system can be described by (11) and satisfy the assumptions (a)-(e), then the characteristic model of the system can be established in the form of a second-order difference equation as

$$\begin{aligned} x(k+1) = f_1(k)x(k) + f_2(k)x(k-1) \\ + g_0(k)u(k) + g_1(k)u(k-1) \end{aligned} . \quad (12)$$

In the dynamic process, under the same input, the error between characteristic model output and the practical plant output is maintained within a permitted small range. In the steady state, the two outputs are equal. The parameters $f_1(k)$, $f_2(k)$, $g_0(k)$, $g_1(k)$ are slow time-varying. If the system is a minimum phase system, (12) can be simplified as

$$x(k+1) = f_1(k)x(k) + f_2(k)x(k-1) + g_0(k)u(k) . \quad (13)$$

According to the Newton's second law, the simplified car-like robot dynamics model can be easily obtained. The assumption (b) and (c) are satisfied. The velocity of the rear wheels and the steering angle of the front wheels are independent from each other. Therefore, the dynamics model can be considered as two systems with single input and single output, which satisfies the assumption (a). In practical situations, the velocity and angular velocity of the car-like mobile robot cannot change suddenly because of inertia. In most cases, all the parameters are continuous and smooth which satisfies (d). And assumption (e) is automatically satisfied.

We use the rear wheel velocity control system as an example to design a dynamics controller, and the steering angle control system of front wheels are the same. The error system is described as

$$e(k+1) = f_1(k)e(k) + f_2(k)e(k-1) + g_0(k)u(k) . \quad (14)$$

The parameters are estimated using least squares algorithm, and we have

$$\begin{aligned} e(k+1) = \hat{f}_1(k)e(k) + \hat{f}_2(k)e(k-1) \\ + \hat{g}_0(k)u(k) + \Delta(k) \end{aligned} , \quad (15)$$

where $e(k) = x(k) - x_r(k)$. $x(k)$ is the actual velocity output, and $x_r(k)$ is the expected velocity. $\Delta(k)$ represents the error of parameter estimation, uncertainty and disturbance.

The controller of system (14) is written as the following form

$$u(k) = u_{\mathrm{eq}}(k) + u_{\mathrm{s}}(k) \,. \tag{16}$$

The adaptive control law is

$$u_{\mathrm{eq}}(k) = -\frac{1}{\hat{g}_0(k)}[\hat{f}_1(k)e(k) + \hat{f}_2(k)e(k-1)] \,. \tag{17}$$

And the sliding mode control law is

$$u_{\mathrm{s}}(k) = \frac{1}{\hat{g}_0(k)}\{(1-qT)e(k) - \varepsilon T \operatorname{sgn}[e(k)]\} \,. \tag{18}$$

Assuming $|\Delta(k)| < f_{\mathrm{m}}$, the parameters should satisfy the following ranges:

$$\begin{cases} T < \dfrac{2|e_0|}{2\varepsilon + q|e_0|} \\ q > 0 \\ \varepsilon > \dfrac{f_{\mathrm{m}}}{T} \end{cases} \,. \tag{19}$$

In (19), sgn(□) represents the sign function. e_0 is the minimum error. T is the sampling period, and ε, q are positive constants.

Theorem 1. Suppose the system can be described as (14), and the control law (16) is used with all parameters satisfying (19), then the system is bounded stable.

Proof. Taking (16) into (15), we have

$$e(k+1) = (1-qT)e(k) - \varepsilon T \operatorname{sgn}[e(k)] + \Delta(k) \,. \tag{20}$$

Considering $[e(k+1)-e(k)]\operatorname{sgn}[e(k)]$, we can obtain

$$\begin{aligned} &[e(k+1)-e(k)]\operatorname{sgn}[e(k)] \\ &= -qT|e(k)| - \varepsilon T + \Delta(k)\operatorname{sgn}[e(k)] \\ &< -\varepsilon T + \Delta(k)\operatorname{sgn}[e(k)] \\ &< -\varepsilon T + f_m \end{aligned} \,. \tag{21}$$

From the ranges of parameters (19), we know $\varepsilon > \dfrac{f_{\mathrm{m}}}{T}$. So $[e(k+1)-e(k)]\operatorname{sgn}[e(k)] < 0$.

Then considering $[e(k+1)+e(k)]\operatorname{sgn}[e(k)]$, we have

$$\begin{aligned} &[e(k+1)+e(k)]\operatorname{sgn}[e(k)] \\ &= (2-qT)|e(k)| - \varepsilon T + \Delta(k)\operatorname{sgn}[e(k)] \\ &= (2-qT)|e(k)| - 2\varepsilon T + \varepsilon T + \Delta(k)\operatorname{sgn}[e(k)] \\ &> (2-qT)|e(k)| - 2\varepsilon T + f_{\mathrm{m}} + \Delta(k)\operatorname{sgn}[e(k)] \\ &> (2-qT)|e(k)| - 2\varepsilon T \end{aligned} \,. \tag{22}$$

From (19), $T < \dfrac{2|e_0|}{2\varepsilon + q|e_0|}$ can be written as $2 - qT > \dfrac{4\varepsilon}{2\varepsilon + q|e_0|} > 0$, and $(2-qT)|e_0| - 2\varepsilon T > 0$.

Therefore,

$$\begin{aligned} &[e(k+1)+e(k)]\operatorname{sgn}[e(k)] \\ &> (2-qT)|e(k)| - 2\varepsilon T \\ &> (2-qT)|e_0| - 2\varepsilon T \\ &> 0 \end{aligned} \,. \tag{23}$$

As a result, under the control law (16), the system satisfies

$$\begin{cases} [e(k+1)-e(k)]\operatorname{sgn}[e(k)] < 0 \\ [e(k+1)+e(k)]\operatorname{sgn}[e(k)] > 0 \end{cases} . \tag{24}$$

And the stability condition of discrete sliding mode control is satisfied [10].

4 Simulation results

Simulation1:

Two car-like mobile robots are expected to keep a certain formation and change to another formation at $t = 60\mathrm{s}$. The mass of load is varying all the time during the simulation.

In the simulation, the wheelbase of the robot is $l = 1.9\mathrm{m}$. The leader robot R_1 moves along a circle trajectory, and the following robot R_2 is controlled using the proposed control method to keep a certain formation with the leader. The initial formation parameters are $\rho = 5\mathrm{m}$ and $\varphi = 0^\circ$. The initial position error is $(x_e, y_e, \theta_e)^{\mathrm{T}} = (5,5,0)^{\mathrm{T}}$. The initial load mass of R_2 is $m_0 = 2000\mathrm{kg}$ and decreases by $5kg$ each second. When $t = 60\mathrm{s}$, the formation parameters are changed to $\rho = 5\mathrm{m}$ and $\varphi = -90^\circ$. The formation, position error and angle error are shown in Figure 4.

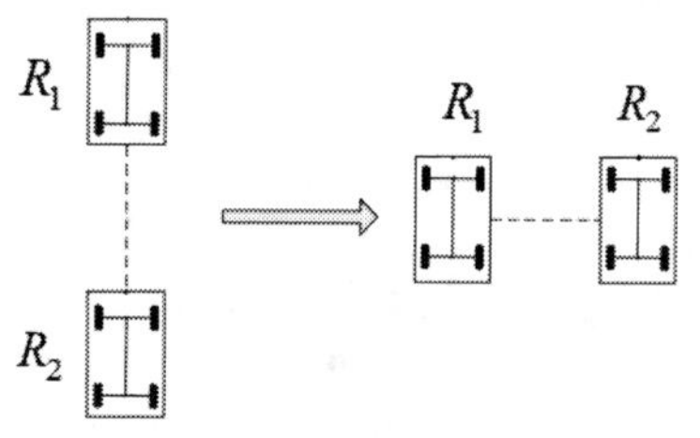

(a) Formation of car-like mobile robots

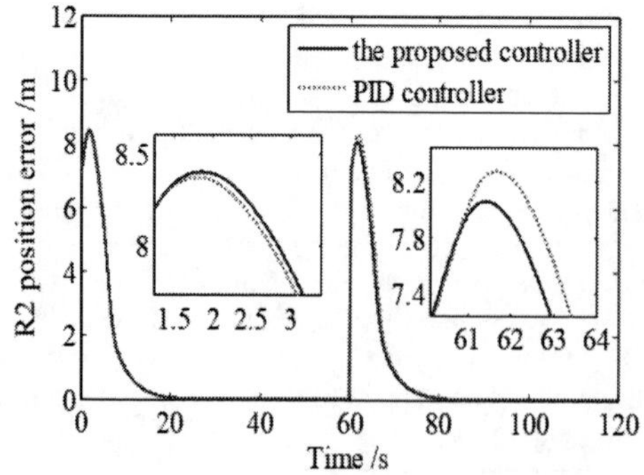

(b) Position error of R_2

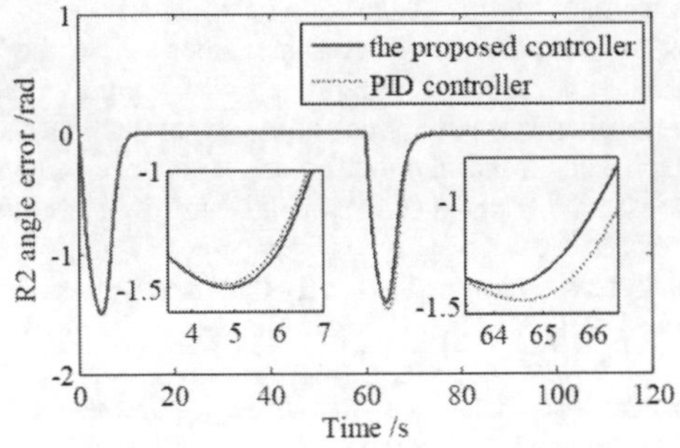

(c) Angel error of R_2

Figure 4. Simulation results using time-varying dynamics parameters.

From (b) and (c) of Figure 4, we can see that when $t < 60\text{s}$, the two controllers both can keep the expected formation with high accuracy. However, when $t > 60\text{s}$, PID controller cannot adjust to the change of dynamics parameters. Its speed of convergence becomes lower. And the proposed controller based on the characteristic model has good adaptive ability which makes it faster converge to the expected formation position. The proposed method has better control performance when the dynamics parameters are time-varying.

Simulation 2:

In practical, the work environments of car-like mobile robots are always rough terrain workspaces which have higher requirements for robot dynamics controller. In this simulation, a compensation dynamics controller is designed for robot formation control. Virtual 3D simulation environments are built in MRDS4 which provides direct debugger interface to MATLAB and has a physics engine PhysX making the simulation more realistic. The simulation environment are shown in Figure 5.

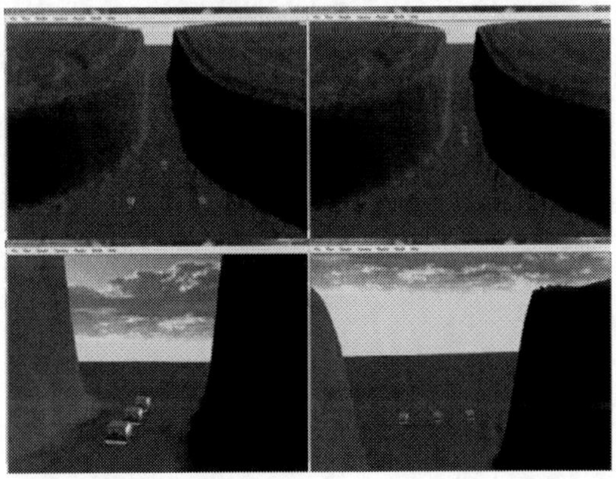

Figure 5. Simulation environment.

In this simulation, the wheelbase of the robot is $l = 1.9\text{m}$. Three robots R_1, R_2, R_3 go through a narrow canyon under expected formations. The initial formation parameters of R_2, R_3 are $\rho = 7\text{m}$, $\varphi_2 = 30^\circ$, $\varphi_3 = -30^\circ$; When the robots are in the canyon area, the formation parameters are $\rho_2 = 5\text{m}$, $\rho_3 = 10\text{m}$ $\varphi = 0^\circ$; After the canyon area, the formation parameters are changed to $\rho = 7\text{m}$, $\varphi_2 = 90^\circ$, $\varphi_3 = -90^\circ$. The formation, trajectories, position errors and angle errors are shown in Figure 6.

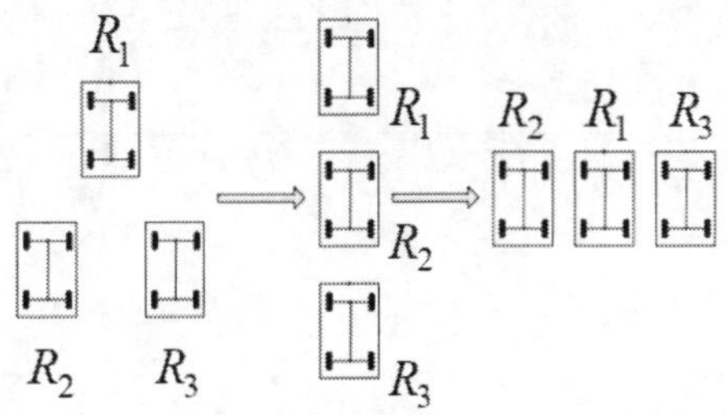

(a) Formation of car-like mobile robots

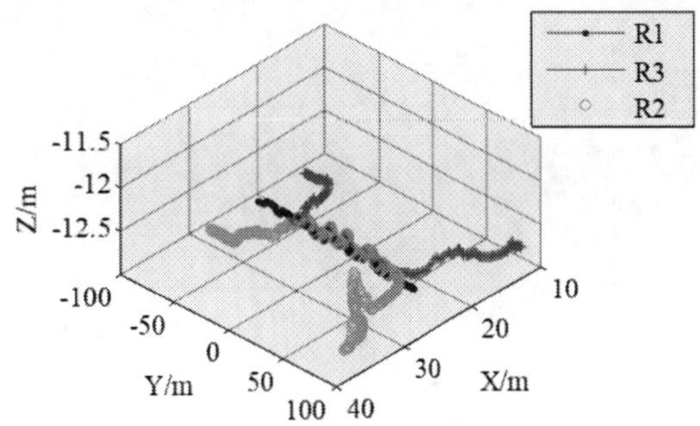

(b) Trajectories of robots

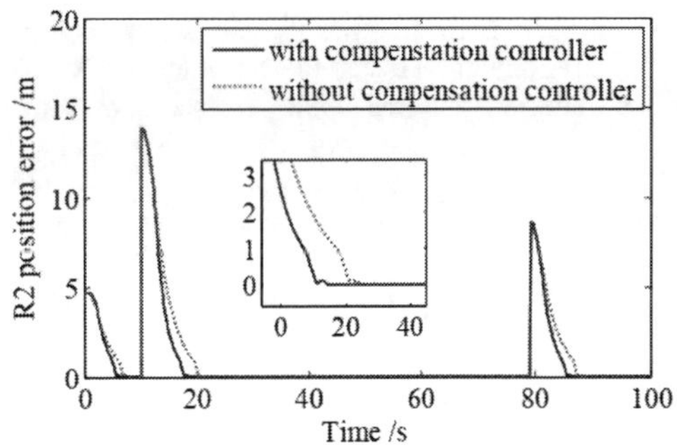

(c) Position error of R_2

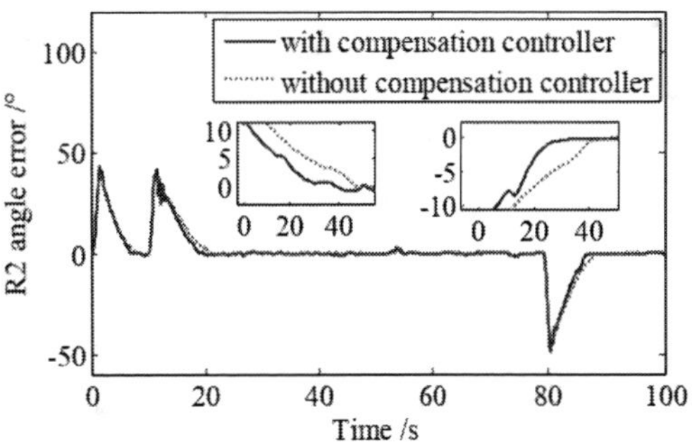

(d) Angle error of R_2

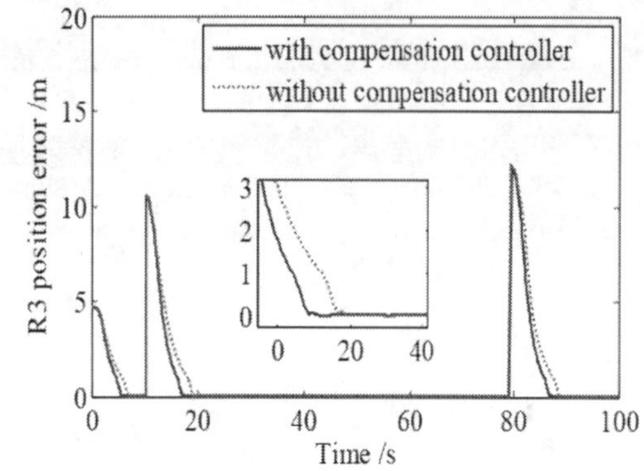

(e) Position error of R_3

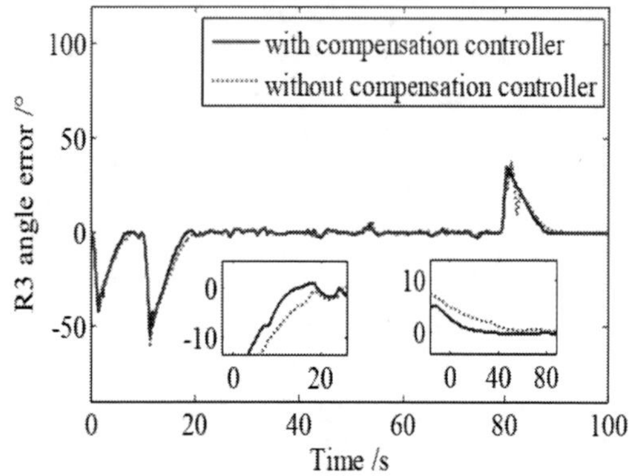

(f) Angle error of R_3

Figure 5. Simulation results in rough terrain environment.

From the simulation results, we know that rough terrain environment affects the accuracy of formation control. The position and angle errors cannot converge very fast when only the original dynamics controller is used. If we add the compensation controller based on characteristic model into the dynamics loop, the robots respond much faster making the convergence time shorter than without compensation.

5 Conclusions

In this paper, a dynamics controller design method based on characteristic model is proposed for the formation control problem of car-like mobile robots. First, we obtain the kinematic error state equations according to the leader-follower method. A kinematics controller is designed and the stability is proved by Lyapunov theory. Then the characteristic model of the dynamics inner loop is established. A sliding mode controller is designed based on the second order discrete model, and the stability of the closed-loop system is analyzed. Finally, two simulations are designed to verify the effectiveness of the proposed method. The proposed method can decrease the response time of robots and improve the control performance when the dynamics parameters are time-varying and the work environment is rough terrain.

Acknowledgment

This work was supported by National Natural Science Foundation of China under Grant No. 61333008, 61673214, 61773211, 61673217 and 61673219.

References

1. C.R. Weisbin, G. Rodriguez, IEEE Rob. Autom Mag. **7**, 25 (2002)
2. G.K. Dey, R. Hossen, M.S. Noor, K.T. Ahmmed, *International Conference on Informatics, Electronics & Vision, Dhaka, Bangladesh* (IEEE, Dhaka Bangladesh, 2013)
3. I.A. Hameed, J. Intell. Robot. Syst. **74**, 965 (2014)
4. H. Wu, J. Hu, Y. Xie, *Characteristic model-based intelligent adaptive control* (China Science and Technology Press, Beijing, 2009)
5. X. Chen, Y. Jia, IET Control Theory Appl. **9**, 1302 (2015)
6. Z. Xu, M. Schroter, N. Dan, L. Ma, K. Schilling, *Proceedings of the 31st Chinese Control Conference* (IEEE, Hefei China, 2012)
7. R. Zhang, S. Li, Q. Chen, Robot, **35**, 651 (2013)
8. Z. Cao, Y. Zhao, Y. Fu, Acta Electron Sin, **40**, 632 (2012)
9. S. Ramaswamy, S. Balakrishnan, American Control Conference (IEEE, Seattle America, 2008)
10. W. Gao, Y. Wang, A. Homaifa, IEEE Trans. Ind. Electron. **42**, 117 (1995)

Research and Implementation of Robot Path Planning Based onVSLAM

Zi-Qiang Wang[1], He-Gen Xu[2], You-Wen Wan[3]

[1]*College of Electronics and Information Engineering, Tongji University,Shanghai, China*
[2] *College of Electronics and Information Engineering, Tongji University, Shanghai, China*
[3] *College of Electronics and Information Engineering, Tongji University, Shanghai, China*

Abstract. In order to solve the problem of warehouse logistics robots planpath in different scenes, this paper proposes a method based on visual simultaneous localization and mapping (VSLAM) to build grid map of different scenes and use A* algorithm to plan path on the grid map. Firstly, we use VSLAMto reconstruct the environment in three-dimensionally. Secondly, based on the three-dimensional environment data, we calculate the accessibility of each grid to prepare occupied grid map (OGM) for terrain description. Rely on the terrain information, we use the A* algorithm to solve path planning problem. We also optimize the A* algorithm and improve algorithm efficiency. Lastly, we verify the effectiveness and reliability of the proposed method by simulation and experimental results.

1 Introduction

Using robots in warehouses can greatly improve the efficiency of warehouse transfer, and the path planning of warehousing and logistics robots is the key to improving efficiency. Path planning based on occupied grid map is a common method of environment for robots, such as Dijkstra[1], A*[2], D*[3] and so on. Dijkstra's main application is to find the shortest path between the start and end point of the map, but the path search is non-heuristic and slow. D* mainly solves the problem of the change of travel cost caused by the dynamic change of the environment. For a static environment such as a warehouse, the A* algorithm can find the shortest path between two points more quickly and efficiently[4-5]. While a warehouse is a static environment, robots need the ability to rapidly map the environment if they need to be deployed quickly to different warehouses.

Monocular vision SLAM technology can make robots to build maps. There are two types of SLAM visual odometer, which are divided into direct method and feature point method [6]. Feature point method mainly includes PTAM[7], ORB-SLAM[8] and other algorithms.Direct methods are mainly LSD-SLAM [9-10], DSO [11-12] and other algorithms. The advantage of the feature point method is that the map drifts small and the loop-closures detection is accurate. However, due to the relatively sparse feature points, it can be used for finding robot localization, difficult to use for robot navigation and path planning, and relativelycalculate slow. Direct method can extract relatively dense environmental features for robot navigation and positioning, and can run in real time on a PC. But the map drift is more terrible than the feature point method, loop-closures detection time is long.

This paper mainly considers how to let the robot quickly perceive the environment and complete the path planning. First of all,direct method SLAM are used to construct the 3D point cloud map of the warehouse environment, then makeoccupied grid map according to barrier in 3D map. Secondly, this paper uses A* algorithm to plan the path in the robot's working environment.

2 Map preparation

2.1Large-scale direct monocular SLAM

Jakob Engel and Daniel Cremers at the Technical University of Munich proposed a monocular SLAM algorithm based on direct method to construct a large-scale, globally consistent environment map called Large-Scale Direct (LSD) Monocular SLAM [9-10]. Vladyslav-Usenko and Jakob Engel implemented a three-dimensional reconstruction of the real-time street environment on the moving cars[13]. Jakob Engel, JorgStückler achieved accurate depth estimation and relatively dense three-dimensional reconstruction on a stereo camera [14]. For static, stable light conditions indoor, warehouse and other environments, the use of LSD-SLAM can make the robot has a precise positioning and map building capabilities.

The algorithm consists of three major components: tracking, depth map estimation and map optimizationas visualized in Fig.1:

1. The tracking component continuously tracks new camera images. That is,it estimates their rigid body pose $\xi \in \mathfrak{se}(3)$with respect to the currentkeyframe, using the pose of the previous frame as initialization.

2. The depth map estimation component uses tracked frames to either refineor replace the current keyframe. Depth is refined by filtering over manyper-pixel, small-baseline stereo comparisons coupled with interleaved spatialregularization as originally proposed in [9]. If the camera has moved too far,a new keyframe is initialized by projecting points from existing, close-bykeyframes into it.

3. Once a keyframe is replaced as tracking reference – and hence its depth mapwill not be refined further – it is incorporated into the global map by themap optimization component. To detect loop closures and scale-drift, asimilarity transform $\xi \in \mathfrak{sim}(3)$ to close-by existing keyframes (includingits direct predecessor) is estimated using scale-aware, direct $\mathfrak{sim}(3)$-imagealignment.

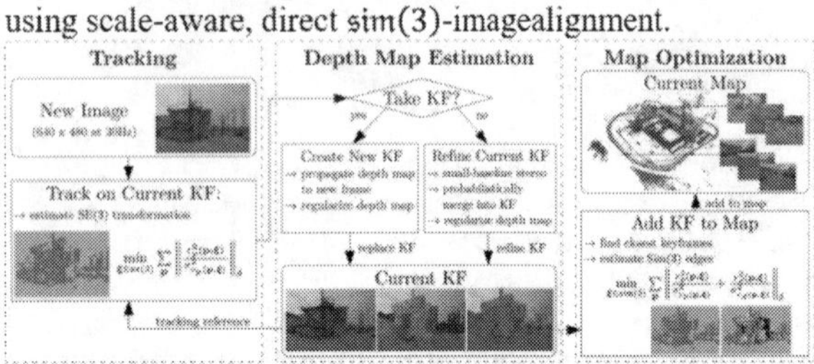

Figure 1. Overviewover the complete LSD-SLAM algorithm

2.2 Occupied grid mappreparation

Semi-dense three-dimensional point cloud data can be obtained from the LSD-SLAM algorithm, and the main obstacles in the indoor environment can be identified, but the three-dimensional data still needs to be converted into a two-dimensional grid map to be able to use the A* algorithm for path planning . Suppose a total of n three-dimensional data points, recorded as a set:

$$A = \{ a_1(x_1, y_1, z_1), \ldots, a_n(x_n, y_n, z_n) \}$$

Specific steps are as follows:

STEP 1:

Filter out point cloud data that formed by objects higher than the robot itself, this objects are hanging on the robot some, such as lights, ceilings, etc., the robot can't reach higher than itself. To ensure that the camera is installed horizontally, according to the pinhole imaging model, the data with the y-axis less than 0 is removed and obtain the point set:

$$B = \{ b\,(x, y, z) |\ b \in A \text{ and } y \geq 0 \}$$

Suppose the number of points left nowis$m$, then B can be described as:

$$B = \{ b_1(x_1, y_1, z_1), \ldots, b_m(x_m, y_m, z_m) \}$$

STEP 2:

$\forall\, b \in B,\ y = 0$, and all points are projected onto the same plane. Then a new set of points is formed:

$$P = \{ p_1(x_1, 0, z_1), \ldots, p_m(x_m, 0, z_m) \}$$

Suppose the number of horizontal grids is H and the number of vertical grids is$V$, then the grid points can be represented as point sets:

$$Q = \{ q\,(h, v) |\ 0 \leq h \leq H, 0 \leq v \leq V, h, v \in \mathbb{N} \}$$

There is a mapping $f: P \rightarrow Q$between$P$and$Q$,seeequations (1).

STEP 3:

After making Pas Q, the grid q_i will containtotal l points internally, as shown in Fig.2(a),blue points are clouds points. Select a threshold T_1, when $l \geq T_1$, $q_i = 1$, which means that the grid can'treach, when $l < T_1$, $q_i = 0$, means the grid can reach, so that you can make into OGM map, as shown in Fig.2(b), the black grid can't reach, white grid isreachable area.

$$f := \begin{cases} X_{max} = MAX\{x_1, x_2, \ldots, x_m\} \\ X_{min} = MIN\{x_1, x_2, \ldots, x_m\} \\ u = \frac{X_{max} - X_{min}}{H} \\ Z_{max} = MAX\{z_1, z_2, \ldots, z_m\} \\ Z_{max} = MAX\{z_1, z_2, \ldots, z_m\} \quad (1) \\ v = \frac{Z_{max} - Z_{min}}{V} \\ p_i(x_i, z_i) \in P, q_i(h_i, v_i) \in Q \\ h_i = \left\lfloor \frac{x_i - X_{min}}{u} \right\rfloor \\ v_i = \left\lfloor \frac{z_i - Z_{min}}{v} \right\rfloor \end{cases}$$

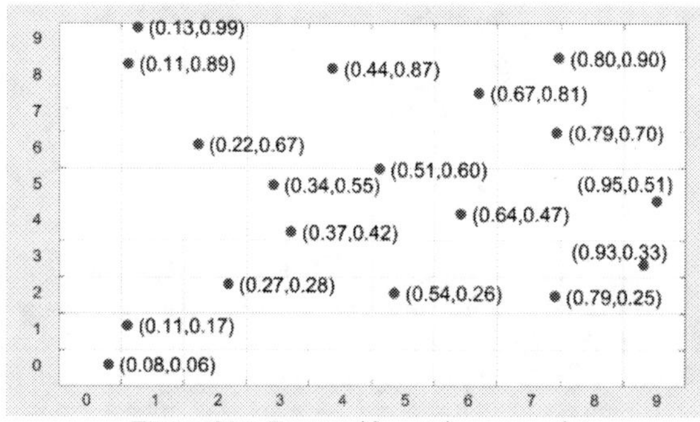

Figure 2(a). Every grid contain some points.

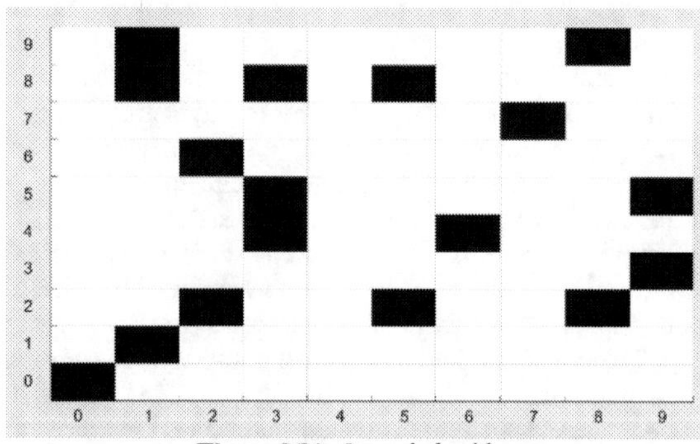

Figure 2(b). Occupied grid map.

2.3 Occupied grid mapexpansion

After obtaining OGM map, the robot itself has a certain size, but in the path planning, it is to use a grid to represent the robot, so the robot may collidesome obstacles. It is necessary to expand the obstacles before proceeding with the path planning, update to the OGM map, and then do the pathplanning. For acertain non-accessible grid, expansion area to the center of the circle, T_2 radius of the circular area.

3 Path planning

3.1 A* algorithm

The A* algorithm is a commonly used heuristic path planning algorithm that can be applied to find the shortest path from the start to the end of a grid map. The algorithm relies on the evaluation function $f(n)$ to measure whether the path is optimal.

$$f(n) = g(n) + h(n) \qquad (2)$$

The evaluation function $f(n)$is an evaluation function of node n, $g(n)$represents the movement cost evaluation of node n from the starting point, which is the movement distance of all the passed parent nodes by adding n to the starting point; $h(n)$is cost to the end point, which is a heuristic value where the Manhattan distance is used to guide the algorithm to find the end point. As shown in Fig.3, the redletter S indicates the start node, the green letter E indicates the end node and the blue node indicates the found path. Algorithm pseudo-code shown in Fig.4.

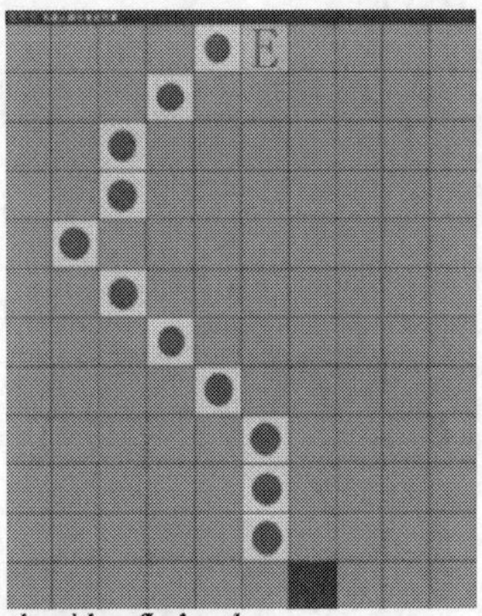

Figure 3. A* algorithm find path.

```
Input  : Location of Start and End node
Output : Father node of End

open       = list containing Start node;
closed     = empty set;
movecost(x;y) = distance from node x to node y.
while End node not in open : do
    i = node with lowest f(i) in open;
    remove i from open;
    add i to closed ;
    count = 0;
    for j = neighbor node of i and not in closed and reachable do :
      count++;
      cost = g(i) + movecost(i;j);
      if j in open and cost < g(j) then:
        remove j from open;
      if j not in open and not in closed then:
        add j into open;
        f(j) = g(j) + h(j);
        set father node of j is i;
   if count == 0 then:
      can't ¯nd path;
      break out;
```

Figure 4. A* algorithm pseudo-code.

3.2 A* algorithm performanceimprovement

We perform three main operations on the open set: the main loop finds the best node in the open set and adds it to the closed set, checks whether it is in the open set while visiting the neighbour node, and inserts the new node into the open set. Generally, the list is used to store the open set, the time complexity of deleting the insertion operation is $O(n)$, and here the binary heap is used to store the open set, which can reduce the time complexity to $O(\log n)$.

4 Simulation and experimental

4.1 Mapping using LSD SLAM

The physical prototype used in this paper is omnidirectional wheel mobile platform. The physical prototype has three 60mm omnidirectional wheels driven by MD36 type 24V DC motor, self-developed CAN bus DC servo motor driver and open UART and CAN motion control interface to the upper computer equipment. Using HTPC industrial computerforimage progressing and obstacle avoidance calculation. Using MT9V034 type camera, global shutter, angle of view 140°, focal length 1.9mm, resolution 640×480 , frame rate 60Fps, complete visual collection; Using RPLIDAR A1 type Lidar, 360° scan, 6mmeasurement radius, the data acquisition frequency of 5.5HZ ,avoid obstacles. The overall electrical connection is shown in Fig.5(a), and the robot entity is shown in Figure Fig.5(b).

Controlling the physical prototype to complete the loop-back motion in the indoor environment, as shown in Fig.6(a), the visual odometer calculates the pose change of $\mathfrak{se}(3)$, and optimizes the map and Open FAB-Map to complete the loopback detection using g^2o . Finally, 639007 3D point cloud data are obtained, and an ASCII version of the polygon file format document (PLY) is saved. The obtained 3D point cloud map and the pose diagram of the physical prototype are shown in Fig.6(b).

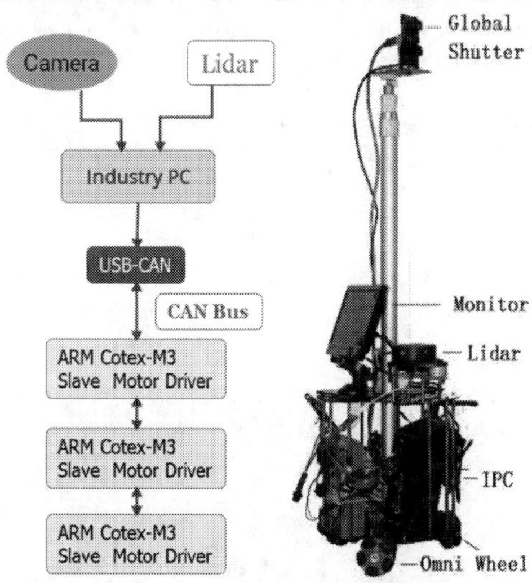

Figure 5 (a). Electrical Connection**(b).** Robot

4.2 Occupied grid mappreparation

After obtaining the PLY file, 364413 points cloud data are obtained according to STEP1 of Section 2.2, and then all the points are projected onto the ground to obtain a scattergram. Fig.7(a) is a path planning software interface running by the industrial computer.The software is based on OpenGL set up to achieve real-time display of 2-D terrain data, the number of horizontal grid is 168, the number of vertical grid is 120, the red point is the point of cloud projection, accounting for one pixel. Progressing the map according to STEP 2 and get the grid map as shown in Fig.7(b). The red area is not accessible and the green area is the accessible area. There are 8007 unacceptable grids and 121153 grids available.

Figure 6 (a). Indoor room**(b).** Point clouds map

Figure 7 (a). Scatter plot**(b).** OGM.**(c).** Expansion OGM

Then the expansion of the grid points is performed, using a Bresenham drawing circle algorithm from computer graphics--using a series of discrete points to approximate the circle and obtaining a grid expansion with a radius of 2 as shown in Fig.7(c). Radius can be dynamically adjusted

4.3 Path Planning Algorithm Simulation

To evaluate the performance and performance of the A* algorithm, the algorithm was tested in the simulation software described above. The author first tests the performance of two data structure A* algorithms at different start node andend node on different size maps of $168 \times 120, 336 \times 240, 672 \times 480$,as shown in Fig.8(a) and (b). The yellow grid indicates the path to the plan. The letter S indicates the start node and the letter E indicates the end node. Fig.9 shows the comparison of the three methods at different fifty pairs of start node andend node. The vertical axis shows the fifty pairsaverage time.Unit issecond. It can be seen that the time cost of A* binary heap is about 1/5 of the A* list, while the Dijkstra algorithm has the longest time, ten times as much as the A* list.

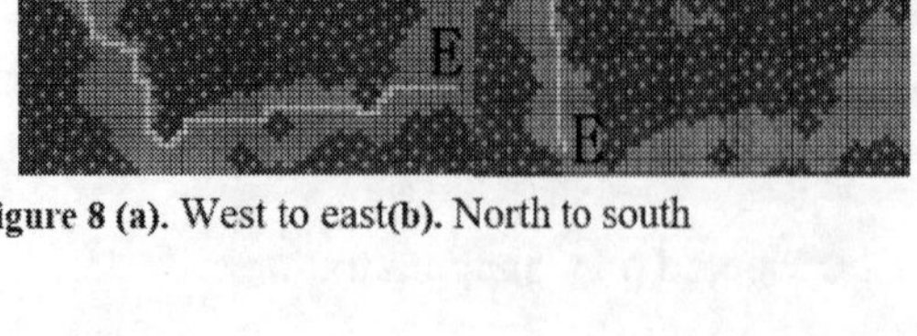

Figure 8 (a). West to east**(b).** North to south

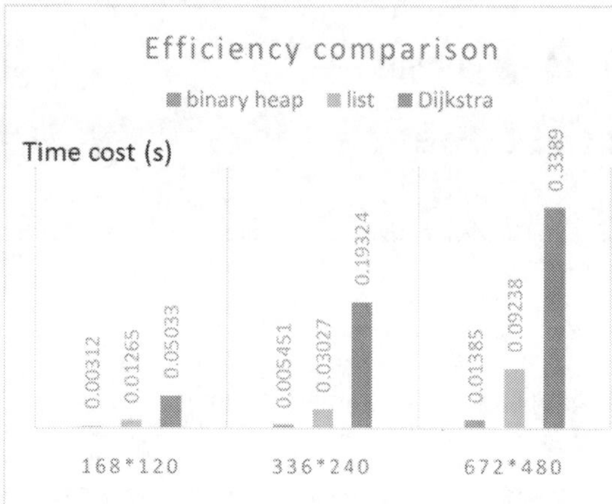

Figure 9. Vertical axis is average of fifty different pairs of start node and end node summary time cost. Horizontal axis is three different maps. Orange is time cost of A* using binary heap data structure, yellow is time cost of A* using list data structure, green is time cost of Dijkstra algorithm.

5 Conclusions

Inthis paper, the grid map preparation method and path planning strategy based on monocular vision simultaneous positioning and map construction are studied.

The environment is reconstructed using the large-scale direct method (LSD) model, which enables the environment to be accurately reconstructedin various scenarios of robots. Based on the environment three-dimensional data, occupancy grid map (OGM) was prepared for the description of the terrain. According to the terrain information, A * algorithm was used for path planning, and the performance of the algorithm was improved. Simulation shows that the proposed method is more than 5 times faster than the traditional method.

Next research will focus on controlling robot driving along the path and fixing robot pose according to VSLAM $\mathfrak{se}(3)$ pose.

References

1. E.W.Dijkstra, *A Note on Two Problems inCinnexion with Graphs. Numerische Mathematik*,1,269-271,(1959).
2. Amit Patel *Introduction to A** ,1-28,(1997-10-16)
3. Ferguson D,StentzA,Using interpolation to improve path planning: the field D algorithm. Journal of Field Robotics, 2006, 23(2) : 79-101.
4. XIN Yu,LIANG Huawei,DU Mingbo,MEI Tao, WANG Zhiling,JIANG Ruhai,An Improved A* Algorithm for Searching Infinite Neighbourhoods. Robot,(20-14),36(05):627-633.
5. LI Chong,ZHANG An,BI Wenhao.Single-Boundary Rectangle Expansion A* Algorithm.Robot,2017,39 (01):46-56
6. XiangGao,TaoZhang,YiLiu,QinruiYan.Fourteen Lectures on Visual SLAM: From Theory to Practice

[M]. Beijing: Publishing House of Electronics Industry, 2017: 130-137
7. Georg Klein, David Murray. Parallel Tracking and Mapping for Small AR Workspaces. In Proc. International Symposium on Mixed and Augmented Reality (ISMAR 2007, Nara)
8. Raúl Mur-Artal and Juan D. Tardós. ORB-SLAM2: an Open-Source SLAM System for Monocular, Stereo and RGB-D Cameras. *IEEE Transactions on Robotics,* vol. 33, no. 5, pp. 1255-1262, (2017)
9. J. Engel, J. Sturm, D. Cremers, Semi-Dense Visual Odometry for a Monocular Camera, J. Engel, J. Sturm, D. Cremers, ICCV (2013)
10. J. Engel, T. Schöps, D. Cr*emers*, LSD-SLAM: Large-Scale Direct Monocular SLAM, ECCV(2014)
11. J. Engel, V. Koltun, D. Cremers.Direct Sparse Odometry , In arXiv:1607.02565, (2016)
12. R. Wang, M. Schwörer, D. Cremers Stereo DSO: Large-Scale Direct Sparse Visual Odometry with Stereo Cameras , In International Conference on Computer Vision (ICCV), (2017)
13. V. Usenko, J. Engel, J. Stueckler, D. Cremers, Reconstructing Street-Scenes in Real-Time From a Driving Car , In Proc. of the Int. Conference on 3D Vision (3DV), (2015)
14. J. Engel, J. Stueckler, D. Cremers,Large-Scale Direct SLAM with Stereo Cameras , In-International Conference on Intelligent Robots and Systems (IROS), (2015)
15. Arren Glover,William Maddern,OpenFABMAP: An Open Source Toolbox for Appearance-based Loop Closure Detection,ICRA,14-18 May (2012)

The Square Root UKF-SLAM Algorithm Based on the Smallest Proportion of Skewness in Single Line Sampling

MengYuan Chen[1 2], GuoWei Qin[1] and Tong Xu[1]

[1]Key Lab of Electric Drive and Control of Anhui Province, Anhui Polytechnic University, Wuhu 241000, China
[2] Department of Precision Machinery and Precision Instrumentation, University of Science and Technology of China, Hefei 230027, China

Abstract. In view of the distortion in the filter gain matrix calculation as well as the high computational complexity and the nonlocal effect of symmetric sampling that exists in the UKF-SLAM algorithm， the square root UKF-SLAM algorithm based on the smallest proportion of skewness in single line sampling was proposed. According to the mended algorithm, the square root of covariance matrix is brought into iteration operation instead of covariance matrix, moreover, the smallest proportion of skewness in single line sampling is utilized for the optimization of sampling strategy. The results of simulation show that the algorithm can effectively improve the position accuracy in robot as well as the estimation accuracy of feature map. Furthermore, the computational complexity is reduced and the algorithm stability is improved.

1 Introduction

In the unknown environment, the robot takes advantage of its own sensors to create feature maps for the environment incrementally. At the same time, the machine corrects its own position that has been created in the map, namely SLAM, a problem about simultaneous localization and mapping in mobile robot. Moreover, a classical algorithm that had been applied to SLAM is the method that dynamic equations and observation equations of robot are expanded by first order Taylor series, and then the Extended Kalman Filter algorithm was used for the optimal estimation of the robot's position and the feature map. This model was first proposed by Smith and Cheeseman [1]. Many scholars optimized and improved it based on EKF-SLAM framework. For example, Lu presented that polar coordinates can be used to compare two adjacent observed values to reduce external interference and improve the algorithm's estimation accuracy as well as robustness [2]; Zhanghai Jiang put forward an improved compression-type EKF-SLAM algorithm to reduce the computational complexity [3]. Zhou Wu proposed a global observation map model to improve the presentation methods [4] of the map. However, the inherent defects, such as the truncation error generated from the linearization procedure and the large amount of calculation from Jacobi matrix, make the EKF-SLAM algorithm unsuited to the large-scale environmental map.

Therefore, Julier et al. pointed out Unscented Kalman Filter(UKF), a new filtering algorithm that can be directly applied to nonlinear systems when they studied how to improve EKF-SLAM. UKF is a nonlinear Gaussian state estimator based on the minimum variance estimation criteria [5]. Similar to the Extended Kalman Filter algorithm, Gaussian random variables are used to approximate the state distribution. But the difference is that UKF approximate the state distribution through deterministic sampled (Sigma points) obtained from some certainty sampling. These sampling points can better describe the true mean and variance of the Gauss random variable. In addition, its precision approximates to the second-order of Taylor while the EKF can only approximates to the first-order accuracy under the same conditions. Besides， in terms of computational complexity, UKF and EKF belong to the same order and do not need to calculate the Jacobian matrix, so UKF is easier to meet our objectives than the EKF [6]. Furthermore, the follow-up work found that the state covariance matrix of UKF algorithm is easy to lose non-negative qualitative and symmetry due to the rounding errors from the computer in the process of data transmission, which will lead to filter divergence [7]. On the contrary, the square root filter and the strategy that use the square root of the covariance matrix instead of covariance matrix itself can solve this problem perfectly, which has been successfully applied to GPS / DR combination [8] and indoor geolocation [9].

In the first place, this article focuses on MATLAB comparative simulation experiments that simulate the process of SLAM which UKF algorithm and the square root UKF algorithm (SR-UKF) apply to. The result shows that the SR-UKF algorithm can significantly improve the accuracy of the robot's pose estimation. However, in terms of SLAM, a kind of system which need high requirements of real-time, it is particularly important to reduce the computational complexity and improve the computational efficiency [10]. To solve the above problems, the SR-UKF algorithm was given a depth study by changing the sampling strategy and using the smallest proportion of skewness in single line sampling instead of symmetric sampling, as well as reducing the computational complexity through reducing the number of points of Sigma, which enables the systems towards the direction of better real-time. Therefore, a square root UKF-SLAM algorithm based on the smallest proportion of skewness in single line sampling has been proposed in this paper.

2 The probability description about SLAM

It assumes that the state vector of a robot is $X_k^v=[x_k^v,y_k^v,\varphi_k^v]^T$; the feature map is $X_k^m=[m_1,m_2,\cdots,m_n]^T$; the observation sequence is $Z_{1:k}=[Z_1,Z_1,\cdots,Z_k]^T$; the control command is $u_{1:k}=[u_1,u_2,\cdots,u_k]^T$. Dynamic and observation equation are respectively supposed as $X_k=f\left(X_{k-1},u_k\right)+w_k$ and $Z_k=h\left(X_k\right)+v_k$, moreover, w_k and v_k are gaussian distribution followed by $N(0,Q_k)$ and $N(0,R_k)$. The mathematical description of SLAM problem is:

$$p(X_k^v,X_k^m\,|\,Z_{1:k},u_{1:k}) \tag{1}$$

$Z_{1:k}$ and $u_{1:k}$ are known conditions. We calculate posterior probability distribution in formula (1) by Baye filter principle to obtain the optimal state estimation. The whole process is divided into two steps, the first step is prediction, the prior probability distribution at k time is obtained by computational model of robot pose $p(X_k^v\,|\,X_{k-1}^v,u_k)$ and posterior probability distribution at k+1 time:

$$p(X_k^v,X_k^m\,|\,Z_{1:k},u_{1:k})=\int p(X_k^v\,|\,X_{k-1}^v,u_k)p(X_{k-1}^v,X_{k-1}^m\,|\,Z_{1:k-1},u_{1:k-1})dX_{k-1}^v \tag{2}$$

The second step is observation update, Z_k is gained from the observation model of system at k time, and the posterior probability distribution is solved by the prior probability distribution.

$$p(X_k^v,X_k^m\,|\,Z_{1:k},u_{1:k})=\eta p(Z_k\,|\,X_k^v,X_k^m)p(X_k^v,X_k^m\,|\,Z_{1:k-1},u_{1:k}) \tag{3}$$

η is the normalization coefficient.

3 SR-UKF-SLAM algorithm and the simulation analysis

3.1 UKF-SLAM algorithm

UT Transformation as the core, kalman filter as the basic framework, UKF take advantage of certainty sampling to get the posterior state distribution from state estimation. Before filtering, the original state vector is augmented to gauss noise variable:

$$X_k^a=[X_{k-1}^{a\ T},w_k^T,v_k^T]^T \tag{4}$$

$$P_k^a=\begin{bmatrix}P_k & 0 & 0\\ 0 & Q_k & 0\\ 0 & 0 & R_k\end{bmatrix} \tag{5}$$

It assumes that the mean is $\hat{X}_{k-1}^a$ and the variance is P_{k-1}^a at k-1 time. The UKF-SLAM algorithm block diagram was shown in Figure 1. The whole algorithm is divided into three parts , including sigma point、forecast period and update phase calculations. Besides, the calculation of sigma point adopts the symmetrical sampling method. $2n+1$ Sigma points, $\xi_i, i=0,1,\cdots,2n$ (n is the system status dimension of X_k^a), whose corresponding first and second order metric are W_i^m and W_i^c respectively. These sampling points are taken into the model of nonlinear system for transition, which helps to get the statistical properties of the state vector.

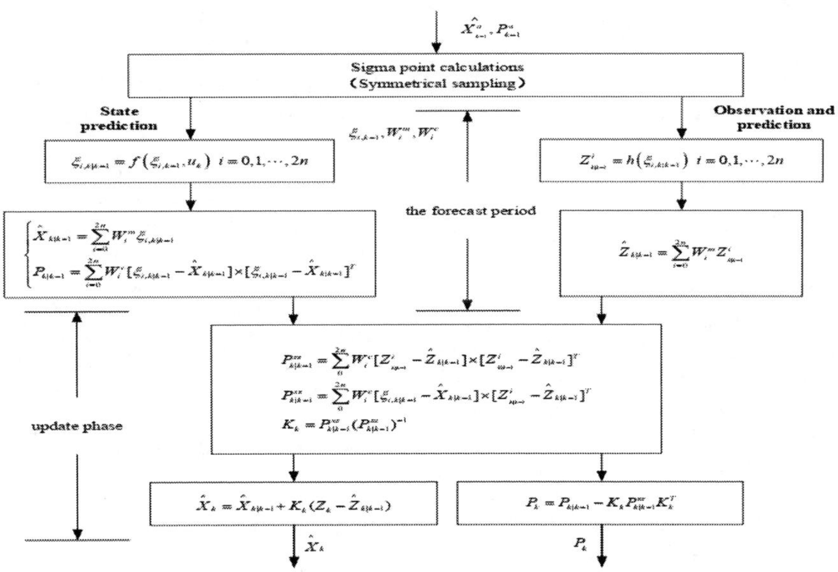

Figure 1. EKF-SLAM algorithm block diagram.

3.2 SR-UKF-SLAM algorithm

In the aforesaid filtering model, the rounding errors of calculation are gradually accumulated with the depth of filtering. In addition, the possibility that the filtering error covariance matrix P_k and the predicting error covariance matrix $P_{k|k-1}$ lose non-negative qualitative and symmetry will increase, which makes the calculation of filtering gain distorted as well as the filter diverging. Therefore, on the basic of UKF-SLAM algorithm, with the update of covariance square root S_k instead of covariance P_k ($P_k = S_k S_k^T$), a UKF-SLAM algorithm based on the square root [11] (SR-UKF-SLAM) is proposed. Furthermore, in the specific process of algorithm, the two calculation methods including QR decomposition and Cholesky factor update are utilized to reduce the computational complexity. As distinguished from the EKF-SLAM algorithm, its optimization process is as follows:

Predict phase:

$$S_{k|k-1} = QR\left\{\left[\sqrt{W_{1:2n}^{c}}\left(\xi_{1:2n,k|k-1}^{a} - \hat{X}_{k|k-1}\right) \quad \sqrt{Q_{k-1}}\right]\right\} \tag{6}$$

$$S_{k|k-1}^{z} = QR\left\{\left[\sqrt{W_{1:2n}^{c}}\left(Z_{1:2n,k|k-1}^{i} - \hat{Z}_{k|k-1}\right) \quad \sqrt{R_{k-1}}\right]\right\} \tag{7}$$

Update phase:

$$S_{k|k-1} = cholupdate\left\{S_{k|k-1}, \xi_{0,k|k-1}^{a} - \hat{X}_{k|k-1}, W_0^c\right\} \tag{8}$$

$$S_{k|k-1}^{z} = cholupdate\left\{S_{k|k-1}^{z}, Z_{0,k|k-1}^{i} - \hat{Z}_{k|k-1}, W_0^c\right\} \tag{9}$$

In order to avoid the matrix inverse operation, back substitution was used to deal with filtering gain K_k . As shown in the formula (10), formula (11)、formula (13) are applied to the iterative update of state vector.

$$K_k = P_{k|k-1}^{xz} / (S_{k|k-1}^{z})^T / S_{k|k-1}^{z} \tag{10}$$

$$\hat{X}_k = \hat{X}_{k|k-1} + K_k(Z_k - \hat{Z}_{k|k-1}) \tag{11}$$

$$U = K_k S_{k|k-1}^{z} \tag{12}$$

$$S_k = cholupdate\left\{S_{k|k-1}, U, -1\right\} \tag{13}$$

3.3 Simulation and Analysis

In order to study the effect of the estimation accuracy from square root transition, the SR-UKF-SLAM and UKF-SLAM algorithm were respectively simulated through open source simulation platform developed by Tim Bailey in MATLAB. The computer in the experiment employs Intel Core-i5 dual-core processor with 2.6 GHz basic frequency and 4 GB internal storage. In the experiment, with 250*200m analog outside circumstance, the robot paraded a circle along 17 path points(“●”) from coordinates (0,0), the running speed of the robot is up to 2 m/s, 35 static feature points (“*”) were randomly distributed in the environment. The simulation results of two algorithms are shown in Figures 2 and 3, the filament represents the ideal trajectory of the robot, the “dotted line” indicates the estimated path of the robot, "+" is the position of feature point, "ellipse" is the error range of the feature point estimation. Figure 2 and figure 3 shows that, during the initial period, the calculation error between UKF-SLAM and SR-UKF-SLAM algorithm became relatively small. However, as time goes on, due to the constantly increased new features, there has been an increasingly deviation gap between estimated and ideal trajectory. and the error ellipse from the features map is also significantly higher than the SR-UKF-SLAM algorithm.

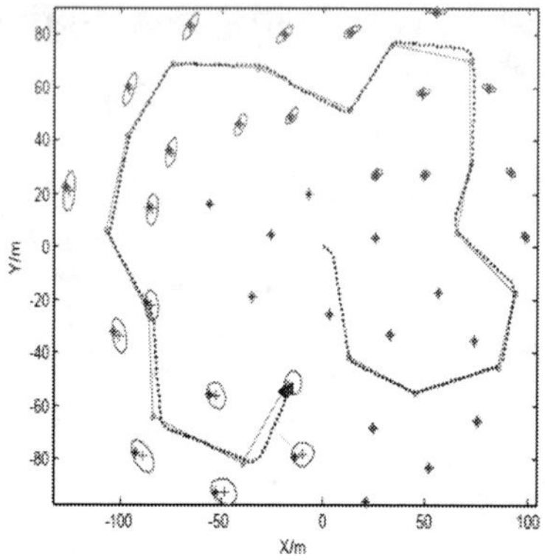

Figure 2. Simulation result of UKF-SLAM.

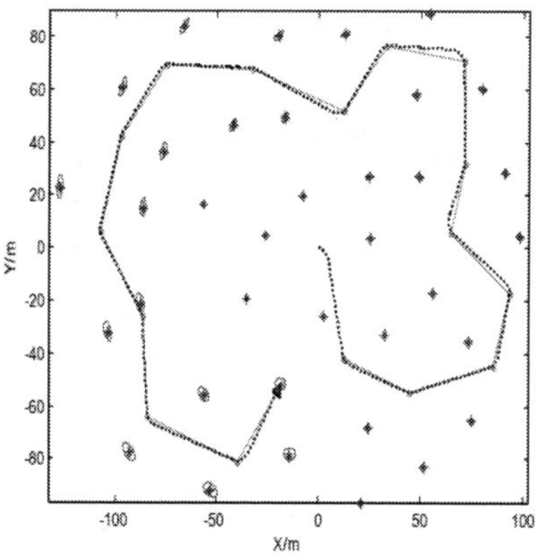

Figure 3. Simulation result of SR-UKF-SLAM.

It took about 330 seconds for robot to explore one lap through average value from multiple experiments. In this paper, the simulation results gained from the first 300s were analysed. (a)、(b)、(c) in figure 4 respectively indicate the calculation accuracy of robot in x and y direction as well as the error comparison of course angle that were obtained from UKF-SLAM UKF-SLAM (as UKF showed in the picture, dotted line) and SR-UKF-SLAM (as SR-UKF showed in the picture, solid line) algorithm. From the picture, the estimating errors of position in robot based under two algorithms became greater, however, in contrast to UKF-SLAM, the estimation error of SR-UKF-SLAM algorithm is relatively stable and always maintain a low level.

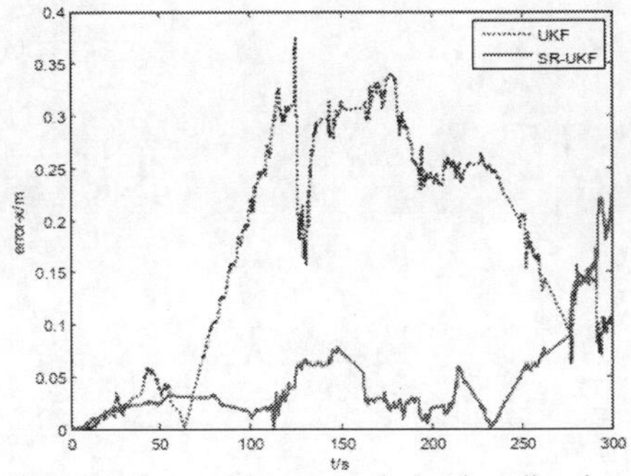

(a) Estimation position error of robot in x direction.

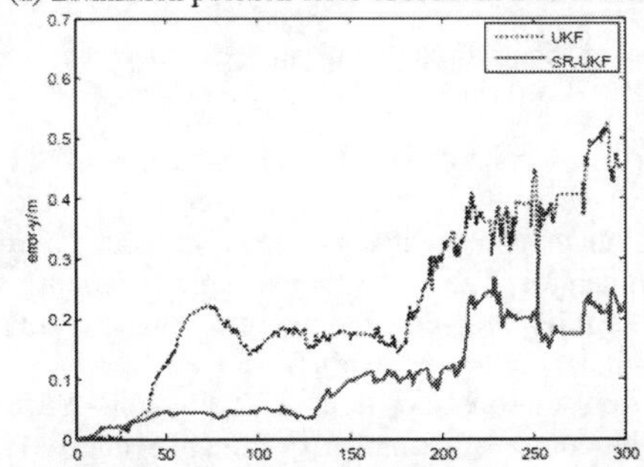

(b) Estimation position error of robot in y direction.

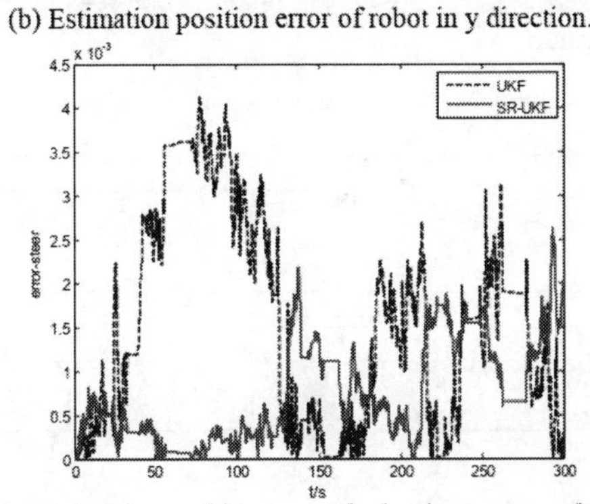

(c) Estimation position error of robot in course angle.

Figure 4. A comparison of estimation error of UKF-SLAM and SR-UKF-SLAM.

4 Simulation experiment and analysis of SPSR-UKF-SLAM algorithm

Although the SR-UKF-SLAM algorithm improves the estimation accuracy of robot-position and features maps to a certain degree, however, in terms of SLAM which requires hard real-time performance, we hope that the computational complexity can be further reduced through reducing the number of sigma points. In the symmetric sampling, the distance from sigma point to $\hat{X}_k^a$ ascends with the increasement of state dimension of X_k^a, which easily causes nonlocal effect and affects the accuracy and stability of the filtering.

The smallest proportion of skewness in single line is a kind of sampling mode that needs minimal points，due to the direct influence on the calculation amount of UKF from the number of sampling points, the strategy is more suitable for the system with higher real-time performance. According to Julier's analysis in the literature [11], at least n+1 sampling points are required to determine the posterior distribution of the system for the n-dimensional state vector distribution. Moreover, the study found that the proportion correction can be a good solution to the non-local effects. Therefore, combined with the advantages of the sampling strategy above, a square root UKF-SLAM algorithm based on the smallest proportion of skewness in single line sampling, namely the SPSR-UKF-SLAM algorithm, was proposed in the process of SLAM，Which helps modify the sampling strategy of SR-UKF-SLAM algorithm, its specific sampling process is as follows:

(1) The weight coefficient of initialization $0 \le W_0^m = W_0^c < 1$.

(2) The first and second order weight coefficient of sigma sampling points are:

$$W_i^m = \begin{cases} \frac{W_0^m}{\alpha^2} + \left(1 - \frac{1}{\alpha^2}\right), & i = 0 \\ \frac{1 - W_0^m}{2^n \alpha^2}, & i = 1,2 \\ \frac{2^{i-2} W_1^m}{\alpha^2}, & i = 3,4,\cdots,n+2 \end{cases} \tag{14}$$

$$W_i^c = \begin{cases} W_0^m + (1 + \beta - \alpha^2), & i = 0 \\ W_i^m, & i \neq 0 \end{cases} \tag{15}$$

In the equation: α ,as a smaller positive number, is used to control the distance from sigma point to the center point $\hat{X}_k^a$ with the range of $\left[10^{-4}, 1\right]$; β is used to describe a priori information about the distribution of random variables., in the case of a Gaussian distribution, the best approach is taking $\beta = 2$.

(3) The initial vector ($L = 1$) is

$$\xi_0^1 = [0], \xi_1^1 = \left[-\frac{1}{\sqrt{2W_1^m}}\right], \xi_2^1 = \left[\frac{1}{\sqrt{2W_1^m}}\right] \tag{16}$$

According to the state dimension of input $L = 2,3,\cdots,n$, the iteration formula of vector is:

$$\xi_i^L = \begin{cases} \begin{bmatrix} \xi_0^{L-1} \\ 0 \end{bmatrix}, & i = 0 \\ \begin{bmatrix} \xi_0^{L-1} \\ -\frac{1}{\sqrt{2W_{L+1}^m}} \end{bmatrix}, & i = 1,2,\cdots,L \\ \begin{bmatrix} 0 \\ -\frac{1}{\sqrt{2W_{L+1}^m}} \end{bmatrix}, & i = L+1 \end{cases} \tag{17}$$

(4) The sigma points obtained from formula (24) is:

$$\xi_{i,k-1}^a = \hat{X}_{k-1}^a + \alpha S_{k-1} \xi_i^L, i = 0,1,2,\cdots,L+1 \tag{18}$$

The simulation experiments was carried out on the optimized algorithm in the same simulation environment, the simulation results shown in Figure 5. With respect to the results shown in Figure 2, the estimated trajectory from robot goes a step further approaching the ideal trajectory in the process of SPSR-UKF-SLAM algorithm, and the error ellipse is further reduced. (a) , (b) , (c)in figure 6 respectively indicate the calculation accuracy of robot in x and y direction as well as the error comparison of course angle that were obtained from SR-UKF-SLAM (as SR-UKF showed in the picture, dotted line) and SPSR-UKF-SLAM (as SPSR-UKF-SLAM showed in the picture, solid line) algorithm. From the picture, SPSR-UKF-SLAM improves the estimation precision of the robot pose state while reducing the computational complexity.

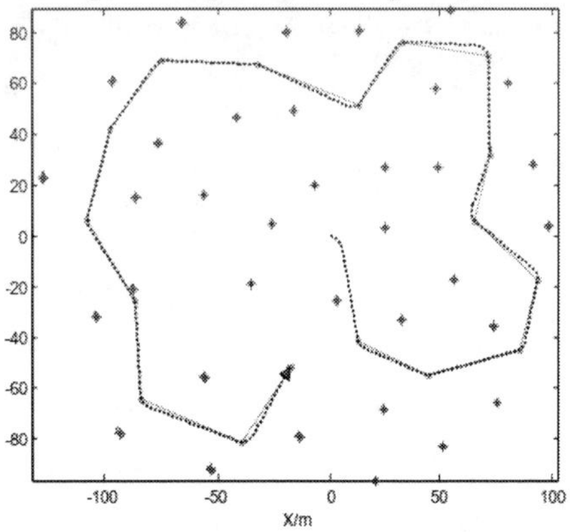

Figure 5. The simulation results of SPSR-UKF-SLAM.

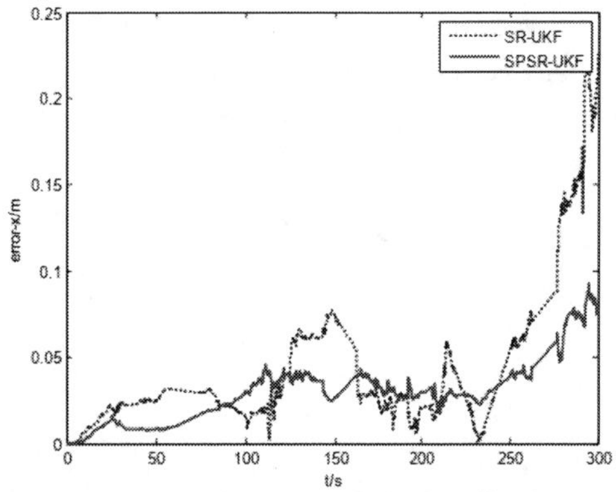

(a) Estimation position error of robot in x direction.

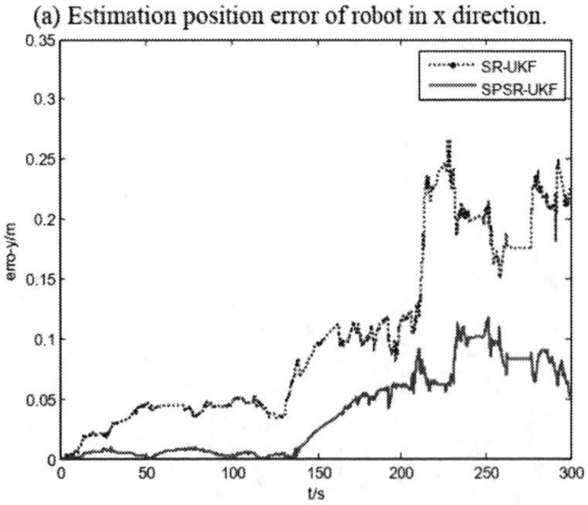

(b) Estimation position error of robot in y direction.

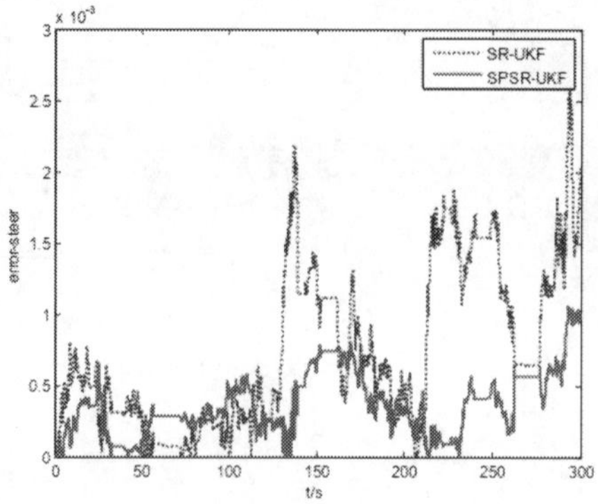

(c) Estimation position error of robot in course angle.

Figure 6. A comparison of estimation error between SR-UKF-SLAM and SPSR-UKF-SLAM.

In order to compare the performance of the three SLAM algorithms, the average errors of the three algorithms are calculated as shown in table 1. From the data in table 1, The estimation accuracy of robot position from SPSR-UKF-SLAM is the highest in three algorithms. Compared with SR-UKF-SLAM algorithm, the errors of robot-position in X direction, Y direction and the course angle are respectively reduced 19.08%, 29.31% and 50.47% under SPSR-UKF-SLAM algorithm.

Table 1. Statistical comparison of algorithm errors.

algorithm	errors		
	x direction /m	y direction /m	course angle /rad
UKF-SLAM	0.1742	0.2284	0.1543
SR-UKF-SLAM	0.0435	0.1020	0.0741
SPSR-UKF-SLAM	0.0352	0.0721	0.0367

5 Conclusion

According to the filter divergence from the rounding errors of caculation that exists in EKF-SLAM algorithm, the covariance square root was utilized for update instead of covariance. Furthermore, in term of the problems such as a great quantity of sampling site, the poor real-time performance and the nonlocal effect that easily produced, the sampling schematism was transformed into the one based on smallest proportion of skewness in single line sampling, from which a more optimized algorithm ,the square root UKF-SLAM algorithm based on the smallest proportion of skewness in single line sampling, was proposed. Moreover, the MATLAB simulation experiments show that, compared with the UKF-SLAM algorithm, the new algorithm can effectively improve the estimation accuracy of the robot's position and the feature maps with its relatively lower computational complexity and greater stability.

Acknowledgments

This work was supported by 2018 Anhui Natural Science Foundation (Bionic SLAM study of rat mixed cells with DGSOM neural network) and 2018 Anhui Excellent top-notch talent cultivation project in Colleges and Universities.

References

1. R Smith, M self, P Cheesseman. Estimating uncertain spatial relationships in robotics [A]. Proc of Conf Uncertainty in Artificial Intelligence[C]. Amsterdam: North -Holland, 1988, 435-461.
2. Taizhi Lv. A new anti-disturbance EKF-SLAM algorithm [J].computer engineering, 2012, 38(21):1-4.
3. Haiqiang Zhang, Lihua Dou, Hao Fang, etc. An improved SLAM algorithm based on compression type EKF [J]. System Simulation, 2009, 21(18):5668-5680.
4. Wu Zhou, Chunxia Zhao, Yaqiang Shen, etc. The study of SLAM based on global observation map model [J]. Robot, 2010, 32(5):647-654.
5. S J Julier, J K Uhlmann. H Durrant-Whyte. A New Method for the Nonlinear Transformation of Means and Covariances in Filters and Estimation [J]. IEEE Transactions on Automatic Control (S0018-9286), 2000, 45(3):477-482.
6. Xinxi Shi, Chunxia Zhao, Jianhui Guo. SLAM framework algorithm of mobile robot based on PF / EKF UKF [J]. Digital Journal.2009, 8:1865-1868.
7. Merwe R V D, Wan E A, The Squre-Root Unscented Kalman Filter for State and Parameter–Estimation[C].International Conference on Acoustics, Speech, and Signal Processing, Utah,2001.
8. Jing Yang, Nanning Zheng. A GPS/DR Integrated Positioning Algorithm Based on SR-UKF. System Simulation [J].2009, 21(3):721-722,742.
9. Lei An, Guoliang Zhang, Wenjun Tan. The Indoor Kalman filter localization algorithm of Robot based on RSSI [J]. Computer Engineering and Application. 2010, 48(8):230-232.
10. Shurong Li, Pengfei Ni. SLAM algorithm based on the square root of UKF [J]. Computer Engineering and Application, 2011, 47(22):209-212.
11. S J Julier, J K Uhlmann. Reduced sigma point filters for propagation of means and covariances through nonlinear transformation: Proc. of American Control Conf[C]. Jefferson City, 2002: 887-892.

Novel Design of a Biaxial and Four-Wheeled Robot Capable of Steering

Guiping Lan[1], Yujun Wang[1], Can Fang[1], Min Yi[1]

[1] *College of Computer and Information Science , Southwest University , BeiBei 400415, China*

Abstract. The paper presents a biaxial and four-wheeled robot, mainly composed of two axle, two circular wheels and two three-leaved wheels. By analyzing the difference between the velocity of the circular wheel and three-leaved wheel in same axle, the steering principle of the difference velocity is proved. The three-leaved wheels are installed by a complementary phase method to ensure the stability of the robot walking. Through the steering principle, the control method of robot' s forward, backward, turn left and turn right is designed. A large number of experimental results show that the robot has the characteristics of high steering efficiency, simple mechanical structure and easy to control.

1 Introduction

In order to the wheeled robot has better mobile efficiency, many scholars have designed various structures for steering. Fu Yi-li, et al [1] proposed a variety of steering methods using omni-directional wheeled mode. Chang jian,et al [2] proposed a shape-shifting robot skid-turning with line configuration method. Gao Jun-xiao et al [3] proposed GA fuzzy neural network algorithm to achieve the steering control of the robot. Yan Yong-bao et al [4] proposed a independent drive skid-steering vehicle. Li Xue-yuan et al [5] proposed a steering method with difference velocity. Xing Jie [6] analyzed differential steering and torque matching of 4-WD mobile robot. Ye Chang-long et al [7] designed an omnidirectional mobile robot with MY wheels. Qian J et al [8] designed and developed an omni-directional mobile robot oriented to an intelligent manufacturing system.Zi B et al [9]designed a winding hybrid-driven cable parallel manipulator. They proposed the mechanical structures or algorithms are too complex and difficult to control.

To cope with these problems, the biaxial and four-wheeled robot is proposed in this paper that was focused on small-size, high efficiency of movement, simple mechanical structure, easy to control and strong capability of steering. This paper is organized as follows: Section 2 describes the mechanical structure of the robot. Section 3 detail the steering principle. The experimental results are shown in Section 5. Finally, Section 6 is the conclusions.

2 Conception of the robot

2.1 Mechanical structure

The paper presents a biaxial and four-wheeled robot, composed of two axles, two transmission mechanisms, two motors, two circular wheels and two three-leaved wheels, as shown in Figure 1. Each end of the axle is mounted with a circular wheel and a three-leaved wheel, respectively. Two circular wheels and two three-leaved wheels are installed on both sides of the body. When each motor drives the axle through the transmission mechanism, the circular wheel and the three-leaved wheel are rotated synchronously. The transmission ratio of the transmission mechanism is 1: 1. The phase difference between two three-leaved wheels is 60°. The radius of the circular wheel and the three-leaved wheel is same. The angle between each leaf of the three -leaved wheel is 60° as shown in Figure 2. Figure 3 is a physical diagram of a biaxial four-wheeled robot.

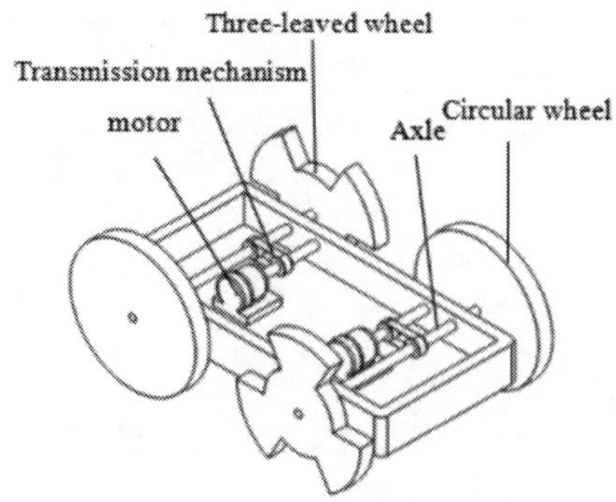

Figure 1. 3D model of the robot

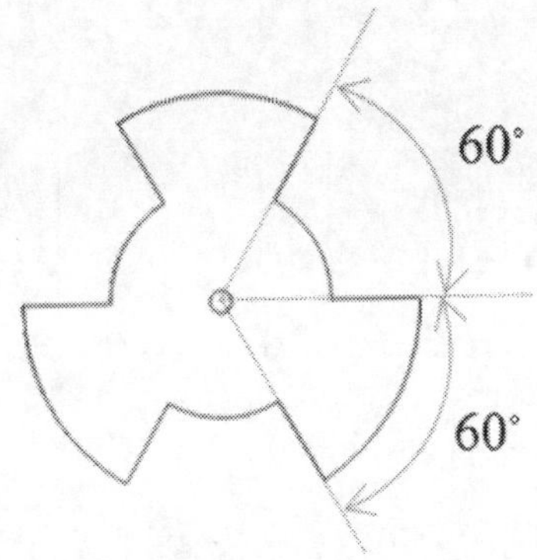

Figure 2. Three-leaved wheel structure

Figure 3. The biaxial and four-wheeled robot

2.2 Control system

In order to achieve high motion efficiency, the design of the control system is essential. Two low-torque motors are used to drive the wheels with encoder. Two motor driving modules can be used to drive the motors. The microcontroller (ATMega 328p) is used to control the motor driving modules. And the MCU receives control commands from the host computer. The block diagram of the robot control system is shown in Figure 4.

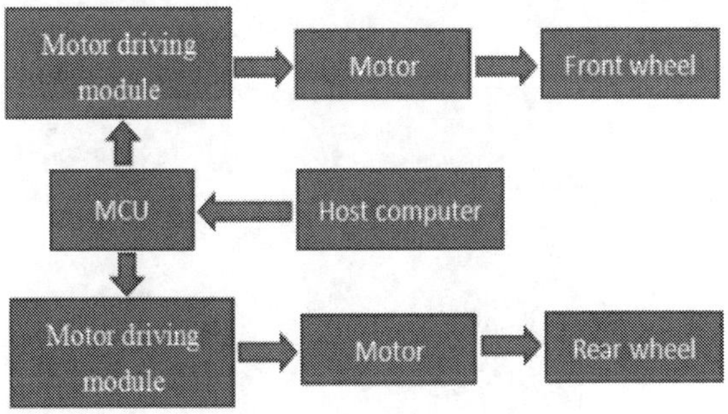

Figure 4. Robot control system

3 Motion analysis

In structured environment, wheeled robots have four main forms of movement, forward, backward, turn left and turn right. The circular wheel and the three-leaved wheel on the same axle have the same angular velocity. Assuming that the time required for the wheels to rotate one round is t, the walking distance of the circular wheel and three-leaved wheel on the ground can be obtained by :

$$S_1 = R\omega t$$
$$S_2 = \frac{1}{2}R\omega t \tag{1}$$

Where S_1, S_2 are the walking distance of the circular wheel and three-leaved wheel on the ground, respectively; R is the radius of the circular wheel and three-leaved wheel. ω is the angular velocity of the circular wheel and three-leaved wheel.

The walking velocity of the circular wheel and three-leaved wheel on the ground can be obtained by :

$$v_1 = \frac{dS_1}{dt} = R\omega$$
$$v_2 = \frac{dS_2}{dt} = \frac{1}{2}R\omega \tag{2}$$

Since the speed of the circular wheel is greater than the speed of the three-leaved wheel, the robot rotates to the side of the three-leaved wheel, as shown in Figure 5. When the motor drives the circular wheel and three-leaved wheel forward, a clockwise torque(τ_1) is generated, as shown in Figure 5 (a). When the motor drives the circular wheel and three-leaved wheel backward, a counterclockwise torque(τ_2) is generated, as shown in Figure 5 (b).

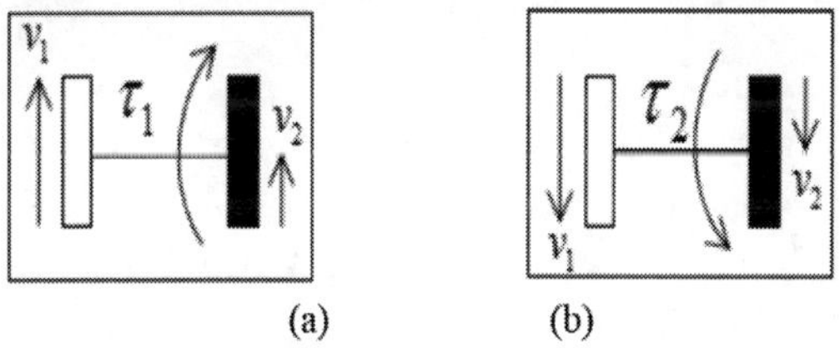

Figure 5. Coaxial steering principle
(White box is circular wheel, black box is three-leaved wheel)

From the steering principle mentioned above, we can get four kinds of biaxial four-wheeled robot movement: forward, backward, turn left and turn right. Assuming that the output power of the two motors is the same, the friction coefficient between the wheels and the ground is the same.

When the robot moves forward (or backward), the front and rear motors simultaneously drive the circular wheels and three-leaved wheels forward (or backward) as shown in Figure 6(a)(b). In the process of movement the following relationship:

$$
\begin{aligned}
v_1 &= v_4 \\
v_2 &= v_3 \\
\tau_f &= \tau_r
\end{aligned} \tag{3}
$$

Where v_1, v_4 are the walking velocity of the front circular wheel and the rear circular wheel on the ground, v_1, v_4 have the same value and direction; v_2, v_3 are the walking velocity of the front three-leaved wheel and the rear three-leaved wheel on the ground. v_2, v_3 have the same value and direction; τ_f, τ_r are the torque of the front axle and the rear axle on the ground. τ_f, τ_r have the same value and the opposite direction.

When the robot turns left, the front motor drives the circular wheel and three-leaved wheel backward and the rear motor drives the circular wheel and three-leaved wheel forward as shown in Figure 6(c). In the process of movement the following relationship:

$$
\begin{aligned}
v_1 &= v_4 \\
v_2 &= v_3 \\
\tau_f &= \tau_r
\end{aligned} \tag{4}
$$

Where v_1, v_4 have the same value and the opposite direction; v_2, v_3 have the same value and the opposite direction; τ_f, τ_r have the same value and the same direction.

When the robot turns right, the front motor drives the circular wheel and three-leaved wheel forward and the rear motor drives the circular wheel and three-leaved wheel backward as shown in Figure 6(d). In the process of movement the following relationship:

$$
\begin{aligned}
v_1 &= v_4 \\
v_2 &= v_3 \\
\tau_f &= \tau_r
\end{aligned} \tag{5}
$$

Where v_1, v_4 have the same value and the opposite direction; v_2, v_3 have the same value and the opposite direction; τ_f, τ_r have the same value and the same direction.

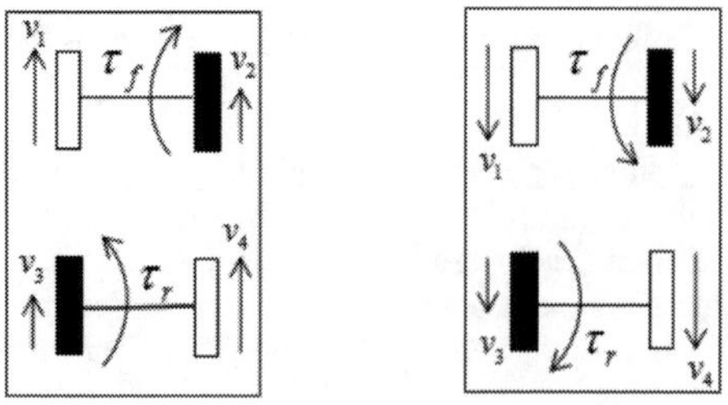

(a) (b)

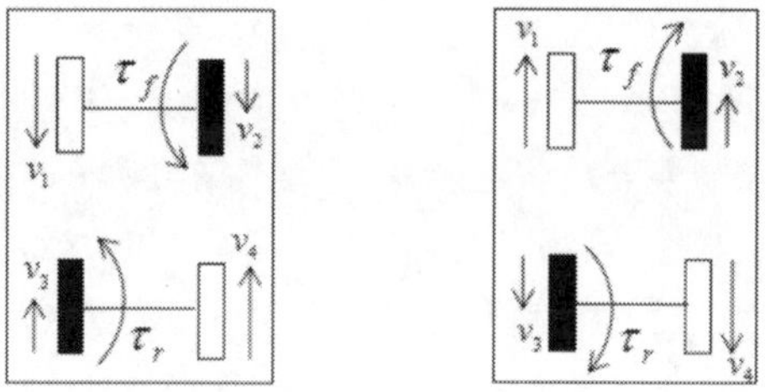

(c) (d)

Figure 6. The four movements of the robot

4 Experiment

In order to verify the steering principle of the robot and test the efficiency of walking, a series of experiments were carried out. In the experiment, four batteries(MCOBEAM / 3400Mah / 3.7V) are used the power of the DSAWbot. MCU is Arduino nano V3.0. The motor drive module uses TSK-BTN7971.Two motors(Germany Faulhaber23420L012CR) can be used to drive the wheels.The size of robot used in this experiment was 310mm by 310mm by 200mm for the length, width and high, respectively. The radius of wheel is 80mm. Figure 6 is the process of the left turn of the robot. Figure 7 is the process of the right turn of the robot. Experiments show that the steering principle proposed in this paper is correct and can achieve the left and right turn of the robot. Experiment result show that the robot can to steer on the ground. In the case of limited power and size, the walking efficiency of the robot are maximized. The relevant experimental results are shown in Table 1.A large number of experimental results show that the robot has the characteristics of high steering efficiency, simple mechanical structure and easy to control as shown in Figure 7 and Figure 8.

(a)

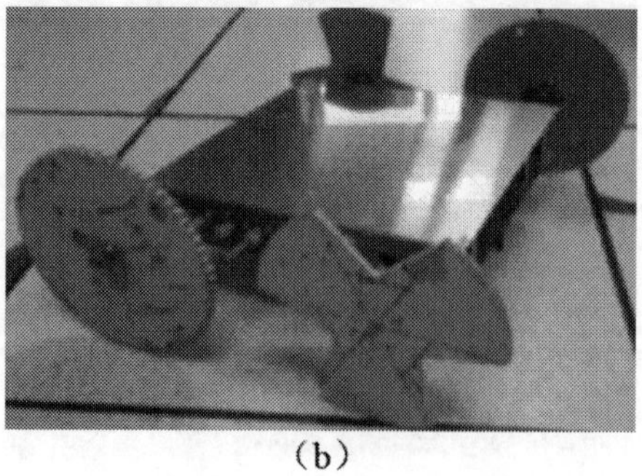

(b)

(c)

Figure 7. Turn left

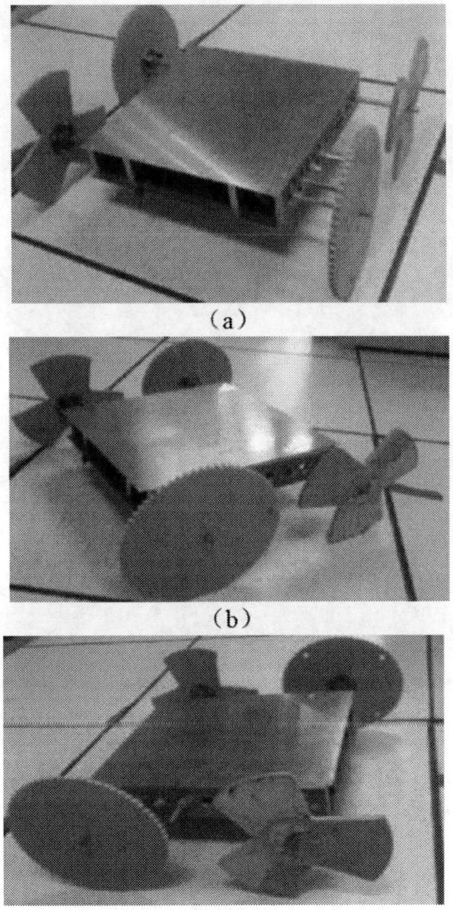

(a)

(b)

(c)

Figure 8. Turn right

Table 1. Specifications of the robot

Parameters	robot
Total weight	8.47kg
load capacity	4.25kg
Speed in structure environment	3.16km/h
Size	310mm*310mm*200mm
Radius	80mm

5 Conclusion

In this paper, a biaxial four-wheeled robot is proposed, which utilizes the phase complementarity between two three-leaved wheels. Each time there are two circular wheels and one three leaved wheel on the ground, which guarantees the stability of the robot during the walking process. By analyzing the difference between the velocity of the coaxial circular wheel and the three-leaved wheel during the movement, the four kinds of motion principle of the wheeled robot are proved, and the principle is further verified by experiment.

(1) Each axle of the three-leaved and circular wheel have the same angular velocity, the robot walking more efficient.

(2) Every moment, the robot has three wheels on the ground, to ensure the stability of walking.

(3) The steering principle of coaxial difference velocity can be used to achieve robot's movement.

(4) The mechanical structure of the biaxial four-wheeled robot is simpler.

In the complex environment, the robot can work autonomically and stably. The robot is highly efficient either as a post-disaster detection and rescue robot or as a service robot.

Acknowledgments

This work is financially supported by "Fundamental Research Funds for the Central Universities"(XDJK2017D070, XDJK2017D063).

References

1. Fu Yi-li, Li Han, Xu He. Research on several kinds of steering mode of OWMR[J]. Manufacturing Automation, 2005, 27(10): 33-37.
2. Chang Jian, Wu Cheng-dong, Li Bin. Shape-shifting robot skid-turning with line configuration[J]. Journal of Mechanical Engineering, 2012, 48(7). 9-15.
3. Gao Jun-yao, Lu Ji-lian. Steering control of wheeled mobile robot using GA fuzzy neural network[J]. Transactions of Beijing Institute of Technology, 2003, 23(2): 176-180.
4. Yan Yong-bao, Zhang Yu-nan, Yan Nan-ming.Simulation study of motion control algorithm for a six-wheel independent drive skid-steering vehicle[J]. Acta Armamentarii, 2013, 34(11):1461-1468.
5. Li Xue-yuan, Zhang Yu, Hu Ji-bin. Skid-steering resistance characteristics of wheeled vehicle[J], Acta Armamentarii. 2013, 32(12): 1433-1438.
6. Xing Jie. Differential steering and torque matching of 4-WD mobile robot[D]. Shanghai: Tongji University, 2008.
7. Ye Chang-long, Li Huai-yong, Ma Shu-gen. Kinematics study on an omnidirectional mobile robot with MY wheels[J]. robot, 2012, 34(2): 144-151.

8. Qian J, Zi B, Wang D, et al. The Design and Development of an Omni-Directional Mobile Robot Oriented to an Intelligent Manufacturing System. Sensors, 2017, 17(9): 2073; doi:10.3390/s17092073
9. Zi B, Sun H, Zhang D. Design, analysis and control of a winding hybrid-driven cable parallel manipulator. Robotics and Computer-Integrated Manufacturing, 2017, 48: 196-208.

Design and Implementation of Image Capture Sentry Gun Robot

Muhammad Kashif[1], Muhammad Arslan[2], Rocky Chakma[3], Farhad Banoori[4], Abdullah Al Mamun[5], Ganantu Lal Chakma[6]

[1, 2, 3, 5] *School Of Automation of Science and Engineering, South China University of Technology, China*
[4] *School of Electronic Information and Engineering, South China University of Technology, China*
[6] *Department of EEE, Atish Dipankar University of Science & Technology, Bangladesh*

Abstract. This paper is about the development and design of semi-autonomous sentry robot using aurdino controller. There is a global issue of that everyone wants to secure themselves from different sorts of danger and to navigate their surroundings. This robot called Sentry Gun, because it consists of a Gun mounted on a stand and an operator can also operate it from a secure location. Sentry Gun is based on latest research, it uses high resolution camera to scan the target area and movement can be detected by using software i.e processing .org As compare to humans it can perform far better while operating in autonomous mode in red zones, it will bring a significant fall in the death toll, if installed outside every check post, law enforcement agencies and borders, to stop illegal intrusions.

1 Introduction

For intelligence based systems it is used for security and defense purposes. These ideas have been taken from the computer game and implement it on our thesis work. The sentry gun model suggested in this paper has two degrees of freedom; one degree allows the gun turret to scan the area along the horizontal axis (azimuth) while the other degree allows the gun turret to move up and down (elevation). These two degrees allow the gun turret to move in any direction in a 2D plane. The most important part of this project was that the aim of the turret should be accurate i.e. the sentry is able to detect and locate intruders effectively [1].The earliest functioning military sentry guns were the close-in weapon systems point-defense weapons for detecting and destroying short range incoming missiles and enemy aircraft first, used exclusively on naval assets, and now also as land-based defenses. Sentry gun robot which can operate both in manual mode and automatic mode to track the moving objects. This gun should be installed on all the borders and outside all the check posts. It has a camera which will enable the operator (sitting in a secure location) to see all the incoming and outgoing peoples, all these incoming and outgoing peoples can be attended from a secure location with the help of Sentry gun Robot. We are mainly focusing on the autonomous mode for all of its calibration and functioning. Robots might also prove vastly superior to humans in the battle.

The idea of implementing Sentry Gun technology is not something novel. Numerous attempts have been made on a commercial level, military level, and even independent level. In 2006, MediaZen in cooperation with Samsung Techwin Inc., Korea University, Sysfo Lab Co. and Sejin System Co., produced the Guard Robots and Intelligent Surveillance Systems to defend the South Korean borders. From 2003 to 2006, Private sector and South Korea's government invested about $5 million on such military robots [2]. This technology has also been installed to protect Korean military bases in Iraq. This system uses visuals and infra-red cameras to distinguish between humans, animals, and trees. It can detect targets up to a 2.5 mile during the day, and about half of it at night. Bob Rudolph is quite a prominent name when we talk about Paintball Sentry Gun. On his website, he has shared the details of his numerous attempts at designing a paintball sentry gun. To date, he has created 9 different prototypes. He has also documented the flaws and strengths of each design and is still actively continuing development. He created Autonomous System last year, LLC as an entity to sell his hardware and software designs as a way to finance this hobby. However, he hasn't stopped publishing information on his latest software and hardware prototypes.

2 Backgrounds

There are many ways to detect motion but particularly in robotics, most general approaches use image processing or Light Detection and Ranging (LIDAR) tracking [3]. These two methods basically require extensive processing power or costly sensors.

MATEC Web of Conferences **160**, 06007 (2018) https://doi.org/10.1051/matecconf/201816006007
EECR 2018

Figure 1. Sentry gun robot

Many military grade solutions are available for this problem but usually, they are very expensive solutions and a low priced solution for this problem makes it more potent. Now the military is gaining a lot of practical experience and work for the interaction of human and robots. Robotics is an evolving branch of technology that deals with the design, construction, operation and application of robots and computer systems for their control, sensory feedback and information processing [4]. It improves security, reduces unnecessary causalities and reduces operational cost.

3 Modes of Operation

It has two modes of operation. They are described below-

3.1 Autonomous mode

In automatic mode sentry gun robot would be standing at a fixed location and automatically aim and fire the target detected by the sensors. In automatic mode it is operated in red alert areas where no one is allowed to go. So it would fire anyone who enters the surveillance area despite the fact whether he is an enemy or an alley.

3.2 Manual mode

In manual mode it is controlled by the operator for shooting the particular targets. It's like a gun in the hand and you decide what to shoot and what not to shoot but the only difference in this scenario is that you are operating it from a safer location. You have a laptop in front of you and you just have to click on the target the gun would automatically calculate the angle, aim at the exact target and shoot it. So there would be little chances of missing the target.

4 Hardware

The main hardware used in our project is Aurdino Uno, Servo motors and JF39 Air soft Gun. Aurdino Uno is used as a processing unit and Servo motors are used to control the vertical and horizontal motion and for pressing the trigger.

4.1 Arduino

In 1975, Arduino was originally assembled up with a purpose of creating such a device that controls student-built interactive projects with less expense comparing with other prototypes available in the market. David Cuartielles and Massimo Banziare the founders. Along with these the Arduino development team also includes: Gianluca Martino, David Cuartielles, Massimo Banzi, David Mellis, and Tom Igoe. Along with providing advantages for students, teachers, interested amateurs, and developers, it also has simplified the work as compare with other microcontrollers.

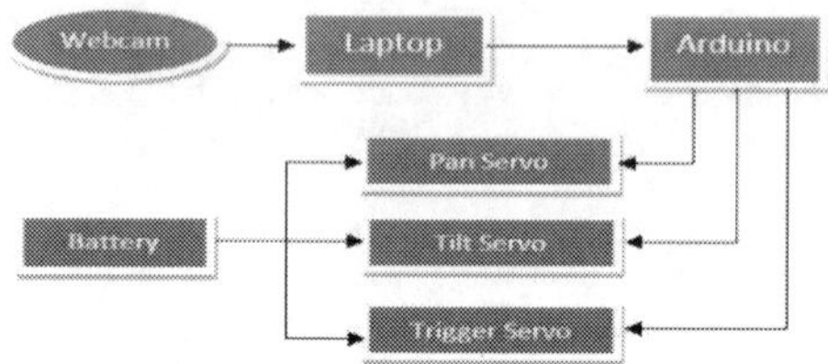

Figure 2. Block Diagram

Basic features of aurdino

i. Low Cost: The plus point of Aurdino board as compared to other microcontroller is that it has relatively less cost.
ii. Cross-platform: Aurdino can work on Linux, Windows, and Macintosh, while others are only for Windows.
iii. Simple and clear programming environment: Aurdino programming is more flexible and easy, even beginners can benefit from it. As it's based on processing environment, so professional programmers are also taking advantage of its features.
iv. Extensible and open source software: Aurdino software was published as an open source tools and is available for extension. All types of source codes and libraries of Aurdino environment are available openly, with which anyone can get used to Aurdino.
v. Extensible hardware: The Aurdino board is based on Atmel's ATMEGA8, ATMEGA328P PU and ATMEGA168 microcontrollers. It can be connected with any type of sensor, and circuits.

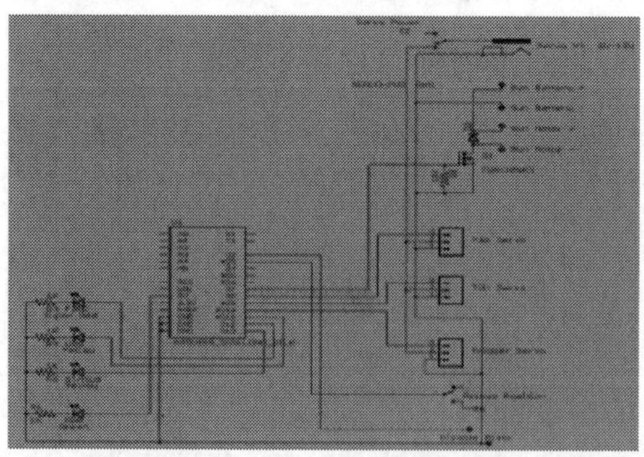

Figure 3. Circuit Diagram

4.2 Selection of Motor

We needed motors that could give our sentry gun a vertical and horizontal motion but the selection became harder and harder when we studied different types of motors

4.2.1 SERVO MOTOR V STEPPER MOTOR

i. Servomotor consumes power only when it rotates to its commanded position, while rest of the time it doesn't consumes any power to maintain its commanded position.

ii. Stepper motors continuously consumes power to hold the commanded position or to move to the commanded position and due to continuous consumption of power it becomes warm.

iii. Servomotors are generally replacing Stepper motors due to its high performance

iv. Stepper motors have built-in output steps.

v. Stepper motor can only work properly with a load that is well within its capacity, otherwise over load may lead to missed steps and as a result we would get positioning errors.

vi. Servo motors contains encoder and controller which is the reason of their high cost, but they improve the performance of the overall system (for all of accuracy, power and speed) relative to the capacity of the basic motor.

5 Interconnection between software and hardware

Webcam is connected with Laptop or C.P.U through serial port and in C.P.U the processing and Aurdino software are executed their respective codes given in appendix A and B. Data from the Webcam is collected through processing Software and is send to Aurdino board through Aurdino software to control the motion of Sentry gun through servo motors.

5.1 Manual Mode

In manual mode the gun is aimed at the target using the mouse cursor, by moving the cursor towards the right side of the screen, the x-axis servo rotates clockwise and while moving the cursor towards the left side of the screen the servo rotates counter clockwise, this servo's shaft is attached to the sentry gun, so when the cursor is moved to the right side the servo moves the gun to the right and vice versa, Similarly when the mouse cursor is moved up the servo will move the gun upwards and when the cursor is moved down the servos will bring the gun downwards. There is another servo attached to the trigger, by clicking or holding the left mouse button the servo will press the trigger and upon releasing the left mouse button the servo will release the tiger. If the gun is in semi mode and you hold the left mouse button the gun will fire only ones but if the gun is in auto mode and you hold the left mouse button the gun will start continuous firing and wouldn't stop until you release the left mouse button.

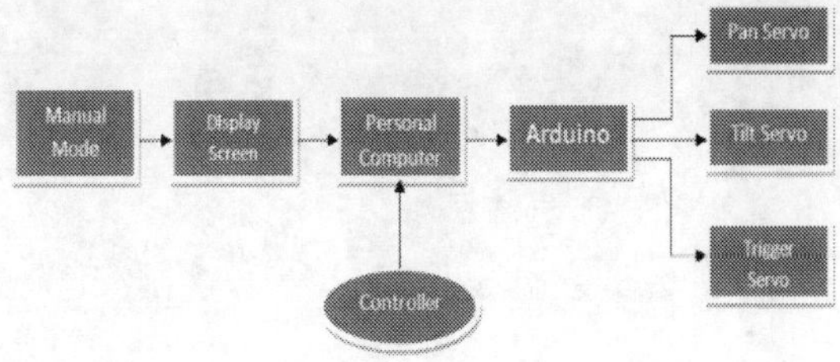

Figure 4. Manual Mode

5.1 Autonomous Mode

In Autonomous Mode, the system could only successfully track and hit 1 to 2 objects in a frame moving at up to approximately 3 to 4 miles per hour (human walking speed). Its accuracy was extremely sensitive to the calibration of the system, but it has about 95% hit rate and can shoot up to twice per second. The sentry gun will not require any user control, instead it will aim and fire at its own discretion driven by the image-processing engine. There are two methods of finding targets. One is by background subtraction, the other by frame differencing. Frame differencing detects only moving targets while Background Subtraction detects moving targets as well as stationary targets. However, in the environment where sentry gun is typically placed, there are often changing conditions, such as shadows gradually shifting ,leaves blowing around and even decoys thrown out before an attack to distract the sentry gun. These will mess up the background subtraction, but they will not affect frame differencing as much. However, choose for yourself which to use, based on the application.

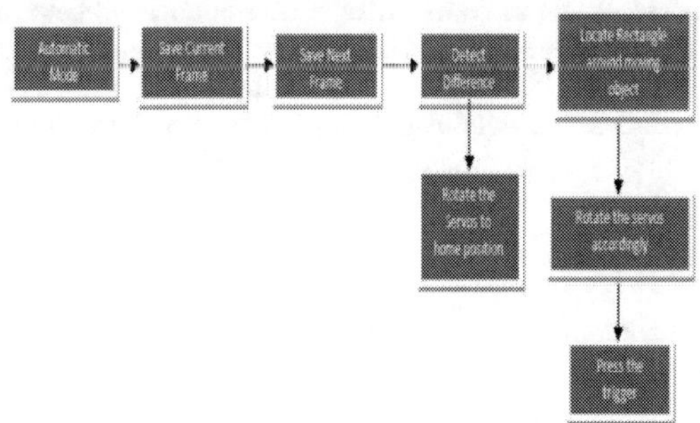

Figure 5. Autonomous Mode

The Tolerance buttons lowers and raise the tolerance of the subtraction both in Frame Differencing and Background Subtraction; it determines how sensitive image-processing should be, to color change. Sensitivity

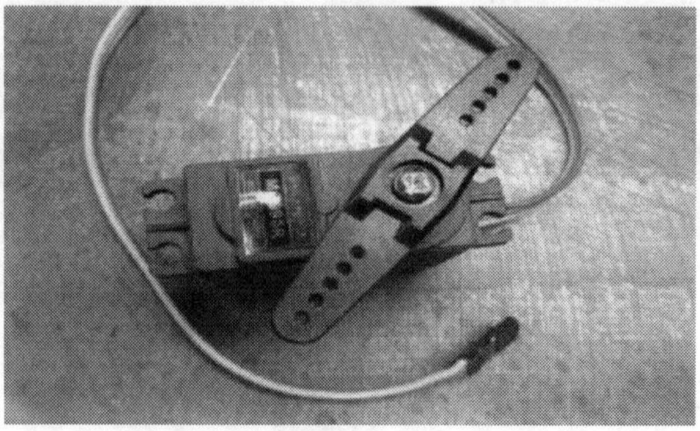

Figure 6. Trigger Servo Motor

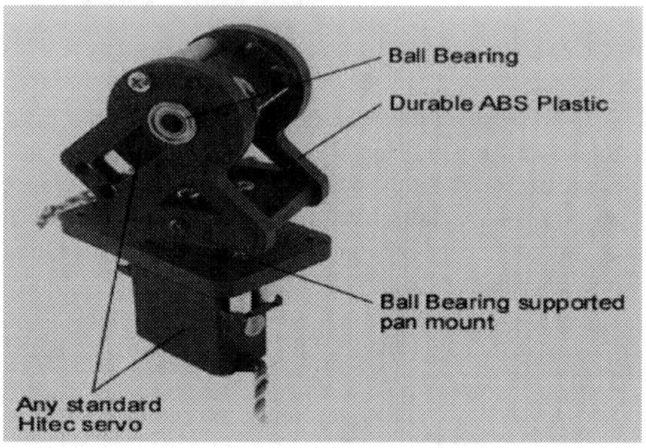

Figure 7. Pan Servo Motor

can also be adjusted by using the min Blob Area variable; it defines the minimum number of pixels a target must have for the gun to fire on it.

When Autonomous Mode is in background subtraction state, clicking on Background button will store the current camera image as a background image and will subtract the next images form this background image. While in Frame Differencing, it does not have any effect on the result. The Frame Differencing speed buttons adjust the sensitivity of the frame differencing state to the motion, by defining how many frames should be used to find motion.

6 Results

Some of the results are given below –

6.1 User controlled mode result

The movement, positioning, and firing of the Sentry Gun in user mode performed as we expected. Because the tilt and pan servos selected could not support large weight, it was necessary to select the weapon around this constraint. DC gear motors could handle more weight than servos but servos allow for precise control of angular positioning, velocity and acceleration.

The grouping of the projectiles fired from the weapon in USER mode was well within an acceptable range when fired within the determined operating range of 25 feet

6.2 Autonomous tracking result

In Autonomous mode the tracking was somewhat disappointing as compared to our original hopes; it not only met our expectations it even surpass them. Image processing is a very CPU intensive task and the laptop selected for the system could only successfully track and hit 1 to 2 objects in a frame moving at up to approximately 3 to 4 miles per hour (human walking speed). Its accuracy was extremely sensitive to the calibration of the system, but it has about an 80% hit rate and can shoot up to twice per second. This performance is not only well within the acceptable range, it exceeded our expectations.

6.3 Constrains

There were Several factors that limited the performance of our Sentry Gun during testing. The low budget webcam we used to perform the Sentry Gun's image-processing was above all other constrains. Even at its maximum 30fps rating, the webcam was not sufficient for high quality real time applications.

The algorithm of image-processing itself, turned out to be the biggest constrain in the Sentry gun's system. In real-time systems it is very difficult and requires a high clock frequency as well as large amounts of memory to perform the task of processing large datasets. Ideally, the image-processing algorithms should be run on a multi-core processer, and the system should be parallelized.

7 Scope of Project

The idea of implementing Sentry Gun technology is not something novel. Numerous attempts have been made on commercial level, military level, and even independent level. In 2006, MediaZen in cooperation with Samsung Techwin Inc., Korea University, Sysfo Lab Co. and Sejin System Co., produced the Guard Robots and Intelligent Surveillance Systems to defend the South Korean borders. From 2003 to 2006, Private sector and South Korea's government invested about $5 million on such military robots. This technology has also been installed to protect Korean military bases in Iraq. This system uses visuals and infra-red cameras to distinguish between humans, animals and trees. It can detect targets up to a 2.5 mile during the day, and about half of it at night. Bob Rudolph is quiet a prominent name when we talk about Paintball Sentry Gun. On his website, he has shared the details of his numerous attempts at designing a paintball sentry gun. To date, he has created 9 different prototypes. He has also documented the flaws and strengths of each design and is still actively continuing development. He created Autonomous System last year, LLC as an entity to sell his hardware and software designs as a way to finance this hobby. However, he hasn't stopped publishing information on his latest software and hardware prototypes.

Conclusions

Use In this paper the development and design of a robotic sentry gun for security applications are discussed by detection as well as engagement of a target. It is a soldier that never sleeps, to hit a target is its first priority .In future these autonomous robots can replace soldiers. The hardware of the this system includes mechanical structure mainly consists of accessible equipment that make it low budget system .The algorithm of image-processing itself, turned out to be the biggest constrain in the Sentry gun's system, Ideally, the image-processing algorithms should be run on a multi-core processer, and the system should be parallelized. It's not the final design we can modify it with so many modifications like with 3D camera and face/object recognition. We can also add voice password recognition. This is an open source project then perhaps this is a good place to work on AI algorithms to improve the gun's performance.

References

1. Husnain Ahmad, Syed Mohsin Ali, Usman Ayub Sheikh, Zain Murtaza, Muhammad Rizwan, "Design and Manufacturing of a Low-Cost 2-DOF Autonomous Sentry Gun" (IEEE Conference, pp.196-201, 2016).
2. Jean Kumagai, "A Robotic Sentry for Korea's Demilitarized Zone", (IEEE Journals & Magazines, vol.44, pp.16-17, 2017).
3. Sam Shue, Dr. Claude Hargrove, Dr. James Conrad, "Low Cost Semi-Autonomous Sentry Robot", (IEEE Conference, pp.1-5, 2012)
4. Et. Karthikrajan, "Multi terrain robotic vehicle with step climbing mechanism", (International Journal of Advanced Research in Basic Engineering Sciences and Technology (IJARBEST).

Design and Implementation of Fire Extinguisher Robot with Robotic Arm

Abdul Waris Memon[1], Juan Du[1], Abdul Haleem Abro[2], Sharmeen Iftikhar Shah[2] and Moazzam Ali Bhutto[3]

[1] *School of Automation Science and Engineering, South China University of Technology, China*
[2]*Department of Electronic-Information Technology, University of Sindh Jamshoro, Sindh, Pakistan*
[3]*School of Electronic Information and Engineering, South China University of Technology, China*

Abstract. Robot is a device, which performs human task or behave like a human-being. It needs expertise skills and complex programming to design. For designing a fire fighter robot, many sensors and motors were used. User firstly send robot to an affected area, to get live image of the field with the help of mobile camera via Wi-Fi using IP camera application to laptop. If any signs of fire shown in image, user direct robot in that particular direction for confirmation. Fire sensor and temperature sensor detects and measures the reading, after confirmation robot sprinkle water on affected field. During extinguish process if any obstacle comes in between the prototype and the affected area the ultrasonic sensor detects the obstacle, in response the robotic arm moves to pick and place that obstacle to another location for clearing the path. Meanwhile if any poisonous gas is present, the gas sensor detects and indicates by making alarm.

1 Introduction

1.1 Fire Extinguisher Robot

A fire extinguisher robot is one that have a fire quencher included to it. By attaching a fire quencher of the robot, the mechanization put off the fire of the affected area. This paper comprises those outline and development of a robot that is able to detect and quench fire. This robot executes the fire extinguish and fire detection idea. It has utilization of fire sensors which detect the affected fired area. A robot put off the recreated tunnel fire, industry, school, home and forests fire. Flame sensors are used for the initial detection of the fire, If the flame is detected then the robot resonances those alerts with the help of ringer which is set on the prototype, the robot actuates an electronic valve discharging sprinkles of water on the fired area. This project helps to develop interests as well as innovations in the fields of robotics while working towards a practical and obtainable solution to save lives and diminish the risk of property destruction. This fire detector robot will play its role, where human fails to detect fire. Fire contenders face dangerous circumstances when smothering blazes furthermore rescuing victims, it may be an unavoidable part and condition for a fire warrior. Done contrast, a robot has capacity and functioned by itself or be controlled from a distance. In short, fire extinguisher robots have decline the need to fire contenders should get under hazardous circumstances [1-2].

1.2 Robotic Arm

This robot also implements removing of obstacles with a robotic arm. Robotic arm is used to pick and place the obstacles from one place to another location. It is used to clear the path and to reach the affected area easily. Obstacles are detected by ultrasonic sensor.

2 Literature Review

2.1 Fire Extinguisher Robot

Home could furnish safety, convenience, and productivity for men in the 21st century. Several functions, frameworks and systems are used to integrate an intelligent home system. A standout amongst the practically significant frameworks may be the fire identification work framework in a smart home system. In the previous literatures, huge numbers masters Look into in the administration robot. Exactly Look into tended to done Creating target-tracking framework for administration robot [3-5], for example, Hisato Kobayashi et al. Recommended a strategy on identify individual towards a self-sufficient versatile watch robot [6]. Moreover, exactly scrutinize tended to in the robot need the ability about fire-fighting. Ruser, furthermore Magori depicted those fire identification with a consolidation for ultrasonic also microwave doppler sensor. Ruser, H. Furthermore Magori, V, "Fire identification with joined together ultrasonic-microwave doppler sensor," in incidents from claiming IEEE ultrasonic Symposium, Vol. 1, 1998, pp. 489-492. Luo also Su use two smoke sensors, two temperature sensors and two fire sensors to identify fired affected area, and finding which sensor will be disappointment utilizing versatile combination strategy.

2.2 Robotic Arm

Robotic manipulators also robotic arms were those exceptionally first robots which were utilized within the industries and they have been in used since 1960s. An extensive number for investigations need centered once robots for making a difference people with physical challenges, to example, those desktop robotic framework Toward Hammel. With the proceeded advancement in registering power, machine vision, Furthermore propelled sensors, researches need began to concentrate on employments for robotic arms over Every day an aggregation Also fill in situations in one task for example, Care-O-bot and the late versatile controller robot PR2 starting with Willow carport. Drug store assignments incorporated capacity oversaw economy from claiming medication and taking care of risky solution.

3 Methodology

Fire extinguisher robot with a robotic arm is divided in to three sections. Such as- Obstacle Detection, Temperature, Flame and Gas Sensors Detection Robot Direction Control.

3.1 Obstacle Detection

The block diagram, flow chart and schematic diagram of obstacle detection are shown in Fig 1, Fig 2 and Fig 3 in which ultrasonic sensor senses the obstacle and sends signal to the arduino (1), arduino generates the command for the robotic arm to clear the path.

Figure 1. Block Diagram of Obstacle Detection

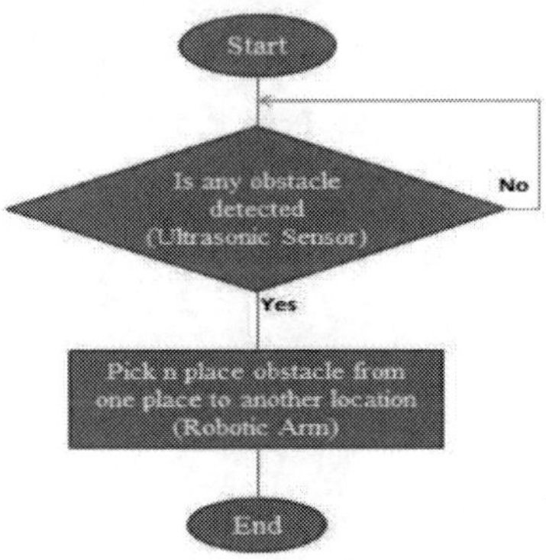

Figure 2. Flow Chart of Obstacle Detection

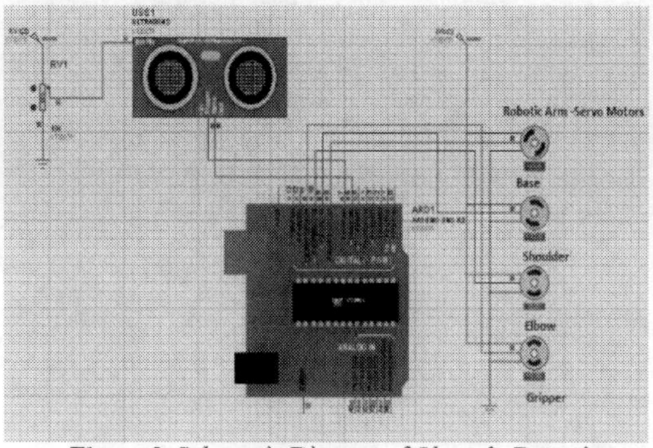

Figure 3. Schematic Diagram of Obstacle Detection

3.2 Temperature, Flame and Gas Sensors Detection

The block diagram and flow chart of flame, gas, and temperature sensors detection are shown in Figure 4 and Figure 5 in which temperature sensor senses the temperature of environment, if the temperature crosses its limit, it activates the flame sensor to sense the fire. Further, flame sensor sends the signal to arduino (2). After that, arduino generates the command for water sprinkler to make the fire off from the desired location.

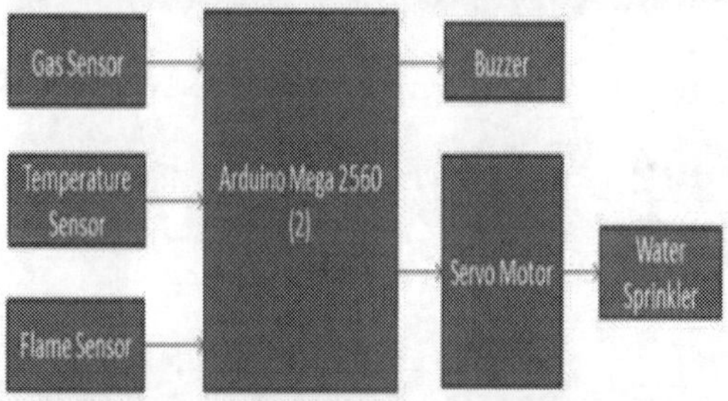

Figure 4. Block Diagram of Temperature, Flame and Gas Sensors Detection

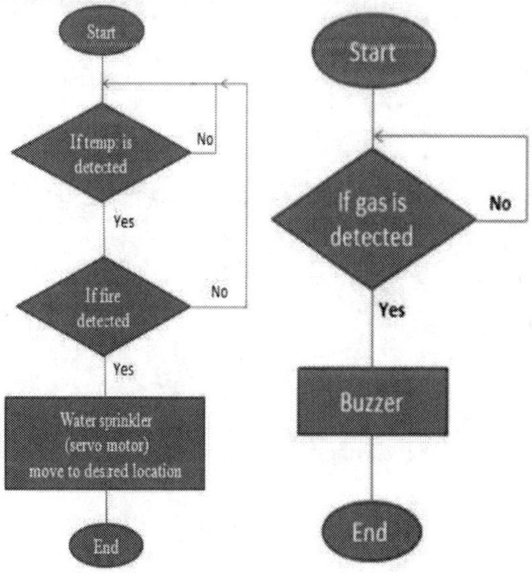

Figure 5. Flowchart of Temperature Sensor Detection and Gas Sensor Detection

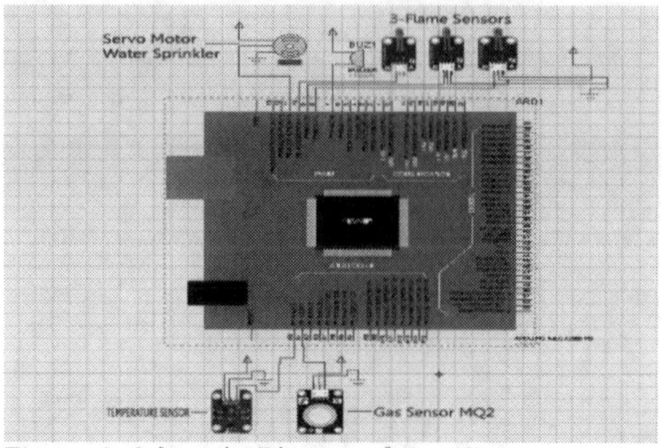

Figure 6. Schematic Diagram of Gas, Temperature Sensor Detection

3.3 Robot Direction Control

The block diagrams of video feed backing and robot movement are shown in Fig 7 (a) and Fig 7 (b) in which android cell phone is connected to the laptop via mobile IP camera application which is connected with its IP address of desired location. Further, RF remote control controls the robot direction through it, robotic car can be moved to any direction.

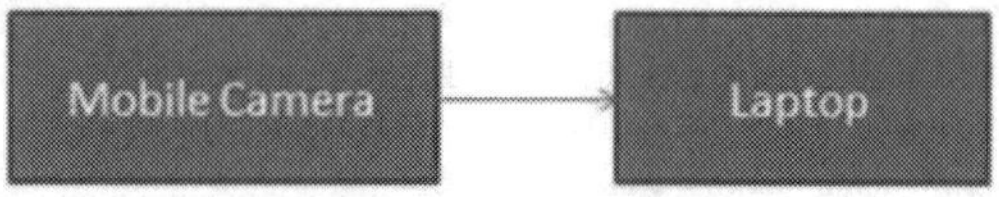

Wi-Fi

Figure 7(a). Block Diagram of Video Feed Backing

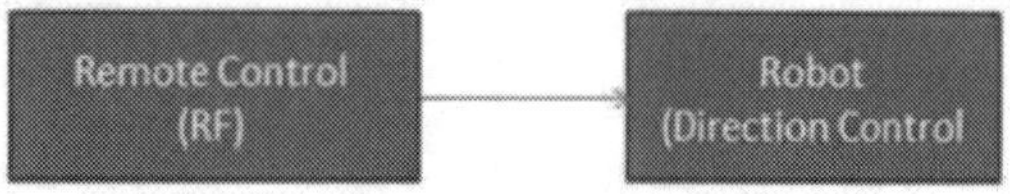

Figure 7(b). Block diagram of Robot Movement

4 Results

4.1 Obstacle Detection

The process of obstacle detection is shown in figure 8 to figure 12, picking and placing the obstacle towards left side for path clearing.

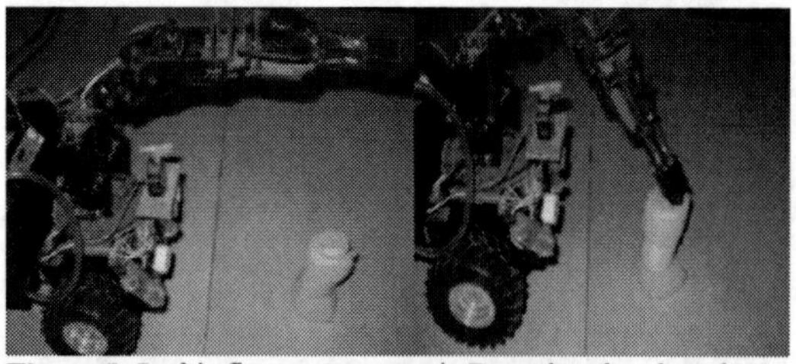

Figure 8. In this figure prototype is Detecting the obstacle & Robotic Arm is moving towards Obstacle

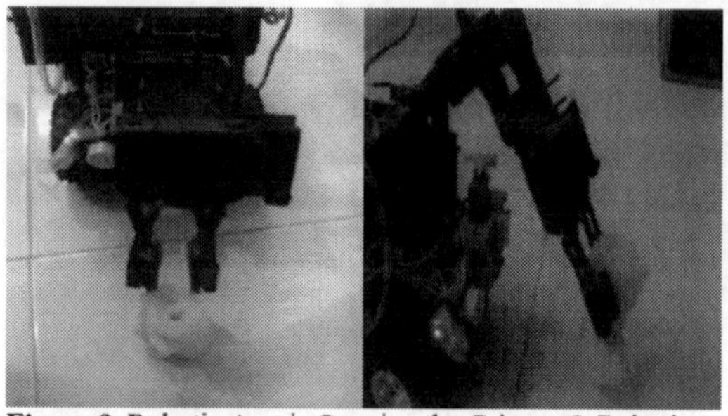

Figure 9. Robotic Arm is Opening the Gripper & Robotic Arm is picking the Obstacle

Figure 10. Robotic Arm is moving the obstacle at Left Side to clear its path

Figure 11. Robotic Arm is dropping the Obstacle

Figure 12. Finally in this image Robotic Arm is closing the Grip & Robotic Arm is coming towards Initial Position

4.2 Temperature, Flame and Gas Sensor Detection

4.2.1 Temperature Sensor Detection

Figure 13 shows the Temperature Sensor Detection. The robot is in sleep mode. It will be switched on if any temperature can be detected by the temperature sensor.

Figure 13. Temperature Sensor Detection

4.2.2 Flame Sensor Detection

The fire sensing process is illustrated from Figure 14 to 16, Figure 14 shows that fire has been detected on left side and extinguished by the water sprinkler. Figure 15 shows that fire is detected at center and extinguished by water sprinkler. Figure 16 shows that fire is detected at right side and extinguished by water sprinkler. When the water sprinkler extinguishes the fire the pump will be stopped automatically.

Figure 14. Fire Sensing At Left Side and Fire Extinguished By the Water Sprinkler

Figure 15. Fire Sensing At Centre & Fire Extinguished By the Water Sprinkler

Figure 16. Fire Sensing At Right Side & Fire Extinguished By the Water Sprinkler

4.2.3 Gas Sensor Detection

Figure 17 shows the detection of gas by gas sensor from 10 cm and 15 cm and beep the buzzer to indicate the presence of poisonous gas.

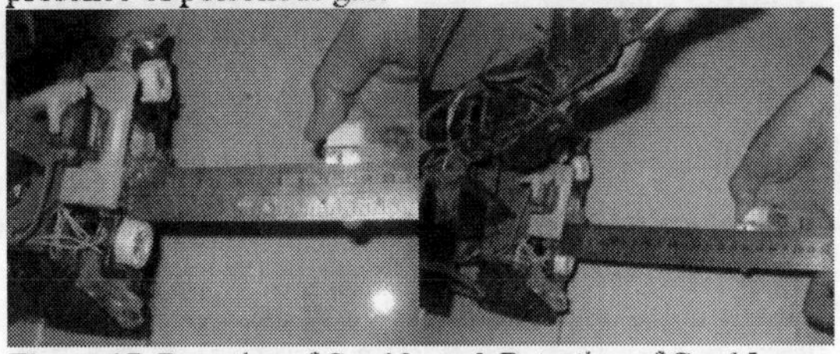

Figure 17. Detection of Gas 10 cm & Detection of Gas 15 cm

4.2.4 Robot Direction Control

An IP camera application is used for video feed-backing, which is shown in Figures 18 to Figure 19. An IP Camera application is used through android cell phone on transmission side and in resulting visualization of location is received on monitoring side.

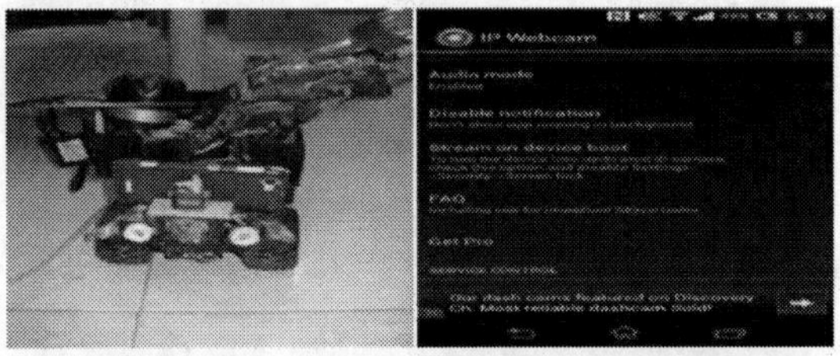

Figure 18. Prototype with IP Camera & IP Webcam Cell Phone Application

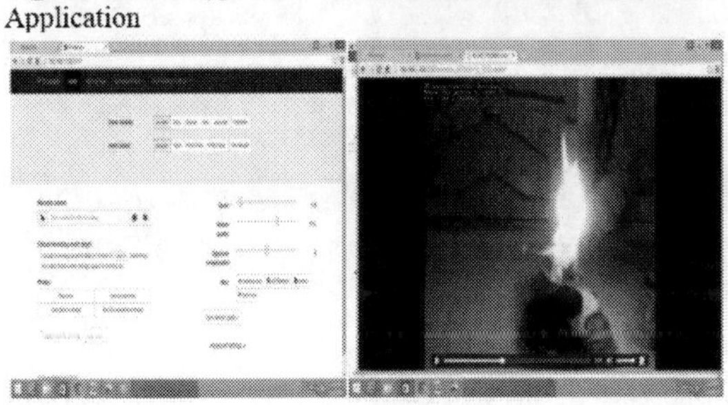

Figure 19. IP Camera on Transmitter & Receiver Side (Laptop) and Video Feed-backing On Receiving Side

5 Conclusion

Fire-fighting robot can be easily and conveniently used and operated automatically when any fire incident occurs in educational, industrial and hospital areas to save human life. Fire-fighting Robot comprises of numerous sensors and motors, and has small in size, less in weight, with rechargeable batteries, in result it requires less space. Prototype provides us greater efficiency to detect the flame, temperature and gas presented in the affected area. The extinguisher robot effectively extinguishes fire before it becomes uncontrollable and gives threat to life. Fire-fighting robot also successfully removes obstacle from the path to another location by using a robotic arm to clear the way. Hence, the robot can play a crucial role in our society.

References

1. Et. Akshay Desmukh, "A survey on fire detector and extinguisher robot," (5th International Conference on Recent Development in Engineering Science, Humanities and Management, pp. 256–264, 16th April 2017), ISBN : 978-93-86171-39-9.
2. Et. Dr. Raju N.Pamnchal, "Fire Fighting Vehicle," International Journal of Engineering Science and Research (IJETSR)," Volume 4, (Issue 9, September 2017), ISSN 2394-3386.
3. Et. Ting-Li Chien, "Detection system of security robot using multisensor fusion algorithms," The Fourteenth International Symposium on Artificial Life and Robotics 2009(AROB 14th 09), " February 5-7, 2009.
4. Et. S.O. Lee, "A stable target-tracking control for unicyle mobile robots," Proceedings of the IEEE/RSJ International Conference on Intelligent Robots and Systems(IROS 2000)," Vol.3, 2000, pp. 1822-1827.
5. L. E. Parker and B. A. Emmons, "Cooperative multi-robot observation of multiple moving targets," Proceedings of the IEEE/RSJ International Conference on Robotics and Automation," Vol.3, 1997, pp. 2082-2089.
6. Et. Kuo L. Su, "Design a low cost security robot applying in family," 2nd International Conference on Autonomous Robots and Agents, (December 13-15, 2004).

Project Supported by: ① Fundamental Research Funds for the Central Universities (Grant No. 2015ZM135) ; ② Science and Technology Project of Nansha, Guangzhou, Guangdong (Grant No. 2015KF012) ; ③ Science and Technology Project of Guangzhou (Grant No. 201707010061); ④ Science and Technology Project (International Cooperation Research) of Nansha, Guangzhou, Guangdong (Grant No. 2016GJ012).

A Micro SLAM System Based on ORB for RGB-D Cameras

Fei Wang[1, 2], Xiaogang Ruan[1, 2], Pengfei Dong[1, 2] and OUATTARA SIE[1, 2]

[1]*Faculty of Information Technology, Beijing University of Technology, Beijing 100124, China*
[2]*Beijing Key Laboratory of Computational Intelligence and Intelligent System, Beijing 100124, China*

Abstract. In this paper, a micro SLAM system based on ORB features for RGB-D cameras has been proposed. With only a RGB-D sensor, this method can be applied in small environment for localization and mapping. Furthermore, the task of 3D reconstruction can also be accomplished by using the approach. The pose graph based on Bundle Adjustment is adopted for reducing the estimation error. In order to further speed up computing to meet the requirement of real-time, we have proposed the piecewise optimization strategy. The approach is evaluated on public benchmark datasets. Compared with several state-of-the-art scheme, this method has proven to work well in these environments.

1 Introduction

Simultaneous Localization and Mapping(SLAM) is the key technology for autonomous navigation of robot. It acquires information about itself and environment by sensors such as laser sensors, cameras and inertial measurement units. Visual SLAM has become a hot research in the field of computer vision and robotics in recent years. Compared with other sensors, RGB-D camera can obtain more abundant environmental information, and help to achieve accurate and robust positioning, and has been widely used recently.

Visual SLAM mainly estimates the motion of the camera through the matching of image features, which is commonly called visual odometry [1]. The commonly used visual features are SIFT[2], SURF[3], ORB[4], etc. The commonly used method is to obtain the feature matching between two adjacent frames, then compute the estimated camera motion by PNP, ICP[5][6] and some other algorithms. The estimation error of camera motion will accumulate over time, leading to more and more inaccurate estimation of motion [7]. So it is difficult to build a globally consistent map. Therefore, after the estimation, it is necessary to optimize the pose and environment map.

In this paper, we build a SLAM system based on the ORB feature for small environments. Figure 1 shows the different environment map using tum data. ORB has the advantages of rotation and scale invariance, while the extraction time is short, which meets the real-time requirements. We use a simple matching method to select the key frame. We use the method of matching the frame to the environment model to create the map and use the bundle adjustment to optimize the camera pose and feature point position. It usually takes a lot of time to optimize the entire map because of the large number of feature points. We proposed a piecewise optimization strategy. For the relocalization problem, we select a number of previous key frames to match the current frame.

The evaluation was conducted on a publicly RGB-D data set. The result shows that our method can achieve higher accuracy and robustness in the estimation of motion and map points positions.

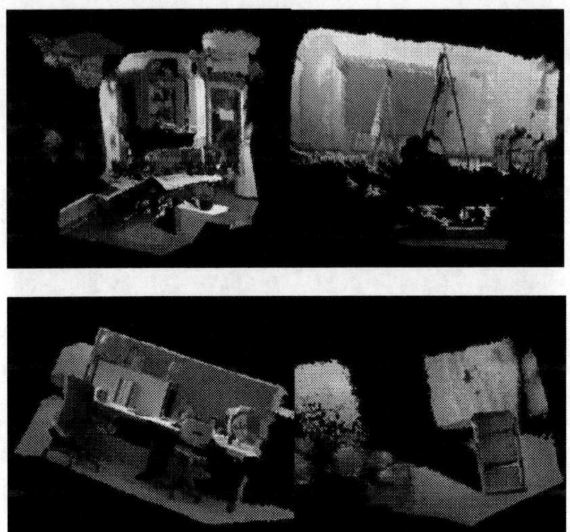

Figure 1. The building map of different environment.

2 Related work

The image feature is a set of information related to the computational task, and the main features used in visual slam are SIFT, SURF, ORB [2,3,4], etc.. SIFT features arediscriminative, because their descriptors are represented by high dimensional vectors, and have rotation invariance, scale invariance, and robustness to noise and illumination variations[8]. The SIFT feature is used in visual SLAM, but the time complexity of SIFT feature is too high because of its high vector dimension. SURF features have scale invariance, rotation invariance, and the speed of the algorithm relative to the SIFT feature is increased by 3 to 7 times. In [9,10,11], SURF was used for the slam system, compared with the SIFT characteristics, the time complexity is reduced. The Euclidean distance between two feature vectors is usually calculated when matching the SIFT and SURF features of two images, which is used as the similarity measure of feature points.

ORB feature is the combination of FAST [12] feature detection operator and BRIEF [13] descriptor, and some improvements are made on the basis of it. The extraction efficiency of ORB is 100 times the SIFT feature, 10 times the SURF feature, which is the biggist advantage of ORB. In [6,14,15,16,17], ORB was used in SLAM system which greatly accelerate the speed of the algorithm. ORB feature matching is based on the Hamming distance of BRIEF binary descriptor as similarity measurement.

Frame to frame alignment will cause large cumulative floating, so there is always error in the pose estimation process. In order to reduce the error caused by frame alignment, the SLAM [6,18,19,20] method based on key frames is proposed. HENRY P, et al. [20] propose that a new key frame is created when the common feature points of the two images are below a certain threshold. KERL C, et al [6] propose a method to select key frames based on entropy, computing similarity entropy ratio for each frame, if the value is less than a predefined threshold, the previous frame is selected as the new key frame, and insert it to the map, which greatly reduces the drift.

In monocular slam, the motion of the camera can be obtained by calculating the fundamental matrix, the essential matrix or the homography matrix, so as to realize initializing. 3D positions of feature points will be obtained by triangulation, and then the feature matching can be done by PNP. In RGB-D images, we can obtain the depth information of the feature points directly through the depth image, and then estimate the camera motion using iterative closest point (ICP) [21].

Error accumulation is inevitable in the process of 3D reconstruction, and closed loop is an important means to eliminate it. The key point of loop closing is to detect whether the camera passed the same location. Williams B, et al. compared the loop closing methods [22] and draw the conclusion that the matching performance of image to image is better than that of map to map and image to map. Eda E D and Drummond T W propose a unified approach to relocation and closed-loop detection, which continually searches for visited locations using dictionary methods based on 16-dimensional SIFT features [23]. Document [24, 25] uses the dictionary method based on the SURF descriptor for loop-closing detection, and SURF feature extraction takes about 400 ms. In the literature [14], the position recognition is carried out by the dictionary method based on the ORB feature. The method can identify the position from different views because of the invariance of rotation and scale.

It is very important for a robot working in complex and dynamic environment to quickly generate a 3-D map, which will play a key role in localization and path planning. Our SLAM system overview is shown in Figure 2. Loop-closing constraint will be added to the map for correction after successful loop closing detection. The optimization of map can be solved by bundle adjustment [26], which is to optimize the all the camera pose and feature points for the same time. However it is difficult to achieve real-time optimization because of the high calculation complexity. An executable method is pose graph optimization. The camera pose is represented as the vertex in the graph, and the error between two frame transformation is the edge. Errors are distributed along the graph, namely evenly distributed over all pose on the graph. The optimization technique of pose graph is used to realize the effective correction of rotation, translation and scale floating [27]. Essential graph is constructed after successful loop detection, and then using pose graph method to executing optimizing[6]. All the key frame are included in the essential graph, while there are less contraint between edges compare to covisibility graph[28]. The overview of our SLAM system is shown in Figure 2.

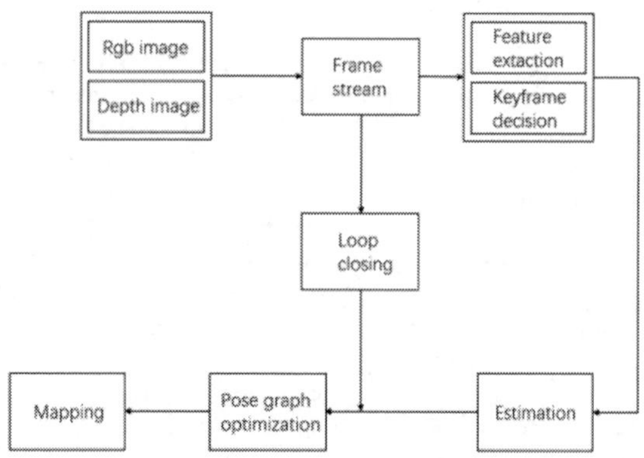

Figure 2. The SLAM system overview in the paper. The pose graph optimized by bundle adjustment using g2o.

3 Approach

3.1 Motion model

The movement of the camera in the three-dimensional space belongs to the rigid body motion and can be decomposed into the rotation R and the translation t. Where R satisfies the following conditions:

$$SO(3) = \{\mathbf{R} \in R^{3\times3} \mid \mathbf{R}\mathbf{R}^T = T, \det(\mathbf{R}) = 1\} \quad (1)$$

The rotation matrix R and translation t will not satisfy the linear relationship when they are using to express continuous rigid motion. Thus, homogeneous coordinates and transformation matrices are introduced:

$$\mathbf{T} = \begin{bmatrix} \mathbf{R} & \mathbf{t} \\ \mathbf{0}^T & 1 \end{bmatrix} \quad (2)$$

$$SE(3) = \left\{ \mathbf{T} = \begin{bmatrix} \mathbf{R} & \mathbf{t} \\ \mathbf{0}^T & 1 \end{bmatrix} \in \mathrm{R}^{4\times4} \mid \mathbf{R} \in SO(3), \mathrm{t} \in \mathrm{R}^3 \right\} \quad (3)$$

The corresponding homogenous coordinate of point in 3-dimension space is defined as $p=(X, Y, Z, 1)$. The coordinates of point p after transformation T are represented as $p'=Tp$.

Since the rigid motion in three-dimensional space has only six degrees of freedom, while there are 12 values in T matrices. So T is a redundant expression. Therefore, we use Lie algebra se(3) ξ corresponding to Lie group SE(3) to express the motion of camera. ξ is a six dimensional vector, and the transformation matrix T can be obtained by the exponential mapping of Lie algebra. The backend optimization problem can be transformed into an unconstrained one using Lie algebra to express the transformation.

3.2 Camera model

The internal matrix of the camera is expressed as follows:

$$\mathbf{K} = \begin{bmatrix} f_x & 0 & c_x \\ 0 & f_y & c_y \\ 0 & 0 & 1 \end{bmatrix} \quad (4)$$

where f_x, f_y are the focal length and c_x, c_y are the primary coordinate of the imaging plane.

For the spatial point $p=(X, Y, Z, 1)$, the corresponding pixel coordinates are denoted as $p_{uv}=(u, v)$, and their corresponding relations are as follows:

$$\begin{cases} u = f_x \dfrac{X}{Z} + c_x \\ v = f_y \dfrac{Y}{Z} + c_y \end{cases} \quad (5)$$

3.3 Feature extraction and matching

We hope that the slam system can track and construct the map in real time, so we select the ORB feature points with high efficiency and invariance of rotation and scale. Feature points generally include key points and descriptors. It is necessary to compute the derivative of pixels when using Harris corners, which will increase the complexity of detection algorithm. The improved fast corner is adopted in ORB. We take two steps to extract the key points. First of all, we extract key points as usual. And then the thresholds are then reduced appropriately in areas with less key points, where there are often lacking in texture features or weak contrast. Thus ensuring that the key points can be evenly distributed over the image. After that, the description of the corresponding key point is computed, using an improved BRIEF descriptor, usually represented by a 128-bit or 256-bit binaries. The detected feature points are shown in Figure 4.

Feature matching is a key step in visual slam, which solves the problem of data association in slam, that is, to determine the relationship between the current landmarks and the front markers. The accurate matching of the descriptors between image to image or image to map can reduce the heavy burden for the subsequent pose estimation and optimization.

However, due to the local characteristics of image features, the situation of mis-matching is widespread and has not been effectively solved for a long time. We can find some mismatched points in Figure 4. The main methods used to remove the wrong match include the feature point distance check and geometric relationship check. The former sets a distance threshold. And the matching point whose distance is longer than the threshold will be regarded as mismatching point. This is because the ORB feature points have various forms of invariance, so the distance of correctly matched feature points is naturally relatively close. The latter randomly selects n pairs, calculates the transformation matrix between these chosen images, and then checks with the remaining pairs. Repeat the process above until finding the best result. The results of removing mismatches are shown in Figure 5.

Figure 3. Feature points detected.

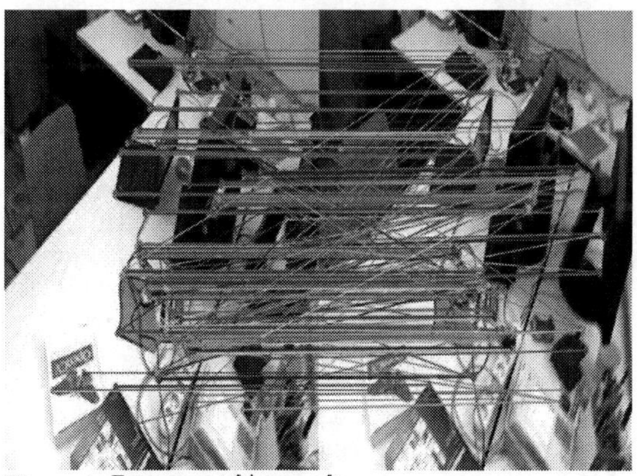

Figure 4. Feature matching results.

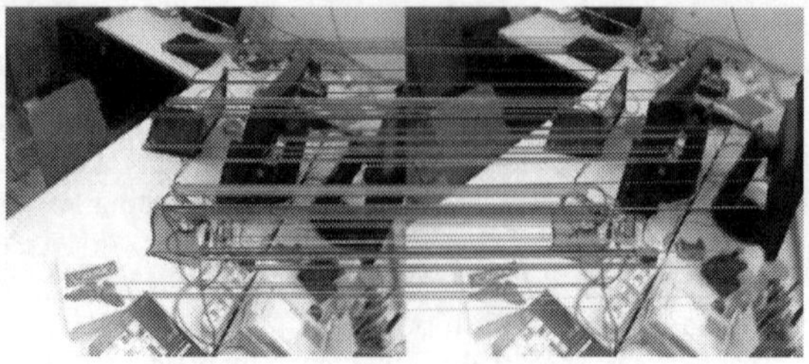

Figure 5. Feature matching results after removing the mismatched key points.

3.4 Pose estimation

We can get the 3D coordinates of the feature points directly, after extracting the feature points. Camera motion can be estimated using ICP algorithm. ICP is called iterative closest point algorithm [17], which is a common method for aligning point clouds of two frames. The algorithm has two key aspects: one is to find the corresponding point pair between the two-frame point cloud, and the other is to compute the transformation matrix that minimizes the distance of the two-frame point cloud according to these corresponding points. The algorithm can accurately calculate the transformation matrix of point clouds of two frame accurately. However, the algorithm is sensitive to initial values. When the initial transformation is not selected properly, the algorithm may fall into local minima. Moreover, when the point cloud is a dense point cloud, because of the large amount of data and the long running time of the algorithm, it can not meet the requirements of real-time. In this paper, the transformation matrix T_t between the current frame F_t and the environmental model M_t is calculated using ICP algorithm, because both of them are sparse feature points, the data volume is small, and therefore with the fast speed. And we assume that the camera moves at constant speed. Between t-2 and t-1, the amount of change in camera movement is $\Delta T = T_{t-1} / T_{t-2}$. In order to calculate the T_t , the initial value of the ICP algorithm is set as $T_{init} = T_{t-1}\Delta T$. The transformation matrix of the camera relative to the global coordinate system is obtained by continuous iteration.

3.5 Bundle adjustment

The transformation matrix obtained by the ICP algorithm is only the initial estimate of the pose. The error of the transformation matrix obtained by ICP algorithm will be larger, when the number of feature points match the current frame and model is small. In addition, there is also possible for ICP algorithm being trapped into local minima. At the same time, with the gradual accumulation of the camera motion error, the posture estimation is becoming more and more inaccurate. Therefore,T is not the optimal pose estimation.

We need to optimize the pose to solve the estimation deviation and accumulation error. In this paper, the bundle adjustment method is used to optimize the pose and map. Ideally, the projection point which is reprojected from the spatial by the transformation matrix obtained by ICP should be coincident with the obsevered point. However, due to the existence of errors in the transformation matrix, it is impossible to completely coincide. Therefore, we optimize the transformation pose by constructing a least squares problem to minimize the re-projection error. Assuming that the 3-D coordinates of the observation point is P, the observed pixel coordinates is P_{uv} and the position posture of the camera is expressed using the Lie algebra ξ . The least squares problem can be constructed as follows:

$$\xi^* = \arg\min_{\xi} \sum \rho(\|P_{uv} - \frac{1}{s} K \exp(\xi) P\|^2) \quad (6)$$

where ρ is a kernel function. In the paper, the Hubel kernel is used.

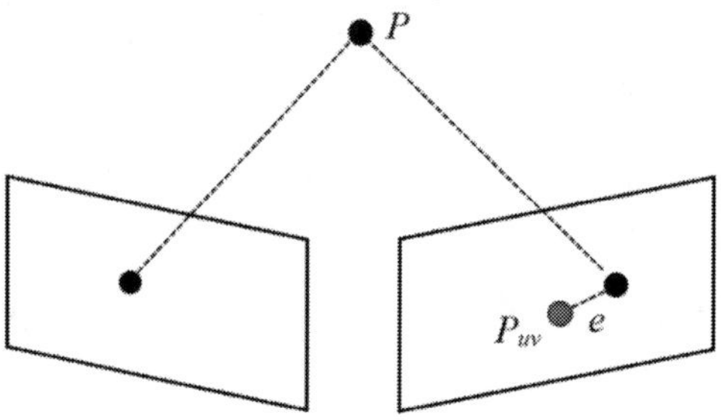

Figure 6. Minimzing the error of reprojection.

The sketch map of reprojection error is shown in Figure 6. We use piecewise optimized strategies in our experiments. Each time K frames will be optimized, that is, the first time from the 1-k frame are optimized and the second time the k+1 -2k frame are optimized, which can further accelerate the optimization speed. And finally performing global optimization.

3.6 Loop closing

Loop closing detection is an important method to eliminate accumulation error. Bag of wors method has been widely used in loop detection because of its effectiveness. The bag of words refers to the technique of converting the contents of an image into a digital vector using a visual dictionary tree. Firstly, the feature of the training image set is extracted, and then the feature descriptor space is clustered by K-means.

We use a k tree to express the dictionary. Similar to hierarchical clustering, it is a direct extension of k-means. Suppose we have N feature points, and we will build a tree with the depth of d and divided into k branch each time. The steps are as follows :

1. In the root node, all samples are clustered into k class by k-means, obtaining the first layer.
2. For each node of the previous layer, the samples belonging to the node are clustered into k class, getting the next layer.
3. And so on, finally get the leaf layer. The leaf layer is called Words.

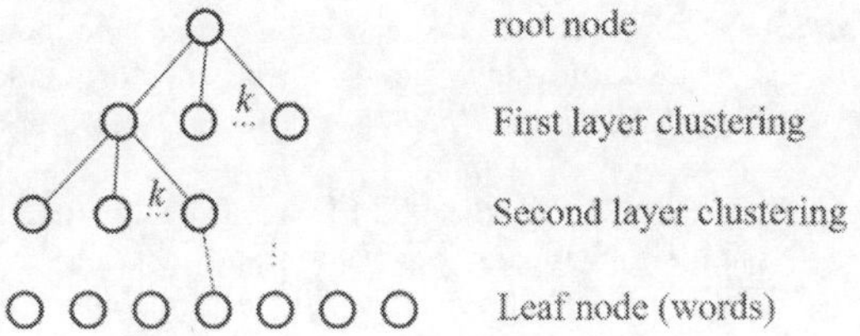

Figure 7. Sketch map of *k* branch tree dictionary.

We will get a *k*-branch, *d*-depth tree that can accommodate k^d words. The k branch tree are shown in Figure 7. When searching for a word corresponding to the given words, we can get the needed one by comparing it with the clustering center of each intermediate node, ensuing the logarithmic level of search efficiency, speeding up the detection.

4 Evaluation

In order to verify the effectiveness of the algorithm and to ensure consistent and comparable experiments in this paper, we use the publicly available Tum data set provided by Technical University of Munich. It contains indoor benchmark data set acquired by RGB-D vision sensor. The motion trajectory of the camera is captured by a high precision motion capture system. The experiment runs on a laptop with Intel Core i7-6700hq CPU.

The constraint between frames can be added through loop detection, further improving the robustness of the system, and the tracking and mapping quality of the whole SLAM is also improved. In this algorithm, loop closing detection is performed after the key frame decision, and the results of the ICP solution are added to the graph optimization as shown in Figure 8.

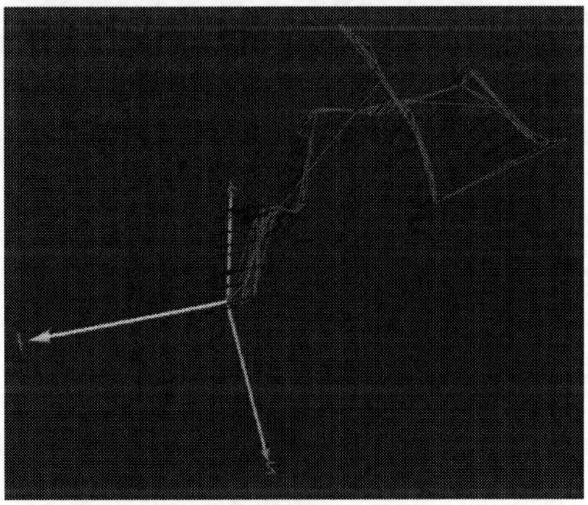

Figure 8. Constraint produced by loop closing.

In order to obtain the change of robot pose, we use the method of ICP to estimate the adjacent pose of the robot by using the 3D-3D feature point constraint, and optimize the position and posture with the help of graph optimization. The results of the experiment are shown in Figure 9 and Figure 10. The experimental trajectory of the robot and the reference trajectory of the data set under different data sets are given, for making the trajectory more intuitive. Where the red line represents the tracjectory computed by the ground truth file, the black line represents the estimated one of the robot by our method. The error of the robot posture can be effectively evaluated by comparing the root mean square error(RMSE). It can be seen that this method can realize the trajectory tracking of robot well. The root mean square error between the real pose and the estimation of the robot's trajectory in the 3-dimensional space is recorded in Table 1.

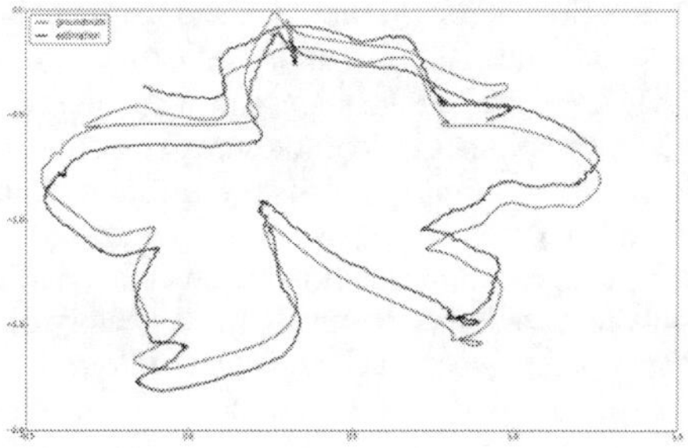

(a) fr1/plant

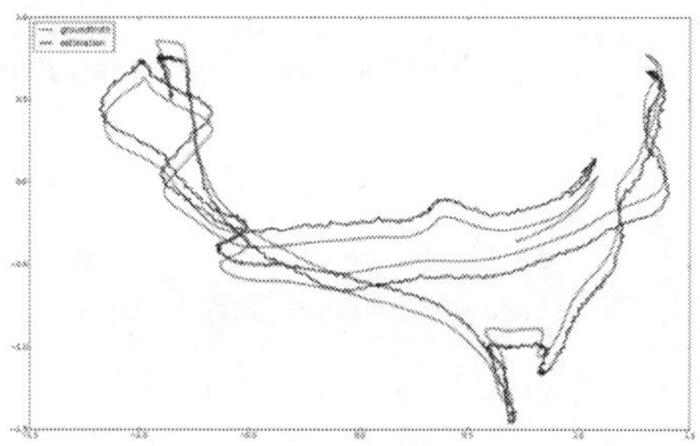

(b) fr1/room

Figure 9. The comparison between groundtruth and estimation.

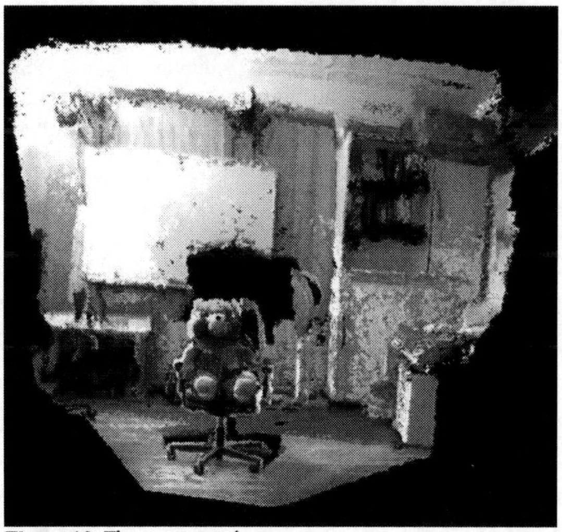

Figure 10. The constructed map.

Table 1. Comparison of RMSE (*m*)

Data set	RGB-D SLAM	DVO SLAM	ORB SLAM2	Our method
fr1/desk	0.026	0.021	0.016	0.023
fr1/desk2	0.048	0.046	0.022	0.042
fr1/room	0.087	0.043	0.047	0.06

5 Conclusion

In this paper, we proposed a method of localization and map building for RGB-D cameras. ORB feature points are adopted for image matching between adjacent frame. And then we use ICP algorithm to estimate the camera pose and 3-D coordinate of feature points. Because there is always drift in pose estimation and map building. It is necessry to reduce the drift at the same time. We take two steps to optimize the location and map. First, we use Bag of words to detect loop closure. Then, we build the pose graph to optimize the pose estimation obtained by ICP.We have proposed a piecewise optimization strategy, which can speed up the optimizing process. And finally, a global optimization is executed. The trajectory of camera and environment map can be obtained at last.

For the next step, we plan to use the method of machine learning to detect loop closing such as convolutional neural networks(CNN), which has been proven to be very effective in the field of image recognition. Further more, we are planning to bulild the environment map with higher quality to realize robot navigation.

References

1. D. Nister, O. Naroditsky, and J. Bergen, "Visual odometry," in *IEEE Conf. on Computer Vision and Pattern Recognition (CVPR)*, 2004
2. D. G. Lowe, "Distinctive image features from scale-invariant keypoints", *International Journal of Computer Vision*, vol. 60, no 2, pp.91-110, 2004.
3. H. Bay, T. Tuytelaars, and L. Van Gool, " Surf : Speeded up robust features", in Computer Vision-ECCV 2006, pp.404-417, Springer, 2006.
4. E. Rublee, V. Rabaud, K. Konolige, and G. Braski, " Orb : an efficient alternative to sift or surf", in *2011 IEEE International Conference on Computer Vision(ICCV)*, pp. 2564-2571, IEEE, 2011.
5. R. A. Newcombe, S Izadi, O. Hilliges, D. Molyneaux, D. Kim, A. J. Davison, P.Kohli, J. Shotton, S. Hodges, and A. Fitzgibbon, "KinectFusion : Real-time dense surface mapping and tracking", in *IEEE Intl. Symp. On Mixed and Augmented Reality(ISMAR)*, 2011.
6. Raúl Mur-Artal, J. M. M. Montiel and Juan D. Tardós. ORB-SLAM: A Versatile and Accurate Monocular SLAM System. *IEEE Transactions on Robotics*, vol. 31, no. 5, pp. 1147-1163, 2015.
7. C. Kerl, J. Sturm, and D. Cremers, "Dense visual SLAM for RGB-D cameras", in *IEEE/RSJ International Conference on Intelligent Robots and Systems (IROS)*, 2013.
8. Ali, A.M., Nordin, M.J., SIFT based monocular SLAM with multi-clouds features for indoor navigation, *TENCON 2010 - 2010 IEEE Region 10 Conference*, 2010, pp. 2326–2331
9. Z. Zhang, Y. Huang, C. Li, Y. Kang, Monocular vision simultaneous localization and mapping using SURF, in *World Congress on Intelligent Control and Automation*, 2008, pp. 1651–1656.
10. Y. Ye, The research of SLAM monocular vision based on the improved surf feature, *International Conference on Computational Intelligence and Communication Net-works*, Hong Kong, China, 2014:344–348.
11. Y. T. Wang, Y. C. Feng, Data association and map management for robot SLAM using local invariant features. *IEEE International Conference on Mechatronics and Automation (ICMA)*, 2013.
12. E. Rosten and T. Drummond. Machine learning for highspeed corner detection. In *European Conference on Computer Vision*, volume I, 2006.
13. M. Calonder, V. Lepetit, C. Strecha, and P. Fua. Brief: Binary robust independent elementary features. In *European Conference on Computer Vision*, 2010.
14. R. Mur-Artal and J. D. Tardos, Fast relocalisation and loop closing in keyframe-based SLAM, in IEEE International Conference on Robotics and Automation (ICRA), 2011, pp. 2564-2571.
15. X. Fen and W. Zhen, "An embedded visual SLAM algorithm based on Kinect and ORB features," in *34th Chinese Control Conference*, July 2015, pp. 6026–6031.
16. G. Xin, X T. Zhang, X.Wang, and J. Song, "A RGBD SLAM algorithm combining ORB with PROSAC for indoor mobile robot", International Conference Computer Science and Network Technology, Harbin, pp. 71-74, 2015.
17. L. Jun, T. Pan, K. Tseng, J. Pan, Design of a monocular simultaneous localisation and mapping system with ORB feature. In *Proceedings of 2013 IEEE International Conference Multimedia and Expo*, San Jose, CA, USA, 15–19 July 2013; pp. 1–4.
18. G. Klein and D. Murray. Parallel tracking and mapping for small AR workspaces. In *Proc. Sixth IEEE and ACM International Symposium on Mixed and Augmented Reality (ISMAR'07)*, Nara, Japan, November 2007. 1
19. G. Klein and D. Murray. Improving the agility of keyframebasedSLAM. In *European Conference on Computer Vision*, 2008.
20. P. Henry, M. Krainin, E. Herbst, X. Ren, and D. Fox. RGB-D Mapping: Using depth cameras for dense 3D modeling of indoor environments. In *the 12th International Symposium on Experimental Robotics (ISER)*, December 2010.
21. P. J. Best and N. D. Mckay, A method for registration of 3-D shapes, *IEEE Trans. Pattern Anal. Mach. Intell. (PAMI)*, vol. 14, no. 2,1992.
22. B. Williams, M. Cummins, J. Neira, P. Newman, I. Reid, and J. D. Tardos, A comparison of loop closing techniques in monocular SLAM, *Robotics and*

Autonomous Systems, vol. 57, no.12, pp. 1188-1197,2009.
23. E. D. Eade and T. W. Drummond. Unified loop closing and recovery for real time monocular SLAM. In *Proc. BMVC*, 2008.
24. D. Galvez-Lopez and J. D. Tardos. Real-time loop detection with bags of binary words. In *Intelligent Robots and Systems (IROS), 2011 IEEE/RSJ International Conference on*, pages 51-58, September 2011.
25. M. Cummins and P. Newman, "Appearance-only SLAM at large scale with FAB-MAP 2.0," The International Journal of Robotics Research, vol. 30, no. 9, pp. 1100–1123, 2011.
26. B. Triggs, P. McLauchlan, R. Hartley, A. Fitzgibbon, Bundle Adjustment - a Modern Synthesis, *LNCS*, Springer Verlag, 1883:298-375, 2000.
27. H. Strasdat, J. M. M. Montiel, and A. Davison, "Scale drift-aware largescale monocular SLAM," in *Proc. of Robotics: Science and Systems(RSS)*, 2010.
28. H. Strasdat, A. J. Davison, J. M. M. Montiel, and K. Konolige, Double window optimisation for constant time visual SLAM, in *IEEE International Conference on Computer Vision (ICCV)*, Barcelona, Spain, November 2011, pp. 2352–2359.

Detecting the Trajectory of Moving Object for Single-Pixel Imaging System

Shao Hui , Wu Dongsheng ,Chen Jie, Li Zhao, Xu Xiaoxue, Zhai Fangbing
School of Electronic and Information Engineering, Anhui Jianzhu University, Hefei, 230601, China

Abstract. In order to get the trajectory of moving object using single-pixel imaging system, an algorithm is proposed. The same pseudorandom masks are employed to illuminate the different time scene. A time weighted sum of the background correction signals is employed to get the trajectory information using compressed sensing (CS) method. In ideal situation, we can obtain other parameters (e.g., speed, orientation) besides the trajectory. However, the reflective intensity of the object can be change due to the reflective angle change caused by the motion in some situations. This will mislead for achieving the speed, orientation parameters. In order to eliminate this effect, a division method is utilized. At last, the computer simulation results prove the effect validity of the proposed algorithm.

1 Introduction

Localization of changes in a scene is an important task in a large number of application and disciplines, such as video surveillance, remote sensing, medical diagnostics, and scientific research [1-5]. Detection of changes is the central step in motion detection and moving object tracking applications. Typically, detection of changes is accomplished by examining the difference between consecutive frames within a sequence of image frames. The location of changes is obtained by subtracting the two consecutive frames that has been widely applied in common imaging system based on CCD detectors. From the numbers of the different images, we can get the trajectory of moving object. Now, there are a lot of methods employed to get the trajectory, such as background subtraction [2,3], optical flow [4] and so on [5].

In recent years, single-pixel imaging system [6-13] has been paid more and more attention and has been successfully applied in many fields, such as three-dimensional imaging [6-9], optical security [10] and so on [11]. Single-pixel imaging system needs to modulate the illumination light intensity, and the reflected light is detected by single-pixel no-scanning detector. Compressed sensing (CS) [14-16] is a new sampling theory developed from the sparse characteristic of the signal. In the far smaller than Nyquist sampling rate, the signal is recovered by CS method. Now, compressed sensing algorithm is largely utilized to recover the object image in single pixel imaging system. Sun [6] employed the different detectors with different visual angles to obtain the two-dimensional images of the object. According to the shape from shading technique, the three-dimensional image of the object is obtained. C.Z. et. al. [7] according the return signal distribution to obtain the three structuration of the object. In this method, the two-dimensional image of a certain range is achieved by the certain-time returned signal. The three-dimensional image can be acquired from these two dimensional images. Photon counting single-pixel lidar is built by the [8] and applied to obtain three-dimensional image and video at 14 frames per second. A background subtraction method [8,9] for object tracking at low light levels is proposed. From the numbers of the object tracking images, we can get the trajectory of moving object. But it will need a lot of CS operations. Trajectory of the interested object is an important subject for analyzing the state and estimating motion parameters in remote sensing, tracking. However, there is no research on trajectory acquisition for single pixel imaging system. In this paper, we proposed a trajectory acquisition method.

2 Compressive detecting the object trajectory

Figure 1 shows the typical single pixel imaging system. Light transmitted by the light source is collimated by the Lens and then illuminate the object. The reflective/transmission light is modulated by the modulation information-system and then collected by the Lens to the single pixel detector. The light emitted from the light source can be continuous, also can be pulsed. In three dimensional imaging applications, the pulse light source is generally used [7,8]. Modulation information-system is a key device. Digital Micro-mirror Device (DMD) [17] which is invented by L. J. Uorbeck in 1987 is often employed in single pixel imaging system for the incoherent light source. The light modulated by the DMD can be described as two elements (0,1) matrix, and we call it as mask. Phase spatial light modulator (SLM) is employed for the coherent light source. In this situation, the mask is computed by the computer program based on the light transmission equations. The whole imaging process requires many different modulations. Figure 2 shows the different masks.

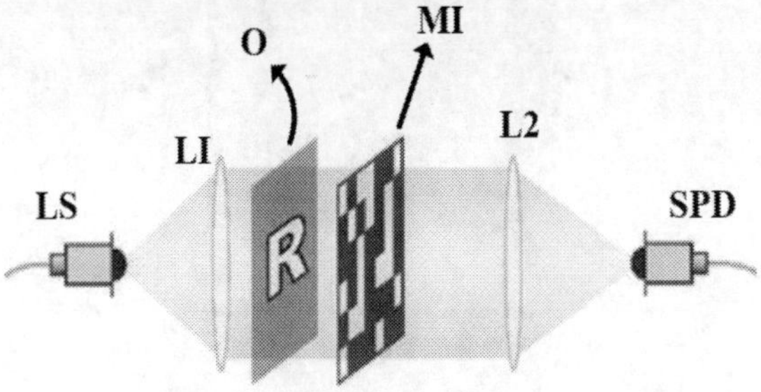

Figure 1. Structure diagram of typical single pixel imaging system. LS: light source; LI: lens; O: object; MI: modulation information-system; L2: lens; SPD: single pixel detector.

Now, we analyze the theory of the single-pixel detector imaging system. Matrix O is used to denote the object information, and the size of O is $N=N_1\times N_2$. Matrix Bn is employed to describe the n-th mask matrix information, and has the same size as O. The detected signal from the single pixel detector in the n-th acquisition denote as in . For the n-th acquisition, we can express as:

Figure 2. Different masks in the imaging process

$$i_n = B_{nl} \cdot O_l \quad (1)$$

Where B_{nl} and O_l show the column vectors of the matrixs B_n and O, respectively. In order to recover the object information, a lot of detected data for different masks are used. The imaging process can be expressed in mathematical form as follows:

$$I = B \cdot O_l \quad (2)$$

where I is the detected intensity vector form the single pixel detector which can be expressed as $I=[i_1, i_2, i_3..., i_M]^T$, symbol T denotes the matrix transpose operation, B denotes the mask matrix as $B=[B_{1l}; B_{2l}; B_{3l}; ...; B_{Ml}]^T$. Because the object has N unknown parameters, the number of the detected data have to greater than or equal to N in order to recover the accurate object information. Otherwise, there are many solutions to the above equation.

Fortunately, CS algorithm offers the way which let us to recover the object information from a smaller number of samples than that of the object information. If the coefficients of a signal in a group of vectors containing a large number of zero elements, this signal is sparse. In fact, most natural signals are sparse mapped in soma basis vectors. The scene is usually complex, so the sparse transform is needed. The equation can be described as:

$$O_l = \Psi \alpha \quad (3)$$

where Ψ is the sparse operator (e.g., wavelet, Fourier, Total Variation operators), α is the sparse signal. We can rewrite the Eq. (2) as

$$I = B \cdot O_l = B \cdot \Psi \alpha \quad (4)$$

Usually CS algorithm is employed to solve l_1-norm minimization problem stated in general as follows:

$$\min \|\alpha\|_1 \text{ such that } I = B \cdot \Psi \alpha \quad (5)$$

According to the above method, we can get the object information.In order to track a moving object by removing static components from a scene, a compressive subtract object tracking method is proposed. The same pseudorandom masks are project to the different time scenes. If the detected signals produce the same correlation value, it would imply that the scene is identical. The opposite situation would reveal information about the changes. The method can be described as:

$$\min \|\alpha\|_1 \text{ such that } \Delta I = B \cdot \Psi \alpha \quad (6)$$

where ΔI is the difference between two detected signals, and equal to $\Delta I=I_l-I_2$. If we use this method to get the trajectory of moving object, a series of operations are needed. First, the Eq. (6) is used to get the change information between the two adjacent time scenes. And then according to the above many change information, computer algorithm is employed to obtain the trajectory of moving object. This treatment is complicated.

In the following section, we will propose a simple method to get the trajectory of the moving object. The same pseudorandom masks are project to the scenes in different times (t_1, t_2, ... , t_n), and we describe the detected singles as (I_1, I_2,..., I_n). Suppose the moving object enter the scene in the moment t_2, the detected singles I_1 regard as the background singles. We can get the summed and weighted summed signals as following:

$$I_S = \Delta I_1 + \Delta I_2 + ... + \Delta I_h \quad (7)$$

$$I_{SW} = \alpha(\Delta I_1) + 2\alpha(\Delta I_2) + ... + h\alpha(\Delta I_h) \quad (8)$$

Where $\Delta I_h = I_h - I_1$ is background correction signals, and the parameter α is time parameter. Using the above Eq. (7) as the detected signal and submitting it in the Eq. (6), the result O_s show the trajectory of the moving object. However, we cannot get other parameters (e.g., speed, orientation) from this result because time information lost. We using a time weighted coefficients in the Eq. (8). In

https://doi.org/10.1051/matecconf/201816007002

ideal situation, the reflective intensity of the object is not change. Using the above Eq. (8) as the detected signal and submitting it in the Eq. (6), we can get the trajectory of the moving object according to gray value of the result Os, and other parameters can also obtained. However, reflective intensity is related to the reflective angle in some situations. The motion of the object caused the change of the reflective angle. At this time, the trajectory can also obtained by the gray value, but the achieved other parameters (e.g., speed, orientation) will take misleading due to the inconsistency of the reflective intensity. Therefore, we need to correct this inconsistency by the intensity normalization. A division method is employed as:

$$T = \frac{O_{WS}}{O_S + \varepsilon} \tag{9}$$

where ε is the regularization parameter. In the next simulation, we will verify the validity of the proposed method to get the trajectory of moving object, and the accurately information (e.g., speed, orientation) can also be acquired.

3 Simulation Experiments

In this section, we will give the results from the two situations:1) object moving in the scene, but the reflective intensity unchanged;2) object moving in the scene, and the reflective intensity change. Figure 3 shows the part of the image frame sequence in the simulation experiment one. Frame 1 is the background. In the next frames, a circular object with uniform reflectance moving into the scene. The number of image frames is twenty. One thousand random intensity distribution patterns are used in this simulation. Using these patterns project on the frames and detecting the signals, the results O_s and O_{ws} can be obtained and are shown in Figure 4. Figure 4(A) is the trajectory of the moving object. However, because of the lack of time information, other parameters (e.g., speed, orientation) are unable to distinguish from the result O_s. This result is consistent with the theoretic analysis. Result O_{ws} contains the time information which is expressed as the gray value. According to this result, we can obtain the orientation parameter. If the distance between two points is known, velocity parameter can also be solved.

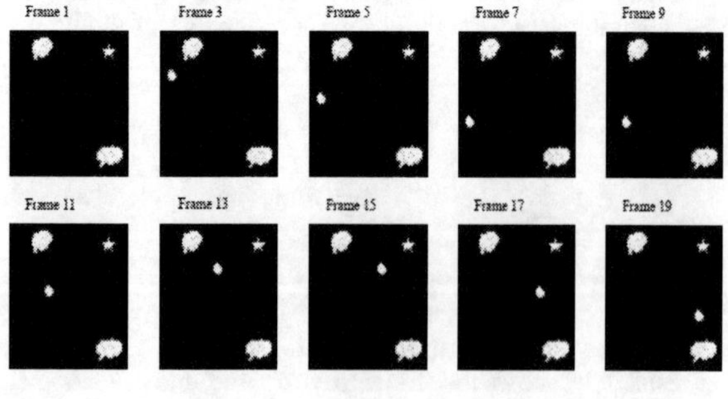

Figure 3. The part of the image frame sequence in the simulation experiment one.

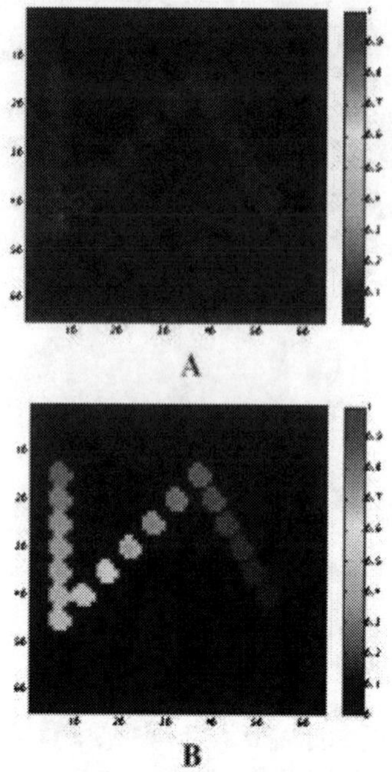

Figure 4. The trajectory of moving object: (A) result from the Eq. (7); (B) result from the Eq. (8).

In order to simulate the change of the object reflectivity intensity, a random coefficient is used in this simulation. The reflectivity intensity of the background assumes constant. The part of the image frame sequence in the simulation experiment two are shown in Figure 5. The other parameters are the same as the first case. Figure 6 (A) and (B) are the results O_s and O_{ws}, respectively. Due to the change of the object reflectivity intensity, the trajectory can obtain, but other parameters (e.g., speed, orientation) cannot achieve. A division method is employed to eliminate the effect of the object reflectivity intensity change. Figure 6(C) shows the final result. The regularization parameter is equal to 0.01. Complete accurate information can be obtained from the final result.

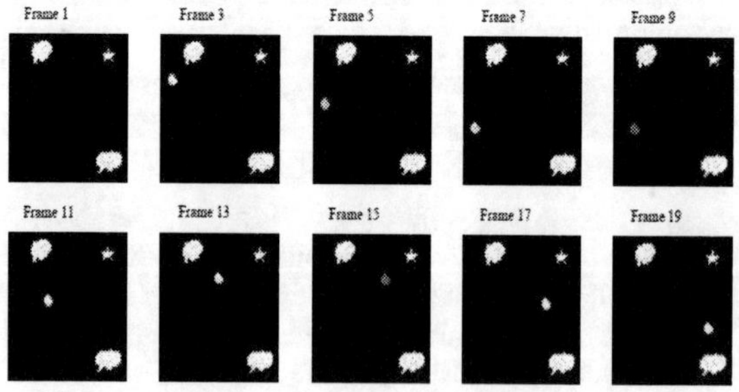

Figure 5. The part of the image frame sequence in the simulation experiment two.

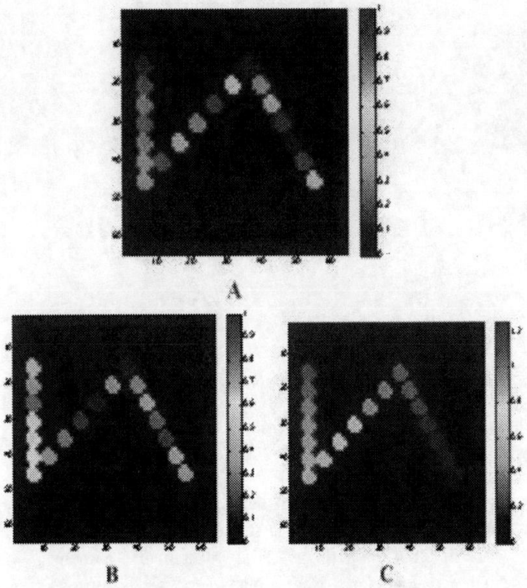

Figure 6. The trajectory of moving object:
(A) result from the Eq. (7);
(B) result from the Eq. (8);
(C) result by the division method Eq. (9).

4 Conclusions

We proposed an algorithm to obtain the trajectory of moving object for single-pixel imaging system. Our approach requires detecting the signals from the different time scenes using the same pseudorandom masks. By the sum of the background correction signals, the trajectory of moving object is obtained using the CS method. Other parameters (e.g., speed, orientation) are also important for analyzing the object state. A time weighted sum of the background correction signals is employed to get the trajectory information. In ideal situation, we can obtain the important parameters besides the trajectory. But in some situation, the reflective intensity of the object can be change due to the reflective angle change caused by the motion. In order to eliminate this effect, a division method is utilized. From the final result, we can obtain the accurate information.

Acknowledgments

This research is supported by Natural Science Foundation of Anhui Province (KJ2016A149, KJ2015ZD14), Research Found for Doctoral Program of Anhui Jianzhu University (2015QD07), National Natural Science Foundation of China (61471002).

References

1. P. M. Lane and M. Cada.Optical Fourier processor and point-diffraction interferometer for moving-object trajectory estimation. Appl Optics **38**, 4306-4315 (1999).
2. A. Elgammal, R. Duraiswami, D. Harwood, and L. S. Davis. Background and foreground modeling using nonparametric kernel density estimation for visual surveillance.P IEEE **90**, 1151-1163 (2002).
3. D. S. Lee. Effective Gaussian mixture learning for video background subtraction. IEEE T Pattern Anal **27**, 827-832 (2005).
4. K. Ma and H. Y. Wang.Region-based nonparametric optical flow segmentation with pre-clustering and post-clustering. IEEE International Conference on Multimedia and Expo, Vol I and Ii, Proceedings, A201-A204 (2002).
5. J. Badenas, J. M. Sanchiz, and F. Pla.Motion-based segmentation and region tracking in image sequences. Pattern Recogn **34**, 661-670 (2001).
6. B. Sun, M. P. Edgar, R. Bowman, L. E. Vittert, S. Welsh, A. Bowman, and M. J. Padgett.3D Computational Imaging with Single-Pixel Detectors. Science **340**, 844-847 (2013).
7. C. Z. Wen, L. Gong, J. Jiao, E. Li, M. Chen, H.Wang, W. Xu, S. Han. Three-dimensional ghost imaging ladar. arXiv:**1301**.5767 (2013).
8. G. A. Howland, D. J. Lum, M. R. Ware, and J. C. Howell. Photon counting compressive depth mapping .Opt Express **21**, 23822-23837 (2013).
9. O. S. Magana-Loaiza, G. A. Howland, M. Malik, J. C. Howell, and R. W. Boyd. Compressive object tracking using entangled photons. Appl Phys Lett 102 (2013).
10. W. Chen and X. D. Chen. Ghost imaging for three-dimensional optical security. Appl Phys Lett **103** (2013).
11. D. Shi, S. Hu, and Y. Wang. Polarimetric ghost imaging. Opt Lett **39**, 1231-1234 (2014).
12. M. Stuff, B. Thelen, J. Garbarino, and N. Subotic. Compressive Sensing and 3-D Radar Imaging. 2012 IEEE Statistical Signal Processing Workshop (Ssp), 540-543 (2012).
13. A. Colaco, A. Kirmani, G. A. Howland, J. C. Howell, and V. K. Goyal. Compressive Depth Map Acquisition Using a Single Photon-Counting Detector: Parametric Signal Processing Meets Sparsity. 2012 IEEE Conference on Computer Vision and Pattern Recognition (Cvpr), 96-102 (2012).
14. W. Chan, K.Charan, D.Takhar, etl.. A single-pixel terahertz imaging system based on compressed sensing. Appl Phys Lett **93** 121105(2008)
15. E. J. Candes, J. Romberg, and T. Tao. Robust uncertainty principles: Exact signal reconstruction from highly incomplete frequency information. IEEE T Inform Theory **52**, 489-509 (2006).
16. Y. Rivenson, A. Rot, S. Balber, A. Stern, and J. Rosen. Recovery of partially occluded objects by applying compressive Fresnel holography. Opt Lett **37**, 1757-1759 (2012).
17. http://www.ti.com/lsds/ti/dlp/.

Short-Term Traffic Flow Prediction of Highway Network Based on Network-Constrained Lasso and NN

Cong Wu[1], Zhaozheng Chen[2] and Xiaofei Li[1]

[1]Engineering Research Center of Wideband Wireless Communication Technology, Ministry of Education, Nanjing University of Posts and Telecommunications, 210003 Nanjing, China
[2]Jiangxi Expressway Networking Management Center, 330036 Nanchang, China

Abstract. Accurate and timely traffic flow prediction is important for the successful deployment of intelligent transportation systems. Most of existing methods have not made good use of information from adjacent sections to analyse the trends of the object section. A new method for traffic flow prediction of highway network, namely network-constrained Lasso (Least absolute shrinkage and selection operator) and Neural Networks, was proposed. Unlike existing methods, our approach incorporated all the spatial and temporal information available in a highway network to carry our short-term traffic flow prediction for the objective section. To capture the spatial correlation of traffic network, the Laplacian matrix was introduced to describe the highway network structure. Subsequently, a network-constrained Lasso method was applied for sparse variable selection. With the extracted historic and real-time data, the back propagation neural networks were implemented to predict traffic flow at different time intervals in future. The experimental results verified that the proposed method could achieve above 90% average accuracy in the 30-minutes speed predictions for 78 road sections.

1 Introduction

Real-time and accurate traffic flow prediction is the successful deployment of advanced traffic control and dynamic route guidance which can reduce road congestion and improve mobility of transportation. Due to the dynamic and stochastic characteristics of traffic flow, short-term prediction has become one of the difficulties of Intelligence Transportation System.

Traffic researchers have conducted substantial studies, e.g., ARIMA[1], time series[2], Kalman[3], Neural Networks (NN)[4, 5] and SVR[6]. Nevertheless, most of them have not achieved satisfactory performance in traffic network, because they didn't make good use of spatio-temporal correlation between multiple sections.

In order to reveal the potential influence of traffic flow changes between sections of highway network, some researchers have made attempts to increase the prediction accuracy. S Sun [7] and E Castillo [8] performed a short-time traffic flow prediction based on Bayesian network, using space structure information of adjacent sections in highway. B Ghosh [9] reported an time-series analysis, which was propitious to analyze the relationship of traffic flow network. In addition, Y Kamarianakis [10]and C Hu [11] have taken use of topology and distance information between multiple section, and performed short-time traffic flow prediction by ARIMA and Hybrid Process Neural Network. However, the prediction algorithm mentioned above were mainly based on prior knowledge and mass calculation. Consequently, those algorithms haven't obtained an ideal prediction result in real-time systems, because of their complexity in describing substantive characteristics of the high dimensional traffic data.

In this paper, the Laplacian matrix was introduced to depict the highway network structure. Subsequently, a network-constrained least absolute shrinkage and selection operator (Lasso) method was implemented as the penalty function for sparse variable selection. As neural network has been proved to do well in complex and nonlinear system, the back propagation (BP) neural networks with the extracted data were used to predict nonlinear traffic flow at different time intervals in future.

2 Variable selection

2.1 Laplacian matrix representation of spatial correlation

There were many influencing factor of traffic flow between sections of highway network. From the perspective of qualitative analysis, the relationship was more significant while a section was adjacent and close to another. With the increase of distance and converged traffic flow from other sections, the relationship was gradually diminished.

In this work, we used a weighted graph $G = (V, E, W)$ to describe the highway network structure, where V is a set of vertices corresponding to the sensors (e.g. loop detectors) of independent road sections, $e(u, v)$ is the set of edges indicating that the section u and v are linked, and $w(u, v)$ is the set of weights of the edges, which represent the spatial correlation governed by the network topology. Note that the value of $w(u, v)$ was calculated according to the number and location of upstream and downstream sections in our work.

In graph theory, Laplacian Matrix L, also known as Admittance Matrix or Kirchohoff Matrix, is a matrix form for graph G. As in [12], the element $L(u, v)$ in the normalized Laplacian matrix L was defined in equation (1), where the vertices degree was defined $d_v = \Sigma_u w(u, v)$.

$$L(u,v)=\begin{cases} 1-w(u,v)/d_u & \text{if } u=v \text{ and } d_u \neq 0 \\ -w(u,v)/\sqrt{d_u d_v} & \text{if } u=v \text{ are adjacent} \\ 0 & \text{otherwise.} \end{cases} \quad (1)$$

2.2 Laplacian matrix representation of spatial correlation

Compared with classical variable selection methods, such as subset selection, the Lasso is computationally feasible for high-dimensional data. The selection procedure is continuous and hence more stable than the subset selection which is discrete[13]. The traditional penalty regression can be formulated as follows in Equation (2), as a combination of regularized least squares and a penalty term[14].

$$\hat{\beta} = \arg\min_{\beta} \left\{ \frac{1}{n}\sum_{i=1}^{n}(y_i - x_i\beta_i)^2 + \sum_{j=1}^{p}\lambda\left|\beta_j\right| \right\} \quad (2)$$

In this work, a network-constrained penalty was constructed by the Lasso penalty and a penalty induced by the Laplace matrix of the graph[15]. A network-constrained penalty function was defined in Equation (3), where $|\beta|_1 = \sum_{j=1}^{p}|\beta_j|$ was the L1-norm.

$$J(\lambda_1,\lambda_2,\beta) = (y - \mathbf{X}\beta)^T(y - \mathbf{X}\beta) + \lambda_1|\beta|_1 + \lambda_2\beta^T L\beta \quad (3)$$

In Equation (3), the tuning parameters λ_1, λ_2 control the amount of regularization for sparsity and smoothness. When $\lambda_2 = 0$, the estimate reduces to the traditional Lasso. It is easy to verify[16] that this matrix L is nonnegative definite and can be decomposed as $L=SS^T$. Equation (3) can be rewritten as Equation (4).

$$J(\lambda_1,\lambda_2,\beta) = (y - \mathbf{X}\beta)^T(y - \mathbf{X}\beta) + \lambda_1|\beta|_1 + \lambda_2\sum_{u\sim v}\left(\frac{\beta_u}{\sqrt{d_u}} - \frac{\beta_v}{\sqrt{d_v}}\right)^2 w(u,v) \quad (4)$$

The network-constrained estimation $\hat{\beta}$ was defined as the minimizer of Equation (4).

$$\hat{\beta} = \arg\min_{\beta}\left(J(\lambda_1,\lambda_2,\beta)\right) \quad (5)$$

As in [17], the network-constrained regularization problem was solved by a similar efficient computation procedure.

3 Prediction model with network-constrained Lasso and NN

As shown in Figure 1, we adopted Laplacian matrix to depict the traffic network topology and spatial correlation, and then applied network-constrained Lasso for variable selection with historical data. After finding the most helpful sections which have great relativity to the objective section, we constructed many BPNNs to predict traffic flow in a region of highway network. In Figure 1, n and m were the number of T, time interval of 5 minutes.

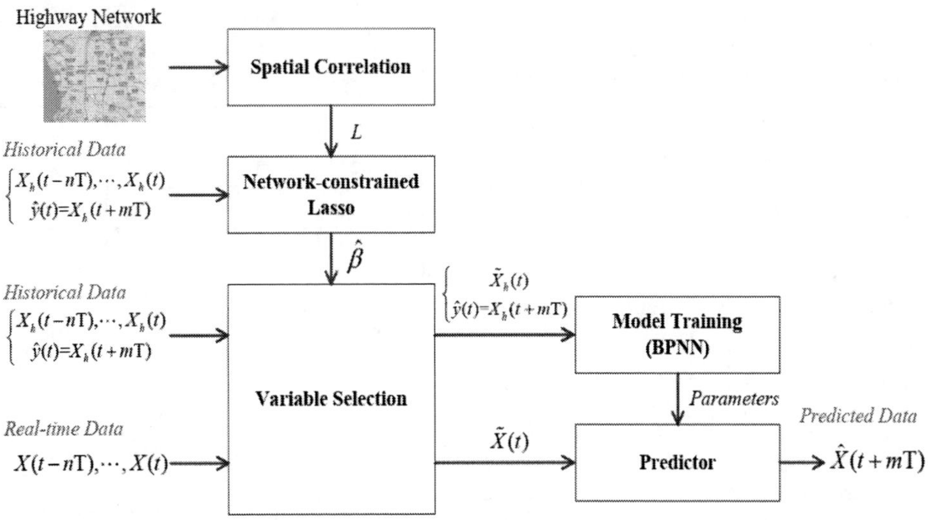

Figure 1. Prediction flowchart with network-constrained Lasso and NN.

4 Application and experimental results

4.1 Data description

The Freeway Performance Measurement System (PeMS) of California collects and processes the data of loop detectors every 30s such as flow, occupancy and speed. This paper focused on one region of Los Angeles and selected the average traffic flow data in 5 minutes' interval from 2013/3/4 to 2013/3/31. A total number of 78 loop detectors in this freeway network were listed in Figure 2, some of which locate near the crossways of traffic network. There were 8064 records in each detector with heavy, medium, and low traffic loads.

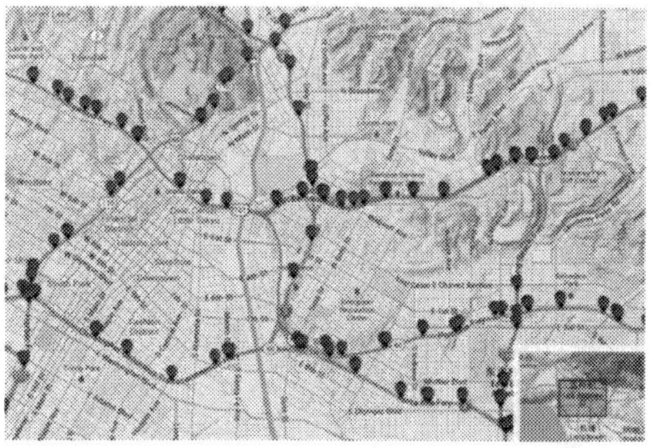

Figure 2. The map of traffic network and loop detectors.

4.2 Comparison of 15-min traffic flow prediction

In this part, we used the average speed data collected in 5 minutes as the input $X(t)$, and predicted average speed $X(t+3T)$ in next 15 minutes with NN and the proposed method. Results were shown in Table 1 and Table 2, corresponding to 15 sections far away from crossways and 15 sections near crossways in the network region. In this paper, Mean Absolute Percent Error (MAPE) was calculated to evaluate the effectiveness of NN and the proposed model.

In Table 1 and Table 2, we could also find that the proposed method proved to be more accurate than NN model for the short-term prediction of average speed. With the proposed method for 15-min prediction, the average accuracy of sections far away from crossways, and sections near crossways was 90.9%, and 93.6%, respectively. Meanwhile, the variable number of the proposed method had a large drop from 78 to 23.0, and 16.3, respectively.

4.3 Comparison of traffic flow prediction from 5-min to 60-min

With the same input X(t), NN model and the proposed method was used to predict 5-min, 10-min, 15-min, 20min, 30-min and 60-min traffic flow. As shown in Table 3, for 78 loop detectors, the average accuracy of the proposed method was 96.7%, 95.3%, 94.3%, 93.2%, 91.2%, and 87.6%, respectively. The maximum average prediction accuracy improvement of the proposed method was up to 4.6% compared with NN.

Table 1. Performance comparison of sections near crossways.

Section	NN		The proposed method	
	Variable Number	MAPE	Variable Number	MAPE
718173	78	16.9%	14	6.3%
773281	78	14.9%	7	7.8%
717264	78	15.0%	2	7.6%
716942	78	15.0%	21	7.7%
763237	78	14.3%	27	10.4%
717047	78	14.4%	53	12.9%
718045	78	17.0%	21	8.6%
760650	78	12.2%	29	7.4%
773024	78	9.9%	15	5.1%
716946	78	12.7%	28	5.9%
716951	78	18.3%	46	10.8%
772501	78	12.5%	43	8.5%
716940	78	14.2%	5	13.5%
716939	78	15.4%	5	13.7%
718166	78	15.9%	29	9.8%
Average	**78**	**14.6%**	**23.0**	**9.1%**

Table 2. Performance comparison of sections far away from crossways.

Section	NN		The proposed method	
	Variable Number	MAPE	Variable Number	MAPE
718421	78	11.6%	24	5.6%
717262	78	6.5%	10	3.1%
717265	78	8.1%	10	6.9%
717268	78	8.7%	14	5.6%
716069	78	6.9%	6	4.6%
716328	78	9.0%	8	4.9%
773062	78	9.6%	10	5.8%
737320	78	7.8%	21	6.6%
717052	78	6.9%	35	5.8%
716943	78	8.3%	8	7.9%
716944	78	17.7%	27	12.0%
771667	78	9.4%	20	4.3%
718371	78	10.0%	12	8.0%
717445	78	13.0%	36	8.6%

717263	78	10.0%	4	6.2%
Average	**78**	**9.6%**	**16.3**	**6.4%**

Table 3. Performance comparison of predicting different time intervals in future.

Prediction	NN		The proposed method	
	Variable Number	MAPE	Variable Number	MAPE
5-min traffic flow $X(t+T)$	78	4.7%	3.0	3.3%
10-min traffic flow $X(t+2T)$	78	6.8%	9.4	4.7%
15-min traffic flow $X(t+3T)$	78	8.8%	20.9	5.7%
20-min traffic flow $X(t+4T)$	78	10.6%	24.3	6.8%
30-min traffic flow $X(t+6T)$	78	12.8%	34.3	8.8%
60-min traffic flow $X(t+12T)$	78	17.0%	44.5	12.4%

4.4 Comparison of traffic flow prediction with historic data

Usually historic data played an important role in traffic flow prediction. In this part, we predicted 15-min traffic flow with real-time data and historic data at previous time intervals, i.e., X(t-T), X(t-2T). From the results in Table 4, it is apparent that high accuracy result was not due to the variable number but owing to spatio-temporal correlation of traffic network.

Table 4. Performance comparison of predicting different time intervals in future.

Prediction	NN		The proposed method	
	Variable Number	MAPE	Variable Number	MAPE
$X(t)$	78	8.8%	20.9	5.7%
$X(t)$, $X(t-T)$	156	9.5%	33.5	5.9%
$X(t)$, $X(t-T)$, $X(t-2T)$	234	10.3%	34.3	5.8%

5. Conclusion

According to large amount of actual data testing and comparison mentioned above, we proposed a traffic flow prediction method for highway networks. The advantage of the proposed method is mainly due to its simplicity and effectiveness. The simplicity enables its wide applications in real-time traffic management systems.

For future work, we will investigate deep learning algorithms with Lasso for traffic flow prediction and apply these algorithms on different public open traffic data sets to verify their effectiveness.

Acknowledgement

This work was supported in part by the Natural Science Foundation of China under Grant 61471206, in part by NUPTSF under Grant NY217153, in part by the Key Research and Development Foundation Project of Jiangsu province under Grant BE2016001-4.

References

1. BM Williams, LA Hoel, J. Transp. Eng. **129**, 6 (2003)
2. T Thomas, W Weijermars, EV Berkum, IEEE T. Intell. Transp. **11**, 1 (2010)
3. J Guo, B Williams, Transport. Res. Rec. **2175**, 1 (2010)
4. X Jiang, H Adeli, J. Transp. Eng. **131**, 10 (2005)
5. KY Chan, TS Dillon, J Singh, E Chang, IEEE T. Intell. Transp. **13**, 2 (2012)
6. M Castro-Neto, YS Jeong, MK Jeong, LD Han, Expert Syst. Appl. **36**, 3 (2009)
7. S Sun, C Zhang, G Yu, IEEE T. Intell. Transp. **7**, 1 (2006)
8. E Castillo, M Nogal, JM Menéndez, SS Cambronero, P Jimenez, IEEE T. Intell. Transp. **13**, 2 (2012)
9. B Ghosh, B Basu, M O'Mahony, IEEE T. Intell. Transp. **10**, 2 (2009)
10. Y Kamarianakis, P Prastacos, Transport. Res. Rec. **1857**, 1 (2003)
11. C Hu, K Xie, G Song, T Wu, *11th International IEEE Conference on Intelligent Transportation Systems* (Beijing, China, 2008)
12. FRK Chung, *Spectral graph theory* (American Mathematical Soc., 1997)
13. Y Kamarianakis, W Shen, L Wynter, Appl. Stoch. Model. Bus., **28**, 4 (2012)
14. S Sun, R Huang, Y Gao, J. Transp. Eng. **138**, 11 (2012)
15. C Li, H Li, Bioinformatics, **24**, 9 (2008)
16. C Li, H Li, *The annals of applied statistics* **4**, 3 (2010)
17. H Zou, T Hastie, R Tibshirani, J. Comput. Graph. Stat., 15, 2 (2006)

The Optimal Degree of Gravity Spherical Harmonic Model Calculation for Gravity Disturbance Compensation in Inertial Navigation

Junbo TIE[1], Meiping WU[1], Juliang Cao[1], Junxiang Lian[1], Shaokun Cai[1] and Lin Wang[1]

[1]*National University of Defense Technology, Changsha 410073, People's Republic of China*

Abstract. In recent years, the significant improvement of inertial navigation, leaves the gravity disturbance as the important factor which affects the accuracy of inertial navigation. This paper focus on the compensation for gravity disturbance with gravity spherical harmonic model, especially the optimal degree of gravity spherical harmonic model with which to calculate the gravity disturbance. The effect of gravity disturbance on inertial navigation is analysed based on the amplitude-frequency response characteristics of inertial navigation error differential equation, then the dominantly influential frequency band of gravity disturbance can be found which is the target of compensation. Combination the dominantly influential frequency band with the spatial resolution of the Earth's gravity spherical harmonic model EGM2008 which is used to calculate the gravity disturbance, the optimal degree can be determined based on an algorithm proposed in this paper. Finally, shipborne inertial navigation experiment confirms the correctness and effectiveness of the proposed algorithm.

1 Introduction

Inertial navigation system (INS) is currently the main means of navigation which measures the vehicle's motion and constantly updates position. INS is based on Newton's Second Law, which combines specific force and gravity into acceleration. The velocity and position can be obtained through integration of acceleration. Specific force can be measured by inertial sensors, three accelerometers of INS, and the gravity information is acquired from the normal gravity model.

The performance of INS depends not only on the quality of inertial sensors, but also on the accuracy of the gravity information [1, 2]. In the limiting situation of drift-free gyroscopes and perfect accelerometers, INS still cannot be error-free, because uncertainties in the normal gravity model will produce errors. In recent years, the significant improvement of INS, especially on inertial sensors, leaves gravity disturbance as the important error source of the navigation solution, particularly for rough topological areas [4, 5]. For future higher precision navigation, the compensation of gravity disturbance has to be considered [10].

Nowadays, with the release of ultra-high degree Earth Gravitational Model, such as Earth Gravitational Model 2008 (EGM2008) [6, 7], the gravity disturbance on the vehicle's trajectory can be calculated directly which can be used to compensate the effect of gravity disturbance on INS. EGM2008 is publicly released by the U.S. National Geospatial Intelligence Agency (NGA) EGM Development Team. This gravitational model is a spherical harmonic model (SHM) and is complete to spherical harmonic degree and order 2159, and contains additional coefficients extension to degree 2190 and order 2519 and the full access to this model's coefficients is provided in [3]. It is because EGM2008 has degree and order up to 2190 that the gravity disturbance can be precisely calculated with this model. However, the high degree and order bring a great computation burden on the embedded navigation computer of INS in real time application. [4] indicates the total operation count for calculating spherical model. To reduce the computation complexity and maintain the compensation accuracy, the interest lies in this paper is to find an optimal degree used in the calculation of EGM2008 instead of the maximum degree 2190.

The paper is organized as follows: in section 2, the effect of gravity disturbance on INS is analysed to determine the dominantly influential frequency band of gravity disturbance. Then the calculation of SHM is introduced in section 3 and the algorithm of determining the optimal degree is proposed in section 4. The shipborne inertial navigation experiment is in section 5, and the conclusion is provided in section 6.

2 Effect of Gravity Disturbance on INS

To find the dominantly influential frequency band of gravity disturbance, the error characteristics of gravity disturbance induced error is investigated. The north channel of INS is analyzed here and the only error source considered is the gravity disturbance, and the error differential equation are as follow [8-10].

$$\delta\dot{\theta} = -\delta v_N / R \tag{1}$$

$$\delta \dot{v}_N = g\delta\theta + \delta g \quad (2)$$

$$\delta \dot{p} = \delta v_N \quad (3)$$

where $\delta\theta$ is the pitch error, δv_N is the north velocity error, R is the mean radius of the Earth, δp is the position error, g is the normal gravitational acceleration, and δg is the input error source, the gravity disturbance.

These error equations can be described with state equation. According to the theory of linear system, the transfer function can be obtained from by the following transformation [11].

$$G(s) = \left[C(sI - F)^{-1} B\right] \quad (4)$$

The transfer function can be obtained as follow

$$G(s) = R / (Rs^2 + g) \quad (5)$$

The amplitude-frequency response of the north channel of INS can be obtained based on the transfer function, equation (5).

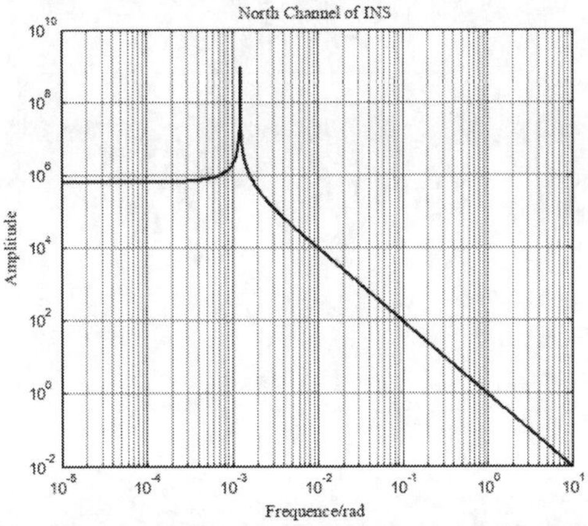

Figure 1. The amplitude-frequency response of the north channel of INS

It can be seen that the response of INS for gravity disturbance is alike a low-pass filter, and the cut-off frequency can be determined according to the -3dB definition, the cut-off frequency $f_{cut\text{-}off}$ is 0.00215 *rad* or 3.42×10^{-4} Hz, and the peak of the amplitude-frequency response is the Schuler frequency of INS. Because of the low-pass characteristic, the high frequency gravity disturbance is filtered out, and only the low frequency gravity disturbance will have a negative effect on the INS. Therefore, the dominantly influential frequency band of gravity disturbance is determined.

$$f_{compen} \le 3.42\times10^{-4} Hz \quad (6)$$

3 Gravity Disturbance and Gravity Spherical Harmonic Model

The gravity disturbance means the difference between the gravity vector calculated through the normal gravity model (NGM) and the actual one, including difference in magnitude and direction [12]. The difference in magnitude is the so-called gravity anomaly, and the difference in direction is the so-called deflection of vertical (DOV) which is the chief cause of the gravity-induced errors [13-15]. As shown in figure 2, DOV is the angle between the true plumb line and the normal line of the reference ellipsoid which is chosen to approximate the Earth's sea-level surface. DOV has a north component x and an east component h, the order of DOV component can maximally reach 100 arcsec which is significantly higher than the modern accelerometer bias, whose typical value is 10 mGal (1 mGal = $10^{-5} m/s^2$) [4], corresponding to a 2 arcsec DOV [3]. The horizontal gravity disturbance, including north component δg_N and east component δg_E, are caused by DOV as follow.

$$\delta g_N = g\cdot\xi \quad \delta g_E = g\cdot\eta \quad (7)$$

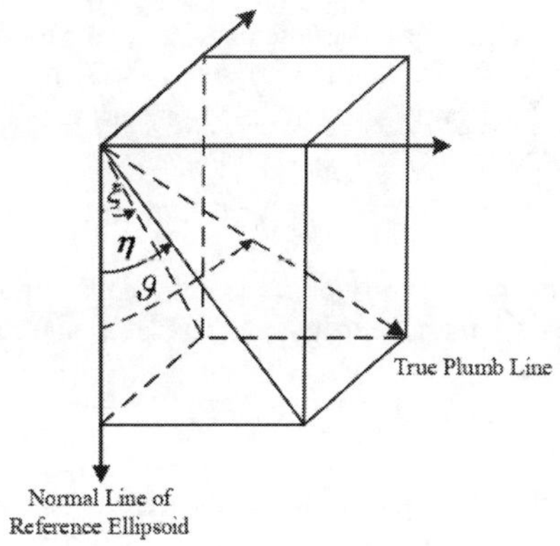

Figure 2. The definition of DOV

The horizontal gravity disturbance can be calculated based on the gravity spherical harmonic model [3, 16]

$$\delta g_N = -\frac{GM}{r^2}\sum_{n=2}^{\infty}\left(\frac{a}{r}\right)^n \sum_{m=0}^{n}\begin{pmatrix}\bar{C}^*_{nm}\cos m\lambda \\ +\bar{S}_{nm}\sin m\lambda\end{pmatrix}\frac{d\bar{P}_{nm}(\cos\theta)}{d\theta} \quad (8)$$

$$\delta g_E = -\frac{GM}{r^2\sin\theta}\sum_{n=2}^{\infty}\left(\frac{a}{r}\right)^n \sum_{m=0}^{n} m\begin{bmatrix}\bar{C}^*_{nm}(-\sin m\lambda) \\ +\bar{S}_{nm}\cos m\lambda\end{bmatrix}\bar{P}_{nm}(\cos\theta) \quad (9)$$

where G is Newton's gravitational constant, M is the mass of the Earth, r is the geocentric radius distance, θ is the polar distance, $\bar{C}^*_{nm}$ and $\bar{S}_{nm}$ are coefficients of the spherical harmonic of degree n and order m, $\bar{P}_{nm}(\cos\theta)$ is the normalized associated Legendre function of the first kind. The Legendre function can be resolved by recursion[17].

4 The Algorithm of Choosing Optimal Degree

In Section 2, the dominantly influential frequency band of gravity disturbance has been determined. However, gravity disturbance is a spatial-domain signal rather than a time-domain signal, and the gravity disturbance signal comes into INS as the position of INS changing on the surface of the Earth, hence the gravity disturbance signal can be transformed from spatial-domain into time-domain with the velocity of INS, and the associated dominantly influential frequency band can also be transformed into spatial domain as

$$f'_{compen} = \frac{f_{compen}}{v} \quad (10)$$

where v is the velocity of INS, and f'_{compen} is the compensation frequency band in spatial-domain, and the corresponding period of the compensation frequency band which is a spatial distance can be determined

$$D_{compen} = \frac{1}{f'_{compen}} = \frac{v}{f_{compen}} \quad (11)$$

According to the property of spherical harmonic function, the spatial distance D_{compen} is determined by the degree of the SHM. Assume that the degree used in SHM calculation is n, then the sphere of the Earth is divided by the Legendre function into n zones, so the spatial resolution of degree n SHM is

$$D = \frac{\pi R}{n} \quad (12)$$

Substituting equation (12) into equation (11), the optimal degree of gravity spherical harmonic model can be obtained.

$$n_{compen} = \frac{\pi R f_{compen}}{v} \quad (13)$$

There are some special cases need to be considered, the first is that the maximum of n_{compen} is 2190 when the used gravity spherical harmonic model is EGM2008. Second, if the velocity is very small, even zero, then assume n_{compen} equal to 2190 is appropriate.

5 The Shipborne Inertial Navigation Experiment

The shipborne experiment instruments consists of a high-precision strapdown inertial navigation system (SINS) as shown in figure 4 and a Novatel® GPS receiver contained in the electric system of SINS.

The inertial sensors used in SINS are high precision ring laser gyros and quartz flexible accelerometers. The sampling rate of SINS is 200 Hz and the sample rate of GPS receiver is 2 Hz, and the GPS data is processed off line with precise point positioning (PPP) technology to obtain position results as the reference in this experiment. The ship was first at moor, then sailed in maneuver, and the experiment time is 24 hours. The DOV values on the sailing trajectories are calculated based on EGM2008, and the position information used to calculate DOV is from GPS solution, and the period of calculating gravity disturbance with SHM is 100 seconds.

There are three types method to deal with the inertial navigation date: processing the inertial navigation data without compensation for gravity disturbance is **type I**. Processing the data with compensation where the degree of SHM is set to 2190 is **type II**, and if the degree of SHM is optimally set based on the proposed algorithm in section 4, that's **type III.**

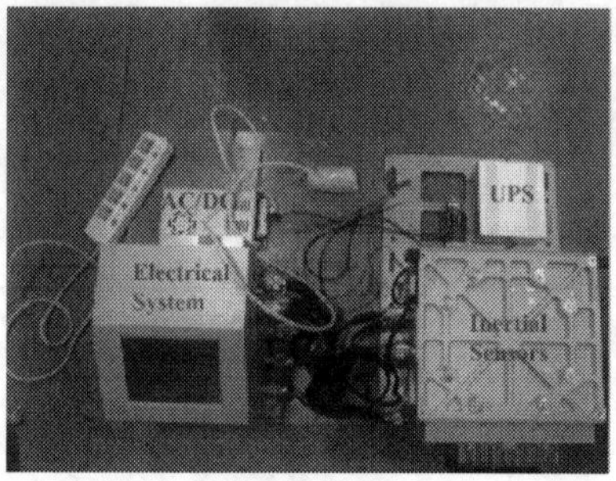

Figure 4. High precision strapdown inertial navigation system

The comparison of north component of gravity disturbance on the trajectory between type II and type III is shown in figure 5, and the comparison of east component of gravity disturbance on the trajectory between type II and type III is shown in figure 6.

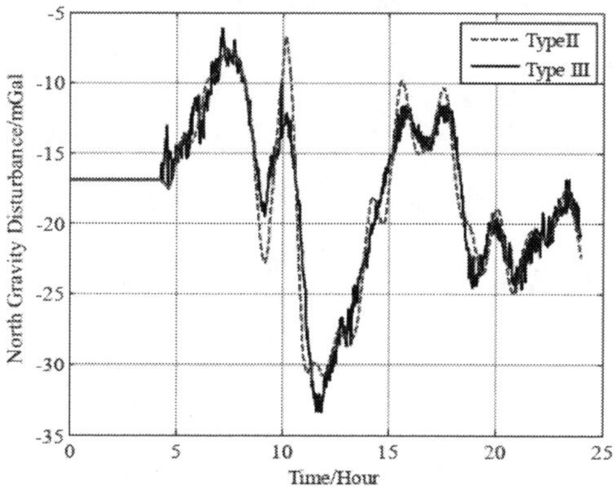

Figure 5. The comparison of north component of gravity disturbance on the trajectory

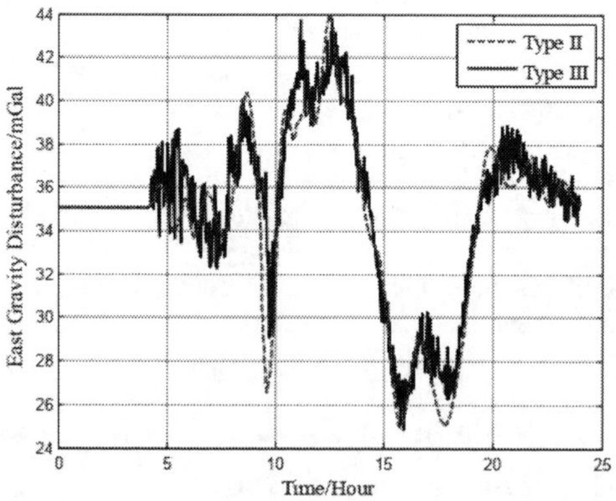

Figure 6. The comparison of east component of gravity disturbance on the trajectory

The comparison of time-consuming between type II and type III is shown in Table 1. With the optimal degree, the time for calculating SHM decreases by 635.8 seconds, about 36.1%. From figure 7, it can be seen that the average one-time time consuming of calculating SHM with full degree 2190 in type II is about 2 second. As for Type III, in the first five hours, the experiment ship is at moor, the velocity almost equals to zero, so the optimal degree is 2190, after that the experiment ship begins to move, and the optimal degree is decreased to 1400 as shown in figure 8, then the time-consuming is shortened to about 1 second.

Table 1. Comparison of time-consuming between type II and type III

Time-consuming Item /Second	Type II	Type III
Total time-consuming	3542.1	2938.3
Time for navigation calculation	1722.5	1748.6
Time for calculating SHM	1762.4	1126.6
Time for initial alignment and others	57.2	63.1
Decrease of total time-consuming	603.8	
Decrease of time for calculating SHM	635.8	

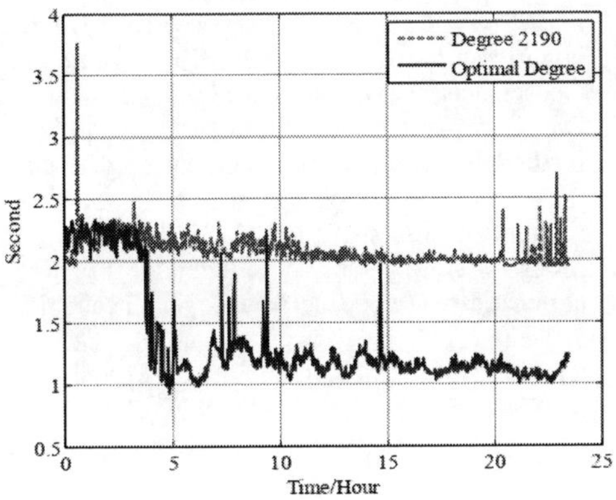

Figure 7. The time-consuming for one-time calculating SHM in type II and type III

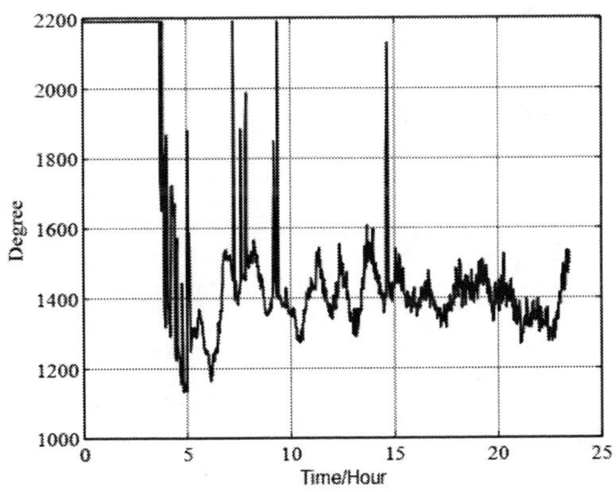

Figure 8. The optimal degree of calculating SHM in type III

The positioning error comparison among the three types are shown in figure 12, it can been seen that the position errors of type II and type III are both less than type I, which means the compensation for gravity disturbance is efficient. Subtracting the position error of type II and type III from Type I, the improvement of positioning accuracy can be obtained, as shown in figure 10, and it can be seen that the precision improvement with optimal degree is almost consistent with the full degree. Attention is needed here, the vibration of precision improvement of type III is obviously smaller than that of type II, this is probably because that there exists some error in the high degree components of the SHM, and these errors is avoided by utilization of the optimal degree.

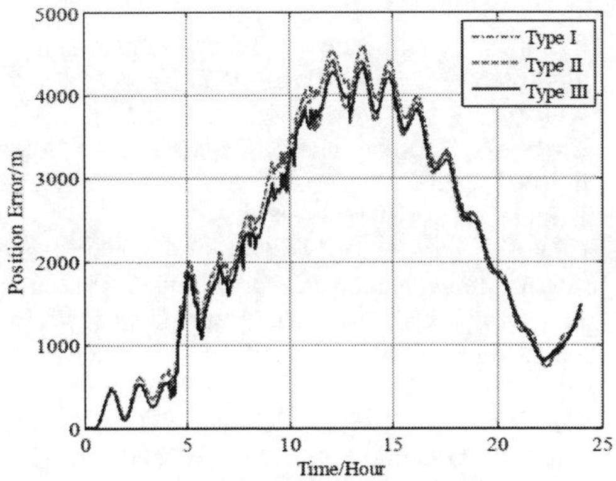

Figure 9. The comparison of position error

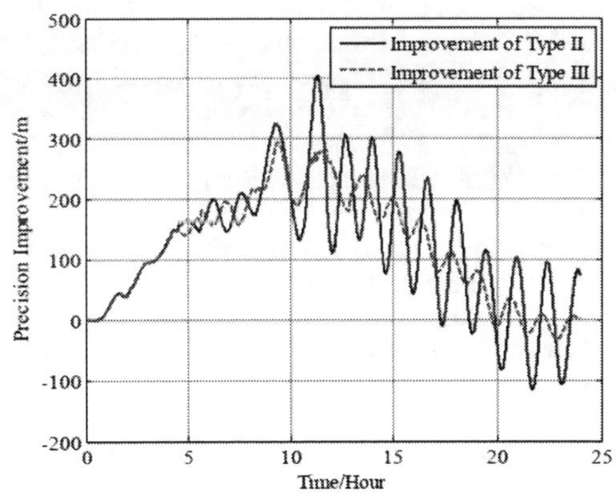

Figure 10. The comparison of position error

6 Conclusion

In this paper, the optimal degree of calculating SHM for gravity disturbance compensation is discussed. First, the effect of gravity disturbance on inertial navigation is analysed based on the amplitude-frequency response characteristics of inertial navigation error differential equation. Second, the dominantly influential frequency band of gravity disturbance is determined which is the target of compensation. Then combination the dominantly influential frequency band with the spatial resolution of the Earth's gravity spherical harmonic model, the algorithm of determining the optimal degree with which to implement the SHM calculation is proposed. Finally, the shipborne inertial navigation experiment confirms the correctness and effectiveness of the proposed algorithm, with the optimal degree the time-consuming of calculating SHM decreases by 36.1%, and the efficiency of DOV compensation is almost consistent with that of full degree.

Acknowledgment

This work is supported by the National Key R&D Program of China, contract No. 2016YFC0303002; and National Natural Science Foundation of China, contract No. 61603401.

References

1. Kwon J H and Jekeli C Gravity requirements for ultra-precise inertial navigation J. Navig. 58 479-92 (2005)
2. David W. Harriman and J. Chris Harrison Gravity-Induced Errors in Airborne Inertial Navigation J.Guidance. 9 419-426 (1986)
3. URL of EGM2008: http://earth-info.nga.mil/GandG/wgs84/gravitymod/egm2008
4. Jing W, Gongliu Y, Xiangyun L and Zhou X Application of the spherical harmonic gravity model in high precision inertial navigation systems Meas. Sci. Technol. 27 095103 (2016)
5. Zhou X, Gongliu Yang, Jing W and Jing L An improved gravity compensation method for high-precision free-INS based on MEC-BP-AdaBoost Meas. Sci. Technol. 27 125007 (2016)
6. Pavlis N. K., Holmes S. A., Kenyon S. C. and Factor J. K. An Earth Gravitational Model to Degree 2160: EGM2008. In Proceedings of the European Geosciences Union General Assembly 2008, Vienna, Austria, 13–18 (2008)
7. Pavlis N. K., Holmes S. A., Kenyon S. C. and Factor J. K. The development and evaluation of the Earth Gravitational Model 2008 (EGM2008) J. Geophys. Res. Solid Earth 117 1-38 (2012)
8. D. Titterton and J. Weston, Strapdown Inertial Navigation Technology. (U.K.: The Institution of Electrical Engineers 2004)
9. Paul D. Groves Principles of GNSS, Inertial, and Multisensor Integrated Navigation Systems (London: Artech House 2008)
10. Jekeli C Precision free-inertial navigation with gravity compensation by an onboard gradiometer Journal of Guidance, Control and Dynamic. 29 704–13 (2006)
11. Chi-Tsong Chen. Linear System Theory Design (Oxford University Press, USA 1998)
12. Wellenhof B H and Moritz H Physical Geodesy 2nd edn (Graz: Springer 2005)
13. Xiao Z., Gongliu Y., Qingzhong C. and Jing W. A novel gravity compensation method for high precise free-ins based on extreme learing machine Sensors. 16 2019 (2016)
14. Ruonan W, Qiuping W, Fengtian H, Tianyi L, Peida H and Haixia L Gravity Compensation Using EGM2008 for High-Precision Long-Term Inertial Navigation Systems Sensors. 16 2177 (2016)
15. Jing W, Gongliu Y An Online Gravity Modeling Method Applied for High Precision Free-INS. Sensors 16 1541 (2016)
16. Jekeli C, Lee J K and Kwon J H 2007 On the computation and approximation of ultra-high-degree spherical harmonic series [J]. Journal of Geodesy, 81 603-615 (2007)
17. Wei Zi-Qing. Recurrence Relations for Fully Normalized Associated Legendre Functions and Their Derivatives and Integrals. Geomatics and Information Science of Wuhan University, 41 27-36 (2016)
18. Jiancheng Fang, LinZhouting Chen, and Jifeng Yao. An Accurate Gravity Compensation Method for High-Precision Airborne POS. IEEE Transactions on Geoscience and Remote Sensing, 52 (2014)
19. Dai D, Wang X, Zhan D and Huang Z. An improved method for dynamic measurement of deflections of the vertical based on the maintenance of attitude reference. Sensors, 14, 16322 (2014)
20. Siouris G M. Gravity modeling in aerospace applications. Aerospace Science and Technology, 13 301-315 (2009)
21. Gleason D. Critical Role of Gravity Compensation in a Stand-Alone Precision INS. DARPA PINS Meeting, Arligton, Virginia (2003)
22. Jekeli C. Gravity on Precise Short-Term, 3-D Free Inertial Navigation. Navigation, 44 347-357 (1997)

A Novel Information Fusion Method for Redundant Rotational Inertial Navigation Systems Based on Reduced-Order Kalman Filter

Lin Wang[1] , Wenqi Wu[1], Guo Wei[1], Jinlong Li[2] and Ruihang Yu[1]

[1]*National University of Defense Technology, 410073 Changsha, China*
[2]*92330 Troops of People's Liberation Army, Production Department, 266102 Qingdao, China*

Abstract. The redundant rotational inertial navigation systems can satisfy not only the high-accuracy but also the high-reliability demands of underwater vehicle on navigation system. However, different systems are usually independent, and lack of information fusion. A reduced-order Kalman filter is designed to fuse the navigation information output of redundant rotational navigation systems which usually include a dual-axis rotational inertial navigation system being master system and a single-axis rotational inertial navigation system being hot-backup system. The azimuth gyro drift of single-axis rotational inertial navigation system can be estimated by the designed filter, whereby the position error caused by that can be compensated with the aid of designed position error prediction model. As a result, the improved performance of single-axis rotational inertial navigation system can guarantee the position accuracy in the case of dual-axis system failure. Semi-physical simulation and experiment verify the effectiveness of the proposed method.

1 Introduction

The modern inertial navigation system (INS) should satisfy not only the high-accuracy but also the high-reliability demands of underwater vehicle during long-endurance sailing trial [1]. Rotational inertial navigation system (RINS) based on optical gyro has been proven to meet the high-accuracy demand of underwater vehicle [2]. In order to satisfy the high-reliability demand, redundant RINS configuration is usually adopted by underwater vehicle [3].

RINS can average out the influence of gyro drifts and accelerometer biases with the aid of gimbals [4-6]. The redundant RINS configuration including a dual-axis RINS and a single-axis RINS is a good scheme which makes trade-off among the accuracy, the reliability and the cost. The dual-axis RINS can average out all the gyro drifts and accelerometer biases, and the performance of it mainly depends on gyro angular random walk [4, 5]. Hence, the dual-axis RINS is with the higher accuracy, and usually as master INS in spite of its lower reliability and higher cost in comparison to single-axis RINS. On the other hand, the single-axis RINS has the advantages of the higher reliability and the lower cost regardless of its poorer performance in comparison to dual-axis RINS, which makes it an excellent backup system for dual-axis RINS [6]. However, information fusion between the two systems is usually ignored, and the systems generally run independently from one another.

Aimed at solving the above problem, an augmented error states Kalman filter is designed to estimate the inertial sensor deterministic errors of the redundant RINSs, whereby a prediction model is established to estimate the position error of single-axis RINS caused by its gyro drifts and accelerometer biases [7]. As a result, the position output from single-axis RINS after position error compensation is still at high level even if the dual-axis RINS fails. However, since all the inertial sensor deterministic errors except for the single-axis RINS azimuth gyro drift can be averaged out, it is not necessary to take them as error states, which will relax computation burden.

In this paper, without using external information, a reduced-order Kalman filter is designed for the redundant RINS configuration to estimate the azimuth gyro drift of single-axis RINS. And on this basis, a prediction model is established to predict the position error of single-axis RINS caused by azimuth gyro drift. After compensation for the position error, the single-axis RINS is comparable in position accuracy with the dual-axis RINS. Finally, semi-physical simulation and experiment are conducted to verify the effectiveness of the proposed method.

2 Reduced-order Kalman filter design

2.1. Error states dynamic model

An augmented error states Kalman filter is designed for redundant RINS configuration in [7], where the attitude error's difference, the velocity error's difference and the position error's difference between the single-axis RINS

1 and the dual-axis RINS 2 as well as the gyro drifts and accelerometer biases are taken as error states. The augmented error state vector is given by

$$\mathbf{x}(t) = [\boldsymbol{\phi}_{12}^n \quad \delta\mathbf{v}_{12}^n \quad \delta\mathbf{p}_{12}^n \quad \boldsymbol{\varepsilon}_1^{b_1} \quad \boldsymbol{\varepsilon}_2^{b_2} \quad \nabla_1^{b_1} \quad \nabla_2^{b_2}]^T \quad (1)$$

with

$$\boldsymbol{\phi}_{12}^n = \boldsymbol{\phi}_1^n - \boldsymbol{\phi}_2^n, \delta\mathbf{v}_{12}^n = \delta\mathbf{v}_1^n - \delta\mathbf{v}_2^n, \delta\mathbf{p}_{12}^n = \delta\mathbf{p}_1^n - \delta\mathbf{p}_2^n \quad (2)$$

$$\boldsymbol{\varepsilon}_1^{b_1} = [\varepsilon_x^{b_1} \quad \varepsilon_y^{b_1} \quad \varepsilon_z^{b_1}]^T, \boldsymbol{\varepsilon}_2^{b_2} = [\varepsilon_x^{b_2} \quad \varepsilon_y^{b_2} \quad \varepsilon_z^{b_2}]^T \quad (3)$$

$$\nabla_1^{b_1} = [\nabla_x^{b_1} \quad \nabla_y^{b_1}]^T, \nabla_2^{b_2} = [\nabla_x^{b_2} \quad \nabla_y^{b_2}]^T \quad (4)$$

where $\boldsymbol{\phi}_{12}^n$ denotes the difference between the attitude error of single-axis RINS 1, i.e., $\boldsymbol{\phi}_1^n$, and that of dual-axis RINS 2, i.e., $\boldsymbol{\phi}_2^n$; $\delta\mathbf{v}_{12}^n$ denotes the difference between the velocity error of single-axis RINS 1, i.e., $\delta\mathbf{v}_1^n$, and that of dual-axis RINS 2, i.e., $\delta\mathbf{v}_2^n$; $\delta\mathbf{p}_{12}^n$ denotes the difference between the position error of single-axis RINS 1, i.e., $\delta\mathbf{p}_1^n$, and that of dual-axis RINS 2, i.e., $\delta\mathbf{p}_2^n$; $\boldsymbol{\varepsilon}_1^{b_1}$ and $\boldsymbol{\varepsilon}_2^{b_2}$ respectively denote the gyro triad drift of single-axis RINS 1 and dual-axis RINS 2 ; $\nabla_1^{b_1}$ and $\nabla_2^{b_2}$ denote the accelerometer biases of the two systems. Note that the error states related with vertical channel are ignored, e.g., the vertical velocity error's difference, the altitude error's difference, and the vertical accelerometer biases. The position error are in the form of latitude error and longitude error. Corresponding error state vector dynamic model can be expressed by

$$\dot{\mathbf{x}}(t) = \mathbf{F}(t)\mathbf{x}(t) + \mathbf{G}(t)\mathbf{w}(t) \quad (5)$$

with

$$\mathbf{w}(t) = [w_{\varepsilon_x}^{b_1}, w_{\varepsilon_y}^{b_1}, w_{\varepsilon_z}^{b_1}, w_{\varepsilon_x}^{b_2}, w_{\varepsilon_y}^{b_2}, w_{\varepsilon_z}^{b_2}, w_{\nabla_x}^{b_1}, w_{\nabla_y}^{b_1}, w_{\nabla_x}^{b_2}, w_{\nabla_y}^{b_2}]^T \quad (6)$$

where $\mathbf{w}(t)$ is the process nosie, and the system matrix $\mathbf{F}(t)$ and process noise matrix $\mathbf{G}(t)$ are given in [1].

In fact, all the inertial sensor deterministic errors except for the azimuth gyro drift of single-axis RINS 1, i.e., $\varepsilon_z^{b_1}$, can be averaged out by rotating the inertial mearuement unit (IMU) of RINS periodically. It is not necessary to take them as error states, so the error state vector in Eq. (1) can be reduced to a reduced-order error state vector as

$$\mathbf{x}_r(t) = [\boldsymbol{\phi}_{12}^n \quad \delta\mathbf{v}_{12}^n \quad \delta\mathbf{p}_{12}^n \quad \varepsilon_z^{b_1}]^T = \mathbf{T}\mathbf{x}(t) \quad (7)$$

with

$$\mathbf{T} = \left[\begin{array}{c|c} \mathbf{I}_7 & \mathbf{0}_{7\times10} \\ \hline \mathbf{0}_{1\times7} & \mathbf{T}_{22} \end{array}\right], \mathbf{T}_{22} = [0 \quad 0 \quad 1 \mid \mathbf{0}_{1\times7}] \quad (8)$$

Corresponding dynamic model of reduced-order error state vector is given by

$$\dot{\mathbf{x}}_r(t) = \mathbf{F}_r(t)\mathbf{x}_r(t) + \mathbf{G}_r(t)\mathbf{w}(t) \quad (9)$$

with

$$\mathbf{F}_r(t) = \mathbf{T}\mathbf{F}(t)\mathbf{T}^T, \mathbf{G}_r(t) = \mathbf{T}\mathbf{G}(t) \quad (10)$$

where the components of $\mathbf{F}_r(t)$ and $\mathbf{G}_r(t)$ are given by

$$\mathbf{F}_r(t) = \mathbf{T}\mathbf{F}(t)\mathbf{T}^T = \left[\begin{array}{ccc|c} \mathbf{F}_1 & \mathbf{F}_2 & \mathbf{F}_3 & \mathbf{C}_{b_1}^n\mathbf{N} \\ \mathbf{F}_4 & \mathbf{F}_5 & \mathbf{F}_6 & \mathbf{0}_{2\times1} \\ \mathbf{0}_{2\times3} & \mathbf{F}_7 & \mathbf{F}_8 & \mathbf{0}_{2\times1} \\ \hline & \mathbf{0}_{1\times7} & & 0 \end{array}\right], \mathbf{N} = \begin{bmatrix} 0 \\ 0 \\ -1 \end{bmatrix} \quad (11)$$

$$\mathbf{F}_1 = \begin{bmatrix} 0 & \frac{v_E \tan L}{R_E + h} + \omega_{ie}\sin L & -\omega_{ie}\cos L - \frac{v_E}{R_E + h} \\ -\frac{v_E \tan L}{R_E + h} - \omega_{ie}\sin L & 0 & -\frac{v_N}{R_N + h} \\ \omega_{ie}\cos L + \frac{v_E}{R_E + h} & \frac{v_N}{R_N + h} & 0 \end{bmatrix}, \mathbf{F}_2 = \begin{bmatrix} 0 & -\frac{1}{R_N + h} \\ \frac{1}{R_E + h} & 0 \\ \frac{\tan L}{R_E + h} & 0 \end{bmatrix} \quad (12)$$

$$\mathbf{F}_3 = \begin{bmatrix} 0 & 0 \\ -\omega_{ie}\sin L & 0 \\ \omega_{ie}\cos L + \frac{v_E}{(R_E + h)\cos^2 L} & 0 \end{bmatrix} \quad (13)$$

$$\mathbf{F}_4 = \begin{bmatrix} 0 & -f_U & f_N \\ f_U & 0 & -f_E \end{bmatrix} \quad (14)$$

$$\mathbf{F}_5 = \begin{bmatrix} \frac{v_N \tan L}{R_E + h} & \frac{v_E \tan L}{R_E + h} + 2\omega_{ie}\sin L \\ -2\omega_{ie}\sin L - 2\frac{v_E \tan L}{R_E + h} & 0 \end{bmatrix} \quad (15)$$

$$\mathbf{F}_6 = \begin{bmatrix} 2\omega_{ie}v_N\cos L + \frac{v_N v_E}{(R_E + h)\cos^2 L} & 0 \\ -2\omega_{ie}v_E\cos L - \frac{v_E^2}{(R_E + h)\cos^2 L} & 0 \end{bmatrix} \quad (16)$$

$$\mathbf{F}_7 = \begin{bmatrix} 0 & \frac{1}{R_N + h} \\ \frac{1}{(R_E + h)\cos L} & 0 \end{bmatrix} \quad (17)$$

$$\mathbf{F}_8 = \begin{bmatrix} 0 & 0 \\ \frac{v_E \tan L}{(R_E + h)\cos L} & 0 \end{bmatrix}, \mathbf{M} = \begin{bmatrix} 1 & 0 & 0 \\ 0 & 1 & 0 \end{bmatrix}$$

$$\mathbf{G}_r(t) = \left[\begin{array}{cc|cc} -\mathbf{C}_{b_1}^n & \mathbf{C}_{b_2}^n & \multicolumn{2}{c}{\mathbf{0}_{3\times4}} \\ \hline \multicolumn{2}{c|}{\mathbf{0}_{2\times6}} & \mathbf{M}\mathbf{C}_{b_1}^n\mathbf{M}^T & -\mathbf{C}_{b_2}^n\mathbf{M}^T \\ \hline \multicolumn{4}{c}{\mathbf{0}_{3\times10}} \end{array}\right] \quad (18)$$

The navigation frame is local level geographic frame (East-North-Up). The body frames of single-axis RINS 1 and dual-axis RINS 2 are b_1 and b_2, respectively. v_E and v_N are true east velocity and north velocity of underwater vehicle, respectively. ω_{ie} is the earth rotation rate. L and h are true latitude and height. R_E and R_N are transverse radius of curvature and meridian radius curvature, respectively. f_E, f_N and f_U are respectively the east, the north and the up component of specific force.

2.2. Observation model

As shown in Figure 1, the velocity outputs and position outputs from single-axis RINS 1 and dual-axis RINS 2 meet the following relationships

$$\tilde{\mathbf{p}}_1^n = \mathbf{p}^n + \delta\mathbf{p}_1^n,\ \ \tilde{\mathbf{p}}_2^n = \mathbf{p}^n + \mathbf{p}_{12}^n + \delta\mathbf{p}_2^n \quad (19)$$

$$\tilde{\mathbf{v}}_1^n = \mathbf{v}^n + \delta\mathbf{v}_1^n,\ \ \tilde{\mathbf{v}}_2^n = \mathbf{v}^n + \boldsymbol{\omega}_{b_1b_2}^n \times \mathbf{p}_{12}^n + \delta\mathbf{v}_2^n \quad (20)$$

where $\tilde{\mathbf{p}}_1^n$ and $\tilde{\mathbf{p}}_2^n$ respectively denote the position output of single-axis RINS 1 and dual-axis RINS 2, $\tilde{\mathbf{v}}_1^n$ and $\tilde{\mathbf{v}}_2^n$ respectively denote the velocity output of them; $\mathbf{p}^n$ and $\mathbf{v}^n$ respectively denote the true position and true velocity of underwater vehicle; $\mathbf{p}_{12}^n$ denotes the lever-arm between the single-axis RINS 1 and dual-axis RINS 2, which can be accurately calibrated in advance. The operator 'δ' denotes the error perturbation of corresponding variables.

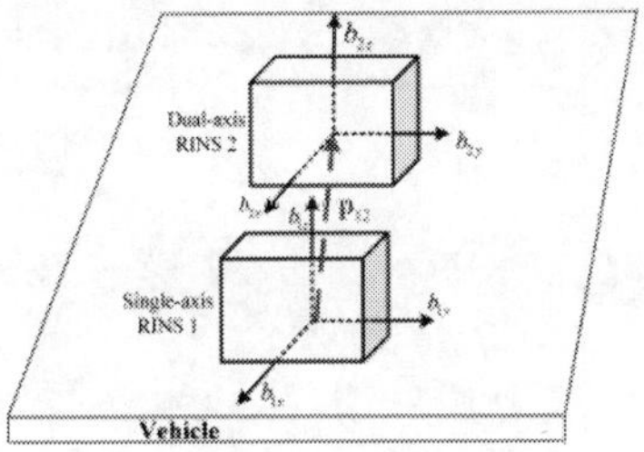

Figure 1. Redundant RINS configuration.

Subtracting the second equation of Eq. (19) and (20) from the first one yields

$$\delta\mathbf{p}_{12}^n = \delta\mathbf{p}_1^n - \delta\mathbf{p}_2^n = \tilde{\mathbf{p}}_1^n - \tilde{\mathbf{p}}_2^n \quad (21)$$

$$\delta\mathbf{v}_{12}^n = \delta\mathbf{v}_1^n - \delta\mathbf{v}_2^n = \tilde{\mathbf{v}}_1^n - \tilde{\mathbf{v}}_2^n \quad (22)$$

where the influence of lever-arm is ignored. Hence, the observation model of reduced-order error state vector is given by

$$\mathbf{z}(t) = \mathbf{H}_r(t)\mathbf{x}_r(t) + \boldsymbol{\upsilon}(t) \quad (23)$$

with

$$\mathbf{z}(t) = [\delta v_{E12}\ \ \delta v_{N12}\ \ \delta L_{12}\ \ \delta\lambda_{12}]^T$$
$$\mathbf{H}_r(t) = \begin{bmatrix} \mathbf{I}_2 & 0_{2\times2} & 0_{2\times3} \\ 0_{2\times2} & \mathbf{I}_2 & 0_{2\times3} \end{bmatrix} \quad (24)$$

where δv_{E12} and δv_{N12} are the east component and north component of $\delta\mathbf{v}_{12}^n$; δL_{12} and $\delta\lambda_{12}$ are the latitude error's difference and longitude error's difference between single-axis RINS 1 and dual-axis RINS 2; $\boldsymbol{\upsilon}(t)$ is observation noise.

2.3 Observability analysis

Since the system $(\mathbf{F}_r(t), \mathbf{H}_r(t))$ satisfies the definition of piece-wise constant system, the stripped observability matrix can be used to analyse its observability [8]. The stripped observability matrix is given by

$$\mathbf{O}(m) = \left[\mathbf{O}_1^T \,\vdots\, \mathbf{O}_2^T \,\vdots\, \cdots \,\vdots\, \mathbf{O}_m^T\right]^T \quad (25)$$

with

$$\mathbf{O}_j = \left[\mathbf{H}_r^T \,\vdots\, (\mathbf{H}_r\mathbf{F}_r)^T \,\vdots\, \cdots \,\vdots\, (\mathbf{H}_r\mathbf{F}_r^{7})^T\right]^T, j = 1,2,\ldots,m \quad (26)$$

where m is the segment.

The observability analysis is conducted in uniform linear motion scenario for simplicity. Substituting Eqs. (11) and (24) into Eq. (25), and calculating the rank of it yields

$$rank\mathbf{O}(1) = 8 \quad (27)$$

which means the reduced-order error state vector is always completely observable.

3 Information fusion method

As shown in Figure 2, with the aid of reduced-order Kalman filter, the azimuth gyro drift of single-axis RINS 1 can be estimated online, whereby the position error caused by it can be predicted. In normal condition, the dual-axis RINS 2 is usually as master INS and outputs position information. Moreover, once the dual-axis RINS 2 fails, the single-axis RINS 1 will keep on outputting position information after position error compensation. As a result, the performance of single-axis RINS 1 is comparable with that of dual-axis RINS in the case of dual-axis RINS 2 failure.

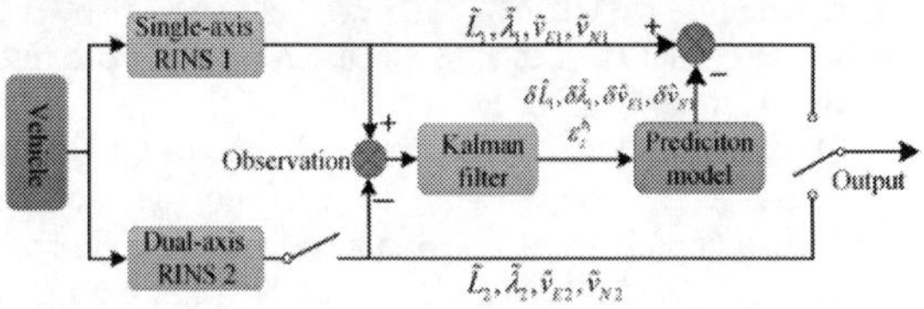

Figure 2. Information fusion method.

The error state vector of single-axis RINS 1 can be expressed as

$$\mathbf{x}_1(t) = [\boldsymbol{\phi}^n\ \ \delta\mathbf{v}_1^n\ \ \delta\mathbf{p}_1^n\ \ \varepsilon_z^{b_1}]^T \quad (28)$$

The dynamic model of above error state vector is given by

$$\dot{\mathbf{x}}_1(t) = \mathbf{A}_1(t)\mathbf{x}_1(t) + \mathbf{G}_1(t)\mathbf{w}_1(t) \quad (29)$$

with

$$\mathbf{A}_1(t) = \mathbf{F}_r(t), \mathbf{G}_1(t) = \mathbf{G}_r(t)$$
$$\mathbf{w}_1(t) = \left[w_{\varepsilon_x}^{b_1}\ \ w_{\varepsilon_y}^{b_1}\ \ w_{\varepsilon_z}^{b_1}\ \ w_{\varepsilon_x}^{b_2}\ \ w_{\varepsilon_y}^{b_2}\right]^T \quad (30)$$

Discretizing the above equation yields the position error prediction model of single-axis RINS 1 as

$$\hat{\mathbf{x}}_1(t_{k+1}) = \boldsymbol{\Phi}_1(t_{k+1}, t_k)\hat{\mathbf{x}}_1(t_k) \quad (31)$$

with

$$\boldsymbol{\Phi}_1(t_{k+1}, t_k) = e^{\mathbf{A}_1\Delta t}, \hat{\mathbf{x}}_1(t_0) = \mathbf{0} \quad (32)$$

where $\hat{\mathbf{x}}_1(t_k)$ is the error state vector prediction at time t_k, and its initial value is set as zero vector; $\boldsymbol{\Phi}_1(t_{k+1}, t_k)$ is the transition matrix; Δt is the discretization step size. Note that only the system error of single-axis RINS 1 caused by azimuth gyro drift can be predicted, and the prediction model does not work on random error caused by white noise.

With the position error prediction value subtracted from the position information output of single-axis RINS

1 as follows, the position accuracy of it can be improved to a certain extent

$$\bar{L}_1 = \tilde{L}_1 - \delta \hat{L}_1,\ \ \bar{\lambda}_1 = \tilde{\lambda}_1 - \delta \hat{\lambda}_1 \\ \bar{v}_{E1} = \tilde{v}_{E1} - \delta \hat{v}_{E1},\ \ \bar{v}_{N1} = \tilde{v}_{N1} - \delta \hat{v}_{N1} \tag{33}$$

where the superscripts '-' and '^' denote the correction output value and prediction value, respectively.

4 Experiment and discussion

4.1 Semi-physical simulation test

The single-axis RINS 1 adopts a rotation scheme which includes four dwell positions, and the dual-axis RINS 2 adopts a 16-phase sequence cycle rotation scheme [5, 6]. The actual inertial sensor noise data reflecting the sensor noise level is added to the simulation data. The gyro angular random walk is less than $0.001°/\sqrt{h}$, and the power spectral density of accelerometer noise is less than $20\mu g/\sqrt{Hz}$. The constant gyro triad drifts of single-axis RINS 1 are 0.003°/h, -0.002°/h and 0.0003°/h, and those of dual-axis RINS 2 are 0.004°/h, -0.005°/h and 0.003°/h. The horizontal accelerometer biases of single-axis RINS 1 are 20 μg and -40 μg, and those of dual-axis RINS 2 are 20 μg and -30 μg. The simulation time is 120 hours. The assumed dual-axis RINS 2 failure moment is the 16th hour. After that moment, the position output is provided by the single-axis RINS 1.

As shown in Figure 3, the azimuth gyro estimate error is within its 3σ bounds, and the estimate accuracy can achieve 0.00005°/h. Figure 4 depicts the predictions of latitude error and longitude error. As shown in the figure, the prediction model works well. Figure 5 shows the single-axis RINS' radial position error comparison before and after error prediction compensation. It is found that the position accuracy of single-axis RINS 1 after error compensation is comparable with dual-axis RINS 2.

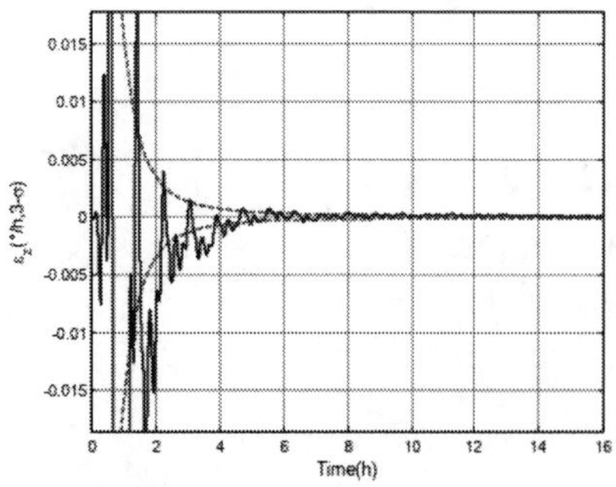

Figure 3. Azimuth gyro drift estimate error and 3σ bounds.

Figure 6 shows the radial position error curves of single-axis RINS, which are respectively corresponding to augmented error states Kalman filter and reduced-order error states Kalman filter. As shown in the figure, the information method based on reduced-order Kalman filter is comparable in performance with the information method based on augmented error states Kalman filter. Hence, even if the dual-axis RINS 2 fails, the position accuracy is still at high level.

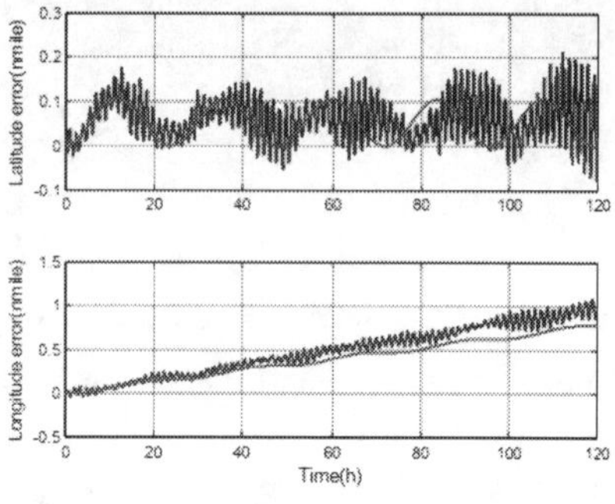

Figure 4. Position error prediction.

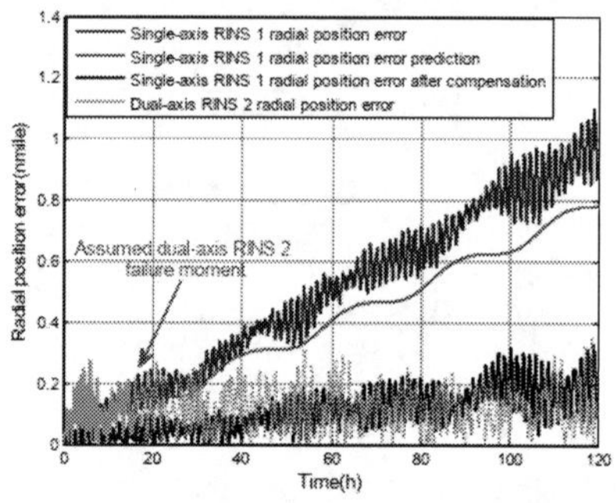

Figure 5. Comparison of position error.

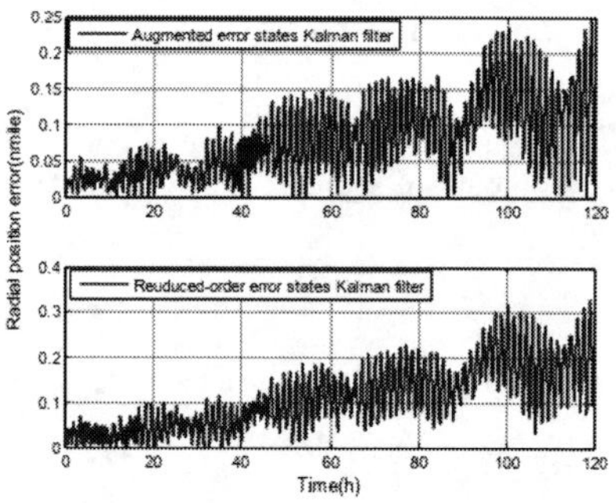

Figure 6. Position error comparison.

4.2 Lab test

Lab static test is conducted to verify the effectiveness of proposed method. After self-alignment, both the systems work in pure inertial navigation mode. The assumed dual-axis RINS 2 failure moment is the 24th hour. After that moment, the single-axis RINS 1 outputs position information with error compensation. Figure 7 depicts the predictions of latitude error and longitude error. As expected, the prediction model works well, especially on longitude error. Figure 8 depicts the normalized radial position error comparison of single-axis RINS. After position error compensation, its accuracy is improved to the same level as dual-axis RINS 2.

https://doi.org/10.1051/matecconf/201816007005

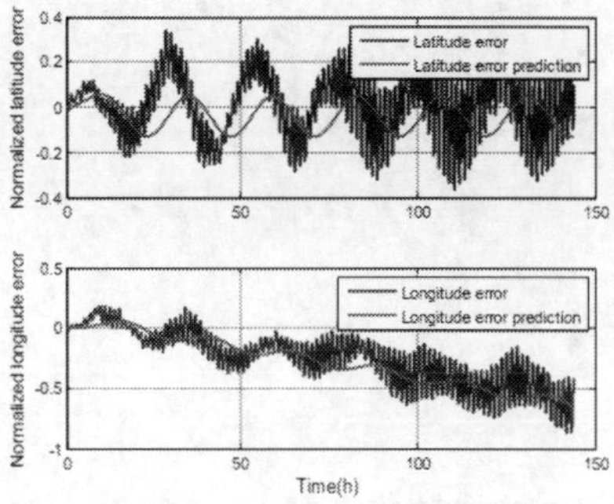

Figure 7. Position error prediction (Lab test).

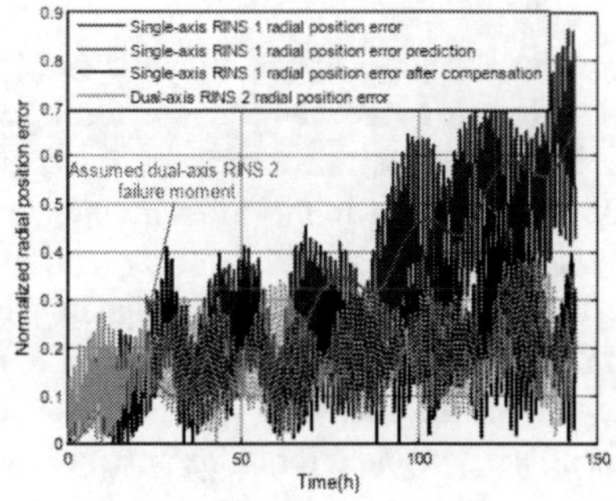

Figure 8. Comparison of position error (Lab test).

4.3 Sea trial

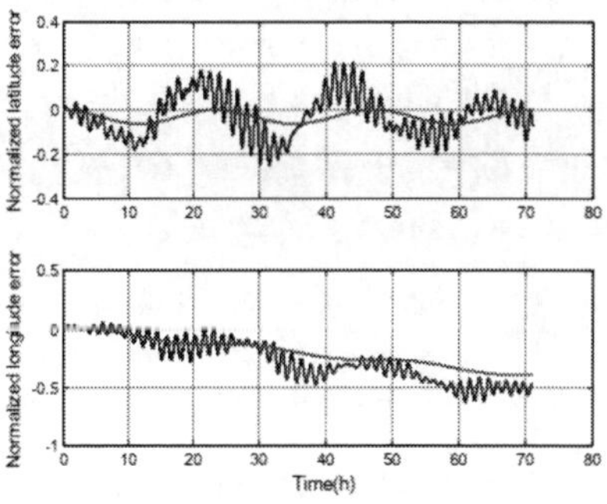

Figure 9. Position error prediction (Sea trial).

A sea trial is conducted to verify the effectiveness of the proposed method. The test time is about 72 hours. Both the systems work in pure inertial navigation mode. GPS position information is used as external reference. The assumed dual-axis RINS 2 failure moment is the 14th hour. Also, after that moment, the single-axis RINS 1 keeps on outputting position information after error compensation. As shown in Figure 9, the position error prediction model works well, especially on longitude error which builds up as time. As depicted in Figure 10, after position error compensation, the position accuracy of single-axis RINS 1 is improved greatly, which is comparable with that of dual-axis RINS 2.

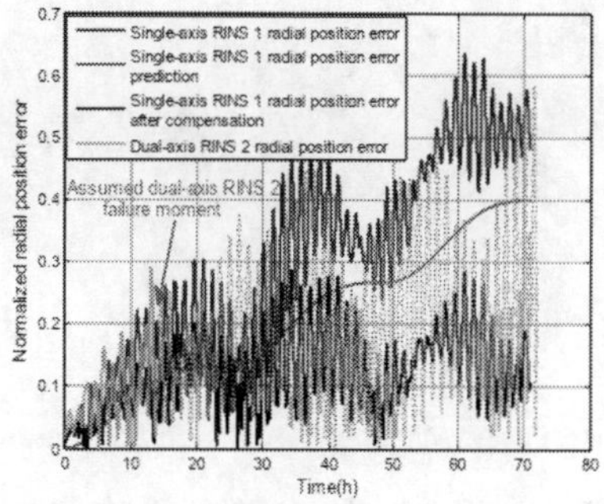

Figure 10. Comparison of position error (Sea trial).

5 Conclusion

A novel information fusion method for redundant rotational inertial navigation systems is proposed. A reduced-order Kalman filter is designed to estimate the azimuth gyro drift of single-axis RINS 1. The position error prediction model of single-axis RINS 1 is established to predict its error, and the corresponding position error compensation method is devised. The reduced-order Kalman filter relaxes the computation burden, and is comparable in performance with the augmented Kalman filter. The test results show that the performance of single-axis RINS 1 is improved to the same level as dual-axis RINS 2, which can still guarantee position accuracy in the case of dual-axis RINS 2 failure.

References

1. D.H. Titterton, J.L. Weston. *Strapdown inertial navigation technology* (IET, 2004)
2. E. Levinson, C.S. Giovanni. Laser gyro potential for long endurance marine navigation, *IEEE Position Location and Navigation Symposium*, **115** (1980)
3. E. Levinson, J. Horst, M. Willcocks. The next generation marine inertial navigator is here now, *IEEE Position Location and Navigation Symposium* (1994)
4. E. Levinson, R. Majure. Accuracy enhancement techniques applied to the marine ring laser inertial navigator (MARLIN), Navigation **34** (1987)
5. B. L. Yuan, D. Liao, S. L. Han. Error compensation of an optical gyro INS by multi-axis rotation. Meas. Sci. Technol. **23** (2012)
6. T. Tucker, E. Levinson. The AN/WSN-7B marine gyroscompass/navigator, *Prodeedings of the 2000 National Technical Meeting of the Institute of Navigation* (2000)
7. L. Wang, W. Q. Wu, G. Wei, X. F. Pan, C. F. Gao. Dual marine INS joint rotation and modulation for cooperative positioning and error parameter estimation, J. Chin. Inertial Technol. **25** (2017)
8. I. Y. Bar-Itzhack, D. Goshen-Meskin. Observability analysis of piece-wise constant systems-part I: theory. IEEE Trans. on Aerosp. Electron. Syst. **28** (1992)

Error analysis in gravity aided navigation

Wei Hongwei[1], Wu Meiping[1], Tie Junbo[1] and Wang Shenquan[2]

[1]College of Mechatronics Engineering and Automation, National University of Defense Technology, Changsha 410073, China,

[2]Manned Space System Research Center，Beijing100094，China,

Abstract. For underwater vehicles, because GPS signals are not always available, it is necessary to use inertial navigation system (INS) for navigation. In order to improve the navigation performance, the gravity value calculated by the gravity field model is introduced into the navigation system to reduce the influence of gravity on the navigation accuracy. However, due to the difference between the actual gravity field and the gravity field model, the navigation error caused by gravity field still exists. In this paper, this error is studied based on the theoretical analysis and simulation. The results show that the accuracy of gravity-aided navigation will decrease sharply when the position error of the underwater vehicles reach 9.6 n miles, because at this moment, the calculation error of vertical deflection is the largest. In the simulation experiment, the position error of the gravity-aided navigation system reach 9.6 n miles in 44 hours. Therefore, for the underwater vehicles navigation system, it is recommended that the vehicles should be corrected once every 44 hours.

1 Introduction

As a basic physical field, gravity field restricts all physical events of the earth itself and its adjacent space [1]. The inertial navigation system is also affected by the earth gravity field, and the navigation accuracy of inertial navigation system is closely related to the value of gravity.

As the accuracy of inertial transducers increase, the influence of gravity becomes the main factor limiting the precision of high accuracy inertial navigation system [2]. Therefore, in inertial navigation system, the influence of gravity field need to be eliminated through reasonable compensation method. In the low-precision inertial navigation system, Equation (1) is often used to calculate the value of gravity, and the influence of gravity field on inertial navigation system is compensated using this value [3].

$$\gamma = \frac{\gamma_0}{\left(1 + h / R_0\right)^2} \tag{1}$$

in which:

$$\gamma_0 = 9.780318\left(1 + 5.3024\times10^{-3}\sin^2 L - 5.9\times10^{-6}\sin^2 2L\right)$$
$$R_0 = \sqrt{R_M R_N} \tag{2}$$

The radius of curvature in meridian R_M and the radius of curvature in prime vertical R_N can be calculated by the following equations:

$$\begin{cases} R_M = R\left(1-e^2\right)/\left(1-e^2\sin^2 L\right)^{3/2} \\ R_N = R/\left(1-e^2\sin^2 L\right)^{1/2} \end{cases} \tag{3}$$

where R is semi-major axis of the reference spheroid; e represents the major eccentricity of the reference spheroid; L is latitude of the inertial navigation system and h is the height above ground.

In the above-mentioned compensation method, it is considered that the direction of the gravity vector coincides with the direction of the reference ellipsoid normal. In fact, the reference ellipsoid norm is not consistent with the direction of the geoid normal, so this compensation method cannot completely eliminate the influence of the gravity field on the inertial navigation system.

The angle between the reference ellipsoid norm and the geoid normal is called the vertical deflection. The results of JIN Ji-hang's research indicate that the position error of the inertial navigation system caused by vertical deflection is about 3 km on the set route [4].

In order to improve the accuracy of the gravity compensation method, Zhao Zhong et al. analyzed the effect of vertical deflection on high accuracy inertial navigation system [5]. Tie Jun-bo et al. improved the method of calculating gravity. He introduced EGM2008 into the inertial navigation system[6]. EGM2008 is Earth's gravitational model with very high resolution, which has been released by the Naional Geospatial-Intelligence Ageny (NGA) Development Team [7]. EGM2008 is complete to spherical harmonic degree and order 2159, and it contains additional coefficients extending to degree 2190 and order 2159.

Based on EGM2008, the horizontal gravity disturbance can be calculated. Introduce the horizontal gravity disturbance into the inertial navigation system, and then the accuracy of inertial navigation positioning can be increased by 120m/4h [6].

The compensation method that using vertical deflection, which is achieved from the horizontal gravity disturbance that calculated using EGM2008, to assist inertial navigation can effectively reduce the error of the navigation system and improve the navigation accuracy. Nevertheless, this compensation method does not eliminate the errors caused by gravity field. In this paper, we will study the error in this compensation method, and analyse the distribution of vertical deflection based on EGM2008. Finally, based on the simulation experiment, the optimal time for the underwater vehicle to be corrected is given.

2 Calculation of vertical deflection

In order to study the distribution of vertical deflection, the value of vertical deflection should be calculated based on EGM2008. The calculation steps are shown in figure 1: First, calculate the horizontal gravity disturbance with EGM2008, and then the vertical deflection is calculated by the relationship between horizontal gravity disturbance and vertical deflection.

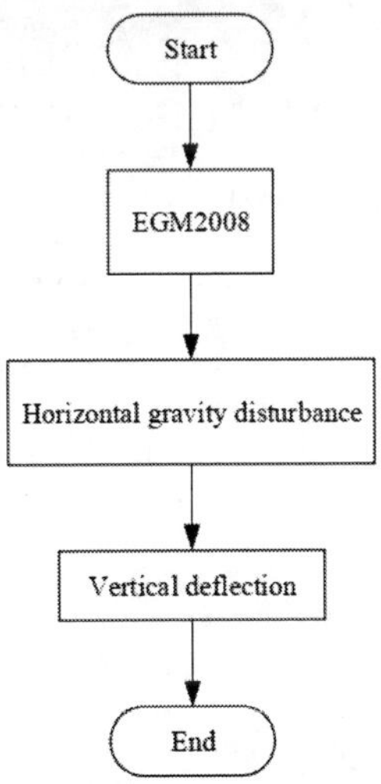

Figure 1. Calculation steps of vertical deflection.

2.1 Relationship between horizontal gravity disturbance and vertical deflection

The prime vertical component of vertical deflection is denoted by η (rad), and the meridional component of vertical deflection is denoted as ε (rad). The meridional component and prime vertical component of gravity disturbances are denoted by δg_N and δg_E, respectively.

There is a relationship between them as follows[6]:

$$\begin{aligned} \delta g_N &= -\gamma \cdot \eta \\ \delta g_E &= -\gamma \cdot \varepsilon \end{aligned} \tag{4}$$

where γ is the normal gravity, which can be calculated by equation (1).

2.2 Horizontal gravity disturbance

Horizontal gravity disturbances can be calculated, using EGM2008 as follows:

The perturbed potential of gravity is denoted by T, since it is a harmonic function. It can be expressed by the following equation:

$$T = \frac{GM}{r}\sum_{n=2}^{N}\left(\frac{a}{r}\right)^n \sum_{m=0}^{n}\left(\bar{C}^*_{nm}\cos m\lambda + \bar{S}_{nm}\sin m\lambda\right)\bar{P}_{nm}(\cos\theta) \tag{5}$$

in which GM and a are fundamental parameters of WGS-84, the values of them are shown in Table 1. $\bar{C}^*_{nm}$ and $\bar{S}_{nm}$ are spherical harmonic coefficients corresponding to the fully normalized Legendre functions, which are provided by EGM2008.

Table 1. Fundamental parameter of WGS-84.

Parameter	Value
a	6378136.3
f	1/298.256415099
GM	$0.3986004415\times10^{20}$
ω_{ie}	7292115×10^{-11}

The relationship between the perturbed potential of gravity and the horizontal gravity disturbance is as follows [8-9]:

$$\begin{aligned} \delta g_N &= -\frac{1}{\rho}\frac{\partial T}{\partial\theta} \\ \delta g_E &= -\frac{1}{\rho\cdot\sin\theta}\frac{\partial T}{\partial\lambda} \end{aligned} \tag{6}$$

in which δg_N and δg_E are northward and eastward horizontal gravity disturbance, respectively.

Substituting equation (5) into equation (6) yields:

$$\begin{aligned} \delta g_N &= -\frac{GM}{r^2}\sum_{n=2}^{\infty}\left(\frac{a}{r}\right)^n\left(\bar{C}^*_{nm}\cos m\lambda + \bar{S}_{nm}\sin m\lambda\right)\frac{d\bar{P}_{nm}(\cos\theta)}{d\theta} \\ \delta g_E &= -\frac{GM}{r^2\sin\theta}\sum_{n=2}^{\infty}\left(\frac{a}{r}\right)^n\sum_{m=0}^{n} m\left(-\bar{C}^*_{nm}\sin m\lambda + \bar{S}_{nm}\cos m\lambda\right)\bar{P}_{nm}(\cos\theta) \end{aligned} \tag{7}$$

Then, the horizontal gravity disturbance can be calculated by equation(7).

3 Analysis of vertical deflection

According to the above methods, the vertical deflection in the range of $10°N-11°N$, $110°E-112°E$ is shown in figure 2-3.

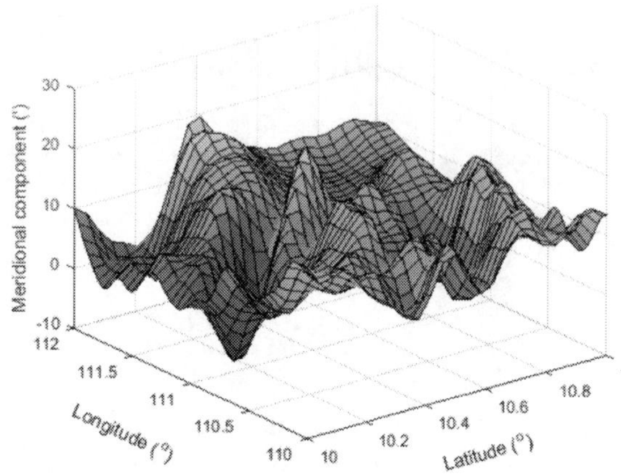

Figure 2. Meridional component of the vertical deflections in arc-seconds

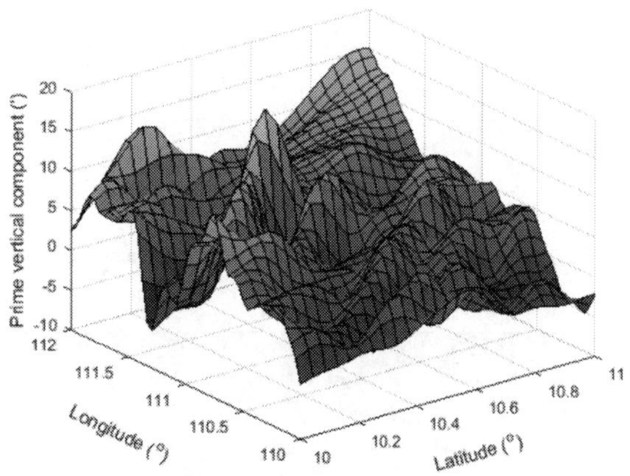

Figure 3. Prime vertical component of the vertical deflections in arc-seconds

Then the vertical deflections in a survey line is analyzed. The range of the survey line is $111°E$, $10°N-11°N$. The vertical deflection of this survey line is shown in figure 4.

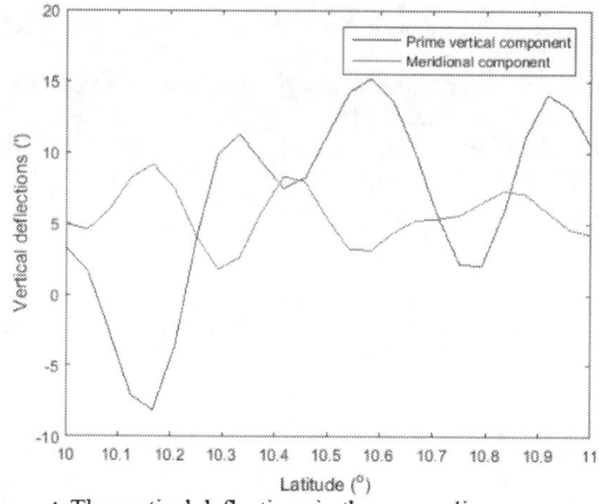

Figure 4. The vertical deflections in the survey line

As shown in figure 4, the blue line represents the prime vertical component in the survey line, and the red line represents the meridional component in the survey line. It can be seen that the periodicity of the two components of the vertical deflection is about 0.32 degrees.

For the survey line of $10.5°N$ and $110°E-112°E$, the conclusion above also holds.

Therefore, it can be concluded that for gravity aided inertial navigation, if the position error reaches 9.6 n miles, then the error of the calculated vertical deflection is maximized. That is to say, if continue using gravity aided navigation, it will lead to greater error. In this case, the gravity aided navigation system needs to be corrected.

4 Simulation

Based on the simulation experiment, the precision of gravity aided navigation is studied in this section. Because of the lack of inertial navigation data of underwater vehicles, a high-precision strapdown inertial navigation system for ship is used to replace the navigation system for underwater vehicle.

In the following simulation experiment, the data of a shipboard laser strapdown inertial navigation system is used as the raw data for simulation; the position calculated by GPS precision single point positioning is used as an accurate value for reference.

Simulation lasted 48 hours, first, the raw data was used to calculate the position of the carrier in the inertial navigation system. In the calculation, it is assumed that the projection of gravity vector in the X axis and Y axis direction of navigation coordinate system is zero. And the projection of gravity vector in Z axis direction is calculated by equation (1) and equation (2). The position error of the inertial navigation system is obtained by comparing the position calculated by the GPS with the position information of the inertial navigation system.

Next, based on the model of earth gravity field EGM2008, the value of vertical deflection was calculated. And then the value of vertical deflection was used to compensate for the influence of horizontal components in the gravity field on the inertial navigation system. By this compensation, the gravity field used in the inertial navigation system will be closer to the real gravity field, and the error will be smaller. Such inertial navigation system is known as gravity assisted inertial navigation system. In the same way, the position obtained by the gravity-assisted inertial navigation system is compared with the position calculated by the GPS to obtain the position error of the gravity-assisted inertial navigation system.

The navigation position errors that using gravity aided navigation and without gravity aided navigation are shown in figure 5. The blue line indicates the position error of the navigation system using gravity aided, and the red line indicates the position error of the navigation system without gravity aided. It can be seen that the position error that using gravity aided navigation is less than the error that without gravity aided navigation.

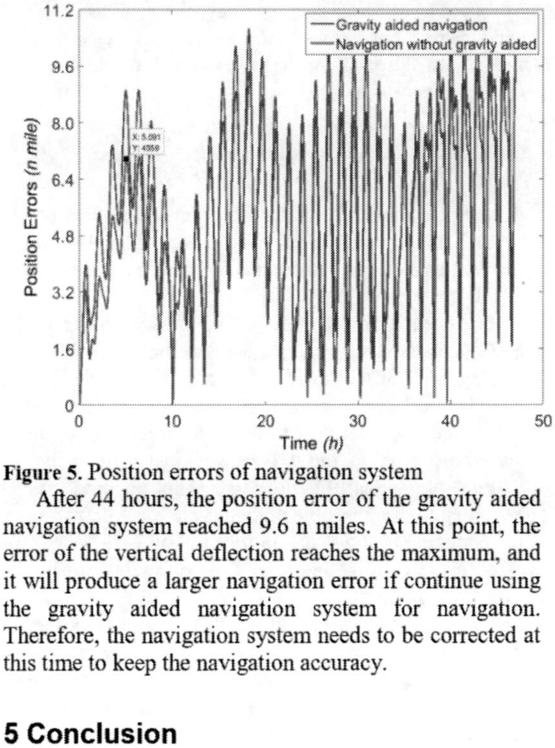

Figure 5. Position errors of navigation system

After 44 hours, the position error of the gravity aided navigation system reached 9.6 n miles. At this point, the error of the vertical deflection reaches the maximum, and it will produce a larger navigation error if continue using the gravity aided navigation system for navigation. Therefore, the navigation system needs to be corrected at this time to keep the navigation accuracy.

5 Conclusion

In this paper, the errors in gravity assisted inertial navigation system for underwater vehicle is analysed. Furthermore, the errors caused by using the gravity field model EGM2008 to calculate the vertical deviation and its influence on inertial navigation system are studied.

Through the above analysis and further simulation experiments, we can get the conclusion that when the position error reaches 9.6 n miles, the error in calculation of gravity reaches the largest. Therefore, the gravity assisted navigation system should be corrected when the position error reaches 9.6 n miles.

According to the results of this paper, we can calculate the optimal time to calibrate the gravity assisted inertial navigation system based on the actual performance of the underwater vehicle. In this way, we can make the underwater vehicle voyage as long as possible under the condition that the navigation accuracy is guaranteed.

With the conclusions of this paper, the navigation accuracy of underwater vehicle can be improved and the rapid growth of the navigation error can be effectively suppressed.

Acknowledgements

We would like to express our gratitude to all those who helped us during the writing of this paper. And this paper is supported by the National Key R&D Program of China (2017YFC0601701).

References

1. K.D. Zhang, *Research on the Methods of Airborne Gravimetry Based on SINS/DGPS*. Changsha, Hunan, P.R.China, National University of Defense Technology, (2007)
2. CSIT. Overview of present state and trends for inertial technology. *Report on Advances in Inertial Technology*, Beijing, 12-15 (2010)
3. D.H. Titterton, *Strapdown Inertial Navigation Technology(2nd edition)*: American Institute of Aeronautics and Astronautics,Inc. (2004)
4. J.H Jin, S.F. Bian, Analysis of inertial navigation system positioning error caused by gravity disturbance. *Geomatics and Information Science of Wuhan University*, **35**(1): 30-32 (2010)
5. Z. Zhao, P. Wang, Analysis and compensation of vertical deflection effect on high accuracy inertial navigation system. *Journal of Chinese Inertial Technology*, **23**(6): 701-705 (2013)
6. J.B. Tie, M.P. Wu, S.K. Cai, K.D. Zhang, J.X. Lian, Gravity disturbance compensation method based on Earth Gravitational Model 2008. *Journal of Chinese Inertial Technology*. (to be published)
7. Nikolaos K. Pavlis, Simon A. Holmes, Steve C. Kenyon, John K. Factor; The development and evaluation of the Earth Gravitational Model 2008 (EGM2008), *Journal of Geophysical Research: Solid Earth (1978-2012)* Volume **117**, Issue B4, April (2012)
8. Rapp, R. H. A FORTRAN program for the computation of gravimetric quantities from high degree spherical harmonic expansions. Report No. **334**. Ohio State University, Dept. of Geodetic Science and Surveying, Columbus, Ohio (1982).
9. Colombo, O. L. Numerical methods for harmonic analysis on the sphere. Report No. **310**. Dept. of Geodetic Science, The Ohio State University, Columbus, Ohio (1981)

Classification of Motor Imagery EEG Based on Sparsification and Non-negative Matrix Factorization

Jing Su , Zuyuan Yang, Haiping Wang and Wei Han

School of Automation, Guangdong University of Technology

Abstract. The analysis of EEG is a hot topic in the area of biomedical signal processing. In this paper, the EEG signals with Mu (μ) rhythm and Beta (β) rhythm are used to solve the motor imagery problem, i.e., the imagery of the left hand and the right hand. The collected raw data is first filtered by FIR band-pass filter, followed by using the maximization of feature difference to increase the sparsity of the matrix. Then, to reduce the redundant information of these features, a non-negative matrix factorization (NMF) method is employed. Due to the usage of the NMF scheme, the obtained factorizations has been better class property. Simulations show that our method achieves higher classification accuracy (more than 91%) than existing results (about 86%).

1 Introduction

Biomedical signal is a weaker low frequency signal in the background of strong noise. It is a natural unstable signal which is emitted by complex living organisms. From the characteristics of the signal, it's detection methods and processing technology are different from the general signal. From the nature of electricity, it can be divided into electrical signals and non-electrical signals, such as ECG, EMG, EEG and other electrical signals. Body temperature, blood pressure, respiration, blood flow, pulse, and heart sounds belong to non-electrical signals[1]. EEG is a kind of bio-electric activity with rhythm and spontaneity. It is closely linked with life. Once the organism dies, the electricity phenomenon disappears[2].

In the practical application of brain-computer interface, there are three kinds of experimental models: motion imagination, steady-state visual evoked(SSV-EP)[3] and P300[4], which SSVEP and P300 are based on the induced pattern of EEG. The essence of the movement imagination of EEG is a spontaneous EEG, in this dynamic state can not feel the muscle movement. Some studies have shown that the area of sensory motor cortex activated by motor imagination is the same as the cortex area of sensory motor brain activated by actual physical exercise [5]. If one does not actively participate in motor activities, feel the processing or imagine such movements and processing[6], one will show signify-cant oscillations in the sensory area of the 8Hz-12Hz and 18Hz-25Hz frequency band EEG. This oscillation, commonly referred to as Mu (μ) rhythms (8Hz-12Hz) and Beta(β) rhythms (18Hz-25Hz), that is generated by the circuit of the thalamic cortex [7]. In the exercise of imagination, ipsilateral hemispheric μ rhythm and β rhythm are inhibited, while the countermeasures of half-brain μ rhythm and β rhythm increases. This phenomenon is event-related desynchronization (ERD) or event-related synchronization (ERS) [8,9].

In recent years, feature extraction methods based on EEG signals have been rapidly developed. These studies mainly use different feature extraction techniques to evaluate different classifiers. Recently, Zhao et al. calculated the energy difference between the μ rhythm and the β rhythm through power spectrum analysis to extract the signal features[10]. This method has the advantages of clear objective and simple calculation and is conducive to the realization of online BCI system. Sun et al. presents a method of online classification for BCI based on common spatial pattern(CSP) for feature extraction and support vector machine as a classifier[11]. Zhang et al. proposed a semi-supervised feature learning method based on deep stack network [12]. This method combines unsupervised learning of restricted boltzmann machine with gradient descent algorithm based on batch mode, which need two stages of pre-training and fine-tuning. However, these algorithms all have some limitations, such as poor classification accuracy or need labels as supervisory information.

Recently, non-negative matrix factorization(NMF) is commonly used to classify EEG signals. Lee et al. exploited non-negative matrix factorization to select discriminative features in the time-frequency representa-tion of EEG[13]. The non-negative matrix factorization method was used to extract attention-related EEG features[14]. In this paper, we propose an unsupervised feature extraction method called sparsification and non-negative matrix factorization. The central idea of the method is that since the characteristics of motor imaging EEG focus on the μ rhythm and β rhythm, the signal is first filtered by a 8-25 Hz band-pass filter and then the sparsity of the matrix, increases by maximizing the feature difference, extracting distinguishable features with matrix decomposition to extract, and finally clustering these features. The main advantage of this paper is that non-negative matrix factorization is an unsupervised learning process that is widely used and does not require known signals as oversight. After the feature difference is maximized, the sparsity of the matrix is increased, the computational complexity of non-negative matrix factorization is reduced, and the efficiency of the algorithm and the accuracy of the classification are improved. Experiments show that we method based on non-negative matrix decomposition to learn the discriminable basis vectors, which is very suitable for offline analysis of motor imaging EEG in many experiments.

The structure of this paper is as follows. In the second part, we introduces how to increase the matrix sparsity, and the background of non-negative matrix factorization, and explain the main idea of the algorithm. The real data experiment results are shown in Section 3, and the conclusions are presented in Section 4.

2 Method

2.1 Sparsification

In the experiment, we distinguish the left-right movement imagination by ERD and ERS. In order to improve the efficiency and classification accuracy of the algorithm, we first increase the signal sparsity by maximization of feature difference.

Suppose the k -th experimentally measured EEG signal is E_k ,

$$E_k = \begin{bmatrix} x_{1k}(1) & x_{1k}(2) & \dots & x_{1k}(t) \\ x_{2k}(1) & x_{2k}(2) & \dots & x_{2k}(t) \\ \vdots & \vdots & \ddots & \vdots \\ x_{nk}(1) & x_{nk}(2) & \dots & x_{nk}(t) \end{bmatrix} \tag{1}$$

It consists of the signal with n channels, the length of time is t , and the sparsification process is as follows:

1. Find the covariance matrix of left and right hand motion imagination respectively C_l and C_r [15]:

$$C_l = \sum_{K \in I_l} E_k^T E_k \qquad C_r = \sum_{K \in I_r} E_k^T E_k \tag{2}$$

If the k -th example is a left-hand imagination movement, $k \in I_l$, otherwise , $k \in I_r$.

2. The eigenvalue decomposition of the covariance matrix is represented as

$$C = C_l + C_r = O_c \Sigma O_c^T \tag{3}$$

where Σ is the eigenvalue diagonal matrix, O_c is the corresponding eigenvector matrix.

3. Construct a whitening matrix, convert the covariance matrix C_l and C_l do the corresponding eigenvalued ecomposition, described by

$$P = \sqrt{\Sigma^{-1}} O_c^T \tag{4}$$

$$\begin{aligned} S_l &= P C_l P^T = O \Sigma_l O^T \\ S_r &= P C_r P^T = O \Sigma_r O^T \end{aligned} \tag{5}$$

According to the principle of simultaneous diagonaliza-tion of the matrix, S_l and S_r have the same eigenvector, it's O , and their sum of corresponding eigenvalues is 1 , which is:

$$S_l + S_r = I \tag{6}$$

where, I is the identity matrix. And so, $M = O^T P$ is the space filter that we want. Since the sum of the eigenvalues is 1, if S_l has the maximum eigenvalue, then the corresponding S_r has the smallest eigenvalue, and vice versa. So distinguishing between these two sets of data can be implemented with the feature vector M . The first and last eigenvectors of the EEG projection into M will give the least squares meaning to distinguish the best features of the two types of EEG. According to the projection matrix, single trial EEG experiment can be decomposed into:

$$V = ME \tag{7}$$

In this way, the characteristics used for classification can be obtained from the V variance of the first and last line.

For example, suppose there are two matrices E_1, E_2 :

$$E_1 = \begin{pmatrix} 1.5 & 1.3 \\ 1.7 & 1.6 \\ 2.0 & 2.0 \\ 1.9 & 1.5 \\ 1.6 & 1.4 \end{pmatrix} \qquad E_2 = \begin{pmatrix} 1.4 & 1 \\ 1.8 & 1 \\ 1.3 & 1 \\ 1.6 & 2 \\ 1.8 & 1 \end{pmatrix}$$

After the sparsification, the projection matrix is obtained

$$|M| = \begin{pmatrix} 5.8402 & 5.0347 \\ 5.1576 & 5.7689 \end{pmatrix}$$

Since the non-negative matrix ensures that all data is non-negative, so add absolute value here. Therefore, the matrix E_1, E_2 after the maximization of feature difference are shown as V_1, V_2 :

$$V_1 = \begin{pmatrix} 2.0554 & 0.0525 \\ 1.6726 & 0.6712 \\ 1.3652 & 1.4684 \\ 3.3600 & 0.9126 \\ 2.1237 & 0.0209 \end{pmatrix} \qquad V_2 = \begin{pmatrix} 0.0759 & 2.1 \\ 1.2287 & 1.3 \\ 0.3716 & 1. \\ 0.9709 & 3.4 \\ 0.7129 & 1.8 \end{pmatrix}$$

It can be seen that, after the maximization of feature difference, the sparsity of the matrix is well increased.

After sparsification of motor imagery EEG signals, the signals of different channels in different experiments were respectively classified into different matrices, and the features were extracted by NMF. Finally we conducted clustering analysis of the features of matrix.

2.2 NMF

Non-negative matrix factorization(NMF) has been introduced as a matrix factorization tool that produces a useful decomposition in the dimensionality reduction of data set. The non-negative matrix factorization can be described as follows:

$$V \approx W \times H \tag{8}$$

$V \in R^{m \times n}$ consists of column vectors each of which corresponds to a different sample. Find out basis matrix $W \in R^{m \times r}$ and weight matrix $H \in R^{r \times n}$ .Each column vector in V has a corresponding representation vector in the H . Each matrix can be represented as a composition of multiple vectors, $V = [v_1\ v_2\ v_3 \ldots v_n]$ and $H = [h_1\ h_2\ h_3 \ldots h_n]$.

We consider the squared Euclidean distance as a discrepancy measure between the V and the $W \times H$, leading to the following cost function:

$$D(V\|WH) = \|V - WH\|^2 = \left(\sum_{i=1}^{m} \sum_{j=1}^{n} V_{ij} - [WH]_{ij} \right)^2 \tag{9}$$

It is shown that the cost function defined by Eq. (9) is a convex function of W and H separately. Therefore, multiplicative and alternating update rules for H and W are derived by gradient decent optimization [16]. The update rules is used to gradient descent algorithm. It is proven by LEE and SEUNG[16] that convergence is guaranteed when step size is chosen as η_H for the update H_{aj} and chosen as η_W are given by Eq.(10).

$$\eta_H = \frac{H_{aj}}{(W^T WH)_{aj}}, \quad \eta_w = \frac{W_{aj}}{(WHH^T)_{aj}} \tag{10}$$

The update rule of matrix H is given by Eq.(11).

$$H_{aj} \leftarrow H_{aj} \frac{(W^T V)_{aj}}{(W^T WH)_{aj}} \tag{11}$$

The update rule of matrix W is given by Eq.(12).

$$W_{ia} \leftarrow W_{ia} \frac{(VH^T)_{ia}}{(VHH^T)_{ia}} \tag{12}$$

Our signal classification process is divided into the following steps:

Step 1: Filtering the original signal X through an 8-25 band-pass filter to obtain the signal of the frequency band related to the characteristic, denoted as E .

Step 2: Sparsification of E by Eqs. (1)-(7), and the matrix V after sparsification was obtained.

Step 3: Divided the V into two matrices, denoted as V_{C3} and V_{C4} .The basis matrix (W_{C3} and W_{C4}) and the weight matrix (H_{C3} and H_{C4}) of the signal are obtained by Eqs. (11) and (12).

Step 4: Clustering of H_{C3} and H_{C4}.

3 Experimental results

In this section, we investigate the use of our proposed NMF algorithm for data clustering. Several experiments are carried out to show the effectiveness of our algorithm for motor imaging EEG classification. For our empirical study, we used one of BCI competition 2003 data sets, which was provided by the department of medical informatics, institute for biomedical engineering, Graz university of technology, Austria [17]. The data set involves left-right imagery hand movements and consists of 140 labeled trials for training and 140 unlabeled trials for test. Each trial has a duration of 9 seconds, where a visual cue (arrow) is presented pointing to the left or the right after 3-second preparation period and imagination task is carried out for 6 seconds. It contains EEG acquired from three different channels (with sampling frequency 128 Hz) C3, Cz and C4. In our study we use only two channels, C3 and C4, because ERD has contralateral dominance and Cz channel contains little information for discriminant analysis.

3.1 Performance evaluation and comparisons

In this paper, we use a sparsification measure based on the relationship between the C3 and the C4 channel. We compare the motor imaging EEG with the maximization of feature difference to those without the maximization of feature difference, as shown in the Fig.1.

Divide the data with the maximization of feature difference into two matrices, V_{C3} contains the signal of the C3 channel of the left and right hand motion imaging signal, V_{C4} contains the signal for C4 channel , V_{C3} and V_{C4}, respectively, through the NMF feature extraction and classification, inferred accuracy of their classification.

The test set in the dataset was extracted and classified separately by CSP and NMF, and compared with the method we proposed, as shown in Table 1.

Table 1. Classification accuracy.

Method Accuracy	Left-Hand	Right-Hand	Average
CSP	85.71%	82.86%	84.29%
NMF	87.14%	85.71%	86.42%
S+NMF	94.28%	88.57%	91.43%

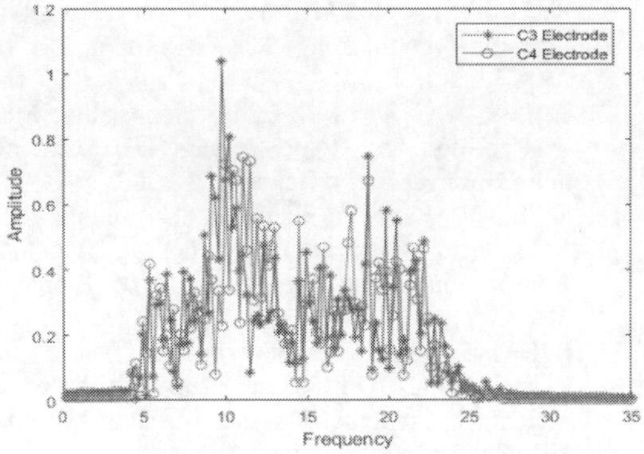

(a) Motor imagery on left-hand

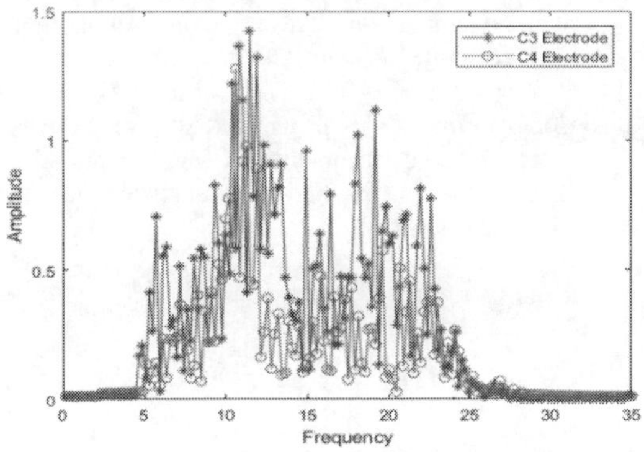

(b) Motor imagery on left-hand after sparsification

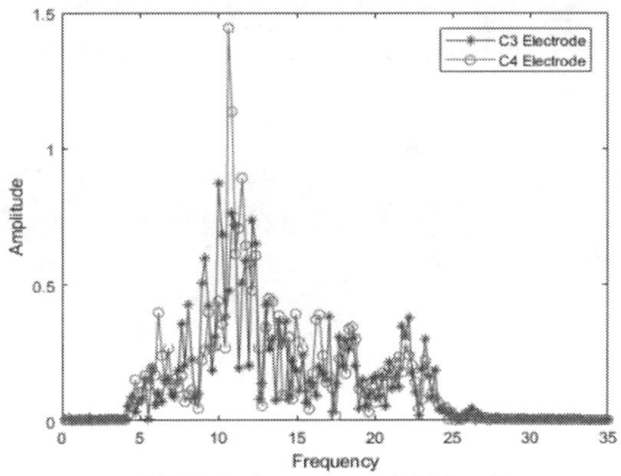

(c) Motor imagery on right-hand

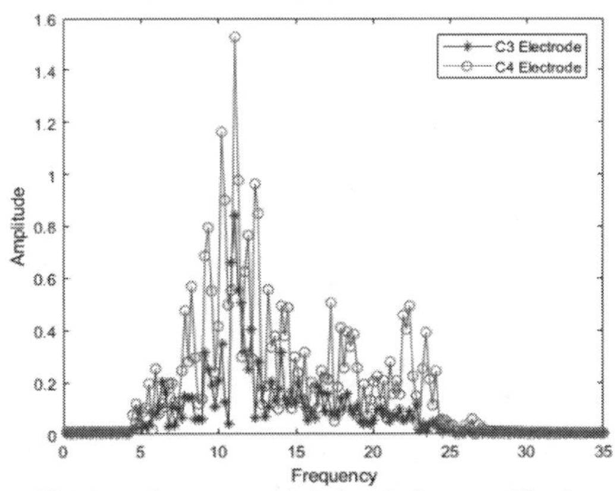

(d) Motor imagery on right-hand after sparsification

Figure 1. The signal after sparsification comparison.

To show the data classification performance, we compare our algorithm with other related methods on BCI competition 2003 data sets. The detailed classification accuracy and comparison shown in Table 1.As we can see, when only CSP and NMF were used, the average classification accuracy of the NMF was 2.13% higher than the CSP. And the accuracy of NMF classification was improved by 5.01% after the sparsification of data.

4 Conclusion

In this paper, we propose a non-negative matrix factorization method to classify motor imagination EEG. Based on the sparse degree of matrix and the selection of dimension will affect the efficiency of the algorithm and the accuracy of the classification. Firstly, the method of maximization of feature difference is used to increase the sparsity of the matrix; Secondly, the data of training set and test set of BCI competition in 2003 data are respectively classified by our method. Experiment shows: the average classification accuracy of test set was 91.43%, the highest single class motion imaging accuracy reached 94.28%.

Acknowledgments

This work was supported in part by the Guangdong program for support of top-notch young professionals under grant 2014TQ01X144, the National Natural Science Foundation of China under grants 61722304, 61333013, the Pearl River S$\&$T Nova Program of Guangzhou under grant 201610010196, the Guangdong Natural Science Funds for Distinguished Young Scholar under grant 2014A030306037.

References

1. A. Karagiannis, P. Constantinou, "Noise-Assisted Data Processing With Empirical Mode Decomposition in Biomedical Signals." IEEE Engineering in Medicine & Biology Society, 2011, 15(1):11-8.
2. M. Boersma, D. Smit, H. Bie, et al. "Network analysis of resting state EEG in the developing young brain: structure comes with maturation." Human Brain Mapping, 2015, 32(3):413-425.
3. E. Kalunga, S. Chevallier, Q. Barthélemy, et al. "Online SSVEP-based BCI using Riemannian geometry". Neurocomputing, 2016, 191:55-68.
4. E. Lukhanina, N. Mel'Nik, N.Berezetskaya, et al. "Co-rrelations between indices of P300 EEG potential, cognitive tests, and variational pulsometry in Parkin-sonian patients." Neurophysiology, 2008, 40(1):39-47.
5. S. Pittaccio, F. Zappasodi, S. Viscuso, et al. "Primary sensory and motor cortex activities during voluntary and passive ankle mobilization by the SHADE orthosis. "Human Brain Mapping, 2011, 32(1):60-70.

6. A. Bamdadian, C. Guan, K. Kai, et al. "Towards improvement of MI-BCI performance of subjects with BCI deficiency." International Ieee/embs Conference on Neural Engineering. IEEE, 2015:17-20.
7. S. Huang, X. Wu."Feature Extraction of EEG Signals Based on Mu/Beta Rhythm Imaging." Guangzhou: School of Bioscience and Engineering, South China University of Technology, 2010.
8. Çınar, Salim, N. Acır . "A novel system for automatic removal of ocular artefacts in EEG by using outlier detection methods and independent component analysis". Pergamon Press, Inc. 2017.
9. D. Toledo, G. Manzano, J. Barela, et al. "Cortical corr-elates of response time slowing in older adults: ERP and ERD/ERS analyses during passive ankle movement." 2016, 127(1):655-663.
10. L. Zhao, H. Shen, S. Cui,et al. "Research of Brain-Computer Interface Technology Based on Labview." IEEE Computer Society, 2008:72-80.
11. H. Sun, Y. Xiang, Y. Sun,et al. "On-line EEG classifica-tion for brain-computer interface based on CSP and SVM." International Congress on Image and Signal Processing. IEEE, 2010:4105-4108.
12. N. Zhang, X. Tang, Q. Liu. "Feature Extraction and Recognition of EEG Based on Semi-supervised Learning." Engineering Science Edition, 2017(s2): 230-237.
13. H. Lee, A. Cichocki, S. Choi. "Nonnegative Matrix Factorization for Motor Imagery EEG Classification." Springer Berlin Heidelberg, 2006:250-259.
14. M. Liu, J. Wang, C. Zheng, et al. Attention-related EEG features extracted by nonnegative matrix factorization [J]. Chinese Journal of Biophysics, 2006,22 (1): 67-72.
15. S. Duan, Y. Shang, L. Pan. "Feature Extraction and Classification of EEG Signals in Multi-Kinematics Imaging." Computer Measurement and Control, 2016,24 (2): 283-287.
16. D. D. Lee, H. S. Seung. "Algorithms for nonnegative matrix factorization." Advances in Neural Information Processing Systems, 2001, 13(6):556--562.
17. B. Blankertz, K. Müller, G. Curio, et al. "The BCI Competition 2003: progress and perspectives in detection and discrimination of EEG single trials." IEEE Trans. biomedical engineering, 2004, 51(6):1044-51.

Maintaining temporal validity of real-time data on non-continuously executing resources

Tian Bai[1], Hong Lu[2] and Juan Yang[2]

[1]Hunan Institute of Science and Technology, College of Computer Science, 414000, Yueyang, China
[2]Wuhan Maritime Communication Research Institute, 430079, Wuhan, China

Abstract. This paper studies the problem of temporal validity maintenance on non-continuously-executing resources. Response time bounds for sensor transaction scheduling are derived in the context of the hybrid extended multiprocessor periodic resource model. Then two deadline and period assignment schemes are proposed to maintain the temporal validity of real-time data. The calculation based scheme (DPA-C) uses the response time bounds to compute deadlines and periods. The check based scheme (DPA-A) assigns deadlines and periods directly. It then checks the feasibility of the assignment based on the response time bounds. Experiments are conducted to evaluate the performance of the proposed schemes. The results show that DPA-C performs better than DPA-A in terms of the scheduling success ratio and the imposed update workload.

1 Introduction

Cyber-physical systems (CPS) are widely used in many application areas to monitor the environmental status and make response to critical events. Typical application areas include flight control, health monitoring and industrial process control. In CPS, the real-time data objects are defined to model the current status of external entities. Each data object is associated with a validity interval which specify the lifetime of the object's values. A real-time data object is temporally valid if its current value's validity interval doesn't expire [1]. It's very important to guarantee the real-time data validity, since otherwise CPS would make wrong control decisions based on the invalid data objects.

In recent years, there has been a number of works on temporal validity maintenance. The More-Less (ML) scheme uses sensor transactions to update the values of real-time data objects periodically [1, 2]. The deferrable scheduling algorithm for fixed priority transactions (DS-FP) adopts the sporadic task model [3, 4]. It defers the release times of transaction instances as much as possible by exploiting the semantics of validity constraints. The earliest deadline first (EDF) scheme was adopted in [5, 6] for scheduling sensor transactions. Han et al. proposed two schemes to maintain validity constraints during mode switches [7]. The problem of co-scheduling sensor transactions and application transactions was studied in [8]. Li et al. studied the temporal validity maintenance problem on partitioned multiprocessors [9].

Current studies for maintaining the temporal validity of real-time data assume the processing resource is continuous available for transaction execution. This assumption, however, is not true in many cases. For example, a complex real-time system usually consists of several subsystems. Some subsystems may run on one physical platform. Each of them can only get a fraction of the resource. As another example, in order to achieve thermal resiliency, real-time systems will change between different power modes dynamically. Running in a low power mode can be viewed as running on a platform with discontinuous resource supply.

In this paper, we study how to maintain temporal validity of real-time data on platforms where the resource supply is not continuous. We derives the response time bounds of sensor transactions under global fixed priority scheduling scheme in the context of the hybrid extended multiprocessor periodic resource model. Based on the response time bounds, we then present two schemes to assign relative deadlines and periods to transactions so that the real-time data validity is properly maintained. The calculation based scheme (DPA-C) uses the bounds to compute deadlines and periods. The check based scheme (DPA-A) makes the assignment directly. It then checks whether the transaction set is feasible under the assignment. Experiments are conducted to evaluate the performance of these two schemes. The results show that compared with DPA-A, DPA-C can achieve higher scheduling success ratio while impose lower update workload.

The rest of the paper is organized as follows. Section 2 presents real-time data model and resource model. Section 3 presents the response time bounds of sensor transactions and two schemes for deadline and period assignment. The experimental results are given in section 4 and in section 5 we give the conclusion.

2 System model

Consider a set of real-time data objects $X = \{X_i\}_{i=1}^{n}$ and a set of sensor transaction $\Gamma = \{\tau_i\}_{i=1}^{n}$. The validity interval of data object X_i is V_i. Sensor transaction τ_i update the value of X_i. The worst-case execution time of τ_i is C_i.

Definition 1 X_i is temporally valid at time t, if the sampling time of X_i's current value $ts(X_i)$ plus V_i is not less than t, that is $ts(X_i)+V_i \geq t$ [1].

The transaction set is executed on a non-continuously-executing resource. In this paper we introduce the hybrid extended multiprocessor periodic resource model (HEMPR). The model is characterized by Ω= (Π, Δ, Θ, m_1, m). It specifies that for a resource that consists of m processors, m_1 processors of it collectively provide Θ unit of resource in every period Π within the deadline Δ ($\Delta \leq \Pi$). The rest of the processors each can guarantee full resource supply. The HEMPR model can be viewed as a combination of the MPR model [10] and the dedicated recourse model (DR). If we set $m_1 = m$ and $\Delta = \Pi$, then we get MPR. If m_1 is set to be zero, then we get DR. Thus our model is more flexible than MPR and DR. Notice that the processors in HEMPR can be either physical or virtual processors. The virtual processors themselves should be allocated to proper physical processors.

There are two approaches to schedule the sensor transactions on multiprocessors: partitioned approach and global approach. In the partitioned approach, a transaction will be allocated to a processor and can be executed only on this processor. In the global approach, transactions are allowed to migrate between different processors during their execution. In this paper the global fixed priority scheduling approach (GFP) is considered. We assume the transactions are sorted according to non-decreasing order of their validity intervals. Higher priorities are assigned to transactions with lower validity intervals. A transaction instance is allowed to be executed on one processor at any time.

3 Temporal validity maintenance on non-continuously executing resources

3.1 Response time bounds

The worst-case resource supply pattern for the HEMPR model can be obtained by making some modifications to the reasoning for MPR model in [10]. In our pattern, the resources provided by m_1 processors in the first period are allocated from time zero. For subsequent periods, the resources are allocated as late as possible without violating the deadline constraint. Then the supply bound function of HEMPR is given by:

$$sbf(t) = \begin{cases} N\Theta + (m - m_1)t + \beta(L) & t \geq a \\ (m - m_1)t & otherwise \end{cases} \quad (1)$$

In (1), $a = \Pi - \lceil \Theta / m_1 \rceil$, $N = \lfloor (t-a)/\Pi \rfloor$, $L = \min\{\Delta,\ t - a - N\Pi\}$. $\beta(L)$ is calculated as follows:

$$\beta(L) = \begin{cases} 0 & L \leq \Delta - \Pi + a \\ \max\{0, \Theta - (\Delta - L + 1)m_1\} + b & otherwise \end{cases} \quad (2)$$

In (2), $b = \Theta - \lfloor \Theta / m_1 \rfloor$.

In order to derive the upper bound of a sensor transaction's response time, we extend Nan's idea of busy period extension that is designed for dedicated resources [11].Consider a transaction instance $\tau_{k,l}$ with release time $r_{k,l}$ and finish time $f_{k,l}$. Let t_o denote the earliest time instant before $r_{k,l}$ such that all resource supply of HEMPR in time interval $[t_o, r_{k,l})$ are consumed by higher priority instances. There are at most $m_1 - 1$ sensor transactions with carry-in instance. Let t denote the length of the interval, then the upper bound of total workload of higher priority transactions in the interval is:

$$W_k(t) = \sum_{i=1}^{n} I_k^{NC}(\tau_i, t) + \sum_{m_1 - 1 \text{ largest}} (I_k^{CI}(\tau_i, t) - I_k^{NC}(\tau_i, t)) \quad (3)$$

The calculation of $I_k^{CI}(\tau_i, t)$ and $I_k^{NC}(\tau_i, t)$ is given by equations (5)-(8) in [11] and is omitted here for brevity.

The next problem is to obtain an upper bound for interference that $\tau_{k,l}$ suffers in $[t_o, f_{k,l})$. Let $I_k(t)$ denote the bound. For a dedicated resource, it is $\lfloor W_k(t)/m \rfloor$. But for HEMPR, it's not the case since the resource supply may not be continuous. In order to obtain $I_k(t)$, we compute the parallel supply functions for HEMPR and then use these functions to calculate the bound. A level-i supply function $Y_i(t)$ is the minimum resource supply in every time interval of length t with parallelism at most i [12]. Based on the worst-case resource supply pattern, we then have:

$$Y_i(t) = \begin{cases} t \cdot i & i \leq m - m_1 \\ sbf(t,i) & otherwise \end{cases} \quad (4)$$

In (4), $sbf(t,i)$ is obtained by replacing m in $sbf(t)$ with i and Θ with $\Theta - (m-i)(\Pi - a - 1) - \max\{b - i, 0\}$. Let $D(t, j)$ denote the duration over an interval of length t during which the time is provided by j processors in parallel, then :

$$D(t,j) = \begin{cases} Y_{m-m_1+b+1}(t) - Y_{m-m_1+b}(t) & j = m \\ Y_{m-m_1+1}(t) - Y_{m-m_1}(t) - D(t,m) & j = m - m_1 + b \\ t - D(t,m) - D(t, m - m_1 + b) & j = m - m_1 \\ 0 & otherwise \end{cases} \quad (5)$$

Then according to [12], $I_k(t)$ is given by:

$$I_k(t) = D(t, m - m_1) I_{\{m\}}(m_1) + \sum_{i=1}^{|S|} \min\{D(t, s(i)), \left\lfloor \max\{(W_k(t) - \sum_{j=1}^{i-1} D(t, s(j))s(j)), 0\} / s(i) \right\rfloor\} \quad (6)$$

In (6), $I_{\{m\}}(m_1)$ is an indicator function to take the case $m = m_1$ into consideration. $S = \{m - m_1, m - m_1 + b, m\} / \{0\}$. $s(i)$ is i-th element of S.

Let R_k denote the minimal value that satisfies the equation below:

$$I_k(t) + C_k = t \quad (7)$$

Then $R_k - r_{k,l} + t_o$ must be an upper bound of the response time of τ_k for this particular t_o. The reason is that $I_k(t)$ is an upper bound of interference on $\tau_{k,l}$ in $[t_o, f_{k,l})$, and there are no unused resource units in $[t_o, r_{k,l})$, so it's not possible to obtain a larger response time than $R_k - r_{k,l} + t_o$. Since $r_{k,l}$ is not earlier than t_o, we can see that R_k is a response time bound of τ_k for arbitrary t_o.

3.2 Deadline and period assignment schemes

In order to maintain the temporal validity constraints, the sum of the relative deadline D_k and the period T_k of τ_k must be no larger than the validity interval V_k. In addition, the transaction set must be schedulable by GFP on HEMPR. In this section two schemes are proposed for period and deadline assignment.

Calculation based scheme (DPA-C) uses the response time bounds derived in subsection 3.1 to calculate deadlines and periods directly. This calculation is carried out from τ_1 to τ_n. For τ_k, if R_k is larger than $V_k/2$, Γ is considered as unschedulable. Since $I_k(t)$ and $W_k(t)$ are both non-deceasing functions of t, R_k can be obtained in an iterative way. That is:

$$R_k^{i+1} = I_k(R_k^i) + C_k \quad (8)$$

In (8), R_k^i denotes the value of R_k in i-th iteration. R_k^{-1} is set to zero. DPA-C is presented in Algorithm 1. Notice that when m_1=m, R_i^0 is set to $\Delta - \Pi + 2a + C_i$ which is the length of the interval without resource supply. For τ_k, the time to compute $W_k(t)$ is $O(n)$, then $I_k(t)$ can be calculated in constant time based on $W_k(t)$ since there are at most three different parallel levels to be considered. There are at most $V_k/2 - C_k$ iterations to get R_k. Let $V_{\max} = \max\{V_i \mid 1 \le i \le n\}$, then the time complexity of DPA-C is $O(n^2 V_{\max})$.

Algorithm 1 Calculation based Assignment
Input: Γ and Ω
Output: deadlines and periods of Γ if Γ is schedulable, otherwise report failure

```
for i = 1 to n
  k=0;
  if m1 < m
    R_i^0 = C_i;
  else
    R_i^0 = Δ − Π + 2a + C_i;
  end
  while true
    if R_i^k > V_i / 2
      Γ is unschedulable, return failure;
    else if R_i^k = R_i^(k−1)
      break;
    end
    compute R_i^(k+1) using (8);
    k = k + 1;
  end
  D_i = R_i^k;
  T_i = V_i − D_i;
end
```

The iterative computation of DPA-C could be time-consuming for transactions with large validity intervals. Different from DPA-C, the check based scheme (DPA-A) adopts a very simple method to assign deadlines and periods. For τ_k, the scheme sets $D_k = \lfloor V_k/(F+1) \rfloor$ and $T_k = V_k - D_k$. F is the scaling factor of DPA-A. Its value should be not less than 1 and is chosen by system designers.

A schedulability condition for τ_k is obtained from (7):

$$I_k(D_k) + C_k \le D_k \quad (9)$$

Here $W_k(D_k)$ is computed using only $I_k^{CI}(\tau_i, D_k)$ to guarantee a safe bound since the iteration is not used. For each transaction, DPA-A checks whether (9) is satisfied. If not, the transaction set is considered as infeasible. Otherwise DPA-A continues to test subsequent transactions. For each transaction, it's not hard to see that the time to check (9) is $O(n)$. So the time complexity of DPA-A is $O(n^2)$. It is lower than that of DPA-C. However, as one can see from the experimental results, this reduction comes with the cost of lower scheduling success ratio and higher imposed workload.

4 Performance evaluation

In this section the results of the experimental evaluation on DPA-C and DPA-A are presented. There are two performance metrics. The first metric is the scheduling success ratio. It's defined as the ratio of the schedulable transaction sets under the particular scheme to the total number of transaction sets. The second metric is the update workload imposed by sensor transactions defined as $\sum_{1 \le i \le n} u_i / ((\Theta + (m - m_1)\Pi)/(m\Pi))$. This definition takes the discontinuity of the resource supply into account.

The parameters and their default settings are given in table 1. The resource deadline of HEMPR is set to be the same as the resource period. The ratio between Θ and $m_1\Pi$ is chosen randomly in [0.7, 1]. The ratio between m_1 and m is chosen randomly in [0.5, 1]. The value of Θ and m1 can then be obtained based on the ratios.

Table 1. Parameters and settings.

Parameters	Meaning	Value
V_i	Validity interval	[100,300]
C_i	Computation time	[5,20]
N_p	Number of processors	[2,8]
Π	Resource period	200

Fig.1 shows the scheduling success ratio of DPA-C and DPA-A when the density of the transaction set is varied from 1.5 to 2.5. The density is defined as the sum of the ratio of a transaction's computation time and its validity interval. In the experiment, N_p is set to 4 and the number of dedicated processors is set to be half of N_p. The ratio between Θ and $m_1\Pi$ is fixed at 0.8. The scaling factor for DPA-A is set to 2. It can be seen that the scheduling success ratio of DPA-C is constantly no lower than that of DPA-A. For example, when the density is about 1.9, almost all transaction sets can't be accepted by DPA-A while DPA-C still can accept any transaction set. The reason is DPA-A uses looser bounds for feasibility test which will lead to higher reject rate of transactions.

Fig.2 shows the update workload generated from two

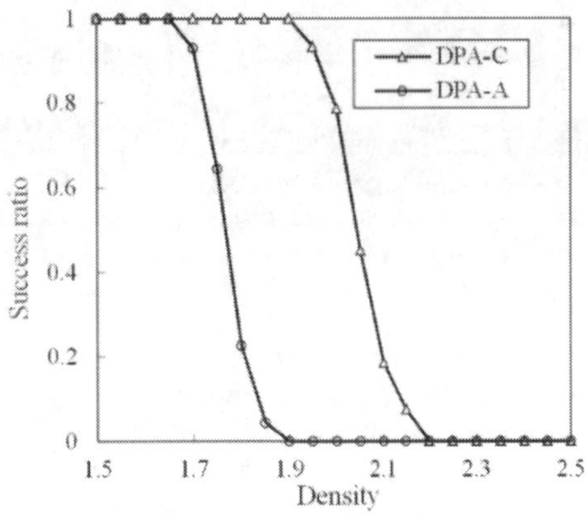

Figure 1. Scheduling success ratio comparison.

schemes. The parameter settings are the same as those described for fig.1. It can be seen that the update workload produced by DPA-C is lower than that of DPA-A. Both schemes' update workload becomes higher when the density increases. For temporal validity maintenance schemes, the update workload is an import metric. Reducing the workload of sensor transactions will leave more resources to other types of transactions, thus the overall system performance is improved.

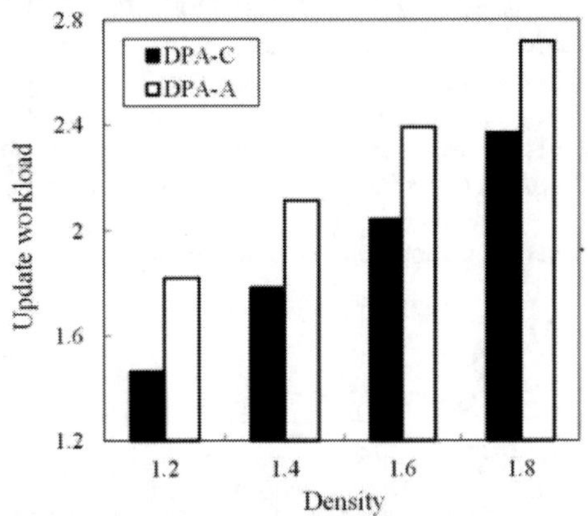

Figure 2. Update workload comparison.

Fig.3 shows the scheduling success ratio of two schemes under different ratios between Θ and $m_1\Pi$, while the density is fixed at 1.75. Other parameter settings remain unchanged. It is observed that the success ratio of DPA-C is constantly higher than that of DPA-A. In addition, both schemes' success ratio will increase when the recourse capacity increases. More experiments are carried out by varying parameter settings. The results are similar to what we have presented here and are omitted.

5 Conclusion

In this paper, we study the problem of maintaining temporal validity on platforms where the resource supply is not continuous. The response time bounds of sensor transactions scheduled by GFP for HEMPR model are derived. Two deadline and period assignment schemes,

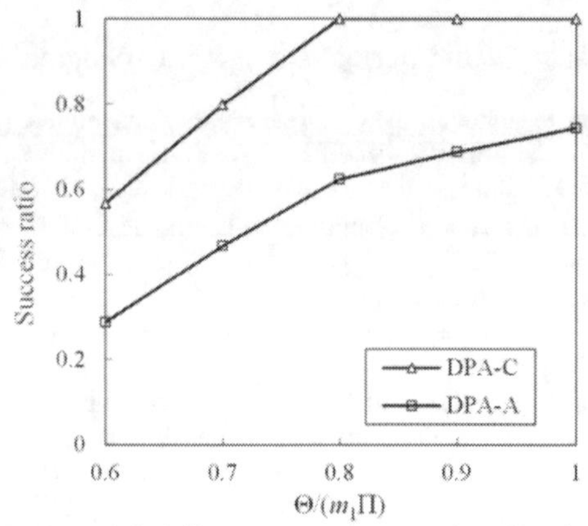

Figure 3. Scheduling success ratio comparison (density=1.75).

named calculation based scheme (DPA-C) and check based scheme (DPA-A), are proposed to maintain the validity constraints. DPA-C calculates deadlines and periods by using the response time bounds. DPA-A makes the assignment directly and then checks whether the assignment is feasible based on the derived bounds. The experimental results show that DPA-C achieves higher scheduling success ratio and imposes lower update workload than DPA-A does. This paper only considers the sensor transaction scheduling problem. In the furfure we will study how to simultaneously schedule sensor transactions and user transactions on non-continuously-executing resources.

References

1. M. Xiong, K. Ramamritham, Deriving deadlines and periods for real-time update transactions, IEEE Trans Comput,**53**(5), 567–583(2004)
2. J. Wang, S. Han, K. Y. Lam, et al., Maintaining data temporal consistency in distributed real-time systems, Real-Time Syst, **48**(4), 387–429(2012)
3. S. Han, D. Chen, M. Xiong, et al., Schedulability Analysis of Deferrable Scheduling Algorithm for Maintaining Real-Time Data Freshness, IEEE Trans Comput, **63**(4), 979–994(2014)
4. M. Xiong, S. Han, D. Chen, et al., Desh: overhead reduction algorithms for deferrable scheduling, Real-Time Syst, **44**(1), 1–25(2010)
5. M. Xiong, Q. Wang, K. Ramamritham, On earliest deadline first scheduling for temporal consistency maintenance, Real-Time Syst, **40**(2), 208–237(2008)
6. J. Li, M. Xiong, V. C. S. Lee, et al., Workload-Efficient Deadline and Period Assignment for Maintaining Temporal Consistency under EDF, IEEE Trans Comput,**62**(6), 1255–1268(2013)
7. S. Han, D. Chen, M. Xiong, et al., Online scheduling switch for maintaining data freshness in flexible real-time systems, *Proceeding of Real-Time Systems Symposium*, IEEE, Washington D.C., 115–124(2009)
8. S. Han, K. Y. Lam, J. Wang, et al., On Co-Scheduling of Update and Control Transactions in Real-Time Sensing and Control Systems: Algorithms,

Analysis, and Performance, IEEE Trans on Knowl Data En, **25**(10), 2325–2342(2013)
9. J. Li, D. Chen, M. Xiong, et al., Workload-Aware Partitioning for Maintaining Temporal Consistency upon Multiprocessor Platforms, *Proceedings of the IEEE 32nd Real-Time Systems Symposium*, IEEE, Vienna, 126–135 (2011)
10. I. Shin, A. Easwaran, I. Lee, Hierarchical Scheduling Framework for Virtual Clustering of Multiprocessors, *Euromicro Conference on Real-time Systems*, IEEE, Washington D.C., 181–190(2008)
11. N. Guan, M. Stigge, W. Yi, et al., New Response Time Bounds for Fixed Priority Multiprocessor Scheduling, IEEE Trans Commun, **51**(2), 387–397 (2010)
12. E. Bini, M. Bertogna, S. Baruah, Virtual Multiprocessor Platforms: Specification and Use, *IEEE Real-time Systems Symposium* , IEEE, Washington D.C., 437–446 (2009)

AUTHOR INDEX

AUTHOR INDEX

AUTHOR INDEX